9 Year Questions: Sorted and Solved

CSIR UGC NET Life Science

THE AUTHORS

Dr. Kailash Choudhary is the Director, Life Science, at Institute for Advanced Studies (IFAS), Pune, Maharashtra. He is actively involved in teaching and research for the last 18 years on different aspects of Lifesciences. Dr. Choudhary was born in Jodhpur and completed his school from Jodhpur followed by Bachelor in Science and then Masters in Botany, in 2000 from Jai Narian Vyas University securing Gold Medal. In June 2006 he completed his Ph.D degree on "Somatic Cell genetics and cloning in Acacia and Vigna Species". In 2004 he joined Lachoo Memorial College, Jodhpur where he served as Assistant Professor in Department of Botany and Biotechnology upto 2016. Dr Choudhary has been the recipient of several honours and awards. He secured All India rank 1 in prestigious CSIR NET JRF exam of Country for 10 consecutive time. He also cleared GATE lifesciences with top rank. He has cleared ARS-NET, ICMR-JRF and Rajasthan SET exams. He is imparting guidance to NET and GATE Lifesciences/Biotechnology aspirants from last 18 years at IFAS Jodhpur, Pune, Hyderabad, Kolkata, online and over 2500 students have qualified various national level exams in stream of life science. He has published around 23 articles, reviews, book and conference chapters, including nearly 20 original research articles in National and International peer reviewed Journals. His areas of interest include Conservation Biology, Plant Physiology and Biotechnology especially Plant Tissue Culture and Science and Society Related Programmes in the context of the Indian Thar Desert Region.

Dr. Ram Prakash Saran is presently working as Assistant Professor in Zoology at Jai Narain Vyas University, Jodhpur (Rajasthan). He did his M.Sc. in Zoology from Jai Narayan Vyas University, Jodhpur (Rajasthan) in the year 2005 with specialization in cell biology. He was throughout first class in studies up to M.Sc. He has cleared many national level exams like UGC-CSIR NET, GATE lifesciences, SET lifesciences. He is engaged in researches on threatened bird species of India with special emphasis on vultures, cytotoxicity of veterinary drugs on avian fauna and trophic transfer of nanoparticles and organic pollutants. At present he is involved in teaching - diversity among animal forms, animal physiology, Immunology, Developmental biology, Animal behaviour and biological evolution.

9 Year Questions: Sorted and Solved

CSIR UGC NET Life Science

Dr. Kailash Choudhary
Director, Life Science,
Institute for Advanced Studies (IFAS),
Pune, Maharashtra

Dr. Ram Prakash Saran
Assistant Professor, Zoology
Jai Narain Vyas University, Jodhpur (Rajasthan)

Published by:

SCIENTIFIC PUBLISHERS (INDIA)
5A, New Pali Road, P.O. Box 91
Jodhpur 342 001, India
E-mail: info@scientificpub.com
Website: www.scientificpubonline.com

Print 2026

ISBN: 978-93-89832-19-8
eISBN: 978-93-89832-20-4

Printed in India

Preface

The manifold purpose of writing this book is to cater the needs of the candidates aspiring for CSIR NET examinations. It is observed that the CSIR NET exam evokes fear in the minds of the students, as student do not know what to prepare and to what extent. The latest trend of CSIR exam is randomized computer based multiple choice questions. For student it becomes very difficult to find out the topic from which question has been asked and in its absence students fails to develop any strategy for the exam. It is imperative for every CSIR exam aspirant to understand the concepts and the pattern of exam.

Through this book an effort has been made to dispel CSIR exam fear. For sure success, a student should approach this exam with a strategy and planning in his/her mind. He/she should face CSIR NET and other lifescience exams any time and at any place with confidence. To help students, we have compiled this book in user-friendly and well-structured manner. Last 18 CSIR exam question are sorted in 13 sections and each section is further divided into sub-sections. This fine shorting of questions will help the student to understand the important topic of the subject, which is most frequent in examinations. The students will also be able to understand a change in trend of the questions over time. This will help them to understand the important topics of the subject and to prepare for exam in systemic planned way, which will enhance their selection chances. The quality of the MCQs asked in CSIR exams is always higher than other examinations. Practice of these questions will also help the students who are preparing for GATE, ICMR, DBT, SET and other exams in lifesceince or related streams.

A word or suggestion from your side may add another feather to the cap in the compilation of questions in this book. The author looks forward to the comments, suggestions and criticism from the students. Constructive suggestions and feedback from users would be highly appreciated, gratefully acknowledged and suitably incorporated.

Contents

SECTION-1

BIOCHEMISTRY

1. Structure of Atoms, Molecules, Bonds, Interactions

1. Which of the following bonds will be most difficult to break? **(MODELPAPER)**

   (a) C–O (b) C–C
   (c) C–N (d) C–S

2. from the following statements, **(DEC 2016)**

   A. Hydrogen, Deuterium and Tritium differ in the number of protons
   B. Hydrogen, Deuterium and Tritium differ in the number of neutrons
   C. Both Deuterium and Tritium are radioactive and decay to Hydrogen and Deuterium, respectively.
   D. Tritium is radioactive and decays to Helium
   E. Carbon–14 decays to Nitrogen–14
   F. Carbon–14 decays to Carbon–13

   Pick the combination with ALL correct statements

   (a) A, B and F (b) B, D and E
   (c) A, C and D (d) C, E and F

3. The interaction energy between atom A and B is 400 kJ/mol^{-1}. The type of interaction between them is **(DEC 2013)**

   (a) pi – pi. (b) covalent.
   (c) ion – dipole. (d) hydrogen bond.

4. On the molar scale which of the following interaction in a nonpolar environment provides the highest contribution to the bio–molecule? **(DEC 2011)**

   (a) van der Walls interaction
   (b) Hydrogen bonding
   (c) Salt bridge
   (d) Hydrophobic interaction

5. Which one of the following interactions plays a major role in stabilizing B–DNA? **(JUNE 2012)**

   (a) Hydrogen bond
   (b) Hydrophobic interactions
   (c) Van der Wall's interactions
   (d) Ionic interactions

6. Which one of the following non–covalent interactions between two non–bonded atoms A and B is most sensitive to the distance between them? **(JUNE 2013)**

   (a) A and B are permanent dipoles and are involved in hydrogen bonding.
   (b) A and B are fully ionized and are involved in salt bridge formation.
   (c) A and B are uncharged and repel each other.
   (d) A and B are uncharged and attract each other.

7. In proteins, hydrogen bonds form as follows: Donor (D)–H–Acceptor (A). Hydrogen bond is more favorable if the angle between D–H and A is **(DEC 2014)**

   (a) $<90^0$ (b) 180^0
   (c) $>180^0$ (d) 120^0

8. The maximum number of hydrogen bonds that can form between H_2N-NH_2 (hydrazine) and water is **(JUNE 2014)**

   (a) 2 (b) 1
   (c) 3 (d) 4

9. Out of the following hydrogen bonding schemes shown by ..., which one corresponds to the weakest hydrogen bond in a given solvent condition? **(DEC 2012)**

   (a) O–H ...O< (b) N–H...H<
   (c) O–H...N< (d) N–H...N<

10. According to Coulomb's law, two point charges 5.5 A^0 apart in vacuum would interact giving rise to energy of 60 kcal/mole. The energy of interaction of the same system in water (dielectric constant of water is 80 at 20°C) is **(DEC 2013 GU)**

   (a) 4800 kcal/mole.
   (b) 60 kcal/mole.
   (c) 0.75 kcal/mole.
   (d) 140 kcal/mole.

11. The interaction energy between two opposite charges separated by 3Å in vacuum is –500 $kJmol^{-1}$. The interaction energy between these two charges in water will be closest to **(DEC 2019)**

(a) –1500 $kJmol^{-1}$ (b) –166 $kJmol^{-1}$

(c) –55 $kJmol^{-1}$ (d) –6 $kJmol^{-1}$

12. H_2O is a liquid whereas H_2S is a gas at 25°C. The reason for this is **(DEC 2013 GU)**

(a) the number of electrons in the outermost electron shells of oxygen and sulfur are different.

(b) lack of hydrogen bonds in H_2O.

(c) the hydrogen–sulfur bond is more polar than the hydrogen–oxygen bond.

(d) lack of hydrogen bonds in H_2S.

13. Predominant interactions between phospholipids that stabilize a biological membrane include **(JUNE 2016)**

(a) hydrogen bonds and covalent interactions.

(b) van der Waal and ionic interactions.

(c) hydrophobic interactions and hydrogen bonding.

(d) covalent and hydrophobic interactions.

14. The free energy ΔG of a dissolved solute **(JUNE 2011)**

(a) Increases with solute concentration.

(b) decreases with solute concentration

(c) is independent of solute concentration.

(d) depends only on temperature.

15. The solubility of gases in water depends on their interaction with water molecules. Four gases i.e. carbon dioxide, oxygen, sulphur dioxide and ammonia are dissolved in water. In terms of their solubility which of the following statements is correct? **(JUNE 2016)**

(a) Ammonia > Oxygen > Sulphur dioxide > Carbon dioxide

(b) Oxygen> Carbon dioxide> Sulphur dioxide > Ammonia

(c) Sulphur dioxide > Oxygen > Ammonia> Carbon dioxide

(d) Ammonia > Sulphur dioxide > Carbon dioxide> Oxygen

16. From the following statements,

A. For a reaction to occur spontaneously the free energy change must be negative

B. The interaction between two nitrogen molecules in the gaseous state is predominantly electrostatic

C. By knowing the bond energies, it is possible to deduce whether the bond is covalent or hydrogen bond

D. Hydrophobic interactions are not important in a folded globular protein.

Pick the combination with ALL WRONG statements. **(DEC 2016)**

(a) A and B (b) B and C

(c) C and D (d) B and D

17. Hydrogen bonds in proteins occur when two electronegative atoms compete for the same hydrogen atom

Donor–H.....Acceptor

The angle 'θ' between donor and acceptor of a hydrogen bond was determined from large number of X–ray structures of proteins, as shown below:

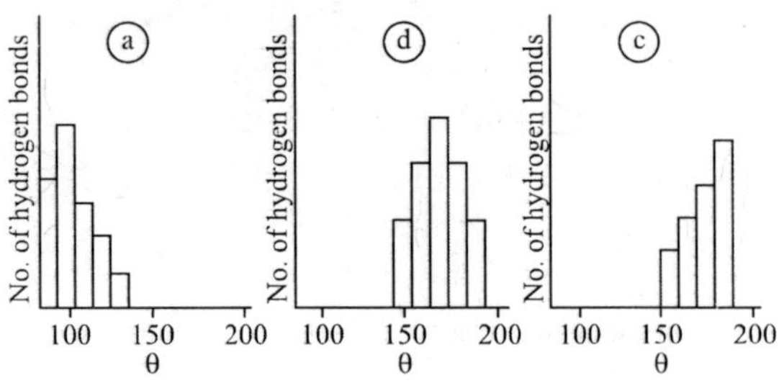

Which one of the distribution of 'θ' was 'observed from the proteins? **(JUNE 2015)**

(a) Only b (b) Only a

(c) Only c (d) a and b

18. There are hydrogen bond donors (D), and acceptors (A). When the mixture (A+D) is transferred from water, designated as (w) to a nonpolar solvent (np), the free energy change (ΔG) of transfer is 6.12 kcal mol^{-1} for the (A+D) mixture. A and D form a hydrogen bond in this nonpolar solvent and ΔG for this process is – 2.4 kcal mol^{-1}. When the hydrogen bonded molecule A–D is transferred from water to the nonpolar solvent ΔG for this process is 0.62 kcal mol^{-1}. The value of ΔG for the formation of hydrogen bond (A+B ←→ A–B) in water is **(MODELPAPER)**

(a) –0.1 kcal mol^{-1}. (b) +3.1 kcal mol^{-1}.
(c) –3.1 kcal mol^{-1}. (d) –1.78 kcal mol^{-1}.

19. If van der Waals interaction is described by the following relation.

$$\Delta G_{\text{Van}} = \frac{A}{r^{12}} - \frac{B}{r^{6}} + \frac{q_1 q_2}{r}$$

Where ΔG_{Van} is the free energy of the van der Waals interaction, A and B are constants, r is the distance between two nonbonded atoms 1 and 2 and q_1 and q_2 are partial charges on the dipoles 1 and 2. In this relation, the parameter A describes **(DEC 2011)**

(a) electron shell attraction
(b) electron shell repulsion
(c) dipole–dipole attraction
(d) dipole–dipole repulsion

20. A protein in 100 mM KCl solution was heated and the observed T_m (midpoint of unfolding) was 60°C. When the same protein solution in 500 mM KCl was heated, the observed T_m was 65°C. What is the most probable reason for this increase in T_m? **(JUNE 2011)**

(a) Hydrophobic interaction is increased and electrostatic repulsion is decreased.
(b) Hydrophobic interaction is decreased and electrostatic repulsion is increased.
(c) Hydrogen–bonding is increased.
(d) van der Waals interaction is increased

21. Values of T_m (midpoint of denaturation), ΔH_m (enthalpy change at T_m) and ΔC_p (constant–pressure heat capacity change) of a protein are measured in a differential scanning calorimeter. $\Delta G_D(T)$, the Gibbs free energy change at any temperature T(K) can be estimated using the following form of the Gibbs–Helmholtz equation with the values obtained from these measurements:

$$\Delta G_D(T) = \Delta H_m\left(\frac{T_m - T}{T_m}\right) - \Delta C_p\left[T_m - T + T\ln\left(\frac{T}{T_m}\right)\right]$$

The stability curve for the protein simulated using the observed thermodynamic values is given below:

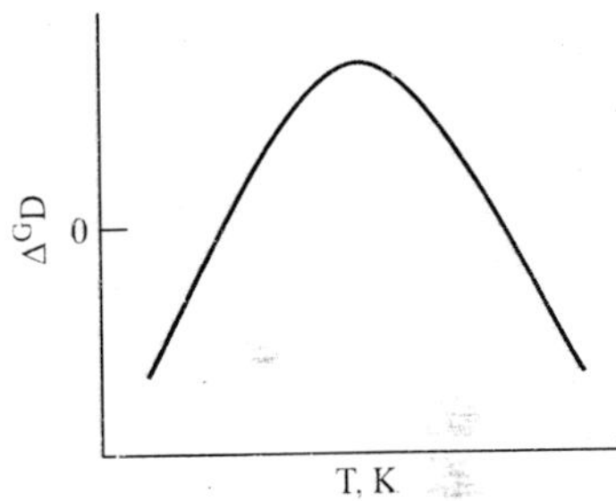

The shape of the stability curve is due to **(JUNE 2011)**

(a) hydrogen–bonding and electrostatic interactions only.
(b) van der Waals and electrostatic interactions only.
(c) only electrostatic interactions.
(d) only hydrophobic interaction.

22. Match the following bonds with their approximate energies: **(DEC 2018)**

(a)	Hydrogen bond	(i)	0.5 Kcal
(b)	Van der walls forces	(ii)	40 Kcal
(c)	Covalent bond	(iii)	80 Kcal
(d)	Ionic bond	(iv)	3 Kcal

(a) (a)–(iv), (b)–(iii), (c)–(ii), (d)–(i)
(b) (a)–(ii), (b)–(i), (c)–(iii), (d)– (iv)
(c) (a)–(i), (b)–(iv), (c)–(ii), (d)– (iii)
(d) (a)–(iv), (b)–(i), (c)–(iii), (d)–(ii)

23. Given below are some physicochemical properties (column X) and their manifestations (column Y). **(DEC 2019)**

X	Y
A. Pauling electronegativity	(i) Charge separation
B. Isolated $\pi\pi$–orbital overlap	(ii) Solvation of atoms
C. Aromaticity	(iii) Restricted rotation
D. Dielectric constant	(iv) Planarity of molecules

Which one of the following is the most appropriate match?

(a) A–i, B–iv, C–ii, D–iii

(b) A–iii, B–ii, C–iv, D–i

(c) A–ii, B–iii, C–iv, D–i

(d) A–iv, B–ii, C–i, D–iii

Answer key

1.(d)	2.(b)	3.(b)	4.(c)	5.(a)	6.(c)	7.(b)
8.(d)	9.(d)	10.(c)	11.(d)	12.(d)	13.(c)	14.(a)
15.(d)	16.(d)	17.(c)	18.(b)	19.(b)	20.(a)	21.(d)
22.(d)	23.(c)					

2. pH, buffer, Colligative properties, Molarity, Kinetics

1. Equal volumes of pH 4.0 and pH 10.0 solutions are mixed. What will be the approximate pH of the final solution? **(JUNE 2019)**

(a) 7.0 (b) 5.0

(c) 6.0 (d) 4.0

2. 100 ml of 0.02 M acetic acid (pK_a = 4.76) is titrated with 0.02 N KOH. After adding some base to the acid solution, the observed pH is 2.76. At this pH degree of protonation is **(MODEL PAPER)**

(a) 0%. (b) 10%.

(c) 90%. (d) 99%.

3. 10 mM acetate buffer (pH 4.00) is diluted one million times with distilled water (pH 7.00). pH of this diluted buffer is: **(JUNE 2014)**

(a) 4.00 (b) 7.04

(c) 8.00 (d) 6.96

(Help: $\log_{10} 10^x = x$; $\log_{10} 1.10 = 0.04$; $\log_{10} 1.01 = 0.004$)

4. 100 ml of 0.1 M sodium acetate solution has a pH of 8.90. To this solution 1000 μl of 1M acetic acid (pKa = 4.76) of pH 2.80 is added. The pH of this mixture will be: **(DEC 2013)**

(a) 8.90 (b) 4.76

(c) 2.80 (d) 5.76

5. A buffer is prepared by mixing 50 ml of 0.20 M acetic acid and 50 ml of 0.20 M sodium acetate. The pH of the buffer is 4.75 which is equal to the value of pKa of the acid. The molarity of the buffer is **(DEC 2013 GU)**

(a) 0.10. (b) 0.15.

(c) 0.20. (d) 0.40.

6. The apparent pH of a fluid is 7.45, where bicarbonate buffer is involved for maintaining its pH. Values of pKa of carbonic acid are 6.15 and 10.45. The molar ratio of [conjugate base]:[acid) is **(JUNE 2013)**

(Hint: antilog 1.3 = 20.0, and antilog 10^3 = 1000)

(a) 1: 20 (b) 20: 1

(c) 1: 1000 (d) 1000: 1

7. The pH of blood of a healthy person is maintained at 7.40 ± 0.05, Assuming that this pH is maintained entirely by the bicarbonate buffer (pKa_1, and pKa_2 of carbonic acid are 6.1 and 10.3, respectively), the molar ratio of [bicarbonate]/ [carbonic acid] in the blood is **(DEC 2011)**

(a) 0.05 (b) 1

(c) 10 (d) 20

8. The ionic strength of a 0.2 M Na_2HPO_4, solution will be **(DEC 2015)**

(a) 0.2 M (b) 0.4 M

(c) 0.6 M (d) 0.8 M

9. The genome of a bacterium is composed of a single DNA molecule which is 10^9 bp long. How many moles of genomic DNA is present in the bacterium? [Consider Avogadro No = 6 × 10^{23}] **(DEC 2015)**

(a) $1/6 \times 10^{-23}$ (b) $1/6 \times 10^{-14}$
(c) 6×10^{14} (d) 6×10^{23}

10. The (OH^-) of 0.1 N HCl solution is **(JUNE 2018)**

(a) 10^{-14} M (b) 10^{-13} M
(c) 10^{-12} M (d) 10^{-7} M

11. Which one of the following statements is true? **(DEC 2018)**

(a) The specific rotation of enantiomers will be identical.
(b) The rate constant of a first order reaction has only time but no concentration units.
(c) The value of pH + pOH depends on temperature.
(d) The bond disassociation energy (kJ/mol) of –C–C – will be greater than –C=C–.

12. Which one of the following graphs best describes the dependence of free energy change (ΔG) of ATP hydrolysis on Mg^{2+} concentration? **(DEC 2019)**

(a)
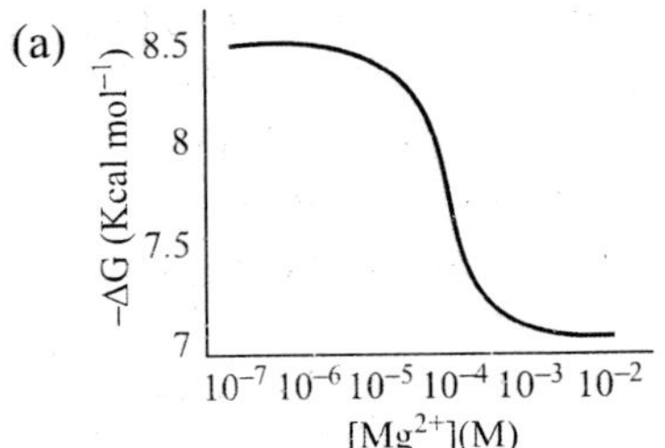

(b)
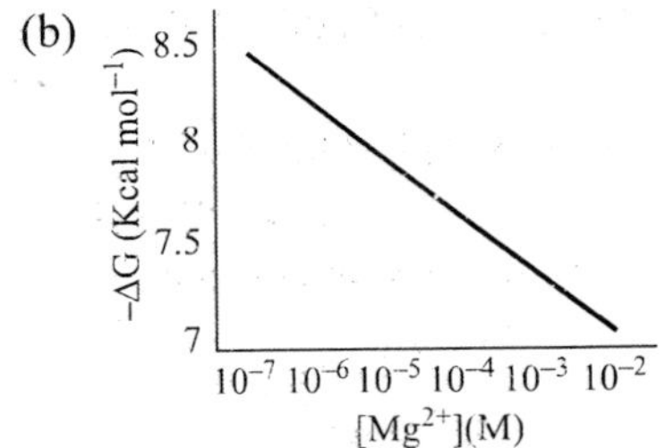

(c)
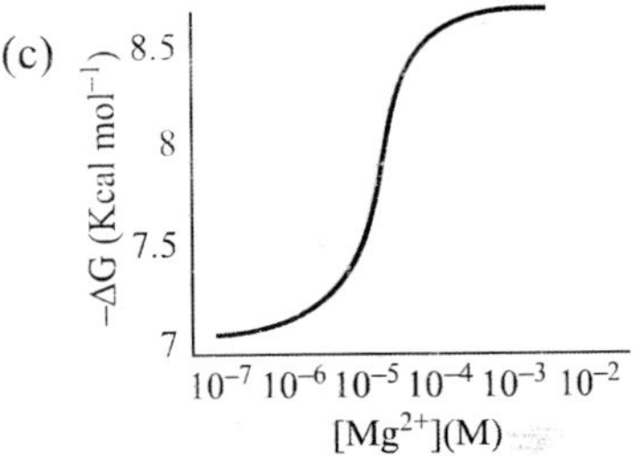

(d)
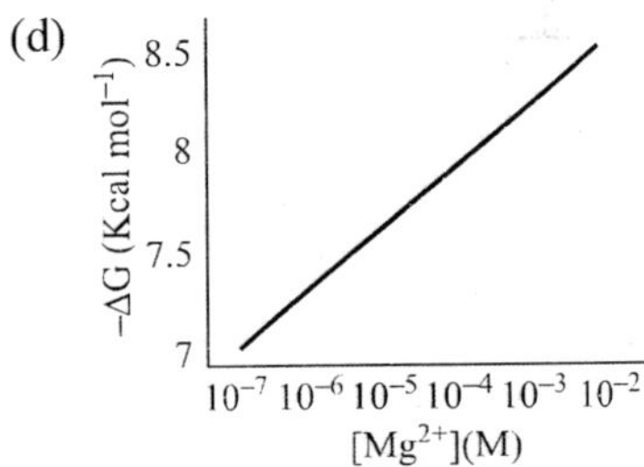

13. Following are statement related to biophysical chemistry of molecules

A. When solute is dissolved in a solvent, the increase in boiling point is independent of the number of particles into which the solute dissociates and dependent only on the molarity of solute.
B. When two atoms are linked by a covalent bond, van der Waals interaction between then contributes substantially to the bond energy
C. If a plot of reactant concentration versus time is not linear, but a plot of 1/reactant concentration versus time is linear, the reaction is second order.
D. Glucose, tyrosine and tryptophan can be easily distinguished by analyzing their UV spectra at equimolar concentration.

Choose the combination with correct statements **(DEC 2019 ASSAM)**

1. A and C 2. B and D
3. A and B 4. C and D

Answer Key

1.(a)	2.(d)	3.(d)	4.(d)	5.(c)	6.(b)	7.(d)
8.(c)	9.(a)	10.(b)	11.(b,c)	12.(a)	13.(d)	

3. Thermodynamics

1. The internal energy of a gas increases by 1J when it is compressed by a force of 1 Newton through 2 metres. The heat change of the system is: **(JUNE 2014)**
   (a) 1 J (b) –1 J
   (c) 2 J (d) –2 J
2. DNA melting temperature (T_m) was found to be 47°C and enthalpy measured at T_m was 0.032 kJ. The entropy change would be: **(DEC 2018)**
   (a) 1×10^{-3} kJ (b) 1×10^{-4} kJ
   (c) 3×10^{-2} kJ (d) 6×10^{-2} kJ
3. ΔG_o for the base pairing of oligonucleotides (n=5) at 300 K is – 18 kJ mol^{-1}. What would be the approximate value of the equilibrium constant K? **(JUNE 2014)**
   (a) 100 (b) 10
   (c) 1000 (d) 1
4. The standard free energy change (ΔG_o) per mole for the reaction A → B at 30°C in an open system is – 1000 cal/mole. What is the approximate free energy change (ΔG) when the concentration of A and B are 100 micromolar and 100 millimolar, respectively? **(JUNE 2016)**
   (a) 3160 (b) 316
   (c) 31610 (d) –3160
5. The Gibbs free energy of binding of a ligand with a protein is determined using calorimetric measurements at 25°C. The value of $\Delta G°$ thus determined is –1.36 kcal/mole. The binding constant for the ligand–protein association is: **(DEC 2012)**
   (a) 1.30×10^{-12} (b) 0.10
   (c) 1.00 (d) 0.97
6. Phosphogluconolactone is added to 0.1 M glucose–1–phosphate (G–6–P). The standard free energy change of the reaction, G–6–P⇄G–1–P is +1.36 kcal/mole at 25°C. The equilibrium concentrations of G–6–P and G–1–P, respectively are **(JUNE 2012)**
   (a) 96 mM, 45 mM (b) 100 mM, 10mM
   (c) 45 mM, 96 mM (d) 10 mM, 100 mM
7. In a human cell, the concentration of ATP, ADP and Pi are 2.25, 0.50 and 0.825mM respectively. The free energy of hydrolysis of ATP at pH 7.0 and 25°C will be (standard free energy change is –30.5 kJ/mol) **(MODEL PAPER)**
   (a) –30.5kJ/mol. (b) –61kJ/mol.
   (c) –15.25 kJ/mol. (d) –52 kJ/mol.
8. The lifetime of a peptide bond in proteins is very large (~1000 years). Which statement below is INCORRECT with respect to stability of the peptide bond? **(DEC 2014)**
   (a) The free energy of hydrolysis is negative
   (b) The free energy of hydrolysis is positive and large
   (c) The energy barrier to be crossed to go to the hydrolyzed state is large
   (d) The peptide bond can be hydrolyzed by 6N HCl at 100°C.
9. Protein stability is represented as **(JUNE 2017)**

   $$\text{Folded} \underset{}{\overset{K_{eq}}{\rightleftharpoons}} \text{Unfolded}$$

   Prior to development of sensitive calorimeters, thermodynamic parameters of processes were determined by following equation

   $$\ln K_{eq} = \frac{-\Delta H''}{R}\left(\frac{1}{T}\right) + \frac{\Delta S''}{R}$$

   ΔH^0 and ΔS^0 are standard changes m enthalpy and entropy, respectively.
   Which one of the following statements is correct for estimating ΔG, ΔH and ΔS?
   (a) Determining the ratio of folded and unfolded protein at 37°C
   (b) Plotting K_{eq} as a function of ΔH
   (c) Plotting K_{eq} against ΔS
   (d) Plotting K_{eq} against temperature
10. From the following statements: **(JUNE 2017)**
    A. In proteins the amino acids that can undergo oxidation are Cys and Met.

B. A tetrasaccharide composed of alternate L and D isomers will not be optically active.

C. The ΔG(Kcal/mol) values for K_{eq} of 0.1, 0.01 and 0.001 are 1.36, 2.72 and 4.09, respectively. It can be concluded that the relationship between ΔG and K_{eq} is parabolic.

D. The oxidation states of Fe in haemoglobin is +2. In cytochrome C, the oxidation states of Fe can be +2 or +3. E. In DNA, the sugar and bases are planar.

F. High–energy bonds hydrolyze with large negative ΔG.

Choose the combination with **ONLY ONE WRONG** statement.

(a) A, E, F (b) B, C, D

(c) C, D, E (d) A, B, C

11. Equilibrium constant (K'eq) of reaction is a ratio of product to substrate concentration. The reaction between (K'eq) and free energy change in a reaction ($\Delta G'$) is follows

$\Delta G' = -RT \ln K'eq$

Reaction A and reaction B have K'eq values of 10 and 100, respectively.

Which of the following statement is correct with respect to $\Delta G'$? **(DEC 2016)**

(a) $\Delta G'$ of A= $\Delta G'$ of B

(b) $\Delta G'$ of A > $\Delta G'$ of B

(c) $\Delta G'$ of B > $\Delta G'$ of A

(d) $\Delta G'$ of A ≈ $\Delta G'$ of B

12. It has been observed that for the DNA double helix melting, the value of ΔH (enthalpy change of denaturation) are 80 and 90 kcal/mole at 70° and 80°, respectively. Assuming that ΔCp (constant–pressure heat capacity change) is independent of temperature, estimate ΔH associated with the denaturation of DNA at 37°C. This estimated value of ΔH (kcal/mole) is **(DEC 2013)**

(a) 27 (b) 37

(c) 47 (d) 57

13. Differential scanning calorimetric study of calf thymus DNA was carried out to measure midpoint of thermal denaturation (T_m). ΔH_m (enthalpy change at T_m) and ΔC_p (constant–pressure heat capacity change). It has been observed that ΔC_p=0, T_m=75.5°C and ΔH_m=50.4 kcal/mole. The Gibbs free energy change at 37°C is **(JUNE 2012)**

(a) 25.5 kcal/mole (b) 2.6 kcal/mole

(c) 0.6 kcal/mole (d) 5.6 kcal/mole

14. Equilibrium constant (K) of noncovalent interaction between two non–bonded atoms of two different groups was measured at 27°C. It was observed that K = 100 M^{-1}. The strength of this noncovalent interaction in terms of Gibbs free energy change is: **(DEC 2011)**

(a) 2746 cal/mole (b) –2746 cal/mole

(c) 247 cal/mole (d) –247 cal/mole

15. The reaction of glutamate and ammonia to glutamine and water has a value of + 14 kJ mol^{-1} for ΔG_o. This is coupled with the ATP reaction (ATP + H_2O →ADP + phosphate). The ΔG_o for this reaction is –30 kJ mol^{-1}. The ΔG_o (KJ mol^{-1}) for the coupled reaction Glutamate + NH_3 + ATP →Glutamine +ADP + Phosphate under equilibrium condition is **(JUNE 2014)**

(a) 16 (b) –44

(c) –16 (d) 44

Answer Key

1.(b)	2.(b)	3.(c)	4.(a)	5.(b)	6.(b)	7.(d)
8.(b)	9.(d)	10.(a)	11.(b)	12.(c)	13.(d)	14.(b)
15.(c)						

4. Enzymes

1. Enzymes accelerate a reaction by which one of the following strategies? **(DEC 2015)**

(a) Decreasing energy required to form the transition state.

(b) Increasing kinetic energy of the substrate.

(c) Increasing the free energy difference between substrate and the product.

(d) Increasing the turn over number of enzymes.

2. Enzymes help to lower the activation energies of reactions by **(DEC 2013)**

(a) covalent interaction with substrates.

(b) binding only with the solvent molecules.

(c) changing reaction equilibria.

(d) forming weak interactions with substrates.

3. From the following statements **(DEC 2017)**

A. Enzymes enhance reaction rates by a factor of 2 to 10.

B. The activation energy for a reaction is lowered by enzymes.

C. The interactions between enzymes and . substrates are hydrogen bonding, hydrophobic and ionic.

D. Substrate concentration does not affect the rate of enzyme catalyzed reactions.

Pick the combination with all **INCORRECT** statements.

(a) A, B (b) B, C

(c) A, C (d) A, D

4. Match the coenzymes in column I serving as transient. Carriers of specific atoms or functional groups in column II **(DEC 2017)**

Column I		**Column II**	
(A)	Coenzyme A	(i)	Aldehyde groups
(B)	Flavin adenine dinucleotide	(ii)	Amino groups
(C)	Pyridoxal phosphate	(iii)	Hydrogen atoms
(D)	Thiamine pyrophosphate	(iv)	Acyl groups

Select correct combinations from the options given below:

(a) A–(iv), B–(iii), C–(ii), D–(i)

(b) A–(iii), B–(iv), C–(i), D–(ii)

(c) A–(i), B–(ii), C–(iii), D–(iv)

(d) A–(ii), B–(i), C–(iv), D–(iii)

5. Some coenzymes that serve as transient carriers of specific chemical groups are shown below **(JUNE 2019)**

	Coenzymes		**Chemical Group transfer**
A	Coenzyme A	(i)	Electrons
B	Flavin adenine Dinucleotide	(ii)	Acyl groups
C	Pyridoxal phosphate	(iii)	Hydride ions
D	Nicotinamide adenine dinucleotide	(iv)	Amino groups

Choose the combination with all correct matches

(a) A – (i); B – (ii); C – (iii); D – (iv)

(b) A – (ii); B – (i); C – (iv); D – (iii)

(c) A – (iii); B – (iv); C – (ii); D – (i)

(d) A – (iv); B – (iii); C – (i); D – (ii)

6. A researcher has isolated a restriction endonuclease that cleaves at only one specific 10 base pair site.

A. Would this enzyme be useful in protecting cells from viral infections, given that a typical viral genome is 5 x 10^4 base pairs long?

B. Restriction endonucleases are slow enzymes with turnover number of 1 s^{-1}. Suppose the isolated endonuclease was faster with turnover numbers similar to those for carbonic anhydrase (10^6 s^{-1}), would this increased rate be beneficial to host cells, assuming that the fast enzymes have similar levels of specificity?

The correct combination of answer is **(JUNE 2011)**

(a) (A) : No (B) : Yes

(b) (A) : No (B) : No

(c) (A) : Yes (B) : No

(d) (A) : Yes (B) : Yes

7. Michaelis and Menten derived their equation using which of the following assumption? **(DEC 2011)**
   (a) Rate limiting step in the reaction is the breakdown of ES complex to product and free enzyme
   (b) Rate limiting step in the reaction is the formation of ES complex
   (c) Concentration of the substrate can be ignored
   (d) Non–enzymatic degradation of the substrate is the major step

8. Following are statements on enzyme kinetics. Choose the correct statement. **(DEC 2019 ASSAM)**
   (a) Sufficiently high concentrations of substrate cannot completely relieve competitive inhibition.
   (b) Sufficiently high concentrations of substrate can relieve non–competitive inhibition.
   (c) Allosteric nature of an enzyme cannot be inferred from a plot of reaction velocity and substrate concentration.
   (d) For an enzyme following Michaelis–Menten kinetics, the initial velocity is determined at the beginning when enzyme–substrate dissociation is insignificant.

9. Michaelis–Menten enzyme kinetics for a simple reaction involving an enzyme (E) and substrate (S) is given by the scheme: **(DEC 2013)**

$$E + S \underset{}{\overset{K_m}{\rightleftharpoons}} ES \underset{}{\overset{K_{cat}}{\rightleftharpoons}} E + P$$

description of K_m, k_{cat} and their relationship is provided in the following statements.

A. Km represents association constant of the ES complex.

B. Km represents the dissociation constant of the ES complex.

C. k_{cat} is the rate constant for the chemical conversion of the ES complex to substrate bound enzyme and product.

D. k_{cat}/K_m is a rate constant that refers to the properties and reactions of the free enzyme and free substrate.

Which of the combinations of above statements is true?

(a) A and C (b) B and D
(c) A and D (d) B and C

10. A is converted to E by enzymes E_A, E_B, E_C, E_D. The K_m values of the enzymes are 10^{-2}, 10^{-4}, 10^{-5} and 10^{-4}, respectively. If all the substrates and products are present at a concentration of 10^{-4} M, and the enzymes have approximately the same V_{max} the rate limiting step will be: **(DEC 2012)**

(a) $C \underset{E_C}{\rightleftharpoons} D$ (b) $D \underset{E_D}{\rightleftharpoons} E$
(c) $A \underset{E_A}{\rightleftharpoons} B$ (d) $B \underset{E_B}{\rightleftharpoons} C$

11. A plot of V/[S] versus V is generated for an enzyme catalyzed reaction, and a straight line is obtained. Indicate the information that can be obtained from the plot. **(JUNE 2012)**
   (a) V_{max} and turnover number K_m can be obtained only from a plot of 1/V versus 1/[S].
   (b) K_m/V_{max} from the slope
   (c) V_{max}. K_m and turnover number
   (d) only K_m and turnover number.

12. Given below is the [P] vs time plot of an enzymatic reaction carried out by the enzyme X

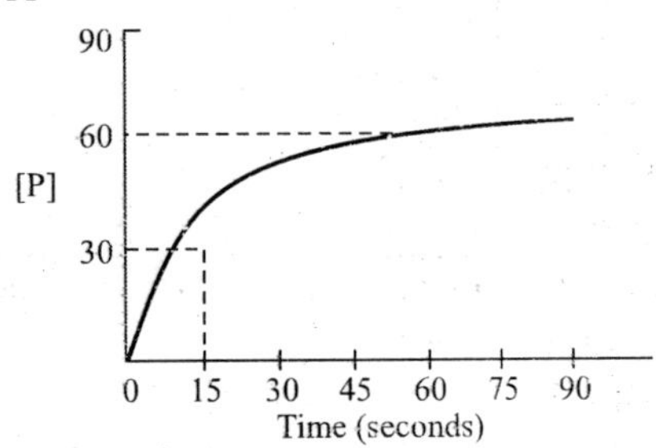

Which one of the following statements is the correct interpretation of the data?**(DEC 2019)**

(a) The K_m and V_{max} of the enzyme X are 15 and 60 units, respectively.

(b) The V_{max} is 60 but the Km cannot be determined

(c) The K_m 15 but the V_{max} cannot be determined

(d) neither the Km nor the V_{max} of the enzyme X can be determined from these data

13. The turnover number and specific activity of an enzyme (molecular weight 40,000 D) in a reaction (V_{max} = 4μmol of substrate reacted/ min, enzyme amount = 2 μg) are (JUNE 2016)

(a) 80,000/min, 2×10^4 μ mol substrate/min

(b) 80,000/min, 2×10^3 μ mol substrate/second

(c) 40,000/min, 1×10^3 μ mol substrate/min

(d) 40,000/min, 2×10^3 μ mol substrate/min

14. A 1% (w/v) solution of a sugar polymer is digested by an enzyme (20μg, MW=200,000). The rate of monomer sugar (MW=400) liberated was determined to have a maximal initial velocity of 10 mg formed/min. The turnover number (min^{-1}) will be (JUNE 2015)

(a) 5×10^4 (b) 2.5×10^{-2}

(c) 4.0×10^{-6} (d) 2.5×10^5

15. A solution of 1% (w/v) starch at pH 6.7 is digested by 15 μg of β–amylase (mol wt 152,000). The rate of maltose (mol wt = 342) had a maximal initial velocity of 8.5 mg formed per min. The turnover number is

(MODEL PAPER)

(a) 0.25×10^5 min^{-1}. (b) 25×10^5 min^{-1}.

(c) 4×10^{-6} min^{-1} (d) 2.5 x 10^4 min^{-1}.

16. Enzyme parameters of four isozymes is given below.

Isozyme	K_m (micromolar)	V_{max}
A	0.1	15
B	1.5	45
C	4.0	100
D	0.01	10

These isozymes are localized in different tissues. In liver the substrate concentrationis 0.2 micromolar. The liver isozyme is likely to be **(DEC 2014)**

(a) A (b) B

(c) C (d) D

17. In the accompanying figure, reaction kinetics of three proteins (a, b, c) is presented. Protein concentrations used to obtain this data are a – 1 mg/ml; b– 4 mg/ml; c – 2 mg/ml.

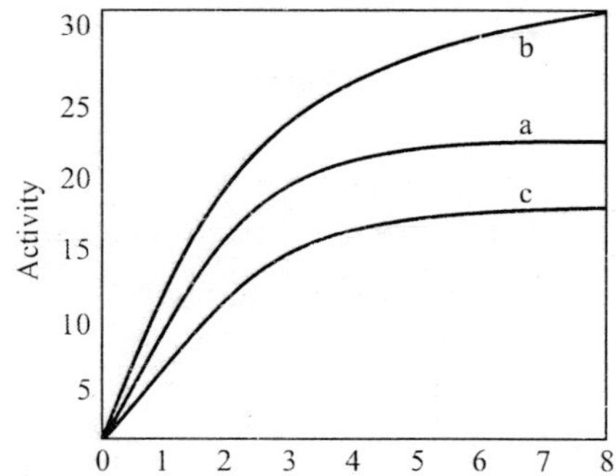

If catalytic efficiency is defined as K_{cat}/K_m, which of the following statements is correct?

(JUNE 2015)

(a) b > c > a (b) a > b > c

(c) a > c > b (d) c > a > b

18. The K_{cat}/K_m values of an enzyme–catalyzed reaction when plotted as a function of pH yielded a bell–shaped curve with a maximum around pH 6.0. Which of the following conclusions best describes the above observation?

(DEC 2013 GU)

(a) A His residue may be important for catalysis.

(b) The enzyme would be highly stable at pH 6.0.

(c) The enzyme would be most active below and above pH 6.0.

(d) The enzyme will require a metal ion as cofactor for catalysis.

19. The following small peptide substrates are used for determining elastase activity and the following data have been recorded.

Substrate	K_m (mM)	K_{cat} (S^{-1})
PAPA↓G	4.02	26
PAPA↓A	1.51	37
PAPA↓F	0.64	18

The arrow indicates the Cleavage site. From the above observations, it appears that:

(A) PAPAF is digested most rapidly.

(B) PAPAG is digested most rapidly.

(C) A hydrophobic residue at the C– terminus seems to be favored..

(D) A smaller residue at the C–terminus seems to be favored.

(E) Elastase always requires a smaller residue at the N–terminus of the cleavage site.

Which of the following is true? (JUNE 2013)

(a) (A), (C), (E) (b) (B), (D), (E)

(c) (E) only (d) (D), (E) only

20. A serine protease was tested for its activity on the following peptide substrates of different lengths and sequences. The obtained kinetic parameters of the protease are shown along with the peptide.

Peptide substrate	K_{cat} (S^{-1})	K_m (mM)
Ac–X–Ala–CO–↓–NH_2	0.01	100
Ac–Y–X–Ala–CO–↓–NH_2	0.10	4.0
Ac–Z–Y–X–Ala–CO–↓–NH_2	8.0	4.0
Ac–Y–X–Val–CO–↓–NH_2	6.0	35.0

Arrow denotes site of cleavage. Based on the above data, the following statements are made:

A. Catalytic efficiency (K_{cat} / K_m) increases with the size of the peptide.

B. Amino acid at the hydrolytic cleavage position of the peptide is critical for binding of the peptide with the protease

C. Catalytic efficiency decreases from three amino acid peptide to four amino acid peptide.

Which of the following combinations of the above statements is correct? **(JUNE 2017)**

(a) A and B (b) A and C

(c) B and C (d) A , B and C

21. The V_{max} and K_m from a Lineweaver–Burk plot of an enzyme reaction where $\frac{1}{v} = 40\ \mu M^{-1}$ min at $\frac{1}{s} = 0$ and $\frac{1}{s} = -1.5 \times 10^2$ mM^{-1} at $\frac{1}{v} = 0$ are **(JUNE 2018)**

(a) 0.025 μM min^{-1} and 0.67×10^{-2}mM

(b) 0.025 μM^{-1} min and 0.67×10^{-2}

(c) 0.025 μM min^{-1} and 1.5×10^2 mM^{-1}

(d) 0.038 μM min^{-1} and 0.67×10^{-2}mM

22. An allosteric enzyme has two heterotropic effectors, X and Y. The allosteric constant, L for the enzyme in the absence of effector molecules is 180. For the X–saturated form, the value of L increases from 180 to 1200, while for Y–saturated form it decreases to 60. What kind of effector molecules are X and Y? **(DEC 2019 ASSAM)**

(a) X and Y both are positive regulators.

(b) X is a negative regulator while Y is a positive regulator.

(c) X is a positive regulator while Y is a negative regulator.

(d) X and Y are not allosteric regulators.

23. Which one of the following statements is **NOT** correct?

(a) Allosteric enzymes do not obey Michaelis–Menten kinetics.

(b) The free–energy change provides information about the spontaneity but not rate of a reaction.

(c) Competitive and non–competitive inhibitions are kinetically indistinguishable.

(d) A K_{cat}/K_m (M^{-1} S^{-1}) of ~2x 10^8 for an enzyme indicates that the value is close to diffusion–controlled rate of encounter.

24. Choose the INCORRECT statement from the following statements made for an enzyme–catalyzed reaction **(JUNE 2019)**

(a) The kinetic properties of allosteric enzyme do not diverge from Michaelis–Menten behaviour.

(b) In feedback inhibition, the product of a pathway inhibits an enzyme of the pathway

(c) An antibody that binds tightly to the analog of the transition state intermediate of the reaction $S \rightarrow P$, would promote formation of P when the analog is added to the reaction.

(d) An enzyme with $K_{cat} = 1.4 \times 10^4\ s^{-1}$ and $K_m = 9 \times 10^{-5}$ M has activity close to the diffusion controlled limit.

25. The hydrolysis of pyrophoshate to orthophosphate is important for several biosynthetic reactions. In *E. coli*, the molecular mass of the enzyme pyrophosphatase is 120 kD and it consists of six identical subunits. The enzyme activity is defined as the amount of enzyme that hydrolyzes 10 μmol of pyrophosphate in 15 minutes at 37^0 under standard assay condition. The purified enzyme has a V_{max} of 2800 units per milligram of the enzyme. How many moles of the substrate are hydrolysed per second per milligram of the enzyme when the substrate concentration is much greater than K_m? **(DEC 2011)**

(a) 0.05 μmol (b) 62 μmol

(c) 31.1 μmol (d) 1 μmol

26. Binding of two ligands to their binding proteins were investigated. Following binding isotherms were obtained.

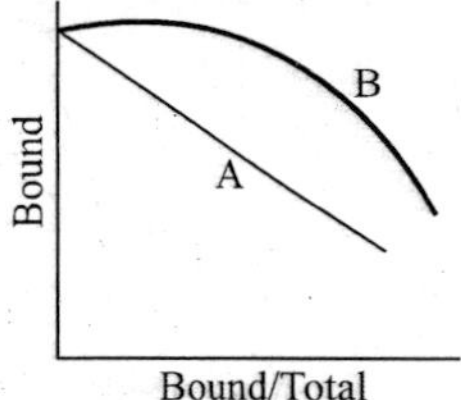

Which of the following statements is correct? **(DEC 2014)**

(a) A is obtained with an oligomeric protein and B is obtained with a monomeric protein

(b) B is obtained with protein with positive cooperativity

(c) A and B were obtained by the same protein at two different temperatures

(d) The profile B is not possible

Answer Key

1.(a)	**2.**(d)	**3.**(d)	**4.**(a)	**5.**(b)	**6.**(b)	**7.**(a)
8.(d)	**9.**(b)	**10.**(c)	**11.**(c)	**12.**(d)	**13.**(a)	**14.**(d)
15.(c)	**16.**(a)	**17.**(b)	**18.**(a)	**19.**(a)	**20.**(a)	**21.**(a)
22.(b)	**23.**(c)	**24.**(a)	**25.**(c)	**26.**(b)		

5. Enzyme Inhibition

1. Reaction products inhibit catalysis in enzymes by **(DEC 2014)**

(a) covalently binding to the enzyme.

(b) altering the enzyme structure

(c) occupying the active site.

(d) form a complex with the substrate.

2. Indicate the **INCORRECT** statement from the following: **(DEC 2017)**

(a) Allosteric enzymes function through reversible noncovalent binding of allosteric modulators or effectors.

(b) Monoclonal antibodies that catalyze hydrolysis of esters or carbonates can be produced.

(c) Enzymes are not inhibited irreversibly by heavy metals such as Hg^{2+}, Ag^+.

(d) Acid phosphatases hydrolyze biological phosphate esters at ~ pH 5.0.

3. Choose the correct statement from the following: **(DEC 2019 ASSAM)**

(a) Iodoacetamide inactivates an enzyme by reaction with a critical serine residue at neutral pH.

(b) Proline racemase causes isomerisation of L–proline to D–proline. Ribose will be an appropriate transition state analog.

(c) Tosyl–1–phenylalanine chloromethyl ketone binds at the active site of chymotrypsinand modifies an essential arginine residue.

(d) $^{2-}O_3PO—CH_2—\underset{\underset{O}{||}}{C}—CH_2—Br$ binds to triose phosphate isomerase at the active site and covalently modifies a glutamic acid residue required for enzyme activity.

4. Penicillin acts as a suicide substrate. Which one of the following steps of catalysis does a suicide inhibitor affect? (JUNE 2016)

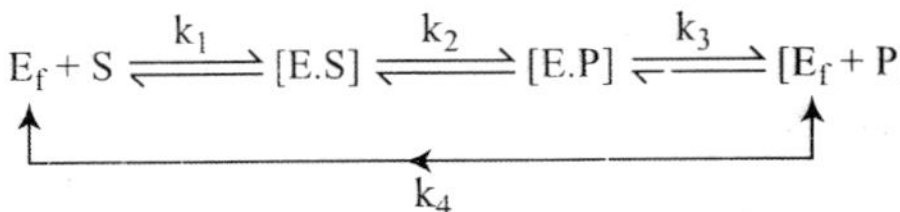

(a) K_1 (b) K_2
(c) K_3 (d) K_4

5. Which one of the following enzyme reaction represents noncompetitive initiation?

(MODEL PAPER)

(a) $E + S \rightleftharpoons ES + E + P$

$ES + I \overset{K_I}{\rightleftharpoons} EI$

(b) $E + S \rightleftharpoons E + S + E + P$

$ES + I \underset{K_I}{\rightleftharpoons} ESI$

(c) $E + S_1 \rightleftharpoons ES_1 \rightleftharpoons E'P_1 \overset{\downarrow P_1}{\rightleftharpoons}$

$E' \overset{\downarrow S_2}{\rightleftharpoons} E'S_2 \rightarrow E + P_2$

(d) $E + S \rightleftharpoons ES \rightarrow E + P$

$E + S + I \overset{K_I}{\rightleftharpoons} EI + S$

$ES + I \underset{K_I'}{\rightleftharpoons} ESI$

$EI + S \rightleftharpoons ESI$

6. For a reversible non–competitive inhibition of an enzyme, choose the plot that you would use to determine *Km:* **(DEC 2018)**

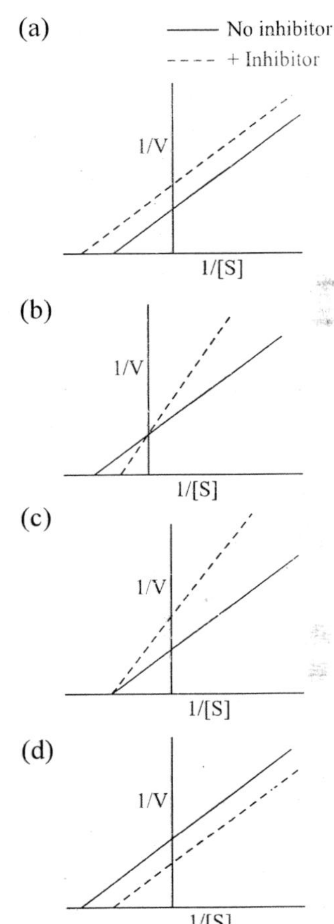

7. A researcher was investigating the substrate specificity of two different enzymes, X and Y, on the same substrate. Both the enzymes were subjected to treatment with either heat or an inhibitor which inhibits the enzyme activity. Following are the results obtained where, a= inhibitor treatment, b = heat treatment, c= control.

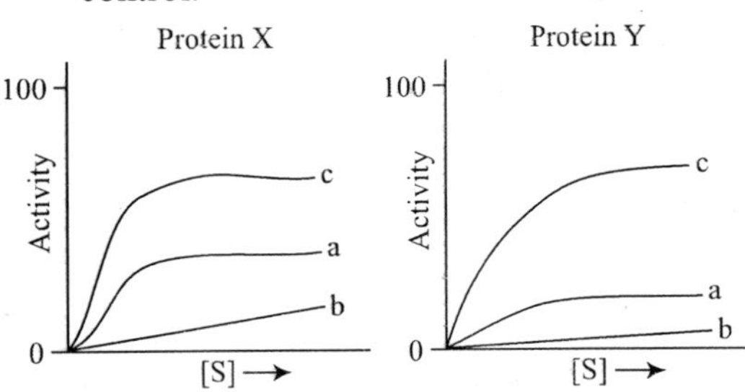

Which of the following statements is correct? **(DEC 2016)**

(a) Only protein X is specific for the substrate, S

(b) Only protein Y is specific for the substrate, S

(c) Both X and Y are specific for the substrate, S

(d) Both X and Y are non–specific for the substrate, S

8. If the product of an enzyme binds to the enzyme–substrate complex to exhibit its activity through a decrease in both K_m and V_{max}, this type of inhibition is called **(DEC 2013 GU)**

(a) competitive inhibition

(b) non–competitive inhibition.

(c) uncompetitive inhibition.

(d) partially–competitive inhibition.

9. An enzyme catalyzed reaction was measured in the presence and absence of an inhibitor for an uncompetitive inhibition, **(DEC 2012)**

(a) only Km is increased

(b) only is Vmax decreased.

(c) both Km and Vmax are decreased

(d) both Km and Vmax are not affected

10. An enzyme follows Michaelis–Menten kinetics. The activity of the enzyme was measured in the presence or absence of a molecule, "X". Given below are the double reciprocal equations for the enzyme activity with or without 'X'.

(i) Without molecule 'X'

$$\frac{1}{v_0} = \frac{K_m}{V_{max}}\left(\frac{1}{[S]}\right) + \frac{1}{V_{max}}$$

(ii) With molecule 'X'

$$\frac{1}{v_0} = \frac{K_m}{V_{max}}\left(\frac{1}{[S]}\right) + \frac{1}{V_{max}}\left(1 + \frac{[X]}{K_X}\right)$$

What kind of molecule is 'X'? Select from the options given below. **(DEC 2019 ASSAM)**

(a) Temperature independent competitive inhibitor

(b) Temperature dependent competitive inhibitor

(c) Non–competitive inhibitor

(d) Uncompetitive inhibitor

Answer Key

1.(c)	2.(c)	3.(d)	4.(c)	5.(d)	6.(c)	7.(c)
8.(c)	9.(c)	10.(d)				

6. Carbohydrates, Lipids, Amino acid, Vitamins

1. The –COOH group of cellular amino acids can form which of the following bonds inside the cell? **(JUNE 2016)**

(a) Ether and ester bonds.

(b) Ester and amide bonds.

(c) Amide and ether bonds.

(d) Amide and carboxylic anhydride bonds.

2. Glucose residues in amylose are linked by **(DEC 2013)**

(a) $\beta 1 \rightarrow 4$ (b) $\alpha 1 \rightarrow 4$

(c) $\alpha 1 \rightarrow 6$ (d) $\beta 1 \rightarrow 6$

3. The sequence of monosaccharides including position and configuration of glycosidic bonds in a glycoprotein is to be determined. Which one of the following methods can be employed? **(MODELPAPER)**

(a) Glycoprotein→removal of oligosaccharides by alkaline hydrolysis →nuclear magnetic resonance analysis of cleaved mixture of oligosaccharides

(b) Two dimensional nuclear magnetic resonance spectroscopic analysis of the glycoprotein

(c) Glycoprotein→release of oligosaccharides with endoglycosidases followed by purification to separate oligosaccharides →enzymatic hydrolysis of purified oligosaccharides with specific glycosidases

→mass spectroscopic analysis of smaller oligosaccharides

(d) Glycoprotein →treat with trypsin followed by MALDI analyses of tryptic peptides

4. Following are structures of stereoisomers of aldohexoses which differ in the stereochemistry

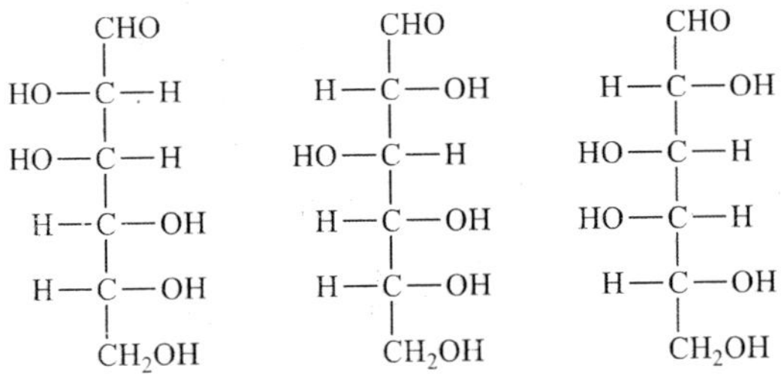

D-Mannose	D-Glucose	D-Galactose
CHO	CHO	CHO
HO—C—H	H—C—OH	H—C—OH
HO—C—H	HO—C—H	HO—C—H
H—C—OH	H—C—OH	HO—C—H
H—C—OH	H—C—OH	H—C—OH
CH_2OH	CH_2OH	CH_2OH

based on above structures, following information was given below:

A. D–glucose and D–mannose are epimers because they differ in the stereochemistry at C–2 position.

B. D–glucose and D–galactose are epimers because they differ in the stereochemistry at C–4 position.

C. D–mannose and D–glucose are epimers because they differ in the stereochemistry at C–3 position.

D. D–galactose and D–glucose are epimers because they differ in the stereochemistry at C–5 position.

Choose one of the correct combinations of above statements: (JUNE 2018)

(a) A and B (b) C and D

(c) B, C and D (d) A and D

5. Which one of the following statements is correct? **(DEC 2015)**

(a) In all L–amino acids, only the C^{α} carbon atom is chiral

(b) Deoxyribose is optically inactive

(c) The specific rotation of sucrose will be the sum of the specific rotations of D– glucose and D–fructose

(d) Phosphatidyl choline isolated from biological membranes is optically active

6. An amino acid contains no ionizable group in its side chain (R). It is titrated from pH 0 to 14. Which of the following ionizable state is not observed during the entire titration in the pH range 0 – 14? **(JUNE 2011)**

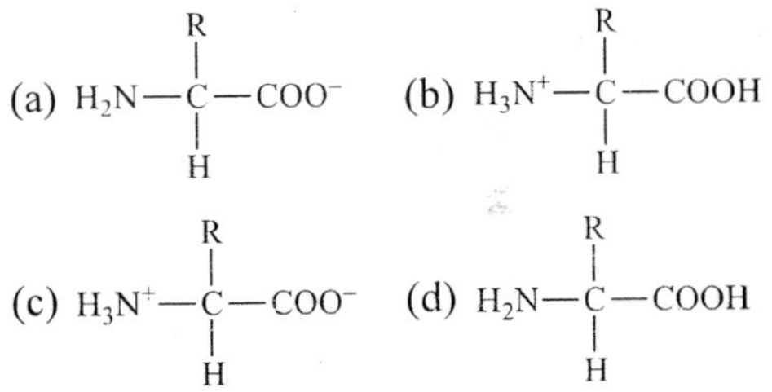

(a) $H_2N—C(R)(H)—COO^-$ (b) $H_3N^+—C(R)(H)—COOH$

(c) $H_3N^+—C(R)(H)—COO^-$ (d) $H_2N—C(R)(H)—COOH$

7. Choose the most appropriate pH at which the net charge is 0 for the molecule from the data shown below. **(DEC 2016)**

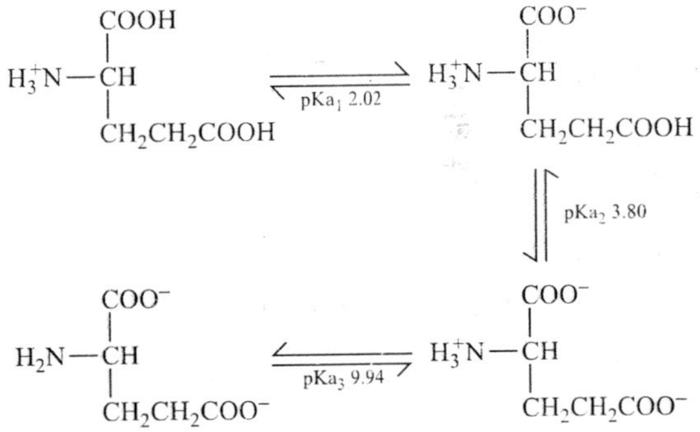

$H_3\overset{+}{N}—CH(COOH)—CH_2CH_2COOH$ ⇌ (pKa₁ 2.02) $H_3\overset{+}{N}—CH(COO^-)—CH_2CH_2COOH$

⇌ (pKa₂ 3.80) $H_3\overset{+}{N}—CH(COO^-)—CH_2CH_2COO^-$

⇌ (pKa₃ 9.94) $H_2N—CH(COO^-)—CH_2CH_2COO^-$

(a) 2.02 (b) 2.91

(c) 5.98 (d) 6.87

8. Which statement best describes the pKa of amino groups in proteins? **(JUNE 2013)**

(a) pKa of α–amino group is higher than the pKa of ε–amino. group.

(b) pKa of α–amino group is lower than the pKa of ε–amino group.

(c) pKa of α–amino group is same as the pKa of ε –amino group.

(d) pKa of α–amino group is higher than the pKa of guanidine side chain of arginine.

9. Choose the correct statement from the following: **(DEC 2017)**

(a) Disulfide bonds in a 20–residue peptide can be formed only if the cysteines are adjacent to each other.

(b) The amino acid isoleucine has only one chiral centre.

(c) Both bases and sugar contribute to chirality of nucleic acids.

(d) The pI of aspartic acid is less than that of glutamic acid.

10. Which one of the following peptides can coexist in both cis– and trans conformation?

(a) Ala–Ala–$CONH_2$

(b) Pro–Gly–$CONH_2$

(c) Asn–Gly–$CONH_2$

(d) Val–Pro–$CONH_2$

11. Which one of the following pair of amino acids are glucogenic and ketogenic in nature? **(JUNE 2018)**

(a) Alanine and Lysine

(b) Lysine and Leucine

(c) Isoleucine and Phenylalanine

(d) Aspartate and Lysine

12. Which one of the following statements is true regarding amino acids? **(DEC 2019)**

(a) Proline has high propensity to form α–helix in globular proteins

(b) Both isoleucine and threonine can exist as diastereomers

(c) Side chain pKa of aspartic acid is more than the side chain pKa of glutamic acid

(d) The ψ dihedral angle of proline is more restricted than the φ dihedral angle

13. Which one of the statements given below is correct? **(DEC 2019 ASSAM)**

1. The Ramachandran plots of Acetyl–L–AlaCONHMe and Acetyl–D–Ala–CONHMe will be identical.
2. Fe is a transition metal but Ca is not.
3. Side chain pKa of L–Glu and D–Glu will not be identical.
4. Hydrogen bonds are fundamentally van der Waals interaction.

14. Match the following vitamins with the corresponding pathological conditions arising from their deficiencies. **(DEC 2019)**

Vitamin	Disease
(i) A	(a) Pernicious anaemia
(ii) B_{12}	(b) Subdermal haemorrhaging
(iii) D	(c) Night blindness
(iv) K	(d) Rickets

(a) i – c; ii – a; iii – d; iv–b

(b) i – c; ii – b; iii – d; iv–a

(c) i – c; ii – a; iii – b; iv–d

(d) i – d; ii – a; iii – b; iv – d

Answer Key

1.(b)	2.(b)	3.(c)	4.(a)	5.(d)	6.(d)	7.(b)
8.(b)	9.(d)	10.(d)	11.(c)	12.(b)	13.(b)	14.(a)

7. Nucleotide and Nucleic acid (DNA/RNA)

1. The conformation of a nucleotide in DNA is affected by rotation about how many bonds? **(MODEL PAPER)**

(a) 4 (b) 6

(c) 7 (d) 3

2. Which nitrogen of adenosine gets protonated if pH of nucleoside is lowered from 7 to 3? **(JUNE 2012)**

(a) N_1 (b) N_3

(c) N_7 (d) N_9

3. In which form of DNA, the number of base pairs per helical turn is 10.5? **(DEC 2011)**

(a) A (b) B

(c) X (d) Z

4. Which one of the following bases has the largest hydrogen bonding possibility? **(DEC 2013)**

(a) Adenine (b) Guanine

(c) Cytosine (d) Uracil

5. Chirality of DNA is due to **(DEC 2014)**

(a) the bases.

(b) base stacking.

(c) hydrogen bonds between bases.

(d) deoxyribose

6. DNA is not hydrolyzed by alkali whereas RNA is readily hydrolyzed. This is due to **(DEC 2014)**

(a) The double helical structure of DNA
(b) The presence of uridine in RNA
(c) Due to features observed in RNA such as stem–loop structures
(d) The presence of 2'–OH group in RNA

7. The B–form of double helix has not been observed in RNA due to **(DEC 2013 GU)**

(a) the presence of U.
(b) the presence of I.
(c) absence of base stacking.
(d) altered conformation of the ribose ring.

8. Various modifications of nucleotides occur in nucleic acids. Which of the following combinations contains at least one modification that does **NOT** occur in nucleic acids? **(JUNE 2017)**

(a) N, N–dimethylguanosine, pseudouridine. 2'O–methyluridine
(b) 2–thiouridine, dihydrouridine, Nisopentenyladenine
(c) 5–methyldeoxycytosine, 5–thiouridine, pseudouridine
(d) dihydrouridine, 4–thiouridine, 2'O–methyluridine

9. The length of DNA comprising of 6.4 x 10^9 base pairs with the base pairs occurring at intervals of 0.34 nm along the DNA helix **(DEC 2013 GU)**

(a) cannot be calculated unless further information is provided.
(b) is ~ 2.18 µm.
(c) ~2.18 m.
(d) depends on base composition.

10. A segment of B–DNA encodes an enzyme of molecular mass 50 kD. The estimated length of this segment in µm would be **(JUNE 2013)**

(a) 0.1547 (b) 0.1547×10^3
(c) 0.4641 (d) 0.4641×10^3

11. The diploid genome of a species comprises 6.4 × 10^9 bp and fits into a nucleus that is 6 µm in diameter. If base pairs' occur at intervals of 0.34 nm along the DNA helix, what is the total length of DNA in a testing cell? **(JUNE 2013)**

(a) 3.0 m (b) 3.5 m
(c) 2.2 m (d) 4.0 m

12. On sequence analysis of a double stranded DNA, the results showed the content of cytosine, C was 20%. What is the amount of A and T put together? **(JUNE 2019)**

(a) 20% (b) 30%
(c) 50% (d) 60%

13. Consider a short double–stranded linear DNA molecule of 10 complete turns with 10.5 bp/turn. The ends of the DNA molecule are sealed together to make a relaxed circle. This relaxed circle will have a linking number of **(DEC 2016)**

(a) 105 (b) 20.5
(c) 10.0 (d) 10.5

14. A plasmid with a linking number (Lk) of 200, topological winding (Tw) number of 200 and a writhing number (Wr) of 0 was transformed into *E. coli*. The plasmid was re–isolated from the culture of the transformant.

The re–isolated plasmid was found to possess the same molecular weight as the original plasmid, but it possessed a writhing number of –5. Following statements are made about this observation.

A. Lk of the re–isolated plasmid would be 195.
B. Lk of the re–isolated plasmid would remain 200.
C. Tw of the re–isolated plasmid would remain 200.
D. Tw of the re–isolated plasmid would be 195

Which one of the following combinations of the above statements is correct representation of the facts? **(JUNE 2018)**

(a) A only (b) A and C
(c) A and D (d) D only

15. Following are three single stranded DNA sequences that form secondary structures.

(a) ATTGAGCGATCAAT

(b) ATTGAGCGATATCAAT

(c) AGGGAGCGATCCCT

Based on their stability, which one is correct? **(JUNE 2015)**

(a) (a) = (b) = (c) (b) (c) > (a) > (b)

(c) (b) > (c) = (a) (d) (b) > (c) > (a)

16. Denaturation profiles of DNA are shown below

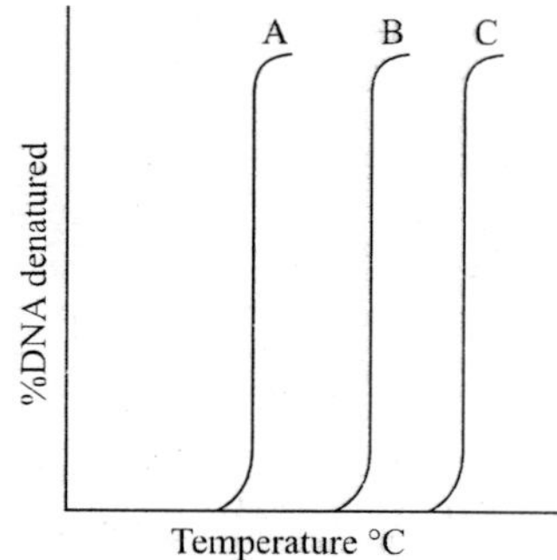

The differences in the profiles arise because **(DEC 2011)**

(a) the DNA is single stranded but of different sizes

(b) A + T content of A > B > C and the DNA are from complex genomes

(c) G + C content of C > B > A in DNA of comparable sizes isolated from simple genomes

(d) G + C content is identical but A + T content in A > B > C in DNA of comparable sizes isolated from simple genomes

17. Following statements were made about the presence of various hydrogen bond acceptor/donor groups in the major groove of B form DNA.

i. 4 and 6 amino groups, of cytosine and adenine, respectively.

ii. 4 and 6 keto groups of thymine and guanine, respectively.

iii. 2 and 6 amino groups of guanine and adenine, respectively.

iv. 2 keto and 4 amino groups of thymine and cytosine, respectively.

Which one of the following options consists of all correct statements? **(DEC 2019 ASSAM)**

(a) (i) and (ii) (b) (ii) and (iii)

(c) (iii) and (iv) (d) (i) and (iv)

18. The following statements are made: **(DEC 2017)**

A. α and β anomers of glucose are interconvertible and the ratio of their abundance is 1:2, respectively.

B. Single chain lipids ($\geq C_{14}$) form micelles and double chain lipids form bilayers in water.

C. Proline is energetically favoured at the C–termini of an α–helix than at the N–termini.

D. Major groove of DNA readily accommodates several common structural motifs in protein than the minor groove.

E. Replacement of a canonical Watson–Crick pairing by Wobble base pairs does not change the surface properties in t–RNA.

Which one of the following combinations is **INCORRECT**?

(a) A and D (b) B and E

(c) C and E (d) B and C

19. The following table lists the two major forms of DNA duplexes, conformation of base attached to the sugar and the nature of major and minor grooves. **(DEC 2017)**

DNA duplex		Properties	
(A)	A form	(i)	Syn conformation of the base to sugar
(B)	B form	(ii)	Anti conformation of the base to sugar
		(iii)	Wide major groove
		(iv)	Narrow maior groove
		(v)	Wide minor groove
		(vi)	Narrow minor groove

Which of the following combination correctly depicts the types of DNA duplexes and their properties (numbered within bracket)?

(a) A – (i) (iii) (vi); B – (ii) (iii) (vi)
(b) A – (i) (iii) (vi); B – (ii) (iii) (v)
(c) A– (ii) (iv) (v); B – (ii) (iii) (vi)
(d) A – (ii) (iv) (v); B – (i) (iii) (vi)

20. Following statements are made on nucleic acid structure:

A. In the B–form of DNA, the sugar pucker is C2' endo.

B. In RNA, the sugar pucker is C3' exo.

C. The wobble base pair is formed between G and A in RNA.

D. A change in the sugar pucker from C2' endo in the B form of DNA to C3' endo alters the width and depth of the major groove.

Which one of the following combinations of above statements is correct? **(JUNE 2018)**

(a) A and C (b) B and D
(c) A and D (d) B and C

21. Sugar puckering in double stranded nucleic acids is exclusively **(JUNE 2019)**

(a) C–2' endo in double stranded DNA
(b) C–3' endo in double stranded DNA
(c) C–2' endo in double stranded RNA
(d) C–3' endo in hybrid duplex with one strand as DNA and other as RNA

22. Following statements are being made about the orientation of the N–glycosidic bond between the base and the sugar in the following DNA duplexes.

A. 'anti' for B form DNA duplexes
B. 'syn' for B form DNA duplexes
C. 'anti' for A form DNA duplexes
D. 'syn' for A form DNA duplexes

Which one of the following combinations of the above statements is correct? **(JUNE 2018)**

(a) A and C (b) B and C
(c) A and D (d) B and D

23. A form and Z form of double stranded DNA differ in the handedness of their helices, nucleotide sequences, and configuration of base to sugar. Based on these properties, which one of the following statements defines a correct combination for A and Z forms of DNA? **(DEC 2019)**

(a) Right handed double helix and anti–configuration for the base to sugar arrangement in A DNA; and left handed double helix with alternating sequence of G and C (as a general pattern), and alternating syn– and anti–configurations for the base to sugar arrangement in the Z DNA.

(b) Right handed double helix and syn–configuration for the base to sugar arrangement in A DNA; and left handed double helix with alternating A and G sequence (as a general pattern),and anti–configurations for base to sugar arrangement in the Z DNA.

(c) Left handed double helix and anti–configuration for base to sugar arrangement in the A for DNA and right handed double helix and syn–configuration for base to sugar arrangement in the Z form DNA.

(d) Left handed double helix and syn–configuration for base to sugar arrangement in the A form DNA and right handed double helix and anti–configuration for the base to sugar arrangement for the Z form DNA.

24. The following statements are made

A. B form of DNA has –10 base pairs/turn and A form of DNA has –2.3Å helix rise per base pair

B. Both the A and B form of DNA have wider major groove and narrow minor groove

C. The crystalline nature of cellulose is brought about by $\alpha(1\rightarrow4)$ linkage between the glucose subunits.

D. The double bonds in natural lipids are always cis, which provides fluidity to the plasma membrane.

Which of the following combinations represent the correct statements? **(DEC 2019)**

(a) A and C (b) B and C
(c) A and D (d) C and D

25. Which one of the following statements on nucleic acids is NOT true? **(DEC 2018)**

(a) The conformation of ribose in DNA is α–2'–deoxy–D–ribofuranose.
(b) Hydrolysis of RNA takes place under alkaline conditions unlike DNA, as the 2'–hydroxyl in RNA acts as a nucleophile in an intramolecular displacement
(c) DNA can occur in different three–dimensional forms.
(d) In DNA, deamination of cytosine to uracil can occur in a non–enzymatic manner.

Answer Key

1.(c)	2.(a)	3.(b)	4.(b)	5.(d)	6.(d)	7.(d)
8.(c)	9.(c)	10.(c)	11.(c)	12.(d)	13.(c)	14.(b)
15.(b)	16.(c)	17.(a)	18.(c)	19.(c)	20.(c)	21.(d)
22.(a)	23.(a)	24.(c)	25.(a)			

8. Ramachandran Plot

1. The peptide unit (C^{α} – C'O – NH – C^{α}) is planer due to **(JUNE 2014)**

(a) restriction around C^{α} – C' bond
(b) restriction around C' – N bond
(c) restriction around N – C^{α} bond
(d) hydrogen bonding between carbonyl oxygen and imino hydrogen of the peptide backbone

2. The area of allowed regions in the Ramachandran map will be least for **(JUNE 2011)**

(a) Gly (b) L–Ala
(c) L–Prol (d) α–methyl L–Valine

3. The ϕ and Ψ values of a β–strand composed of all D–amino acids will mainly occupy which quadrant in the Ramachandran plot? **(DEC 2017)**

(a) upper left (b) upper right
(c) lower left (d) lower right

4. Acetyl–$(Ala)_{18}$–$CONH_2$ exists in α–helical conformation in solution. Most of the backbone dihedral angles (φ, ψ) will be . **(DEC 2014)**

(a) –60°, –30°
(b) 60°,30°
(c) –60°,–30°(50%) and 60°,30° (50%)
(d) –80°, 120°

5. Choose the correct statement about peptides in Ramchandran plot. **(DEC 2016)**

(a) Peptides that are unstructured will have all the backbone dihedral angles in the disallowed regions.
(b) It is not possible to conclude whether two peptide adopts entirely helix or entirely beta sheet confirmation.
(c) The occurrence of beta turn conformation formation in a peptide can be deduced.
(d) The sequence of a peptide can be deduced.

6. An α–helix in a peptide or protein is characterized by hydrogen bonds and characteristic dihedral angles.

Choose the right combination. **(JUNE 2011)**

(a) Hydrogen bonding between the amide CO of residue i and amide NH of residue i + 4. Dihedral angles in the region ϕ = –50°, ψ = –60°.
(b) Hydrogen bonding between the amide NH of residue i and amide CO of residue i + 4. Dihedral angles in the region of ϕ = –50°, ψ = –60.
(c) Hydrogen bonding between the amide CO of residue i and amide NH of residue i + 4. Dihedral angles in the region of ϕ = – 50°, ψ = +60
(d) Hydrogen bonding between the amide CO of residue i and amide NH of residue 1 + 3. Dihedral angles in the region of ϕ = –50°, ψ = –60

7. Given below are statements related to protein structures:

A. The dihedral angles of an amino acid X in Acetyl–X–N–Methyl amide in the

Ramachandran plot, occur in very small but equal areas in the left and right quadrants. It can be concluded that X is not one of the 20–coded amino acids.

B. The dihedral angles of a 20–residue peptide are represented in the Ramachandran plot. It is possible to conclude that the peptide does not have a proline.

C. Two proteins can have a similar fold even if they do not share significant similarity in their primary structure.

D. On denaturation of a protein by urea, the interactions that would be disrupted are ionic bonds and van der Waal's interaction but not disulfide bonds.

Choose the combination with **ALL CORRECT** answers: **(JUNE 2017)**

(a) A, B, C (b) A, C, D

(c) B, C, D (d) A, B, D

8. The following are four statements on the peptides/proteins conformation: **(DEC 2016)**

A. Glycine has a largest area of conformationally allowed space in the Ramachandran plot of Φ and Ψ

B. A 20–residue peptide that is acetylated at the N– terminus and amidated at the C–terminus has $\Phi = -60^0\ (\pm 5)$, $\Psi = -30^0(\pm 5)$ for all the residues. It can be concluded that conformation of the peptide is helix–turn–strand

C. The allowed values of Φ, Ψ for amino acids in a protein are not valid for short peptide

D. A peptide Acetyl–A_1 –A_2–A_3–A_4–$CONH_2$ (A–A_4 are amino acids) adopts well defined β–turn. The dihedral angles of A_2 and A_3 determined the type of β– turn

Choose the combination of correct statements.

(a) A and B (b) B and C

(c) A and D (d) C and D

9. The exact backbone dihedral angles in a folded protein can be obtained by **(DEC 2015)**

(a) deconvolution of its circular dichroism spectra obtained at different pH and temperature

(b) estimating the number of protons that exchange with deuterium on treating the protein with D_2O

(c) forming fibres of the protein and analyzing the fibre diffraction pattern

(d) analysis of the crystal structure of the protein obtained by X–ray diffraction at high resolutions

10. In a 30–residue peptide, the dihedral angles ϕ/ψ have been determined by one or more methods. When their values are examined in the Ramachandran plot, it is **(JUNE 2015)**

(a) not possible for ϕ/ψ values to be distributed in the helical as well as beta sheet region.

(b) possible that the ϕ/ψ values are all in the helical region although circular dichroism spectral studies indicate beta sheet conformation.

(c) possible to conclude that the peptide is composed of entirely D–amino acids.

(d) not possible to conclude if the peptide is entirely helical or entirely in beta sheet conformation.

11. Following are statements on β–turns:

A. All the 20 coded amino acids have equal propensity to form β –turns.

B. Pro cannot occur in β–turns.

C. Pro–Gly sequence strongly favours β–turns.

D. In Asn–Gly β–turns, Asn can have positive φ, Ψ values.

Choose the combination with all correct statements **(DEC 2018)**

(a) B,D (b) A,C

(c) A,D (d) C, D

12. Indicate which one of the following statements about nucleic acids and protein structures is correct. **(JUNE 2016)**

(a) Hydrogen bonding between the bases in the major and minor grooves of DNA is absent
(b) Both uracil and thymine have a methyl group but at different positions.
(c) The backbone dihedral angles of α–helices and β–sheets are very similar. Only the hydrogen bonding pattern is different.
(d) A β–turn is formed by four amino acids. The type of β–turn is determined by the dihedral angles of the second and third amino acid.

13. Which one of the following statements on 'protein conformation is NOT true? **DEC 2018)**
(a) Dihedral angles of side–chains in amino acids are depicted in the Ramachandran plot.
(b) Infrared spectroscopy can be used to deduce hydrogen bonding in peptides.
(c) Three dimensional structures of protein composed of 100 amino acids can be obtained by nuclear magnetic resonance spectroscopy
(d) Globular proteins have α–helical and β–sheet components

14. Which one of the statements on protein conformation, detailed below is INCORRECT? **(JUNE 2019)**
(a) L– amino acids can occur in Type I' β–turns where φ, ψ are positive
(b) A peptide rich in proline is unlikely to adopt α– helical structure
(c) Proline residues have high propensity to occur in β– turns
(d) The dihedral angles φ, ψ of amino acids in unfolded proteins are exclusively positive

Answer Key

1.(b)	2.(d)	3.(d)	4.(a)	5.(c)	6.(a)	7.(b)
8.(c)	9.(d)	10.(c)	11.(d)	12.(d)	13.(a)	14.(d)

9. Protein Folding– Secondary and Tertiary Structure

1. In an alpha helical polypeptide, the backbone hydrogen bonds are between **(JUNE 2015)**
(a) NH of n and CO of n + 4 amino acids
(b) CO of n and NH of n + 3 amino acids
(c) CO of n and NH of n + 4 amino acids
(d) NH of n and CO of n + 3 amino acids

2. Consider a 51–residue long protein containing only 100 bonds about which rotation can occur. Assume that 3 orientations per bond are possible. Based on these assumptions, how many conformations will be possible for this protein? **(JUNE 2012)**
(a) 3^{100} (b) 100^3
(c) 3^{51} (d) $51 \times 100 \times 3$

3. Which one of the following is unfavorable for protein folding? **(JUNE 2014)**
(a) Hydrophobic interaction
(b) van der Waals interaction
(c) Conformational entropy
(d) Hydrogen bonding

4. Which one of the following is the most appropriate statement regarding folded proteins? **(JUNE 2014)**
(a) Charged amino acid side chains are always buried.
(b) Charged amino acid side chains are seldom buried.
(c) Non–polar amino acid side chains are seldom buried.
(d) Tyrosine residues are always buried.

5. A protein has 30% alanine. If all the alanines are replaced by glycines, **(JUNE 2013)**
(a) helical content will increase.
(b) β–sheet content will increase.
(c) there will be no change in conformation
(d) the alanine–substituted protein will be less structured than the parent protein.

6. A protein is composed of leucine, isoleucine, alanine, glycine, proline, one lysine, one arginine and two cysteines connected by a disulfide bond. Conformational analysis indicates that the protein has elements of helix and beta structure. The protein is most likely. **(JUNE 2014)**

(a) a non–specific protease.

(b) not an enzyme

(c) a lipase.

(d) a flippase

7. Phosphorylation of ADP to ATP occurs through energy metabolism, comprising oxidative phosphorylation or substrate level phosphorylation or photo–phosphorylation (in plants). ATP can also be formed from ADP through the action of adenylate kinase. Crystal structure determination of adenylate kinase shows that the C–terminal region has the sequence *–val–asp–asp–val–phe–ser–gln–val–cys–thr–his–leu–asp–thr–leu–lys*. What can be a possible conformation of the sequence? **(JUNE 2012)**

(a) A helix that is not amphipathic

(b) Amphipathic helix

(c) Leucine zipper helix

(d) Beta helix

8. Following statements are made related to protein structure

A. The hydrogen bonding patterns between the CO and NH groups are n → n + 3 in α–helix; n → n + 4 in 3_{10} helix and n → n + 5 in π helix.

B. In a β turn, there are 10 atoms between the hydrogen bond donor and acceptor.

C. In a γ turn, there are 6 atoms between the hydrogen bond donor and acceptor.

D. Parallel sheets have evenly spaced hydrogen bonds, which bridge the strands at an angle.

Which one of the following combinations of above statements is correct? **(JUNE 2018)**

(a) A and C (b) A and B

(c) C and D (d) B and D

9. Out of the statements mentioned below **(DEC 2017)**

A. L–threonine and L–*allo*–threonine interact identically with plane polarized light.

B. van der Waals' interactions are always attractive.

C. Poly (pro) II–helix is not stabilized by intermolecular hydrogen bonds.

D. The folding of a protein is associated with an overall positive change in free energy and negative change in entropy.

E. Lysine acetylation on histone is associated with loosening of the histone complex from DNA.

Which of the following combinations is **CORRECT**?

(a) A and C (b) B and D

(c) C and E (d) D and E

10. Following are statements related to peptide/protein conformation.

A. The circular dichroism spectra of collagen and a protein in α–helical conformation will be identical.

B. The allowed region for the dihedral angles φ, ψ in Gly, spans a large area in the Ramachandran map. This can be drastically reduced by substituting the two hydrogens with methyl groups.

C. Proline has high frequency of occurrence in β–turns.

D. In a β–hairpin structure, the dihedral angles φ, ψ of amino acids flanking the β–turn region will be –60°, –30°, respectively.

Choose the combination with both INCORRECT statements. **(DEC 2019 ASSAM)**

(a) B and C (b) A and B

(c) A and D (d) C and D

11. A researcher has developed a program to evaluate the stability of a protein by substituting each amino acid at a time by the other 19 amino acids. For a protein, a researcher has observed the following changes

in stability upon substitution of amino acids in loops, helices, sheets, protein core and on the protein surface.

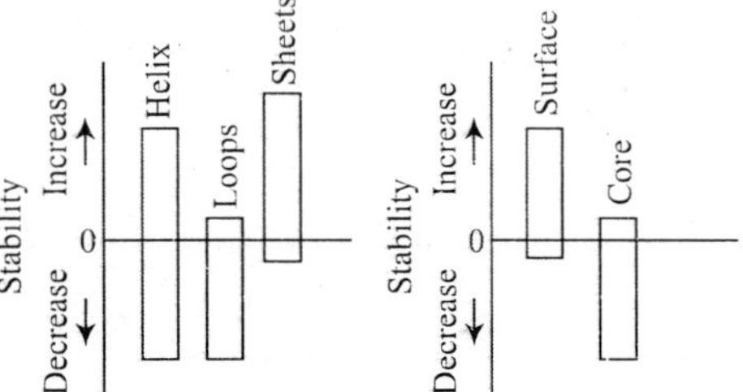

Substitutions in

(a) loops are more tolerant

(b) sheets are more tolerant

(c) core is less tolerant,

(d) helices are less tolerant

(e) surface is more tolerant

Which of the above statements are correct?

(JUNE 2016)

(a) a and c (b) c and d

(c) b and e (d) a and b

12. Thermodynamics of protein folding is depicted as a free energy funnel below. **(JUNE 2019)**

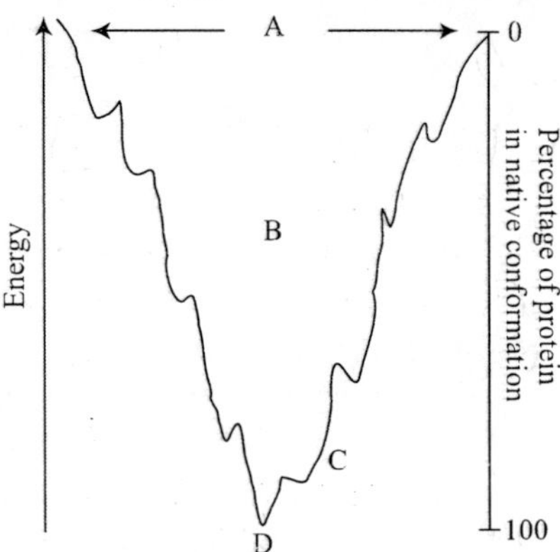

Gives below are regions in the diagram (Column X) and their representation (Column) Y

X	Y
A	(i) Native structure
B	(ii) Structure with highest entropy
C	(iii) Molten Globule
D	(iv) Discrete folding intermediates

(a) A – (ii); B – (iii); C – (iv); D – (i)

(b) A – (i); B – (ii); C – (iii); D – (iv)

(c) A – (iii); B – (iv); C – (ii); D – (i)

(d) A – (iv); B – (i); C – (ii); D – (iii)

Answer Key

1.(c)	2.(a)	3.(c)	4.(b)	5.(d)	6.(b)	7.(b)
8.(d)	9.(c)	10.(c)	11.(c)	12.(a)		

10. Proteins: Electrophoresis/ Chromatography

1. A researcher investigated the a set of conditions for a protein with an isoelectric point of 6.5 and also bind to calcium. This protein was subjected to four independent treatments: (i) pH 6.4, (ii) 10% glycerol, (iii) 10mM $CaCl_2$, (iv) 40% ammonium sulphate. This was followed by centrifugation and estimation of protein in supernatant. The results are depicted in the graph below:

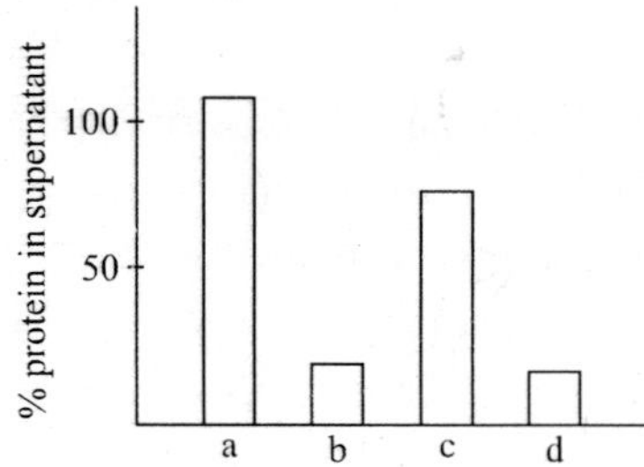

Which of the following treatments best represents the results shown in the graph?

(DEC 2016)

(a) a = ammonium sulphate, b = glycerol, c = pH 6.4, d = $CaCl_2$

(b) a = $CaCl_2$, b = glycerol, c = ammonium sulphate, d = pH 6.4

(c) a = pH 6.4, b = $CaCl_2$, c = ammonium sulphate, d = glycerol

(d) a = $CaCl_2$, b = pH 6.4, c = glycerol, d = ammonium sulphate

2. Glycerol is added to protein solutions to stabilize the preparations by **(JUNE 2017)**

(a) increasing the viscosity of solution
(b) stabilizing the pH
(c) preferential hydration of proteins
(d) interacting and neutralising the surface charges on the proteins

3. A mixture of two proteins was subjected to following three chromatographic columns:
(a) Cation exchange,
(b) Size exclusion (Sephadex 100) and
(c) Reverse phase
Following elution profiles were obtained

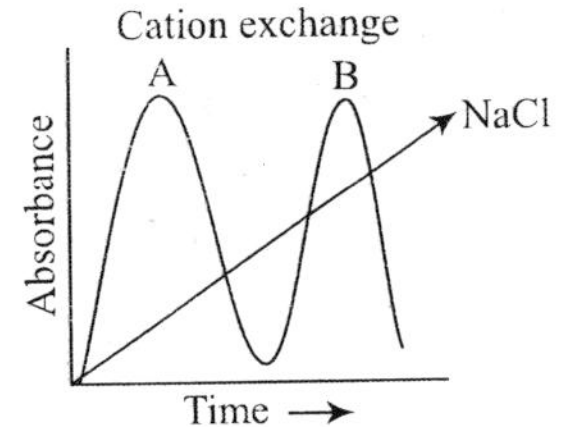

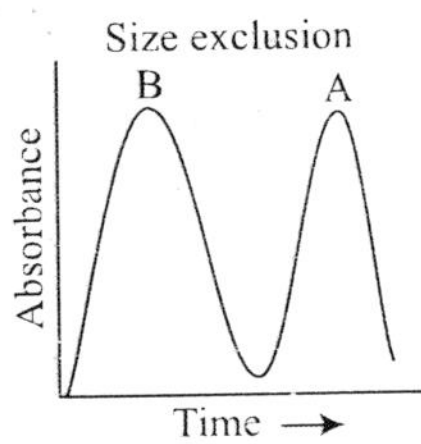

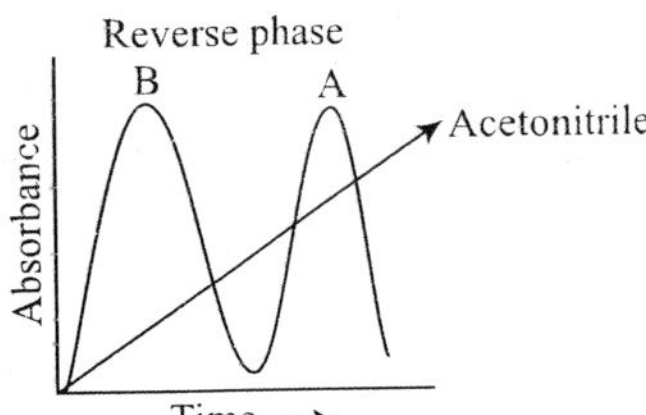

Which of the following statements is correct? **(DEC 2015)**
(a) A is larger and more hydrophobic than B.
(b) B is more anionic and more hydrophobic than A.
(c) A is more hydrophobic and smaller than B.
(d) A is more cationic and smaller than B.

4. In isoelectric focussing experiments, proteins are separated on the basis of their **(JUNE 2014)**
(a) relative content of only positively charged residues.
(b) relative content of only negatively charged residues.
(c) relative content of positively and negatively charged residues.
(d) mass to charge ratios

5. Analysis of a homotetrameric protein and a double stranded DNA (that had been incubated in standard buffer) on native gels revealed that they migrated true to their physical states (tetrameric nature of the protein and double stranded nature of the DNA). Following hypotheses were made for the effect of adding high salt to the incubation mix and subsequent analysis on native gels.
A. The protein would migrate as a homotetramer and DNA in double stranded form.
B. The protein would migrate as a monomer but DNA in double stranded form.
C. The protein would migrate as a homotetramer but the DNA in single stranded form.
D. The protein would migrate as a monomer and the DNA in single stranded form.
Which of the following combination of hypotheses is most likely? **(DEC 2019)**
(a) A and B (b) B and C
(c) C and D (d) A and D

6. A small fraction of clear cellular lysate was run on an isoelectric focusing gel (IEF) to purify a particular protein, which showed a number of sharp bands corresponding to different pI value. The protein of interest has a pI of 5.2. Therefore, the band corresponding to pI 5.2 was cut, eluted with appropriate buffer and subjected to SDS PAGE, which showed 3 distinct bands. One of the following inferences CANNOT be drawn from the above observations? **(JUNE 2014)**
(a) Several different proteins having same pI may be present at the single band on IEF gel.
(b) SDS–PAGE showed 3 distinct bands which may represent molecular mass of a different proteins.
(c) The protein of interest may be composed of 3 subunits.
(d) As the IEF gel showed a distinct band corresponding to pI 5.2, which is the pI of the protein of interest, the protein is composed of a single subunit.

7. Ribonuclease A (RNase A) consists of 124 amino acid residues. RNase A is cleaved by subtilisin (a protease) in a quantitative fashion at Ala 20 – Ser 21 peptide bond generating two fragments of 20 and 104 residues respectively. However, gel–filtration chromatography of the subtilisin–treated reaction mixture yields a single peak that elutes at the same position of the full–length RNase A. Notably, under identical chromatographic conditions as above, similar size fragments (20, 104, 124 residues) are seen to separate well. Which of the following is the most likely explanation for this observation? **(DEC 2013 GU)**
   (a) The smaller fragment is formed in vanishingly small amounts and cannot be detected.
   (b) The smaller 20–residue fragment is unstable and degraded.
   (c) The two fragments are non–covalently bound despite the cleavage of 20–21 peptide bond.
   (d) The two fragments are linked by disulfide bonds after subtilisin cleavage.

8. The following observations are made on a 30–residue polypeptide **(DEC 2019)**
   (a) Unordered structure is observed in water but a helical conformation is observed in medium of low dielectric constant.
   (b) The peptide is resistant to degradation by proteases.
   (c) Red blood cells are lysed by the peptide.
   (d) β–mercaptoethanol has no effect on peptide structure.

   Which of the following statements can be correctly attributed to the above observations?
   (a) The peptide is entirely composed of D–amino acids and is amphipathic.
   (b) The peptide is entirely composed of L–amino acids and is not amphipathic.
   (c) The peptide is rich in disulphide bonds between D–cysteines.
   (d) The peptide is entirely composed of L–aromatic amino acids.

9. A multimeric protein when run on SDS gel showed 2 bands at 20 kDa and 40kDa. However, when the protein was run on a native gel, it showed a single band at 120 k Da. The native form of the protein would be **(JUNE 2019)**
   (a) homotrimer (b) heterotetramer
   (c) heterodimer (d) heterotrimer

10. Mouse IgG is left either intact (left lane A,B,C, D) or digested with papain or pepsin or treated with β–mercaptoethanol (β–ME) and run on non–reducing SDS–PAGE and stained with Coomassie blue. In a separate experiment, papain–digested products are immunoblotted with an anti–idiotypic monoclonal antibody. Following four profiles are attributed to each of these treatments. **(DEC 2012)**

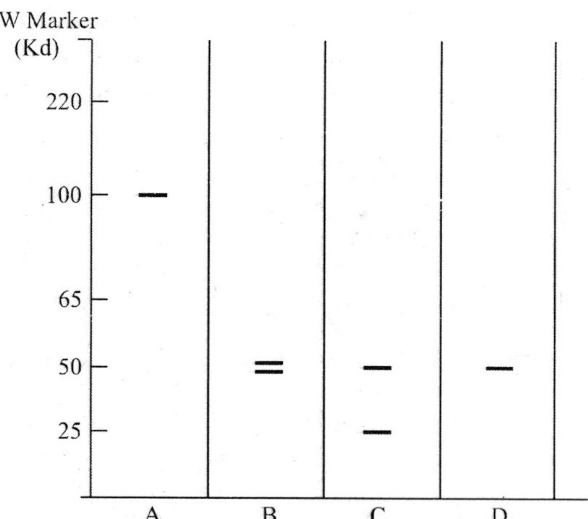

   Which one of the following possibilities is correct?
   (a) A (pepsin), B (papain), C (β–ME), D (papain, followed by anti–idiotype immunoblot)
   (b) A (papain), B (pepsin), C (papain, followed by anti–idiotype inununoblot) D. (β–ME)
   (c) A (papain, followed by anti–idiotype immunoblot), S (papain), C (pepsin) D (β–ME)
   (d) A (β–ME). B (papain), C (pepsin), D (papain, followed by anti–idiotype immunoblot)

11. The molecular mass of a protein determined by gel filtration is 120 kDa. When its mass is determined by SDS–PAGE with and without β–mercaptoethanol, it is only 60 kDa. What is the most probable explanation for these observations? (DEC 2012)
    (a) Protein is a dimer in which two identical chains are cross–linked by disulphide bond(s).
    (b) Protein is a monomer of molecular mass 60 kDa but it is excluded from the gel matrix due to strong repulsion between the gel matrix and the protein.
    (c) Protein is most likely to be composed of two sub units having identical molecular mass.
    (d) Protein is a monomer but it is nicked into half its size by SDS.

12. Four single amino acid mutants (a to d) of a protein in the epitope–region of a monoclonal antibody *X* were made and expressed in *E. coli*. The lysates from the four *E. coli* cultures expressing these four proteins were run or an SDS–PAGE gel and subsequently transferred to nitrocellulose membrane and Western blotted using a monoclonal antibody *X* raised against the wild type protein. The results are presented in the figure below.

a b c d wt

Monoclonal

Load control

The four single mutation, upon sequencing, were found to be Valine (V) to Alanine (A); Glycine (G) to Proline (P); Alanine (A) to Aspartic acid (D) and isoleucine (I) to leucine (L). Which one of the following statements is correct? **(DEC 2015)**
    (a) b is due to V→A and c is due to G→P
    (b) b is due to G→P and d is due to V→A
    (c) d is due to I→L and a is due to A→D
    (d) c is due to V→A and a is due to I→L

13. A protein undergoes post–translational modifications. In an experiment to identify the nature of modifications, the following experimental results were obtained.
    A. Protein moved more slowly in an SDS PAGE.
    B. Isoelectric focusing (IEF) showed that there was no change in the pI.
    C. Mass spectrometric analysis showed that the modification was on serine

    The modification that the protein undergoes is likely to be **(JUNE 2015)**
    (a) phosphorylation
    (b) glycosylation
    (c) ubiquitinization
    (d) ADP–ribosylation

14. Your have isolated spleen cells and purified your desired enzyme from the spleen cell extract by a sequence of steps involving ammonium sulfate fractionation and gel chromatography. You found an increase in the specific activity of the enzyme in the purified fraction compared to cell free extract and made the following speculations:
    A. Cell free extract contained inhibitor bound enzyme
    B. Structure of the enzyme was modified during the course of purification
    C. Purification steps gradually removed the inhibitors
    D. The final fraction contained concentrated enzyme preparation

    Which one of the following combination is appropriate? **(DEC 2013 GU)**
    (a) A, B and D (b) A, C and D
    (c) A, B and C (d) B, C and D

15. A biochemist purifies a new enzyme, generating the following purification table.

S. No	Procedure	Total protein (mg)	Activity (units)
i.	Crude Extract	20,000	4,000,000
ii.	Salt precipitation	5,000	3,000,000
iii.	Ion–exchange chromatography	200	800,000
iv.	Affinity chromatography	50	750,000
v.	Size–exclusion chromatography	45	675,000

The most effective purification step is

(MODEL PAPER)

(a) iv. (b) iii.
(c) v. (d) ii.

Answer Key

1.(d)	**2.**(c)	**3.**(c)	**4.**(c)	**5.**(a)	**6.**(d)	**7.**(c)
8.(a)	**9.**(b)	**10.**(a)	**11.**(c)	**12.**(a)	**13.**(b)	**14.**(a)
15.(a)						

11. Protein sequencing: Edman & Mass spectroscopy

1. The sequence of the peptide KGLITRTGLIKR can be unequivocally determined by

(JUNE 2016)

(a) Only Edman degradation.
(b) Amino acid analysis and MALDI MS/MS mass spectrometry.
(c) MALDI MS/MS mass spectrometry.
(d) MALDI mass spectrometry after treatment of the peptide with trypsin.

2. Consider the structureless oligopeptide R–G–P–S–T–K–M–P–E–Y–G–S–T–D–Q–S–N–W–H–F–R. The number of bonds that will be cleaved by trypsin and chymotrypsin treatments separately, are: **(JUNE 2014)**

(a) 1, 2 (b) 2, 2
(c) 2, 3 (d) 1, 3

3. Which peptide bond (s) marked as a, b, c, d and e will be broken when the following oligopeptide is treated with trypsin at pH 7.0?

(DEC 2012)

Lys—(a) —Arg—(b) —Pro—(c) —Lys—(d) —Arg—(e) —Gly

(a) a, b, d, e (b) b, d, e
(c) d, e (d) d

4. The oligopeptide, with F–A–R–P–M–T–S–R–P–G–F is treated with trypsin, chymotrypsin and carboxypeptidase–B. Apart from original, the number of fragments obtained will be

(JUNE 2012)

(a) 4 (b) 3
(c) 2 (d) 0

5. Tryptic digestion of a heptapeptide produced a dipeptide and a pentapeptide. Chymotryptic cleavage of the same heptapeptide yielded a tripeptide and a tetrapeptide. The tripeptide was found to be composed of two Thr (T) and one Lys (K) residue. The tetrapeptide contained one Ala (A), one Met (M), one Arg (R) and one Tyr (Y) residue. Which of the following sequence fits the heptapeptide?

(DEC 2013 GU)

(a) AKMYTTR (b) ARMYTTK
(c) AYMKTTR (d) MYARTKT

6. Tryptic digest of a heptapeptide (built from 3 lysine (K), 2 alanine (A), 1 tyrosine (Y) and 1 phenylalanine (F) yielded tri and tetrapeptide. Which of the following is the correct sequence of the heptapeptide? **(DEC 2013)**

(a) KAYAKFK (b) YKAAFKK
(c) KYKAAKF (d) KYAAKFK

7. In order to determine the primary structure of an octapeptide, amino acid composition was detennined by acid hydrolysis (A). The intact oligopeptide was treated with carboxypeptidase (B), chymotrypsin (C), trypsin (D) and CNBr (E). The peptides were separated in each case and acid hydrolysis was carried out for B – E. Following results were obtained (the brackets represent mixtures of amino acids in each fragment):

A. (2Ala, Arg, Lys, Met, Phe, 2Ser)

B. (Ala, Arg, Lys, Met, Phe, 2Ser) and Ala

C. (Ala, Arg, Phe, Ser), (Ala, Lys, Met, Ser)

D. (Ala, Arg), (Lys, Phe, Ser), (Ala, Met, Ser)

E. (Ala, Arg, Lys, Met, Phe, Ser), (Ala, Ser)

Which one is the correct sequence of the presence of this protein upon oligopeptide?

(JUNE 2013)

(a) Arg–Ala–Ser–Lys–Met–Phe–Ser–Ala

(b) Arg–Ala–Ser–Lys–Phe–Met–Ser–Ala

(c) Ala–Arg–Ser–Phe–Lys–Met–Ser–Ala

(d) Ala–Arg–Phe–Ser–Lys–Met–Ser–Ala

8. An unknown peptide was isolated from the leaf of a medicinal plant and purified to homogeneity. The peptide did not yield any sequence when subjected to Edman degradation. However, tryptic digest of the peptide produced a unique sequence. The mass of the intact peptide was found to be 18 Da less than that obtained from the trypsin treated sample. The possible interpretation of the above experimental results could be that the

A. the N–terminus of the peptide was blocked by acetylation or methylation.

B. the peptide was cyclic and contained at least one internal arginine or lysine residue.

C. the peptide was cyclic and contained a lysine or arginine residue at the C terminus.

D. the peptide was cyclized by peptide bond formation between α–amino group and α–carboxyl group.

Which of the above statements is true?

(DEC 2013)

(a) A and D (b) A and B

(c) B and C (d) B and D

9. A 30–residue peptide was treated with trypsin and the tryptic peptides were separated by HPLC. Four peaks A, B, C and D were obtained. Peptides corresponding to A, B, C and D were reduced and alkylated selectively at cysteine residues. The sequences obtained from A, B, C and D after reduction and alkylation were: A, AEK; B, C(S–alkyl)EPGYR and WC(S–alkyl)SPPK; C,C(Salkyl) EHFR and C(S–alkyl)GGK; D, C(S–alkyl)EAFC(S–alkyl)L. The sequence of the 30–residue peptide is **(JUNE 2011)**

(a) AEKCEPGYRCEHFRWCSPPKCGGKCEAFCL

(b) AEKCEPGYRCEHFRWCSPPKCGGKCEAFCL

(c) AEKCEPGYRCEHFRWCSPPKCGGKCEAFCL

(d) AEKCEPGYRCEHFRWCSPPKCGGKCEAFCL

10. The electrospray ionization spectrum of a mixture of two peptides show peak with m/z values 301, 401, 501 and 601. The molecular weight of the peptides are? **(DEC 2016)**

(a) 1200 and 1250 (b) 1200 and 1500

(c) 1350 and 1500 (d) 1220 and 1350

11. The electro spray ionization spectrum of a highly purified protein shows multiple closely spaced peaks. This most likely arises due to

(JUNE 2014)

(a) degradation of the protein during recording of the spectrum.

(b) the presence of multiple conformations.

(c) multiple charged species of the protein.

(d) extensive aggregation of the protein.

12. In which of the following techniques does molecular fragmentation offer clues to the covalent chemical structures of biomolecules?

(DEC 2012)

(a) MALDI–TOF MS spectrometry

(b) MALDI–TOF MS/MS spectrometry

(c) ESI–TOF MS mass spectrometry

(d) LC–couples ESI–TOF MS mass spectrometry

13. The MALDI spectrum of a peptide shows a peak at *m/z* corresponding to 3600. When the ESI spectrum is recorded, peaks at *m/z* corresponding to 721, 904 and 1801 were obtained. When the MALDI MS/MS spectrum was recorded, large number of peaks with *m/z* less than 3600 were observed. The spectral data indicate that the peptide is **(JUNE 2011)**

(a) highly impure.

(b) pure with molecular mass of 3600 and partial sequence of the peptide can be determined.

(c) highly unstable and degrades rapidly.

(d) degraded under condition employed for recording ESI spectrum.

14. The MALDI mass spectrum of a peptide gave a single peak with M/z of 2000. The ESI mass spectrum of the same peptide gave multiple peaks. These observations indicate that **(JUNE 2018)**

(a) degradation has occurred while acquiring ESI mass spectrum

(b) multiple charged species of the same compound are observed in the ESI spectrum

(c) the sample is impure

(d) ESI induces polymerization of the peptide

15. Highly purified peptides Pl, P2 and P3 were subjected to MALDI mass spectral analysis. The following observations were made:

P1: Showed a m/z of 16 more than the expected value.

P2: Showed a m/z of 80 more than the expected value. MS/MS spectra of the peptide resulted in a precursor ion with m/z 98 less than the expected m/z.

P3: Showed a m/z that was double the expected value. [Note: z = + 1 for all the mass spectra.] Which one of the options given below comprises all correct interpretations? **(DEC 2018)**

1. P1: Cys is oxidized; P2: has undergone oxidation at multiple Met residues; P3: is a non–covalent dimer.
2. P1; Met is oxidized; P2: is phosphorylated; P3: is a covalent dimer.
3. P1: Met is oxidized; P2: multiple Cys are oxidized; P3 is a covalent dimer.
4. P1: Cys is oxidized; P2: phosphorylated and oxidized at Met; P3: is a noncovalent dimer.

16. Given below are names of techniques (Column A) and their characteristic features/applications (Column B):

	Column A		Column B
A.	Hybridoma technology	(i)	Separation of technology proteins according to charge
B.	MALDI–TOF	(ii)	Identification of protein complexes in cells
C.	Ion–exchange chromatography	(iii)	Production of identical antibodies
D.	Co–immunoprecipitation	(iv)	Determination of molecular weight of proteins and/or peptides

Which one of the following represents a correct match between Column A and Column B: **(DEC 2018)**

(a) A – (ii) ; B – (iii) ; C – (iv) ; D – (i)

(b) A – (iii) ; B – (iv) ; C – (i) ; D – (ii)

(c) A–(iv); B–(i); C–(ii); D–(iii)

(d) A – (ii) ; B – (iv) ; C – (i) ; D – (iii)

17. From statements on protein structure and interactions detailed below, indicate the correct statement **(JUNE 2016)**

(a) The concentration of a tryptophan containing protein can be determined by monitoring the fluorescence spectrum of the protein.

(b) A peptide with equal number of Glu and Lys amino acids can show multiple charged species in its electrospray ionization mass spectrum.

(c) The circular dichroism spectrum of a protein shows predominantly helical conformation. Analysis of its two dimensional NMR spectrum shows predominantly β–structure.

(d) Binding constant can be determined by two interacting molecules by the technique of surface plasmon resonance only if there is strong hydrophobic interactions between them.

Answer Key

1.(a)	2.(a)	3.(d)	4.(d)	5.(b)	6.(c)	7.(c)
8.(d)	9.(c)	10.(b)	11.(c)	12.(b)	13.(b)	14.(b)
15.(b)	16.(b)	17.(b)				

12. Circular Dichroism spectroscopy

1. A and B are two enantiomeric helical peptides. Their chirality can be determined by recording their **(JUNE 2016)**

(a) CD spectrum.

(b) UV spectrum.

(c) Edman sequencing

(d) Fluorescence spectrum.

2. Two 18–residue helical peptides A and B are enantiomers. They can be distinguished by **(DEC 2014)**

(a) recording their MALDI mass spectrum.

(b) hydrolysis followed by amino acid analysis

(c) sequencing by Edman's method.

(d) examining their circular dichoraism spectra

3. An optical measurement of protein is taken both before and after digestion of the protein by a protease. In which of the following spectroscopic measurement the signal change, i.e., before v/s after protease treatment, could be the maximum? **(DEC 2016)**

(a) Absorbance at 280 nm

(b) Circular dichromism

(c) Absorbance at 340nm

(d) Fluorescence value

4. Denaturation of a highly helical protein having disulfide bridges and two phenylalanines can be monitored as a function of temperature by which one of the following techniques? **(JUNE 2017)**

(a) Recording circular dichroism spectra at various temperatures

(b) Monitoring the absorbance at 214 nm at various temperatures

(c) Estimating the –SH content during heat denaturation

(d) Monitoring the ratio of absorbance at 214 nm and at 250 nm at various temperatures

5. Which one of the following statements regarding proteins is **CORRECT**? **(DEC 2017)**

(a) $n \rightarrow \sigma^*$ transition requires less energy than $n \rightarrow \pi^*$ transition and can be monitored by mass spectrometry.

(b) $n \rightarrow \pi^*$ transition requires more energy than $\pi \rightarrow \pi^*$ transition and can be monitored by UV–V is spectroscopy.

(c) $n \rightarrow \pi^*$ transition requires less energy than $n \rightarrow \sigma^*$ transition and can be monitored by CD spectroscopy.

(d) $\sigma \rightarrow \sigma^*$ transition requires less energy than $n \rightarrow \pi^*$ transition and can be monitored by CD spectroscopy

6. The amino acid alanine has high propensity to occur in helical conformation. The Circular dichroism spectrum of an equimolar mixture of two 20–residue peptides, one composed of only L–alanine and the other only D–alanine is recorded in the region of 185–250 nm. Which one of the following will be observed? **(JUNE 2013)**

(a) No signal: as the chiroptical properties of the two peptide will cancel out.

(b) Bands with only negative ellipticity; as helix formed by D–Ala peptide will be unstable

(c) Bands with only positive ellipticity; as both the peptides will form right handed helices

(d) Bands with identical negative and positive ellipticity.

7. The circular dichroism spectrum of a polypeptide composed of 50 amino acids does not show any signal in the region of 185–250 nm. The reason is: **(DEC 2013 GU)**

(a) the polypeptide is composed of only achiral amino acids

(b) there is no helix or β sheet conformation, only beta–turns are present

(c) the polypeptide is in random conformation

(d) the % of helix and β sheet are equal

8. The structure of a protein is known from X–ray diffraction studies which gave 30% α– helix, 50% β–sheet and 20% random coil. Circular dichroism (CD) measurements gave 50% α–helix, 40% β–sheet and 10% random coil. What could NOT be a possible explanation for these observations. **(MP)**

(a) Protein structure in the crystal is different from that in the solution.

(b) CD analysis for structural components is not appropriate for this protein.

(c) Contributions from other chromophores also contribute to the CD spectrum of the protein

(d) Protein contains high content of disulphide bonds.

9. Two homologous proteins were isolated from a psychrophile (P) and a thermophile (T). The purified proteins were subjected to denaturation, protease digestion and circular dichroism (CD).

Following observations were made:

A. The CD spectra of P and T proteins are identical

B. Their amino acid composition is 95% identical

C. T and P are equally susceptible to proteolysis in the presence or absence of reducing agent

D. T has higher midpoint of thermal denaturation than P

The reason for enhanced stability in T is due to **(DEC 2014)**

(a) Altered secondary structure

(b) Increased number of disulfides in T

(c) Increase in water of hydration

(d) Increase in number of salt bridges

10. Which one of the following statements is correct? **(DEC 2015)**

(a) Electrospray ionization mass spectrum of a compound can be obtained only if it has a net positive charge at pH 7.4

(b) Helical content of a tryptophan containing peptide can be obtained by examining the fluorescence spectrum of tryptophan

(c) The occurrence of beta sheet in a protein can be inferred from its circular dichroism spectrum

(d) The chemical shift spread for a compound is more in its ^{1}H NMR spectrum as compared to its ^{13}C NMR spectrum

11. A 26–residue peptide composed of alanine and leucine shows a circular dichroism (CD) spectrum characteristic of α–helix at 50°C in 5 mM phosphate buffer at pH 7.4. Deconvolution of the spectrum indicates 60% α–helical and 40% random conformation. When the peptide solution is cooled gradually to 25°C, and the CD spectra are recorded at different temperatures, the most likely observation will be that **(DEC 2013)**

(a) the % helical content will decrease and % random conformation will increase.

(b) the % helical content will increase and % random conformation will decrease.

(c) there will be transition from α–helix to β– sheet.

(d) there will be transition from α–helix to β–hairpin.

12. A protein polypeptide chain exists in α–helical conformation in a solvent and it has a value of

–30,000 deg cm^2 dmol^{-1} for the mean residue ellipiticity at 222 nm ($[\theta]_{222}$) in the temperature range 20–50 ^{0}C. On raising the temperature above 50^{0}C, $[\theta]_{222}$ increases and reaches a value of –2,000 deg cm^2 dmol–1 at 70^{0}C and the value of $[\theta]_{222}$ remains unchanged above 70^{0}C. The observed value of $[\theta]_{222}$ is –14,000 deg cm^2 dmol–1 at 60^{0}C. If one assumes that the heat–induced denaturation is a two–state process, the fraction of α– helix at 60^{0}C is

(MODEL PAPER)

(a) 0.40 (b) 0.43
(c) 0.50 (d) 0.57

13. At 25°C values of $[\theta]_{222}$, the mean residue ellipticity at 222 nm, are – 33,000 and –3,000 deg cm^2 dmol^{-1} for a polypeptide existing in α–helical (α) and β–structure (β), respectively. If this polypeptide undergoes a two–state heat–induced $\alpha \rightarrow \beta$ transition, and a value of $[\theta]_{222}$ = –18,000 deg cm^2 dmol^{-1} is observed at 60°C, then this observation leads to the conclusion that the α helix conversion to β–structure:

(DEC 2013)

(a) 40% (b) 50%
(c) 55% (d) 60%

14. Poly–L–lysine exists in pure α–helix, β–sheet and random coiled conformation depending upon the solvent conditions. Tile values of mean residue ellipticity at 220 nm ($[\theta]_{220}$) are –35,700, –13,800 and +3,900 deg cm^2 dmol^{-1} for α–helix, β–sheet and random coil conformations of this polypeptide, respectively. The polypeptide exists in α–helix conformation at pH 10.8 and 25^0 C. Addition of urea leads to a two state transition between α–helix and random coil conformation. It has been observed that $[\theta]_{222}$ of the polypeptide is –14800 degcm^{2}dmol^{-1}in the presence of 6M urea.

The percentage of the polypeptide in α–helix conformation is: **(JUNE 2013)**

(a) 37 (b) 41
(c) 47 (d) 50

15. Following observations are–made regarding a peptide sequence.

– The peptide is inert to Ellman's reagent. However, on reacting with β–mercaptoethanol, the peptide gives a positive Ellman's test.
– The peptide sequence gives a broad minimum around 211 nm in the CD spectrum.
– With increasing concentration of the peptide, the melting temperature of the peptide increases.
– On treating the peptide with D_2O, half the total number of amides get exchanged.

Which one of the following statements is correct? **(DEC 2017)**

(a) It is an α–helical peptide that undergoes aggregation
(b) It is an α–helical disulfide bridged peptide that undergoes aggregation
(c) It is an β–hairpin peptide, which is stabilized by a disulfide bridge
(d) The peptide is composed of an α–helix and β–sheet connected by a disulfide bridge

16. A polymer is synthesized from an achiral amino acid. Conformation of the polymer can be investigated by the following techniques.

A. Fibre diffraction
B. Nuclear magnetic resonance spectroscopy
C. Circular dichroism spectroscopy.
D. Differential scanning calorimetry

Choose the combination which would indicate that the polymer adopts a helical conformation

(DEC 2017)

(a) A, C (b) B, D
(c) A, B (d) C, D

17. Haemoglobin has characteristic circular dichroism (CD) peaks C in the far–UV, near UV and Soret regions. Contribution to near–UV CD comes entirely from **(DEC 2013)**

(a) aromatic amino acid residues.
(b) heme group.
(c) heme and aromatic amino acid residues.
(d) peptide bonds and aromatic amino acid residues

18. Cytochrome–c has only one tryptophan residue (W) which is buried. The protein in cacodylate buffer (pH 6.0) is excited at 280 nm, and its emission spectrum measured in the range of 300–450 nm. The same measurement was repeated on the protein in the buffer containing 6M guanidine hydrochloride. It was observed that there is an increase in the intensity of the emission spectrum of the guanidine hydrochloride treated cytochrome–c. The most probable reason for this increase is: **(JUNE 2014)**

(a) W is near a hydrophobic patch present in the unfolded protein.
(b) W is near heme in the native protein
(c) W is near carboxylate amino acid side chains in the native protein.
(d) W is in a polar pocket in the native protein.

19. Researcher is investigating structural changes in protein by following tryptophan fluorescence and by Circular dichorism. Fluorescence and CD spectra of pure protein were obtaining the absence of any treatment (a), in the presence of 0.5 molar urea (b), upon adding acrylamide, a quencher of tryptophan (c) and upon heating (d) the data are shown below.

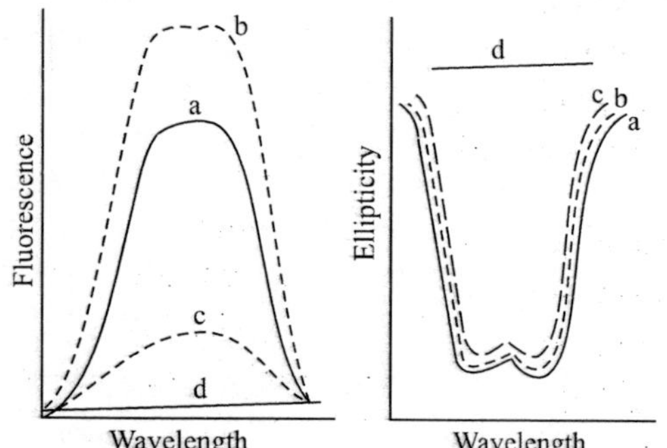

Which one of the following statements is correct? **(DEC 2016)**

(a) CD is more sensitive to structural changes than fluorescence
(b) Fluorescence is more sensitive to structural changes than CD
(c) Both the methods are equally responsive to structural changes
(d) Acrylamide alter the secondary structure of the protein

20. From the following statements,

A. Surface plasmon resonance can be used to determine binding constants only in the range of $10^2 - 10^3$ M.
B. *de novo* sequencing is not possible by mass spectral methods.
C. The position of hydrogen atoms in proteins is not directly determined by X–ray, diffraction.
D. Circular dichroism and nuclear magnetic resonance spectroscopy do not give the same information on protein structure.

Choose the option with all correct statements. **(JUNE 2018)**

(a) A, B, C (b) A, C, D
(c) B, D (d) C, D

21. Protein conformational dynamics **CANNOT** be determined by which one of the following techniques **(JUNE 2018)**

(a) NMR spectroscopy
(b) Differential scanning calorimetry
(c) Mass spectroscopy
(d) Fluorescence microscopy

Answer Key

1.(a)	**2.**(d)	**3.**(b)	**4.**(a)	**5.**(c)	**6.**(a)	**7.**(a)
8.(b)	**9.**(d)	**10.**(c)	**11.**(b)	**12.**(b)	**13.**(b)	**14.**(c)
15.(c)	**16.**(c)	**17.**(c)	**18.**(b)	**19.**(b)	**20.**(d)	**21.**(b)

13. Carbohydrate Metabolism and TCA

1. What is the effect of sudden increase in the levels of ATP and citrate on an erythrocyte undergoing glycolysis? **(DEC 2017)**
   (a) It inhibits glycolysis.
   (b) It stimulates glycolysis.
   (c) The rate of glycolysis remains unaltered.
   (d) The rate of glycolysis increases gradually.

2. Given below are statements that may or may not be correct.
   A. Fructose 2, 6– biphosphate is an allosteric inhibitor of phosphofructokinase–I.
   B. The TCA cycle intermediates, succinate and oxaloacetate can both be derived from amino acids.
   C. A diet rich in cysteine can compensate for a methionine deficient diet in humans.
   D. dTTP for DNA synthesis can be obtained from UTP.
   E. In the fatty acid biosynthetic pathway, the carbon atom from HCO_3^- in the synthesis of malonyl CoA is not incorporated into palmitic acid.

   Choose the option that represents the combination of all the CORRECT statements. **(JUNE 2017)**
   (a) A, B , C and E (b) B, D and E
   (c) A, D and E (d) Only B and C

3. The following are the statements about pyruvate kinase (PK). **(DEC 2015)**
   A. ATP is an allosteric inhibitor of PK
   B. Fructose 1,6 bisphosphate is an activator of PK
   C. ADP is an allosteric inhibitor of PK
   D. Alanine is an allosteric modulator of PK

   Which of the above statement(s) are true?
   (a) A, B, C (b) A, B, D
   (c) B, C, D (d) only A

4. The following are some statements regarding glycolysis: **(DEC 2018)**
   A. Glycolysis is not regulated by pyruvate kinase.
   B. Lactate can be an end product of glycolysis.
   C. Glycolysis cannot function anaerobically.
   D. In erythrocytes, the second site in glycolysis for ATP generation can be bypassed.

   From the above, choose the combination with both **INCORRECT** statements:
   (a) A and B (b) B and D
   (c) C and D (d) A and D

5. The glycolysis and citric acid cycles are important pathways to generate energy in the cell. Given below are statements regarding the production of ATP.
   A. Electrons released during the oxidative steps of glycolysis and citric acid cycle produce 10 molecules of NADH and 2 molecules of, $FADH_2$ per molecule of glucose.
   B. Electrons released during the oxidative steps of glycolysis and citric acid cycle produce 20 molecules of NADH and 4 molecules of F ADH_2 per molecule of glucose.
   C. The coenzymes produced are oxidized by electron transfer chain.
   D. The conversion of ADP and Pi to ATP takes place in the intermembrane space of mitochondria.

   Which one of the following combinations of above statements is correct? (JUNE 2015)
   (a) A and B (b) B and C
   (c) C and D (d) A and C

6. AsO_4^{3-} is chemically very similar to PO_4^{3-} and could be used as an alternate substrate by phosphate requiring enzymes. When 1,3–bisphospho–glycerate is converted to 3–phosphoglycerate, A TP is generated. However, ATP is not formed when 1–arseno–3–phosphoglycerate is converted to 3–phosphoglycerate. The reason for this could be **(DEC 2013 GU)**

(a) The bond between arsenate and carboxylic acid can be hydrolyzed only at 50°C.

(b) 3–phosphoglycerate is formed non–enzymatically due to rapid hydrolysis of the bond between arsenate and carboxylic acid group.

(c) Enzymes cannot hydrolyze the bond between arsenate and carboxylic acid.

(d) The bond between arsenate and carboxylic acid can be hydrolyzed only by conversion of GDP to GTP.

7. What are A, B and C in the following reactions? **(DEC 2013)**

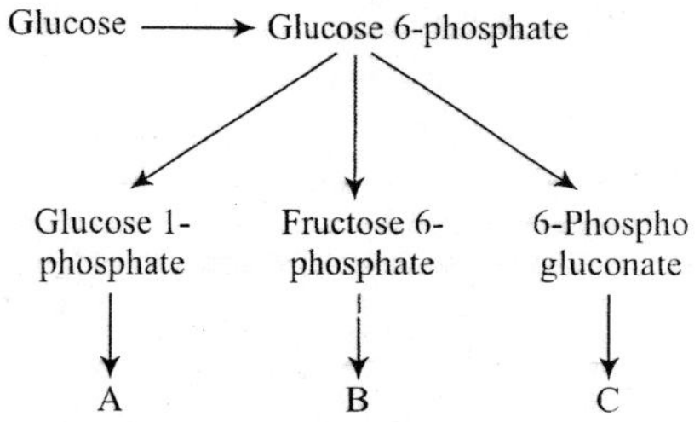

(a) Pyruvate, ribose 5–phosphate, glycogen.

(b) Ribose 5–phosphate, glycogen, pyruvate.

(c) Glycogen, pyruvate, ribose 5–phosphate.

(d) Glycogen, citrate, ribose 5–phosphate.

8. Which one of the following enzymes is NOT a part of pyruvate dehydrogenase enzyme complex in linkage step of glycolysis and TCA cycle? **(JUNE 2015)**

(a) Pyruvate dehydrogenase.

(b) Dihydrolipoyl transferase.

(c) Dihydrolipoyl dehydrogenase.

(d) Dihydrolipoyl oxidase.

9. During respiration, which of the following processes occur only inside mitochondria and not cytoplasm? **(DEC 2012)**

(a) Glycolysis and the pentose–phosphate pathway.

(b) Glycolysis and the citric acid cycle.

(c) The citric acid cycle and oxidative phosphorylation.

(d) Glycolysis and oxidative phosphorylation.

10. The citric acid cycle in respiration yields: **(DEC 2012)**

(a) 1 GTP, 3 NADH, 1 $FADH_2$ 2 CO_2

(b) 2 GTP, 2 $FADH_2$, 6NADH, 2CO_2

(c) 4GTP, 6NADH, 4$FADH_2$, 2CO_2

(d) 32 GTP, 2 NADH, 4 $FADH_2$, 4 CO_2

11. The energy–rich fuel molecules produced in the TCA cycle are **(JUNE 2017)**

(a) 2 GTP, 2 NADH and 1 F ADH_2

(b) 1 GTP, 2 NADH and 2 $FADH_2$

(c) 1 GTP, 3NADH and 1 $FADH_2$

(d) 2 GTP and 3 NADH

12. Indicate the names of the following molecules **(JUNE 2016)**

(A)
COO^-
|
CH_2
|
$HO—C—COO^-$
|
CH_2
|
COO^-

(B)
COO^-
|
CH_2
|
$CH—COO^-$
|
$HO—CH$
|
COO^-

(C)
COO^-
|
CH_2
|
CH_2
|
$C=O$
|
COO^-

(D)
COO^-
|
$C=O$
|
CH_2
|
COO^-

(a) A= isocitrate, B = α–ketoglutarate, C = oxaloacetate, D = citrate

(b) A= citrate, B = isocitrate, C = α – ketoglutarate, D = oxaloacetate

(c) A= isocitrate, B = citrate, C = α – ketoglutarate, D = oxaloacetate

(d) A= citrate, B = isocitrate, C = oxaloacetate, D = α – ketoglutarate

13. The molecule citrate, isocitrate, malate and succinate take part in the citric acid cycle. Identify the structure with correct name: **(DEC 2019 ASSAM)**

(a) $^-OOC—CH_2—C(COO^-)(OH)—CH_2—COO^-$: Isocitrate

(b) $^{-}OOC-CH_2-\underset{H}{\overset{COO^-}{C}}-\underset{OH}{\overset{H}{C}}-COO^-$: Citrate

(c) $^{-}OOC-CH_2-\underset{H}{\overset{OH}{C}}-COO^-$: Malate

(d) $^{-}OOC-CH_2-CH_2-\overset{O}{\overset{||}{C}}-COO^-$: Succinate

14. The following reactions are part of the citric acid cycle. The numbers in parenthesis indicate the number of carbon atoms in each molecule.

Isocitrate(6) $\xrightarrow{A}$ α–ketoglutarate(5) $\xrightarrow{B}$ succinyl CoA(d) $\xrightarrow{C}$ succinate(d) $\xrightarrow{D}$ fumarate(d)

Which of the following sequences of the reaction systems A→D is correct? **(JUNE 2012)**

(a) $NAD^+ \rightarrow NADH+H^+, \rightarrow NAD^+, CO_2 \rightarrow NADH+H^+$, GDP, $CO_2 \rightarrow$ GTP, FAD ,iP $\rightarrow FADH_2$

(b) $NAD^+ \rightarrow NADH+H^+$, NAD^+, $CO_2 \rightarrow NADH+H^+$, ADP, $CO_2 \rightarrow$ATP, FAD, iP $\rightarrow FADH_2$

(c) $NAD^+ \rightarrow NADH+H^+$, FAD^+, $CO_2 \rightarrow FADH_2$, ADP $\rightarrow$ ATP, NAD^+, iP $\rightarrow NADH+H^+, CO_2$

(d) $NAD^+ \rightarrow NADH+H^+, FAD^+, CO_2 \rightarrow FADH_2$, GDP→GTP,$NAD^+$,iP $\rightarrow NADH+H^+, CO_2$

15. In the following transformations,

Succinate + X ⇌ fumarate

Succinate + Y ⇌ oxaloacetate

Malate + Y ⇌ oxaloacetate

Pyruvate + Z ⇌ oxaloacetate

the correct combination of X, Y, Z is: **(DEC 2019 ASSAM)**

(a) X=FAD, Y=NAD, Z=ATP

(b) X= ATP, Y=FAD, Z= NAD

(c) X=NAD, Y=FAD, Z=ATP

(d) X=NAD, Y= ATP, Z=FAD

16. The first step in glycogen breakdown releases glucose units as **(JUNE 2019)**

(a) glucose 6– phosphate

(b) glucose 1– phosphate

(c) glucose

(d) glucose and glucose 6– phosphate

Answer Key

1.(a)	2.(b)	3.(b)	4.(d)	5.(d)	6.(b)	7.(c)
8.(d)	9.(c)	10.(a)	11.(c)	12.(b)	13.(c)	14.(a)
15.(a)	16.(b)					

14. Lipids, Amino acids and Nucleotide Metabolism

1. Which of the following ii NOT true for cholesterol metabolism? **(JUNE 2016)**
   (a) HMG–CoA reductase is the key regulator of cholesterol biosynthesis.
   (b) Biosynthesis takes place in the cytoplasm.
   (c) Reduction reactions use NADH as cofactor.
   (d) Cholesterol is transported by LDL in plasma.
2. Sting of a bee causes pain, redness and swelling. Melittin is a major peptide in bee venom. Melittin is a membrane binding peptide that is involved in activating phospholipases in the membrane. The possible target phospholipase that is activated by melittin is **(JUNE 2015)**
   (a) Phospholipase C to generate inositol phosphates.
   (b) Phospholipase A_2 to generate arichidonic acid.
   (c) Phospholipase D to generate 1', 3'– inisitol.
   (d) Phospholipase A_1 to generate palmitic acid
3. Biosynthesis of tyrosine is detailed below:
   Shikimic acid –A→ shikimic acid–5–phosphate –B →C→ chorismic acid → prephenic acid→D → transaminase →tyrosine.
   Identify A, B, C and D **(DEC 2011)**

(a) ATP, phosphoenolpyruvic acid, 3–enolpyruyl shikimIc acid–5–phosphate, p–hydroxyphenylpyruvic acid

(b) GTP, pyridoxal phosphate, 3–enolpyruvyl shikimic acid–5phosphate, phenylpyruvic acid

(c) NADP, 3–phosphohydroxypyruvic acid, 3–enolpyruvic shikimic acid–5–phosphate, p–hydroxyphenylpyruvic acid

(d) ATP, 3–phosphohydroxypyruvic acid, 3–enolpyruvyl shikimic acid–5–phosphate, pyridoxylphosphate

4. Choose the correct answer from the following statements on biosynthesis. **(DEC 2018)**

(a) In the biosynthesis of palmitate, all the carbon atoms are derived from activated malonate.

(b) The amino acids Met, Thr, Lys, Ile, Val and Leu are biosynthesized from oxaloacetate and pyruvate in most bacteria.

(c) Alanine is a major precursor for the biosynthesis of porphyrin.

(d) Tryptophan is converted to L–DOPA in the biosynthesis of epinephrine.

5. The first few steps of the tryptophan biosynthetic pathway in plants along with the available mutants is provided below

$$\text{Chlorismate} \xrightarrow{trp\,4} \text{Anthranilate} \xrightarrow{trp\,1}$$

$$\text{PR} - \text{anthranilate} \xrightarrow{pai} \text{CDRP}$$

Column X represents genotypes and Column Y summarizes the anthranilate fluorescence phenotype in UV light.

Column X	Column Y
Genotype	**Phenotype in UV Light**
A–wild type	i. Strong fluorescence
B–*trp*1	ii. No fluorescene
C–*trp*1; *trp* 4	iii. Weak fluorescence
D–*pai*	

Which of the following combinations represent the correct match between column X and column Y? **(DEC 2019 ASSAM)**

(a) A–i, B–ii, C–i, D–iii

(b) A–ii, B–i, C–ii, D– iii

(c) A–ii, B–i, C–iii, D–i

(d) A–iii, B–i, C–iii, D–ii

6. The inborn error of amino acid metabolism, alkaptonuria. is due to the lack of one of the following enzymes: **(JUNE 2019)**

(a) Fumaryl acetoacetate hydrolase

(b) α–keto acid decarboxylase

(c) Hornogentisate oxidase

(d) p–hydroxyphertylpyruvate dehydroxylase

7. A cell line deficient in salvage pathway for nucleotide biosynthesis was fed with medium containing ^{15}N labelled amino acids. Purines were then extracted. Treatment with which one of the following amino acids is NOT likely to produce ^{15}N labelled purines? **(DEC 2015)**

(a) Aspartic acid (b) Glycine

(c) Glutamine (d) Aspartamine

8. In the biosynthesis of purine: **(DEC 2016)**

6 C, 7 N, 1N, C^5, C^8, 2C, C^4, N 3, N 9

(a) All N atoms, C4 and C5 from Aspartic acid

(b) N1 is from aspartic acid; N3 and N9 are from Glutamine side– chain; N7, C4 and C5 are from Glycine

(c) N1 is from Aspartic acid; N3 from Glutamine side– chain ; N9 from N attached to Cα of Glutamine; N7, C4 and C5 from Glycine

(d) N1 is from Glutamine; N3 from Glutamine side– chain; N9 from N attached to Cα of Glutamine; N7, C4 and C5 from Glycine

9. Precursors of the atoms in the purine skeleton are **(JUNE 2011)**

6 C, 7 N, 1N, C^5, C^8, 2C, C^4, N 3, N 9

(a) N1, Asp; C2 and C8, formate; N3 and N9 guanidine C4, C5 and N7, Gly; C6, CO_2.

(b) N1, Asp; C2 and C8, citrate; N3 and N9, amide nitrogen of Gln, C4,,C5 and N7; Gly; C6, CO_2.

(c) N1, Asp; C2 and C8, formate; N3 and N9 amide nitrogen of Gln, C4, C5 and N7, Gly; C6, CO_2.

(d) N1, Glu; C2 and C8, acetate; N3 and N9, amide nitrogen of Asn; C4, C5 and N7, Gly; C6, CO_2.

Answer Key

1.(c)	2.(b)	3.(a)	4.(b)	5.(b)	6.(c)	7.(d)
8.(b)	9.(c)					

15. Red–Ox reaction, ETC and ATP synthesis

1. Three electron acceptors 'X', 'Y' and 'Z' have redox potential (E_0') of +0.15V, +0.05V and –0.1V, respectively.

   For a reaction

   $B + 2H + 2e^- \rightarrow BH_2 \quad E_0' = +0.05V$

   Which of these three electron acceptors are appropriate? **(JUNE 2017)**

   [useful equation: $\Delta G_0' = -nFE_0'$]

   $\Delta G_0'$ = free energy change;

   n = number of electron

   F = Faraday constant

   (a) X and Y (b) Only X

   (c) Y and Z (d) Only Z

2. Excess oxygen consumed after a vigorous exercise is **(DEC 2016)**

   (a) to pump out lactic acid from muscle

   (b) to increase the concentration of lactic acid in muscle

   (c) to reduce dissolved carbon dioxide in blood

   (d) to make ATP for gluconeogenesis

3. Proton motive force during oxidative phosphorylation is generated in mitochondria by **(DEC 2014)**

   (a) exchanging protons for sodium ions

   (b) pumping protons out into intermembrane space

   (c) pumping hydoxyl ions into the mitochondria

   (d) hydrolysis of ATP

4. Coupling of the reaction centers of oxidative phosphorylation is achieved by which one of the following? **(DEC 2015)**

   (a) Making a complex of all four reaction centers.

   (b) Locating all four complexes in the inner membrane.

   (c) Ubiquinones and cytochrome C.

   (d) Pumping of protons.

5. The respiratory chain is relatively inaccessible to experimental manipulation in intact mitochondria. Upon disrupting mitochondria with ultrasound, however, it is possible to isolate functional sub mitochondrial particles, which consist of broken cristae that have resealed inside out into small closed vesicles. In these vesicles the components that originally faced the matrix are now exposed to the surrounding medium. This arrangement helps in studying electron transport and ATP synthesis because: **(JUNE 2012)**

   (a) it is difficult to manipulate the concentration of small molecules(NADH, ATP, ADP, Pi) in the matrix of intact mitochondria

   (b) in broken cristae ,the enzymes and other molecules responsible for electron transport are more active

   (c) intact mitochondria are more unstable than broken cristae

   (d) purification of intact mitochondria is not possible

6. In a mitochondrial respiration experiment, a researcher observed the following profile of oxygen consumption upon addition of following compounds at times I, II and III.

(a) ADP + Pi
(b) Dinitrophenol, an uncoupler
(c) Oligomycin, an ATPase inhibitor
(d) Cyanide
(e) Succinate

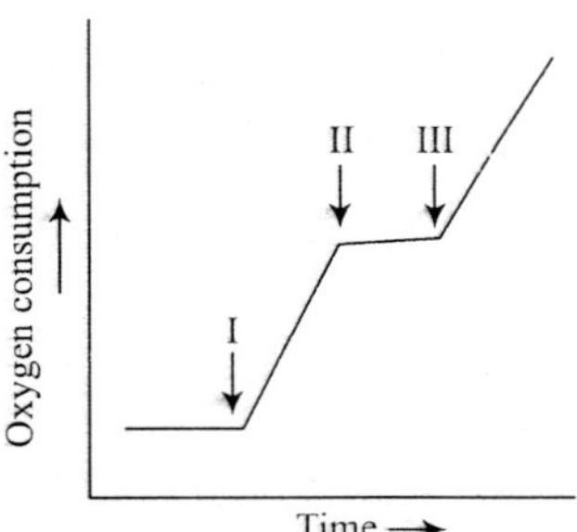

Which of the following describes the profile appropriately? **(JUNE 2016)**

(a) I – b; II – d; III – e
(b) I – a; II – d; III – c
(c) I – a; II – e; III – c
(d) I – a; II – c; III – b

7. Rotenone is an inhibitor of the electron transport chain. The addition of rotenone to cells results in which of the following? **(JUNE 2017)**
   (a) Generation of mitochondrial reactive oxygen species and block in ATP generation.
   (b) Block in ATP generation but no generation of reactive oxygen species.
   (c) Generation of reactive oxygen species but no block in ATP generation.
   (d) Permeabilization of the inner membrane to compounds which are usually not able to traverse the membrane.

8. A practical class was going on where the students were demonstrating ATP synthesis *in vitro* using active mitochondria. Some students added one of the following to their tubes
   A. Dinitrophenol (DNP), an uncoupler
   B. Mild acidification of the medium
   C. Glutilferone, that permeabilizes both the membranes
   D. An outer membrane permeable H^+ quencher compound, Elila

   In which one of the above, ATP synthesis will be detected? **(DEC 2015)**

   (a) A (b) B
   (c) C (d) D

9. Match the chemical agents that interfere in oxidative phosphorylation process with their respective mode of action.

	Column I		Column II
(A)	Antimycin A	(i)	Inhibits Fo component of ATP synthase
(B)	Oligomycine	(ii)	Disrupts inner mitochondrial membrane potential
(C)	Valinomycine	(iii)	Prevent electron transport from Fe/S cluster to ubiquinone
(D)	Rotenone	(iv)	Blocks electron transfer from cytochrome b to cytochrome C_1
		(v)	Inhibits adenine nucleotide translocase

Choose the correct combination from below: **(DEC 2017)**

(a) A–(ii), B–(iv), C–(v), D–(iii)
(b) A–(iv), B–(i), C–(ii), D–(iii)
(c) A–(i), B–(iii), C–(ii), D–(v)
(d) A–(v), B–(ii), C–(i), D–(ii)

10. From the following statements:
    A. Biosynthesis of proteins and nucleic acids from precursors results in production of chemical energy in the form of ATP, NADH, NADPH and $FADH_2$,
    B. The spontaneity of a reaction in cells does not depend whether ΔG_0 for the reaction is positive or negative.
    C. Both oxidative phosphorylation and photo–phosphorylation involve oxidation of H_2O to O_2
    D. Only chemical potential energy contributes to proton motive force in mitochondria.

Which one of the following combinations represents all INCORRECT statements? **(JUNE 2018)**

(a) A, B, C (b) B, C, D
(c) A, B, D (d) A, C, D

11. Which of the following proteins acts as an energy transducer? **(MODELPAPER)**
(a) G–protein. (b) Bacteriorhodopsin.
(c) Hemoglobin. (d) Heat shock protein.

Answer Key

1.(a)	**2.**(d)	**3.**(b)	**4.**(c)	**5.**(a)	**6.**(d)	**7.**(a)
8.(b)	**9.**(b)	**10.**(d)	**11.**(b)			

SECTION-2

CELL BIOLOGY

1. Membrane Structure and Lipid Rafts

1. Most common type of phospholipid in the cell membrane of nerve cells is (DEC 2011)
   (a) phosphatidylcholine
   (b) phosphatidylinositol
   (c) phoshphatidylserine
   (d) sphingomylein

2. Phosphotidyl serine, an important component of biological membrane. is located in (JUNE 2012)
   (a) the outer leaflet but flip flop to inner
   (b) both leaflets of plasma membrane
   (c) middle of the bilayer
   (d) the inner leaflet but flip flops to outer leaflets under specific conditions

3. Glycophorin having one highly hydrophobic domain is able to span a phospholipid bilayer membrane only (DEC 2015)
   (a) once (b) twice
   (c) thrice (d) four times

4. If one of the two fatty acyl chains is removed from– the phosphoglyceride by hydrolysis in solution, such phospholipids will form: (DEC 2018)
   (a) liposomes
   (b) micelles
   (c) phospholipid bilayer
   (d) symmetric phospholipid bilayer

5. Which of the following predicted property of lipid bilayers would result if the phospholipids had only one hydrocarbon chain instead of two? (MODELPAPER)
   (a) The bilayers formed would be much less fluid.
   (b) The diameter of the head group would be much larger than the acyl chain and would tend to form micelles rather than bilayers.
   (c) The bilayers formed would be much more fluid.
   (d) The bilayers would be more permeable to small water–soluble molecules.

6. The gel to liquid crystalline phase trasition temperature in phosphatidyl choline (PC) lipids compose of dioleoyl (DO), dipalmitoyl (DP), disteroyl (DS) and palmitoyl oleoyl (PO) fatty acids in increasing order will be (DEC 2016)
   (a) DOPC > DPPC > POPC > DSPC
   (b) DSPC > DPPC > POPC > DOPC
   (c) DPPC > DSPC > DOPC > POPC
   (d) POPC > DPPC > DOPC > DSPC

7. The gel to liquid crystalline transition temperature (Tm) of phospholipids is dependent on the fatty acid composition. Considering this, Tm of (JUNE 2013)
   (a) all the phospholipids will be identical.
   (b) DPPC will be lowest and DOPC will be highest.
   (c) POPC and DOPC will be identical and lower than DMPC or DPPC.
   (d) DOPC will be lowest and DPPC will be highest.

8. When bacteria growing at 20°C are warmed at 37°C, they are most likely to synthesize membrane lipids with more (JUNE 2015)
   (a) short chain saturated fatty acids
   (b) short chain unsaturated fatty acids
   (c) long chain saturated fatty acids
   (d) long chain unsaturated fatty acids

9. The membrane lipid molecules assemble spontaneously into bilayers when placed in water and form a closed spherical structure known as (DEC 2011)
   (a) Lysosome (b) Peroxisome
   (c) Liposome (d) Endosome

10. Major disadvantage of using liposome as targeted drug delivery vehicle is that (JUNE 2012)
   (a) It get internalized by phagocytosis inside lysosome
   (b) It is very unstable and has low shelf life
   (c) It get intercalated in cell membrane
   (d) Its drug entrapment efficiency is very low

11. The membrane phospholipid structures in bacteria and archaea differ. Which one of the following correctly states the differences between the two? **(DEC 2019)**

(a) The bacterial membrane phospholipid consist of D–glycerol linked to hydrophobic chains (tails) with ester bonds whereas those of the archaeal membranes consist of L–glycerol linked to hydrophobic tails through ether bonds.

(b) The bacterial membrane phospholipid consist of L–glycerol linked to hydrophobic chains (tails) with ester bonds whereas those of the archaeal membranes consist of D–glycerol linked to hydrophobic tails through ether bonds.

(c) The bacterial membrane phospholipid consist of D–glycerol linked to hydrophobic chains (tails) with ether bonds whereas those of the archaeal membranes consist of L–glycerol linked to hydrophobic tails through ester bonds.

(d) The bacterial membrane phospholipid consist of L–glycerol linked to hydrophobic chains (tails) with ether bonds whereas those of the archaeal membranes consists of D–glycerol linked to hydrophobic tails through ester bonds.

12. Both sphingomyelin and phosphoglycerides are phospholipids. Which one of the following statements is NOT correct? **(JUNE 2016)**

(a) While one has a fatty acid tail attached via an ester bond, in another, the fatty acid tail is attached via an amide bond.

(b) The hydrophilicity of both is dependent on the phosphate group and other head groups attached to the phosphate group.

(c) Only one of them may contain a carbon–carbon double bond (C=C).

(d) Both may have choline as head group.

13. Molecules primarily responsible for the formation of lipid raft are **(JUNE 2014)**

(a) phosphatidyl serine and phosphatidyl choline.

(b) phosphatidyl inositol and cholesterol.

(c) glycosylphosphatidyl inositol and cholesterol.

(d) sphingolipids and cholesterol.

14. Lipid rafts are rich in both sphingolipids and cholesterol. Cholesterol plays a central role in raft formation since lipid rafts apparently do not form ill its absence. Why do you think cholesterol is essential for the formation of lipid rafts? **(DEC 2015)**

(a) Cholesterol decreases the mobility of sphingolipids in the lipid bilayer.

(b) Large head groups of sphingolipids repel each other in presence of cholesterol.

(c) Cholesterol interacts with fatty acid tails in the membrane.

(d) The planar cholesterol molecules are postulated to fill the voids that form underneath the large head groups of the sphingolipids.

15. Lipid rafts are involved in signal transduction in cells. Rafts have composition different from rest of the membrane. Rafts were isolated and found to have cholesterol to sphingolipid ratio of 2:1. The estimated size of the raft is 35 nm^2. If the surface areas of cholesterol is 40 A^0 and sphingolipid is 60 A^0, how many cholesterol and sphingolipids are present in one raft?

(JUNE 2015)

(a) 50 cholesterol; 25 sphingolipid

(b) 200 cholesterol: 100 sphingolipid

(c) 40 cholesterol : 20 sphingolipid

(d) 20 cholesterol: 10 sphingolipid

16. Sphingolipids (S) and cholesterol (C) molecules of the lipid bilayer aggregate into multiple tiny rafts instead of a single large one. Considering that size of a lipid raft depends on the affinity of S and C for one another and other lipids in the membrane, choose the option that best describes this property. **(DEC 2017)**

(a) S and C bind to one another tightly and independent of any other lipid molecules.

(b) S and C bind to one another with same affinity as they bind to other lipid species of the membrane.

(c) Sand C bind to one another with high affinity under the influence of some cytoskeletal elements.

(d) S and C have slightly higher affinity than other 'lipid molecules of the membrane and are in dynamic equilibrium with their free forms.

17. Ability of a membrane protein to span the lipid bilayer strictly depends on the presence of **(JUNE 2018)**

(a) Zinc finger domain

(b) α– helices

(c) parallel β sheet

(d) antiparallelβ sheet

18. Match the enzymes in Column A with their respective biological function in Column B:

	Column A	Column B
A	Lipases	i. Catalysis of ATP–dependent translocation of the amino phospholipids, phosphatidyl ethanol amine and phosphatidylserine from the extracellular to the cytosolic leaflet of the plasma membrane
B	Flippases	ii. Catalysis of ATP–dependent trans location of plasma membrane phospholipids from the cytosolic to the extracellular leaflet.
C	Floppases	iii. Catalyze hydrolysis of triacylglycerols.
D	Scramblase	iv. Catalyze the movement of any membrane phospho–lipid across the bilayer down its concentration gradient.

Choose the correct combinations of answers from the options given below: **(JUNE 2018)**

(a) A – iii, B – i, C – ii, D – iv

(b) A – i, B – iii, C – iv, D – ii

(c) A – iv, B– ii, C – i, D– iii

(d) A– ii, B– iv, C – iii, D – i

Answer Key

1.(d)	**2.**(d)	**3.**(a)	**4.**(b)	**5.**(b)	**6.**(b)	**7.**(d)
8.(c)	**9.**(c)	**10.**(a)	**11.**(a)	**12.**(c)	**13.**(d)	**14.**(d)
15.(a)	**16.**(d)	**17.**(b)	**18.**(a)			

2. Applied Questions– Membrane Proteins, FRAP

1. A membrane associated protein is composed of seven "α–helices", with each helix containing 19 hydrophobic residues. While treating the membrane with all kinds of proteases, a major portion of this protein remains intact. Treatment with high salt (till 1.5M NaCl) and buffer with pH 5.0 failed to dissociate this protein from the membrane.

Predict the most appropriate nature and orientation of this protein in the membrane. **(JUNE 2017)**

(a) Peripheral glycoprotein

(b) Integral protein with seven membrane spanning regions

(c) Peripheral protein with both N and C – terminals remain exposed to outer surface of the cell membrane

(d) Peripheral protein with both N and C – terminal remain exposed to cytosolic surface of the cell membrane

2. Labelling of membrane spanning domain of any integral membrane protein in a given plasma membrane vesicle (without disrupting its structure) is successfully carried out by **(JUNE 2016)**

(a) immunochemical methods.

(b) metabolic labelling with radioisotopes.

(c) hydrophobic photoaffinity labelling.

(d) limited proteolysis followed by metabolic labelling.

3. Glycophorin of red blood cell (RBC) membrane spans the membrane only once and the N–terminal is projected extracellularly and the C–terminal is exposed to the cytosolic side with the help of antibodies (labelled with fluorophors) against N–terminal and C–terminal peptides, orientation of glycophorin across membrane can be verified. Which one of the following statements is correct?

(JUNE 2016)

(a) Intact RBC can be labelled with C–terminal antibody.

(b) Permeabilized RBC can be labelled with C–terminal antibodies as well as N–terminal antibodies.

(c) Intact RBC cannot be labelled with N–terminal antibodies.

(d) Inside out ghost of RBC can be labelled with N–terminal antibodies.

4. It is well established that "Band 3" protein of red blood cell membrane is solely responsible for Cl^- transport across membrane. A lysine group in the Cl^- binding site of "Band 3" is crucial for this event. Keeping this in mind what is the most appropriate way to load and retain a small anionic fluorescent probe (x) inside the red blood cells (RBCs) suspended in phosphate buffered saline (PBS), pH 7.4.

(JUNE 2016)

(a) Incubate the RBCs with x in phosphate buffered saline (PBS, pH 7.4) at 37°C for 30 min.

(b) Incubate the RBCs with x in PBS at 4°C for 30 min.

(c) Incubate the RBCs with x in Hepes sulphate buffer (pH 7.4) at 37°C for 30 min.

(d) Incubate the RBCs with x in Hepes sulfate buffer (pH 7.4) at 37°C for 30 min followed by treatment with a NH2 group modifying agent (covalent modification).

5. One population of large unillamellar vesicles (LUV), labeled with a flurophore attached to phosphatidyl ethanolamine is mixed with a population of unlabelled LUV in the ratio of 1:5.

A protein is added to these lipid vesicles. Considerable enhancement in fluorescence is observed. This is due to **(DEC 2013 GU)**

(a) distribution of the flurophore between the vesicles as a result of membrane fusion.

(b) translocation of a population of flurophore to the unlabelled vesicles due to vesicle aggregation.

(c) extrusion of the flurophore from the vesicles into buffer.

(d) formation of micelles from the vesicles.

6. A new protein "X" has been discovered in erythrocyte plasma membrane. The following observation are made about "X".

(a) It can be extracted from disrupted erythrocyte membranes into a concentrated salt solution (NaCl), and it can be cleaved into fragments by proteinase K.

(b) Treatment of erythrocytes with proteinase K followed by disruption and extraction of membrane components yields intact X.

(c) However, treatment of erythrocyte "ghosts" (which consist of just plasma membranes, produced by disrupting the cells and washing out the hemoglobin) with proteinase K followed by disruption and extraction yields extensively fragmented "X".

What do these findings indicate about the location of "X" in the plasma membrane? Identify the correct answer. **(DEC 2013 GU)**

(a) Integral membrane protein with both N– and C– terminal exposed outside the cell.

(b) Peripheral outer membrane

(c) Integral membrane protein

(d) peripheral inner membrane

7. In an experiment, red blood cells were subjected to lysis and any unbroken cells were removed by centrifugation at 600g.

The supernatant was taken and centrifuged at 100,000g, the pellet was extracted with 5M NaCl and again centrifused at 100,000g. Which of the following protein will be present in supernatant? **(DEC 2016)**

(a) Band3
(b) Glycophorin
(c) G protein– coupled receptor
(d) Spectrin

8. ^{125}I–labelled diaminofluorene (DAF) is a well known photoactivable hydrophobic probe of plasma membrane integral protein. To determine the approximate length and number of hydrophobic domain in any integral membrane protein, a controlled experiment (following standard protocol) is carried out. In order to ascertain the aforesaid aspects indicate the correct combination of experimental protocols from the following choices: **(JUNE 2014)**

(a) Intact membrane was allowed to interact with DAF and unincorporated DAF measured.
(b) Intact membrane was allowed to interact with DAF, lysed and total protein precipitated with TCA and amount of radioactivity incorporated in the total proteins in the TCA precipitated fraction measured.
(c) Intact membrane was allowed to interact with DAF, then membrane was solubilized with detergent, digested with proteases (limited proteolysis) run on SDS–PAGE followed by autoradiography.
(d) Intact membrane was allowed to interact with DAF followed by complete proteolysis, SDS–PAGE and finally auto-radiography.

9. Phosphatidyl serine (PS) is mostly located in the inner bilayer of plasma membrane of red blood cells (RBCs). You have to prove this fact about PS by an experiment. You are provided with PS–specific lytic enzymes (PSE) and other reagents needed. Identify the correct sequence of experiments to be carried out to settle this issue. **(DEC 2012)**

(a) RBCs → inside out vesicles→ PSE → Thin Layer Chromotography (TLC)
(b) RBCs → right side out vesicles → TLC →PSE
(c) RBCs → PSE → Inside out vesicles →TLC
(d) RBCs → PSE → TLC → Inside out vesicles

10. You are studying the binding of proteins to the cytoplasmic face of cultured liver cells and have found a method that gives a good yield of inside–out vesicles from the plasma membrane. Unfortunately, your preparations are contaminated with variable amounts of right–side–out vesicles. Nothing you have tried avoids this contamination. Somebody suggests that you pass the vesicles over an affinity column made of lectin coupled to Sepharose beads. What is the rational of this suggestion? **(JUNE 2011)**

(a) Right–side–out–vesicles wIll be lysed by lectin coupled to Sepharose beads.
(b) Right–side–out–vesicles will simply bind to the lectin coupled Sepharose beads.
(c) Lectin will bind to the carbohydrate residues present only on the inside out vesicles.
(d) Lectin will bind to only glycoproteins and glycolipids present on the inside–out vesicles.

11. A typical animal cell (nucleated) membrane contains glycolipids and glycoproteins in the plasma membrane. To determine its topological distribution, "lectin" is used as a probe. The following interactions may be the basis of the probing method: **(MODELPAPER)**

A. Protein–protein interaction
B. Protein–sugar interaction
C. Protein–lipid interaction
D. Protein–sterol interaction.

The appropriate answer is

(a) Only (A)
(b) Only (B)
(c) Only (D)
(d) All of (A), (B) and (C).

12. α–bungarotoxin binds to acetylcholine receptor (AChR) protein with high specificity and prevents the ion–channel opening. This interaction can be exploited to purify AChR from membrane using: **(JUNE 2017)**

(a) Ion–exchange chromatography
(b) Gel filtration chromatography
(c) Affinity chromatography
(d) Density gradient centrifugation

13. When the cholera toxin (protein of Mr 90,000 Da) gains access to the human intestinal tract, it binds tightly to specific receptors in the plasma membrane of the epithelial cells lining the small intestine, causing membrane bound adenylyl cyclase to undergo prolonged activation resulting in extensive loss of H_2O and Na^+. Pretreatment of the epithelial cells with various phospholipases and proteases failed to inhibit the binding of cholera toxin to its receptor and the fluid loss but treatment with exoglycosidase, prior to binding, significantly reduces these effects. Which of the following molecule could be the receptor for this toxin? **(JUNE 2017)**

(a) Phosphatidyl choline
(b) Sodium–potassium ATPase
(c) Ganglioside
(d) Chloride–bicarbonate exchanger

14. Lateral diffusion of proteins in membrane can be followed and diffusion rate calculated by **(JUNE 2016)**

(a) Atomic force microscopy
(b) Scanning electron microscopy
(c) Transmission electron microscopy
(d) FRAP

15. Using FRAP (Fluorescence Recovery After Photo–bleaching) techniques, diffusion coefficient of three integral membranes proteins M_1, M_2 and M_3 in a kidney cell is calculated as 1 μm/s, 0.05 μm/s and 0.005 μm/s, receptively. Considering fluid–mosaic nature of biological membrane and relationship of structural organization of integral membrane protein with diffusion coefficient, which protein(s) will have highest number of integral membrane domain? **(DEC 2013)**

(a) M_2 and M_3 (b) M_2 only
(c) M_3 only (d) M_1 and M_3

16. Fluorescence recovery after photo bleaching (FRAP) is a method to estimate the diffusion of molecules in a membrane. Fluorescently labelled in a membrane such as

(i) a receptor tagged with green fluorescent protein (GFP)
(ii) a receptor labelled with GFP which interacts with cytoskeleton
(iii) a labelled lipid
(iv) a labelled protein that binds to the membrane surface

are photo bleached and the recovery profiles (a–d) were obtained to estimate their diffusion coefficients

The following data were obtained.

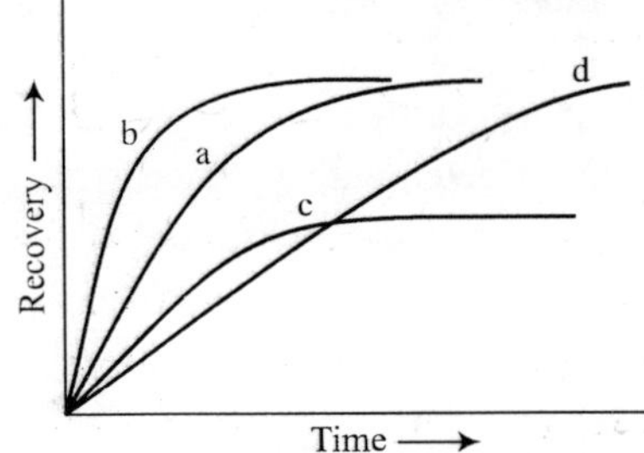

Which one of the combination is correct? **(DEC 2016)**

(a) a = i; b = ii (b) b = iii; a = iv
(c) c = iii; d = iv (d) d = ii; b = i

17. To investigate the dynamic nature of two unrelated centrosome–localized GFP–tagged proteins (GFP–A; GFP–B), a team of scientists conducted fluorescence recovery after photo bleaching (FRAP) experiment. The FRAP profile of these two proteins is given below:

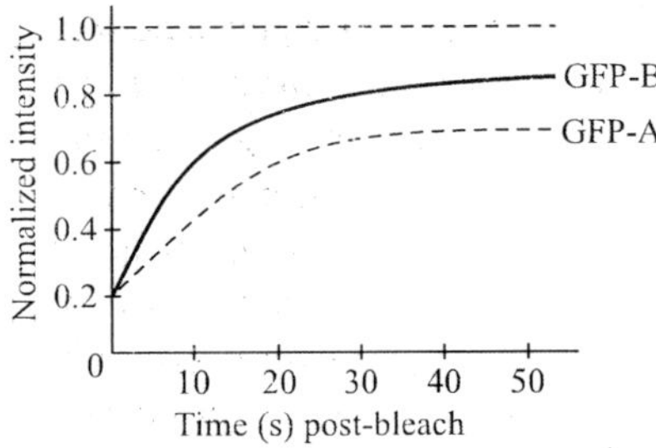

The following statements for this FRAP analysis were made

A. GFP–B shows faster exchange rate than GFP–A

B. GFP–A shows faster exchange rate than GFP–B

C. GFP–A has more immobile fraction than GFP–B

D. GFP–B has more immobile fraction than GFP–A

Which of the above mentioned statements for GFP–A and GFP–B are correct? **(DEC 2019)**

(a) A and C (b) A and D

(c) B and C (d) B and D

18. An integral membrane protein (P) has been identified as a cell surface protein of hepatocytes and assigned to bind to hepatitis B virus (HBV) and promote its entry into cytosol. Upon binding to HBV particles, the C–terminal of P interacts with F–actin in the cytosol and in turn, helps in the entry of the HBV particles. P was successfully cloned and expressed in animal cells in culture wherein its N– terminal is exposed on the surface while the C–terminal resides in the cytosol. The recombinant protein P so expressed retains its complete structure and function. From the list of experiments given below, which one of the experiments will you perform to show that C– terminal of the protein P via interacting with F–actin helps in HBV entry? **(JUNE 2018)**

(a) Incubating radiolabelled HBV with hepatocytes in culture and follow up its association with F–actin by immuno–precipitation analysis using anti–F–actin antibody.

(b) Incubating radiolabelled HBV with hepatocytes over–expressing the C–terminal mutant of P and repeat the rest of the experiment as in "1"

(c) Incubating radiolabelled HBV with hepatocytes over–expressing the N–terminal mutant of P and repeat the rest of the experiment as in "1".

(d) Using wild type P as well as C–terminal mutant of P and their individual over–expression in a heterologous cell line (completely devoid of endogenous P protein) and then repeat experiment as in "1".

19. Two liposome preparations ("X" and "Y") are made using basic lipid composition as phosphatidylcholine (PC) and cholesterol (Chol). In "Y" a ganglioside (asialo–GM_1) is incorporated during the preparation besides PC and Chol. In an attempt to find out the localization of asialo–GM_1, in the membrane bilayer of "Y" (taking "X" as a negative control) and considering liposome as a true depiction of lipid bilayer structure of cellular membrane, following reagents are provided as probes:

A. Phospholipase A

B. Galactose binding lectin

C. Exoglycosidase

D. Cyclodextrin

Choose the most appropriate reagent(s) from the above list to ascertain the localization of asialo–GM_1– **(JUNE 2018)**

(a) Only A (b) Both C and D

(c) Both B and C (d) Both A and D

20. A group of scientists performed an experiment where the)' artificially fused mouse cells with monkey cells. The resulting fused cells were labelled with fluorescently tagged antibodies against mouse and monkey surface receptor proteins, X and Y respectively. At the time 0 minute just after fusion events, two receptors were confined to their own half in the heterokaryon. However, such surface receptors (X and Y) intermixed on the cell surface after 60 minutes. Which one of the given statements correctly reflects the outcome of the experiment? **(JUNE 2018)**

(a) The proteins in cytoplasm are in a dynamic state.

(b) The proteins on the membrane surface are in a dynamic state.

(c) Surface membrane proteins exchange with the cytosolic proteins.

(d) Membrane surface proteins are in a static phase.

21. Detergents at low concentration generally do not denature proteins and are thus used for extracting proteins in their folded and active form. For isolation of Porins, an *E. coli* membrane protein, which one of the following purification approaches will be most appropriate? **(DEC 2018)**

(a) Use of low concentration of non–ionic detergent without salt.

(b) Use of low concentration of ionic detergent.

(c) Use of salt solution containing non– ionic detergent.

(d) Use of salt solution containing ionic detergent.

Answer Key

1.(b)	**2.**(c)	**3.**(b)	**4.**(d)	**5.**(a)	**6.**(d)	**7.**(d)
8.(c)	**9.**(a)	**10.**(b)	**11.**(b)	**12.**(c)	**13.**(c)	**14.**(d)
15.(c)	**16.**(b)	**17.**(a)	**18.**(d)	**19.**(c)	**20.**(b)	**21.**(c)

3. Membrane Transport and potential

1. Cystic fibrosis transmembrane conductance regulator (CFTR) is known to control the transport of which ion? **(JUNE 2015)**

1. Ca^{2+} 2. Mg^{2+}

3. HCO_3^- 4. Cl^-

2. Which of the following statement regarding membrane transport is FALSE? **(DEC 2014)**

(a) Polar and charged solutes will not cross cell membranes effectively without specific protein carriers.

(b) Each protein carrier will only bind and transport one (or a few very similar) type of solute.

(c) Sugars such as glucose are always transported by active transport rather than facilitated diffusion carrier.

(d) Ions are typically transported by special proteins that form membrane channels.

3. At pH 7.0, tryptophan crosses a lipid bilayer at about one thousandth the rate of the closely related substance indole. The correct explanation for this observation at pH 7.0 is: **(DEC 2013 GU)**

(a) Tryptophan bears a net negative charge

(b) Tryptophan bears a net positive charge

(c) Tryptophan bears a positive and a negative charge

(d) Tryptophan cannot bind to its transporter in the lipid bilayer

4. *E. coli* proliferates faster on 'glucose' than it does on 'lactose' because lactose is **(DEC 2013)**

(a) taken up more slowly than glucose

(b) not hydrolyzed by *E. coli.*

(c) taken up faster than glucose.

(d) toxic to the cells.

5. Out of the list given below, which is the correct order of increasing lipid bilayer permeability? **(DEC 2013)**

(a) N_2 > Ethanol> H_2O> Glucose > Ca^{+2} > RNA

(b) H_2O > Glucose> Ethanol > N_2 > Ca^{+2} > RNA

(c) Ca^{+2} > RNA > N_2 > Ethanol > H_2O > Glucose

(d) Ethanol> RNA > Ca^{+2} > H_2O> Glucose >N_2

6. Acetylcholine is a potent neurotransmitter, which is released from the neurons. After release theydiffuse across the synaptic cleft and combine with nicotinic acetylcholine receptor molecules in the membrane of the postsynaptic cell. The interaction of acetylcholine with the nicotinic acetylcholine receptor produces large transient increase in the permeability of the membrane to specialized ions resulting in signal transduction for nerve impulse. Acetylcholine receptor is a **(DEC 2019)**

(a) ligand–gated cation channel

(b) ligand–gated anion channel

(c) voltage–gated cation channel

(d) voltage–gated anion channel

7. Acetylcholine receptor is an archetype for: **(DEC 2012)**

(a) Ligand–gated ion channel

(b) ATPase dependent voltage–gated ion channel

(c) ATPase dependent Ca^{2+}–gated ion channel

(d) ATPase independent voltage gated ion channel

8. ATP– binding cassette (ABC) transporters **(JUNE 2012)**

(a) all are P–glycoprotein

(b) found only in eukaryotes

(c) are both membrane spanning transporter domain that recognizes substrate and ATP–binding domain

(d) affect translocation by forming channels

9. Na^+–K^+ ATPase is a tetramer of 2α and 2β subunits. On which of the following subunits are the Na^+ and K^+ binding sites present? **(JUNE 2011)**

(a) both on α

(b) both on β

(c) Na^+ on α and K^+ on β

(d) Na^+ on β and K^+ on α

10. The Na^+/ K^+ ATPase pumps found on the plasma membrane of most animal cells. A mutation in the intrinsic phosphorylation site of the pump is most likely to affect **(JUNE 2019)**

(a) the outward movement of Na^+ only.

(b) inward movement of K^+ only

(c) both the inward and outward movement of K^+ and Na^+.

(d) has no effect on pump activity but affects its stability

11. ATP–driven pumps hydrolyze ATP to ADP and phosphate and use the energy released to pump ions or solutes across a membrane. There are many classes of these pumps and representatives of each are found in all prokaryotic and eukaryotic cells. Which of the following statements about these pumps is NOT correct?

(a) P–type pumps are multipass transmembrane proteins which phosphorylate themselves during pumping and involve in ion transport.

(b) F–type pumps normally use the H^+ gradient across the membrane to drive the synthesis of ATP.

(c) V–type pumps normally use voltage gradient for transport of small molecules.

(d) ABC transporters primarily pump small molecules across cell membrane **(DEC 2012)**

12. Cystic fibrosis(CF) trans membrane conductance regulator (CFTR) protein is known to be a cAMP– dependent Cl^- channel. CF patients (with mutant CFTR proteins) show reduced Cl^- permeability and as a result exhibit elevated Cl^- level in sweat. To prove this, CFTR proteins (both wild type and mutant) are inserted in a model membrane (liposome) and Cl^- transport is followed with radioactive Cl^-.it is known that topology of CFTR in membrane is important for its function. Despite no proteolytic degradation of denaturation of CFTR proteins, wild type CFTR failed to transport Cl^- in liposome.

Which of the following is the correct explanation of this? **(JUNE 2012)**

(a) CFTR protein gets mutated during insertion in liposomes.

(b) CFTR protein loses affinity with Cl^- ions.

(c) CFTR protein gets wrongly inserted in liposomes.

(d) CFTR protein loses channel forming property in liposomes.

13. When a membrane is depolarized to a voltage value more positive than the threshold voltage, it leads to the generation of **(DEC 2013)**

(a) Donnan potential.

(b) Action potential.

(c) resting potential.

(d) electrochemical potential.

14. KCI (100 mM) was entrapped inside large unilamellar vesicles. A diffusion potential across the bilayer can be generated by diluting with buffer containing **(DEC 2012)**

(a) 100 mM KCl and a protonophore.

(b) 100 mM NaCl and a protonophore.

(c) 100 mM KCl and a K^+– specific ionophore.

(d) 100 mM NaCl and a K^+– specific ionophore

15. Which channel is responsible for initiating muscle contraction through depolarization of muscle cell membrane? **(MODELPAPER)**

(a) Na^+ channel (b) K^+ channel

(c) Ca^{2+} channel (d) Mg^{2+} channel

16. A nerve impulse or action potential is generated from transient changes in the permeability of the axon membrane to Na^+ and K^+ ions. The depolarization of the membrane beyond the threshold level leads to Na^+ flowing into the cell and a change in membrane potential to a positive value. The K^+ channel then opens allowing K^+ to flow outwards ultimately restoring membrane potential to the resting value, The Na^+ and K^+ channels operate in opposite directions because. **(DEC 2011)**

(a) there is an electrochemical gradient growth generated by proton transport

(b) there is a difference in Na^+ and K^+ concentrations on either side of the membrane

(c) Na^+ is a voltage–gated channel, whereas K^+ is ligand–gated

(d) Na^+ is dependent on ATP whereas K^+ is not

17. Which one of the listed below is a P–type ion transporter? **(JUNE 2018)**

(a) Mg^{+2} and Fe^{+2}

(b) Mg^{+2} and Fe^{+3}

(c) Mg^{+2} and Cl^-

(d) Na^+–K^+, Ca^{+2} and H^+

18. The cell maintains a high concentration of protons inside the lysosome because of **(JUNE 2018)**

(a) antiporter in the lysosomal membrane

(b) ATP–powered proton pump in the lysosomal membrane

(c) facilitated diffusion proton channel in the lysosomal membrane

(d) facilitated diffusion proton uniporter in the lysosomal membrane

19. Phosphoenol pyruvate : sugar phosphotransferase system (PTS) transports a variety of sugars into bacteria. In *E. coli,* PTS consists of EI, EII (EIIA, EIB, and EIIC), and Hpr. During this process the sugar molecule is phosphorylated by direct transfer of phosphate group from **(DEC 2019)**

(a) EI–P (b) EIIA–P

(c) EIIB–P (d) Hpr–P

20. The cell membrane of neuron maintains intracellular conditions that differ from those of the extracellular environment. Such difference in intra– and extracellular conditions are critical to the function of the nerve cell as the nerve cell membrane resembles a charged capacitor.

Assuming the electric field (E) across a parallel–plate capacitor is uniform and if membrane thickness is 7 nm and potential difference across the membrane is –60 mV, calculate E of the membrane. **(JUNE 2018)**

(a) 6×10^5 V m^{-1} (b) 7×10^5 V m^{-1}

(c) 8.6×10^6 V m^{-1} (d) 6.6×10^6 V m^{-1}

Answer Key

1.(d)	2.(c)	3.(c)	4.(a)	5.(a)	6.(a)	7.(a)
8.(c)	9.(a)	10.(c)	11.(c)	12.(c)	13.(b)	14.(d)
15.(a)	16.(b)	17.(d)	18.(?)	19.(b)	20.(c)	

4. Protein Sorting

1. Which of the following is NOT an example of trasmembrane transport between different subcellular compartments? **(DEC 2016)**
   (a) Transport from cytoplasm into lumen of the endoplasmic reticulum
   (b) Transport from endoplasmic reticulum to the Golgi complex
   (c) Transport from stroma into thylakoid space
   (d) Transport from mitochondrial inter-membrane space into the mitochondrial matrix

2. Which one of the following describes the primary function of flipases? **(DEC 2016)**
   (a) Help in increasing lipid protein interaction in the outer leaflet of the bilayer
   (b) Move certain phospholipids from one leaflet of the membrane to another
   (c) Localize more negatively charged membrane proteins in the lipid bilayer
   (d) Cause uncoupling of v–SNARES and t–SNARES after fusion of incoming vesicle with target membrane

3. Glycolipids and sphingomyelin are produced by the addition of sugars or phosphorylcholine to ceramide on cytosolic and luminal surfaces respectively, of the Golgi apparatus. Finally, after such modifications, these molecules are located on the outer half of the plasma membrane. What key events are responsible for such localization? **(JUNE 2015)**
   (a) Membrane fusion only.
   (b) Action of Flippase and membrane fusion.
   (c) Action of only Flippase.
   (d) Flip flop of these molecules in the golgi membrane catalyzed by proton pump

4. Membrane–bound and free ribosomes, are structurally identical, but differ only at a given time in terms of association with **(JUNE 2014)**
   (a) acetylated proteins
   (b) glycosylated proteins
   (c) phospholipids
   (d) Nascent proteins

5. Signal Recognition Particle (SRP) plays an essential role in protein import in the endoplasmic reticulum (ER). In mammalian cells, SRP is a rod–like ribonucleoprotein complex containing six protein subunits and one RNA molecule. SRP stalls protein translation by blocking:
   A. elongation of the polypeptide chain
   B. mRNA loading onto the ribosomes.
   C. assembly of 60S and 40S ribosome particles.
   D. binding of the initiation factors.

   Which one of the following represents the correct statement/s? **(DEC 2019 ASSAM)**
   1. A only 2. A and B
   3. B and C 4. A and D

6. Fluorescently tagged protein was used to study protein secretion in yeast. Fluorescence was observed in:
   (a) the Golgi
   (b) the secretory vesicles
   (c) the rough ER.

   Which of the following describes the best sequence in which these events occur? **(DEC 2016)**
   (a) a → b → c (b) b → c → a
   (c) c → a → b (d) c → b → a

7. Which one of the following activities is NOT involved in protein folding in the endoplasmic reticulum? **(DEC 2018)**

(a) Peptidylprolylisomerase

(b) Protein disulphide isomerase

(c) Protein glycosylation

(d) Protein ubiquitination

8. Which one of the following statements correctly applies to proteins which are translated on the rough endoplasmic reticulum? **(DEC 2015)**

(a) Cytoplasmic proteins which are targeted to the nucleus in response to hormone stimuli.

(b) Proteins targeted to lysosomes, plasma membrane and cell exterior.

(c) Proteins which are targeted to the nucleus through endoplasmic reticulum lumen as the lumen is in direct connection with the inter membrane space of the nucleus.

(d) All proteins which get targeted to peroxisomes

9. The tetrapeptide "KDEL" is well known as a retrieval signal of several newly synthesized proteins. This process is mediated through specific receptor –'– KDEL interaction. Any single amino acid change in this tetrapeptide is not allowed in terms of its binding its receptors and its subsequent retention in specific organelle whereas, secretory proteins are devoid of such tetrapeptide. From this observation indicate the localization of the receptor of this tetrapeptide: **(JUNE 2014)**

(a) Plasma membrane

(b) Golgi

(c) Endoplasmic reticulum.

(d) Golgi and ER both

10. You are following the intracellular sorting of an integral plasma membrane protein in a living cell, in culture. You have decided to probe this protein by metabolic labeling technique with ^{35}S–methionine (pulse–chase technique).

After one cycle of division, the cells were treated with a potent inhibitor of protein biosynthesis and processed for subcellular fractionation. In which of the following fractions will you expect the immunoprecipitation with a specific antibody? **(JUNE 2013)**

(a) Only cytoplasm.

(b) Only plasma membrane.

(c) Both endoplasmic reticulum and plasma membrane.

(d) Only secretory vesicles and endoplasmic reticulum

11. You are studying a protein that inserts itself into a model membrane (liposomes) during a reconstitution process. The protein has an N–terminal, 18–amino acid hydrophilic segment that is located on the outside of the membrane, a 10–amino acid hydrophobic transmembrane segment flanked by negatively and positively charged amino acids, and a C–terminal domain that resides inside the lumen (as depicted below in the form of a cartoon).

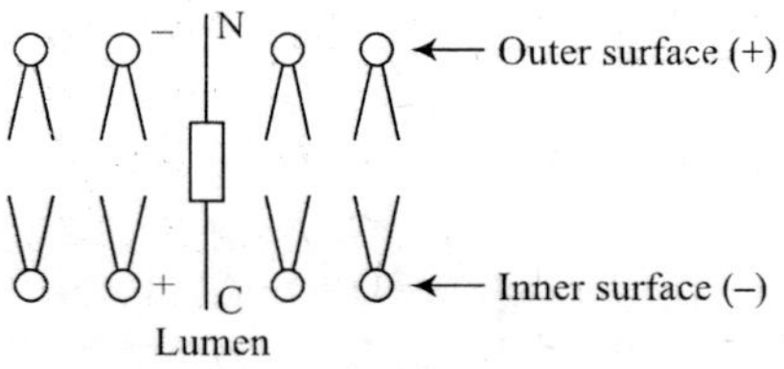

For proper reconstitution of the protein, which of the following strategies will be appropriate? **(JUNE 2014)**

(a) Increase the number of negatively charged amino acids in the N–terminal

(b) Increase the number of positively charged amino acids in the C–terminal

(c) Removal of positively charged amino acids from the C–terminal

(d) Increase the length of hydrophobic segment

12. Membrane proteins are synthesized on endoplasmic reticulum and transported to various oraganelle. One hypothesis for membrane protein sorting is hydrophobicity matching i.e., the proteins with a shorter transmembrane portion would partition into thinner membranes. You are given the following three observations:

A. It was found that transmembrane portion of proteins in Golgi membranes are short than those in plasma membranes

B. Presence of cholesterol increases the thickness of the bilayer

C. The phospholipid composition of Golgi and plasma membranes are same

Which of the following statements is correct?

(DEC 2015)

(a) Proteins in plasma membrane have longer transmembrane portion than proteins in Golgi membranes

(b) Proteins in Golgi membranes have longer transmembrane portion than proteins in plasma membranes

(c) Proteins of both Golgi and plasma-membranes have same length of transmembrane portion

(d) Cholesterol is more in Golgi membrane than in plasma membrane

13. In Tay–Sachs disease, accumulation of glycolipids occurs especially in nerve cells. These cells are greatly enlarged with swollen lipid–filled endosomes and the children with this disease die at a very early stage. Such condition occurs due to a specific defect in:

(JUNE 2017)

(a) Specific lysosomal enzyme that catalyzes a step in the breakdown of gangliosides

(b) Sorting of an enzyme that adds a phosphate group at 6th position of mannose in all acid hydrolases

(c) One of the Rab proteins involved in recycling of vesicles

(d) v–SNARE molecules which cause abnormal vesicle tethering and docking and affect vesicle fusion with lysosomes

14. In order to study intracellular trafficking of protein 'A', it was tagged with GFP (A–GFP). Fluorescence microscopy showed that A–GFP co–localizes with LAMP1. In the presence of bafilomycin A, an inhibitor of H^{+}– ATPase, A–GFP does not co–localize with LAMP1. Instead, it localizes with LC3 puncta. Which one of the following statements is true?

(DEC 2016)

(a) A–GFP targets to the ER in the absence of bafilomycin A.

(b) Autophagy is required for trafficking of A–GFP to lysosome.

(c) Bafilomycin A facilitates targeting of A–GFP to ER.

(d) Bafilomycin A facilitates targeting of A–GFP to mitochondria.

15. A null mutation is created in a gene which is responsible for specific phosphorylation at 6th carbon position of mannose on acid hydrolases occurring in cis–Golgi. The following statements are given towards explaining the effect of this mutation–:

A. The lysosomes will be devoid of lysosomal enzymes

B. Lysosomal enzymes will be secreted out

C. Lysosomal enzymes will get localized in cytoplasm

Which statement or combination of statements will explain the effect of mutation if the acid hydrolases in the mutation not get degraded?

(DEC 2014)

(a) A and C (b) B and C

(c) C only (d) A and B

16. The principal pathway for transport of lysosomal hydrolases from the trans Golgi network (pH 6.6) to the late endosomes (pH 6.0) and the recycling of M6P (mannose 6 phosphate) receptors back to the Goigi depends on the pH difference between those two compartments. From what you know about M6P receptor binding and recycling and the pathways for delivery of material to lysosomes, predict what would happen if the pH in late endosomes was raised to 6.6? **(JUNE 2013)**

(a) M6P will bind to hydrolases but will not release the hydrolases in the late endosomes.

(b) M6P will bind to hydrolases and will release the hydrolases in the late endosomes.

(c) At higher endosomal pH, the receptor would not release the hydrolase and could not be recycled back to the trans Golgi network.

(d) M6P will be degraded at higher pH.

17. Presence of the nuclear localization signal (NLS) in a steroid receptor indicates that the receptor resides **(DEC 2012)**

(a) on the nuclear membrane.

(b) within the nucleus.

(c) on the cell membrane.

(d) in the cytosol.

18. Thousands of proteins that are synthesized in the cytoplasm are imported into the nucleus through the nuclear pore complex (NPC) every minute. These proteins contain Nuclear Localization Signal (NLS) that direct their selective transport into the nucleus. This nuclear import requires

A. A small monomeric G–protein Ran.

B. A nuclear transport receptor that interacts with the NLS on a cargo protein.

C. A GTPase activating protein (GAP) bound to the chromatin tethered to the nuclear membrane

D. A Guanine Exchange Factor (GEF) bound to the chromatin inside the nucleus.

Which one of the following options represents all correct statements? (DEC 2019)

(a) A, C, D (b) A, B, D

(c) B, C only (d) A, C only

19. Given below are organelles (column X) and properties associated with the organelles (column B)

	Column I		Column II
(A)	Lysosomes	(i)	Anterograde transport from ER to Golgi
(B)	*cis*–Golgi	(ii)	Clathrin – coated vesicles
(C)	*trans* – Golgi	(iii)	Cop I vesicle budding
(D)	Cop II vesicles	(iv)	Mannose–6–phosphate receptor
(E)	Endocytic vesicles	(v)	Protein aggregate for secretion

Choose the option that matches the organelles with the most appropriate property.

(DEC 2017)

(a) A – (iv); B – (iii); C – (v); D– (i); E – (ii).

(b) A– (v); B – (iv); C – (i); D – (v); E – (ii)

(c) A – (iii) ; B – (v); C – (i); D – (iv); E – (ii)

(d) A– (iv); B – (v); C – (ii); D – (i); E – (iii)

20. Following is the domain organization of three proteins that are targeted to the mitochondria.

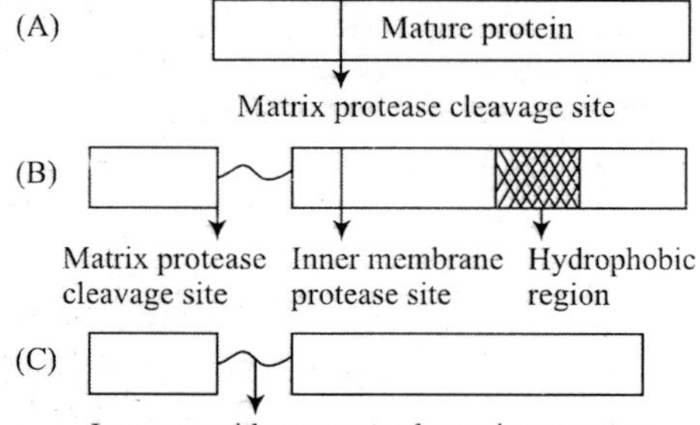

Based on the domain organization in the above figure and assuming the left box to be having the mitochondrial sorting signal, predict the most likely sub–compartment of the mitochondria in which the protein will be found. **(DEC 2015)**

(a) A in matrix; B in inner membrane; C in inter–membrane space

(b) A in inner membrane; B in inter–membrane space; C in outer membrane

(c) A and B are in matrix; C in outer membrane

(d) A in matrix; B and C are in inter–membrane space

21. Mitochondrial proteins are first fully synthesized as mitochondrial precursor proteins in the cytosol and then translocated into mitochondria through signal sequences. Which one of the following is true for mitochondrial precursor proteins? **(DEC 2013 GU)**

(a) They remain unfolded in the cystosol.

(b) They form a folded structure.

(c) They do not form amphiphilic α–helix.

(d) They form β–pleated sheets.

22. Which one of the following proteins is most likely to be found in the inter–membrane space of mitochondria? A protein containing **(DEC 2018)**

(a) an N–terminal matrix targeting sequence followed by hydrophobic stop–transfer anchor sequence

(b) an N–terminal matrix targeting sequence followed by a cleavable hydrophobic sequence that blocks complete translocation

(c) a protein with multiple internal sequences that are recognized by Tim 22 complex

(d) a protein with an outer membrane localization sequence followed by a matrix targeting signal

23. A newly identified sequence was experimentally tested by in vitro transport assay using a radiolabelled protein containing the sequence to test import into mitochondria. Transport assay was done for a' short time with or without membrane potential and after the assay, the mitochondria were either treated or not treated with proteinase K. At the end of the assay the mitochondria were pelleted and total protein of the pellet was isolated and separated on SDS–PAGE and autoradiographed. A representative auto–radiogram is shown below.

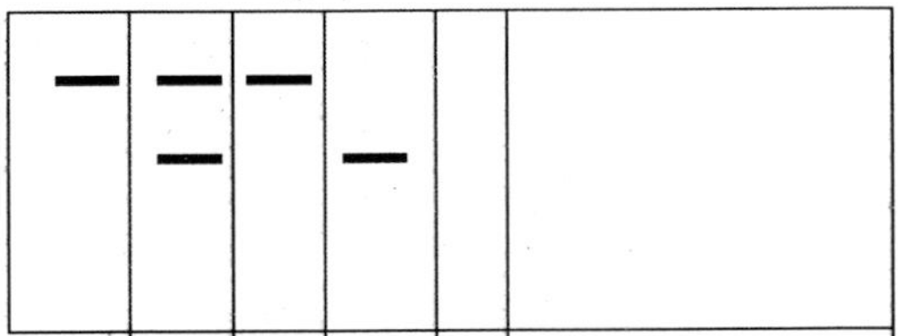

Lysate					
	+	+	+	+	Mitochondria
	+	–	+	–	Ψ Membrane potential
	–	–	+	+	proteinase K

Based on this experimental data, which of the following statements is NOT correct: **(DEC 2014)**

(a) The protein goes into the matrix

(b) Not all of the added protein was imported

(c) The protein requires membrane potential for import

(d) The protein is associated with the outer mitochondrial membrane

24. Table below shows the list of organelles (column A) AND the signals (Column B) that target proteins to the organelle.

	A		B
a	Lysosome	(i)	Stretch of amino acid sequence rich in Lys and Arg residues
b	Mitochondria	(ii)	C–terminal tripeptide
c	Nucleus	(iii)	N–terminal amphipathic helix rich in Lys and Arg
d	Peroxisome	(iv)	Mannose–6–Phosphate

Choose the option that shows all correct matches **(JUNE 2019)**

(a) A – (ii); B – (iii); C – (iv); D – (i)

(b) A – (ii); B – (iv); C – (iii); D – (i)

(c) A – (iv); B – (iii); C – (i); D – (ii)

(d) A – (iv); B – (iii); C – (ii); D – (i)

25. During receptor–mediated endocytosis, apolipoprotein B on the surface of a LDL particle binds to the LDL–receptor present in coated pits containing clathrin. The receptor–LDL complex is internalized by endocytosis, trafficked to lysosomes and the LDL–receptor is finally recycled. A patient reports with familial hypercholesterolemia. This could be due to **(JUNE 2011)**

(a) mutation in the LDL molecule.

(b) defect in LDL–receptor recycling.

(c) mutation in the LDL–receptor.

(d) defect in cholesterol binding with its receptor.

26. Patients suffering from familial hypercholesterolemia (FH) are mostly homozygous for the defective gene and have profoundly elevated levels of serum cholesterol. The reason may be that the gene for highly specific receptor for LDL is either defective or missing in these patients. In an experiment, cells were taken from both normal individual and homozygote (FH) subjects, incubated in buffer with ^{125}I–labeled LDL in presence or absence of excess unlabeled LDL for various time periods and then ^{125}I–labeled LDL bound to cells was measured. Which of the following is the best fit graph for the above experiment? **(JUNE 2014)**

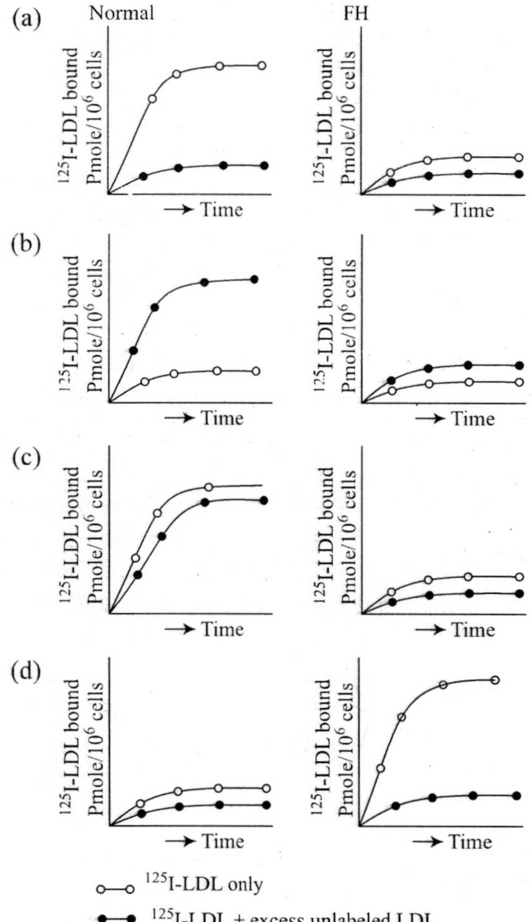

27. During receptor–mediated endocytosis, ligand first binds with cell surface receptor, then traffic through Rab5 positive early endosomal compartment. Finally, it moves to the Lamp1 positive lysosomes via Rab7 positive late compartment. In order to understand the trafficking of ligand A in epithelia cells, cells were allowed to internalize ligand A for various period of times at 37°C. Finally, cells were stained with anti–ligand antibody and probed with secondary antibody labelled with Alexa–Red fluorescence dye. Same cells were also co–stained with anti–Rab5, anti–Rab 7 or anti–Lamp1 antibody and probed with appropriate secondary antibody labelled with Alexa–green fluorescence dyes. Cells were viewed in confocal microscope and observations are

(I) 5 min internalize ligand (Red) in cells are colocalize with anti Rab5 antibody but not with anti–Lamp1 antibody and

(II) 90 min internalized ligand are colocalized with anti–Lamp1 antibody but not with anti–Rab5 antibody.

The following conclusions could be arrived at from the above observations.

A. Ligand A travels to early endosomal compartment by 5 min.

B. Ligand A travels to early endosomal compartment by about 90 min.

C. Ligand A travels to lysozome by about 5 min.

D. Ligand A travels to lysozyme by about 90 min.

Identify the correct inferences.

(MODEL PAPER)

(a) (A) and (B) (b) (B) and (D)

(c) (A) and (D) (d) (D) and (C)

Answer Key

1.(b)	**2.**(b)	**3.**(b)	**4.**(d)	**5.**(a)	**6.**(c)	**7.**(d)
8.(b)	**9.**(d)	**10.**(c)	**11.**(d)	**12.**(a)	**13.**(a)	**14.**(b)
15.(d)	**16.**(c)	**17.**(d)	**18.**(b)	**19.**(a)	**20.**(a)	**21.**(a)
22.(b)	**23.**(d)	**24.**(c)	**25.**(c)	**26.**(a)	**27.**(c)	

5. Cell wall and Cell Organelles

1. A researcher is studying the subcellular localization of a particular protein 'X' in an animal cell. The researcher performs successive centrifugation at increasing rotor speed. The researcher starts spinning the cellular homogenate at 600g for 10 min, collects the pellet, spins the supernatant at 10,000 g for 20 min, collects the pellet, spins the supernatant at 100,000g for 1 hour, collects both the pellet and the final supernatant. On subjecting various pellets and the final supernatant to Western blotting with anti–protein–X antibody, the protein X is, observed to be maximally expressed in pellet after centrifugation at 10,000 g. Based on the above observation, what will be the most likely localization of protein X. **(DEC 2015)**
   (a) Nucleus (b) Ribosomes
   (c) Mitochondria (d) Microsomes
2. A student is asked to isolate and purify various subcellular organelles from a liver cell sample. Following sucrose density gradient centrifugation technique, and taking utmost care mitochondrial fraction is found to be contaminated with nuclear fraction. Which technique out of the choices given below should be appropriate to obtain a clean mitochondrial fraction? **(MODELPAPER)**
   (a) Gel filtration
   (b) Percoll density gradient
   (c) Immuno–magnetic separation
   (d) Pulse field electrophoresis
3. Which one of the following pairs is **NOT** matched correctly? **(JUNE 2014)**
   (a) Glycocalyx – adherence
   (b) Fimbriae – motility
   (c) Pili – conjugation
   (d) Peptidoglycan – cell wall
4. Which one of the following statements is **NOT** true? **(JUNE 2017)**
   (a) Beta–oxidation of long chain fatty acids occurs in mitochondria
   (b) Fatty acid biosynthesis occurs in peroxisomes
   (c) Peroxisomes utilize H_2O_2 to oxidize a variety of substrates
   (d) Peroxisomes import their repertoire of proteins using sorting signals
5. Which of the following pairs of subcellular compartments is likely to have same pH and electrolyte composition? **(JUNE 2013)**
   (a) Cytosol and lysosomes.
   (b) Cytosol and mitochondrial inter membrane space.
   (c) Cytosol and endosome.
   (d) Mitochondrial matrix and inter membrane space.
6. Metachromatic leukodystrophy (MLD) is caused by a deficiency of arylsulfatase A and affects the CNS. MLD is **(JUNE 2017)**
   (a) a lysosomal storage disorder
   (b) a disease due to dysfunctional mitochondria
   (c) caused by loss of the myelin sheath
   (d) caused by a defect in proteins of the nuclear envelope
7. Select a cellular body which is **NOT** a part of the nuclear bodies: **(DEC 2017)**
   (a) P–bodies
   (b) Nucleolus
   (c) Cajal bodies
   (d) Interchromatin granule clusters

Answer Key

1.(c)	2.(b)	3.(b)	4.(b)	5.(b)	6.(a)	7.(a)

6. Cytoskeleton and its role in motility

1. Bacterial actin homologs contribute to the cell shape by serving as a scaffold to direct the synthesis of the peptidoglycan cell wall. Which of the following is/are actin homolog/sin bacteria? **(DEC 2019 ASSAM)**
   (a) MreB and Mbl (b) FtsZ
   (c) DnaK (d) DnaJ

2. Cells are physically linked to one another and to extracellular matrix through their cytoskeleton and this impart strength and rigidity of tissues and organs. Most of the animal cells have three types of cytoskeletal filaments, which are listed in Column A. The possible functions are listed in Column B.

Column A	Column B
A. Intermediate filaments	i. Determine the shape of cell surface and are necessary for cell locomotion
B. Microtubules	ii. Maintain the position of membrane–enclosed organelles and provide intracellular transport
C. Actin filaments	iii. Provide mechanical strength of a cell

Which one is the correct match? **(DEC 2019)**

(a) A–i; B–ii; C–iii (b) A–ii; B–i; C–iii
(c) A–iii; B–ii; C–i (d) A–iii; B–i; C–ii

3. Given below are cytoskeleton filaments (Column X) and their component/associated proteins (Column Y)

Column X		Column Y	
A	Actin	i	Formin
B	Microtubule	ii	Arp 2/3 complex
C	Intermediate filaments	iii	EB1
		iv	Lamin A
		v	Xmap 215

Choose the option that matches cytoskeleton filament (Column X) with the most appropriate protein (Column Y) **(DEC 2019 ASSAM)**

(a) A–i, A–ii; B–iii, B–V; C–iv
(b) A–iii, A–v; B–ii, B–i; C–iv
(c) A–iv, A–V; B–i, B–ii; C–iii
(d) A–ii, A–v; B–i, B–iv; C–v

4. Beating of cilia is regulated by **(JUNE 2015)**

(a) actin (b) myosin
(c) cofilin (d) nexin

5. In an attempt to study the transport of secretory vesicles containing insulin along microtubules in cultured pancreatic cells, how would treatment with "colcemid" affect the transport of these vesicles? **(JUNE 2015)**

(a) Colcemid induces polymerization of microtubules, which in turn would activate vesicular transport.
(b) Polymerization of microtubules is inhibited by colcemid, which in turn would inhibit the transport of secretory vesicles.
(c) Colcemid inhibits the vesicular trafficking through inactivation of v–SNARE protein.
(d) Colcemid activates t–SNARE, proteins and in turn activates vesicular transport.

6. To understand the microtubule–dependent motor activity of a freshly purified motor protein from budding yeast, a researcher set up microtubule based gliding assay. In such an assay where microtubules are fluorescently tagged at its (+) end, the researcher observed that this motor protein moves the microtubule in the direction of its (+) end as shown below: **(DEC 2017)**

The newly identified motor protein is

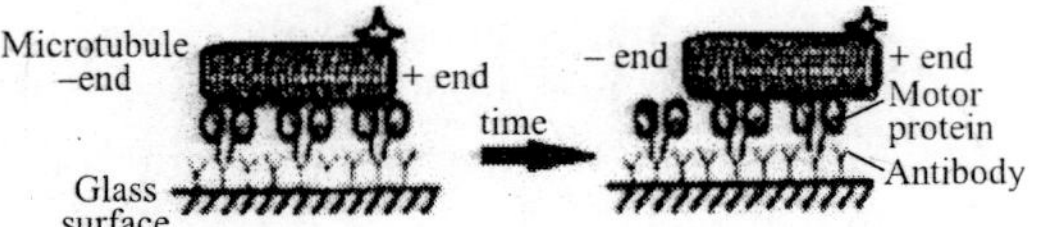

(a) Dynein
(b) Myosin
(c) Kinesin 1
(d) Either Dynein or Kinesin 1

7. In a neuron, proteins and membranes are primarily synthesized in the cell body. These materials must be transported down the axon to the synaptic region using microtubules in an anterograde fashion. Such axonal transport is directed by **(DEC 2019)**

(a) Dynein
(b) Kinesin I
(c) Dynein and Kinesin I
(d) Myosin

8. The erythrocyte membrane cytoskeleton consists of a meshwork of proteins underlying the membrane. The principal component spectrin has α, β subunits which assemble to forms tetramers. The cytoskeleton is anchored to the membrane through linkages with the trans–membrane proteins band 3 and glycophorin C. The cytosolic domain of band 3 also serves as the binding site of glycolytic enzymes such as glyceraldehyde 3–phosphate dehydrogenase.

Analysis of the blood sample of a patient with haemolytic anemia shows spherical red blood cells. The patient carries **(DEC 2011)**

(a) a mutation in glycophorin C

(b) a mutant spectrin with increased tetramerization propensity

(c) mutant β spectrin defective in αβ dimerization ability

(d) mutant glyceraldehyde 3–phosphate dehydrogenase

9. Cytochalasins are the drugs that inhibit actin polymerization into microfilaments. Colchicine blocks tubulin subunit interaction. Which of the following combinations is true for the effect of cytochalasin and colchicine on mammalian cell division? **(DEC 2013 GU)**

(a) Cytochalasin inhibits mitosis and colchicine inhibits cytokinesis.

(b) Cytochalasin inhibits cytokinesis and colchicine inhibits mitosis.

(c) Both cytochalasin and colchicine inhibit mitosis.

(d) Both cytochalasin and colchicine inhibit cytokinesis.

10. Regarding microtubule assembly and disassembly during cell division, which will be the most appropriate answer?**(JUNE 2013)**

(a) Once formed, kinetochore microtubules depolymerize at the plus ends throughout mitosis.

(b) Once formed, kinetochore microtubules polymerize at the plus ends throughout mitosis.

(c) Kinetochore microtubules polymerize at their plus ends up to anaphase, at which point they begin to depolymerize.

(d) Kinetochore microtubules polymerize at their minus ends up to cytokinesis, at which point they depolymerize.

11. A single protofilament of microtubule grows at the speed of 2 μm/min. Considering that there is no catastrophe in the microtubule nucleation and the size of the tubulin unit is of the order of 5 nm, how many tubulin units are added to the growing microtubule per minute? **(DEC 2018)**

(a) 400 (b) 1600

(c) 3200 (d) 5200

12. Eg5 is a well–studied protein in Xenopus. To understand the function of Eg5 in mammalian cells, a team of researchers treated mammalian cells during late G2 phase with Eg5–inhibitor. The following diagrams represent images of mitotic cells.

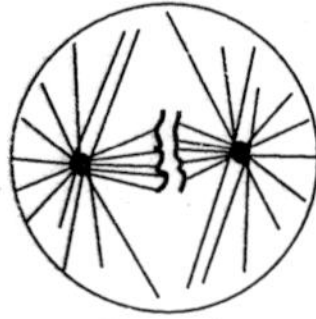

Control

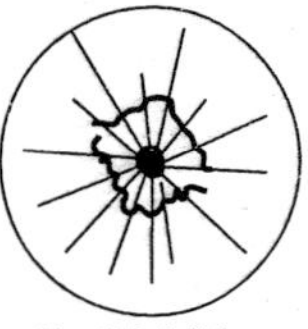

Eg 5 inhibitor

Based on the above observation, what function might be attributed to Eg5 during mitosis? **(JUNE 2018)**

(a) Eg5 inhibits actin dynamics.

(b) Eg5 can activate GPCR signalling.

(c) Eg5 has motor activity.

(d) Eg5 can impact mitochondrial dynamics.

13. An investigator expresses a GFP–fused protein that localizes to the outer membrane of Golgi apparatus. Upon visualising GFP signal in the fluorescence microscope, it was noted that GFP is pericentrosomal in its localization (Fig A). Treatment of such GFP expressing cells with a newly identified drug disrupted the Golgi into small vesicles(Fig B).

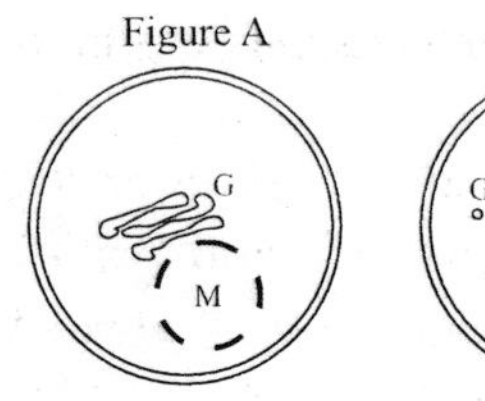

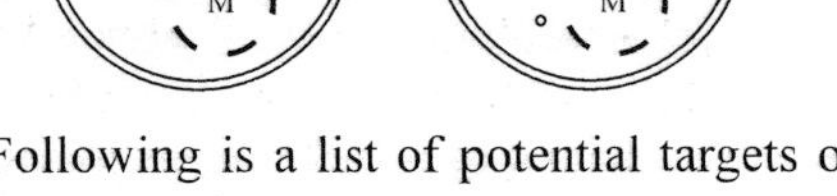

Following is a list of potential targets of the drug:

A. Dynein complex B. Myosin

C. Microtubules D. Dicer

Choose the combination with all correct targets: **(DEC 2018)**

(a) A, Band D only (b) B and D only

(c) A and C only (d) A only

14. Measurements of the rate of actin treadmilling *in vivo* show that it can be several timeshigher than can be achieved with pure actin in vitro. The tread milling in vitro can be enhanced by providing

A. Profilin that binds G–actin on the site opposite the nucleotide binding cleft.

B. co–filin binds specifically to the ADP containing F–actin and destabilizes the actin filament.

C. buffer with ATP and low levels of cations.

D. buffer with ADP and low levels of cations.

Which of the above statements are correct? **(JUNE 2019)**

(a) A and D (b) A and C

(c) C and D (d) A and B

15. Listeria is a food–borne pathogen that causes mild gastro–intestinal symptoms. To move from one host–cell to another, it polymerizes actin into a comet tail like structure. Listeria can assemble host–cell actin at its rear end because:

A. Listeria has on its surface a protein called Act A

B. Listeria can activate Arp 2/3 complex

C. Listeria has on its surface γ–tubulin

D. Listeria has on its surface myosin II motor

Which one of the following options represents all correct statements? **(DEC 2019)**

(a) A and D (b) A and B

(c) A and C (d) C and D

Answer Key

1.(a)	**2.**(c)	**3.**(a)	**4.**(d)	**5.**(b)	**6.**(a)	**7.**(b)
8.(c)	**9.**(b)	**10.**(c)	**11.**(d)	**12.**(c)	**13.**(c)	**14.**(d)
15.(b)						

7. Genome Organization

1. Which one of the following statements about chromatin is **NOT** true? **(DEC 2017)**

(a) DNA winds approximately 1.65 times around the nucleosomes

(b) H2A–H2B bind to both the entry and exit ends of DNA in nucleosomes

(c) Covalent modification of histones influence chromatin compaction

(d) Non–histone proteins are part of mitotic chromosomes

2. Which one of the following statements is NOT true about nucleosomal organization of core particle? **(JUNE 2019)**

(a) The physical structure of DNA is altered in the middle of the core particle.

(b) In the core particle, DNA is organized as flat super helix with 1.65 turns around the histone octamer.

(c) While forming 30nm fibers, generally 6 nucleosomal per turn organize into a two start helix.

(d) The N–terminal histone tails in a core particle are strictly ordered and exit from the nucleosomes between turns of the DNA.

3. CENP–A, a histone variant associated with nucleosomes in the centromeric region, replaces the histone subunit **(DEC 2013 GU)**

(a) H1. (b) H2A.

(c) H3. (d) H4.

4. Assuming that the histone octamer forms a cylinder 9 nm in diameter and 5 nm in height and that the human genome forms 32 million nucleosomes, what fraction (approximately) of the volume of nucleus (6 μm diameter) is occupied histone octamers? **(JUNE 2012)**

(a) 1/22 (b) 1/11

(c) 10/21 (d) 10/11

5. What will happen if histones are depleted from a metaphase chromosome and viewed under a transmission electron microscope?

(DEC 2014)

(a) 30nm chromatin fibres will be observed.

(b) 10 nm chromatin fibres will be observed.

(c) A scaffold and huge number of loops of DNA fibres will be observed

(d) A huge number of loops of DNA fibres without scaffold will be observed.

6. Which one of the following statements is true about human chromosomes? **(DEC 2018)**

(a) The chromosomes that have highest gene density generally localize towards the centre of the nucleus.

(b) The chromosomes that have highest gene· density generally localize near the nuclear periphery to facilitate rapid transport of the nascent transcripts.

(c) The centromeres of different chromosomes tend to cluster together at the centre of the nucleus.

(d) Chromosomal positioning in the nucleus is absolutely random.

7. Chromosome organization becomes clearer from a series of biochemical, electron microscopic and X–ray crystallographic studies. When interphase chromatin is isolated in low salt buffer and observed under EM, 11 nm bead on string organization is seen. Interphase chromatin directly observed under EM shows 30 nm fibre. When histones are depleted from metaphase chromosome and visualized under EM, it shows a huge number of very large' loops associated with scaffold. Following interpretations can be made from these:

A. 11 nm fibre is formed when nucleosomes are brought closer by scaffold.

B. 30 nm interphase chromatin is formed by zig–zag organization of nucleosomes of 11 nm fibre.

C. 30 nm fibre makes a solenoid packing to form the metaphase chromosome.

D. 30 nm fibre gets organized into loops due to SARs getting associated with scaffold proteins and coming closer.

The correct combination of interpretations is:

(JUNE 2014)

(a) A and D (b) A and C

(c) A and B (d) B and D

8. The nucleosome is the fundamental subunit of chromatin in eukaryotes. Following statements are made about nucleosome:

A. Generally, a typical nucleosome contains around 200bp of DNA and two copies of each histone (H2A, H2B, H3 and H4)

B. 146 bp length of DNA per nucleosome core particle is strictly maintained across the organisms

C. The histone octamers are not conserved during/after replication, however, $H3_2$–$H4_2$ tetramers are.

D. Variants have been identified for all core histones except histone H3

E. While wrapping around the core histones, the structure of DNA is altered at the middle of the nucleosome core particle and exhibits increased number of base pairs per turn

Which one of the following combination of statements is most appropriate? **(DEC 2019)**

(a) A, C and E (b) A, B and D

(c) B, D and E (d) A, C and D

9. Eukaryotic genomes are organized into chromosomes and can be visualized at mitosis by staining with specific dyes. Heat denaturation followed by staining with Giemsa produced alternate dark and light bands. The dark bands obtained by this process are mainly **(JUNE 2011)**

(a) AT –rich and gene rich regions.
(b) AT –rich and gene desert regions.
(c) GC–rich and gene rich regions.
(d) GC–rich and gene desert regions.

10. Following observations were made about variations among genomes of eukaryotic organisms:

A. Single nucleotide polymorphisms are the numerically most abundant type of genetic variants
B. Both. interspersed and tandem repeated sequences can show polymorphic variation
C. Mitotic recombination between mispaired repeats causes change in copy number and generates minisatellite diversity in population
D. Smaller variable segments in the genome can be identified by paired end mapping technique

Select the option with all correct statements **(JUNE 2019)**

(a) A, B, C (b) A, C, D
(c) B, C, D (d) A, B, D

Answer Key

1.(b)	**2.**(d)	**3.**(c)	**4.**(b)	**5.**(c)	**6.**(a)	**7.**(d)
8.(a)	**9.**(c)	**10.**(d)				

8. Transposons

1. Transposons can be primarily categorized into two types, DNA transposons and retrotransposons. Given below is some information regarding the above

A. Eukaryotic DNA transposons excise themselves from one place in the genome and integrate into another site.
B. Retrotransposons are RNA sequences that are, first reverse transcribed into cDNA and then integrate into the genome.
C. Retrotransposons move by a copy and paste mechanism through an RNA intermediate.
D. As DNA transposons move via a cut and paste mechanism, there can never be an increase in the copy number of a transposon.

Which of the statement(s) is/are true? **(JUNE 2016)**

(a) A and C (b) B and D
(c) B only (d) D only

2. Given below are some statements about prokaryotic and eukaryotic mobile genetic elements or transposons.

A. Most mobile genetic elements in bacteria transpose via an RNA intermediate.
B. Most mobile genetic elements in bacteria are DNA.
C. Mobile genetic elements in eukaryotes are only retrotransposons.
D. Both, RNA and DNA transposons are found in eukaryotes.

Choose the correct combination. **(JUNE 2015)**

(a) A and C (b) B and C
(c) A and D (d) B and D

3. Hybrid dysgenesis in Drosophila is caused by P–elements. Which one of the following crosses between different cytotypes will lead to dysgenesis? (JUNE 2015)

(a) M–cytotype ♀ x M–cytotype ♂
(b) M–cytotype ♀ x P–cytotype ♂
(c) P–cytotype ♀ x M–cytotype ♂
(d) P–cytotype♀ x P–cytotype ♂

4. Following statements were made while describing the characteristics of transposable elements:

A. Most transposons use a common mechanism in which DNA is nicked to generate blunt ends for its subsequent incorporation.

B. Homologous recombination between the repeats of a transposon may result in precise or imprecise excision.

C. P elements are transposons which show differential splicing pattern in germline and somatic cells.

D. In a transposon, cis–acting mutants that may prevent its transposition are generally located near the ends.

E. An insertion sequence (IS) is a transposon, which is always flanked by long (more than 30 bases) direct repeats known as Long Terminal Repeats (LTR).

Which one of the following combinations of statements is correct? **(DEC 2019 ASSAM)**

1. A, C, E
2. B, C, D
3. C, D, E
4. A, B, D

5. In an experiment on transposition in an eukaryotic system, an intron was cloned within a transposable element and allowed to transpose from a plasmid to genomic DNA. The intron was found to be absent in the transposable element in its new location. It is **(JUNE 2013)**

(a) not a case of transposition.
(b) a case of replicative mode of transposition.
(c) a case of conservative mode of transposition.
(d) a retroposon.

6. An intron was cloned within a transposable element. Absence of the intron following transposition of the element, will indicate that it **(JUNE 2018)**

(a) follows conservative mode of transposition
(b) follows replicative mode of transposition
(c) is a retrotransposon
(d) is an insertion element

7. Which of the following statements is NOT true for transposable element system? **(MODEL PAPER)**

(a) It consists of both autonomous and non–autonomous elements.
(b) Dissociation elements are autonomous in nature.
(c) Transposase is transcribed by the central region of autonomous elements.
(d) Certain repeats in the genome remain fixed even after the element transposes out.

8. You have discovered a new transposon, TnX, and would like to identify its mode of replication. A heteroduplex of the TnX sequence is made with a few mismatches and introduced into bacteria. The newly transposed genomic loci are sequenced. You find that the sequence of the transposon matched exactly with one of its parent strands. This suggests that **(DEC 2017)**

(a) TnX transposes by conservative transposition mechanism.
(b) TnX transposes using a site–specific recombination mechanism.
(c) single–strands of the duplex are inserted.
(d) TnX transposes by replicative mechanism.

9. It is known that there is a large difference in the DNA content between two organisms with similar developmental complexity. This is due to large differences in the number of **(JUNE 2018)**

(a) transposable elements, repetitive DNA and coding sequences
(b) transposable elements and repetitive DNA
(c) introns and coding sequences
(d) introns and transposable elements

Answer Key

1.(a)	2.(d)	3.(b)	4.(b)	5.(d)	6.(c)	7.(b)
8.(d)	9.(b)					

9. Mitosis and meiosis cell

1. How many kinetochores are present in a human cell at mitosis? **(DEC 2019 ASSAM)**

(a) 46 (b) 23
(c) 92 (d) 84

2. The key determinant of the plane of cytokinesis in mammalian cells is the position of **(JUNE 2015)**

(a) chromosomes (b) central spindle
(c) centrioles (d) pre–prophase band

3. The site of the division plane during cytokinesis of animal cells is determined **(JUNE 2019)**

(a) by position of nucleus
(b) by the central spindle
(c) randomly
(d) by the pre–prophase band

4. The mammalian oocyte prior to sperm entry is arrested at what stage of cell division? **(JUNE 2015)**

(a) Prophase of mitosis
(b) Prophase of meiosis I
(c) G1 phase of mitotic cell cycle
(d) Metaphase of meiosis II

5. All cytosolic proteins have nuclear export signals that allows them to be removed from nucleus when it reassembles after **(JUNE 2012)**

(a) meiosis
(b) Mitosis
(c) both mitosis and meiosis
(d) DNA replication

6. In contrast with plant cells, the most distinctive feature of cell division in animal cells is **(DEC 2011)**

(a) control of cell cycle transition by protein kinase
(b) enzyme responsible for DNA replication
(c) ubiquitin–dependent pathway for protein degradation
(d) pattern of chromosome movement

7. Which of the following statements about meiosis is NOT true? **(DEC 2014)**

(a) Kinetochores of sister chromatids attach to opposite poles in Meiosis I.
(b) Kinetochores of sister chromatids attach to opposite poles in Meiosis II.
(c) Chiasma is formed in Prophase I.
(d) Homologous chromosomes are segregated in Meiosis I.

8. Rec 8 is a meiosis specific cohesin that maintains centromeric cohesion between sister chromatids in meiosis I. Which of the phenotypes listed below would you predict will be manifested in a rec8Δ. yeast? **(JUNE 2015)**

(a) Only low viability of dyads.
(b) Improper reduction division and low viability of tetrads.
(c) Improper equational division and low viability of dyads.
(d) Low tetrad viability with no effect on reduction division.

9. The frequency of cells in a population that are undergoing mitosis (the mitotic index) is a convenient way to estimate the length of the cell cycle. In order to measure the cell cycle in the liver of the adult mouse by measuring the mitotic index liver slices are prepared and stained to easily identify cells undergoing mitosis. It was observed that only 3 out of 250 cells are found to be undergoing mitosis.

Assuming that M phase lasts 30 minutes, calculate the approximate length of the cell cycle in the liver of an adult mouse? **(DEC 2015)**

(a) 76 hours (b) 50 hours
(c) 42 hours (d) 21 hours

10. Chromatin condensation is driven by protein complexes, called condensins which are members of a family of "structural maintenance of chromatin" (SMC) proteins that play a key role in the organization of eukaryotic chromosomes. Condensins along with another family of SMC proteins called cohesins significantly contribute to chromosome segregation during mitosis.

If the cells are treated with an inhibitor of cdk 1 phosphorylation immediately before. the cells. enter M phase. which of the. following statements is most likely to be true? **(JUNE 2015)**

(a) Sister chromatids are held together by condensins along the entire length of the, chromosome.

(b) Sister chromatids are held together by cohesins along the entire length of the chromosome.

(c) Sister chromatids are held together by condensins and attached to each other only, at the centromere.

(d) Sister chromatids are held together by condensins and attached to each other only at the telomere.

Answer Key

1.(c)	2.(b)	3.(b)	4.(d)	5.(b)	6.(d)	7.(a)
8.(b)	9.(c)	10.(b)				

10. Cell Cycle and Regulation

1. If the total DNA content of a cell at interphase is 4C, what will be the DNA content of the cell when it is at Anaphase II? **(DEC 2013 GU)**

(a) 2C (b) C

(c) ½ C (d) ¼ C

2. Cells in S– phase of the cell cycle were fused to cells in the following stages of cell cycle: (a) G_1 phase, (b) G_2 phase, (c) M phase these cells were then grown in the medium containing tritiated thymidine. The maximal amount of freshly labelled DNA is likely to be obtained in S– phase cells fused with **(DEC 2016)**

(a) G_1 phase cells

(b) G_2 phase cells

(c) M phase cell

(d) Both G_1 and G_2 phase cells

3. Following observation was made when a mammalian cell in one phase of cell cycle was fused with a cell in another phase of cell cycle:

A. Fusion of a cell in G1 phase with S phase caused the G1 nucleus to enter S phase.

B. Upon fusion of a G2 cell with an S phase cell, G2 cell does not enter S phase.

C. Upon fusion of a G1 cell with G2 cell, G1 nucleus enters G2 phase.

D. Fusion of an S phase cell with a M phase causes the S phase cell to immediately enter mitosis.

Choose the combination with all correct statements. **(DEC 2018)**

(a) A, B, C (b) A,C, D

(c) B,C,D (d) A, B, D

4. Mitotic cyclin–CDK activity peaks in M phase. This is because **(DEC 2016)**

(a) Mitotic cyclin is synthesized only in M phase.

(b) Threshold level of mitotic cyclin accumulates only in late G_2.

(c) Cyclin subunit is activated by phosphorylation only in M phase.

(d) The kinase subunit is activated by dephosphorylation only in M phase.

5. Given below are events in the cell cycle.

(a) Phosphorylation of lamin A, B, C

(b) Phosphorylation of Rb (Retinoblastoma protein)

(c) Polyubiquitination of securin

(d) Association of inner nuclear membrane proteins and nuclear pore complex proteins with chromosomes.

Which one of the following reflects the correct sequence of events in the mammalian cell cycle? **(DEC 2015)**

(a) a→b→c→d (b) b→c→d→a

(c) c→a→b→d (d) b→a→c→d

6. Which one of the following statements about M–phase cyclin is correct? **(DEC 2019 ASSAM)**

(a) Cyclin synthesis and destruction is essential for cell cycle progression

(b) Synthesis of new cyclin, but not destruction is essential for cell cycle progression

(c) Cyclin and the corresponding kinase(s), both exhibit cyclic expression pattern during cell cycle

(d) Cyclic pattern of cyclin protein is regulated by the controlled translation of stored cyclin transcripts, and not dependent on fresh transcription.

7. Which of the cyclins have/has essential functions in S–phase of cell cycle? **(JUNE 2011)**

(a) A–type.

(b) B–type.

(c) D–type.

(d) Both B– and D–types.

8. During eukaryotic cell division, metaphase to anaphase transition is regulated by degradation of **(DEC 2017)**

(a) Cyclin B1

(b) CDK 1

(c) Aurora A kinase

(d) Polo–like kinase

9. In metazoan cell cycle, metaphase to anaphase transition is regulated by the activity of **(DEC 2018):**

(a) Cdkl/cyclinB (b) APC/C

(c) Cdc25 (d) Wee1

10. Many cancers carry mutant p53 genes, while some cancers have normal p53 genes. p53 activates p21 (Waf1) which inhibits G 1/S–Cdks, and phosphorylation of the retinoblastoma protein (Rb). Cancers with normal p53 genes could **(DEC 2011)**

(a) express non–functional form of Rb

(b) express high levels of p53–deubiquitinases

(c) express inactive forms of G1/S–cdks

(d) express inactive forms of G1/S cyclins.

11. Methotrexate is an anti cancer compound which targets dihydrofolate reductase (DHFR). It acts as **(DEC 2013 GU)**

(a) a competitive inhibitor of DHFR and is therefore cytotoxic during S phase of the cell cycle.

(b) a non–competitive inhibitor of DHFR and is therefore cytotoxic during G1 phase of the cell cycle.

(c) an activator of DHFR and is therefore cytotoxic during S phase of the cell cycle.

(d) an activator of DHFR and is therefore helps in the proliferation of cancer cells.

12. During cell cycle progression from G_1 to S, cyclin D–CDK4 phosphorylates Rb and reduces its affinity for E2F. E2F dissociates from Rb and activates S–phase gene expression. Overexpression of protein 'A' arrests G_1 phase progression. Which of the following statement is true? **(DEC 2016)**

(a) 'A' inhibits Rb–E2F interaction

(b) 'A' inhibits CDK4 activity

(c) 'A' phosphorylates E2F

(d) 'A' degrades Rb

13. Cell cycle checkpoints are surveillance mechanisms that ensure order and fidelity of events of the cell cycle. Given below are some of the checkpoint proteins and their functions.

A	Mad2	(i)	Prevention of Cdc14 activation
B	Tem 1	(ii)	Prevention of Cdc20 activation
C	ATM,ATR	(iii)	Inhibition of $p21^{CIP}$
		(iv)	Inhibition of separase action

Match the checkpoint protein with its function. **(DEC 2017)**

(a) A – iv, B – i and ii, C – iii

(b) A – ii and iv, B – i, C – iii

(c) A – ii, B – i and iii, C – iv

(d) A – ii and iv, B – iii, C– i

14. If a proteasome inhibitor is added to synchronously cycling human cells in G2 phase which one of the following events is likely to happen? **(DEC 2014)**

(a) Induce re–replication of DNA
(b) Arrest cells in G2 phase
(c) Arrest cells in anaphase
(d) Block chromatin condensation

15. During mitosis in a eukaryotic cell incorrect kinetochore attachments are corrected by a system of trials and errors. The tension–based mechanism acting at the kinetochore is linked with ensuring that sister chromatid pairs are properly oriented at the spindle. Which one of the following kinases is tethered to the inner kinetochore and phosphorylates the microtubule attachment site, including the Ndc 80 complex to regulate sister chromatids attachments in a bi–orientation manner? **(DEC 2019 ASSAM)**

(a) Aurora A kinase
(b) Aurora B kinase
(c) Aurora C kinase
(d) C–Jun N–terminal kinase

16. Toward the end of M phase the concentration of M phase cyclins drops precipitously and remains low until the next cell cycle. In this case how M–Cdk activity coordinates the activities of Cdh1–anaphase promoting complex (APC)? **(DEC 2013 GU)**

(a) M phase cyclins are destroyed by proteasome digestion following ubiquitination by Cdh1– APC
(b) M phase cyclins are not destroyed by proteasome digestion following ubiquitination by Cdh1– APC
(c) Cdh1–APC never influences M phase cyclins
(d) M phase cyclins are influenced by Cdh1–APC following fatty acylation

17. During cell cycle, entry in the S–phase is tightly regulated. This is possible because:

A. APC/C promotes ubiquitination of S–phase cyclins and mitotic cyclins, marking them for proteolyses at the mitotic exit.
B. Cyclin B1 helps in the activation of S–phase CDKs only in late G1.
C. As mitotic CDK activity declines in late mitosis, cdc14 phosphatase activates APC/C by dephosphorylating Cdh1, thus promoting formation of APC/C^{cdh1}.
D. Securin keeps S–phase cyclins in inactive state till late G1.

Which one of the options represents all correct statements? **(DEC 2019)**

(a) A and B
(b) A and C
(c) B and C
(d) B and D

18. During many important cell processes, many proteins need to undergo degradation to culminate a part of the process. For example, during cell cycle, cycling proteins need to be degraded to allow the cells to exit mitosis. This is achieved by selective ubiquitination of cyclin followed by its degradation by proteasomes. The specific protein factor that is involved in this process is called Anaphase Promoting Complex (APC). APC is possibly a protein which is known as **(MODELPAPER)**

(a) E1 enzyme.
(b) E2 enzyme.
(c) E3 enzyme.
(d) Protease

19. Maturation–promoting factor (MPF) controls the initiation of mitosis in eukaryotic cells. MPF kinase activity requires cyclin B. Cyclin B is required for chromosome condensation and breakdown of the nuclear envelope into vesicles. Cyclin B degradation is followed by chromosome decondensation, nuclear envelope reformation and exit from mitosis. This requires ubiquitination of a cyclin destruction box motif in cyclin B. RNase–treated Xenopus egg extracts and sperm chromatin were, mixed. MPF activity increased with chromosome condensation and nuclear envelope breakdown. However, this was not followed by chromosome decondensation and nuclear envelope reformation because

(DEC 2011)

(a) RNase contamination persisted in the system
(b) cyclin B was missing from the system
(c) ubiquitin ligase had been overexpressed
(d) cyclin B lacking the cyclin destruction box had been overexpressed

20. Budding yeast cells that are deficient for Mad2, a component of the spindle–attachment–check point, are killed by treatment with benomyl, which causes microtubles to depolymerise. In the absence of benomyl, however, the cells are perfectly viable. Which explanation out of the following is able to justify this observation?

(JUNE 2011)

(a) In the absence of benomyl, the majority of spindles forms normally and the spindle–attachment checkpoint (Mad2) plays no role.
(b) In the presence of be no my I, the majority of spindles form normally and Mad2 plays critical role in cell survival.
(c) Other than the role in cell survival, microtuble depolymerization affects oxidative phosphorylation in the absence of Mad2.
(d) Benomyl also affects protein synthesis in the absence of Mad2.

21. In *Schizosaccharomyces pombe*, the recessive ($cdc2^r$) and dominant ($cdc2^D$) mutants have opposing phenotypes. While $cdc2^D$ produces abnormally small cells, $cdc2^r$ produces abnormally long cells.

Some possible explanations are given below.

A. $cdc2^D$ may lack interaction with WEE1
B. $cdc2^r$ may not interact with CDC13 kinase
C. $cdc2^D$ may not interact with CDC25 phosphatase
D. $cdc2^r$ cells may be deficient in interaction with either CDC25 or WEE1

Which combination of the above statements is correct? **(JUNE 2017)**

(a) A and B only (b) A, C and D only
(c) B and C only (d) A, B and C only

22. In eukaryotes, a specific cyclin dependent kinase (CDK) activity is required for the activation of loaded helicases to initiate replication. On the contrary, this CDK activity inhibits the loading of helicases onto the origin of replication. Considering the fact that during each cycle, there is only one opportunity for helicases to be loaded on to origins and only one opportunity for these helicases to be activated, which one of the following graphs best depicts this CDK activity in G_1 and S phases of cell cycle? **(DEC 2016)**

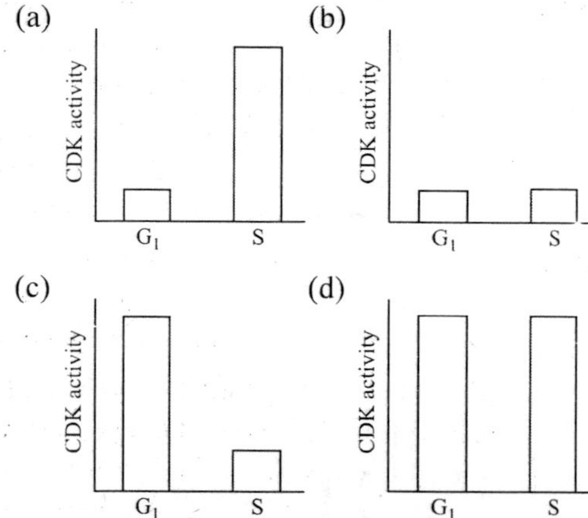

23. Cyclins are the regulatory subunits and cyclin–dependent kinases (CDKs) are the catalytic subunits. Following diagram represents the involvement of cyclins and CDKs in various stages of cell cycle:

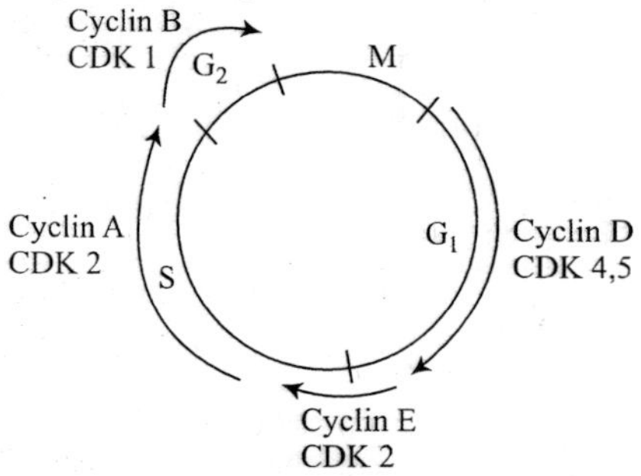

If we knock down cyclin D in a cell by shRNA, which one of the following graphs correctly represents the level of CDK2 activity?

(JUNE 2014)

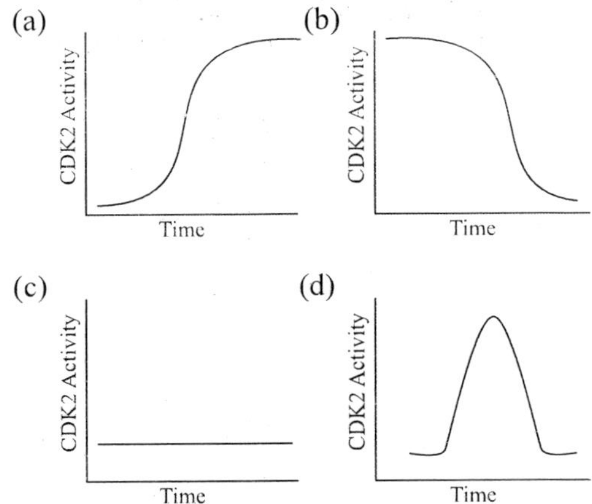

24. Cell cycle is regulated by various cyclins and cyclin dependent kinases (CDK). On receiving mitotic stimuli, cyclin D, the first cyclin expressed, binds with existing CDK4 to form the active cyclin D–CDK4 complex. This in turn phosphorylates retinoblastoma protein (Rb) which activates E2f to further activate the transcription of various downstream cyclins. In a particular cell type there is a mutation in Rb such that it cannot be phosphorylated. What will be the correct expression pattern of cyclin E in these cells after mitotic stimulation?

(DEC 2013)

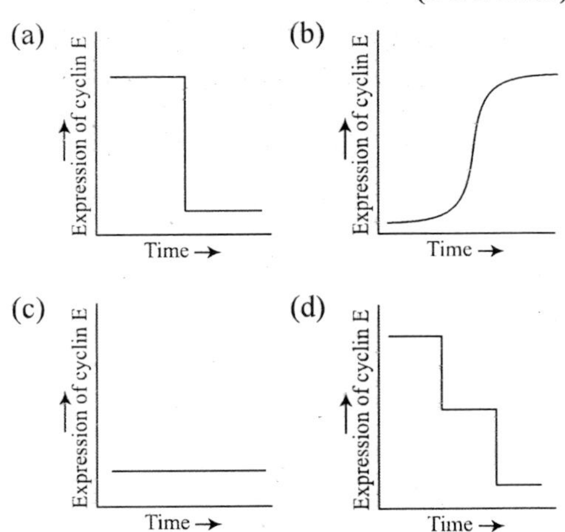

25. Hoechst 33342 is a membrane–permanent dye that fluoresces when it binds to DNA through intercalating process. If a population of cells is incubated briefly with Hoechst dye and stored in flow cyclometer the cells display various levels of fluorescence in different phases of cell cycles as shown in figure below (marked as X,Y,Z)

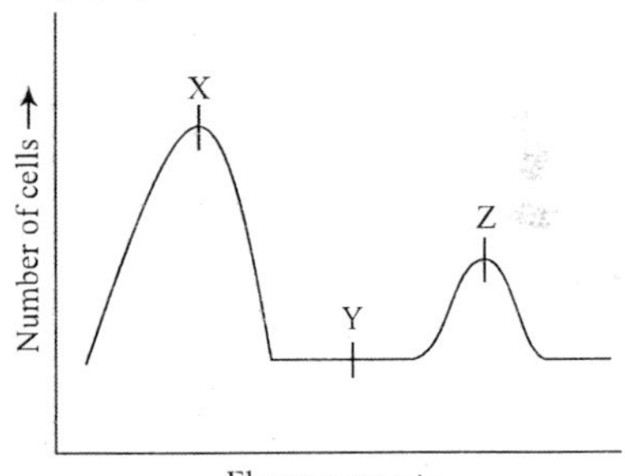

Which of the following is correct?

(JUNE 2012)

(a) X is G_1, Y is G_2+M and Z is S
(b) X is G1, Y is S and Z is G2+M
(c) X is S, Y is G_2+M and Z is G_1
(d) X is S, Y is G_1 and Z is G_2 +M

26. A population of proliferating cells is stained with a DNA binding fluorescence dye so that the amount of fluorescence is directly proportional to the quantity of DNA. Amount of DNA in each cell is measured by flow cytometry. The number of cells with a given DNA content is plotted and following statements were made:

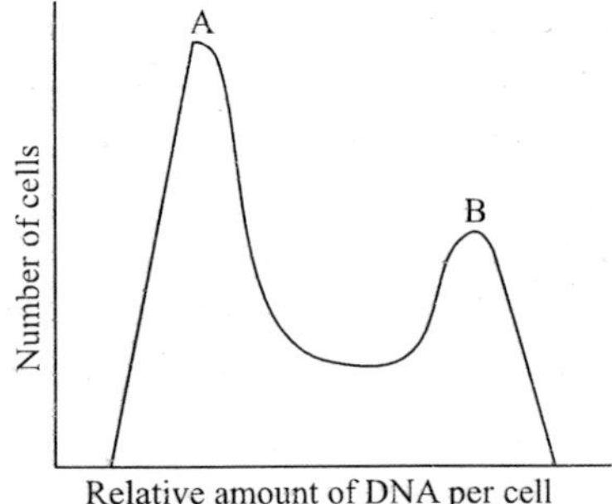

A. Peak A contains the cells of G1 phase
B. Peak B contains the cells of G2 phase
C. Peak A contains the cells of G2 phase
D. Peak B contains the cells of G1 phase

Which of the above mentioned statement is correct? **(MODELPAPER)**

(a) A and B (b) C and D
(c) B and D (d) A and C

27. To assess the impact of a newly identified drug when added to culture of subconfluent HeLa cells, a researcher analyzes the fluorescence activated cell sorting (FACS) profile of untreated (– Drug) versus treated(+ Drug) cells.

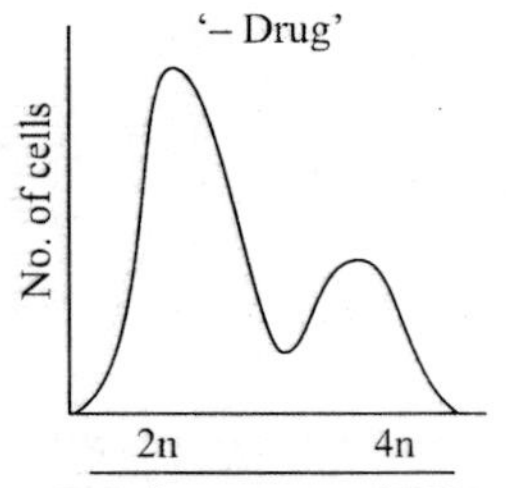

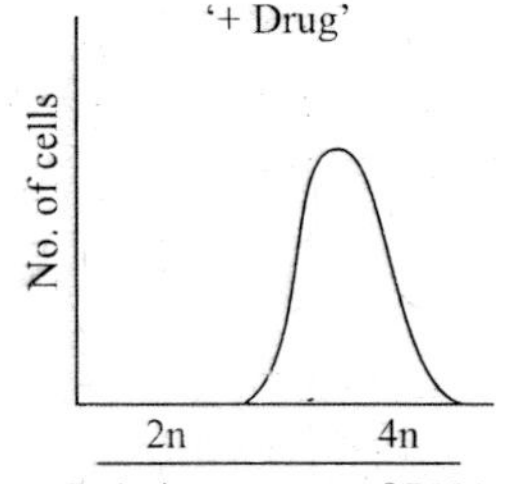

Based on the FACS profile shown above, this drug inhibits **(DEC 2017)**

(a) G_1 phase of the cell cycle
(b) S phase of the cell cycle
(c) G2/M phase of the cell cycle
(d) G_0 phase of cell cycle

28. Cell that grow and divide in a medium containing radioactive thymidine covalently incorporate the thymidine into their DNA during S phase. Consider a simple experiment in which cells are labelled by a brief (30 minutes) exposure to radioactive thymidine. The medium is then replaced with one containing unlabeled thymidine and the cells grow and divide for some additional time.

At different time points after replacement of the medium, cells are examined under a microscope. Cells in mitosis are easy to recognize by their condensed chromosomes and the fraction of mitotic cells that have radioactive DNA can be estimated by autoradiography and plotted as a function of time after the thymidine labeling as in the figure below:

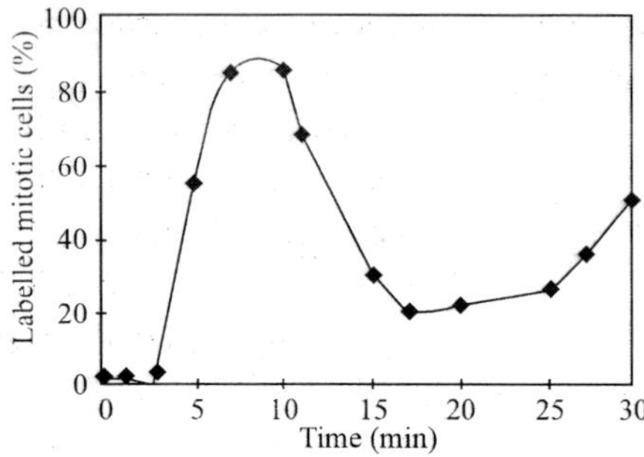

The rise and fall of the curve is because: **(JUNE 2013)**

(a) initial rise of the curve corresponds to cells that were just finishing DNA replication when radioactive thymidine was added (S phase).
(b) the peak of the curve corresponds to cells in M phase.
(c) the rise in curve after 20 min corresponds to cells in apoptotic phase.
(d) the fall in curve after 1 0 min indicates the cells exiting M phase.

29. In a given experiment the cells were labeled for 30 minutes with radioactive thymidine. The medium was then replaced with that containing unlabelled thymidine and the cells wire grown for additional time. At different time points after replacement of medium the fraction of mitotic cells were analysed. Based on the results obtained, the above figure was drawn which shows the percentage of mitotic cells that are labeled as a function of time after brief incubation with radioactive thymidine.

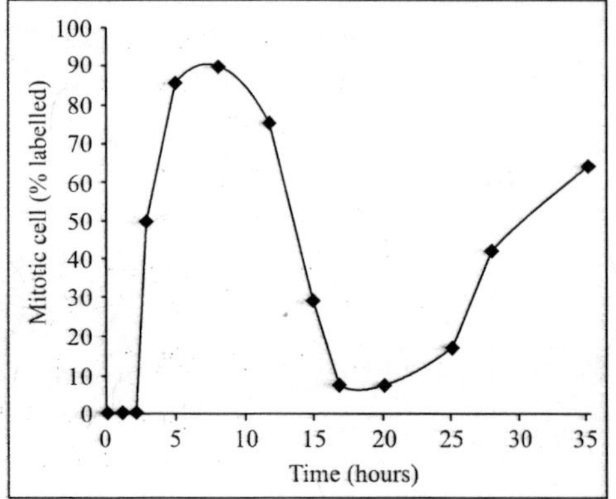

Considering the above experiment, the following statements were made:

A. Cells in the S–phase of the cell cycle during the 30 minute labeling period contain radioactive DNA.

B. It takes about 3 hours before the first labeled mitotic cells appear.

C. The cells enter the second round of mitosis at t_{30} hours.

D. The total length of the cell cycle is about 27 hours with G_1, being more than 15 hours.

Which of the combination of above statements is correct? **(DEC 2012)**

(a) A and B (b) B and C

(c) C and D (d) A and D

30. During cell cycle regulation in eukaryotes, there are post–translational modifications of protein factors, which act as switches for different phases of cell cycle. A cell population of yeast was transfected with gene for wee 1 kinase (modifies cdc2 protein). Assuming that the transfection efficiency was 50% only, which of the following graphical representation of the results is most appropriate? **(JUNE 2012)**

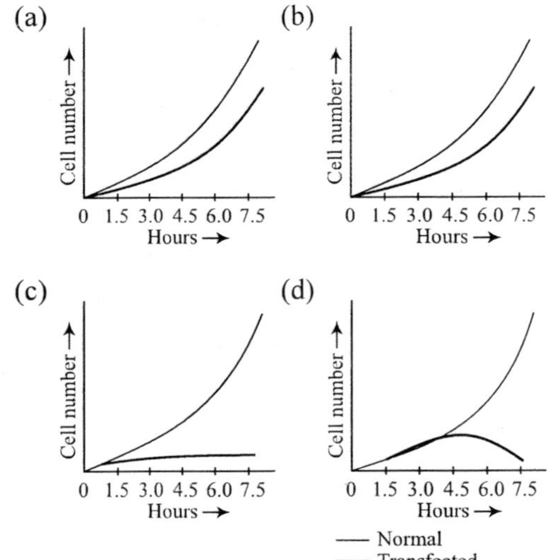

31. Each origin of replication is activated only once. This is achieved because **(JUNE 2014)**

(a) pre–replicative complex can only form in G1 and replication can only be initiated when pre–replication complex is disassembled at the beginning of S–phase.

(b) replication can only be initiated when pre–replication complex is intact.

(c) replication can only be initiated when unphosphorylated Rb is present.

(d) pre–replicative complex can only form in S–phase

32. In eukaryotic cells, replication initiation from a replication origin occurs only once per cell cycle and S– phase CDKs play a vital role in the regulation of DNA replication. In budding yeast a protein complex known as the origin recognition complex (ORC) is associated with DNA replication origin during G1; however origins fire only once at the beginning of S–phase. DNA replication does not start in G1 because:

A. MCM helicases are inactive in G1.

B. Spindle checkpoint is active in G1.

C. DNA polymerase is not recruited in G1.

D. ORC and initiation factors Cdc6 and Cdt1 do not recruts MCM helicases to the site of replication initiation in G1.

Which of the above statements are correct? **(JUNE 2019)**

(a) A and B (b) A and C

(c) B and C (d) B and D

33. In human cells, centrioles assemble centrosomes, and centrosomes are essential for microtubular nucleation. Centriole duplication is tightly coupled to the cell cycle progression. To discover which genes are crucial for centriole duplication, an investigator will conduct:

A. RNAi–based functional genomics screen in human cells to uncover novel genes involved in centriole duplication.

B. Experiments to generate constitutive knockout human cell lines.

C. Chemical based screen in human cell lines.

D. RNA sequencing followed by generating constitutive knockout human cell lines.

Choose the combination with correct statements: **(DEC 2019 ASSAM)**

(a) A and B (b) A and C

(c) B and C (d) B and D

Answer Key

1.(b)	**2.**(a)	**3.**(d)	**4.**(d)	**5.**(d)	**6.**(a)	**7.**(a)
8.(a)	**9.**(1/2)	**10.**(a)	**11.**(a)	**12.**(b)	**13.**(*)	**14.**(c)
15.(b)	**16.**(a)	**17.**(b)	**18.**(c)	**19.**(d)	**20.**(a)	**21.**(a)
22.(a)	**23.**(c)	**24.**(c)	**25.**(b)	**26.**(a)	**27.**(c)	**28.**(a)
29.(a)	**30.**(a)	**31.**(a)	**32.**(b)	**33.**(b)		

11. Microbial Growth, Physiology and Stress

1. The dye used in Gram staining is (JUNE 2014)

(a) Rhodamine (b) Methylene blue.

(c) Giemsa (d) Crystal violet.

2. A fixed smear of a bacteria culture is subjected to the following solutions in the order listed below and appeared red,

(a) Carbolfuchsin (heated)

(b) Acid–alcohol

(c) Methylene blue

Bacteria stained by this method can be identified as **(DEC 2011)**

(a) Non–acid fast *E. coli*

(b) Acid fast *Mycobacterium sp.*

(c) Gram–positive *E. coli*

(d) Gram–negative *Mycobacterium sp.*

3. An organism that has peroxidase and superoxide dismutase but lacks catalase is most likely an **(DEC 2012)**

(a) Aerotolerant anaerobe.

(b) Aerotolerant aerobe.

(c) Obligate anaerobe.

(d) Facultative anaerobe.

4. The table given below lists organisms (column A) and characteristic features (column B).

A	B
(a) Caulobacter	(i) Multicellular fruiting body
(b) Myxobacteria	(ii) Endospore
(c) Methylotroph	(iii)Non–free living, Penicillin resistant
(d) *Bacillus subtilis*	(iv) Immortal stalk cells
(e) Mycoplasma	(v) Can use formate, cyanide and carbon monoxide as a source of carbon

Choose the option that correctly matches organisms with their characteristic features. **(JUNE 2017)**

(a) (a)–(i);(b)–(v);(c)–(iv);(d)–(ii);(e)–(iii)

(b) (a)–(iv);(b)–(i);(c)–(v);(d)–(ii);(e)–(iii)

(c) (a)–(iv);(b)–(v);(c)–(i);(d)–(iii);(e)–(ii)

(d) (a)–(ii);(b)–(i);(c)–(v);(d)–(iv);(e)–(iii)

5. Major stimulus for spore formation in bacteria is **(JUNE 2012)**

(a) Nutrient limitation

(b) heat stress

(c) Cold stress

(d) pH stress

6. For bacterial growth, a single cell elongates in size before it divides into two, in a process called binary fission. During cell growth, **(DEC 2019)**

(a) new peptidoglycan synthesis is required along with the hydrolysis of bonds linking the old peptidoglycan chains.

(b) new peptidoglycan synthesis is required but no hydrolysis of the old peptidoglycan occurs.

(c) the old peptidoglycan is completely degraded and replaced with the newly synthesized longer polymer.

(d) newly synthesized peptidoglycan is utilized to deposit a new layer of the peptidoglycan in the cell wall.

7. A bacterial culture undergoing balanced growth requires a first–order reaction. In other words, the rate of increase in bacteria at any particular time is proportional to the number or mass of bacteria present at that time. If 'N' is the number of cells at any given time 't and 'μ' is "growth rate constant" this process can be expressed mathematically as

(MODEL PAPER)

1. $\frac{dN}{dt} = \mu N.$ 2. $\frac{dt}{dN} = \mu N.$

3. $\frac{d\mu}{dt} = N.$ 4. $\frac{dN}{dt} \times N = \mu.$

8. During the exponential phase of growth, if *No*, *Nt*, and *n* define the initial population number, population number at time *t*, and the number of generations in time *t*, respectively, then

(DEC 2019)

(a) *Nt* = *No x 2n* (b) *No* = *Nt/2*

(c) $Nt/No = 2^n$ (d) $No/Nt = 2^n$

9. A rapidly growing bacterial species such as E. coli exhibits a typical phase of growth cycle in liquid nutrient broth (Iagphase → log phase → stationary phase → death phase).

If a bacterial culture has starting density of 10^3 cells/ml has a lag time of 10 minutes and a generation time of 10 minutes, what will the cell density be at (cells/ml) 30 minutes?

(JUNE 2013)

(a) 6.0×10^3 (b) 2.0×10^3

(c) 3.0×10^3 (d) 4.0×10^3

10. The Scatter plot of growth rate and growth yield for 100 random environment isolates of bacteria is shown below

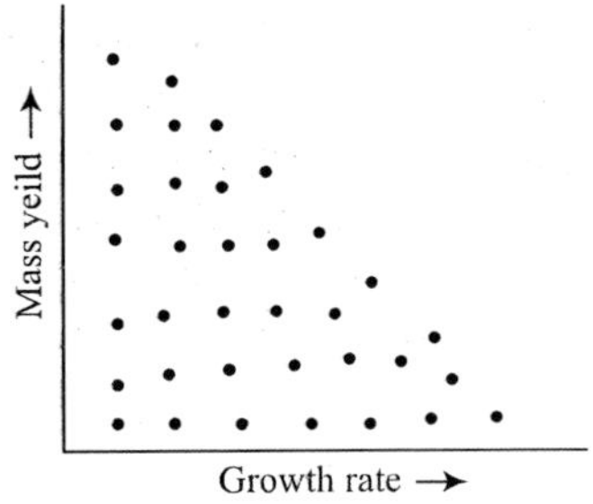

Which of the following can be inferred from data? **(JUNE 2012)**

(a) The two parameters are not related.

(b) Growth rate is inversely proportional to growth yield.

(c) Growth yield is negatively correlated with growth rate.

(d) High growth rate cannot be accompanied by high growth yield.

11. The overall length of the cell cycle can be measured from the doubling time of a population of exponentially proliferating cells. The doubling time of a population of mouse L cells was determined by counting the number of cells in samples of culture at various times. What is the overall length of the cell cycle in mouse L cells? **(JUNE 2011)**

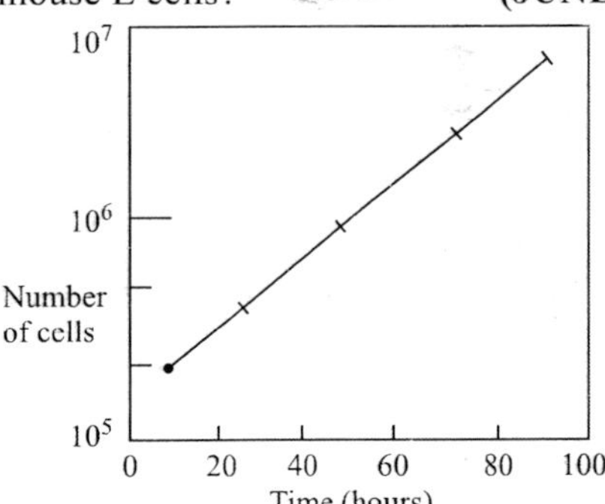

(a) 30 h (b) 20h

(c) 10 h (d) 40 h

12. To test whether bacteria with enhanced toluene degradative abilities could be created for cold environment, a TOL (toluene–degrading) plasmid from a mesophilic bacterial strain was transferred by conjugation into a facultative psychrophile. The psychrophile was able to degrade salicylate (SAL) but donot toluene. The recombinant strain carried the introduced TOL– plasmid and its own SAL plasmid. The results are as follows:

Optimum Growth in Presence of						
	X			Y		
Temp (°C)	A	B	C	A	B	C
37°C	–	–	–	+	–	–
25°C	–	–	–	+	–	–
10°C	–	+	+	–	–	+
0°C	–	+	+	–	–	+

Identify A, B, C, X and Y: **(JUNE 2014)**

(a) A. mesophile B. psychrophile C. transformant X. Toluene Y. Salicylate

(b) A. mesophile B. psychrophile C. transformant X. Salicylate Y. Toluene

(c) A. psychrophile B. transformant C. mesophile X. Salicylate Y. Toluene

(d) A. transformant B. psychrophile C. mesophile X. Toluene Y. Salicylate

13. Addition of antibiotic cephalexin to growing *E. coli* cells lead to filamentation of the cells, followed by lysis. Cephalexin is an inhibitor of **(DEC 2016)**

(a) Protein synthesis

(b) DNA synthesis

(c) Peptidoglycan synthesis

(d) RNA polymerase

14. A bacterial culture was in log phase in the following figure. At time x, an antibacterial compound was added to the culture.

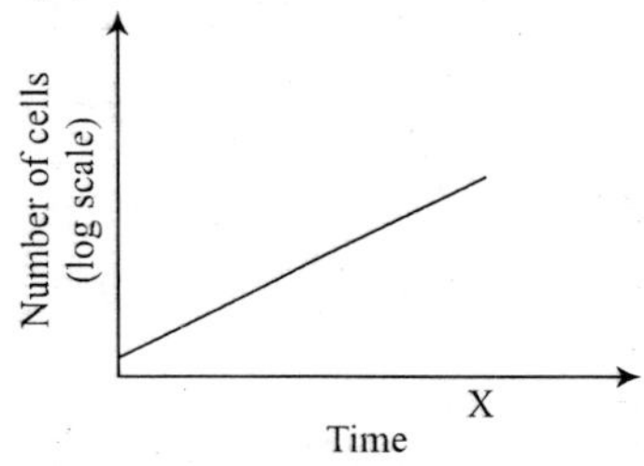

Which of the following lines in the growth curve represents the antibacterial activities of the compound? **(DEC 2013)**

(a)

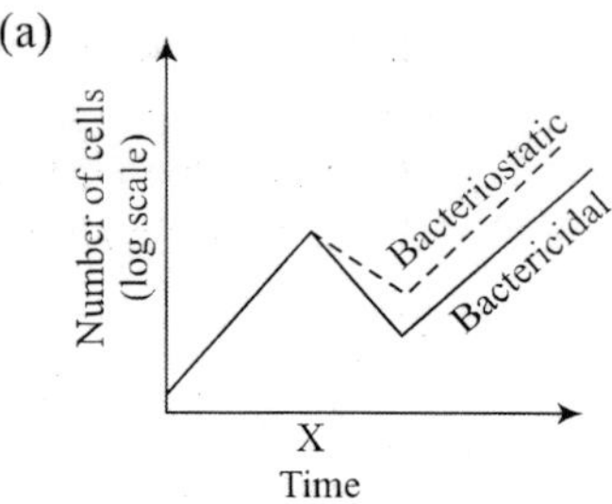

(b)

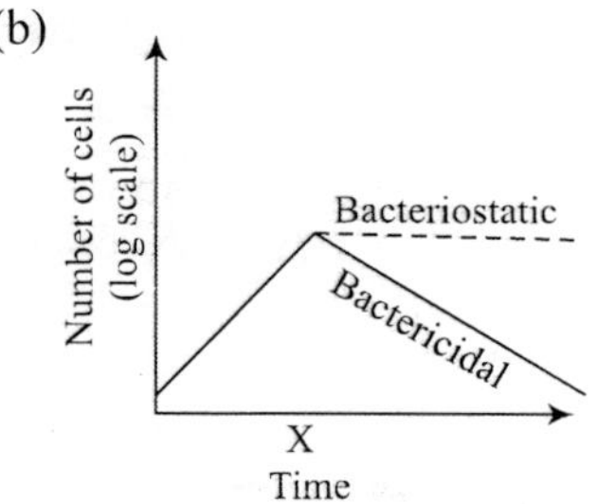

(c)

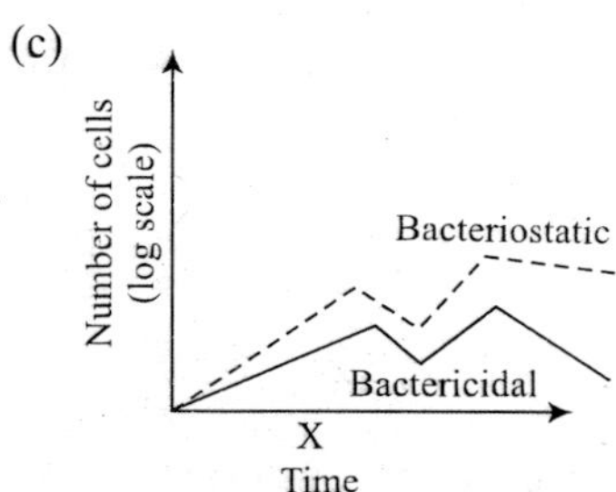

(d)

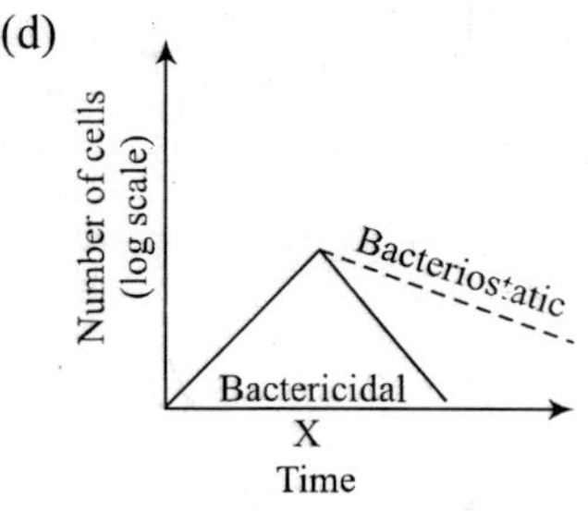

Answer Key

1.(d)	2.(d)	3.(a)	4.(b)	5.(a)	6.(a)	7.(a)
8.(c)	9.(d)	10.(d)	11.(b)	12.(b)	13.(c)	14.(b)

SECTION-3
MOLECULAR BIOLOGY

1. DNA Replication

1. In bacteria chromosomal DNA replication stops at **(DEC 2013)**
   (a) one specific locus.
   (b) several specific loci.
   (c) a single locus, randomly.
   (d) from several loci, randomly.
2. Origin of replication usually contains **(JUNE 2013)**
   (a) GC rich sequences.
   (b) both AT and GC rich sequences.
   (c) no particular stretch of sequences.
   (d) AT rich sequences.
3. During DNA replication, events at the replication fork require different types of enzymes having specialized functions except **(DEC 2012)**
   (a) DNA polymerase III.
   (b) DNA gyrase.
   (c) DNA ligase.
   (d) DNA glycosylase.
4. In E. coli, the major DNA unwinding protein helicase, DnaB translocates in (DEC 2013 GU)
   (a) 5' $\rightarrow$ 3' direction along the lagging strand template.
   (b) 3' $\rightarrow$ 5 'direction along the lagging strand template.
   (c) 5' $\rightarrow$ 3' direction along the leading strand template.
   (d) 3' $\rightarrow$ 5' direction along the leading strand template.
5. You have labelled DNA in a bacterium by flowing cells in medium containing either ^{14}N nitrogen or the heavier isotope, ^{15}N. Furthermore, you have isolated pure DNA from these organisms, and subjected it to CsCl density gradient centrifugation leading separation of light (^{14}N) and heavy (^{15}N) forms of DNA to different locations in the centrifuge tube. In the next experiment, bacteria were regrown first in medium containing ^{15}N, so that all the DNA made by cells will be in heavy form. Then these cells were transferred to medium containing only ^{14}N and allowed the cells to divide for one generation. DNAs were extracted and centrifuged as above in the CsCl gradient. A hybrid DNA band was observed at a positions located between and equidistant from the ^{15}N and ^{14}N DNA bands. Based on the above observation, which one of the following conclusions is correct **(DEC 2015)**
   (a) Replication of DNA is conservative
   (b) Replication of DNA is semi-conservative
   (c) Replication of DNA is dispersive
   (d) Replication by rolling circle mode
6. *E. coli* cells were grown in N^{15} medium for several generations and then shifted to normal medium for one generation. If the DNA isolated from the culture would be centrifuged on a CsCl equilibrium density gradient, the result will be (MODEL PAPER)
   (a) a single band of double helix DNA consisting of one strand with N^{14} and another with N^{15} label.
   (b) single band of double helix DNA consisting of N^{14} and N^{15} in both the strands.
   (c) two bands containing double helix DNA each containing both N^{14} and N^{15} label.
   (d) two bands containing single stranded DNA one with N^{14} and other with N^{15} label.
7. The semi-conservative nature of DNA replication was established by Meselson and Stahl in their classic experiment with bacteria. They grew bacteria in N^{15}-NH_4Cl containing medium, washed and then incubated in fresh medium with N_{14}- containing compounds and allowed to grow for three generations. CsCl density gradient centrifugation of isolated DNA established the nature of semiconservative DNA replication. The pictorial representation below shows the position of differentially labeled DNA in CsCl density gradient.

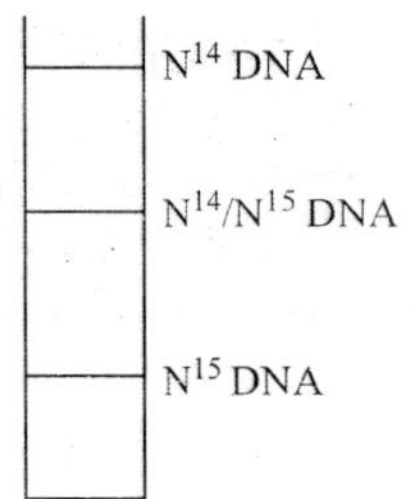

Had the DNA replication been conservative, what would have been the pattern?

(DEC 2012)

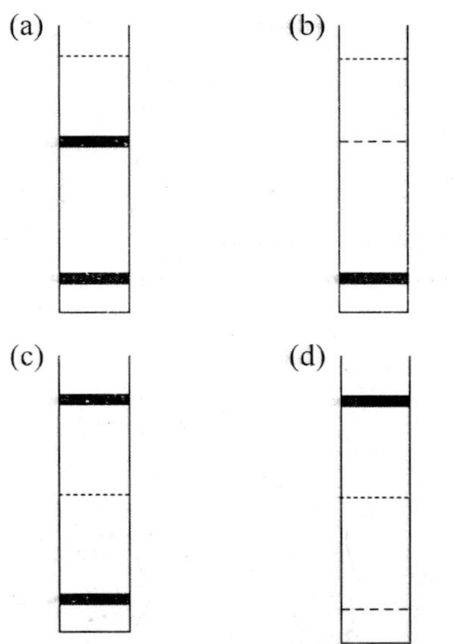

8. In semi-conservative mode of DNA replication two parental strands unwind and are used for synthesis of new strands following the rule of complimentary base pairing. Synthesis of complimentary strands require that DNA synthesis proceeds in opposite direction, while the double helix is progressively unwinding and replicating in only one direction.one of the DNA strands is continuously synthesised in the same direction as the advancing replication fork and is called leading strand whereas the other strands is synthesised discontinuously in segments and is referred to as lagging strands. These short fragments made discontinuously are labelled as okazaki fragments. These okazaki fragments need to be matured into continuous DNA strand by which one of the following combination of enzymes?

(JUNE 2012)

(a) DNA Pol III and DNA ligase
(b) DNA pol I and DNA ligase
(c) DNA pol II and DNA ligase
(d) DNA gyrase and DNA ligase

9. In *E. coli,* though DNA polymerase I (Pol I) plays an essential role in the replication process, it is not the major polymerase. Instead, the enzyme responsible for advancement of replication fork is Pol III. From the four DNA structures (A, B, C and D) given below, students made several interpretations about the shorter arm being extended by Pol I and/or Pol III.

A. PO4-5′ ——— $3'PO_4$
HO-3′ ——— $5'PO_4$

B. HO-3′ ——— $5'PO_4$
HO-5′ ——— 3′-OH

C. PO_4-5′ ——— 3′OH
HO-3′ ——— $5'PO_4$

D. PO_4-5′ ——— 3′-OH
HO-3′ ——— $5'PO_4$

Which one of the interpretation written below is correct? **(DEC 2017)**

(a) A will be extended by Pol III but not by Pol I.
(b) Neither B nor C will be extended by either Pol I or Pol III.
(c) C will be extended by both Pol I and Pol III.
(d) D will be extended only by Pol I, but not by Pol III.

10. The following statements were made about bacterial DNA gyrase, which introduces negative supercoiling in DNA.

A. If gyrase activity is inhibited, there is an increase in replication initiation at the origin.
B. If gyrase activity is inhibited, there is a decrease in replication initiation at the origin.

C. Gyrases are usually ATP-dependent.

D. When gyrase introduces negative supercoils, it decreases, the linking number of a DNA loop by two.

Which of the above statement(s) is/are NOT true? **(DEC 2013 GU)**

(a) C and D (b) Only B

(c) B and C (d) Only A

11. Although ribonucleoside triphosphates (rNTPs) are present at approximately 10-fold higher concentıation than deoxyribonucleoside triphosphates (dNTPs) in the cell but they are incorporated into DNA at a rate that is more than 1000-fold lower than dNTPs. This is because **(DEC 2015)**

(a) DNA polymerase cannot discriminate between dNTPs and rNTPs. But as soon as rNTPs are incorporated in the DNA chain, they are hydrolyzed due to the presence of 2'-OH group.

(b) DNA polymerase cannot discriminate between dNTPs and rNTPs. But IIS soon as rNTPs are incorporated in the DNA chain, they are excised by the proof reading activity of DNA polymerase.

(c) DNA polymerase efficiently discriminates between rNTPs and dNTPs, because its nucleotide binding pocket cannot accommndate a 2'-OH on the incoming nucleotide.

(d) DNA polymerase cannot discriminate between rNTPs and dNTPs. Since the rate of transcription in cell is 10^6 times faster than replication, it cannot compete with RNA polymerase for rNTPs.

12. In a cell free extract containing DNA polymerase I, Mg^{2+}, dATP, dGTP, dCTP and dTTP (^3H), the following DNA molecules were added:

a. Single stranded closed circular DNA molecule containing 824 nucleotides.

b. Single stranded closed circular DNA molecule having 1578 nucleotides base paired with a linear single standard DNA molecule of 824 nucleotides having a free-3'-OH group.

c. Double stranded linear DNA molecule containing 1578 nucleotides having free-3'-OH group at both ends.

d. Double stranded closed circular DNA molecule having 824 nucleotides.

The rate of DNA synthesis was measured by incorporation of ^3H thymidine in the DNA molecule and expressed as the percentage of DNA synthesis relative to total DNA input.

Which one of the following graphs represents the correct result? **(DEC 2013)**

(a)

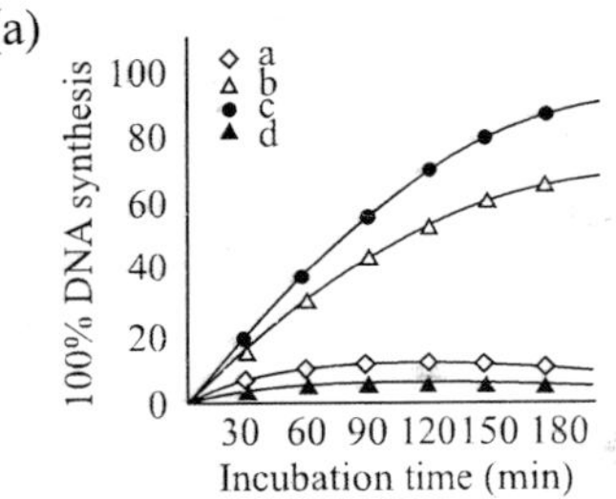

(b)

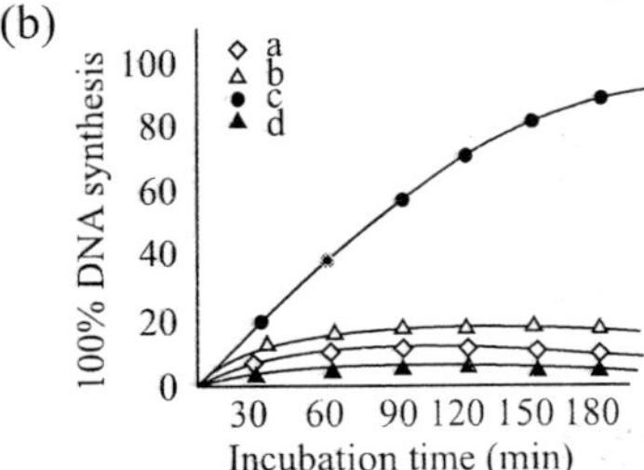

(c)

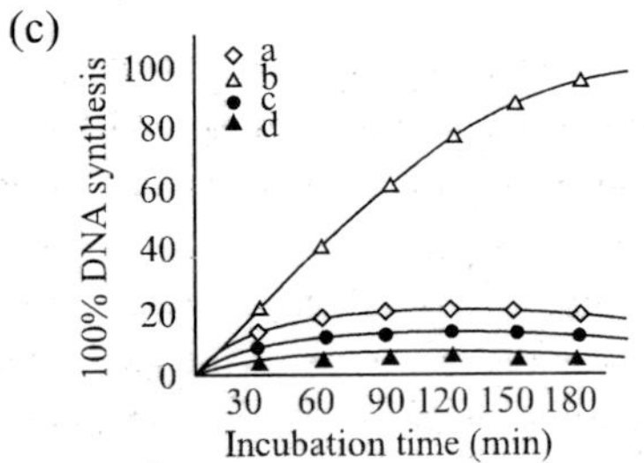

(d)

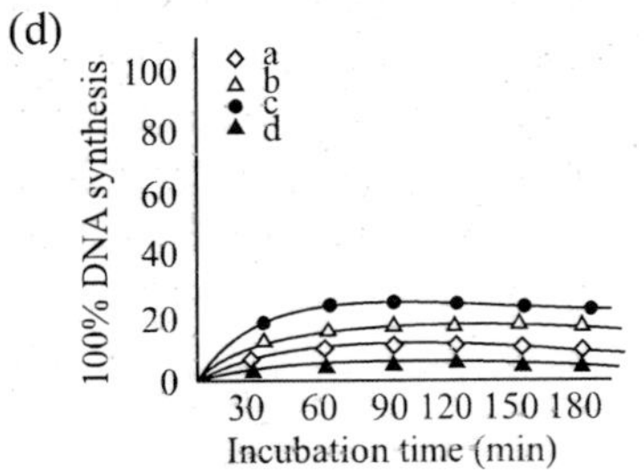

13. Double stranded DNA replicates in a semi –conservative manner. In an in vitro DNA synthesis reaction, dideoxy CTP and, dideoxy CMP were individually added in excess (in separate reaction tubes) in addition to dNTPs and other necessary reagents. Rate of DNA Synthesis was measured by incorporation of ^{3}H-thymidine. The four graphs drawn below represent the rate of DNA synthesis in two separate reaction tubes.

Which of the following graphs represents the expected data? **(JUNE 2012)**

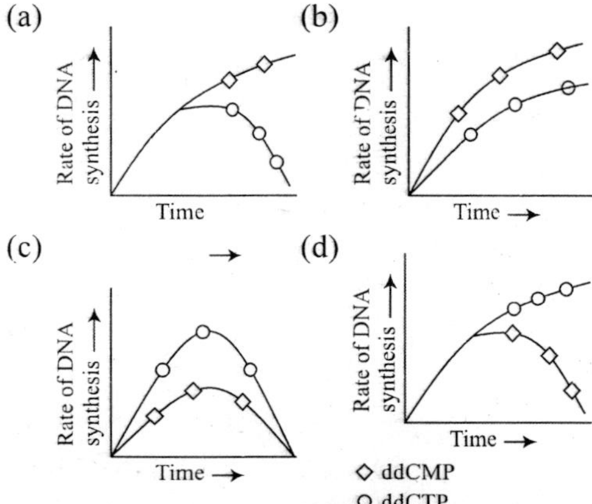

14. Copying errors occurring during replication are . corrected by the proof reading activity of DNA polymerases that recognize incorrect bases **(JUNE 2015)**

(a) at the 5' end of the growing chain and remove them by 5'→3'exonuclease activity.

(b) at the 3' end of the growing chain and remove them by 5'→3'exonuclease activity

(c) at the 3' end of the growing chain and remove them by 3'→5'exonuclease activity

(d) at the 5' end of the growing chain and remove them by 3'→5'exonuclease activity

15. While replicating DNA, the rate of misincorporation by DNA polymerase is 1 in 10^5 nucleotides. However, the actual error rate in the replicated DNA is 1 in 10^9 nucleotides incorporated. This is achieved mainly due to **(DEC 2011)**

(a) spontaneous excision of misincorporated nucleotides

(b) 3'→5' proofreading activity of DNA polymerase

(c) termination of DNA polymerase at misincorporated sites

(d) 5' →3' proof reading activity

16. In all organisms, it is critical that replication initiation be tightly controlled to ensure that chromosome number and cell number remain appropriately balanced. Given below are several statements regarding regulation of replication in *E. coli.*

A. Hemimethylation and sequestration of oriC (origin of replication) by a protein called SeqA prevents initiation of replication.

B. Availability of DnaA protein is an important requirement for initiation of replication.

C. The ratio of ADP : ATP is important as high level of ADP is required for initiation of replication.

D. Recruitment of Hda protein by sliding clamp inhibits ATP hydrolysis required for initiation of replication.

Which of the above statements are **NOT** true? **(DEC 2017)**

(a) A and B (b) B and C
(c) C and D (d) D and A

17. It takes 40 minutes for a typical *E. coli* cell to completely replicate its chromosome. Simultaneous to the ongoing replication, 20 minutes of a fresh round of replication is completed before the cell divides. What would be the generation time of *E. coli* growing at 37°C in complex medium? **(DEC 2015)**

(a) 20 minutes (b) 40 minutes
(c) 60 minutes (d) 30 minutes

18. *E. coli* takes 40 min. to duplicate its genome using a bi-directional mode of replication. If *E. coli* were to use uni-directional mode of replication to synthesize a full copy of DNA complementary to just one of the strands of the genome, it would take **(JUNE 2018)**

(a) 40 min (b) 80 min

(c) 20 min (d) 60 min

19. In *E. coli* grown under nutrient rich conditions, replication of entire genome takes about 40 min., yet it can divide every 20 min. This is so because: **(JUNE 2016)**

(a) While *E. coli* divides every 20 min, equal transfer of genetic material occurs only in the alternate rounds of cell divisions.

(b) A second round of genome replication begins before the completion of first round of replication, and by the time cell is ready to divide, two copies of the genome are available.

(c) Genome replication cell division are not coordinated with each other.

(d) During cell division, only one of the strands of the genome whose synthesis can be achieved in 20 min, is transferred to the daughter cell.

20. Two experiments were performed. In the first one, Okazaki fragments were prepared from a replicating cell of *E. coli* grown in the presence of ^{32}P. In the other, the two strands of *E. coli* chromosome were separated into a H strand and L strand, immobilized onto a nitrocellulose membrane and hybridized with the Okazaki fragments prepared in the first experiment.

Which one of the following options correctly describes the observation? **(JUNE 2016)**

(a) Okazaki fragments will hybridize to only H strand

(b) Okazaki fragments will hybridize to only L strand

(c) Okazaki fragments will hybridize with both H and L strands

(d) Because the H and L strands have been prepared from different cultures of *E. coli*, the Okazaki fragments will hybridize to neither

21. In eukaryotic replication, helicase loading occurs at all replicators during (DEC 2015)

(a) G_0 phase (b) G_1 phase

(c) S phase (d) G_2 phase

22. The figure below shows the structure of a replication fork.

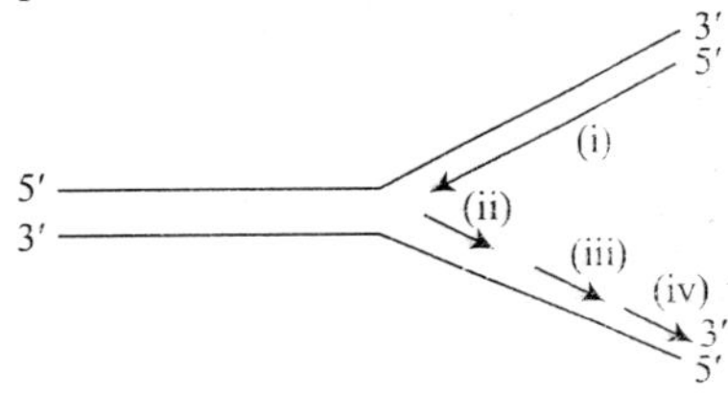

Based on this information following statements are made:

A. (i) represents the leading strand while (ii), (ii) and (iv) represent the Okazaki fragments.

B. Among the Okazaki fragments, synthesis of (iv) occurs prior to the synthesis of (iii) and (i)

C. Among the Okazaki fragments, synthesis of (ii) occurs prior to the synthesis of (iii) and (iv).

Which one of the following options represents the correct statement(s)? **(DEC 2019)**

(a) A only (b) B only

(c) A and B (d) A and C

23. In the diagram below, the dotted line marks the point of initiation of bidirectional replication

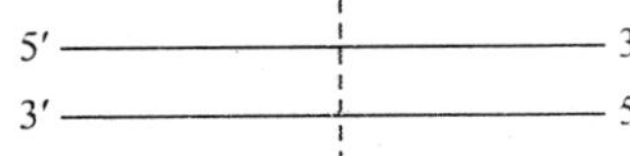

A. On the right side of the dotted line, leading strand synthesis occurs using the upper strand as the template.

B. On the right side of the dotted line, leading strand synthesis occurs using the lower strand as the template.

C. A ligase deficient (lig^-) mutant would affect replication of the upper strand on the left side of the dotted line.

D. A ligase deficient (lig^-) mutant would affect replication of the lower strand on the left side of the dotted line.

Which one of the following options represents the combinations of the correct statements?

(DEC 2019)

(a) A and D (b) B and C

(c) B and D (d) A and C

24. Which one of the following statements about DNA replication is INCORRECT?

(DEC 2019 ASSAM)

(a) During DNA replication, when adjacent bidirectional forks converge, the lagging strand will meet the leading strand of the same template strand

(b) Mispaired nucleotide at the 3'-OH end of the primer strand triggers the 3-5' exonucleolytic proof reading activity

(c) In a replication bubble moving bidirectionally, the same parental DNA strand cannot serve as a template for both the lagging and leading strand synthesis

(d) DNA replication involves a RNA-DNA chimeric molecule

25. For *Escherichia coli* chromosomal DNA replication, which one of the following statements is true? **(JUNE 2019)**

(a) DNA polymerase I is the main polymerase required for DNA replication

(b) DNA polymerase I though identified originally by Kornberg as the one responsible for replication, is not important for the DNA replication process

(c) Requirement of DNA polymerase I is in the context of removal of RNA primer needed for DNA synthesis, and then fill in the same with DNA equivalent

(d) DNA polymerase I is the primary enzyme for error prone DNA synthesis in response to SOS.

26. During replication, the RNA primer is degraded by the 5' - 3' exonuclease activity of

(DEC 2014)

(a) RNAseH

(b) FEN-1 (flap endonuclease 1)

(c) Topoisomerase II B

(d) DNA polymerase

27. During replication, RNaseH removes all of the RNA primer except the ribonucleotide directly linked to the DNA end. This is because

(JUNE 2019)

(a) it can degrade RNA and DNA end. From their 5' end.

(b) it can only cleave bonds between two ribonucleotides.

(c) it can degrade RNA and DNA from their 3' end.

(d) activity of RNaseH is inhibited by the presence of duplex containing both strands as DNA.

28. In a human cell line, a large fraction of double-strand DNA breaks are repaired by non-homologous end joining (NHEJ). An inhibitor of FLAP endonuclease will affect

(JUNE 2019)

(a) recruitment of DNA–dependent kinase

(b) gap trimming

(c) DNA unwinding

(d) pairing of micro-homology regions.

29. In eukaryotic replication, priming of DNA synthesis and removal of RNA primer is catalyzed by **(JUNE 2017)**

(a) DNA Pol α and PCNA, respectively.

(b) DNA Pol α and FEN1, respectively.

(c) DNA Pol δ and FEN1, respectively.

(d) DNA Pol ε and PCNA, respectively.

30. Eukaryotic DNA polymerase a has tightly associated primase activity but moderate processivity. DNA polymerase ε and δ are highly processive but lack primase activity. Given below are four statements about leading and lagging strand synthesis in eukaryotes. Which one is true? **(JUNE 2015)**

(a) Both leading and lagging strands are synthesized by DNA polymerase α. Moderate processivity is essential to maintain fidelity of replication.

(b) Entire leading and lagging strands are synthesized by ε and δ. Eukaryotic replication is primer independent process.

(c) Only the lagging strand synthesis needs primer and synthesized by DNA polymerase α.

(d) Primers for both the strands are synthesized by DNA polymerase α followed by "Polymerase switching" with ε and δ

31. Following are certain statements related to eukaryotic DNA replication:

A. The genome of multicellular animals contain many potential origins of replication.

B. During early development, when embryos are undergoing rapid cell divisions, origin sites are uniformly activated.

C. "Pulse-chase" technique is used to label sites of DNA replication.

D. The rate of elongation of different DNA chains during genome replication varies drastically.

Which one of the following combinations of above statements is correct? **(DEC 2014)**

(a) A, B and C (b) A, C and D

(c) B, C and D (d) A, B and D

32. The following statements are made on DNA replication:

A. Replication fork is a branch point in a replication 'eye' or 'bubble'.

B. A replication bubble contains two replication forks.

C. DNA replication is continuous according to the interpretation made by Okazaki.

D. Multiple priming events are required for both leading and lagging strands to initiate DNA synthesis.

Which one of the following is the correct combination? **(JUNE 2014)**

(a) A and B (b) B and C

(c) C and D (d) A and C

33. The high processivity of DNA polymerases during replication is because of their association with proteins called sliding clamps. For loading onto DNA, sliding clamps require sliding clamp-loaders. Which of the following is true about sliding clamp-loaders, after they successfully load sliding clamps?

(DEC 2013 GU)

(a) With sliding clamps, the processivity of DNA polymerase is very high, hence sliding clamp- loaders cannot compete and fall off.

(b) Binding of DNA with clamp/clamp-loader complex is followed by dissociation of the clamp-loader from the clamp, but it remains attached to the PT (Primer-template) junction.

(c) Once recruited, sliding clamp-loaders remain associated with sliding clamps during replication.

(d) Binding of DNA with the clamp/clamp-loader complex is following by ATP hydrolysis and subsequent release of the sliding clamp loader.

34. In a genetic assay, randomly generated fragments of yeast DNA were cloned into a bacterial plasmid containing gene '*X*' essential for yeast viability on minimal media. The recombinant plasmid was used to transform a yeast strain deficient in recombination and lacking '*X*' gene.

Transformants, which survive on minimal media and form colonies should essentially have: **(DEC 2018)**

(a) Yeast centromeric sequence which ensures integrity of the plasmid after transformation.

(b) Enhancers for the essential gene missing in the transformed strain.

(c) A sequence similar to bacterial origin of replication

(d) Yeast autonomous replicating sequence.

35. Telomerase, a RNA- protein complex which completes the replication of telomeres during DNA synthesis, is a specialized **(DEC 2016)**

(a) RNA dependent DNA polymerase

(b) DNA dependent DNA polymerase

(c) DNA dependent RNA polymerase

(d) RNA dependent RNA polymerase

36. Telomerase, a protein RNA complex, has special reverse transcriptase activity that completes replication of telomerase during DNA synthesis. Although it has many properties similar to DNA polymerase, some of them are also different. Which one of the following properties of telomerase is different from that of DNA-polymerase? (DEC 2016)

(a) Telomerase requires a template to direct the addition of nucleotide

(b) Telomerase can only extent a 3'-OH end of DNA

(c) Telomerase does not carry out lagging strand synthesis

(d) Telomerase acts in processive manner

37. In order to study the role of telomeres in DNA replication, genetically engineered mice were prepared, where the gene for telomerase RNA was knocked out. When cells from these knock out mice were taken and cultured in vitro, they proliferated even after 100 cell divisions which is quite unlikely in the case of human cells. Which of the following is the correct reason? **(JUNE 2013)**

(a) Human and mice are fundamentally different with respect to their requirements for telomerase enzyme in the context of DNA replication.

(b) In vitro, mice DNA becomes circular due to end to end chromosome fusion and does not require telomerase for DNA end replication.

(c) Mice have very long stretch of telomere DNA sequence compared to that of human.

(d) In vitro, mice DNA replication does not require the removal of RNA primers.

38. When circular plasmids having a centromere sequence are transformed into yeast cells, they replicate and segregate in each cell division. However, if a linear chromosome is generated by cutting the plasmid, at a single site with a restriction endonuclease, the plasmids are quickly lost from the yeast. It is known that genes on the plasmids are lost because of the instability of the chromosome ends. What could be done so as to restore its stability and can be inherited? **(JUNE 2015)**

(a) Methylation of adenine residues of the plasmid.

(b) Complexing the plasmid ends with histone proteins.

(c) By incorporating telomere sequences to the end of plasmid.

(d) By incorporating acetylated histone proteins to the plasmid ends.

39. As topoisomerases play an important role during replication, a large number of anticancer drugs have been developed that inhibit the activity of these enzymes. Which of the following statements is NOT true about topoisomerases as a potential anticancer drug target? **(JUNE 2016)**

(a) As cancer cells are rapidly growing cells, they usually contain higher level of topoisomerases.

(b) The transient DNA breaks created by topoisomerases are usually converted to permanent breaks in the genome in the presence of topoisomerase targeted drugs.

(c) As cancer cells often have impaired DNA repair pathways, they are more susceptible-towards topoisomerase targeted drugs.

(d) The drugs which specifically target topoisomerases, usually do not affect normal fast growing cells.

40. Following statements have been made about removal of supercoiling produced by DNA unwinding at the replication fork:

A. Accumulation of supercoils is the result of DNA helicase activity during unwinding of DNA

B. Problem of DNA supercoiling is valid only for circular chromosomes of bacteria and not for the linear chromosomes

C. Supercoiling of DNA is removed by topoisomerases by breaking either one or both strands of DNA on the unreplicated DNA in front of replication fork.

D. Both topoisomerase I and topoisomerase can remove positive super-coiling during replication

Which one of the following options has all correct statements? **(JUNE 2019)**

(a) A, B and C (b) A B and D

(c) A, C and D (d) B, C and D

41. In an *in vitro* experiment using radio-labeled nucleotides, a researcher is trying to analyze the possible products or DNA replication by resolving the products using urea-polyacrylamide gel electrophoresis. In one experimental set up RNase H was added (Set 1), while in another set no RNase H was added (Set 2),

The possible observations of this experiment could be

A. There is no difference in the mobility of labelled DNA fragments between the Set I and Set 2

B. There is distinct difference in the mobility of the newly synthesized labeled DNA fragments between Set I and Set 2

C. The mobility of the newly synthesized labelled DNA fragments in case of Set 1 is faster as compared to the Set 2

D. The mobility of the newly synthesized labelled DNA fragments in case of Set 1 is slower as compared to the Set 2

Which of the following combinations represent correct observations? **(DEC 2011)**

(a) A and B (b) B and C

(c) A and D (d) B and D

42. A bacterial population has a plasmid with copy number 'n'. It was observed that on an average in one out of $2^{(n-1)}$ cell divisions, there was spontaneous plasmid curing. It was inferred from the observation that:

A. Each cell division does not have equal probability of plasmid curing.

B. There is no evidence for any mechanism of plasmid segregation in the two daughter cells.

C. Plasmid distribution to daughter cells is random.

D. Each plasmid has an equal chance of being in either of the two daughter cells.

Which of the combinations of above statements is true? **(DEC 2012)**

(a) A and B (b) B and D

(c) Only A (d) B, C and D

43. Reverse transcriptase has both ribonuclease and polymerase activities. Ribonuclease activity is required for **(DEC 2011)**

(a) the synthesis of new RNA strand

(b) the degradation RNA strand

(c) the synthesis of new DNA strand

(d) the degradation of DNA strand

44. *E. coli* DNA ligase catalyses formation of a phosphodiester pond between the adjoining 3' hydroxyl, and the 5' phosphoryl ends in DNA duplexes. The energetic need for this reaction is met by the hydrolysis of NAD^+ to NMN^+ and AMP in a three-step reaction. Following statements are being made about the mechanism of this reaction.

(i) AMP is linked to the 5' phosphoryl end of the nicked DNA.

(ii) Adenylyl group of NAD^+ is transferred to the $^{\varepsilon\varepsilon}$-amino group of Lys in DNA ligase to form a phosphoamide adduct.

(iii) DNA ligase catalyses the formation of a phosphodiester bond by the nucleophilic attack of the 3' hydroxyl group onto the phosphate and releases AMP.

Based on the statements made above, identify the correct sequence of the reaction steps.

(JUNE 2018)

(a) (i)-(ii)-(iii) (b) (i)-(iii)-(ii)
(c) (ii) -(i)-(iii) (d) (iii)-(i)-(ii)

Answer Key

1.(a)	**2.**(d)	**3.**(d)	**4.**(a)	**5.**(b)	6.(a)	7.(c)
8.(b)	**9.**(b)	**10.**(d)	**11.**(c)	**12.**(c)	**13.**(a)	**14.**(c)
15.(b)	**16.**(c)	**17.**(a)	**18.**(b)	**19.**(b)	**20.**(c)	**21.**(b)
22.(c)	**23.**(c)	**24.**(c)	**25.**(c)	**26.**(b)	**27.**(b)	**28.**(b)
29.(b)	**30.**(d)	**31.**(a)	**32.**(a)	**33.**(d)	**34.**(d)	**35.**(a)
36.(c)	**37.**(c)	**38.**(c)	**39.**(d)	**40.**(c)	**41.**(b)	**42.**(d)
43.(b)	**44.**(c)					

2. DNA damage and repair mechanisms

1. The fidelity of replicative base selection can be reduced by a factor of 10^2 when the repair of DNA synthesis involves **(JUNE 2011)**

(a) AP endonuclease.
(b) ABC exinuclease.
(c) DNA photolyase
(d) TLS DNA polymerase.

2. DNA repair, synthesis and recombination are intimately connected and inter dependent. An apparent commonality between processes of DNA replication and repair in the enzymatically catalyzed synthesis of DNA polynucleotide segments, which can be assembled with preexisting polynucleotides, leading to repair or replication. Synthesis of these polynucleotide segments is catalyzed by a group of enzymes DNA-dependant DNA polymerases. In the case of *E.coli*, DNA polymerase has been isolated in three distinct forms whereas five main types of polymerase have been isolated from mammalian cells. All the polymerases synthesize polynucleotides only in the 5' →3' direction. If polynucleotide chains could be elongated in 3→ 5' direction, the hypothetical growing 5' terminus, rather than the incoming nucleotide, would carry a triposphate that is unsuitable for further elongation.

The 3' →5' exonuclease activity is not associated with all the polymerases and only present in

A. All *E. coli* DNA polymerases
B. Pol I, Pol IV, Pol V, Pol α, Pol β.
C. Pol I, Pol III, Pol δ, Pol ε
D. Pol I, Pol IV, Pol α, Pol ε.

The correct statements are **(MODELPAPER)**

(a) (A) and (B). (b) (A), (B) and (C).
(c) (C) only (d) (A), (C) and (D).

3. The mismatch-repair activity of *E. coli* repairs mis-incorporated bases which is not removed by the proofreading activity of DNA polymerase. However, while doing so, it has to decide which strand of the DNA is newly synthesized and which one is parental. Mismatch repair system does it by which one of the following ways? **(DEC 2015)**

(a) It recognizes nearby GATC sequence.
(b) It recognizes any nearby palindromic sequence.
(c) It recognizes a specific repetitive sequence.
(d) It recognizes the hemi-methylated GATC sequence nearby.

4. Some errors occur during DNA replication that are not corrected by proof reading activity of DNA polymerase. These are corrected by specialized repair pathways. Defect in the activities of some of the following enzymes impair this process.

A. DNA polymerase III and DNA ligase
B. AP endonuclease and DNA glycosidase
C. MutS and Mut L
D. Rec A and Rec F

Defect in which of the above enzymes impair the process? **(JUNE 2016)**

(a) A, B, and C (b) D and B
(c) A and D (d) A and C

5. The 'Uvr ABC' repair mechanism is involved in repairing **(DEC 2013)**
(a) missing bases.
(b) strand break.
(c) cross linked strands.
(d) DNA damage caused by "bulky" chemical adducts.

6. The complex responses to different types of DNA damage in both prokaryotes and eukaryotes fall into three main categories:
(i) damage bypass
(ii) damage reversal
(iii) damage removal
Many repair proteins are isolated like
(a) DNA methyl transferase
(b) DNA glycosylase
(c) DNA polymerase IV
Which one of the following represents the correct combination? **(DEC 2013)**
(a) (i) - (a), (ii) - (b), (iii) - (c)
(b) (i) - (b), (ii) - (c), (iii) - (a)
(c) (i) - (c), (ii) - (a), (iii) - (b)
(d) (i) - (c), (ii) - (b), (iii) - (a)

7. A 6.4 Kb plasmid DNA has two restriction endonuclease sites, *HindIII* and *EcoRI.* Complete double digestion of the plasmid with both the enzymes yields two fragments of 3.1 and 3.3 Kb.
In order to study DNA repair process, a G:T mismatch was introduced in one strand of *HindIII* site and the damaged plasmid was incubated in a reconstituted repair system containing all the factors and enzymes required for repair. If the efficiency of the repair system is 50%, which one of the following band patterns on agarose gel will be obtained after treating the repaired

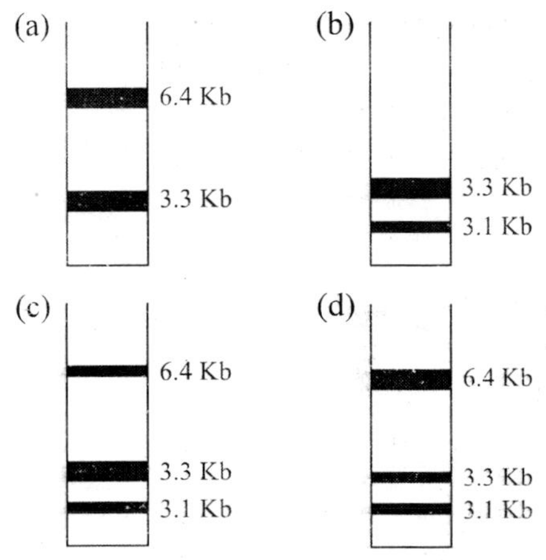

8. A meiotic recombination does not involve **(DEC 2013 GU)**
(a) formation of Holliday junction.
(b) newly replicated identical DNA duplexes.
(c) gene conversion.
(d) crossover and non-crossover.

9. Error-free repair of double strand breaks in DNA is accomplished by **(JUNE 2016)**
(a) non-homologous end-joining.
(b) base excision repair.
(c) homologous recombination.
(d) mismatch repair.

10. A single-strand nick in the parental DNA helix just ahead of a replication fork causes the replication fork to break. Recovery from this calamity requires **(DEC 2011)**
(a) DNA ligase
(b) DNA primase
(c) site-specific recombination
(d) homologous recombination

11. Repair of double strand breaks made during meiosis in the yeast *Saccharomyces cerevisiae* **(MODEL PAPER)**
(a) occurs mostly by non-homologous end joining.
(b) occurs mostly using the sister chromatid as a template.

(c) occurs mostly using the homologous chromosome as a template

(d) is associated with a high frequency of mutations.

12. The homologous genetic recombination is a DNA repair process referred to as recombination repair. Which one of the following statements is INCORRECT for recombination repair? **(JUNE 2014)**

(a) DNA polymerase III stalls at the site of the damage.

(b) DNA polymerase III leaves a gap in the daughter strand.

(c) The gap is filled by recombination between complementary parent strand homologous to daughter strand and the gapped daughter strand.

(d) Homologous recombination process is catalyzed by topoisomerase II.

13. In *E. coli*, *rec*A gene is involved in recombination as well as repair and *dna*B gene is involved in unwinding of DNA double strands during replication. Which of the following statement is/are correct about *Rec* A and *Dna* B?

A. Mutation in *E. coli rec*A gene is lethal.

B. *E. coli* with mutated *dna*B gene does not survive.

C. *Dna* B after uncoiling DNA double strands, prevents further reannealing at the separated strands.

D. *Rec* A gene is involved in SOS response and helps DNA repair.

The correct options are: **(DEC 2012)**

(a) Band C (b) A and B

(c) B and D (d) A and C

14. Which one of the following proteins catalyzes branch migration activity in Holliday junction? **(DEC 2019 ASSAM)**

(a) Spo11 (b) RuvB

(c) Zip1 (d) RuvC

15. Following statements are made about double-strand break repair (DSBR) model of homologous recombination:

A. Process of DSBR recombination is triggered by introducing a double-strand break in a DNA duplex

B. In a process known as 3-end resection, the exonucleases along with a DNA helicase degrade one strand on either side of the break and generates 3'-single stranded termini

C. One strand of the donor duplex is displaced due to formation of heteroduplex DNA and generates a displacement loop (D-Loop)

D. Branch migration allows the point of crossover to move in 5'→3' direction of recipient strand

E. Completion of DSBR recombination may generate either crossover recombinant or non-recombinant product

Which one of the following combination of statements is correct? **(DEC 2019)**

(a) A, B and D (b) A, C and E

(c) A, C and D (d) A, B and E

16. Uracil containing plasmid was constructed and was used in transformation into the wild type (*ung*$^+$) and uracil-N-glycosylase mutated (*ung*$^-$) *E. coli* cells and scored for transformants in the presence of appropriate antibiotics. Which one of the following statements correctly describes the experimental outcome? **(JUNE 2018)**

(a) *ung*$^+$ cells will have fewer transformants compared to *ung*- cells.

(b) *ung*$^-$ cells will give fewer transformants compared to *ung*cells.

(c) No transformants will be obtained in *ung*$^-$ cells as uracil excision repair willnot occur and the plasmid would not replicate.

(d) Presence of uracil in DNA is unnatural and the plasmid DNA with uracils in it will not produce transformants in either *ung*$^+$ or *ung*$^-$cells.

17. A guanine (G) base in DNA is susceptible to modification to 8-oxo-G because of oxidative damage. If the modified G residues are not removed by a DNA glycosylase (*Mut M* or *Fpg*) prior to replication, an adenine (A) may be incorporated against 8-oxo-G. To avoid mutations, the cells have another DNA glycosylase called *Mut Y*. However, when the gene coding for *Mut Y* is deleted in *Escherichia coli*, the strain survives. This observation suggests that in *E. coli*

(DEC 2019 ASSAM)

(a) *MutY* is not responsible for the excision of 'A' incorporated against 8-oxo-G
(b) Alternate DNA repair pathways may substitute for the repair of A incorporated against 8-oxo-G
(c) Incorporation of 'A' against 8-oxo-G does not cause mutation in *E. coli*
(d) The 8-oxo-G:A base pair in DNA distorts its structure in such a way that, during the next round of replication a 'G' is incorporated against A in the 8-oxo-G:A mispair.

Answer Key

1.(d)	2.(c)	3.(d)	4.(d)	5.(d)	6.(c)	7.(d)
8.(b)	9.(c)	10.(d)	11.(c)	12.(d)	13.(c)	14.(b)
15.(b)	16.(a)	17.(b)				

3. Prokaryotic Transcription Process

1. In bacteria, heat-shock response is primarily controlled by **(JUNE 2015)**

(a) Sigma S (σ^S) (b) Sigma 32 (σ^{32})
(c) Sigma E (σ^E) (d) Sigma 70 (σ^{70})

2. Presence of sigma factor (σ70) in Escherichia coli facilitates **(DEC 2019 ASSAM)**

(a) Sequence specific localization of RNA polymerase on the promoter regions of genes
(b) Phosphorylation of C-terminal tail of RNA Polymerase
(c) Interaction of β-subunit of RNA polymerase with α -subunit
(d) Interaction of ώ -subunit with β β'α2, complex of RNA polymerase

3. Which one of the following statements is NOT correct? **(DEC 2018)**

(a) Together with proteins, rRNA provides a site for polypeptide synthesis.
(b) All DNA molecules are unbranched polymers of nucleotides.
(c) DNA is synthesized in a 5'-3' direction while RNA synthesis occurs in a 3'- 5' direction.
(d) A tRNA anticodon may pair with more than one codon.

4. σ subunit of *E. coli* RNA polymerase DOES NOT **(JUNE 2013)**

(a) initiate transcription and fall off during elongation.
(b) increase affinity of the core enzyme to the promoter.
(c) binds to DNA, independent of the core enzyme.
(d) ensures specificity of transcription by interacting with the core enzyme.

5. One of the mechanisms used by bacteria for adaptation to changed environment is altering transcription of their genes. In this regard, which one of the following responses is NOT found in bacteria? **(DEC 2017)**

(a) A gene with two different promoters for expression in different conditions
(b) Use of different sigma factors for transcription of genes
(c) Expression of alternate α, β and β subunits
(d) Expression of anti-sigma factors

6. Rho factor is involved in termination of transcription in prokaryotes. Genetic manipulations indicate that Rho-dependent termination requires the presence of a specific recognition sequence on the newly synthesized RNA upstream of the termination site. The

recognition sequence must be on the nascent RNA rather than the DNA, as demonstrated by Rho's inability to terminate transcription in the presence of pancreatic RNAse. The essential features of this termination site have not been fully elucidated. Construction of synthetic termination sites indicates that it consists of 80 to 100 nts that lack a stable secondary structure and contain multiple regions that are rich in C and poor in G. Which of the following is/ are suggested by the above observation?

(MODEL PAPER)

(a) Rho factor attaches to nascent RNA at its recognition sequence and then migrates along the RNA in the $5' \rightarrow 3'$ direction until it encounters an RNAP paused at the termination site and disrupts the transcriptional complex by acting as an RNA-DNA helicase.

(b) Rho unwinds the RNA-DNA duplex forming the transcription bubble, thereby releasing the RNA transcript.

(c) Rho factor attaches to the RNA at its recognition sequence while RNA is in the RNA-DNA hybrid condition.

(d) There may be other factors and hence Rho factor does not need to unwind the RNA-DNA hybrid to release the transcript.

7. Actinomycin D inhibits the process of transcription in both prokaryotic and eukaryotic organisms. The following statements are made about actinomycin D- mediated inhibition of transcription:

A. Actinomycin D inhibits transcription from a double stranded DNA template by either *E. coli* or yeast RNA polymerases.

B. Actinomycin D inhibits transcription from a single stranded RNA template by eukaryotic viral RNA polymerases.

C. Actinomycin D inhibits transcription from single stranded $\Phi \times 174$ DNA template by *E. coli* RNA polymerase, immediately after viral DNA entry.

D. Actinomycin D inhibits transcription from double stranded RNA template by eukaryotic RNA polymerase II.

Which of the combinations of the above statements is a true representation of the mechanism of actinomycin D mediated inhibition? **(JUNE 2018)**

(a) A only (b) A, B and D

(c) A, B and C (d) A and D only

Answer Key

1.(b)	2.(a)	3.(c)	4.(c)	5.(c)	6.(a)	7.(a)

4. Eukaryotic Transcription Process

1. α-Amanitin is a fungal toxin which inhibits eukaryotic RNA polymerases. The three eukaryotic RNA polymerases show differential sensitivity to this toxin. Which one of the following order (higher to lower) is correct in respect of sensitivity towards α-amanitin?

(DEC 2015)

(a) RNA POL III> RNA POL II > RNA POL I

(b) RNA POL II > RNA POL III > RNA POL I

(c) RNA POL I> RNA POL III > RNA POL II

(d) RNA POL II > RNA POL I > RNA POL III

2. α-amanitin inhibits **(DEC 2011)**

(a) only RNA polymerase I

(b) only RNA polymerase II

(c) only RNA polymerase III

(d) all RNA polymerases

3. In eukaryotes, precursors of micro RNAs (miRNAs) and small interfering RNAs (siRNAs) are usually synthesized by

(JUNE 2017)

(a) RNA Pol I and III, respectively

(b) RNA Pol III and I, respectively

(c) Only RNA Pol I

(d) Only RNA Pol II

4. Which one of the following statements is generally true about RNA polymerase II? **(DEC 2018)**

(a) It is dedicated to transcribing RNA from a single transcription unit, generally a large transcript· which is then processed to yield three types of ribosomal RNA.

(b) It transcribes varieties of small non- coding RNAs which are expressed in all cell types.

(c) It generally synthesizes various types of mRNAs and small non-coding RNAs.

(d) It is exclusively involved in synthesis of rRNA and tRNA.

5. Which of the following are NOT transcribed by RNA polymerase II? **(DEC 2016)**

(a) miRNA and some snRNA

(b) miRNA and snoRNA

(c) mRNA and snoRNA

(d) tRNA and 5S rRNA

6. Small nucleolar RNAs used to process and chemically modify rRNAs are called **(DEC 2014)**

(a) ScaRNAs (b) SiRNAs.

(c) SnoRNA (d) SnRNAs

7. Enlisted below are different types of RNAs produced in the cell (Column A) and their functions (Column B), but not in the same order.

Column A	Column B
A. Sn RNAs	(i) turn off gene expression by directing degradation of selective mRNAs.
B. si RNAs	(ii) regulate gene expression by blocking translation of selective mRNAs.
C. mi RNAs	iii) function in a variety of processes including splicing of pre-mRNA
D. Sno RNAs	(iv) used to process and chemically modify rRNAs.

Choose the correct combination **(DEC 2015)**

(a) A-(iv), B-(ii), C-(i), D-(iii)

(b) A (iii), B-(i), C-(ii), D-(iv)

(c) A-(iv), B-(i), C-(ii), D-(iii)

(d) A- (iii), -B-(ii), C-(i), D-(iv)

8. Genetic studies demonstrated that TBP mutant cell extracts are deficient in transcription of genes from all three promoters viz. class I, II and III. Following statements describe characteristic features of TBP.

A. TBP is considered as an universal basal transcription factor.

B. TBP is not required for transcription of archaeal genes.

C. TBP is involved in recognizing TATA box.

D. TBP operates at all promoters regardless of their TATA content.

Which of the following combinations is NOT correct? **(JUNE 2011)**

(a) A and D (b) C and D

(c) B and D (d) A and C

9. A highly specific inhibitor that targets the phosphorylation activity of TFIIH is added to an in vitro transcription reaction. Which one of the following steps is most likely to be affected? **(DEC 2017)**

(a) Binding of RNA polymerase to promoter sequence

(b) Promoter clearance

(c) Recruitment of TFIID

(d) Open promoter complex formation

10. In order to study the transcription factor TFIIH, it was cloned from a large number of human subjects. Surprisingly, the subjects having mutation in TFIIH, also showed defects in their DNA repair system.

Given below are the explanations:

A. DNA damage is always associated with transcription inhibition.

B. TFIIH has no role in DNA repair.

C. In mammalian system, TFIIH plays an active role in transcription coupled DNA repair process.

D. Because of mutation in TFIIH, transcription initiation is inhibited and incompletely synthesized mRNAs remain attached to the template DNA leading to DNA damage.

Choose the correct answer. **(JUNE 2015)**

(a) A and B (b) C only

(c) B and D (d) D only

11. Following statements were made with respect to transcription in eukaryotes:

A. RNA polymerase III synthesizes mRNAs in the nucleoplasm

B. The target promoter for RNA polymerase III is usually represented by a bipartite sequence downstream of the transcription start site.

C. The assembly factors TFIIIA and TFIIIC assist the binding of the positioning factor TFIIIB at the precise location.

D. TFIIIB is the last factor that joins the initiation complex.

E. Phosphorylated Ser residues in the C-terminal domain (CTD) of RNA polymerase II serve as binding sites for mRNA processing enzymes.

Which one of the following options represents the correct combination of the statements?

(DEC 2019)

(a) A, B and C (b) B, C and E

(c) B, D and E (d) A, C and E

12. A non-enzymatic viral protein X was found to be inducing a cellular gene promoter activity. Although no in vitro DNA binding activity could be identified with X protein, it was found to be co-recruited on the cellular promoter along with a cellular transcription factor in vivo. Which one of the following statements seems to be the best interpretation of the above findings? **(JUNE 2013)**

(a) X is a DNA-binding protein.

(b) X physically interacts with the transcription factor.

(c) X modifies the chromatin for transcription activation.

(d) X is a chaperone.

13. In mammals CG rich sequences are usually methylated at C, which is a way for making genes for silencing. Although the promoters of housekeeping genes are often associated with CpG islands yet they are expressed in mammals. Which one of the following best explains it? **(DEC 2016)**

(a) Methylation of cytosine does not prevent the binding of RNA Pol II with the promoter, so housekeeping genes are expressed.

(b) During housekeeping gene expression, the enzyme methyl transferase is temporarily silenced by mi-RNA, thus shutting down global methylation.

(c) Unlike within the coding region of a gene, CG rich sequences present in the promoters of active genes are usually not methylated.

(d) As soon as the Cytosine is methylated in the promoter region, the enzymes of DNA repair pathways remove the methyl group, thereby ensuring gene expression.

14. Scientists usually find difficulty in identifying the exact transcription termination site in eukaryotes because **(JUNE 2017)**

(a) Immediately following termination of transcription, the 3' end is polyadenylated

(b) The 3' end is generated by cleavage prior to actual termination of transcription

(c) Poly A binding proteins present at 3' end of transcript hides the termination site

(d) 3' end of transcript is complexed with 5' end for initiation of translation

15. Regulatory elements for expression of ribosomal RNA genes resides in the

(JUNE 2012)

(a) transcribed spacer region
(b) non-transcribe spacer region
(c) 5' flanking region of individual ribosomal genes
(d) internal regions within the genes

16. tRNA genes of a Gram positive bacteria were sequenced. While some of the genes possessed sequence corresponding to the -CCA end of the tRNA, others did not. Interestingly, even the genes that lacked the sequence corresponding to the -CCA end of the tRNA were found to code for a functional tRNA. This is because **(DEC 2017)**

(a) the -CCA end in the tRNA is not essential for its function.
(b) the genes that do not possess sequence corresponding to the -CCA ends of the tRNA are repaired prior to their transcription.
(c) the -CCA end is added to the tRNA during their maturation.
(d) the primary transcripts of the genes lacking sequence corresponding to the - CCA end are subjected to g-RNA mediated editing prior to their maturation.

Answer Key

1.(b)	2.(b)	3.(d)	4.(c)	5.(d)	6.(c)	7.(b)
8.(c)	9.(b)	10.(b)	11.(b)	12.(b)	13.(c)	14.(b)
15.(b)	16.(c)					

5. 5' Capping and poly adenylation

1. The 5' Cap of RNA is required for the **(JUNE 2011)**

(a) stability of RNA only.
(b) stability and transport of RNA.
(c) transport of RNA only.
(d) methylation of RNA.

2. Eukaryotic mRNA are modified to possess a 5' cap structure. Which one of the following is an INCORRECT statement about the function of the 5' cap structure? **(JUNE 2019)**

(a) It protects the mRNA from 5'→3' exo-ribonuclease attack.
(b) It facilitates splicing of the nascent transcripts
(c) It protects the transcripts from degradation by RNAse III family enzymes.
(d) It facilitates attachments to 40S subunit of ribosome

3. Eukaryotic mRNAs have an enzymatic appended cap structure consisting of a 7-methylguanosine residue joined to the initial 5' nucleotide of the transcripts. Given below are a few statements regarding capping.

A. Capping protects the mRNA from degradation by 5' -exoribonucleases.
B. During capping, the α-phosphate is released from the 5'-end of the nascent mRNA.
C. Phosphorylation mediated conformational change in carboxyl terminal domain (CTD) of RNA Pol II enables its binding with capping enzymes.
D. During capping, a 5' -5' triphosphate bond is formed between the β-phosphate of the nascent mRNA and α-phosphate of GTP.

Which of the above statement(s) is/are **INCORRECT**? **(JUNE 2017)**

(a) C only (b) B only
(c) A and B (d) C and D

4. HeLa cell extract was used to study transcription of a gene '*X*' having six introns. RNA Pol II complex containing all associated proteins was isolated from actively transcribing system and subjected to proteome analysis.

Results showed the presence of both splicing and capping enzymes in the complex. When transcription elongation was inhibited by flavopiridol, polymerase complex contained only capping enzymes. When phosphorylation of the CTD domain of Pol II was inhibited by a kinase inhibitor, the complex contained neither splicing nor capping enzymes.

From these results, following conclusions were made:

A. Transcription of gene X is coupled to mRNA capping.
B. Transcription elongation is coupled to splicing.
C. Phosphorylation of CTD is required for the recruitment of capping and splicing enzymes.
D. Both capping and splicing of mRNAs occurs simultaneously.

Identify the correct set of conclusions:

(DEC 2012)

(a) A, B and C (b) B, C and D
(c) C, D and A (d) D, A and B

5. An eukaryotic cell undergoing mRNA synthesis and processing was incubated with ^{32}P labelled ATP, with the label at the β-position. Where do you think the radioactive isotope will appear in the mature mRNA? **(JUNE 2016)**

(a) ^{32}P will not appear in the mature mRNA under any circumstances because β and γ phosphates are released during transcription.
(b) Phosphate groups of the phoshodiester backbone of the mRNA will be uniformly labelled as only γ phosphates are released during transcription.
(c) ^{32}P will appear at the 5' end of the mRNA if only it has "A" as the first nucleotide.
(d) No ^{32}P will appear in the mature mRNA because the 5'-terminal phosphate of an "A" residue will be further removed during the capping process.

6. You are working with an in vitro eukaryotic transcription system, which produced both capped and uncapped mRNAs. You incubated these mRNAs with mammalian cell nuclear extract and then quantified the different products as shown below. Which of the following graphs correctly represents the expected result? **(JUNE 2013)**

(a)

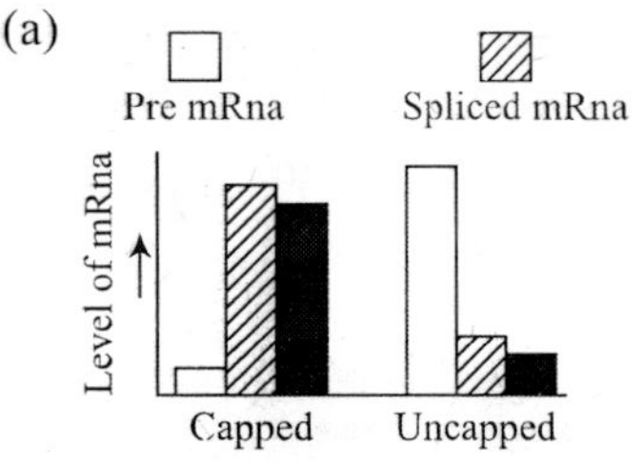

(b)

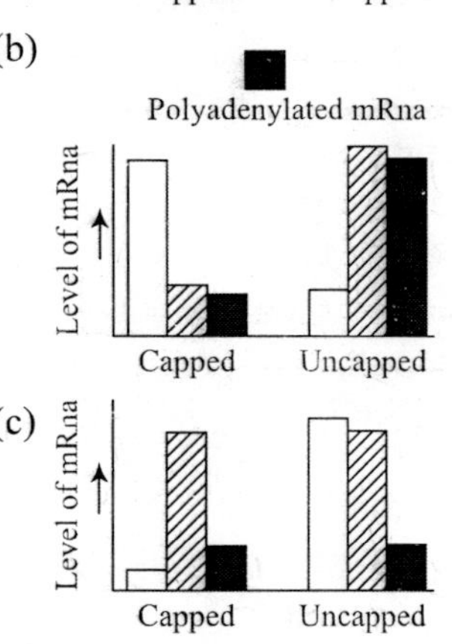

(d)

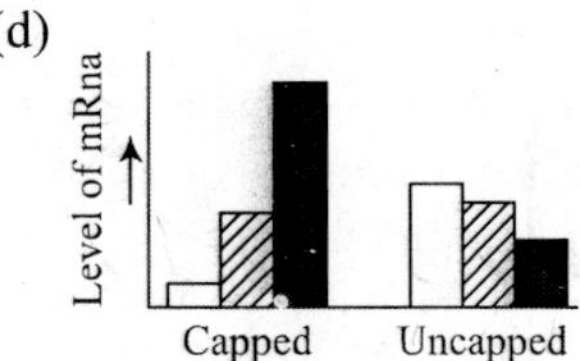

7. The 3'end of most eukaryotic mRNAs is defined by the addition of a polyA tail- a processing reaction called polyadenylation. The addition of poly A tail is carried out by the enzyme Poly(A) Polymerase. Given below are few statements about this process:

A. Poly(A) Polymerase is a template independent enzyme.
B. Poly(A) Polymerase catalyses the addition of AMP from dATP to the 3' end of mRNA.
C. Poly(A) Polymerase is a RNA-template dependent enzyme
D. Poly(A) Polymerase catalyzes the addition of ADP from ATP to the 3' end of mRNA.

E. Poly (A) Polymerase catalyzes the addition of AMP from ATP to the 3' end of mRNA.

F. Poly (A) Polymerase catalyzes the addition of AMP from dADP to the 3' end of mRNA.

Which of the following combination is true?

(JUNE 2015)

(a) B and C (b) C and D
(c) A and E (d) C and F

Answer Key

1.(b)	2.(c)	3.(b)	4.(a)	5.(c)	6.(a)	7.(c)

6. RNA editing, splicing, trans-splicing and transport

1. Which of the following processes does not take place in the 5′→3′ direction?

(MODEL PAPER)

(a) DNA replication (b) Transcription
(c) Nick translation (d) RNA editing

2. RNA editing, a post transcriptional process, is achieved with the help of guide RNA (g RNA) which one of the following statement about the process is NOT true? **(DEC 2016)**

(a) g RNA dependant RNA editing happens in the kinetoplast of DNA.
(b) g-RNA is involved in chemical modification of t-RNA.
(c) This process involves insertion or deletion of uridines.
(d) Sequences edited once may be re-edited using second g-RNA.

3. With an intention to identify the genes expressed in an organism at specific stage of development, mRNAs were isolated from the given organism, cDNAs were synthesized, Cloned in a suitable vector and sequenced. A few of the cDNA sequences showed no matches with the genomic DNA sequence. Further, it was observed that these sequences were U-rich and found to be in stretches dispersed along the sequence. The following may be possible reasons for appearance of such RNA:

A. Splicing
B. Alternate splicing
C. Trans-Splicing
D. Guide RNA mediated introduction of Us involving endonuclease, terminal-U-transferase and RNA ligase
E. Deaminations converting C to U

Which of the following is the most appropriate reasons? **(JUNE 2015)**

(a) A and C (b) B and D
(c) C, D and E (d) D only

4. cDNA clones of one of the gene product of an organism was sequenced. The sequence was searched for the homology in the genomic sequence available in the database. The sequence did not match with any of the genes or any of the complete exon of a gene in the database.

The following causes may explain the sequence changes in the transcript:

A. Alternate splicing
B. Errors occurring in the process of transcription
C. Trans-splicing
D. RNA editing

Which of the above cause or combination of causes may explain such result?

(DEC 2013 GU)

(a) A and D (b) Only D
(c) Only C (d) B and C

5. A 1257 bp genomic DNA sequence of a prokaryotic gene was cloned under a strong constitutive promoter along with a suitable polyA signal and used for development of transgenic tobacco plants.

Molecular analysis revealed the presence of three types/lengths of transgene derived mRNAs: 555 bp, 981 bp and 1257 bp - in the leaves of transgenic plants.

The following statements were proposed to explain the above results.

A. The three mRNAs represent alternatively spliced transcripts due to the presence of putative intronic sequence in the gene.

B. The gene sequence was characterized by the presence of potential polyadenylation signals that resulted in premature termination of transcription.

C. Expression of full-length transcripts (1257 bases) was lethal to the transformed cells.

D. The transgenic plants were chimeric in nature and comprised of a mix of transformed and untransformed cells.

Which of the following combinations of the above statements would correctly explain the obtained results? **(JUNE 2018)**

(a) A and C (b) B and D

(c) A and B (d) C and D

6. Pre-mRNAs are rapidly bound by snRNPs which carry out dual steps of splicing that removes the intron and joins the upstream and downstream exons. The following statements describe some facts related to this event

A. Almost all introns begin with GU and end with AG sequences and hence all the GU or AG sequences are spliced out of RNA.

B. U2 RNA recognizes important sequences at the 3' acceptor end of the intron.

C. The spliceosome uses ATP to carry out accurate removal of intron

D. An unusual linkage with 2' OH group of adenosine within intron form 'Lariat' structure.

Which of the following combinations is correct? **(DEC 2011)**

(a) A and B (b) B and C

(c) C and D (d) D and A

7. In order to ensure that only fully processed mature mRNAs are allowed to be exported to cytosol, pre-mRNAs associated with snRNPs are retained in the nucleus. To demonstrate this, an experiment was performed where the gene coding a pre-mRNA, with a single intron was mutated either at 5'or 3' splice sites or both the splice site. Given below are a few possible outcomes: **(DEC 2016)**

A. Pre-mRNA having mutation at both the splice sites will be retained in the nucleus because of the presence of bound snRNPs

B. Pre-mRNA having mutation at both the splice sites will be exported to cytosol.

C. Pre-mRNA mutated at either 3' or 5' splice sites will be retained in the nucleus because of the presence of bound snRNPs

D. Pre-mRNA mutated at either 3' or 5' splice sites will be exported to cytosol because of the absence of bound snRNPs

Choose the correct combination of possible outcomes:

(a) B and C (b) A and D

(c) B and D (d) A and C

8. Which one of the following statements about mRNA export from the nucleus is INCORRECT? **(DEC 2019 ASSAM)**

(a) Cellular pre-mRNAs with introns are generally not exported out of nucleus

(b) Dbp5 is an RNA helicase that releases mRNA from the transport complex

(c) ATP hydrolysis is required for mRNA transport across the nuclear pore complex

(d) Phe-Gly proteins of the nuclear pore complex are not involved in mRNA export

9. An intron in a yeast reporter gene carries a mutation in the splice site branch point (UACUAAC to UACA*AAC). To suppress the mutation, a library of point mutants of snRNAs was introduced into the mutant strain. The suppressor is most likely to have a point mutation in: **(DEC 2018)**

(a) U1 snRNA (b) U2 snRNA

(c) RNaseP (d) U6 snRNA

10. During maturation process of some RNA molecules, formation of a 2'- 5' phosphodiester bond takes place.

Following statements are made about this phenomenon.

A. Spliceosome mediated removal of intronic sequences occurs through the formation of a 2'- 5' phosphodiester bond.

B. Removal of group II introns occurs through the formation of 2'- 5' phosphodiester bond.

C. Enzymatic removal of introns from the yeast tRNA precursors involves 2'-5' phosphodiester bond formation.

D. RNaseP mediated 5'- end maturation of tRNA precursors involves formation of a 2' - 5' phosphodiester bond.

Which one of the following combinations of the statements is a true representation? **(JUNE 2018)**

(a) A only (b) A and D

(c) A and B (d) C and D

11. Following statements were made about the post-transcriptional processing of RNA in eukaryotes.

A. Soon after transcription initiation, RNA polymerase II pauses -30 nucleotides downstream from the site of initiation until the Cap structure is added to the 5' end of the nascent pre- mRNA

B. The 5' splice sites are functionally divergent whereas the 3' sites are functionally equivalent.

C. In addition to helping in recognition of the splice sites, the exon definition also functions as a splicing regulator by allowing pairing and linking of adjacent 5 and 3' splice sites.

D. The intron definition mechanism applies only to the larger introns (above 500 nucleotides length) and assists in achieving alternate splicing.

E. The splicing reactions carried out in vitro have revealed that the first and second trans-esterification reactions are reversible.

Which one of the following combination of statements is correct? **(DEC 2019)**

(a) A, B and D (b) B, C and D

(c) B, D and E (d) A, C and E

12. In type II splicing **(JUNE 2015)**

(a) a 'G-OH' from outside makes a nucleophilic attack on 5'-P of first base of intron,

(b) a free 2'-OH of an internal adenosine makes a nucleophilic attack on 5'-P of first base of intron

(c) A 3'-OH of an internal adenosine makes a nucleophilic attack on 5'-P of first base of intron

(d) the hydrolysis of last base of exon is carried out by U2/U4/U6

13. A DNA segment was cloned into the active site region of *lacZ* gene and the recombinant plasmid introduced into lac Z^- strain of *E. coli* and plated on a medium containing X-gal. The colonies showed blue color.

Which one of the following statements is correct? **(DEC 2018)**

(a) The nature of the cloned DNA segment need not be special as cloning of any DNA in *lacZ* will result in disruption of its reading frame and production of blue colour on X-gal plates.

(b) The cloned DNA segment could be a Group 1 intron whose removal from the precursor *lac Z* transcript in *E. coli* results in production of mature *lac Z* mRNA which can then produce active Lac Z protein.

(c) The cloned sequence is likely to be *lacy* sequence which is naturally a part of *lac* operon in *E. coli.*

(d) The cloned sequence is likely to be an anti-terminator sequence which allows full length transcription of *lacZ.*

14. Total RNA was isolated separately from cytosol and nuclei of human cells growing in a cell culture. Each sample was mixed with a purified denatured fragment of a DNA corresponding to a large intron of a house keeping gene and incubated under renaturating condition.

Given below are the statements made about the outcome of the experiment.

A. RNA isolated from nuclei will form RNA-DNA duplexes because of the presence of introns in the primary RNA.

B. Cytosolic RNAs usually will not form RNA-DNA duplexes.

C. Both cytosolic and nuclear RNA will not form RNA-DNA duplexes as transcription and splicing occur simultaneously.

D. Cytosolic RNA will form RNA-DNA duplexes because unspliced cytosolic RNAs are exceptionally stable.

(DEC 2014)

Which of the above statement(s) is/are most likely to be true?

(a) C only (b) A and D

(c) A and B (d) Only D

15. Infection with certain viruses inhibits small nuclear RNA (snRNA) processing in eukaryotic host. This favours the expression of viral genes in host because **(DEC 2013 GU)**

(a) inhibition of snRNA destabilizes host mRNA thus favouring viral protein synthesis.

(b) inhibition of snRNA interferes with splicing of host mRNA and prevent antiviral protein synthesis.

(c) inhibition of snRNA inhibits host mRNA transport thus favouring viral mRNA translation.

(d) snRNA inhibits only ViruS mRNA processing. Therefore inhibition of snRNA favours viral mRNA translation.

16. Synthesis of normal hemoglobin requires coordinated synthesis of a α globin and β globin, Thalassemias are genetic defects perturbed in this coordinated synthesis. Patients suffering from deficiency of β globin chains (β-thalasscmia) could also be due to mutations affecting the biosynthesis of β globin mRNA.

The following statements describe the genesis of non-functional β globin leading to β-thalassemia,

A. Mutation in the promoter region of the β globin gene

B. Mutation in the splice junction of the β globin gene

C. Mutation in the intron 1 of the β globin gene

D. Mutations towards the 3' end of the β globin gene that codes for polyadenylation site.

Which of the following combinations is correct? **(DEC 2011)**

(a) A, B and D (b) A, B and C

(c) B, C and D (d) C, D and A

17. In Trypanosomes, a 35 base leader sequence is joined with several different transcripts making functional mRNAs. The leader sequence is joined with the other RNAs by **(DEC 2016)**

(a) A specific RNA ligase

(b) The process of trans–splicing

(c) A nucleophilic attack caused by free guanine nucleotide

(d) A nucleophilic attack caused by 2' OH of an internal A present in the leader sequence

18. Leader sequence in some of the protozoan parasites is transcribed elsewhere in the parasite genome and gets joined with several transcripts make the functional mRNA. The joining of the two transcripts occur by the process of **(DEC 2014)**

(a) alternate splicing.

(b) trans splicing

(c) ligation

(d) RNA editing.

19. In recent years, genome-wide transcription study using high throughput sequence analysis has revealed some novel results that include:

(i) presence of RNA polymerase in both intra- and intergenic regions of the genome

(ii) existence of non-coding RNAs generated from mRNA coding genes.

(iii) existence of sense-and antisense transcripts generated from the promoter and untranslated region of many annotated genes.

Possible interpretation of the above results are:

A. RNA polymerase can loosely bind to any part of the genome but its affinity becomes strong only when it reaches the promoter.

B. Binding of RNA polymerase to non-promoter regions of the genome leads to the generation of various non-coding regulatory RNAs.

C. Non-coding RNAs are generated from mRNA coding genes due to aberrant transcription initiation and termination.

D. Sense and antisense transcripts are generated from the promoter and untranslated regions of protein coding genes by a novel mechanism of bidirectional transcription.

Identify the correct combination of the above interpretations: **(DEC 2013)**

(a) A and B (b) B and D

(c) A and D (d) B and C

20. Which one of the following statements related to transcription and processing of mRNA is INCORRECT? **(DEC 2019)**

(a) During prokaryotic transcription, DNA binding properties of RNA polymerase are altered by sigma factor

(b) In eukaryotic transcription, synthesis of rRNA, mRNA and some small RNAs occurs by RNA polymerases I, II and III, respectively

(c) Splicing observed in tRNA involves successive/sequential cleavage and ligation reactions while pre-mRNA splicing proceeds through lariat formation

(d) mRNA with premature stop codons are degraded by Nonsense-Mediated Decay (NMD) and mRNAs without an in-frame stop codon get accumulated and translated in the cytoplasm.

21. Following observations were made about mRNA degradation pathway in the eukaryotic organism:

A. A sequence independent endonuclease attack initiates degradation pathway for all mRNAs

B. Degradation of the majority of mRNAs is deadenylation dependent

C. Initially, the poly(A) tail is shortened by the PAN2/3 complex, followed by a rapid digestion by CCR4-NOT complex

D. mRNA decay occurs only by 3 to 5' exonuclease activity

E. The 3' to 5' mRNA degradation step is catalyzed by the exosome complex

Which one of the following options represents the combination of correct statements? **(DEC 2019 ASSAM)**

(a) B, C, E (b) B, D, E

(c) A, B, D (d) A, C, E

Answer Key

1.(d)	2.(b)	3.(d)	4.(b)	5.(c)	6.(d)	7.(a)
8.(d)	9.(b)	10.(c)	11.(d)	12.(b)	13.(b)	14.(c)
15.(b)	16.(a)	17.(b)	18.(b)	19.(b)	20.(d)	21.(a)

7. Promoter Deletion Assay/EMSA

1. A promoter deletion study was done in order to determine the binding sites for a transcription factor on the promoter, which is activated on treatment with the drug 'X'. The following constructs were made-

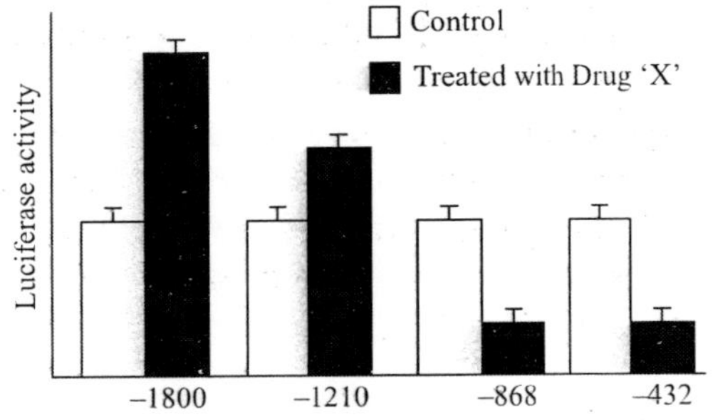

Luciferase assay revealed the following results

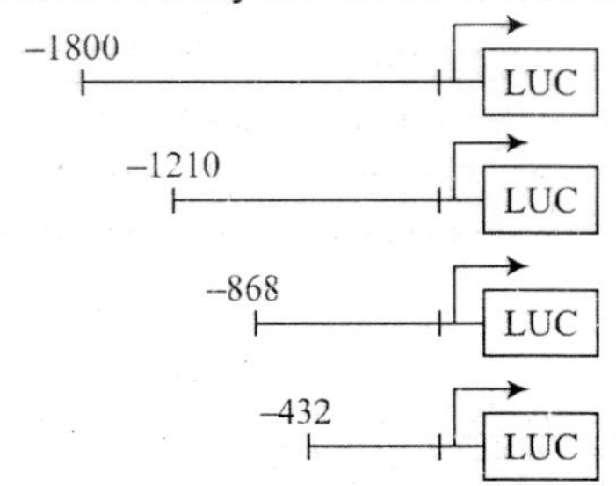

The following statements can be made.

A. Region between -1800 and -1210 contains a binding site for the activator.

B. Region between -868 and -1210 contains a binding site for a repressor.

C. Region between -868 and -432 contains a binding site for a repressor.

D. Region between -1210 and -868 contains a binding site for the activator.

Which of the above is/are true? **(DEC 2014)**

(a) A and C (b) B and C

(c) A and D (d) B only

2.

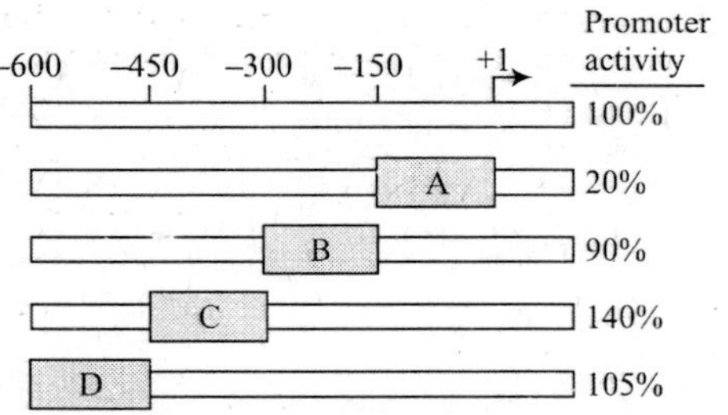

In order to identify the regulatory regions of a novel promoter sequence shown above, four 150 bp deletion constructs were made in a luciferase reporter system as indicated above in boxes A to D. After transfection, the observed level of promoter activity (%) as analyzed by luciferase assay of all the constructs is indicated in the right of the figure. Identify the best correct combination of regions in the options given below that indicate the presence of a positive and a negative regulatory elements respectively. **(JUNE 2014)**

(a) B and D (b) A and C

(c) A and D (d) A and B

3. Deletion analysis of a promoter region of a gene was carried out to identify the regulatory elements in it. In the figure below, the filled boxes denote the areas of deletion and the observed activities (in arbitrary units) of the promoter are as shown.

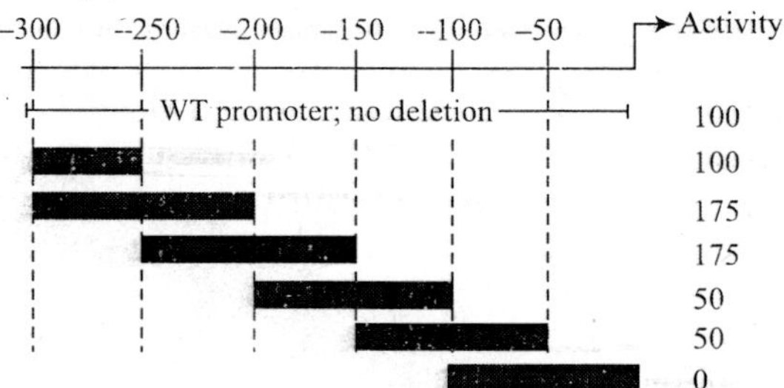

Based on the observations, following statements were made:

A. The region between-100 and -50 houses a positive regulatory element.

B. The region between -200 and -250 houses a negative regulatory element

C. The region between -150 and -200 houses a positive regulatory element.

Which one of the following options represents the correct interpretation of the data?

(DEC 2019)

(a) Both A and B (b) A only

(c) B only (d) Both B and C

4. A researcher was working with three proteins, A, B and C which may have potential roles in gene expression. In order to validate the hypothesis, EMSA (electrophoretic mobility shift assay) was performed. The purified proteins were allowed to bind with a labelled DNA and the results obtained after autoradiography as shown below.

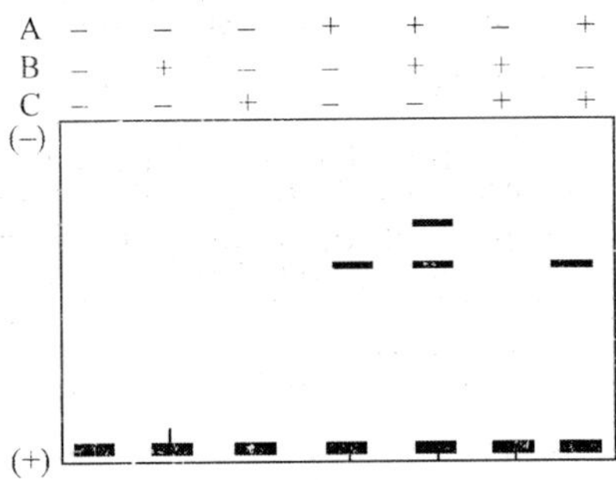

The following interpretations were made **(JUNE 2017)**

(i) Protein A possesses the DNA binding motif

(ii) Protein B possesses the DNA binding motif

(iii) Protein B binds to DNA-protein A complex

(iv) Protein-C binds to DNA only when protein A is bound

Choose the correct combination of interpretations.

(a) (i) and (iv) (b) (i) and (iii)

(c) (ii) and (iii) (d) (iii) and (iv)

5. Run off transcription assays were performed to establish the specificity of three novel sigma factors for their promoters. Result of the experiments are shown below:

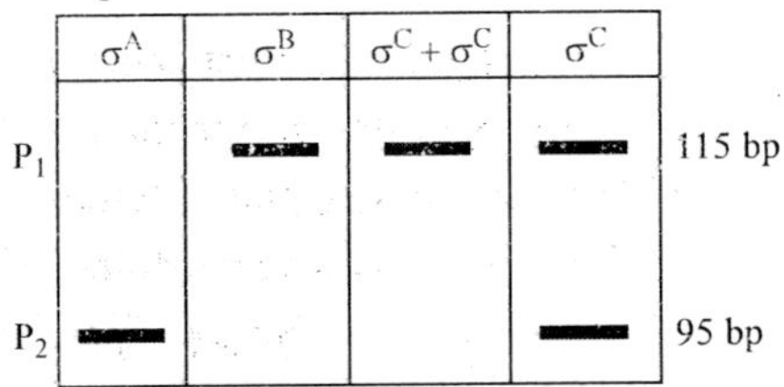

Following inferences were made from these results:

A. σ^A initiates transcription from P2 and σ^B from P1

B. σ^C can initiates transcription from both promoters.

C. σ^B Prevents initiation of transcription from P2

D. σ^A Initiates transcription from P1

Choose the option that correctly interprets that results?

(a) A, B and C only (b) A ,B only

(c) C and D only (d) B, C and D only

6. A researcher wanted to identify the enhancer sequences of a newly discovered gene. Shown below are the relevant regions of some of the reporter constructs the researcher designed to identify the enhancer.

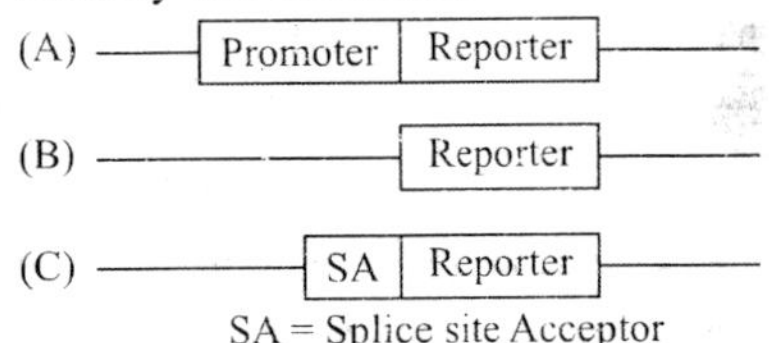

Which of the above constructs can be used to identify the enhancer? **(DEC 2018)**

(a) A only (b) B only

(c) Both A and C (d) C only

Answer Key

1.(c)	2.(b)	3.(c)	4.(b)	5.(a)	6.(a)
3	2	3	2	1	1

8. Transfer RNA and Amino-Acyl t-RNA

1. The different arms in the tRNA structure are shown in Column A. The specific signatures associated with the different arms are shown in Column B. **(JUNE 2019)**

	COLUMN A		COLUMN B
A.	Acceptor arm	(i)	Dihydrouridine
B.	Anticodon Arm	(ii)	8 bp stem and CCA sequence
C.	TψC arm	(iii)	5 bp stem
D	D - arm	(iv)	Pseudo-uridine

Choose the correct matches from the following:

(a) A – (ii); B – (iv); C – (i); D – (iii)

(b) A – (i); B – (iii); C – (iv); D – (ii)

(c) A – (iv); B – (i); C – (ii); D – (iii)

(d) A – (ii); B – (iii); C – (iv); D – (i)

2. Following observations were made about tRNA.

A. tRNA encoding genes cluster at specific regions of the human genome

B. 5' end of the mature tRNA is generated by RNAse P mediated cleavage

C. Modified bases may confer increased stability to tRNA

D. Modified bases of tRNA do not affect the pattern of wobble pairing

E. Suppressor tRNAs compete with the release factors to read the termination codons

Which one of the following combinations of statements represents correct observations?

(DEC 2019 ASSAM)

(a) A, C, E (b) B, E, D
(c) B, C, E (d) A, B, D

3. Aminoacyl tRNAs are escorted to the ribosome by the elongation factor **(JUNE 2017)**

(a) EF-Ts (b) EF-G
(c) EF-Tu (d) eEF-2

4. The specificity of tRNA recognition by a aminoacyl tRNA synthetase that is intrinsic to the tRNA molecule lies on **(DEC 2012)**

(a) acceptor stem.
(b) anticodon loop.
(c) acceptor stem and anticodon loop.
(d) D-arm

5. The challenges faced by aminoacyl tRNA synthetases in selecting the correct amino acid is more daunting than its recognition of the appropriate tRNA. In case of amino acids with similar structures like valine and isoleucine, this challenge is met by the enzyme possibly through its

A. catalytic pocket.

B. editing pocket.

C. anticodon loop.

D. acceptor arm.

Choose the correct set from the following:

(DEC 2012)

(a) A and B (b) A and C
(c) B and D (d) B and C

6. Each amino acyl-tRNA synthetase is precisely able to match an amino acid with the tRNA containing the correct corresponding anticodon., Most organisms have 20 different tRNA synthetases, however some bacteria lack the synthetase for charging the tRNA for glutamine ($tRNA^{Gln}$) with its cognate amino acid. How do these bacteria manage to incorporate glutamine in their proteins? Choose the correct answer. (JUNE 2016)

(a) Glutamine is not present in the newly synthesized bacterial protein. Post translational modification converts glutamate to glutamine at the required sites.

(b) In these, bacteria, the aminoacyl tRNA synthetase specific for tRNA glutamate ($tRNA^{Glu}$) also charges $tRNA^{Gln}$ with glutamine.

(c) In these bacteria, the aminoacyl tRNA synthetase specific for $tRNA^{Glu}$ also charges $tRNA^{Gln}$ with glutamate. A second enzyme then converts the glutamate of the charged $tRNA^{Gln}$ to glutamine.

(d) In these bacteria, the aminoacyl tRNA synthetase charges $tRNA^{Glu}$ with either glutamate or glutamine according to their requirement during protein synthesis

7. Aminoacyl-tRNA synthetases are very specific for aminoacylation of tRNAs with the correct cognate amino acids. However, there is a possibility of a mismatch between the tRNA and its cognate amino acid. This error is corrected by the inherent proof-reading activity of the aminoacyl-tRNA, synthetase. In case of two very similar amino acids, namely valine and isoleucine, isoleucyl-tRNA synthetase employs the following possible approaches for an error free aminoacylation

A. It removes an incorrect amino acid by hydrolyzing the aminoacyl-AMP linkage following first reaction step.

B. It is activated for proof-reading activity, leading to breakage of the bond between the wrong amino acid and tRNA.

C. It has an intrinsic ability to recognize the structural difference between amino acids leading to abortive elimination of the non-cognate amino acid.

D. It gets sequestered in the second step with the wrong amino acid, and that freezes the aminoacylation process.

Which of the following combinations is correct? **(JUNE 2014)**

(a) A and B (b) A and D

(c) B and D (d) C and D

8. Aminacyl tRNA synthetases face two important challenges:

(i) They must recognize the correct set of tRNAs for a particular amino acid.

(ii) they must charge all of these isoaccepting tRNAs with the correct amino acid.

Both of these processes are carried out with high fidelity by the following possible mechanisms:

A. The discrimination ability resides predominantly at the acceptor stem of the tRNAs.

B. The specificity is contributed by the anticodon loop in tRNAs.

C. The specificity is embedded in the amino acyl synthetase at the 'N' terminus.

D. The specificity is contributed by the variable loop of the tRNA.

Which of the following is correct? **(DEC 2013)**

(a) A and B (b) A and C

(c) B and C (d) A and D

9. Amino acid selenocysteine (Sec) is incorporated into polypeptide chain during translation by: **(DEC 2011)**

(a) charging of Sec into $tRNA^{ser}$ followed by incorporation through serine codon

(b) charging of serine into $tRNA^{ser}$ followed by modification of serine into selenocysteine and then incorporation through serine codon

(c) charging of Sec into $tRNA^{sec}$ and then incorporation through selenocysteine codon

(d) charging of serine into $tRNA^{ser}$ followed by modification of serine into selcnocysteine and then incorporation through a specially placed stop codon

10. Presence of selenocysteine in proteins in *E. coli* is a consequence of: **(DEC 2018)**

(a) post-translational modification of cysteine present in special structural regions of the proteins by SelB and SelC.

(b) post-translational modification of serine present in special structural regions of the proteins by SelB and SelC.

(c) aminoacylation of a special tRNA ($tRNA^{SeCys}$) by serine tRNA synthetase with serine followed by further modification of the attached serine to selenocysteine followed by its transport to the ribosome by SelB

(d) aminoacylation of a special tRNA ($tRNA^{SeCyt}$) by serine tRNAsynthetase with selenocysteine followed by its transport to ribosome by SelB.

11. Many organisms encode only 18 aminoacyl-tRNA synthetases (aaRS). These organisms lackaars that use Asn or Gln (as one of the substrates) for direct aminoacylation of the $tRNA^{Asn}$ and $tRNA^{Gln}$, respectively. Which one of the following statements represents the correct option? **(DEC 2019)**

(a) The organisms lacking AsnRS and InRs lack Asn and Gln in their proteins.

(b) In these organisms, selected Asp and Glu residues in the proteins are post-translationally modified by a regulated mechanism

(c) In these organisms, the $tRNA^{Asn}$ and $tRNA^{Gln}$ are first aminoacylated by AspRS and GluRS, respectively, and then the

Asp and Glu attached to the tRNAs are modified to Asn and Gln, respectively.

(d) In these organisms, the precursors of mRNAs that encode AspRS and GluRS are alternatively spliced to generate AsnRs and GinRs.

Answer Key

1.(d)	2.(c)	3.(c)	4.(c)	5.(a)	6.(c)	7.(a)
8.(a)	9.(d)	10.(c)	11.(c)			

9. Translation Process

1. Bacterial ribosomes consist of 30S and 50S ribosomal subunits. The translating monosome has a sedimentation value of **(DEC 2019 ASSAM)**
   (a) 70S because a fixed set of the ribosomal proteins (totalling to a value of ~10S) are removed when 30S and 50S subunits interact with each other
   (b) 70S because the interaction between the two subunits (30S and 50S) excludes some surface area decreasing the overall resistance of movement through the medium
   (c) 80S because the monosome consists of one subunit of 30S and one subunit of 50S
   (d) 50S because the sedimentation of the combined monosome is determined by the sedimentation of the large subunit

2. During each cycle of chain elongation in translation, how many conformational changes does the ribosome undergo that are coupled to GTP hydrolysis? **(JUNE 2015)**
   (a) Zero (b) One
   (c) Two (d) Three

3. What is the minimum number of NTPs required for the formation of one peptide bond during protein synthesis? **(JUNE 2011)**
   (a) One (b) Two
   (c) Four (d) Six

4. Which one of the following statements about eukaryotic translation is NOT true?
   In eukaryotic translation, **(DEC 2014)**
   (a) ribosome binding site on mRNA is called Kozak consensus sequences.
   (b) initiator tRNA is $tRNA^{f\text{-}met}$
   (c) initiator amino acid is methionine.
   (d) translocation factor is eEF2.

5. During protein synthesis in prokaryotes, the peptidyl transferase activity required for peptide bond formation is due to **(JUNE 2014)**
   (a) ribosomal protein L26.
   (b) 16S ribosomal RNA.
   (c) 23S ribosomal RNA.
   (d) aminoacyl tRNA.

6. Which of the following names is appropriate for the sequence 5' –*G*/ANNAUG -3' in a mammalian mRNA? **(DEC 2012)**
   (a) Shine-Dalgarno sequence
   (b) Kozak sequence
   (c) Internal ribosome entry sites
   (d) Translation termination site

7. It has been observed that in 5-10% of the eukaryotic mRNAs with multiple AUGs, the first AUG is not the initiation site. In such cases, the ribosome skips over one or more AUGs before encountering the favourable one and initiating translation. This is postulated to be due to the presence of the following consensus sequence (s):
   A. CCA CC AUG G
   B. CCG CC AUG G
   C. CCG CC AUG C
   D. AAC GG AUG A
   Which of the following sequence sets related to the above postulations is correct? **(JUNE 2011)**
   (a) A and B (b) A and C
   (c) C and D (d) B and D

8. Presence of an internal ribosomal entry site (IRES) in mRNA **(JUNE 2012)**

(a) Inhibit its translation

(b) promotes its post transcriptional processing

(c) has no impact on its translation

(d) promotes its translation under adverse conditions

9. For continuation of protein synthesis in bacteria, ribosomes need to be released from the mRNA as well as to dissociate into subunits. These processes do not occur spontaneously. They need the following possible conditions:

A. RRF and EF-G aid in this process

B. An intrinsic activity of ribosomes and all uncharged tRNA are required

C. IF -1 promotes dissociation of ribosomes

D. IF-3 promote dissociation of ribosome

Which of the following sets is correct? **(DEC 2011)**

(a) A and D (b) A and B

(c) A and C (d) B and D

10. Presence of circular mRNAs for a specific protein in an eukaryotic cell reflects a rapid rate of synthesis of that protein. Following mechanisms are suggested:

A. eIF-4G and PABP promote this process through 5'-3' interaction of mRNA.

B. ribosomes are less active in recognizing circular mRNA.

C. PABP and eIF-4A promote this process.

D. ribosomes can reinitiate translation without being disassembled.

Which of the following is correct? **(JUNE 2011)**

(a) A and D (b) B and D

(c) A and C (d) B and C

11. During elongation step or protein synthesis, translocation moves the mRNA and the peptidyl t-RNA by one codon through the ribosome. Translocation in E. coli involves GTP and EF-G. However, in vitro translocation can take place independent of GTP and EF-G. Based on these observations, the following hypotheses can be made:

(A) The molecular mechanism of translocation in vitro is completely different from that in vivo.

(B) Translocation activity is independent of GTP hydrolysis.

(C) Translocation activity is completely dependent on GTP and EF-G.

(D) Translocation activity is inherent in ribosomes, however, the rate of translocation in vivo is enhanced significantly in presence of GTP and EF-G

Which one of the following combinations is correct? **(JUNE 2013)**

(a) only (D) (b) (A) and (C)

(c) (A) and (B) (d) (C) and (D)

12. During translation in prokaryotes, when ribosomes reach the termination codon, the termination codon is recognized by the class I release factors (RF1 or RF2) leading to the release of the polypeptide. A second class II release factor (RF3) facilitates the termination process. Which of the following statements regarding the mechanism of action of the release factors is INCORRECT? **(JUNE 2018)**

(a) Class I release factors decode the stop codons while the RF3 is a GTPase that stimulates recycling of the class I release factors.

(b) Free RF3 has a higher affinity for GTP than GDP

(c) RF1 and RF2 share a conserved segment of 'GGQ' sequence which is essential for the polypeptide release.

(d) RF1 and RF2, individually possess another stretch of tripeptide sequences which are involved in the recognition of the termination codons.

13. A protein has 4 equally spaced trypsin sensitive sites which results in peptide fragments A_1, A_2, A_3, A_4 and A_5 upon digestion with trypsin.

Peptides A_2 and A_5, represents N-terminal and C-terminal fragments respectively. Now you are asked to synthesis this protein. At time t = 0 you added all the 20 amino acids labelled with ^{14}C and initiated the synthesis. At time t = 4, full length protein is synthesized. If you stop the synthesis of the protein in time t = 1 and digest the protein with trypsin, which peptide will have maximum ^{14}C label than others?

(DEC 2015)

(a) A_3 (b) A_1
(c) A_4 (d) A_2

Answer Key

1.(b)	2.(c)	3.(c)	4.(b)	5.(c)	6.(b)	7.(a)
8.(d)	9.(a)	10.(c)	11.(a)	12.(b)	13.(d)	

10. Genetic Code

1. Which of the following statement about selenocysteine, the 21st amino acid present in several enzymes is true? **(DEC 2013 GU)**
   (a) Selenium is enzymatically transferred to cysteine residue only after the full functional protein gets synthesized .
   (b) selenocysteine is encoded by the same codon/codons specific for cysteine with the help of a special tRNA molecule.
   (c) selenocysteine is encoded by the stop codon UGA with the help of a special tRNA molecule.
   (d) A special pool of selenocysteine exists in the cell from where it is incorporated in the polypeptide chain by AUG with the help of a special tRNA molecule.
2. A synthetically prepared mRNA contains repetitive AU sequences. The mRNA was incubated with mammalian cell extract which contains ribosomes, tRNA s and all the factors required for protein synthesis. Assuming no initiation codon is required for protein synthesis, which of the following peptides will most likely be synthesized? **(JUNE 2012)**
   (a) A single peptide composed of the same amino acid sequence
   (b) Two peptide with alternating sequence of two amino acids
   (c) A single peptide with alternating sequence of three amino acids
   (d) Three different peptides each sequence composed of a single amino acid.
3. In gene regulation, Open Reading frame (ORF) implies **(DEC 2011)**
   (a) interviening nucleotide sequence in between two genes
   (b) a series of triplet codons not interrupted by a stop codon
   (c) a series of triplet codons that begins with a start codon and ends with a stop codon
   (d) the exonic sequence of a gene that curresponds to 5' UTR of the mRNA and thus does not code for the protein
4. A deletion of three consecutive bases in the coding region of a gene cannot result in **(MODEL PAPER)**
   (a) deletion of a single amino acid without any other change in the protein.
   (b) replacement of two adjacent amino acids by a single amino acid.
   (c) replacement of a single amino acid by another without any other change in sequence of the protein.
   (d) production of a truncated protein.
5. What would be the tripeptide produced by translation of the transcript produced by the following DNA sequence? **(DEC 2018)**
   3'-AAGTA CTCT-5'
   (a) Arg-Phe-Trp (b) Arg - Leu - Gly
   (c) Thr- Lys –Ser (d) Phe - Met – Arg
6. Imagine the following RNA sequence is translated in the mammalian cytosol and mitochondria. RNA sequence:

AUG AUA CUG UGA CUU AGG CUC UAA

Following are some putative peptide sequences

	Cytosol		Mitochondria
(a)	Met-Ile-Leu	(i)	Met-Ile-Leu-Trp-Leu-Arg- Leu
(b)	Met-Ile-Leu-Trp-Leu-Arg- Leu	(ii)	Met-Met-Leu-Trp-Leu

Find out the correct combination of peptides made in the cytosol and mitochondria. **(DEC 2017)**

(a) (a) (i) (b) (a) (ii)
(c) (b) (i) (d) (b) (ii)

Answer Key

1.(c)	2.(b)	3.(c)	4.(c)	5.(d)	6.(b)

11. Post-Translational Event

1. Which one of following modification of proteins is co-translational? **(JUNE 2017)**
   (a) Palmitoylation
   (b) Myristoylation
   (c) Famesylation
   (d) Addition of cholesterol
2. Which of the following is NOT a post-translational modification in a mammalian system? **(DEC 2013)**
   (a) Palmitoylation (b) glycosylation
   (c) peptidylation (d) phosphorylation
3. In prokaryotes, the initiatior t-RNA is first charged with a methionine, followed by the addition of a formyl group to the methionine by the enzyme Met-tRNA transfonnylase.
   Given below are several statements in this context.
   A. All prokaryotic proteins have formyl methionine at their amino terminal end.
   B. Defomylase removes the formyl group from the amino terminal methionine.
   C. All prokaryotic proteins have methionine at their amino terminal end.
   D. Aminopeptidases often remove the amino terminal methionine.
   E. Aminopeptidases remove amino terminal formyl methionine.
   Which of the above statement(s) are most likely to be true? **(DEC 2015)**
   (a) A only (b) B and C
   (c) E only (d) B and D
4. In bacteria, N-formyl methionine is the first amino acid to be incorporated into a polypeptide chain. Accordingly, one would think that all bacterial proteins have a formyl group at their amino terminus and the first amino acid is methionine. However, this is not the case, because of the following possible reasons.
   A. Deformylase removes the formyl group only during or after the synthesis of the polypeptides.
   B. Aminopeptidase removes only the amino terminal methionine.
   C. Aminopeptidase removes the amino terminal methionine as well as one or two additional amino acids.
   D. Deformylase removes the formyl group as well as amino terminal methionine and adds one or two amino acids to it.
   Choose the combination of correct answers from the following: **(DEC 2012)**
   (a) B and C (b) A and B
   (c) A and C (d) A and D
5. In a form of stress response, bacteria synthesize a group of proteins called stress proteins (or heat shock proteins) such as DnaK, DnaJ, GroEL, GroES, and GrpE. DnaK is an ATP binding protein, which attaches to the newly synthesized polypeptide in conjunction with DnaJ. Which one of the following statements correctly states a step in the subsequent process of protein folding? **(DEC 2019)**
   (a) The affinity of DnaK to the polypeptide increases upon hydrolysis of the ATP to ADP.

(b) DnaJ is an exchange factor that replaces ADP with ATP in DnaK

(c) ATP hydrolysis is required for the phosphorylation of DnaJ

(d) ATP hydrolysis is required for the phosphorylation of GrpE

Answer Key

1.(b)	2.(c)	3.(d)	4.(c)	5.(a)		

12. Translational Inhibitor

1. Chloramphenicol is a "broad-spectrum" antibiotic which inhibits protein synthesis in prokaryotes. Given below are a few statements regarding the mode of action of chloramphenicol.

   A. Chloramphenicol inhibits the peptidyl-transferase activity of ribosomes.

   B. Chloramphenicol can be used to treat moderate to severe infections, because mitochondrial ribosomes are not sensitive to chloramphenicol.

   C. Chloramphenicol binds to one of the domains of 23S rRNA

   D. Chloramphenicol competes for binding with the E-site tRNA

   Which of the following options describes correctly the mechanism of action of chloramphenicol? **(JUNE 2017)**

   (a) B and D only (b) A and C only

   (c) A, C and D (d) B, C and D

2. Puromycin is an antibiotic used to inhibit protein synthesis. Given below are few statements about the antibiotic.

   A. It enters the E-site of the ribosome where it prevents the release of deacetylated tRNA after the action of peptidyl transferase.

   B. It blocks the translocation process by binding to the translocation factor EF-G.

   C. Puromycin resembles the initiatior tRNA, $tRNA_i^{f\text{-}met}$ and binds exclusively to the P-site.

   D. It resembles the aminoacyl tRNA and binds to the A-site of the ribosome.

   E. Puromycin inhibits only prokaryotic protein synthesis.

   F. Puromycin inhibits both prokaryotic and eukaryotic protein synthesis.

   Which of the above statement(s) is/are true? **(DEC 2014)**

   (a) A and E (b) B only

   (c) D and F (d) C and E

3. An antibiotic that resembles the 3' end of a charged tRNA molecule is: **(DEC 2015)**

   (a) Streptomycin (b) Sporsomycin

   (c) Puromycin (d) Tetracycline

4. When one isolates ribosomes from bacterial lysate, apart from 70S, 50S and 308 ribosomal subunits, one also finds a small population of 100S, 130S and 150S sub-units. EDTA dissociates these larger ribosomal subunits into 50S and 30S, suggesting that they have both the subunits. Upon addition of cations they reassociate into 70S, but none of the other forms could be detected. What is the reason for not obtaining the >70S forms? **(JUNE 2015)**

   (a) The effects of EDTA cannot be reversed by the addition of cations.

   (b) 190S, 130S and150S are modified form of ribosomes that are irreversibly damaged by EDTA.

   (c) 100S, 130S etc. represent polysome that cannot be reassembled denovo without other cellular components.

   (d) They are obtained as an experimental artifact in - preparations of ribosomes

5. Polysome profiling of cells treated with three hypothetical translation inhibitors is shown in the plot below. These three inhibitors are

   (i) CHP – leaky inhibitor of translation

   (ii) LTM – arrests ribosomes at the initiation codon

   (iii) PTM – inhibits ribosome scanning

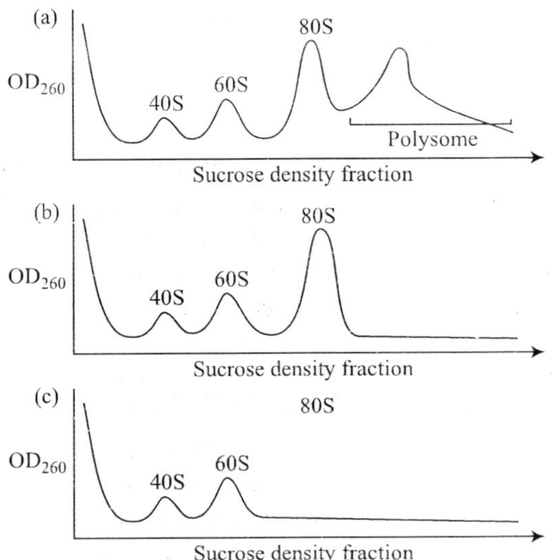

Match the polysome profile to the inhibitor **(DEC 2016)**

(a) (i) – a; (ii) – b; (iii) – c
(b) (i) – b; (ii) – c; (iii) – a
(c) (i) – c; (ii) – b; (iii) – a
(d) (i) – a; (ii) – c; (iii) – b

6. Ribosomes prepared from a bacterium were fractionated by sucrose density gradient centrifugation (panel i) to separate the 30S, 50S and 70S populations. When the ribosome preparation was incubated individually with either elongation factor-G (EF-G), or a newly identified protein X, Or GTP, the profile remained unchanged. Likewise, no changes were seen in the profile when the ribosomal preparation was incubated with EF-G + GTP or protein X + GTP. However, when the ribosomal preparation was incubated with protein X, EF-G and GTP together, it resulted in a change of profile which showed a decrease of the 70S peak area and increase in the peak areas for 30S and 50S (panel ii).

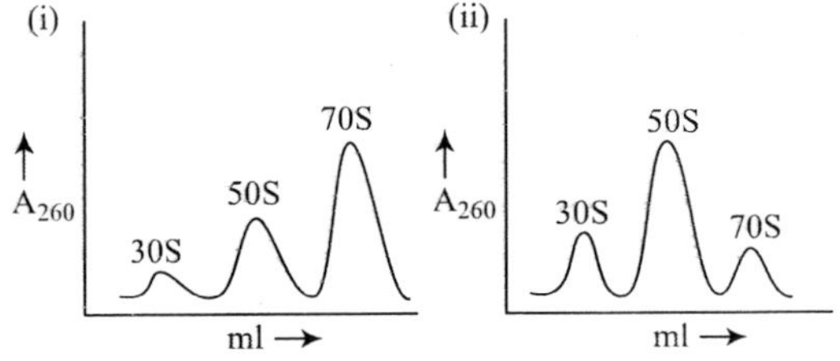

Choose the option that defines a correct conclusion from the observations. **(DEC 2019)**

(a) Protein X is an anti-association factor which functions in the presence of EF-G and GTP
(b) Protein X is a dissociation factor which functions in the presence of EF-G and GTP
(c) Protein X binds GTP
(d) EF-G is known to bind GTP, hence it can be concluded that the effect of GTP is through EF-G and protein X does not bind GTP.

Answer Key

1.(b)	**2.**(c)	**3.**(c)	**4.**(c)	**5.**(a)	**6.**(b)

13. Regulation of gene expression in Phage/virus

1. Viral gene expression after T_3 bacteriophage infection is controlled by: **(DEC 2012)**

(a) Repressor molecule.
(b) Modification of RNA polymerase
(c) Slow injection of nucleic acid.
(d) DNA polymerase.

2. Bacteriophage T4 infects *E. coli* and injects its DNA inside the cell. The transcription of viral genes occurs in three stages: immediate early, early and late. All the promoters on viral genome are available, but the control takes place at the level of **(JUNE 2013)**

(a) promoter strength.
(b) modification of host RNA polymerase.
(c) synthesis of new polymerases.
(d) turn over rate of RNA synthesis.

3. Bacteriophage T4 transcribe the viral genes immediately after infection using **(DEC 2013 GU)**

(a) host RNA polymerase.
(b) viral RNA polymerase.
(c) modified host RNA polymerase.
(d) RNA transcriptase.

4. A culture of an *E. coli* strain that is lysogenic for phage lambda is grown at 32°C. Induction of the prophage from the host chromosome will occur when the culture is exposed to **(JUNE 2011)**

(a) 40°C.

(b) ultra violet radiation.

(c) infra red radiation.

(d) wild type *E. coli* culture

5. During development and differentiation, there is a dynamic programme of differential expression of sets of genes. In bacteria, phage infections are among the simplest examples of developmental process. Typically, only a subset of the phage genome, offer referred to as immediate early genes, are expressed in the host immediately after phage infection. As time passes, early genes start to be expressed, and the immediate early genes and bacterial genes are turned off. In the final stage of phage infection, the early genes give way to late genes. One of the simplest way it is achieved is through

A. expression of cascade of σ factors

B. expression of new RNA polymerases

C. expression of different holoenzymes

D. expression of different transcription factors

The correct reasons are **(MODEL PAPER)**

(a) (A), (D) (b) (A), (C), (D)

(c) (A), (B), (D) (d) (A), (B), (C)

6. Why lysogenic cycle is more beneficial to a virus than lytic cycle under certain circumstances? **(JUNE 2012)**

(a) The Lysogenic cycle prevent local extinction of host while still retaining infectious potential

(b) By integrating with the bacterial chromosomes, the genetic instructions for the virus become refreshed after one or more replication events during binary fission

(c) Lysogenic infection cycles donot harm their host cells, so they can produce virus particles indefinitely

(d) Lysogeny causes more mutations to occur in the virus, creating more variants upon which natural selection can operate

7. In the lysogenic λ-phage **(DEC 2013)**

(a) both CI and Cro are on.

(b) both CI and Cro are off

(c) CI is on while Cro is off.

(d) CI is off while Cro is on.

8. Which one of the following λ proteins acts both as an activator and repressor of transcription? **(DEC 2017)**

(a) cI protein (b) N protein

(c) cII protein (d) Q protein

9. Which one of the following regulatory proteins can act as a positive and negative regulator on binding to the same DNA elements? **(DEC 2019)**

(a) Lac repressor (LacI)

(b) Lambda (CI) repressor

(c) Ara C protein (AraC)

(d) Tip repressor (TrpR)

10. Bacteriophage λ is a temperate bacteriophage and has two modes in its life cycle, lysogenic and lytic. Several genes are involved in these two processes like *N, cI, cII, cIII, Q, int, xis,* etc.

Which one of the following diagrams represents the control mechanism correctly? **(DEC 2013)**

(a)
N → cII, cIII, Q
cII, cIII → (+) ve → cI, INT (Lysogeny)
Q (Lytic)

(b)
N → cII, cIII, XIS, Q (Lytic)
cII → cI → INT (Lysogeny)

(c)

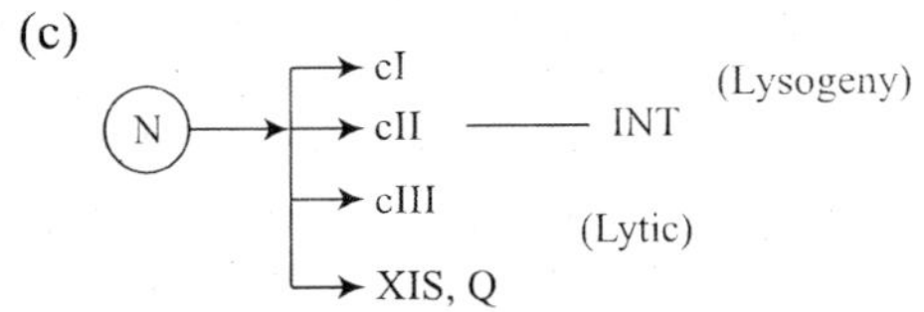

(d)

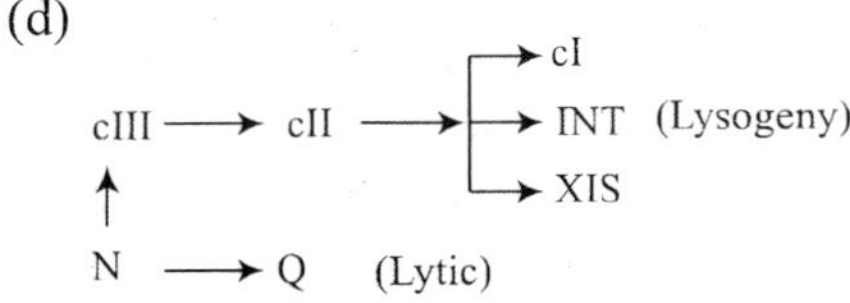

11. Bacteriophage λ genetic circuit may be represented as follows:

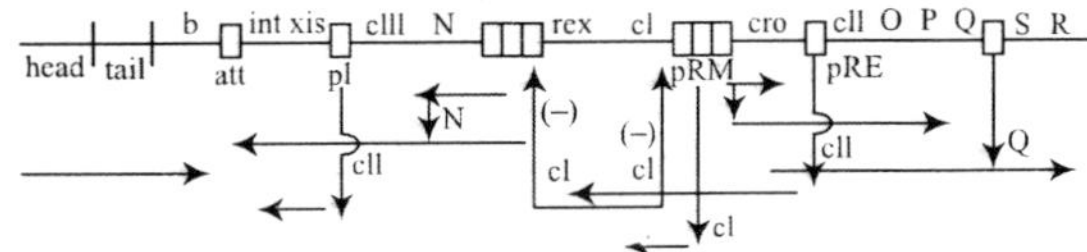

The control of gene expression occurred during the phage infection may be described as follows: **(MODELPAPER)**

A. N and Q protein act as antiterminator

B. CI acts only as repressor

C. CII act as a activator for pRE

D. CI act as positive and negative regulator

Which one of the statements are correct?

(a) (A), (B) and (C).

(b) (B), (C) and (D).

(c) (A) and (D) only.

(d) (A), (C) and (D).

12. Bacteriophage λ is a temperate phage. Immediately after infection, viral specific mRNAs for N and Cro proteins are expressed followed by early mRNAs. At the commitment phase, either lytic cycle starts with the expression of genes for head tail, and lytic proteins or lysogenisation cycle begins with the expression of repressor and integrase genes. During induction of lysogens both INT and XIS proteins are needed along with host factors. Out of the four processes below, some govern Integration of viral genome and its excision?

A. Repression of transcription

B. Retroregulation

C. Rearrangement of viral genome

D. Repression of translation

Identify the correct set of combination:

(DEC 2012)

(a) A and B (b) B and C

(c) C and D (d) D and A

13. Bacteriophage λ, has two modes in its life cycle, lytic and lysogenic. In the lysogenic mode, the expression of all the phage genes are repressed while the expression of repressor gene switches between on and off –position depending on the concentration of repressor. The following statements are made:

(DEC 2011)

A. Repressor may act both as a positive regulator and negative regulator

B. Expression of repressor gene, cI is independent of the expression of cII and cIII genes.

C. Mutation of cI gene will cause it to it to form clear plaques on both wild type *E. coli* and *E. coli* (λ^+)

D. Mutation at operators, O_L and O_R will allow the phage to act as virulent phage.

The correct statements are

(a) A and B (b) B and C

(c) C and D (d) D and A

14. Insertion lambda vectors are used to create cDNA libraries, In some insertion vectors, the site of insertion is within the cI gene. The, recombinants which are cI⁻ can then be screened by plating on *E. coli* hfl. The gene hfl encodes a protease that controls lytic-lysogeny decision through which mechanism?

(JUNE 2011)

(a) Hfl protease degrades cI, thereby promoting lytic pathway

(b) Hfl protease degrades cIII and so cIII cannot interact with CII, lysogenic pathway is thus preferred.

(c) Hfl protease degrades cII, and therefore cI synthesis cannot be established, thus lytic pathway is preferred.

(d) Hfl protease degrades INT, the protein involved in phage integration and the lytic pathway is initiated.

15. The lambda (λ) and P22 phages are two related lambdoid bacteriophages. A recombinant lambda phage (λ^{Mut}) was derived from the wild type lambda phage (λ^{WT}) by replacing its CI repressor gene and the CI binding sites with those from the P22 phage. Both the λ^{WT} and the λ^{Mut} were used independently to infect Escherichia coli strain over-producing λ^{WT} CI repressor. Following outcomes were surmised

(i) Infection with λ^{WT} will lyse the E. coli used

(ii) Infection with λ^{WT} will invariably establish lysogeny in the E. coli used

(iii) Infection with λ^{Mut} will lyse the E. coli used

(iv) Infection with λ^{Mut} will invariably establish lysogeny in the E. coli used.

Which combination of the above statements is correct? **(JUNE 2017)**

(a) (i) and (ii) (b) (ii) and (iii)

(c) (iii) and (iv) (d) (iv) and (i)

Answer Key

1.(b)	**2.**(b)	**3.**(a)	**4.**(b)	**5.**(b)	**6.**(a)	**7.**(c)
8.(a)	**9.**(b)	**10.**(a)	**11.**(d)	**12.**(b)	**13.**(d)	**14.**(c)
15.(b)						

14. Control of gene expression: LAC OPERON

1. The lac operon in *E. coli*, is controlled by both the lac repressor and the catabolite activation protein CAP. In an *in vivo* experiment with *lac* operon, the following observations are made:

A. cAMP levels are high

B. Repressor is bound with allolactose

C. CAP is interacting with RNA polymerase

Which one of the following conclusions is most appropriate based on the above observations? **(JUNE 2012)**

(a) Glucose and lactose are present.

(b) Glucose is present and lactose is absent

(c) Both are absent

(d) Glucose is absent and lactose is present

2. *E. coli* is being grown in a medium containing both glucose and galactose. On depletion of glucose, expression of β-galactosidase will **(JUNE 2016)**

(a) remain unchanged

(b) increase

(c) decrease

(d) initially decrease and then increase

3. *E. coli* cells are growing in a medium containing lactose but no glucose. Indicate the expression status of the lac operon if glucose is added to the culture. **(DEC 2013 GU)**

(a) Decreased expression

(b) Increased expression

(c) No changes in expression level.

(d) Initial rapid increase followed by sharp drop

4. A culture medium contains two carbon sources, one is preferred carbon source (glucose) and the second is a non-preferred source (lactose). Which one below is correct regarding the nature of growth curve of *E. coli* cultured in this medium? **(DEC 2015)**

(a) Growth curve will be same as when grown in presence of only glucose.

(b) Growth curve will be same as when grown in presence of only lactose.

(c) A lag phase will be observed between the two exponential phases.

(d) Two lag phases will be observed between the two exponential phases.

5. A bacterial strain can use carbohydrates and hydrocarbons as growth substrates. The strain uses glucose following a minimal lag period after inoculation, regardless of the other carbohydrates and hydrocarbons in the growth medium. The following observations were also made.

A. In the absence of glucose, lactose is used after a lag period of about three times as long as the lag period for glucose utilization.

B. The presence of hydrocarbons does not affect the lag period for the utilization of lactose.

C. The utilization pattern for all hydrocarbons is similar to that of lactose.

D. Branched hydrocarbons are not immediately utilized if straight chain hydrocarbons are initially present.

Which one of the following specific regulatory mechanisms is consistent with the above observations related to carbohydrate and hydrocarbon utilization? **(DEC 2013)**

(a) Diauxie

(b) End point repression

(c) Catabolite repression

(d) Transcription attenuation

6. Lac repressor inhibits expression of genes in *lac*-operon whereas purine biosynthesis is repressed by the *Pur* repressor. The two proteins have 31% identical sequences and have similar three-dimensional structures. The gene regulatory properties of these proteins differ in relation to

A. binding of small molecules to the repressor.

B. presence of recognition sites on the genome.

C. oligomeric nature of the repressor.

D. DNA binding property.

The correct statements are **(JUNE 2011)**

(a) A and B (b) A, B and C

(c) A and C (d) B, C and D

7. Two *E. coli* cultures A and B are taken. Culture A was earlier grown in the presence of optimum concentration of gratuitous inducer IPTG. Both the cultures are now used to inoculate fresh medium containing sub-optimal concentration of gratuitous inducer. It was observed that culture B was unable to utilize lactose, whereas culture A did so efficiently. The reason behind this is **(JUNE 2011)**

(a) pretreatment with IPTG has resulted in a mutation as a result of which lac operon is constitutively expressed

(b) IPTG has made the cell membrane more porous to small molecu les and so lactose is taken up more efficiently by A as compared to B.

(c) in culture A, lactose permease was induced to a high level, during pretreatment with IPTG, which allowed the preferential uptake of lactose.

(d) in culture A, IPTG activated a receptor which bound lactose more efficiently, thereby triggering a signal.

8. *E coli* was grown in three different experimental conditions. In one, it was grown in medium containing glucose as carbon source; in the second in medium containing both glucose and galactose; and in third was infected with phage. . Match the curves shown below to the treatment **(JUNE 2016)**

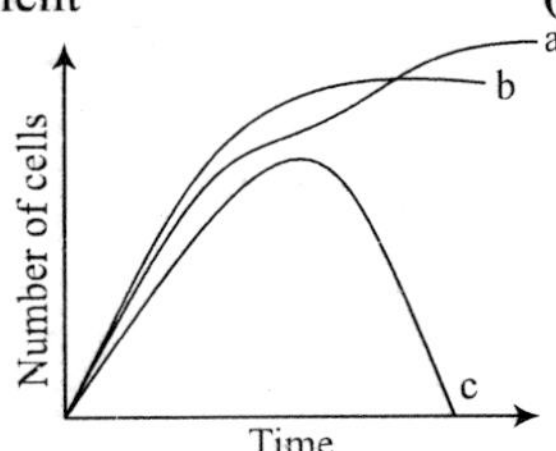

(a) a is grown in glucose; b is grown in glucose and galactose; c is infected with phage

(b) a is grown in glucose and galactose; b in glucose; c is infected with phage.

(c) a is infected with phage; b is grown in glucose and galactose; c in glucose

(d) a is infected with phage; b is grown in glucose; c In glucose and galactose

9. In context to lac operon, if two bacterial strains P, and P2 with the genotypes $O^{C}I^{+}Z^{-}$ and $O^{+}I^{-}Z^{+}$ respectively, were used to produce mero-diploid daughter strain D, which one of the following statements correctly predicts the expression of Z gene (β-galactosidase activity) in all the three strains? (O^{+}, I^{+} and Z^{+} denote the wild type allele of the respective genes).

(JUNE 2015)

(a) P1 - No expression; P2 – constitutive expression; D - Inducible expression.

(b) P1 - No expression; P2 – constitutive expression; D – constitutive expression.

(c) P1 - No expression; P2 – Inducible expression; D - Inducible expression.

(d) P1 - Inducible expression; P2 – constitutive expression; D – Inducible expression.

10. Mutants of *lac Y* (Y^{-}) gene of *E. coli* do not synthesize the lactose permease protein. The following statements refer to the behaviour of lac Y^{-} mutants under different experimental conditions.

A. No synthesis of β- galactosidase when Y' cells are induced with lactose.

B. Synthesis of β- galactosidase when cells are induced with lactose.

C. No synthesis of β- galactosidase when cells are induced with IPTG.

D. Synthesis of β- galactosidase when cells are induced with IPTG.

E. The cells induced with IPTG cannot grow in the presence of TONPG (TONPG is a compound, whose uptake is mediated by lactose permease and cleaved by β-galactosidase to release a toxic compound).

F. Cells induced with IPTG Can grow in the presence of TONPG.

Which combination of the above statements is correct? **(DEC 2012)**

(a) A, D and F (b) B, C and E

(c) A, C and F (d) A, C and E

11. With reference to lac operon, what will be the phenotype of an *E. coli* strain having a genotype $I^{-} O^{+} Z^{+} Y^{-} / F' \, I^{+} O^{c} Z^{-} Y^{+}$? (DEC 2012)

(a) Constitutive for both β-galactosidase and lac permease

(b) Inducible for both β-galactosidase and lac permease.

(c) Inducible for β-galactosidase and constitutive for lac permease.

(d) Constitutive for β-galactosidase and inducible for lac permease.

12. A merodiploid strain of *E. coli* with the genotype $F^{+}O^{c}Z^{-}Y^{+} / O^{+}Z^{+}Y^{+}$ was constructed. The activity of β-galactosidase enzyme was measured in this strain upon following treatments.

A. no induction

B. induction with *n* moles of IPTG

C. induction with *n* moles of lactose

D. induction with *n* moles of lactose in the presence of n moles of glucose

(JUNE 2017)

Which one of the following graphs depicts the expected trends in β-galactosidase activity under the four different conditions?

(a)

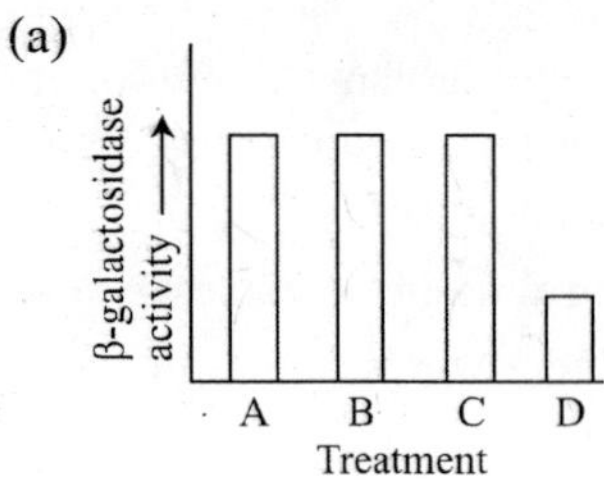

(b)

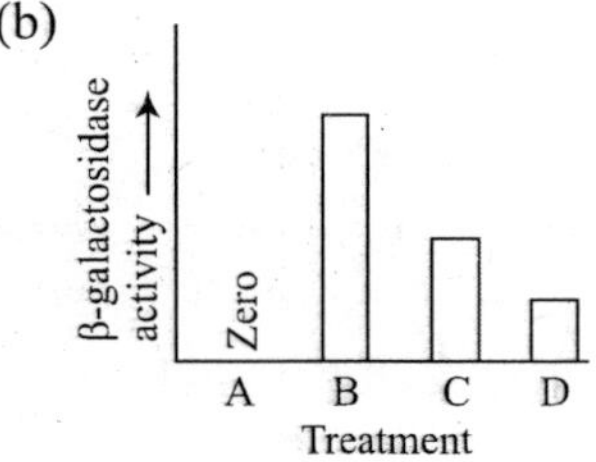

(c)

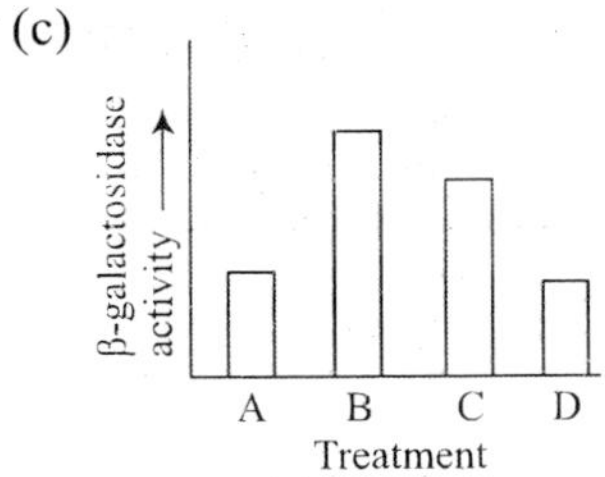

(d)

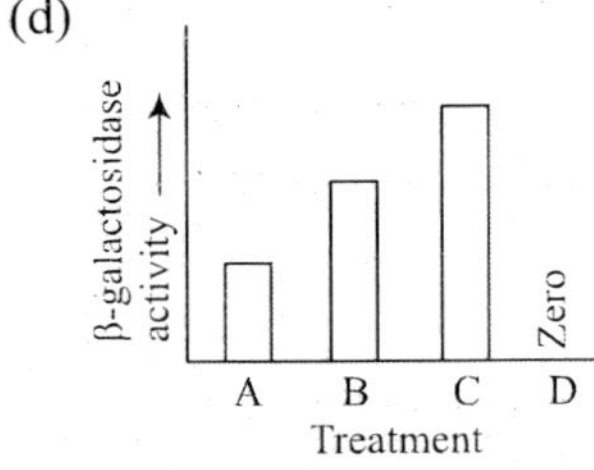

13. A hypothetical operon involved in the synthesis of an amino acid 'X' is 'ON' (transcribing) in the presence of low levels of 'X' and 'OFF' (not transcribing) in presence of high level of 'X'. The symbols a, band c (in the table below) represents a structural gene for the synthesis of X (X- synthase), the operator region and gene encoding the repressor- but not necessarily in that order. From the following data, in which superscripts denote wild type or defective genotype, identity which are the genes for X-synthase, operator region and the repressor.

Strain	Genotype	X-synthase activity in the presence of	
		Low level of 'X'	High level of 'X'
1.	$a^- b^+ c^+$	Detected	Detected
2.	$a^+ b^+ c^-$	Detected	Detected
3.	$a^+ b^- c^-$	Not Detected	Not Detected
4.	$a^+ b^+ c^+$/ $a^- b^- c^-$	Detected	Not Detected
5.	$a^+ b^+ c^-$/ $a^- b^- c^+$	Detected	Not Detected
6.	$a^- b^+ c^+$/ $a^+ b^- c^-$	Detected	Detected

The respective genes for 'X'- synthase, the operator region and repressor are:

(DEC 2015)

(a) a, b, c (b) c, a, b

(c) b, c, a (d) b, a, c

14. E. coli mutants isolated from a genetic screen showed following classes of mutations

A. Point mutations in *lacI*

B. Deletions immediately downstream of the transcription start site of the lacZYA mRNA

C. Duplications of part or whole of lacY

D. Duplications of part or whole of Lac A

Choose the option which is likely to result in constitutive expression of the lac operon?

(JUNE 2019)

(a) Both A and B (b) Both B and C

(c) Both C and D (d) Only A

Answer Key

1.(d)	**2.**(b)	**3.**(a)	**4.**(c)	**5.**(c)	**6.**(a)	**7.**(c)
8.(b)	**9.**(a)	**10.**(a)	**11.**(c)	**12.**(c)	**13.**(d)	**14.**(a)

15. Trp Operon

1. A mutant Trp repressor (TrpR^m) that cannot bind tryptophan is isolated. In this mutant

A. Trp operon is constitutively expressed.

B. Trp operon is super-repressed.

C. Trp repressor does not bind to the operator.

D. only Trp leader sequence is transcribed.

Which one of the following options represents all correct statements? **(DEC 2019 ASSAM)**

(a) A and C only (b) B and C only

(c) A and D only (d) A only

2. You have created a fusion between the trp operon, which encodes the enzymes for trptophan biosynthesis, under the regulatory control of the lac operator. Under which of the following conditions will trptophan synthase be induced in the strain that carries the chimeric operator fused operons? **(JUNE 2013)**

(a) Only when both lactose and glucose are absent.

(b) Only when both lactose and glucose are present.

(c) Only when lactose is absent and glucose is present.

(d) Only when lactose is present and glucose is absent.

3. The tryptophan operon in *E. coli* is an example of **(DEC 2013 GU)**

(a) inducible operon, controlled by a repressor protein and attenuation.

(b) repressible operon, controlled by a repressor protein and attenuation.

(c) inducible operon, controlled by an activator protein and aporepressor protein.

(d) repressible operon, controlled by a repressor protein and anti-termination.

4. Deletion of the leader sequence of *trp* operon of *E. coli* would result in **(MODELPAPER)**

(a) decreased transcription of *trp* operon.

(b) increased transcription of *trp* operon.

(c) no effect on transcription.

(d) decreased transcription of *trp* operon in the presence of tryptophan.

5. *E. coli* cells are growing in a medium with glucose as the sole carbon source. Tryptophan is suddenly added. The cells continue to grow, and divide every 30 minutes. What will be the status of tryptophan synthase activity in the cells over time if the trp mRNA remains stable **(DEC 2013 GU)**

(a) Low (b) High

(c) No change (d) No activity

6. Attenuation is a mechanism involved in the regulation of tryptophan operon in *E. coli*. When tryptophan levels are high in the cell, region 2 of the *trpL* is blocked from pairing with region 3. This allows the pairing of region 3 and 4 leading to the formation a rho-independent termination.

What would be the structure of the *trpL* region in *E. coli* cells where protein synthesis has been inhibited? **(JUNE 2014)**

(a) Region 2 pairs with region 3 allowing transcription of the structural genes.

(b) Region 1 and 2 will pair, allowing 3 and 4 to pair leading to attenuation.

(c) There is no pairing in the *trpL* region and transcription of structural gene occurs.

(d) Region 2 and 3 will pair leading to attenuation.

7. The following graph represents the expression of tryptophan synthetase (TS) in E. coli cells in absence (□) or presence (■) of tryptophan in the medium

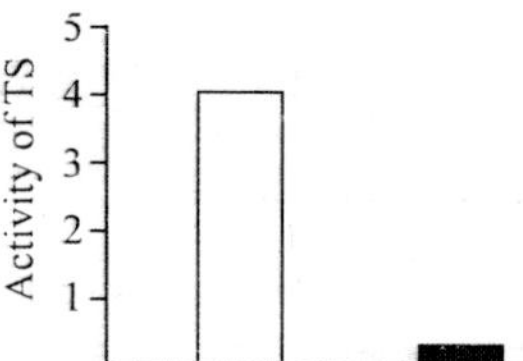

If the two trp codons in the leader sequence of trp operon is mutated to ala, which of the following graphs will best represent activity of TS in *E. coli* cells grown in the absence (□) or presence (■) of tryptophan? **(DEC 2014)**

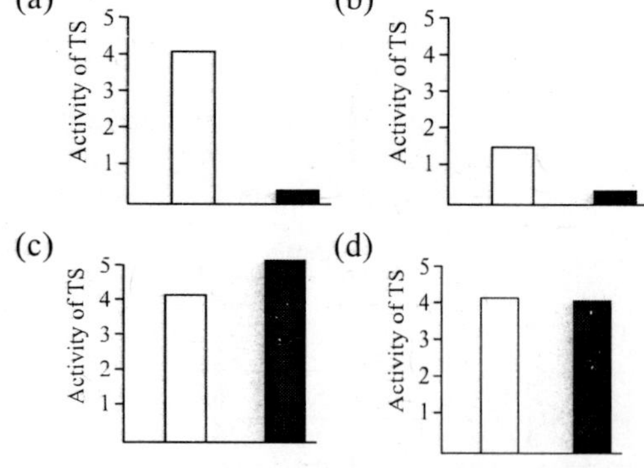

8. A construct (shown below) was generated to access the activity of bidirectional promoter (A and B) which has a common regulatory DNA element (X). The construct was used to transform *E. coli*.

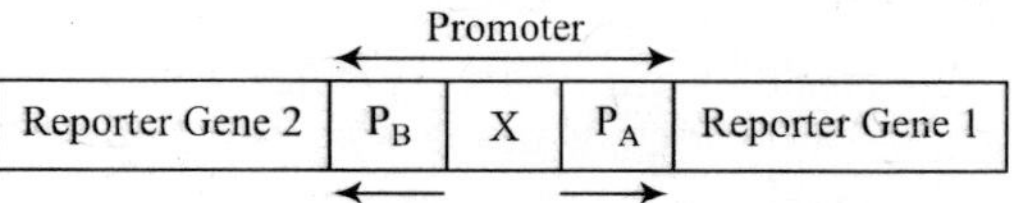

The activity of the promoter (as shown in the graph below) is recorded in the presence of, increasing levels of an *E. coli* encoded regulatory protein which binds to DNA element X.

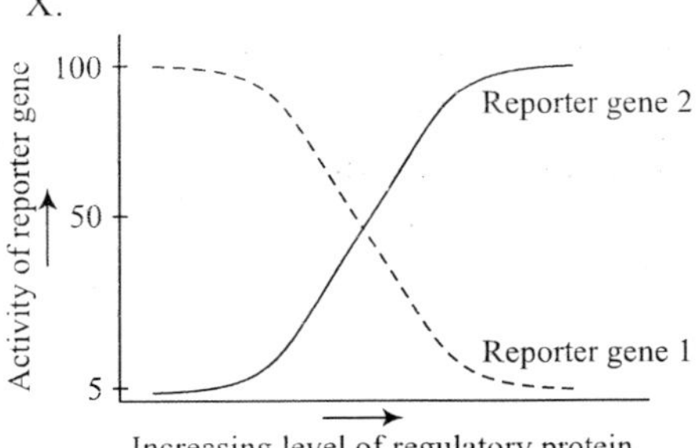

The above experiment is repeated in a mutant *E. coli* with mutation which abolishes binding of the regulatory protein to element X. Which one of the following graphs best depicts the activity of the reporter genes in the mutant strain? **(DEC 2017)**

(a)

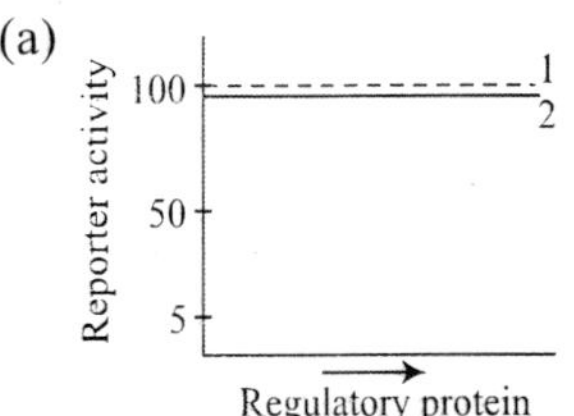

(b)

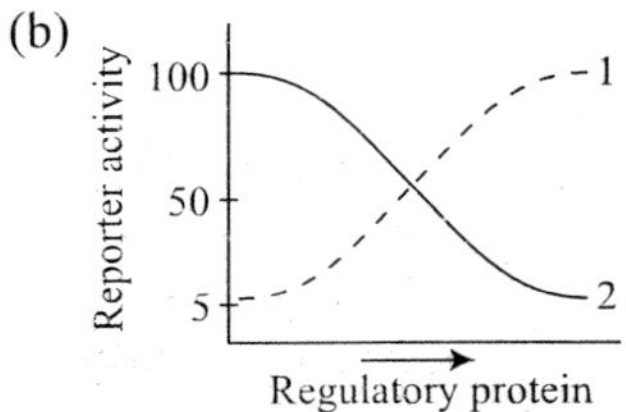

(c)

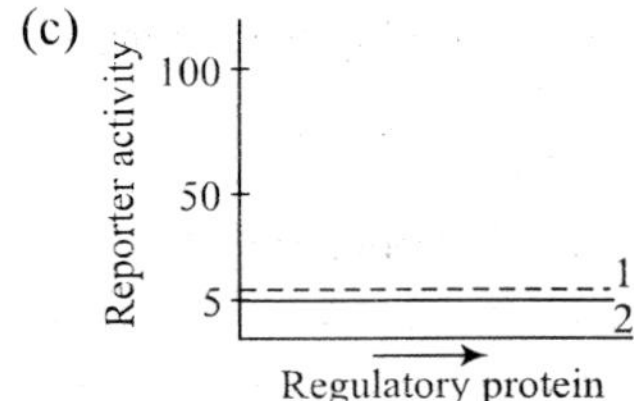

(d)

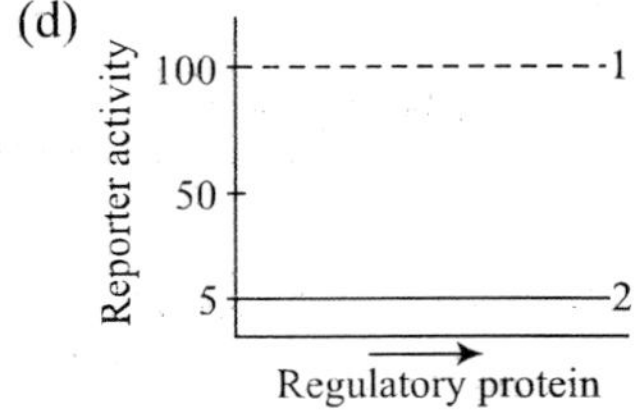

9. Transcriptional regulation of *trp* operon by tryptophan involves binding of tryptophan to **(JUNE 2018)**

(a) the repressor protein and inhibition of transcription by its interaction with the operator region.

(b) RNA polymerase and inhibition of transcription.

(c) the repressor protein leading to structural changes and its degradation by proteases.

(d) the repressor protein leading to its interaction with the sigma subunit and inhibition of transcription.

10. In a strain of *E. coli,* a fusion between the *lac* and *trp* operon took place and the new locus structure is shown below. The strain lacks the wild-type *trp* operon.

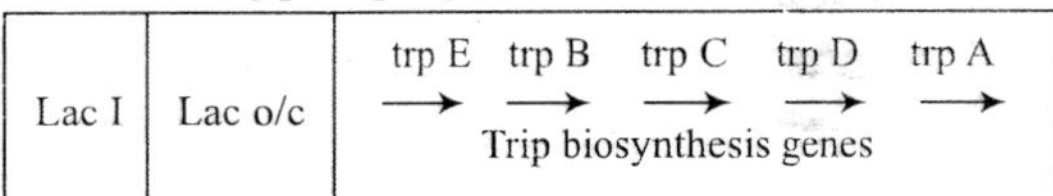

Given below are some of the potential scenarios:

A. Tryptophan will be synthesized in a medium containing lactose and tryptophan.

B. Tryptophan synthesis will be repressed in a medium containing glucose.

C. Tryptophan synthesis will take place only in the absence of sufficient tryptophan in the medium.

Choose the option that correctly describes the behaviours of the fusion operon. (JUNE 2019)

(a) A and B (b) A and C

(c) C only (d) B and C

Answer Key

1.(a)	2.(d)	3.(b)	4.(b)	5.(b)	6.(b)	7.(b)
8.(d)	9.(a)	10.(a)				

16. Control of gene expression in eukaryotes

1. In response to a drug, changes in protein levels were examined in a cell line. A pulse chase experiment was performed using ^{35}S labeled methionine. In comparison, to untreated samples, the following observations were made: few minutes after stimulation, protein X accumulates and this was followed by reduction in protein Y and Z. The correct interpretation of these observations would be: **(JUNE 2015)**
   (a) Protein X is a protease which degrades Y and Z.
   (b) Protein X is a transcriptional repressor that controls expression of Y and Z.
   (c) Expression of protein X and loss of Y and Z are unrelated.
   (d) Information is not sufficient to distinguish between the three possibilities stated above.
2. You are studying a transcriptional activator which is phosphorylated at serine residue that inactivates its nuclear localization signal. Which one of the following will be true if the serine residue in the transcription factor will be replaced by alanine? **(DEC 2013 GU)**
   (a) No phosphorylation of the transcription factor and constitutive activation of the target gene expression.
   (b) No phosphorylation of the transcription, factor and target gene expression will be inhibited.
   (c) No change in phosphorylation and target gene expression.
   (d) Increased phoshorylation of the transcription factor with concomitant increase of the target gene expression
3. The expression of a hypothetical gene was analyzed by Northern and Western blot hybridizations under control and induced condition. The results are summarized below:

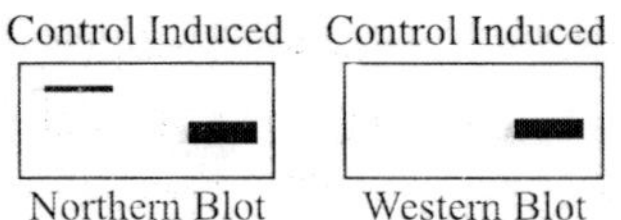

   Expression of genes can be regulated by:
   A. control at transcription initiation
   B. alternative splicing
   C. control of translation initiation
   D. protein stability

   Which of the above regulatory mechanisms can explain the observations shown in the figures? **(DEC 2014)**
   (a) Only B (b) Only A and B
   (c) Only B and C (d) A, B, C and D
4. Locus control region (LCR) lies far upstream from the gene cluster and is required for the appropriate expression of each gene in cluster. LCR regulates expression of globin genes in the cluster through the following ways.
   A. LCR interacts with promoters of individual genes by DNA looping through DNA-binding proteins.
   B. The LCR-bound proteins attract chromatin-remodeling complexes including histone-modifying enzymes and components of the transcription machinery.
   C. LCR acts as an enhancer for global regulation of gene cluster and does not regulate individual genes.
   D. LCR participates in covering inactive chromatin to active chromatin around the gene cluster.

   Choose the correct set of combinations. **(DEC 2012)**
   (a) A and B (b) A and C
   (c) B and C (d) B and D
5. A reporter cell line with stably integrated retroviral promoter-luciferase construct was transfected with an expression vector for a cellular protein. The protein seems to regulate the activation of retroviral promoter as analyzed by luciferase activity assay. Which

one of the following techniques will you use to show "*in vivo*" recruitment of the cellular protein on the integrated retroviral promoter? **(JUNE 2012)**

(a) Electrophoretic mobility shift assay
(b) RNAse protection assay
(c) DNAse hypersensitivity assay
(d) Chromatin immunoprecipitation assay.

Answer Key

1.(d)	2.(a)	3.(d)	4.(a)	5.(d)

17. mOR pathway and Translation regulation

1. Insulin and other growth factors stimulate a pathway involving a protein kinase mTOR, which in its turn augments protein synthesis. mTOR essentially modifies protein(s) which in their unmodified form act as inhibitors of protein synthesis. The following proteins are possible candidates:

A. eEF-1 B. eIF-4E-BP1
C. eIF-4E D. PHAS-1

Which of the following sets is correct? **(DEC 2011)**

(a) A and B (b) B and D
(c) A and C (d) B and C

2. During mitogenic stimulation, cells proliferate at a higher rate and it is primarily determined by an enhanced rate of protein synthesis. Among other mechanisms, MAP kinase pathway of signal transduction is involved in this. Global protein synthesis may be regulated by many mechanisms involving various steps of protein synthesis, namely, initiation, elongation and termination. Thus, many protein factors may be involved in the same. In the above process (mitogenic stimulation) the following factors are the portable targets.

A. eIF -2 B. eEF-1
C. S6 kinase D. eIF-4E BP

The correct answer is **(MODELPAPER)**

(a) A+B (b) C+D
(c) D+A (d) B+D

3. One of the cellular events that TOR, a kinase, positively regulates is the rate of rRNA synthesis. TOR regulates the association of a transcription factor to a Pol I subunit. When TOR is inhibited by the drug rapamycin, the transcription factor dissociates from Pol I. A yeast strain is engineered, which expresses a fusion of the transcription factor and the Pol I subunit. The level of rRNA synthesis is monitored in these cells using pulse labelling following rapamycin addition for the times indicated below. The transcript profile of rRNA observed for the wild type cells is given below:

t(min): 0 20 40

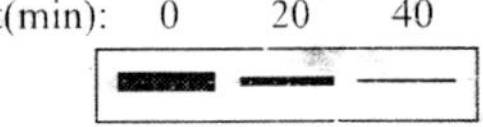

Identify the pattern expected in the engineered strain. **(JUNE 2016)**

(a) (b)
(c) (d)

4. During heat shock, mammalian cells shut down global protein synthesis while inducing heat shock proteins (Hsps). The possible molecular regulation(s) that could explain the phenomenon are:

A. mRNA of all proteins, except those of Hsps, undergoes degradation during heat shock.

B. Cap-dependent translation of most mRNAs is affected during heat shock due to denaturation of cap binding protein, eIF-4E.

C. Translation initiation of Hsp mRNAs takes place through their internal ribosome entry sites (IRES)

D. Hsp mRNAs are abundant during heat shock and thus they compete out other mRNAs for ribosome binding and translation.

Which of the following sets is correct? **(DEC 2013)**

(a) A and D (b) B and C
(c) C and D (d) A and D

5. The cap binding protein (eIF4E), which is involved in the global regulation of translation, is highly regulated in eukaryotic cells. In an experiment, a researcher transfected mammalian cells with (eIF4E) gene for its overexpression. Due to this, the cells will undergo **(JUNE 2013)**
   (a) apoptosis.
   (b) neoplastic transformation.
   (c) no change.
   (d) differentiation.
6. During eukaryotic protein synthesis, stress conditions result in activation of specific kinases leading to phosphorylation of a key translation initiation factor that inhibits protein synthesis from a large number of cellular mRNA. Which one of the following factors is the target of the kinase? **(DEC 2017)**
   (a) eIF4E (b) eIF4G
   (c) eIF2α (d) Gcn4
7. Phosphorylation of eIF2 α subunits (at Ser 51) leads to **(JUNE 2018)**
   (a) inactivation of Met-tRNA$_i$, binding activity of eIF2B.
   (b) sequestration of eIF2B because of tight binding between eIF2 and eIF2B.
   (c) degradation of eIF2B.
   (d) enhanced guanine exchange activity of eIF2B.
8. Phosphorylation of the α-subunit of eIF2 at Ser 51 position in *Saccharomyces cerevisiae* leads to sequestration of eIF2B, guanosine exchange factor. This phenomenon is **(JUNE 2019)**
   (a) known to activate translation of the capped mRNAs in the cytosol
   (b) known to activate translation of many key mRNAs possessing short ORFs(uORFs) in the mRNA sequence that proceed the main ORF
   (c) an essential requirement for translation of IRES containing mRNAs.
   (d) an essential requirement for the transport of mature mRNAs out of the nucleus

Answer Key

1.(b)	2.(b)	3.(b)	4.(b)	5.(b)	6.(c)	7.(b)
8.(b)						

18. Role of chromatin in gene expression

1. Histone deacytalase (HDAC) catalyses the removal of acetyl group from N-terminal of histones. Which amino acid of histone is involved in this process? **(JUNE 2016)**
   (a) Lysine (b) Arginine
   (c) Asparagine (d) Histidine
2. Histone acetylase and chromatin remodeling complexes are recruited to specific regions of chromatin by **(JUNE 2014)**
   (a) gene activator proteins.
   (b) specific promoter sequence.
   (c) phosphorylation of histone acetylase.
   (d) dephosphorylation of chromatin remodeling complexes.
3. After activation of a promoter by the DNA binding activity of a transcription factor, a co-activator is recruited at the region targeted for transcription which in turn creates a binding site for a chromatin remodeling complex. Which one of the following activities of the co-activator is responsible for the recruitment of chromatin remodeling complex? **(JUNE 2014)**
   (a) Histone deacetylase activity.
   (b) Histone methyl transferase activity.
   (c) Histone acetyl transferase activity.
   (d) DNA methyl transferase activity.
4. Prior to transcription, chromatin changes from an inactive state to an active state by various factors in a stepwise manner. Which one of the

following is involved in the initial step during activation of a chromatin loop? **(JUNE 2014)**

(a) HMG14.

(b) SSB

(c) DNA polymerase III

(d) Topoisomerase I.

5. Phosphorylation of serines as well as methylation and acetylation of lysines in histone tails affect the stability of chromatin structure above the nucleosome level and have important consequences for gene expression. The resulting changes in charge are expected to affect the ability of the tails to interact with DNA because **(JUNE 2013)**

(a) DNA is negatively charged.

(b) DNA-histone interaction is independent of net charge.

(c) phosphorylation of serine increases DNA-histone interaction.

(d) methylation and acetylation of lysine increases DNA-histone interaction.

6. DNA methylation plays an important role in transcription regulation in vertebrates. There is an inverse correlation between the level of DNA methylation in the vicinity of a gene and its transcription rate, whereas there is a direct correlation between histone acetylation and increased transcription. β-thalassemia is a common genetic impairment of hemoglobin β-chain synthesis in humans. If these patients can synthesize hemoglobin-F instead of hemoglobin chain in its place, they would be notably benefited. Administration of 5-azacytidine to thalassemia patients increases hemoglobin-F level in erythrocytes and thus benefit the patients.

Which one of the following statements about 5-azacytidine is NOT correct? **(JUNE 2013)**

(a) Cells exposed to 5-azacytidine incorporate it into DNA in place of cytidine.

(b) 5-azacytidine decreases DNA methylation.

(c) 5-azacytidine promotes histone acetylation.

(d) 5-azacytidine does not promote gene expression.

7. In eukaryotic chromatin, 30 nm fiber (solenoid) can open up to give rise to two kinds of chromatin. In one type (A), the promoter of a gene within the open chromatin is occupied by a nucleosome whereas in the other (B), the promoter is occupied by histone H1. The following possibilities are suggested.

A. The gene in (A) is repressed.

B. The gene in (B) is repressed.

C. The gene in (A) is active.

D. The gene in (B) is active.

Which of the following sets is correct? **(JUNE 2011)**

(a) A and D (b) B and C

(c) B and D (d) C and D

8. A gene producing red pigment was placed near centromeres of fission yeast and thus subjected to position effect variegation and produced white colonies, A screen for mutants that increased the red pigment production was undertaken. Which of the following genes, when mutated, is likely to produce this genotype? **(DEC 2014)**

(a) histone deacetylase

(b) Histone acetylase

(c) RNA polymerase II

(d) TATA binding factor

9. Genes translocated to the heterochromatic regions of chromosomes are silenced. In *S. pombe*, a translocation event was detected wherein a gene of interest was translocated to the centromere region and is silenced. Mutagenesis leading to loss of function of the following target genes was done to allow expression of the gene of interest from its new locus.

A. Mutation in histone deacetylase (Clr3).

B. Mutation in histone acetyltransferase (HAT-8).

C. Mutation in histone H3 lysine 9 methyl transferase (Clr4).

D. Loss of Dicer, an RNA processing enzyme.

Which of the above events could allow the expression of this gene from the centromeric region? **(JUNE 2018)**

(a) A, B and C (b) A, C and D

(c) B and C only (d) A and C only

10. Chromatin remodelling leading to histone modifications is invariably required for regulation of transcriptional activity of eukaryotic genes. The residues in histone tails may be methylated or acetylated. The nature of these modifications **(DEC 2019 ASSAM)**

(a) is identical on all the nucleosomes of a gene

(b) varies between the nucleosomes present in the transcription initiation and elongation regions

(c) is always fixed for the nucleosomes in the promoter regions of the genes that are transcriptionally active

(d) in the transcription initiation regions is identical for the genes that are sensitive to nuclease activity

11. The amino acid side chains of the four histones in the nucleosome are subjected to remarkable variety of post-translation modifications such as phosphorylation, acetylation and methylation. Which one of the following post-translational marks on histone tails is usually associated with transcriptional repression? **(DEC 2019)**

(a) Acetylation of H3K9

(b) Methylation of H3K9

(c) Acetylation of H4K5

(d) Phosphorylation of H3S10

12. Following statements are made about chromatin remodelling in human cells:

A. Local chromatin conformation may play more important role than the local DNA sequence of the promoter.

B. Histones in nucleosome can undergo many different covalent modifications, which in turn, alter the chromatin architecture locally.

C. Chromatin remodelling is a developmentally regulated passive process which does not require ATP.

D. Several histone variants exist, which replace the standard histones in specific types of chromatin.

Select the option that has the combination of all correct answers. **(JUNE 2019)**

(a) A, C, D (b) A, B, C

(c) A, B, D (d) B, C, D

13. Readers of histone modifications include: **(DEC 2019)**

(a) SUN domain proteins

(b) BAG domain proteins

(c) PAS domain proteins

(d) TUDOR domain proteins

14. In human, protein coding genes are mainly organized as "exons" and "introns". There are intergenic regions that transcribe into various types of non-coding RNA (not translating into protein). Some introns may harbor also transcription units, which are **(DEC 2011)**

(a) always other protein coding genes

(b) protein coding gene and RNA coding genes

(c) always RNA coding genes

(d) pseudo genes

15. In an experiment, intact chromatin was isolated and digested with micrococcal nuclease in independent tubes for 30 min, 1 h, 2h, and 4h. Further, the DNA was purified from each tube, separated on agarose gel and Southern hybridization was performed with rRNA gene probe and a centromeric DNA probe. Which one of the following patterns of signal intensity from both of the probes is likely to be obtained following Southern hybridization? (DEC 2018)

(a) With increasing time, compared to centromeric probe, a rapid increase in signal intensity of rRNA gene probe was observed.

(b) With increasing time, compared to centromeric probe, a rapid decrease in signal intensity of rRNA gene probe was observed.

(c) Irrespective of incubation period, both probes produced identical band intensities.

(d) Treatment with micrococcal nuclease would instantly degrade the DNA, hence, no hybridization signal would be obtained· in any of the samples.

Answer Key

1.(a)	2.(a)	3.(c)	4.(d)	5.(a)	6.(d)	7.(b)
8.(a)	9.(b)	10.(b)	11.(b)	12.(c)	13.(d)	14.(b)
15.(b)						

19. Gene silencing - RNA interference

1. Which of the following statement is NOT true about small interfering RNA (siRNA)? **(JUNE 2012)**

   (a) siRNA has 21-25 nucleotide sequence with 2 nucleotide overhang at 3' end

   (b) siRNA is processed by RNA protein complex RISC

   (c) siRNA is often induced by viruses

   (d) siRNA does not generally act at the level of transcription.

2. Small RNAs with internally complementary sequences that form hairpin-like structure, synthesized as percursor RNAs and cleaved by endonucleases to form short duplexes are called **(JUNE 2011)**

   (a) snRNA (b) mRNA

   (c) tRNA. (d) miRNA.

3. RNA interference is mediated by both siRNA and miRNA. Which one of the following statement about them is NOT true? **(JUNE 2016)**

   (a) Both siRNA and miRNA are processed by DICER.

   (b) Both siRNA and-miRNA usually guide silencing of the same genetic loci from which they originate.

   (c) miRNA is a natural molecule while siRNA is either natural or a synthetic one.

   (d) miRNA, but not siRNA is processed by Drosha.

4. MicroRNAs (miRNAs) have recently been shown to play a significant role in the fine tuning of gene expression. Some miRNAs induce gene silencing by binding to mRNAs and inducing inhibition of translation. On the other hand, there are miRNAs that bind to mRNAs and activate their degradation. The following characteristics can be applicable to miRNAs that inhibit mRNA translation:

   A. miRNA is partially complementary to region of target mRNA in the 3' UTR.

   B. miRNA always base pairs with mRNA around an AU-rich sequence.

   C. miRNA base pairs with mRNA through 6-7 nucleotides at its 5' end referred to as "seed sequence" as well as few additional bases elsewhere.

   D. miRNA is always partially complementary to the 5' UTR of the target mRNA.

   Choose the correct option from the following: **(JUNE 2014)**

   (a) A and B (b) A and C

   (c) C and D (d) A and D

5. siRNAs and miRNAs are used for achieving gene silencing. Although, major steps are similar there are distinct differences in the key players of the two processing pathways. Following statements relate to some characteristic features of gene silencing.

   A. Both siRNAs and miRNAs are processed by cytoplasmic endonuclease Dicer.

   B. 'Drosha' is needed for processing miRNAs and precursor siRNAs.

C. Both siRNAs and miRNAs show association with Argonaute protein.

D. Both the silencing pathways involve RISC complex.

Which of the following combinations is NOT correct? **(JUNE 2011)**

(a) A and C (b) C and D

(c) B only (d) D and A

6. Virus induced gene silencing (VIGS) is a process that takes advantage of the RNAi mediated antiviral defence mechanism. Which one of the following ultimately guides siRNA to degrade the target transcript (mRNA)? **(JUNE 2014)**

(a) dsRNA

(b) ssRNA

(c) RNA Induced Silencing (RIS) Complex

(d) dsRNA binding protein

7. mRNA of a gene was depleted in human cells using siRNA that arrest cells in the G_2 phase of the cell cycle. In order to test whether the G_2 arrest is due to an off-target or all on-target effect of siRNA mediated mRNA depletion, an investigator can:

A. re-introduce an ectopic copy ofthe gene coding for the wild-type mRNA and protein

B. re-introduce an ectopic copy of the gene that is different in its mRNA sequence at the siRNA target site but encodes for the same protein

C. re-introduce an ectopic copy of the gene that codes for-different mRNA and protein

D. utilize few more siRNAs targeting different regions of the mRNA in question

Choose the combination with correct statements: **(DEC 18)**

(a) A, B, C only (b) C and D only

(c) B and D only (d) B, C, D only

8. In an experiment, the student has infected mammalian host cell with cytoplasmic RNA virus. The virus growth was monitored by measuring the intracellular viral RNA at different time intervals. It was observed that viral RNA titre progressively went down with time, particularly 12 hours post infection. Following are few possibilities which can explain this observation.

A. The virus infection triggered upregulation of miRNAs that might have down-regulated the host factor critical for viral RNA replication.

B. The virus might encode miRNAs that regulate (inhibits) its own replication.

C. One of the viral proteins inhibits replication of the viral RNA to restrict rapid proliferation.

D. Viral RNA goes to nucleus with time and thus not detectable in the cytoplasm 12 hour post infection.

Which one of the following options has all correct statements? **(JUNE 2019)**

(a) A, B and C (b) A, C and D

(c) A. B and D (d) B, C and D

9. Irrespective of the chromosomal configuration, a single X chromosome remains active in all diploid human somatic cell lines. Which one of the following mechanisms best accounts for the above phenomenon? **(JUNE 2019)**

(a) A maternally inherited X–chromosome is developmentally programmed to remain active by avoiding DNA methylation.

(b) Chromosome specific expression and binding of rox1 to one of the X-chromosomes protects it from Xist mediated silencing.

(c) The T-six gene produces just enough of the Xist antisense RNA to block one Xic locus.

(d) A cell produces just enough of the blocking factor to block one Xic locus.

Answer Key

1.(b)	2.(d)	3.(b)	4.(b)	5.(c)	6.(c)	7.(c)
8.(a)	9.(d)					

SECTION-4

CELL COMMUNICATION SIGNALING AND IMMUNOLOGY

1. Host parasite interaction

1. Entry of enveloped viruses into its host cells is mediated by: **(JUNE 2016)**
   (a) Only endocytosis
   (b) Both endocytosis and phagocytosis
   (c) Both endocytosis and membrane fusion
   (d) Only pinocytosis

2. Influenza virus (IV), a well known enveloped animal virus, enters its host cells through membrane fusion process catalyzed by haemagluttinin (HA) protein inside endosomes at 37°C. HA is localized in the lipid bilayer membrane of the IV as an integral membrane protein and is responsible for binding and fusion of IV membrane with the endosomal membrane of host cells. Upon binding, IV is internalized into host cells through receptor mediated endocytosis followed by fusion of the IV membrane with endosome membrane catalyzed by HA. In a situation, if we wish to fuse IV membrane with its host cells (deficient in endocytosis) at the plasma membrane, mention the correct condition out of the following: **(JUNE 2016)**
   (a) Pre-treat IV in pH 5.0 followed by its binding and fusion with host cells at pH 7.4 and 37°C.
   (b) Allow the IV to bind and fuse with host cells at pH 7.4 and 37°C.
   (c) IV and host cells are allowed to bind and fuse at pH 5.0 and 37°C.
   (d) IV is subjected to incubation at 60°C for 30 minutes and allowed to bind and fuse with host cells, at pH 5.0 and 37°C.

3. One highly pathogenic DNA virus enters into the host cells by endocytosis replicates in the nucleus followed by cell lysis. You have drugs at you disposal that block
   A. acidification of vesicles.
   B. mitochondrial transport.
   C. nuclear export.
   D. exocytosis.

   Identify the right combination to prevent the infection. **(DEC 2013)**
   (a) A and B (b) B and D
   (c) A and C (d) A and D

4. Which of the following a food borne toxin **(DEC 2016)**
   (a) Tetanus toxin (b) Botulinum toxin
   (c) Cholera toxin (d) Diptheria toxin

5. Which one of the following is **NOT** a bacterial disease? **(JUNE 2017)**
   (a) Tuberculosis (b) Typhoid
   (c) Tetanus (d) Small pox

6. With which protein of *Yersinia* would integrin proteins of mammalian cells interact for internalization? **(JUNE 2011)**
   (a) Pilin (b) Fimbrin,
   (c) Invasin (d) Adherin

7. Following are the list of the pathogens (column A) and the unique mechanisms they employ for invading the immune response (column B). **(Dec 2016)**

A	B
(a) *Trapanosoma brucei*	(i) Capable of employing unusual genetic processes by which they generate extensive variations in their surface glycoproteins (VSG)
(b) *Plasmodium falciparum*	(ii) Capable of continually undergoing maturational changes in transformation to two different forms which allow the organism to change its surface molecules
(c) *Haemophilus influenza*	(iii) Capable of invading immune response by frequent antigenic changes in its hemagglutinin and neuraminidase glycoproteins

   Which of the following is the correct match between the organism and their respective mechanism to evade immune response?
   (a) a – (i), b – (ii), c – (iii)
   (b) a – (ii), b – (iii), c – (i)

(c) a – (iii), b – (i), c – (ii)
(d) a – (i), b – (iii), c – (ii)

8. Which of the following pathogens adopt the strategy of changing the antigens expressed on their surfaces for evading immune response? **(DEC 2013 GU)**
(a) *Leishmania donovani*
(b) *Trypanosoma brucei*
(c) *Plasmodium falciparum*
(d) *Toxoplasma gondii*

Answer Key

1.(c)	2.(c)	3.(c)	4.(b)	5.(d)	6.(c)	7.(a)	8.(b)

2. Cell Signaling

1. Which of the following signaling molecules enters the cell to initiate its action? **(MODEL PAPER)**
(a) Transferrin (b) Insulin
(c) Glucagon (d) Thyroxine

2. The second messenger, which opens calcium ion pores in endoplasmic reticulum and plasma membrane is **(JUNE 2017)**
(a) Diacylglycerol
(b) cAMP
(c) Phosphatidyl inositol biphosphate
(d) Inositol triphosphate

3. Which of the following is NOT a second messenger? **(DEC 2015)**
(a) Cyclic GMP
(b) Diacylglycerol
(c) Inositol triphosphate
(d) Phosphatidyl inositol

4. Some T lymphocytes respond to antigen stimulation by synthesizing a growth factor that causes T cell proliferation thereby increasing the responsive T lymphocytes resulting in amplification of the immune response. This is an example of **(DEC 2014)**
(a) endocrine signaling.
(b) paracrine signaling.
(c) autocrine signaling
(d) cyclic signaling.

5. Which one of the following combinations is the correct pairing of ligands with their receptors? **(JUNE 2016)**

(i)	FGF	(a)	Patched
(ii)	Hedgehog	(b)	Frizzled
(iii)	Wnt	(c)	Receptor tyrosine kinase

(a) i-c, ii -.a, iii- b
(b) i-a, ii - c, iii - b
(c) i - b, ii - c, iii – a
(d) i - c, ii - b, iii – a

6. Which one of the following lipid-soluble hormones can interact with a cell surface receptor? **(DEC 2019)**
(a) Progesterone (b) Estradiol
(c) Thyroxine (d) Prostaglandin

7. Following are the list of some cellular receptors (Column X) and with possible functional characteristics (Column Y):

Column X	Column y
A. Asialoglyco protein receptor	(i) Intracellular high affinity receptor which upon binding with ligand acts as transcription factor and binds to DNA
B. Transferrin receptor	(ii) Extracellular receptor which upon binding with ligand is subjected to endocytosis Receptor-ligand complex accumulates in acidic endosomes/ CURL (compartment of uncoupling of receptor and ligand), delivers the cargo, receptor-ligand complex recycles back to cell surface and the ligand disassociates.

C. Steroid receptor	(iii) Extracellular receptor which upon binding with ligand is subjected to endocytosis. Receptor-ligand complex accumulates in acidic endosomes/ CURL, ligand gets dissociated, transfers to lysosomes and the receptor recycles back to cell surface.

Which one of the following is the correct match? **(JUNE 2019)**

(a) A – (i); B – (ii); C – (iii)
(b) A – (ii); B – (iii); C – (i)
(c) A – (iii); B – (ii); C – (i)
(d) A – (i); B – (iii); C – (ii)

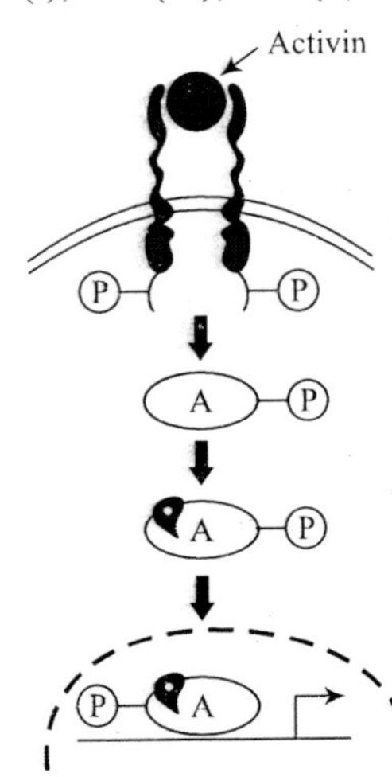

8. In the above signaling cascade, which of the following molecules is denoted by 'B'? **(DEC 2016)**

(a) STAT 5 (b) SMAD 6
(c) GSK 3 β (d) SMAD 4

9. In the table below Column A lists ligands and Column B lists classes of receptors

Column A	Column B
A. Serotonin	i. Binding activates a G-protein which activates or inhibits an enzyme that generates a specific second messenger and opens ion channel
B. Interferons	ii. Binding causes receptor monomers to dimerize. Dimeric receptor then interacts with and activates one or more cytosolic tyrosine protein kinases
C. Glycine	iii. Binding changes the conformation of the receptor so that specifications flow through it
D. Insulin	iv. Binding leads to activation of intrinsic tyrosine kinase activity

Which one is the correct match? **(DEC 2019)**

(a) A-i B-ii: C-iii; D-iv
(b) A-i; B-iii C-iv; D-i
(c) A-iii, B-iv, C-i; D-ii
(d) A-iv, B-i; C-ii; D-iii

10. Hormone act by producing/activating a variety of effectors intracellularly. Below are given a variety of effectors in column 'X' and hormones in column 'Y

Column X	Column Y
A. Inositol triphosphate (IP3)	(i) Leptin
B. cGMP	(ii) IGF-1
C. cAMP	(iii) Oxytocin
D. Receptor Kinase	(iv) Somatostatin
E Associated Kinase	(V) ANF

Which one of the following combinations of effector and the specific hormone is correct? **(DEC 2019)**

(a) A (i) and B (ii) (b) B (iv) and A (iii)
(c) C (iv) and D (ii) (d) E (ii) and C (i)

11. A and B Two cell surface receptors, A and B with a single binding site specifically bind with their respective ligands, X and Y. In the table below are the values for the association constant, K_a of the respective ligand-receptor interactions. K_a is also called the affinity constant.

Receptor	Ligand	Ka
A	X	1×10^8
B	Y	1×10^{11}

Based on these values, which one of the following statements is INCORRECT?

(DEC 2019 ASSAM)

(a) K_d for B-Y binding is smaller than that for A binding

(b) K_1 for A-X binding is lesser than that for B-Y binding

(c) K_{-1} for B-Y binding is smaller than that for A-X binding

(d) K_{-1}/K_1 for B-Y binding is higher than that for A-X binding(K_1 is the forward rate constant, K_{-1} is the reverse rate constant and K_d is the dissociation constant)

Answer Key

1.(d)	2.(d)	3.(d)	4.(c)	5.(a)	6.(d)	7.(c)
8.(d)	9.(a)	10.(b)	11.(d)			

3. G Protein Coupled Receptor

1. The second messenger cAMP, synthesised by adenylyl cyclase transduces a wide variety of physiological signals in various cell types in mammalian cells. Most of the diverse effects of cAMP are mediated through activation of protein kinase A (PKA), also called cAMP dependent protein kinase. Which of the following statements regarding PKA is **NOT** correct? **(DEC 2016)**

   (a) Inactive PKA is a tetramer consisting of two regulatory (R) subunits and two catalytic (C) subunits.

   (b) Each R subunit binds the active site in a catalytic domain and inhibits the activity of the catalytic subunits.

   (c) Each R subunit has two distinct cAMP binding sites and binding of cAMP occurs in a cooperative fashion.

   (d) Binding of cAMP to R subunit causes a conformational change resulting in binding to site other than catalytic site causing strengthening of binding to C subunit activating its kinase activity.

2. G protein-linked receptors are trans-membrane proteins of **(JUNE 2011)**

   (a) single-pass. (b) three-pass

   (c) five-pass. (d) Seven-pass.

3. Which one of the following statements about receptor - enzyme is FALSE? **(JUNE 2016)**

   (a) A receptor - enzyme has an extracellular ligand binding domain, a transmembrane domain and an intracellular catalytic (enzyme) domain.

   (b) Many types of receptor enzymes are found in animals.

   (c) The signal transduction pathways of receptor - enzyme involve phosphorylation cascades.

   (d) Receptor - enzymes interact directly with intracellular G-proteins.

4. Which one of the following events NEVER activates the G-protein coupled receptor for sequestering Ca^{2+} release? **(JUNE 2015)**

   (a) Interaction of bind in to sperm receptors.

   (b) Activation of Frizzled by Wnt.

   (c) Cortical reaction blocking polyspermy

   (d) DNA synthesis and nuclear envelope breakdown.

5. G-protein coupled receptors (GPCR) consist of three protein subunits α, β and γ. In unstimulated state, α-subunit is GDP bound and GPCR is inactive. When GPCR gets activated, it acts like guanine nucleotide exchange (GEF) factor and induces α-subunit to release its bound GDP allowing GTP to bind in its place. In order to regulate G-protein activity by regulating GDP/GTP concentration, α subunit acts as **(DEC 2015)**

   (a) GTPase

   (b) GDP kinase

   (c) cGMP-specific phosphodiesterase

   (d) cAMP-specific phosphodiesterase

6. In cells having G protein coupled receptor, inhibition of protein kinase A by siRNA technology led to diminished transcription of

androgen binding protein (ABP) and CREB protein. Addition of cAMP, which is a second messenger, will lead to (JUNE 2013)

(a) increased transcription of ABP.

(b) increased phosphorylation of CREB protein.

(c) no change in transcription level.

(d) increased ATPase activity of G_α subunit.

7. Glucose is mobilized in muscle when epinephrine activates $G_{\alpha s}$. In an experiment in which muscle cells were stimulated with epinephrine, glucose mobilization was observed even after withdrawal of epinephrine. This could be **(JUNE 2011)**

(a) due to the presence of a cAMP phosphodiesterase inhibitor.

(b) very low rates of cyclic AMP formation.

(c) due to the presence of a cAMP phosphodiesterase activator.

(d) due to the absence of protein kinase A.

8. G protein-coupled receptors (GPCRs) are used to detect and respond to many different types of signals, including neurotransmitters, hormones involved in glycogen and fat metabolism and even photons of light.

Which one of the following statements regarding GPCR is INCORRECT? **(JUNE 2018)**

(a) GPCRs are a large family with a common structure of seven membrane spanning α helices.

(b) GPCRs are coupled to trimeric G proteins comprising three subunits α, β and γ.

(c) The Gα subunit is a GTPase switch protein that alternates between an active ('on') state with bound GTP and an inactive ('off') state with GDP.

(d) The 'on' form gets bound to β and γ subunits and activates a membrane bound effector like adenylyl cyclase, phospholipase C or ion channel.

9. Which one of the following is not a G-protein coupled receptor? **(DEC 2019)**

(a) Epinephrine receptor

(b) Transferrin receptor

(c) Glucagon receptor

(d) Thyroid stimulating hormone receptor

Answer Key

1.(d)	2.(d)	3.(d)	4.(d)	5.(a)	6.(c)	7.(a)
8.(d)	9.(b)					

4. Receptor Tyrosine Kinase and NF-kB Signaling

1. Release of intracellular calcium is due to **(DEC 2013 GU)**

(a) absence of PIP_2 hydrolysis.

(b) rapid dephosphorylation of IP_3

(c) retention of DAG in the cytosol.

(d) activation of IP_3 receptors.

2. The platelet-derived growth factor (PDGF) is a dimer consisting of two identical polypeptides. On binding to its cell surface receptor, it induces dimerization of the receptor by binding to two receptor molecules. If the intracellular C-terminal domain of the receptor is truncated, which one of the following is most likely to occur? **(DEC 2013 GU)**

(a) Both ligand binding and receptor dimerization will be hampered.

(b) Ligand binding will take place followed by consequent signal transduction.

(c) Ligand binding will be hampered.

(d) Ligand binding will take place but no downstream signaling will be triggered.

5. In a type of signal transduction pathway, ligand binding to a receptor triggers activation of a receptor-associated kinase. This kinase may be an intrinsic part of the receptor protein or tightly bound to the receptor. Receptors in which the tyrosine kinase is an intrinsic part of it's polypeptide chain are called the

receptor tyrosine kinase (RTK). Which one of the following statements regarding RTK is **INCORRECT**? **(DEC 2017)**

(a) All RTKs have three essential components: an extracellular domain containing ligand binding site, a transmembrane domain and a cytoplasmic segment that includes a domain with protein tyrosine kinase activity.

(b) Most RTKs are monomeric and ligand binding to the extracellular domain induces formation of receptor dimers.

(c) All cytokine receptors belong to RTKs and cytokine binding activates tyrosine kinase and receptor dimerization.

(d) Ligand binding to RTK leads to autophosphorylation of the protein tyrosine kinase in the cytoplasmic domain. The activated kinase then phosphorylates several tyrosine residues in the receptor's cytoplasmic domain.

4. Epidermal growth factor (EGF) is needed for growth of almost all cells. EGF receptor is a transmembrane protein having an extracellular ligand-binding domain, a transmembrane domain and a cytosolic domain of protein tyrosine kinase (PTK). Binding of EGF to the receptor activates PTK resulting in activation of transcription factors through intracellular transducers. In cell type A, much of the extra cellular ligand-binding domain is deleted by proteases such that cytosolic domain of PTK becomes constitutively active whereas cell type B is having normal EGF receptor. What will be the best-fit graph for the growth of the cultures of cell type A and B in complete medium in presence (+) and absence (-) of EGF? **(DEC 2012)**

(a)

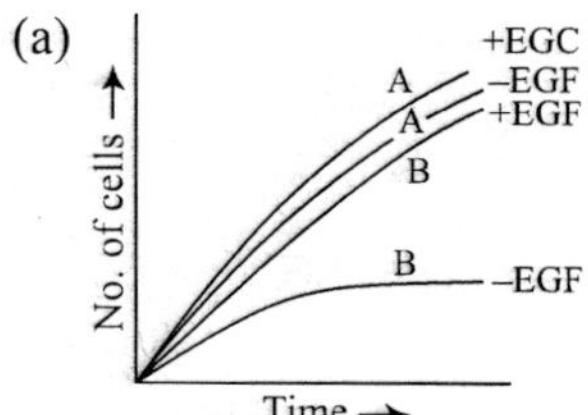

(b)

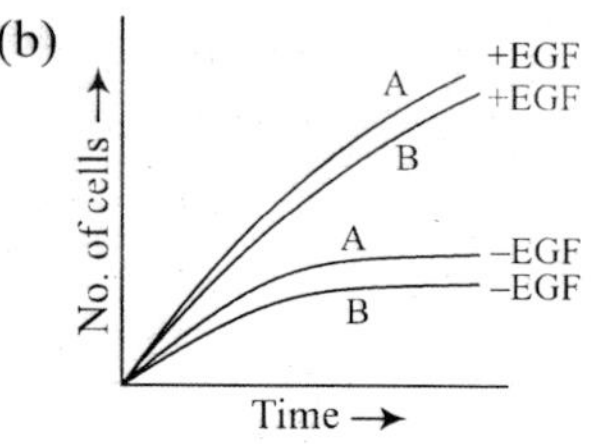

(c)

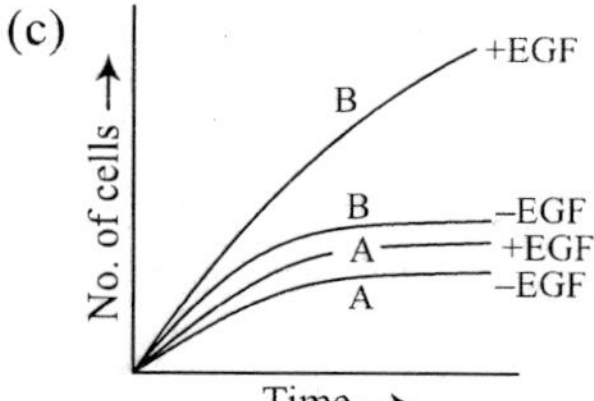

(d)

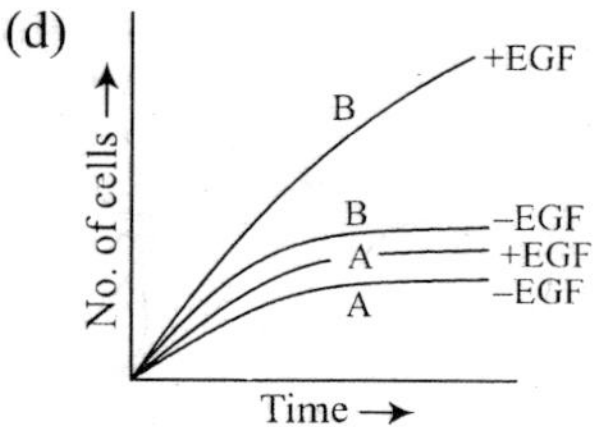

5. Proteins with cytoplasmic domains having tyrosine kinase activity do **NOT** act as receptors for **(DEC 2017)**

(a) Epidermal growth factor (EGF)

(b) Platelet-derived growth factor (PDGF)

(c) Insulin

(d) Transferrin

6. NF-kB is a transcription factor that plays a pivotal role in immune response and inflammatory response. In its resting state, NF-kB is rendered inactive in the cytosol by bound with IKB. When cell is treated with TNF-α, IkB gets phosphorylated by IK and subsequently degraded by preteasome. NF-KB thus becomes free and trans locates to the nucleus and gets activated. In an experiment, a cell is subjected to the following treatment:

A. TNF-α

B. TNF-α + proteasome inhibitor

C. siRNA for IkB + TNF-α D. siRNA for IKK + TNF-α

What will be the best-fit graph for NF-kB activity if it is measured by NF-KB luciferase assay? **(DEC 2013 GU)**

(a)

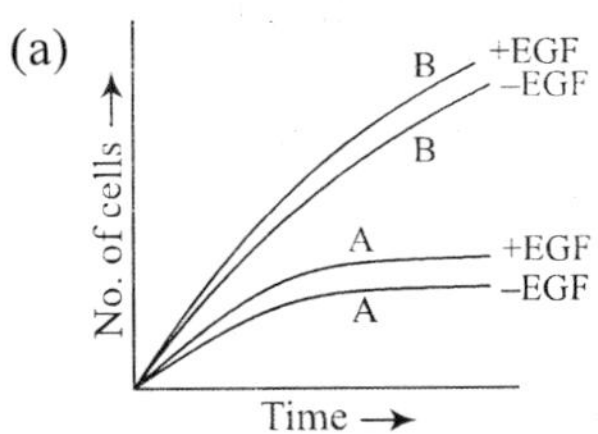

(b)

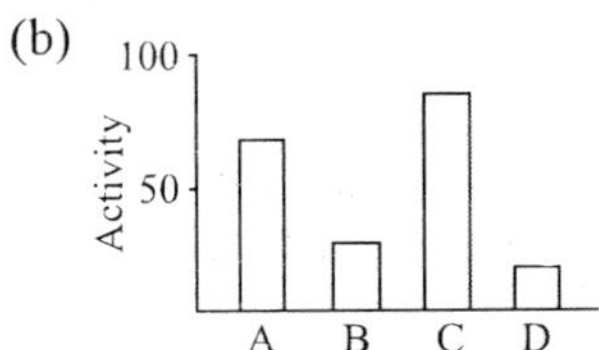

(c)

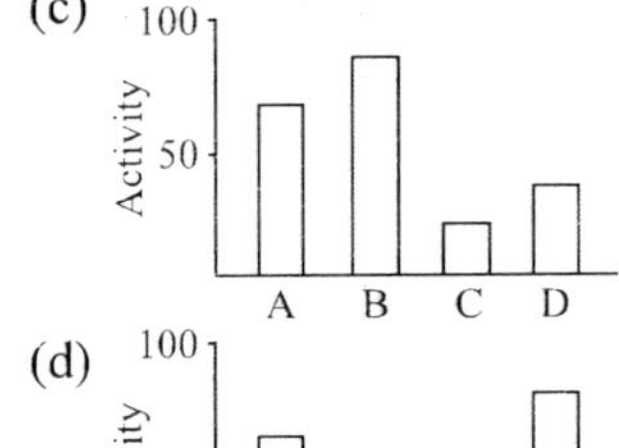

(d)

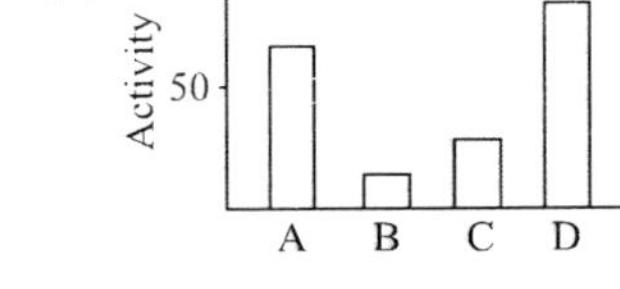

7. Which of the following statement is INCORRECT in relation to treatment of pre-B cell with phorbol ester? **(JUNE 2012)**
   (a) Phorbol esters activates NF-kB for translocation into nuclues
   (b) Phorbol esters activate protein kinase C
   (c) Phorbol ester leads to phosphorylates of NF-kB
   (d) Phorbol ester remove the inhibitor from inactive NF-kB complex in the cytoplasm

Answer Key

1.(d)	2.(d)	3.(c)	4.(a)	5.(d)	6.(a)	7.(c)

5. Nuclear Receptor

1. Which one of the following combinations must be present in a steroid receptor that is located in the cytoplasm? **(JUNE 2015)**
   (a) Nuclear export sequence (NES), leucine zipper
   (b) NES, zinc finger motif
   (c) Nuclear localization sequence (NLS), zinc finger motif
   (d) NLS, leucine zipper

2. Two steroid hormone receptors X and Y both contain a ligand binding domain. Using recombinant DNA technology, a modified hybrid receptor H is prepared such that it contains the ligand binding domain of X and DNA binding domain of Y. three sets of cells overexpressing receptors X, Y and H were then treated separately either with hormone X or with hormone Y. assuming that there is no cross– reactivity, which one of the following graphs best represent the receptor – ligand binding in each case? **(Dec 2016)**

(a)

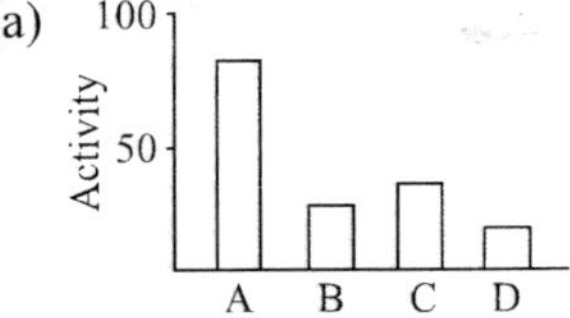

(b)

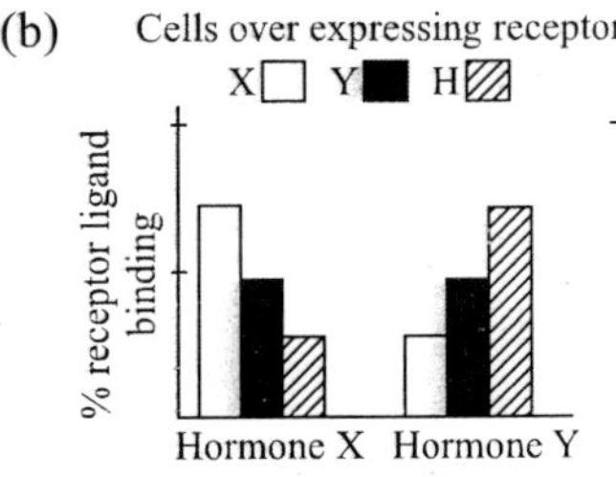

(c)

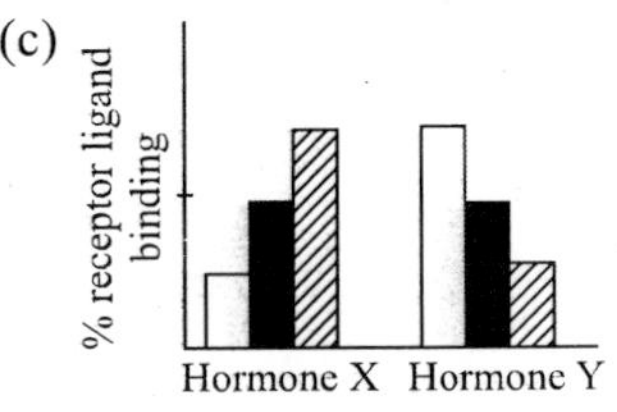

(d)

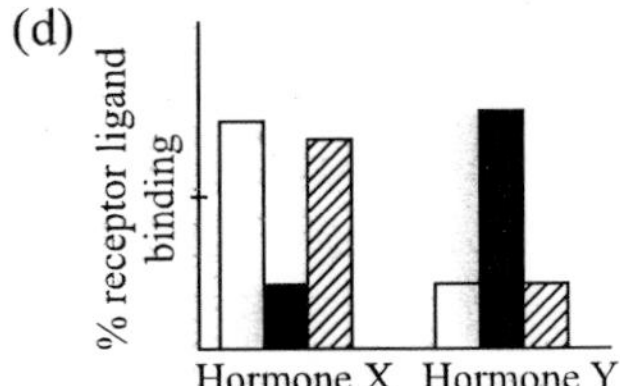

3. Heat shock protein binding region is located in which one of the following domains of glucocorticoid receptor protein?

(DEC 2019 ASSAM)

(a) DNA-binding (b) Ligand-binding
(c) N-terminal (d) Hinge region

4. Which one of the following state about the nuclear receptor super family is NOT true?

(DEC 2015)

(a) The receptors are always cytosol where they remain associated with heat-shock proteins and have variable ligand binding domains in the N- terminal region.

(b) The receptors have characteristic repeat of the C4 zinc-finger motif

(c) The receptors are either homodimeric or heterodimeric and in the absence their hormone ligand, the hetero-dimeric receptors repress transcription, when bound to their response elements

(d) The receptors have a unique N-terminal region of variable length and may contain a nuclear localization signal between the DNA-and ligand-binding domains

5. Based on the structural regions of a nuclear receptor shown in the diagram, the following predictions were made.

A/B C D E F
NH_3^+ COO^-

A. Region F is responsible for binding to ligands and contains two zinc finger-like binding motifs

B. Receptors with A/B domains generally associate with chaperones and do not bind to DNA.

C. Region E indicate that receptors associate with chaperones which protect the nuclear hormone receptors.

D. Region C contains the P-box and the D-box required for dimerization of the receptor and creates contact with DNA phosphate backbone.

Which one of the following is true?

(JUNE 2011)

(a) A and B (b) B and C
(c) B and D (d) C and D

6. A pharmacy student designed a drug to specifically target the receptors for retinoic acid in order to prevent stem cell differentiation. After in vitro trial, the investigator found that the cells underwent differentiation and the drug seemed to be ineffective. The following reasons were given by the student

A. The size of the drug exceeded the size of molecules that could cross the membrane

B. The drug was small in size but hydrophobic in nature

C. The drug did not bind to its receptors

Which of the above could be the probable reason for drug ineffectiveness? **(DEC 2014)**

(a) Only C (b) A and C
(c) A, B and C (d) Only B

7. Cultured animal cells were transfected with expression vector encoding either β-galactosidase (β-gal) alone or expressing a fusion protein of β-gal and glucocorticoid receptor (GR). After transfection, cells were kept in presence or absence of Dexamethasone. Immunofluorescence with a labeled antibody specific for β-gal was used to detect the expressed protein in cytoplasm or nucleus of transfected cells. Possible results of the experiments are:

A. Expression of β-gal alone in the cytoplasm in both absence or presence of Dexamethasone.

B. Expression of β-gal-GR in the cytoplasm in the absence of Dexamethasone.

C. Expression of β-gal alone in the nucleus both in the presence or absence of Dexamethasone.

D. Expression of β-gal-GR in the nucleus in presence of Dexamethasone.

E. Expression of β-gal alone in both cytoplasm and nucleus in presence or absence of Dexamethasone.

F. Expression of β -gal-GR in both cytoplasm and nucleus in presence of Dexamethasone.

Choose the correct combination of results from the following options. **(JUNE 2014)**

(a) B, C and D (b) A, B and D

(c) B, D and E (d) A, B and F

Answer Key

1.(c)	2.(c)	3.(b)	4.(a)	5.(d)	6.(b)	7.(b)

6. Bacterial Signaling/chemotaxis

1. A bacterial response regulator turns on gene A in its phosphorylated form. The amount of "A" shows a sharp and sleep rise at a threshold concentration of the signal sensed by the cognate sensor. This is most likely due to **(DEC 2011)**

(a) increased phosphatase activity of the sensor at the threshold concentration

(b) decreased phosphorylation of the response regulator by the sensor

(c) cooperativity in binding of the response regulator to the target gene

(d) A negative feedback in gene A expression

2. The bacterial flagellar motor is a multi-protein complex. Rotation of the flagellum requires movement of protons across the membrane facilitated by a multi-protein complex. The flagellar motor proteins combine to create a proton channel that drives mechanical rotation. In a screen for mutants, some non-motile ones were selected. These could have **(DEC 2011)**

(a) Mutations in tubulin and actin proteins

(b) Mutations in kinesin proteins

(c) Mutated H^+-ATPase

(d) Mutations in the charged residues lining the ridge of the FliG subunit

3. Bacterial chemotaxis response is mediated by histidine-kinase-associated receptors that activate a two-component signalling pathway which enables chemotaxis receptors to control the flagellar motors. When bacteria move towards attractant, they produce smooth swimming by rotating flagella counter-clockwise, whereas when bacteria move away from repellent, they produce increased tumbling by rotating flagella clockwise.

Which of the following characteristics regarding chemotaxis receptor is NOT true? **(DEC 2018)**

(a) The receptors are dimeric transmembrane proteins that bind specific attractants and repellents on the outside of the plasma membrane.

(b) The cytoplasmic tail of the receptor is stably associated with a histidine kinase CheAyia an adapter protein CheW.

(c) The receptor and its associated proteins are all clustered at one end of the cell.

(d) The binding of an attractant increases the activity of the receptor whereas binding of a repellent decreases the activity.

Answer Key

1.(c)	2.(d)	3.(d)

7. Applied Question-Signaling

1. A ligand recognizes two different cell surface receptors, A and B, on the same cell type. Receptor A, after binding with the ligand is internalized along with the ligand whereas receptor B, after binding with the ligand, initiates tyrosine kinase activity of the intracellular domain. One particular disease is associated with the loss of receptor-mediated signal transduction of the ligand. Different observers inferred that the disease may be resulted due to

A. loss of binding affinity of receptor A due to mutation in the extracellular domain.

B. loss of binding affinity of receptor B due to mutation in the extracellular domain.

C. mutation in the tyrosine kinase domain rendering it inactive.

D. mutation in the intracellular domain rendering it incapable of endocytosis.

Which combination of the above inferences do you think is appropriate for the cause of the disease? **(DEC 2013)**

(a) A and B (b) B and C
(c) C and D (d) A and D

2. Binding of a ligand to a cell-surface receptor activates an intracellular signal transduction pathway through the sequential activation of four protein kinases. In the human cell line A, these kinases are held in a signaling complex by a scaffolding 'protein whereas in another cell line B/these kinases are freely diffusible. Which one of the following possibilities do you think is NOT correct? **(JUNE 2013)**

(a) Speed of signal transduction will be higher in cell A.
(b) Possibility of cross-linking with other signal transduction pathways will be lesser in cell A.
(c) Possibility of signal amplification will be higher in cell A.
(d) Potency of spreading signal through other signaling pathways will be higher in cell B.

3. Four different receptors viz, A, B, C and D bind to same ligand 'X'. In order to determine when receptors has the highest ligand binding affinity following experiment was carried out. Cells were transfected with green fluorescent protein (GFP)-tagged receptors (A, B, C and D) individually, then incubated with red fluorescent protein (RFP) tagged ligand 'X' and subjected to FACS. Following are the results

RFP	Receptor A		Receptor B		Receptor C		Receptor D	
	24%	5%	52%	19%	19%	52%	5%	24%
	52%	19%	24%	5%	5%	24%	19%	52%

GFP

Which receptor has highest binding affinity for ligand 'X'? **(JUNE 2012)**

(a) A (b) B
(c) C (d) D

4. An investigator discovers a new receptor for a known ligand and wanted to identify the binding partner of the receptor i.e. its co-receptor. The anti-receptor antibody is not available but anti GFP-antibody is available. Which one of the following strategies is most likely to identify the co-receptor? **(JUNE 2013)**

(a) The GFP-receptor fusion protein is expressed in a cell line and analyzed by LC-MS/MS.
(b) The GFP-receptor fusion protein is expressed in a cell line and the cells positive for GFP were sorted out, lysed and run on a polyacrylamide gel.
(c) The GFP-receptor protein is coated on ELISA plate, followed by ELISA with anti-.GFP antibody.
(d) The receptor is cloned as a fusion protein of GFP and expressed in stimulated cells. The immuno-precipitated complex obtained by anti-GFP antibody was analyzed by LC-MS/MS.

5. Two protein kinases, K_1 and K_2 function sequentially in regulating intracellular pathway in response to extracellular signal. The following observations are made:

(i) Response is observed even in the absence of extracellular signal when a mutation permanently activates K_1.
(ii) Response is observed even in the absence of extracellular signal when K1 contains an activating mutation and K_2 with inactivating mutation.
(iii) No response in the cells is detected even in the presence of extracellular signal when both kinases are inactivated by mutation.
(iv) No response is observed when K_2 contains an activating mutation

Which one of the following is correct?

(MODEL PAPER)

(a) K_1 activates K_2 (b) K_2 activates K_1

(c) K_1 inhibits K_2 (d) K_2 inhibits K_1

6. In an organism, expression of gene 'X' is induced in the presence of a hormone genetic analysis showed that the hormonal signal is transduced through two proteins *Let1* and *Let(b)* The expression of gene 'X' was studied in lines over expressing (OE) the active Let proteins, knock out (KO) of the Let proteins or combination of both. Results of expression of gene 'X' in presence of the hormone is summarized below:

Lines	Expression of Gene 'X'
WT	++
Let1 OE	++++
Let2 OE	++++
Let1 KO	—
Let2 KO	--
Let1 KO Let1 KO	++++

Based on the above, which one of the following pathways best fits the observation made?

(DEC 2015)

(a)

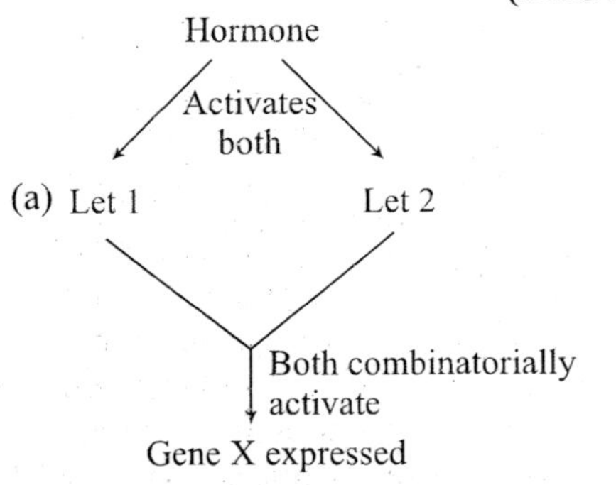

(b) Hormone $\xrightarrow[\text{activates}]{}$ Let 1 $\xrightarrow[\text{activates}]{}$ Let 2 $\xrightarrow[\text{activates}]{}$ gene 'X'

(c) Hormone $\xrightarrow[\text{activates}]{}$ Let 2 $\xrightarrow[\text{activates}]{}$ Let 1 $\xrightarrow[\text{activates}]{}$ gene 'X'

(d) Hormone $\xrightarrow[\text{activates}]{}$ Let 1 $\xrightarrow[\text{activates}]{}$ Let 2 $\xrightarrow[\text{Let 2 is a repressor}]{}$ gene 'X'

7. In a laboratory experiment it was observed that both 'Virus A' and 'Virus B' could infect a mammalian host cell, when infected individually. Interestingly, if the cell were first infected with Virus A (with large MOI), Virus B failed to infect the same cell. If the Virus B (with large MOI) is added first followed by Virus A, both the virus can infect the cells. However, infection with 'Virus A' was found to be in lesser extent. Considering X and Y are the receptors/co-receptors which may be involved for the virus entry, following are few possibilities that can explain the observation.

A. 'Virus A' uses 'X' as receptor and Y as coreceptor.

B. 'Virus B' uses exclusively 'Y' as receptor for entry:

C. Both 'Virus A' and 'Virus B' need X as receptor.

Choose the option with all correct statements.

(JUNE 2019)

(a) A, B and C (b) A and B

(c) B and C (d) A and C

Answer Key

1.(b)	2.(c)	3.(c)	4.(d)	5.(d)	6.(b)	7.(b)

8. Regulation of hematopoiesis

1. Four groups of mice were studied for the factor required for mast cell generation: IL-3 deficient, GM-CSF-deficient, G-CSF-deficient and erythropoietin-deficient. In which mice, mast cell generation is most likely to be deficient? **(JUNE 13)**

(a) IL-3-deficient.

(b) GM-CSF-deficient.

(c) G-CSF deficient.

(d) Erythropoietin-deficient.

2. Mouse bone marrow cells were fractionated to derive stem cell antigen-1^+ (Sca-1^+) cells. These cells were cultured with interleukin-3, or granulocyte-macrophage colony stimulating factor, or macrophage-colony stimulating factor, or granulocyte colony stimulating factor. Most numerous and varied colonies were obtained in the culture stimulated with **(JUNE 2011)**

(a) Interleukin-3

(b) Granulocyte-macrophage colony stimulating factor.

(c) Macrophage-colony stimulating factor.

(d) Granulocyte-colony stimulating factor

3. In bone marrow, stem cells are committed to different lineages. Factors that stimulate colonies of these different lineages are interleukin-3 (multi-CSF), granulocyte-macrophage colony stimulating factor (GM-CSF) and granulocyte or macrophage colony-stimulating factor (G-CSF or M-CSF). In a mouse deficient in GM-CSF the number of hematopoietic cells will be altered. Which one of the following is correct? **(JUNE 2013)**

(a) Mast cells will be normal in number while granulocytes and macrophages will be deficient in number.

(b) Granulocytes count will be normal but not of macrophages.

(c) Macrophage number will remain unaltered.

(d) Mice will be deficient in all the three cell types

4. Mouse erythroleukemia (MEL) cells are used as an in vitro cell culture model for understanding erythropoiesis. The cells are arrested at the stage of pro-erythroblast due to transformation. These cells could be induced by heme to differentiate further so as to synthesize hemoglobin. The most probable molecular mechanism for this could be that heme may suppress and/or downregulate an endogenous heme-regulated inhibitor (HRI) kinase, an inhibitor of globin synthesis. This downregulation in turn promotes differentiation. To validate this hypothesis which of the following approaches is NOT appropriate? **(JUNE 2013)**

(a) Transfect MEL cells with HRI kinase gene.

(b) Knockdown HRI kinase gene in MEL cells.

(c) Determine the rate of protein synthesis in situ as a function of differentiation.

(d) Measure HRI kinase activity as a function of differentiation.

Answer Key

1.(a)	2.(a)	3.(a)	4.(c)

9. Cell-Cell and Cell-Matrix Interaction

1. The extracellular matrix (ECM) is a complex combination of secreted proteins that is involved in holding cells and tissues together. The components of ECM form a network by binding to each other and communicate with cells by binding to adhesion receptors on the cell surface. ECM comprises mainly two classes of macromolecules, proteoglycans and very high molecular weight large proteins. Which one of the following statements regarding ECM constituents is INCORRECT? **(JUNE 2018)**

(a) Proteoglycans are a subset of secreted or cell surface-attached glycoproteins containing covalently linked specialized polysaccharide chains called glycosamino-glycans (GAGs).

(b) GAGs are long branched polymers of specific repeating disaccharides of sialic acid and glucose or galactose.

(c) Major types of GAGs present in ECM are heparan sulphate, chondroitin sulphate, dermatan sulphate, keratan sulphate and hyaluronan.

(d) Major types of large proteins present in ECM are collagen, laminin, elastin and fibronectin.

2. Several types of molecules including the transmembrane glycoproteins can function as matrix receptors and co-receptors. However, the principal receptors on animal cells for binding most extracellular matrix proteins are the integrins. **(DEC 2017)**

Which of the following statements is **NOT** true for integrins?

(a) Integrins are transmembrane linker proteins that link to the cytoskeleton.

(b) An integrin molecule is composed of two non-covalently associated glycoprotein subunits α and β. Both subunits span the cell membrane, with short intracellular C-terminal tails and large N-terminal extracellular domains.

(c) The extracellular portion of the integrin dimer binds to specific carbohydrate residues in extracellular matrix proteins or to ligands on the surface of other cells.

(d) The intracellular portion binds to a complex of proteins that form a linkage to the cytoskeleton.

3. Which of the following is NOT a cell adhesion protein? question **(DEC 2015)**

(a) Cadherin

(b) Selectin

(c) Immunoglobulin (Ig) superfamily

(d) Laminin

4. The extracellular matrix contains a number of non-collagen proteins that typically have multiple domains, each with specific binding sites for other matrix molecules and cell surface receptors. These proteins therefore contribute to both organizing the matrix and helping cells attach to it. The most well characterized matrix protein of this kind is fibronectin. Which one of the following characteristics is NOT TRUE for fibronectin? **(JUNE 2019)**

(a) It is a large glycoprotein found in all vertebrates and important for manycell-matrix interactions.

(b) It is composed of three polypeptides that are disulfide bonded into a crosslink structure.

(c) In human genome, there is only one fibronectin gene containing about 50 exons, but the transcripts can be spliced in different ways to produce many different fibronectin isoforms.

(d) Fibronectin binds to integrin through an RGD motif. Even very short peptide containing RGD sequence can inhibit attachment of cells to fibronectin matrix.

5. Given below are a list of some extracellular matrix (ECM) proteins in column A and their characteristics in column B, but not in correct order:

	A		B
A.	Fibronectin	(i)	Trimeric protein made from three polypeptides which can twist together into a special triple helix
B.	Laminin	(ii)	Heterotrimeric protein comprising α, β and γ chains and many of them are large, cross shaped protein
C.	Nidogen	(iii)	Dimers of two similar polypeptides linked at their C-termini by two disulphide bonds and contain RGD sequence for binding of certain integrins
D.	Collagen	(iv)	A rod-like molecule also called enactin that cross-links with many ECM proteins and also stabilize basal laminae

Which one of the following is the most appropriate match? **(JUNE 2017)**

(a) A - i, B - ii, C - iii, D - iv

(b) A - ii, B - iii, C - iv, D - i

(c) A- iii, B - ii, C- iv, D- i

(d) A - iv, B - i, C - ii, D - iii

6. Which of the following is an intracellular anchor protein? **(DEC 2012)**

(a) Vitronectin (b) Integrin

(c) Vinculin (d) Elastin

7. Integrin molecule link extracellular matrix (ECM) to the actin cytoskeleton of cell. Integrin binds to which of the following ECM macromolecules? **(JUNE 2012)**

(a) Laminin (b) Collagen
(c) Fibronectin (d) Vitronectin

8. Which of the following molecules is involved in Ca^{2+} -dependent cell-cell adhesion? **(JUNE 2011)**

(a) Calmodulin (b) Cadherin
(c) N-CAM (d) Calpain

9. Which one of the following is NOT an extracellular matrix protein? **(DEC 2013)**

(a) Fibronectin (b) Vitronectin
(c) Laminin (d) Cyclin

10. The cylindrical channels in gap junctions are made of **(DEC 2013)**

(a) connexin. (b) collagen.
(c) fibronectin. (d) N-CAM.

11. Which one of the following is a type of intercellular junction in animal cells? **(JUNE 2014)**

(a) Middle lamella (b) Plasmodesmata
(c) Desmosomes (d) Glycocalyx

12. Glycosaminoglycans are usually linked to proteins to form proteoglycans. Which of the following is NOT a proteo-glycan? **(DEC 2014)**

(a) Hyaluron (b) Aggrecan
(c) Betaglycan (d) Syndecan-l

13. Which one of the following glycosaminoglycans is generally NOT covalently attached to protein as proteoglycans? **(DEC 2019 ASSAM)**

(a) Chondroitin sulphate
(b) Dermatan sulphate
(c) Heparan sulphate
(d) Hyaluronan

14. Physical attachment between cells important in imparting strength in tissues. Various physical cell junctions in vertebrate epithelial tissues are classified according to their primary functions. Enlisted below in column A is the major function of a particular junction and column B enlists cell junctions, but the same order. **(DEC 2015)**

	Column A		Column B
A.	Seals gap between epithelial cells	(i)	Desmosomes
B.	Connects actin filament bundle in one cell with that in the next	(ii)	Hemi-desmosomes
C.	Connects intermediate filaments in one cell to those in the next cell	(iii)	Tight Junction
D.	Anchor intermediate filaments in a cell to extracellular matrix	(iv)	Adherens iunction

Choose the correct combination:

(a) A-(i), B-(ii), C--(iii), D-(iv)
(b) A-(ii), B-(iii), C-(iv), D-(i)
(c) A-(iii), B-(iv), C-(i), D-(ii)
(d) A-(iv), B-(i), C-(ii), D-(iii)

15. Which one of the following permits the rapid diffusion of small, water-soluble molecules between the cytoplasm of adjacent cells? **(JUNE 2018)**

(a) Tight junctions
(b) Anchoring junctions
(c) Adherens junctions
(d) Gap junctions

16. In animals, four separates families of cell – cell adhesion proteins are listed in Column A and their functional characteristics are given in Column B **(Dec 2016)**

A	B
(a) Integrin	(i) Lectins that mediates a variety of transient, cell- cell adhesion interactions in the blood stream

(b) Cadherin	(ii) Contains extracellular Ig-like domain and are mainly involved in the fine tuning of cell-cell adhesive interaction during development and regeneration.
(c) Ig-superfamily	(iii) Mediates Ca^{2+}-dependent strong homophilic cell – cell adhesion.
(d) Selectin	(iv) Transmembrane cell adhesion proteins that acts as a extracellular matrix receptor

Which one of the following is the correct combination?

(a) a – (i), b – (ii), c – (iii), d – (iv)
(b) a – (ii), b – (iii), c – (iv), d – (i)
(c) a – (iii), b – (iv), c – (i), d – (ii)
(d) a – (iv), b – (iii), c – (ii), d – (i)

17. Following are a list of extracellular matrix proteins (Column A) along with their functional characteristics (Column B): **(DEC 2018)**

Column A		Column B	
A	Connexin	(i)	The chief endothelial cell proteins that are recognized by the white blood cell integrins and member of immunoglobulin (Ig) super family.
B	Plasmodesmata	(ii)	Cell surface carbohydrate binding proteins that mediate a variety of transient cell-cell adhesion interactions in the bloodstream.
C	ICAM	(iii)	Four-pass transmembrane protein which is the major constituent of gap junctions in forming a continuous aqueous channel.
D	Selectin	(iv)	It is the only class of intercellular junctions in plants that directly connect the cytoplasm of adjacent cells.

Which one of the following is the correct match?

(a) A - i; B-iv; C-iii; D-ii
(b) A - ii; B-iii; C-iv; D-i
(c) A-iii; B-iv; C-i; D-ii
(d) A- iv; B-i; C-ii; D-iii

18. Which one of the following statements about cell-cell interactions is NOT true?

A. Cadherins are transmembrane linker proteins which carry out Ca^{2+}-mediated adhesion between adjacent cells.
B. Integrins are transmembrane adhesion proteins that mediate hemophilic adhesion through actin and intermediate filaments.
C. Selectins are cell surface lectins that mediate a variety of transient, cell-cell adhesion interactions in the bloodstream.
D. ICAMs (intracellular cell adhesion molecules) and VCAMs (vascular cell adhesion molecules) are members of immunoglobulin (Ig) superfamily

(DEC 2014)

(a) Only A (b) Only B
(c) Both C and D (d) Both A and D

19. Physical attachment between cell to cell or between cells to extracellular matrix imparts rigidity to a tissue. Which of the following characteristic is NOT true for cell junction? **(DEC 2013 GU)**

(a) Desmosomes are cell-cell anchoring junctions connecting intermediate filament in one cell with that in the next cell.
(b) Hemidesmosomes are cell matrix anchoring junctions connecting actin filament in one cell with that in the next cell.
(c) Gap junctions in animals and plasmodesmata in plants are channel forming junctions.
(d) Occluding junctions seal the gaps between cells to make the cell sheet into an impermeable barrier.

20. In a tissue,cells are bound together by physical attatchment between cell to cell oe between cell to extra cellular matrix. Following are some of the characterstics of cell junctions:

A. Aderens junctions are cell-matrix anchoring junctions connecting actin filament in one cell with that of matrix

B. Desmosomes are cell-matrix anchoring junctions connecting actin filament to extra cellular matrix

C. gap junctions are channel forming junctions allowing passage of small water soluble molecules from cell to cell.

D. Tight junctions are occluding junctions, which seal gap between two cells

E. Hemidesmosomes are cell-matrix anchoring junctions connecting intermediate filament in one cell to extra cellular matrix.

Which of the following combinations of statements is NOT correct? **(JUNE 2012)**

(a) (A) and (B) (b) (A) and (C)
(c) (C) and (D) (d) (D) and (E)

21. Junctions which tether cytoskeletal filaments inside the cell are known as **(DEC 2017)**

(a) anchoring junctions
(b) occluding junctions
(c) channel- forming junctions
(d) signal - relaying junctions

22. Physical attachment of cells in a tissue or organ is mediated by different types of junctions, one of which is anchoring junctions. This includes both cell-cell adhesions and cell-extracellular matrix adhesions. The central role of anchoring junctions is played by transmembrane adhesion proteins of which the major ones are cadherin and integrin. **(DEC 2019 ASSAM)**

Which one of the following statements regarding cadherin and integrin is NOT correct?

(a) Generally, cadherin mediates attachment of cell to cell whereas integrin mediates attachment of cells to extracellular matrix
(b) Cadherin molecules of a specific subtype on one cell bind to cadherin molecules ofthe same or closely related subtype on adjacent cells
(c) Both cadherin and integrin are composed of two non-covalently attached glycoproteins ubunits which span the cell membrane
(d) Integrin molecules generally act as extracellular matrix receptor

23. A researcher was studying a protein 'X' which has been observed to move across cells when an extracellular electrical stimulus is provided. An artificial peptide 'P' was prepared which resembles the structure of connexins and competitively inhibits connexon formation. Which one of the following statements will best explain the fate of protein 'X' if the cells are treated with peptide 'P' and then electrical stimufus is provided. (JUNE 2015)

(a) X fails to move across cells due to improper formation of tight junctions.
(b) X fails to move across cells due to improper formation of gap junctions.
(c) X moves freely across cells as before.
(d) X fails to move across cells due to improper formation of desmosome.

24. Collagen is the most prevalent extracellular matrix protein. **(JUNE 2014)**

Which of the following is NOT true for collagen?

(a) Collagen is composed of triple helix consisting of two α polypeptide chain and one β polypeptide chain wound around one another in a rope-like structure.
(b) Glycine accounts for almost one third of the amino acids within collagen molecule.
(c) Ascorbate is essential for collagen formation required for hydroxylation of proline.
(d) Individual collagen polypeptide chains are synthesized on membrane-bound ribosomes with N-terminal signal sequences for directing them to ER lumen.

25. Collagens are the most abundant component of the extracellular matrix. In order to maintain normal physiological processes like wound healing, bone development, etc., which one of the following type of enzymes is MOST important? **JUNE 2015**

(a) Peptidases (b) Proteases

(c) Amylase (d) Lipases

Answer Key

1.(b)	2.(c)	3.(*)	4.(b)	5.(c)	6.(c)	7.(*)
8.(b)	9.(d)	10.(a)	11.(c)	12.(a)	13.(d)	14.(c)
15.(d)	16.(d)	17.(c)	18.(c)	19.(b)	20.(a)	21.(a)
22.(c)	23.(b)	24.(a)	25.(b)			

10. Cell-Cell Signaling

1. Which one of the following is a group of signalling molecules that act as morphogens during development of an organism and its effects are mediated through the receptor Patched and its binding partner Smoothened? **(DEC 2017)**

(a) Hedgehog protein (b) Notch protein

(c) Wnt protein (d) β-catenin

2. Which of the following is most likely to be activated in the non-canonical Wnt pathway? **(DEC 2013 GU)**

(a) Disheveled and GSK-3.

(b) RhoA and PLC.

(c) GSK-3 and RhoA.

(d) Disheveled and APC.

3. Intracellular transport and cytoskeletal organization of a cell is regulated by nucleotide exchange of different small molecular weight GTPases of Ras super family. Overexpression of which of the following GTPase modulates the actin-cytoskeleton of HeLa cells? **(DEC 2011)**

(a) Ran in GDP bound form

(b) Ran in GTP bound form

(c) Rho in GTP bound form

(d) Rho in GDP bound form

4. Of the following signaling processes which one is NOT involved in cellular movement or cytoskeletal changes? **(JUNE 2015)**

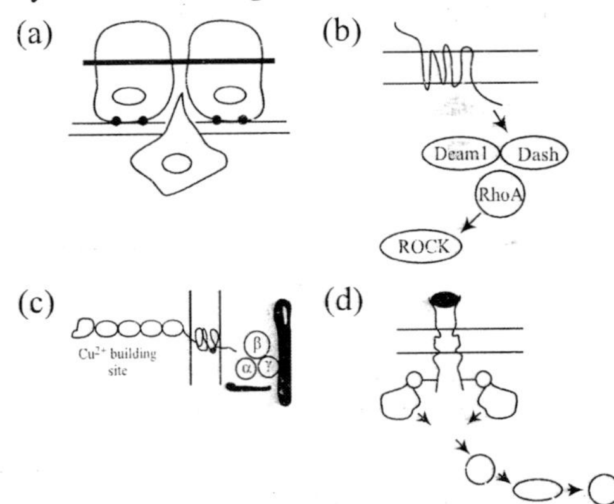

5. Activation of the Wnt signal transduction pathway is extremely important during early development. Of the various pathways, which one of the following is most likely to induce cytoskeletal changes, like cell shape and movement? **(DEC 2014)**

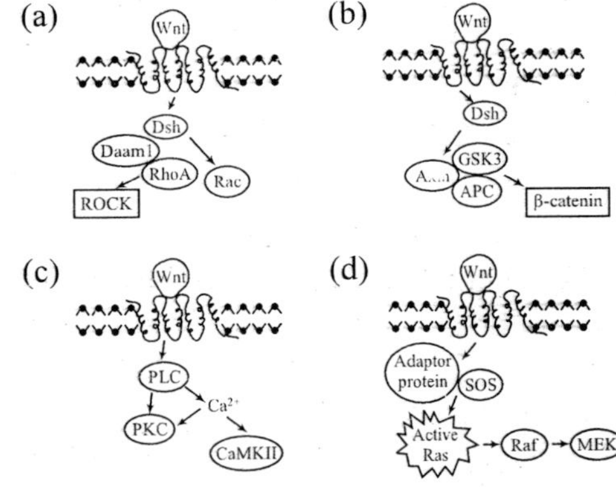

6. Which one of the following graphs represents the relative expression of proteins of Wnt-activated signaling cascades involved during development of an embryo? **(JUNE 2014)**

(a)

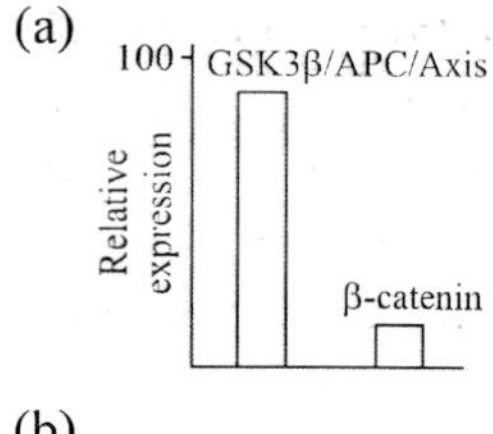

(b)

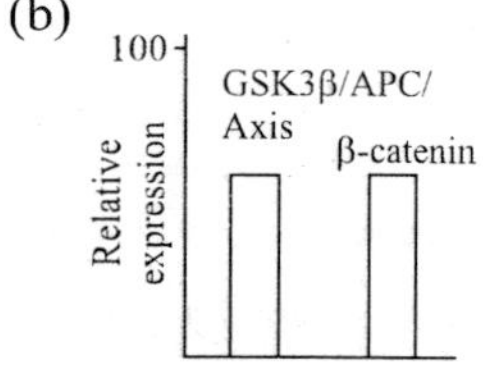

(c)

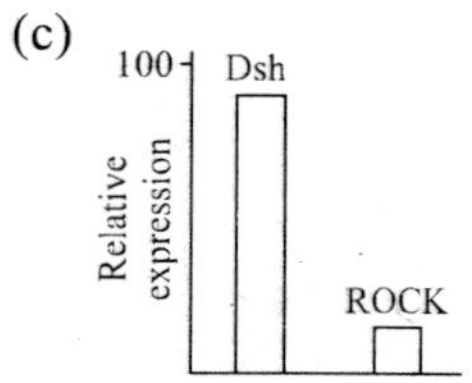

(d)

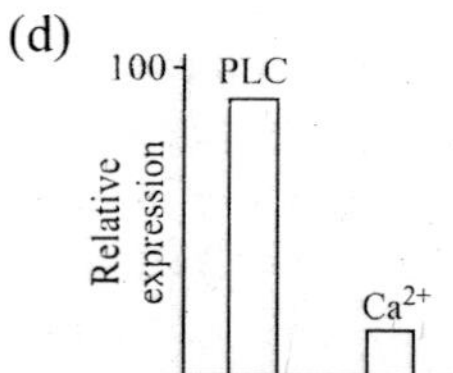

7. Preventing the blocking action of Patched protein leads to activation of Cos-2, which dissociates itself from microtubules, activates Ci/Gli which binds to CBP (CREB - binding protein) and promotes transcription of target genes. Which one of the following treatment of cells will mostly prevent Ci/Gli activated transcription in the cells? **(JUNE 2017)**
   (a) Small molecules which target Frizzled.
   (b) Azepine, an inhibitor of γ-secretase.
   (c) Cyclopamine, which binds to heptahelical bundle of Smoothened.
   (d) Cdk blockers, which negatively regulate TGFβ-induced growth.

8. Cadherins mediate Ca^{2+}-dependent cell-cell adhesion and play an important role in embryonic development by changing the adhesive properties of cell. Aggregation of nerve cells to form an epithelium is correlated with the appearance of N-cadherins on cell surface and vice versa. N-CAM (neural cell adhesion molecules) belongs to Ig-SF (immunoglobulin super family) and involved in fine tuning of adhesive interaction.
   In order to see the effect of mutations of N-cadherin and N-CAM, two sets of mice were generated. Set A - mice with mutation in N-cadherin and set B - mice with mutation in N- CAM. Which of the following results is most likely to occur? **(JUNE 2016)**
   (a) Mice of both set A and set B will die in early development.
   (b) Mice of set A will die in early development but mice of set B will develop normally and show mild abnormalities in the development of nervous system.
   (c) Mice of Set A will show mild abnormalities in the development of nervous system whereas mice of set B will die early in development.
   (d) Mice of both set A and set B develop normally as other cell adhesion molecules will compensate for the mutations.

9. Temporal expression of N-cadherin is extremely important during early development of the mammalian embryos. Accordingly, which one of the following statements about N-cadherin is true? **(DEC 2018)**
   (a) Injection of N-cadherin antibodies just prior to condensation of mesenchymal cells will aid cartilage formation.
   (b) Presence of N-cadherin just prior to condensation will facilitate nodule formation and development of the limb skeleton.
   (c) The border between the nervous system and skin will form properly only if epidermal cells are experimentally made to express N-cadherin.
   (d) Expression of N-cadherin is redundant during separation of neural and epidermal precursor cells.

10. Vascular endothelial (VE)-cadherin is an important cell adhesion molecule for endothelial cells. Endothelial cells that are unable to express VE-cadherin still can adhere to one another via N-cadherin (neural cadherin), but these cells do not survive. Which of the following is the most appropriate reason for this? **(DEC 2013)**

(a) N-cadherin uses VE-cadherin as co-receptor for adhesion.

(b) VE-cadherin acts as co-receptor for VEGF (vascular endothelial growth factor) mediated signal transduction in endothelial cells.

(c) VE-cadherinis important for desmosome formation and interaction of intermediate filaments.

(d) Loss of VE-cadherin impairs Ca^{2+} homeostasis of vascular endothelial cells leading to their death.

11. Which one of the following inactivates the serine/threonine protein kinase, mTOR, related to cell growth in mammalian system? **(DEC 2018)**

(a) Rifamycin (b) Rapamycin

(c) Erythromycin (d) Chloramphenicol

Answer Key

1.(a)	2.(b)	3.(c)	4.(d)	5.(a)	6.(a)	7.(c)
8.(b)	9.(b)	10.(b)	11.(b)			

11. Cancer

1. Which one of the following best defines an oncogene? **(JUNE 2016)**

(a) An oncogene never codes for a cell cycle protein, which pJomotes cell proliferation.

(b) Oncogenes are always involved in inherited forms of cancer.

(c) An oncogene codes for a protein that prevents a cell from undergoing apoptosis.

(d) An oncogene is a dominantly expressed mutated gene that renders a cell advantageous towards survival.

2. Which of the following events will NOT usually lead to transformation of a normal cell into a cancer cell **(DEC 2016)**

(a) Gain of function of oncogenes

(b) Loss of function of tumor suppressors

(c) Gain of function of genes involved in nucleotide excision repair

(d) Loss of function of pro-apoptosis related genes

3. Which of the following matches of oncogene-protein product is NOT correct?**(JUNE 2012)**

(a) *erb*A→Thryoid hormone receptor

(b) *erb*B → Epidermal Growth Factor receptor

(c) *ras*→Guanine nucleotide binding protein with GTPase activity

(d) *fos*→Platelet derived growth factor receptor

4. Out of the following matches of oncogenes with the proteins that each specifies, which one is incorrect? **(DEC 2012)**

(a) erbA - thyroid hormone receptor.

(b) erbB - epidermal growth factor receptor.

(c) ras- guanine-nucleotide binding protein with GTPase activity.

(d) fos - platelet-derived growth factor.

5. A tumour suppressor protein **(DEC 2013)**

(a) is one whose function brings about regression of a tumour

(b) one where mutations are shown to cause or are associated with tumours.

(c) is inactivated by oncogenes.

(d) inhibits the progression of the cell cycle by phosphorylating cyclins.

6. The main difference between normal and transformed cells **(JUNE 2015)**

(a) immortality and contact inhibition

(b) shorter generation time and cell mobility

(c) apoptosis and tumour suppressor gene hyperfunction.

(d) inactivation of oncogenes and shorter cell cycle duration

7. The mutation in an oncogene falls under which of the following classes? **(DEC 2015)**

(a) Loss of function mutation

(b) Frame shift mutation

(c) Gain of function mutation

(d) Dominant negative mutation

8. The mode of action of the anticancer drug methotrexate is through its strong competitive inhibition on **(MODELPAPER)**

(a) dihydrofolate reductase

(b) thymidine synthase.

(c) thymidine kinase.

(d) adenylate cyclase.

9. When adenoma is converted to metastatic adenocarcinoma, which of the following combination of proteins is almost certainly to be degraded? **(JUNE 2013)**

(a) Type IV collagen and laminin.

(b) Fibronectin and β_2 integrin.

(c) Metalloprotease and serine protease.

(d) Elastin and selectin.

10. Conversion of proto-oncogene to oncogene may involve the following processes:

A. mutation in coding sequence

B. gene amplification

C. chromosome rearrangement

D. mutation in non-coding sequence

Which one is appropriate? **(MODEL PAPER)**

(a) A, B and C (b) B, C and D

(c) A, C and D (d) All

11. Cancer is often believed to arise from stem cells rather than fully differentiated cells. Following are certain views related to the above statement. Which one of the following is NOT correct? **(JUNE 2016)**

(a) Stem cells do not divide and therefore require fewer changes to become a cancer cell.

(b) Cancer stem cells can self-renew as well as generate the non-stem cell populations of the tumor.

(c) Teratocarcinomas prove tumors arise from stem cells without further mutations.

(d) Stemness genes can often function as oncogenes.

12. Two classes of genes - proto-oncogene and tumor suppressor gene usually contribute to the development of cancer. Following are some of the statements regarding both the genes.

A. Proto-oncogenes result in the development of cancer by gain-of-function mutation whereas tumor suppressor gene leads to cancer development by loss-of-function mutation.

B. Proto-oncogenes result in development of cancer by loss-of-function mutation whereas tumor suppressor gene leads to cancer development by gain of-function mutation.

C. Mutation in both the alleles of a protooncogene is required for induction of cancer whereas mutation in one of the two alleles in tumor suppressor gene is sufficient for promoting tumorigenesis.

D. Mutation in one of the two alleles in proto-oncogene is sufficient for induction of cancer whereas mutation in both the alleles of a tumor suppressor gene is required for promoting tumorigenesis.

Which combinations of the above statements are true for both the genes? **(DEC 2017)**

(a) A and B (b) A and C

(c) A and D (d) B and C

13. Oncogenes and tumor suppressor genes are termed as cancer-critical genes. Increasingly powerful tools are now available for systematically searching the DNA or mRNAs of cancer cells for either significant mutations or altered expression.

To identify independently an oncogene or a tumor suppressor gene, which of the following would be the most convincing tests to be used?

(JUNE 2012)

(a) Transgenic mice that overexpress the candidate oncogene and knockout mice that lack candidate tumour suppressor gene.

(b) Transgenic mice that overexpress the candidate tumor suppressor gene and knockout mice that lack candidate oncogene.

(c) Transgenic mice that overexpress the candidate oncogene and tumor suppressor gene.

(d) Knockout mice that lack the candidate oncogene and tumor suppressor gene.

14. A western blot analysis after treating cancer cells with a prospective anti-cancer drug is shown below:

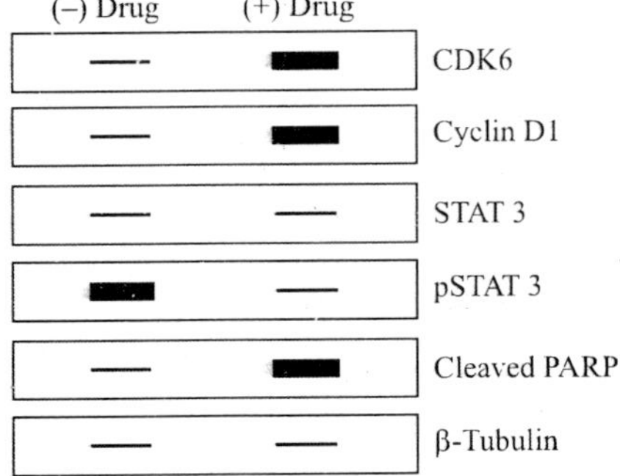

The following assumptions were made:

A. The drug may have arrested the growth of cells at the G1 phase.

B. The drug targeted the JAK-STAT signalling pathway.

C. The drug led to apoptosis of the cells.

D. Drug-induced apoptosis was through the extrinsic or mitochondrial-independent pathway.

Which one of the following combination is correct? **(DEC 2018)**

(a) Only B and D (b) A, B and C

(c) Only A and B (d) B, C and D

15. A student treated cancer cells with an anticancer drug and perform western blot analysis. Which one of the following blots is the best representation under the control (C) and treated (T) samples? **(Dec 2016)**

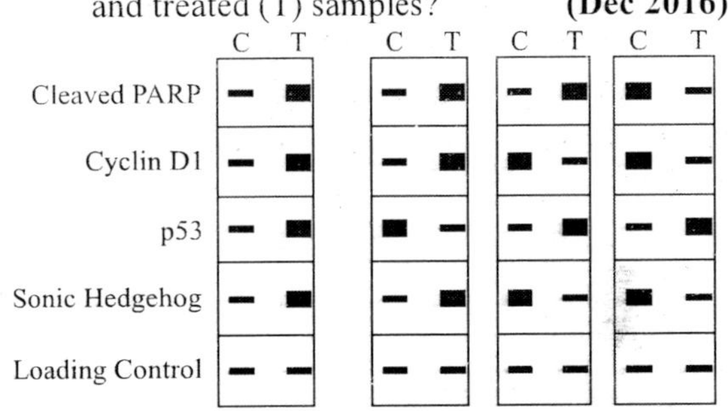

16. In an experiment it was observed that a protein was upregulated in cancer tissues (compared to control tissues) that showed correlation with disease progression. Following are a few possibilities, which can explain the above observation.

A. A mutation could be located in the 3'UTR of the corresponding mRNA at a miRNA binding site.

B. A mutation changes the conformation of the protein, resulting in its better stability.

C. A mutation in the corresponding mRNA promotes ribosome read-through of the termination codon resulting in increased synthesis of the protein.

D. A mutation in the corresponding mRNA increased the stability of the RNA due to change in secondary structure.

Which one of the following combinations represents the most likely explanations? **(JUNE 2019)**

(a) A, B and C (b) B, C and D

(c) C, D and A (d) A, B and D

17. The following statements have been proposed for a cancer cell. **(JUNE 2014)**

A. Binding of p53 with MDM2, a ubiquitin E3 ligase; is a precondition for cancer progression.

B. Phosphorylation of a tyrosine residue in the C-terminus of human c-Src is essential for cell invasion and motility.

C. Loss of function of both alleles of a tumor Suppressor gene prevents metastasis.

D. Dimerization of C-myc-Max leads to enhanced cell proliferation.

Which of the combinations of the above statements is correct?

(a) A and B (b) C and D
(c) A and D (d) B and C

18. Cellular level of tumour suppressor protein *p53* is maintained by the ubiquitin ligase protein, *Mdm2*. Over expression of *Mdm2* was found to convert a normal cell into cancer cells by destabilizing *p53* .Another protein $p19^{ARF}$ inhibits the activity of *Mdm2* thus stabilizing *p53*. Loss of $p19^{ARF}$ function also converts normal cells into cancer cells. Based on the above information, which one of the following statements is correct? **(DEC 2015)**

(a) Both *MDM2* and $p19^{ARF}$ are oncogenes.
(b) Both *MDM2* and $p19^{ARF}$ are tumour suppressor genes.
(c) *MDM2* is an oncogene and $p19^{ARF}$ is tumour suppressor gene.
(d) $p19^{ARF}$ is oncogenes and *MDM2* is a tumour suppressor gene.

19. A patient with breast cancer was given a dose of radiation along with chemotherapy and was apparently cured of the tumor. After five years, a tumor was noticed in the patient's lungs, but the doctors confirmed that it was derived from cells of the mammary gland.

The following possibilities were suggested by the doctor. **(JUNE 2015)**

A. Bacterial infection, after radiation, led to development of the tumors in the lungs.

B. Migration of residual chemo-resistant cells from the mammary gland resulted in tumors in the lungs.

C. Epithelial-to-mesenchymal transition had occurred in the lungs.

D. Cells in the lungs were induced to become a tumor after chemotherapy and from factors secreted by mammary cells:

Which of the following is correct?

(a) B and D (b) Only B
(c) A and B (d) A and C

20. After successive surgery and chemotherapy, the tumor of a breast cancer patient subsided. However after almost 5 years, the tumor relapsed in a more aggressive manner and did not respond to the conventional chemotherapy delivered earlier. The following postulations were made.

A. Chemoresistant cells were persisting within the tumor even after therapy.

B. A population of quiescent cells existed, which under favourable conditions, transformed to new tumor cells.

C. High ABC (ATP-Binding Cassette) transporter expressing cells persisted in the breast during chemotherapy.

D. Breast tumor cells which may have migrated to other tissues, returned to the breast Immediately after chemotherapy was terminated.

Which of the above combination of statements is true? **(DEC 2012)**

(a) A and D (b) A, B and C
(c) Only B (d) B and D

21. While testing the effect of several potent anti-cancer compounds on cycling human oral cancer cells, a student observed that a major percentage of cells showed dose-deep cell death after 12 hours of drug treatment. However the remaining cells repopulated the culture dish once the compounds were removed and the cells were cultured in complete medium. The student made the following assumptions:

A. Not all cells were equally affected by the compounds as they were not synchronized before treatment.

B. The compound selectively killed cells which were in G_0 phase.

C. The cancer stem cells were impervious to the effects of the compounds and therefore repopulated the culture.

D. The cancer cells differentiated into a mesenchymal phenotype and grew in fresh culture medium containing inhibitors of epithelial-to-mesenchymal transition (EMT).

Which one of the following combination of assumptions would best justify the results? **(JUNE 2019)**

(a) B and C (b) A and C
(c) B and D (d) A and B

22. Tumor cells were isolated from a breast cancer patient. These cells were injected into nude mice and they were divided into four groups. Group 1 received EGF receptor-conjugated with methotrexate; Group 2 received transferrin receptor-conjugated with methotrexate. Group 3 received mannose receptor-conjugated with methotrexate; Group 4 received same amount of the free drug. In which of the following cases tumorigenic index would be minimum? **(DEC 2011)**

(a) Free drug
(b) EGF receptor-conjugated drug
(c) Transferrin receptor-conjugated drug
(d) Mannose receptor-conjugated drug

23. Cancer causing genes can be functionally classified into mainly three types: (i) genes that induce cellular proliferation, (ii) tumor suppressor genes, (iii) genes that regulate apoptotic pathway. Epstein-Barr virus that causes cancer by modulating apoptotic pathway, contains a gene having sequence homology with which of the following genes? **(JUNE 2011)**

(a) bax (b) bcl-2
(c) p53 (d) caspase-3

24. An important role of Fas and mediate elimination of tumor cells by killer lymphocytes. In a study of 35 primary lung and colon tumors, half the tumors were found to have amplified and overexpressed a gene for a "secreted protein" that binds to Fas ligand. The main reason for survival of these tumor cells by this "secreted Fas- ligand binding protein" may be attributed to its **(DEC 2013)**

(a) decoy receptor activity.
(b) anti-proliferative activity.
(c) cellular defense activity against cytotoxic killing.
(d) anti-contact inhibition activity.

25. A patient with ER^+/PR^+ breast cancer was cured with a drug 'T', whereas a second patient, did not respond to 'T'. Which one of the following is the best therapy that you should suggest for the second patient? **(DEC 2014)**

(a) Surgery, followed by HER-2/neu targeted drug
(b) A drug that target triple negative (ER^- PR^- $HER\text{-}2^-$) breast cancer
(c) Radiation, followed by drug 'T'
(d) Surgery, followed by radiation only.

26. Which of the following cellular communications shown below will override the process of normal development and lead to cancer? **(DEC 2012)**

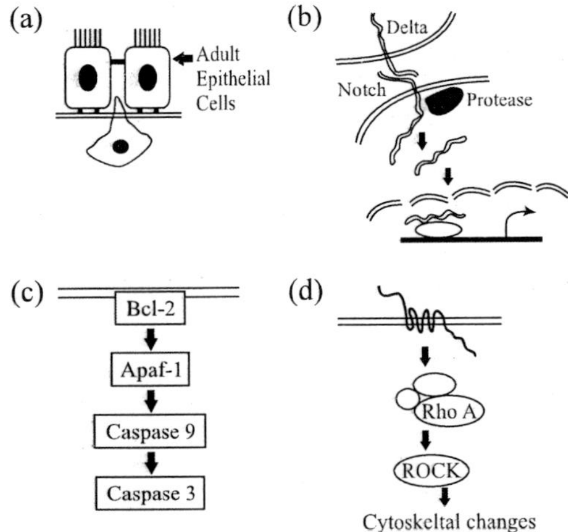

(a) B and C (b) A and C
(c) A and D (d) B and D

27. Virus infects a particular cell type, integrates its genome into a site that contains a proto-oncogene, transforms the cell and increases the level of a protein 'X', which increases cellular proliferation. A compound 'P' is known to increase the level of tumor suppressor proteins in that cell type whereas a compound 'Q' helps in stimulating a protein Z .that can bind to 'X' rendering it inactive. Which one of the following graphs correctly represents the mode of action of 'P' and 'O'? **(JUNE 2016)**

(a)

% cellular proliferation

P	–	+	–	+
Q	–	–	+	+

(b)

% cellular proliferation

P	+	+	–	+
Q	–	–	+	+

(c)

% cellular proliferation

P	–	+	–	+
Q	+	–	+	+

(d)

% cellular proliferation

P	–	+	–	+
Q	–	–	+	+

28. Two important features which aid the development of a tumor and its metastasis are epithelial-to-mesenchymal transition and angiogenesis. A student tested four cell lines to determine their invasiveness and proliferation capability by checking the expression of VEGF-A, TWIST and Cyclin D1. Which one of the following figures is most likely to exhibit the characteristics of a highly metastatic cancer cell? **(DEC 2017)**

(a)

Protein expression (Arbitrary units)
100
50
VEGF-A twist cyclin D1

(b)

Protein expression (Arbitrary units)
100
50
VEGF-A twist cyclin D1

(c)

Protein expression (Arbitrary units)
100
50
VEGF-A twist cyclin D1

(d)

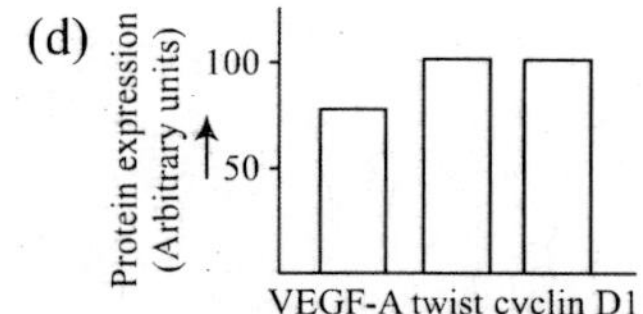

29. In what respect does the genome of slow-acting retroviruses differ from those of transducing viruses? **(DEC 2019)**

(a) They cannot activate nearby cellular proto-oncogenes after integration into the genome of the host cell
(b) They lack an oncogene
(c) They exclude mouse mammary tumor viruses
(d) They have acquired mutations during acquisition of an oncogene

30. Oncogenic viruses could have either DNA or RNA genomes. Listed below are some oncogenic viruses (Column A), their genome types (Column B) and the cancers caused by these viruses (Column C).

A	B	C
(a) Hepatitis B	(i) DNA	(x) Burkitt's lymphoma
(b) Epstein-Barr Virus	(ii) RNA	(y) T cell leukemia
(c) HTLV		(z) Hepatocellular carcinoma

Find out the correct combination. **(DEC 2017)**

(a) (a) (i) (x), (b) (ii) (y), (c) (ii) (z)
(b) (a) (ii) (y), (b) (i) (z), (c) (ii) (x)
(c) (a) (ii) (y), (b) (ii) (z), (c) (i) (y)
(d) (a) (i) (z), (b) (i) (x), (c) (ii) (y)

31. Cervical cancer-causing Papilloma virus produces two oncoproteins E6 and E7 which are responsible for interfering with cell cycle regulation by **(JUNE 2018)**

(a) inactivating Rb and p53, respectively
(b) modulating p53 and pRb, respectively
(c) binding to cyclin D1 and CDK4
(d) activating expression of p21

32. Present-day cancer treatment uses many approaches. Beyond surgery and radiation treatment, which are most often employed in cases of larger, more discrete tumors, drug therapies can be used to target residual tumor cells and to attack dispersed cancers. Chemotherapies by anti-cancer drugs are mostly aimed at blocking DNA synthesis and cell division. A list of anti-cancer drugs is given in column A, their chemical nature in column B and their mechanism of action in column C. **(JUNE 2018)**

	A		B		C
i	Metho-trexate	a	Podophyllotoxin	I	Inhibits formation of tetra hydrofolate
ii	Etoposide	b	Pirimi-dine analogue	II	Inhibit thymidylate synthase
iii	5- fluro uracil	c	Alkaloid	III	Interferes with breakdown of microtubules required for cell division
iv	Paclitaxel	d	Folic acid analogue	IV	Forms stable complex with DNA and topo- isomerase II affecting re-ligation of DNA strands

(a) i-a-I, ii-b-II, iii-c-III, iv-d-IV
(b) i-b-II, ii-a-III, iii-d-I, iv-c-IV
(c) i-c-III, ii-d-IV, iii-a-II, iv-b-I
(d) i-d-I, ii-a-IV, iii-b-II, iv-c-III

33. Match the following tumor cell origin with their nomenclature. **(DEC 2018)**

	TUMOR CELL ORIGIN		NOMENCLATURE
A	Muscle cell	a	Carcinoma
B	Germ cell	b	Sarcoma
C	Epithelial cell	c	Lukemia
D	White blood cell	d	Teratocarcinoma

(a) A - a, B - c, C - d, D - b
(b) A - d, B - a, C - b, D - c
(c) A - c, B - b, C – d,D - a
(d) A - b, B - d, C - a, D - c

Answer Key

1.(d)	**2.**(c)	**3.**(d)	**4.**(d)	**5.**(b)	**6.**(a)	**7.**(c)
8.(a)	**9.**(a)	**10.**(a)	**11.**(a)	**12.**(c)	**13.**(a)	**14.**(b)
15.(c)	**16.**(d)	**17.**(c)	**18.**(c)	**19.**(b)	**20.**(b)	**21.**(b)
22.(b)	**23.**(b)	**24.**(a)	**25.**(b)	**26.**(b)	**27.**(c)	**28.**(d)
29.(b)	**30.**(d)	**31.**(b)	**32.**(d)	**33.**(d)		

12. Apoptosis and Autophagy

1. *Ced-9* gene appears to be a binary switch that regulates cell survival and apoptosis in nematodes. Considering that CED-9 protein can bind to and inactivate CED-4, which of the following would lead to apoptosis? **(JUNE 2012)**

(a) Activation of *Ced-9* gene
(b) Loss of function of CED-3
(c) Loss of function of *Ced-9* gene
(d) Loss of function of CED-4

2. Apoptosis during early development is essential for proper formation of different structures. In C. elegans, apoptosis is accentuated by ced-3 and ced-4 genes, which in turn are negatively regulated by ced-9 and eventually Egl-1. When compared to mammals, functionally similar homolog has been identified. Accordingly, which one of the following statements is NOT correct? **(DEC 2016)**

(a) CED-4 resembles Apaf-1
(b) CED-9 resembles Bcl-XL
(c) CED-3 resembles caspase-3
(d) CED-4 resembles caspase-9

3. The cell death pathway in *C. elegans* can be schematically represented as:

ced9 --I ced4 $\rightarrow$ ced3

Based on the above, which one of the following statements is TRUE? **(DEC 2013)**

(a) A loss-of-function allele of ced9 would lead to survival of cells that normally die.
(b) A loss-of function allele of ced9 would lead to excessive cell death.
(c) A gain-of-function allele of ced9 would lead to excessive cell death.
(d) Neither loss or gain-of-function of ced9 would make any change to the cell death pathway.

4. Following are list of some proteins

A. BCL-2 B. BCL-XL
C. A1 D. BAX

Which of the protein(s) is/are **NOT** anti-apoptotic? **(JUNE 2017)**

(a) D Only (b) C only
(c) A and B only (d) B and D only

5. You have isolated a short chain polypeptide from a bacteria which upon administering to animal cells in culture is demonstrated to get transported to mitochondria and to cause mitochondrial membrane disruption. This peptide is being considered for its therapeutic effects on cancer and is to be tested for its anticancer properties in the following models of the disease.

A. Cancer with Apaf-l mutation
B. Cancer where Bax genes are inactivated
C. Cancer where Bak genes are inactivated

Which of the following is likely to provide best effect (i.e. induce apoptosis) **(DEC 2014)**

(a) Only A (b) B and C
(c) A and B (d) A and C

6. Given below is a list of some proteins known to be associated with apoptosis, their sub-cellular localization (but not in correct order) and possible role in apoptosis.

Proteins		**Localisation**		**Role in apoptosis**	
A	Effector caspase	a.	Cytosol	(i)	Promotes
B	Apaf - 1	b.	Cytosol mitochondria	(ii)	Inhibits
C	Bax	c.	Cytosol nucleus		

Choose the right combination which matches the proteins with their correct localization and role in apoptosis. **(JUNE 2017)**

(a) A – a - ii, B – b - ii, C – c - i
(b) A – c - i, B – a - ii, C –b - ii
(c) A - b - i, B - c - ii, C - a – i
(d) A – c - i, B – a - i, C – b - i

7. A particular type of cancer cell undergoes apoptosis by both extrinsic and intrinsic pathways when treated with a chemotherapeutic agent X. Caspase 8 and Caspase 9 are the initiator caspases associated with extrinsic and intrinsic pathways respectively. Now, if caspase 9 is silenced in the cancer cell by shRNA transfection, what will be the best fit graph for apoptosis scenario in the cancer cell when treated with agent X? **(DEC 2012)**

(a) ☐ Cell after transfection (b) ▨ Cell before transfection

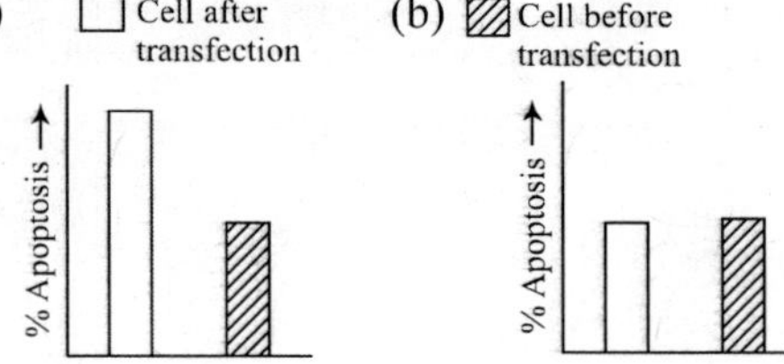

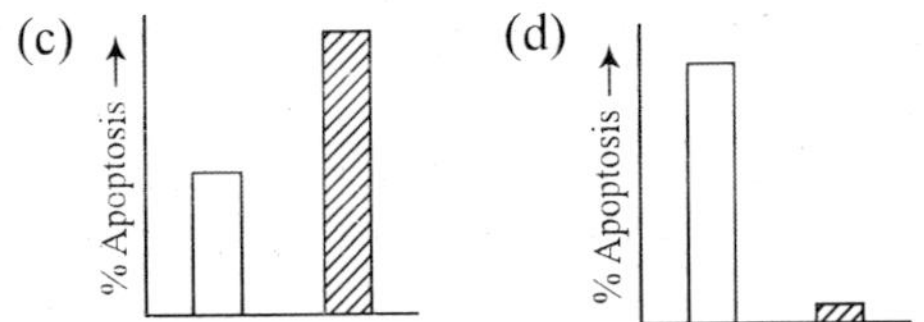

8. The initiator caspase responsible for promoting cell death after specific activation of TRADD is **(DEC 2019 ASSAM)**
   (a) Caspase 1 (b) Caspase 8
   (c) Caspase 7 (d) Caspase 5

9. Cells undergo apoptosis by two distinct and inter-connected pathways: extrinsic and intrinsic. Extrinsic pathway is activated by extracellular ligand binding to cell surface death receptors. Whenever an apoptotic stimulus activates intrinsic pathway, the pro-apoptotic Bax and Bak proteins become activated and induce the release of cytochrome C from mitochondria leading to caspase cascade activation resulting in apoptosis.

   In cell A, cytochrome C is introduced by microinjection whereas in cell B, cytochrome C is introduced by microinjection but Bax and Bak are inactivated. What will be the most appropriate apoptotic response type in both cells? **(JUNE 2013)**

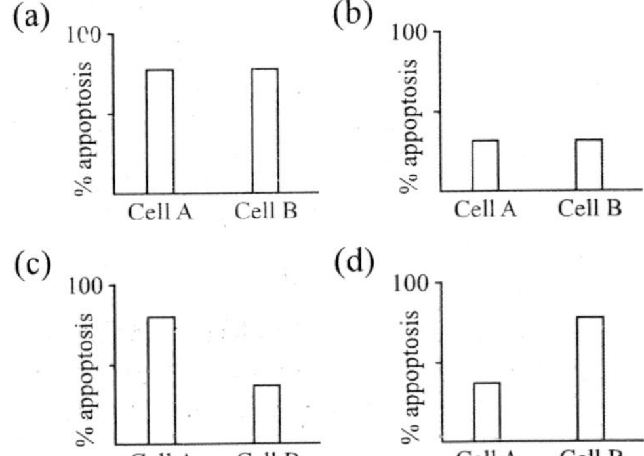

10. A protein X is kept inactive state in cytosol as complexed with protein Y. Under certain stress stimuli, Y gets phosphorylated resulting in its proteasomal degradation. X becomes free, translocates to the nucleus and results in the transcription of a gene which causes cell death by apoptosis. Stress stimuli were given to the following four different cases

    Case A: Protein Y has a mutation such that phosphorylation leading to proteasomal degradation does not occur.

    Case B: Cells are transfected with a gene which encodes for a protein L that inhibits the translocation of protein Y to the nucleus.

    Case C: Cells are transfected only with a empty vector used to transfect the gene for protein L

    Case D: Cells are treated with Z-VAD–FMK, a broad spectrum caspase inhibitor

    Which of the following graphs best describes the apoptotic state of the cells in the above cases? Y-axis represents % apoptotic cells. **(Dec 2016)**

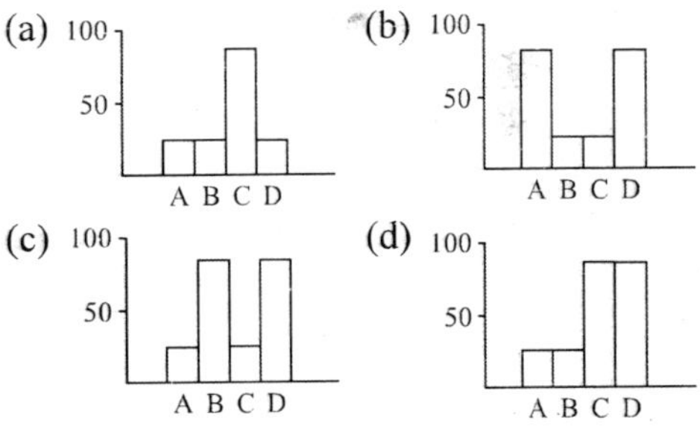

11. During apoptosis, phosphatidyl serine (PS) usually present in the inner leaflet of the plasma membrane flips to the outer membrane. Annexin V is a protein that binds to PS, Using this as a tool, we identifiy the apoptic cells from necrotic and normal cell populations by FACS using FITC-tagged Annexin V. Propidium iodide (PI) is used to stain the nucleus which generally identifies necrotic and late apoptic cells. In which area of the plot you should get early apoptic cells by FACS analysis? **(DEC 2012)**

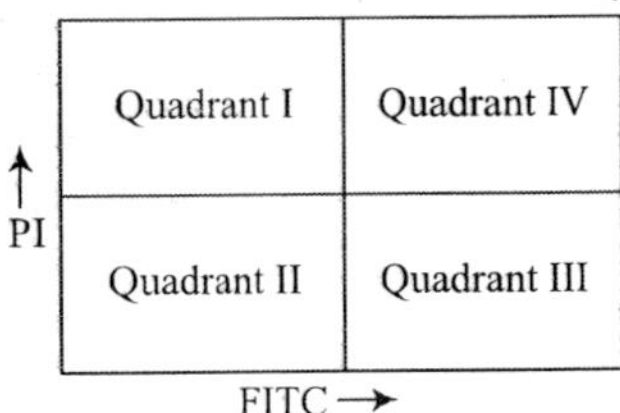

(a) Quadrant I (b) Quadrant II
(c) Quadrant III (d) Quadrant IV

12. Which one of the following assay systems can specifically detect apoptotic cells? **(JUNE 2018)**
   (a) Tetrazolium dye (MTT) based colorimetric assay
   (b) FITC - annexin V based FACS analysis
   (c) ^{51}Cr release assay
   (d) Trypan blue exclusion assay

13. The following intracellular event occurs in a cell that is subjected to conditions of starvation.

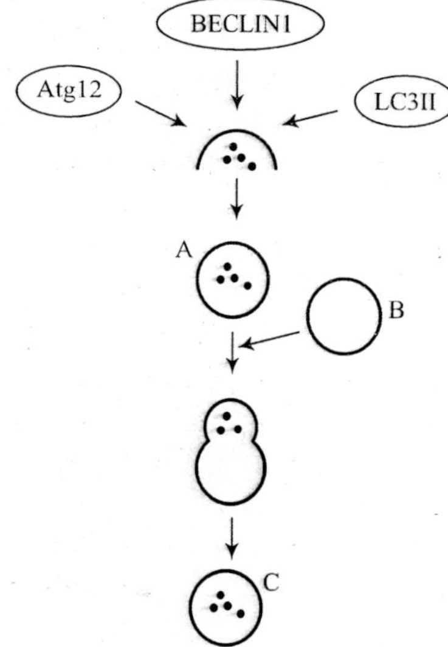

Which one of the following statements correctly represents the event shown above? **(JUNE 2018)**
   (a) The cell is undergoing apoptotic cell death with the help of lysosomes (A).
   (b) The cell is undergoing autophagy by formation of autophagolysosomes (C).
   (c) The cell is undergoing necroptosis.
   (d) The cell is undergoing autophagy and fusion occurs between lysosome (B) and autophagolysosome (A).

14. Some statements regarding the process of autophagy are given below:
   A. Autophagy occurs when cells contain aggregated proteins
   B. Autophagosomes fuse with any organelles
   C. Autophagosome is a single membrane structure
   D. Autophagosomes fuse with lysosomes to form autophagolysosomes

   Which one of the following combination of the above statements is correct? **(DEC 2019)**
   (a) A and B (b) B and C
   (c) C and D (d) D and A

15. Which one of the following best describes death-upon-detachment? **(JUNE 2019)**
   (a) Necroptosis (b) Anoikis
   (c) Extravasation (d) Metastasis

Answer Key

1.(c)	2.(d)	3.(b)	4.(a)	5.(b)	6.(d)	7.(a)
8.(b)	9.(a)	10.(a)	11.(c)	12.(b)	13.(b)	14.(d)
15.(b)						

13. Antigens and immunogenicity

1. There are three substances A, B and C. Given below are the pattern of immunological responses in rabbits when **(JUNE 2015)**
   (i) A is administered along with C,
   (ii) B is administered along with C and
   (iii) A is conjugated with B and administered along with C.

(a) (b)

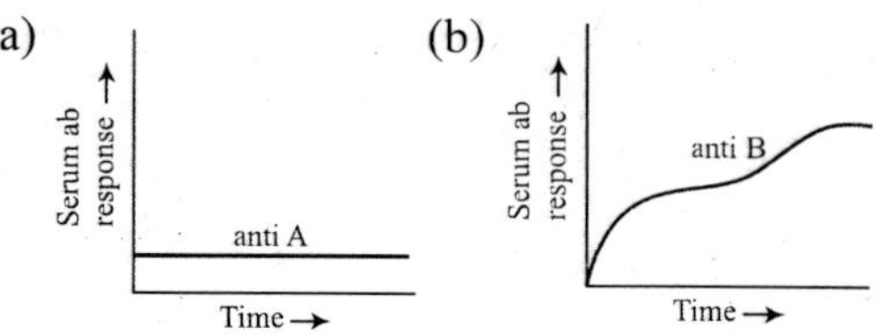

(c)

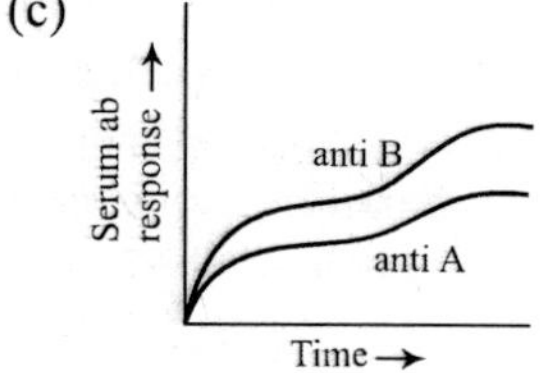

Which one of the following is the correct identification?

(a) A-protein, B-hapten, C-adjuvant

(b) A-hapten, B-protein, C- adjuvant

(c) A-protein, B- adjuvant, C-hapten

(d) A-hapten, B- adjuvant, C-protein

2. A large protein of a pathogenic bacterium has been enzymatically digested to generate a mixture of peptides ranging in size from 3 to 8 amino acids in length. Peptide mixtures were then administered in experimental animals to generate peptide-specific antibodies. In order to develop diagnostics for the bacteria, the antisera were used Western blotting to detect bacterial antigen. Western blotting failed despite the use of a wide range of antisera concentrations. What is the most probable cause of the problem? **(JUNE 2012)**

(a) Peptide-specific antibody mixture is unstable.

(b) Peptide-specific antibodies were not generated as adjuvant was not administered

(c) Peptide-specific antibodies were not generated as they were not coupled to a protein carrier

(d) Peptide-specific antibodies could not recognize the bacterial antigen.

3. A mouse was primed with trinitrophenyl-lipopolysaccharide (TNP-LPS) whereas another mouse was primed with TNP-keyhole limpet hemocyanain (TNP-KLH). After three weeks, these mice were sacrificed and splenic cells were fractionated to B cells and T cells. B cells from TNP-LPS primed mice were co-cultured with T cells from TNP-LPS or TNP-KLH-primed mice. Similarly, B cells from TNP-KLH primed mice were co-cultured with the T cells from TNP-LPS or TNP-KLH-primed mice.

So, we have four co-cultures:

A. $B^{TNP\text{-}LPS}$ x $T^{TNP\text{-}LPS}$

B. $B^{TNP\text{-}LPS}$ x $T^{TNP\text{-}KLH}$

C. $B^{TNP\text{-}KLH}$ x $T^{TNP\text{-}LPS}$

D. $B^{TNP\text{-}KLH}$ x $T^{TNP\text{-}KLH}$

Among these co-cultures, where do you except the highest IgG production? **(JUNE 2012)**

(a) A (b) B

(c) C (d) D

4. You are given a group of four mice. Each mouse is immunized with keyhole limpet hemocyanin or azobenzene arsonate or lipopolysaccharide or dextran. Four weeks later, sera were collected from these mice and antigen-specific IgG1 and IgG2a ELISA were performed. Only one of the mice showed positive response. It was **(DEC 2011)**

(a) keyhole limpet hemocyanin-primed mouse

(b) azobenze arsonate-primed mouse

(c) lipopolysaccharide-primed mouse

(d) dextran-primed mouse

5. A set of neonatal mice are divided into four groups. Group I neonates were not primed with any antigen. Group 2 neonates were primed with KLH. Group 3 neonates were primed with KLH but thymectomized. Group 4 neonates were KLH-primed, thymectomized, but reconstituted with KLH- specific $CD4^+$ T cells. All these mice, when grown adult, were challenged with KLH and the anti-KLH IgG antibody was measured in sera.

Which of the following is the collects order of magnitude of antibody response? (JUNE 2013)

[$>$ = greater than, $\geq$ = greater than or equal to]

(a) Group 1 > Group 2 > Group 3 > Group 4

(b) Group 2 > Group 1 > Group 3 ≥ Group 4

(c) Group 2 > Group 3 > Group 1 > Group 4

(d) Group 4 > Group 1 > Group 2 ≥ Group 3

Answer Key

1.(b)	2.(c)	3.(d)	4.(a)	5.(d)

14. Structure and Function of Antibody

1. Which one of the following human serum immunoglobulins takes part in classical complement fixation pathway? **(DEC 2017)**
   (a) IgD (b) IgE
   (c) IgA (d) IgG
2. Which class of immunoglobulins will increase in case of a chronic infection? **(MODELPAPER)**
   (a) IgA (b) IgG
   (c) IgM (d) IgE
3. Which one of the following immunoglobulin isotypes plays a primary role in protecting against pathogens that invade through the gut or respiratory mucosa? **(DEC 2019 ASSAM)**
   (a) IgE (b) IgM
   (c) IgD (d) IgA
4. The secondary antibodies routinely used for the detection of primary antibodies in western blotting experiment are **(DEC 2016)**
   (a) anti-allotypic (b) anti-idiotypic
   (c) anti-isotypic (d) anti-paratypic
5. If you run a pentavalent IgM through SDS-polyacrylamide gel electrophoresis, how many bands you are supposed to get by Western blotting using alkaline phosphatase conjugated secondary antibody? **(DEC 2014)**
   (a) Five (b) Four
   (c) Three (d) One
6. There are various subclasses of antibodies found in body fluids and body secretion. Many different functions may be attributed to these subclasses. Given below in column I is major functions I different subclasses and column II consists of the name of subclass. **(DEC 2015)**

Column I		Column II	
A	Binds to macrophages by F_c	(i)	IgA
B	Binds to mast cells and basophils	(ii)	IgD
C	First B cell receptor	(iii)	IgE
D	No major specific function known other than antigen binding	(iv)	IgG
E	Protector of mucous membrane	(v)	IgM

Select the correct combination:
   (a) A - (i), B - (ii), C -(iii), D - (iv), E-(v)
   (b) A- (ii), B-(iii), C-(iv), D-(v), E-(i)
   (c) A - (iii), B - (iv), C -(v), D -(i), E- (ii)
   (d) A - (iv), B - (iii), C- (v), D - (ii), E-(i)
7. Polyclonal antibodies are raised against bovine serum albumin in rabbit. Subsequently IgG in the antiserum is purified and digested with either pepsin or papain. Out of the following possibilities, which one is correct? **(MODEL PAPER)**
   (a) Pepsin-digested antibodies cannot precipitate the antigen
   (b) Papain-digested antibodies cannot precipitate the antigen
   (c) Pepsin digestion will produce two Fab molecules
   (d) Pepsin-digested antibodies will lose all interchain disulfide bonds
8. Immunoglobulins have therapeutic applications in cancer treatment, infection clearance and targeted drug delivery. For this reason, immunoglobulins are briefly cleaved by the enzyme pepsin. Following are some of the statements regarding the brief digestion of irnmunoglobulin by pepsin.
   (i) $F(ab)_2$ fragment is generated which retains the antigen binding activity.
   (ii) $F(ab)_2$ fragment having antigen binding activity and the crystallisable Fc fragment are generated.
   (iii) The fragment generated on incubation with a proper antigen forms a visible precipitate.
   (iv) The fragment generated is incapable of forming a visible precipitate on incubation with a proper antigen.

Which of the above statements are correct?

(JUNE 2016)

(a) (i) and (ii) (b) (i) and (iii)
(c) (i) and (iv) (d) (ii) and (iii)

9. In order to precipitate a particular protein by its specific antiserum, it was found that the protein formed cross-linked lattice with specific polyclonal antiserum but failed to precipitate with specific monoclonal antiserum. Which of the following would accurately justify the reason for this behavior? **(DEC 2014)**

(a) The protein has multiple copies of the same epitope specific for the monoclonal antibody.

(b) The protein has multiple distinct epitopes-but each has single copy.

(c) There is tota! absence of epitopes in the protein.

(d) The protein has multiple copies of different epitopes

Answer Key

1.(d)	2.(b)	3.(d)	4.(c)	5.(d)	6.(d)	7.(b)
8.(b)	9.(b)					

15. Antibody diversity and monoclonal antibodies

1. A technician wanted to make rabbit antiserum specific for mouse IgG. The technician injected rabbit with purified mouse IgG but obtained antiserum which reacted strongly with each of the other mouse isotypes. Which of the procedures mentioned below will allow him to make antiserum specific for IgG only?

(DEC 2014)

(a) Injecting rabbit with purified $F(ab)_2$' region of the IgG antibody.

(b) Injecting rabbit with purified heavy chain of IgG antibody

(c) Injecting rabbit with purified light chain of the IgG antibody.

(d) Injecting rabbit with purified F(ab)' region of the IgG antibody.

2. A potentially valuable therapeutic approach for killing tumor cells without affecting normal cells is the use of immunotoxins. Immunotoxins constitute monoclonal antibodies against tumor cells conjugated to lethal toxins. Which of the following molecular approaches do you think is NOT appropriate for generating tumor cell-specific immunotoxin that will not kill normal cells?

(a) Cell surface receptor binding polypeptide chain of toxin molecule should he replaced by monoclonal antibodies against a particular tumor cell type.

(b) Constant region Fc domain of tumor cell specific monoclonal antibody should be replaced by ligation of toxins.

(c) Variable region $F(ab)_2$, domain of tumor cell-specific monoclonal antibody should be replaced by ligation of toxins.

(d) Inhibitor polypeptide chain of toxin should be conjugated to $F(ab)_2$, domain of tumor cell specific monoclonal antibody.

(DEC 2012)

3. A potentially valuable therapeutic approach for killing tumour cells without affecting normal cells is the use of immunotoxins. Immunotoxins consist of particular cell-specific monoclonal antibodies coupled to lethal toxins. Which of the following molecular approaches is NOT appropriate for killing tumor-cells?

(DEC 2013)

(a) Cell surface receptor binding polypeptide chain of toxin molecules should be replaced by monoclonal antibodies which are specific for a particular tumor cell.

(b) Constant region Fc domain of tumor cell-specific monoclonal antibody should be replaced by toxin molecules.

(c) Variable region F(ab) domain of tumor cell-specific monoclonal antibody should be replaced by toxin molecules.

(d) Inhibitor polypeptide chain of toxin molecules should be conjugated to the F(ab) domain of monoclonal antibody tumor-specific monoclonal antibodies.

4. What phenotype would you predict for a mutant mouse lacking one of the genes required for site-specific recombination in lymphocytes? **(JUNE 2015)**

(a) Decrease in T cell counts
(b) Immunodeficient
(c) Increase in T cell counts
(d) Increase in B cell counts

5. Monoclonal antibodies can be modified for better research and therapeutic applications. Several such approaches are mentioned below (Column A) along with their possible applications (Column B).

	Column A		Column B
i.	Binding site of the original mouse mAb are placed onto the Fe regions of human antibodies	a	Abzymes
ii.	Antibodies are modified by conjugation to toxins designed to kill cells to which the antibody will bind	b	Reduce side effect of xenogenic antibodies in immunotherapy
iii.	Generation of mAb that specifically bind and stabilize the transition state of a chemical reaction	c	Immunotoxins

Which one of the following options represents a correct combination of terms in Column A and Column B? **(DEC 2018)**

(a) i-a; ii-b; iii-c (b) i-b; ii-a; iii-c
(c) i-c; ii-b; iii-a (d) i-b; ii-c; iii-a

6. Monoclonal antibodies (mAb) can be potentially used as therapeutic agents. The major advantage is that they can specifically target aberrant cells. However, there is a practical difficulty. Monoclonals are raised in mouse and therefore it is expected that an immune reaction will develop if these are injected into humans. It is therefore necessary to 'humanize' monoclonal antibody by

(a) expressing the genes for the mAb in cultured human cells and isolating the mAb from these cells
(b) replacing the Fc region of a mAb with one derived from a human IgG
(c) replacing C_L and C_H regions of the Mab with that obtained from a human IgG.
(d) taking a human IgG and replacing the CDRs by those derived from the mouse mAb **(JUNE 2011)**

7. In the case of monoclonal antibody production by hybridoma technology, myeloma cells used lack the enzyme hypxanthine - guanine phosphoriboxyl transferase(HGPRT) such that fused cells can only survive when selected on hypoxanthine-aminopterin-thymdine (HAT). What is the role of aminopterin in this medium? **(JUNE 2013)**

(a) To be used as cell cycle inhibitor of myeloma cells.
(b) To block the pathway for nucleotide synthesis.
(c) To facilitate fusion of Myeloid B cells and antibody producing B cells.
(d) To facilitate production of antibody producing B cells

8. Activation-induced cytidinedeaminase (AID) is the key mediator of somatic hypermutation, gene conversion and class-switch recombination. In order to ascertain the role of AID in class-switch recombination, immune response against a target antigen was compared between AID knock-out mice (AID-/-) with that of mice retaining a functional copy of the AID gene (AID+/-). Development of IgM and IgG antibodies against the target antigen was then measured following successive immunization and plotted graphically. Which one of the following is the most appropriate representation of the experiment?

(JUNE 2018)

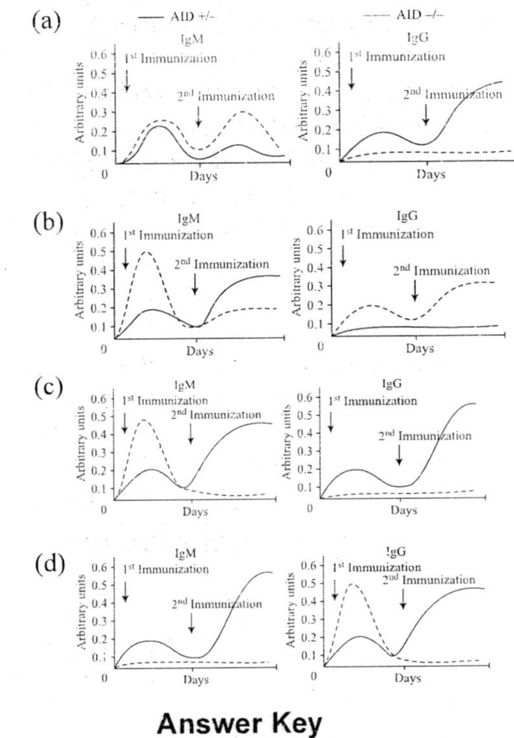

Answer Key

1.(b)	2.(c)	3.(c)	4.(b)	5.(d)	6.(d)	7.(b)
8.(a)						

16. MHC molecules, antigen presentation

1. The major histocompatibility complex (MHC) is referred to as the human leukocyte antigen (HLA) complex in humans and as the H-2 complex in mice. In an experiment, H-2^k mice were primed with the lymphocytic choriomeningitis virus (LeMV) to induce cytotoxic T lymphocytes (CTLs) specific for the virus. Spleen cells from this LCMV-primed mouse were then added to target cells of the same (H-2^k) or different H-2 haplotypes (H-2^b) that were intracellularly radiolabelled with ^{51}Cr and either infected or not infected with LCMV. CTL mediated killing of target cells were then measured by the release of ^{51}Cr into the culture supernatant (Cr release assay). In which of the following cells, ^{51}Cr will be released into the culture supernatant? **(JUNE 2017)**
   (a) H-2^k target cells
   (b) H-2^k LCMV-infected target cells
   (c) H-2^b target cells
   (d) H-2^b LCMV-infected target cells

2. Ten different mouse strains were primed with whole Keyhole limplet hemocyanin (KLH). KLH was broken into 10 peptides for *in vitro* stimulation. The splenocytes from ten different primed mouse strains were re-stimulated with each of these 10 peptides and responsiveness to these were measured *in vitro*. It was found that each of these mouse strains had responded to one of the peptides. When the peptide 3 responder was mated with peptide 4 responder, the splenocyte of F_1 offsprings responded to both the peptide. Which of the following is most appropriate **(JUNE 2012)**
   (a) Mouse strains responding to peptide 3 or peptide 4 have different MHC haplotypes
   (b) Mouse strains responding to peptide 3 or peptides 4 have either of these T cell receptors
   (c) Mouse strains responding to peptide 3 or peptide 4 cannot process KLH
   (d) Mouse strains responding to peptide 3 or peptide 4 did not express MHC class I molecule

3. In an experiment peritoneal macrophages were isolated from strain A of guinea pig. These cells were then, incubated with an antigen. After the antigen pulsed macrophages processed the antigen and presented it on their surface, these were mixed with T cells from (i) strain A or (ii) strain B (a different strain of guinea pig) or (iii) F1 progeny of strain A X B. T cell proliferation was measured in response to antigen pulsed macrophages. T cells of which strain of guinea pig will be activated? **(JUNE 2016)**
   (a) Strain A only
   (b) Strain B only
   (c) Strain A and F1 progeny
   (d) Strain B, and F1 progeny

4. Following are some of the characteristics of MHC class I and class II molecules except one which is applicable only for MHC class I. Identify the appropriate statement. **(JUNE 2016)**
   (a) They are expressed constitutively an all nucleated cells.
   (b) They are glycosylated polypeptides with domain structure. .
   (c) They are involved in presentation of antigen fragments to cells.
   (d) They are expressed on surface membrane of B cells.
5. Major histocompatibility complex (MHC) molecules are encoded by a cluster of genes called MHC locus. There are several reasons why an MHC molecule on the surface of a cell is important. (DEC 2017). Which one of the following reasons is **INCORRECT**?
   (a) To display self-class I to demonstrate that the cell is normal and healthy.
   (b) To display foreign-peptide in class I to show that the cell is infected and to engage with T helper cells.
   (c) To display a self-peptide in class I and II to test developing T cells for autoreactivity.
   (d) To display a self-peptide in class I and II to maintain tolerance to self-proteins.

Answer Key

1.(b)	2.(a)	3.(c)	4.(a)	5.(b)

17. Humoral and cell-mediated immune responses

1. CD 19 is a marker for? **(JUNE 2012)**
   (a) B-cells (b) T-cells
   (c) Macrophage (d) Natural Killer cells
2. Which one of the following statements regarding B cell receptor (BCR) and T cell receptor (TCR) is not true? **(Dec 2016)**
   (a) TCR is membrane bound and does not appear as soluble form as does the BCR
   (b) Unlike BCR, most of the TCR are not specific for antigen alone but for antigen combined with MHC
   (c) In order to activate signal transduction, BCR associates itself with Ig-α/Ig-β whereas TCR associates with CD3
   (d) Antigen binding interaction of BCR is much weaker than TCR
3. Cytotoxic T cells express **(DEC 2015)**
   (a) CD8 marker and are class II MHC restricted
   (b) CD4 marker and are class I MHC restricted
   (c) CD4 marker and are class II MHC restricted
   (d) CD8 marker and are class I MHC restricted
4. Which one of the following statements regarding clonal selection hypothesis is NOT CORRECT? **(DEC 2018)**
   (a) Mature B lymphocytes bear lg receptors on their cell surface and all receptors on a single B cell have variable specificity for antigen.
   (b) On antigen stimulation, B cell matures, migrates to lymphoid organs and replicates. Its clonal descendents bear the same receptor as parental B cell and secrete antibodies with identical specificity.
   (c) After immune response, more B cells bearing receptors will remain in the host and act as memory cells for mounting enhanced secondary response.
   (d) B cells with receptors for self-antigens are deleted during embryonic development.
5. Given below are some properties that are applicable for T cell receptors:
   A. It is associated with a multi-component signal transducing complex, CD3, on the membrane
   B. It is divalent in nature

C. It contains domains which are similar to the immunoglobulin fold structure

D. It exhibits diversity generated by somatic mutation

Which one of the following combinations represents correct statements?

(DEC 2019 ASSAM)

(a) A and B (b) A and C

(c) C and D (d) B and D

6. Which one of the following cells generally does **NOT** secrete IFN-γ? **(JUNE 2017)**

(a) $CD8^+$ T cells (b) T_H1 cells

(c) NK cells (d) T_H2 cells

7. T_H2 response is generated and maintained mainly by which of the following pair of cytokines? **(JUNE 2011)**

(a) IL-4 and IL-10. (b) IL-12 and IFN-γ.

(c) IFN-γ and TNF-α. (d) IL-2 and IL-12.

8. When a primary antibody response and cell-mediated response were induced by a protein in its native conformation, the secondary antibody response could be induced only by the native antigen whereas the secondary cell-mediated immune response could be induced by the native antigen as well as by the denatured antigen. This discrepancy may only be attributed to **(DEC 2013 GU)**

(a) T cell memory

(b) Antigen recognition

(c) Antigen processing

(d) B cell memory

9. Macrophages were collected from BALB/c mice, $CD40^-$ deficient mice, $CD86^-$ deficient mice and ICAM-1-deficient mice. These macrophages were co-cultured with LCMV peptide-specific T cells in presence of the LCMV peptide for three days. The cells were recovered and co-cultured with BALB/c-derived macrophages in presence of the peptide. During the last twelve hour of the co-culture, ^{3}H-thymidine was added to the cultures. The cells were harvested and ^{3}H-thymidine incorporation was assessed. The highest incorporation was observed in

(JUNE 2011)

(a) BALB/c macrophage-T cell co-culture.

(b) CD40-deficient macrophage-T cell co-culture.

(c) CD86-deficient macrophage-T cell co-culture.

(d) ICAM-1-deficient macrophage-T cell co-culture.

10. Following are the experimental observations made on treatment of B cells:

A. Anti-immunoglobulin (anti-Ig) antibody treatment results in B cell apoptosis.

B. Anti-Ig plus CD40 ligand treatment results in B cell proliferation.

C. Anti-Ig plus CD40 ligand plus IL-4 treatment results in B cell proliferation and switching to IgG1.

D. Anti-Ig plus IL4 treatment results in less B cell proliferation but switching to IgE.

From the above observation, which one of the following is the correct interpretation for the role of CD40 in B cell function? **(DEC 2012)**

(a) Induce death of B cells.

(b) Rescue B cells from death and Ig class switch to IgG1.

(c) Inducing Ig class switch to IgE.

(d) Induce Ig class switching to both IgG1 and IgE and inhibit B cell proliferation.

11. A BALB/c mouse was thymectomized on the first day after birth (mouse 1) whereas another was thymectomized on day 7 after birth (mouse 2). A third mouse underwent the same operation on day 21 after birth. After 56 days, sera were prepared from these mice and also from control mice, which had sham operation. The sera were checked for anti-DNA antibodies. Which one of the following observations is the most plausible? (DEC 2013)

(a) Both mouse 1 and mouse 2 had anti-DNA antibodies but mouse 3 did not have anti-DNA antibodies.

(b) Only mouse 1 had anti-DNA antibodies.
(c) Only mouse 3 had anti-DNA antibodies.
(d) Only the control mice had anti-DNA antibodies.

12. CD4$^+$ T cells are co-cultured with macrophages in the presence of immobilized anti-CD3 antibody under four different conditions:

(JUNE 2014)

A. Interleukin (IL)-4 plus anti-IFN γ antibody,
B. IL-12 and anti-IL-4 antibody,
C. Transforming growth factor (TGF)-β,
D. TGF-β and IL-6

for three rounds to induce T-helper cell differentiation identifiable by the cytokines they express predominantly. Which one of the following is the most likely combination of predominant cytokine expression in these cultures?

(a) (A) IL-4, (B) IFN- γ, (C) IL-10, (D) IL-17
(b) (A) IFN- γ, (B) IL-4, (C) IL-17, (D) IL-10
(c) (A) IL-17, (B) IL-4, (C) IFN- γ, (D) IL-10
(d) (A) IL-17, (B) IL-10, (C) IL-4, (D) IFN- γ

13. Four groups of one-day old female BALB/c neonates had received the following treatments:

(JUNE 2014)

A. Epidermal cells from CS7BL/6 male plus anti β2 microglobulin antibody.
B. Epidermal cells from C57BL/6 male plus antibodies to β2 microglobulin, CD40 ligand, CD80 and CD86.
C. Epidermal cells from C57BL/6 female plus anti-CD80 antibody,.
D. Epidermal cells from C57BL/6 female mice.

When these BALB/c neonates grew six weeks old, they received skin transplant from C57BL/6 male mice. Transplantation rejection time varied between these four groups. Starting from the fastest to the slowest rejection, which one of the following is the most likely order?

(a) D > C > A > B (b) A > B > C > D
(c) B > A > D > C (d) D > B > C > A

14. Dendritic cells (DC) from BALB/c mice were treated with IL-10 or with IFN-γ. Similarly, dendritic cells from β2-microglobulin-deficient mice were also treated with IL-10 or with IFN-γ. These cells were co-cultured with CD8$^+$ T cells from hen egg lysozyme (HEL)-specific T cell receptor transgenic mice in presence of the HEL peptide. Five days later, CD8$^+$ T cells were assayed for target cell lysis. Which one of the following combinations will have the highest target cytotoxicity? **(JUNE 2013)**

(a) DC $(BALB/c)^{IL-10}$ x CD8$^+$ T
(b) DC $(BALB/c)^{IFN-\gamma}$ x CD8$^+$ T
(c) DC $(\beta2\text{-microglobulin-deficient})^{IL-10}$ x CD8$^+$ T
(d) DC $(\beta2\text{-microglobulin-deticient})^{IFN-\gamma}$ x CD8$^+$T

15. Strain A mice were crossed with strain B mice and first generation F1 mice were obtained, i.e.$(A \times B)_{F1}$. A scientist then implanted thymectomized and irradiated $(A \times B)_{F1}$ mice with a B-type thymus and then reconstituted the animals immune system with an intravenous infusion of $(A \times B)_{F1}$ bone marrow cells. The chimeric mice were infected with lymphocytic choriomeningitis virus (LCMV) and the spleen T cells were then tested for their ability to kill LCMV-infected target cells from the strain A or strain B mice. Which one of the following is the correct outcome of the experiment?

(DEC 2019)

(a) LCMV-infected target cells from strain A only will be killed
(b) LCMV-infected target cells from strain B only will be killed
(c) LCMV-infected target cells from both strain A and B will be killed
(d) Neither cells from strain A nor from strain B will be killed

16. If a disease caused by an intracellular pathogen is associated with host anti-inflammatory response, which one of the following may lead to an effective therapeutic approach?

(JUNE 2018)

(a) Treatment with TGF-β
(b) Treatment with macrophage activating agent
(c) Depletion of IFN-γ from the system
(d) Treatment with IL-4 and IL-10

17. Many cytotoxic T lymphocytes initiate killing of target cells via delivery of molecules that could induce target-cell damage directly. Which one of the following is the most appropriate? **(DEC 2018)**
(a) Interferon γ (b) Peroxynitrite
(c) Lysozyme (d) Granzyme

18. There are number of specific T cell surface molecules involved in various functions of adaptive immune response. Column X represents a list of T cell surface molecules and Column Y with the possible functional characteristics:

Column X	Column Y
A. T cell receptor	(i) binds to CD40 on B cells and APCs and triggers activation of APCs and activation/ differentiation of B cells.
B. CD 28	(ii) binds to MHC class I molecules and restricts T cytotoxic cells to recognizing only peptide presented on MHC class I
C. CD8	(iii) binds to B7-1,2 or CD80/86 on B cells and APCs, which triggers T cell activation.
D. CD154	(iv) consists of two polypeptide chains a and α and β and some consist of polypeptide chains γ and δ.

Which of the following option has all correct matches? **(JUNE 2019)**
(a) A – (i); B – (ii); C – (iii); D – (iv)
(b) A – (iv); B – (i); C – (ii); D – (iii)
(c) A – (iii); B – (iv); C – (i); D – (ii)
(d) A – (iv); B – (iii); C – (ii); D – (i)

Answer Key

1.(a)	**2.**(d)	**3.**(d)	**4.**(a)	**5.**(b)	**6.**(d)	**7.**(a)
8.(c)	**9.**(a)	**10.**(b)	**11.**(b)	**12.**(a)	**13.**(a)	**14.**(b)
15.(b)	**16.**(b)	**17.**(d)	**18.**(d)			

18. The complement system and Toll like receptors

1. Which one of the following is **NOT** true for alternative pathway of complement activation? **(JUNE 2018)**
(a) Alternative pathway uses the same membrane-attack complex as the classical pathway
(b) Alternative pathway does not require antigen-antibody interactions.
(c) Alternative pathway produces C3 by the same route as the classical pathway
(d) Certain microbial surfaces have physico-chemical properties that may result in activation of alternative pathway

2. In order to prove that liposome can serve as a model membrane (mimicking cellular plasma membrane) and can be used as a target for complement-mediated immunolysis, an experiment as below is designed. To initiate such experiment, hapten-conjugated liposomes are made and loaded with umbelliferyl phosphate (UMP; hydrolysed product of UMP is umbelliferone and is fluorescent). Such loaded, hapten-conjugated liposomes in 10 mM Tris. buffered saline, pH 7.4 were mixed with anti-hapten antibodies and fresh guinea pig serum (as a source of complement) to induce immunolysis of liposomal membrane. To quantify only the membrane lysis component which of the assay sequences below is MOST appropriate? **(DEC 2014)**
(a) Mixture is ultracentrifuged and the supernatant reacted with alkaline phosphatase and fluorescence measured.

(b) Mixture is sequentially reacted with phospholipase and alkaline phosphatase followed by fluorescence measurements

(c) Mixture is directly subjected to fluorescence measurement

(d) Mixture is treated with Triton X-100 and fluorescence measured

3. In a plaque-forming cell assay, antigen specific B-cell numbers are assessed. In this assay, antigen coated sheep red blood cells (SRBCs) are lysed by the hapten specific B cells by complement-mediated cytotoxicity. In an assay that tried to enumerate the TNP-specific B-cells elicited in TNP-KLH-primed mice, no plaques were formed despite the presence of antigen-specific antibody producing B cells. Which of the following is NOT the reason for the absence of plaques? **(JUNE 2014)**

(a) The source of complement has anti- TNP antibody. .

(b) The SRBC were stored for too long.

(c) The B cells were contaminated with LPS or lectins.

(d) The complement had anti-KLH antibody.

4. Toll-like receptors (TLR) present in mammalian macrophages are recognized by types of macromolecules that are not present in vertebrates but are present in certain groups of microbial pathogens. When these pathogens infect macrophages, TLR signalling is stimulated. Following are the list of macromolecules in column A and types of TLR in column B.

	A		B
(i)	Lipopolysaccharide (LPS)	a.	TLR3
(ii)	Flagellin	b.	TLR4
(iii)	Double stranded RNA	c.	TLR5
(iv)	Unmethylated CpG dinucleotides	d.	TLR9

Which of the following is the best possible match of the pathogenic ligand with their corresponding TLR? **(JUNE 2017)**

(a) (i) - a, (ii) - b, (iii) - c, (iv) - d

(b) (i) - b, (ii) - a, (iii) - d, (iv) - c

(c) (i)- b, (ii)- c, (iii)- a, (iv) - d

(d) (i)- c, (ii) - d, (iii) - b, (iv) – a

5. Toll-like receptor 4 is associated with responsiveness to LPS, an that causes lethal endotoxic shock. The mice deficient In Toll-like receptor 4 and BALB/c mice were injected with Escherichia coli. In addition, some BALB/b mice were also injected with the same bacteria alone or with anti-interleukin-10 (IL-10) antibody. The mice resistant to the lethal effect of the bacteria were: **(JUNE 2011)**

(a) BALB/b mice receiving the bacteria.

(b) BALB/b mice receiving the bacteria and the anti-IL-10 antibody

(c) Mice deficient In Toll-like receptor

(d) BALB/c mice receiving the bacteria.

Answer Key

1.(c)	2.(a)	3.(d)	4.(c)	5.(c)

19. Hypersensitivity and autoimmunity

1. Toxic shock is caused by **(DEC 2011)**

(a) toxins produced by some bacteria

(b) Excessive stimulation of a large proportion of T cells by bacterial superantigens

(c) abnormal cytokine production by B cells

(d) excessive production of immunoglobulins

2. Immediate hypersensitivity reactions are associated with **(DEC 2013)**

(a) IgG (b) IgE

(c) IgM (d) IgA

3. Which one of the following is NOT related to immediate hypersensitivity reactions? **(JUNE 2014)**

(a) Mast cell degranulation results in histamine-mediated allergic reactions.

(b) Reaginic antibodies trigger allergic reactions.

(c) Granulomatous reaction is a key to contain infection.

(d) Anaphylactic reaction is triggered primarily by IgE

4. In a type I hypersensitivity- mediated asthmatic response, which one of the following is thought to contribute significantly to the prolonged bronchospasm and build-up of mucous seen in asthmatics? **(JUNE 2019)**

(a) Thromboxane (b) Leukotriene

(c) TGFβ (d) Chondroitin

5. A patient undergoes liver transplantation and during the course of post-operative treatment, becomes susceptible to infection. The patient can be treated in two different modes and can have alternative outcomes. Which of the following statements is correct?**(JUNE 2011)**

(a) Treatment with immunostimulatory drugs reducing the infection but rejecting the transplant.

(b) Treatment with immunostimulatory drugs reducing the infection and retaining the transplant.

(c) Treatment with antibiotics reducing the infection but retaining the transplant.

(d) Treatment with antibiotics reducing the infection but rejecting the transplant.

6. Graft rejection does not involve **(DEC 2011)**

(a) erythrocyte

(b) T cells

(c) macrophages

(d) polymorphonuclear leukocytes

7. Mismatch of blood in parents many result in erythroblastosis fetalis in a new born. Match the correct cause (left column) and usual treatment (right column). **(MODEL PAPER)**

Commonest cause	Used treatment
A. Mother Rh(+) and father Rh(-)	C. replacement of neonate's blood with Rh(-) blood.
B. Mother Rh (-) and father Rh (+)	D. replacement of neonate's blood with Rh(+) blood.

(a) A and C (b) A and D

(c) B and C (d) B and D

8. Gram negative bacteria, *Klebsiella pneumoniae*, upon infecting humans, results in severe septic shock after a few hours of infection. Which of the following is not true for this type of infection? **(JUNE 2013)**

(a) Cell wall endotoxins cause overproduction of cytokines.

(b) Septic shock can be treated by anti- TNFα antibodies.

(c) Recombinant bacterial proteins can be used for the treatment of septic shock.

(d) Recombinant TNFα receptor antagonist can be used for the treatment of septic shock.

9. Which of the following is considered to be a combined B- and T-cell deficiency? **(JUNE 2013)**

(a) Ataxia-telangiectasia

(b) Swiss type agammaglobulinemia

(c) Wiskott-Aldrich syndrome

(d) Bruton's agammaglobulinemia

Answer Key

1.(b)	2.(b)	3.(c)	4.(b)	5.(c)	6.(a)	7.(c)
8.(c)	9.(b)					

20. Mycobacterium tuberculosis and other bacteria

1. Which one of the following pathogens cannot survive and replicate within phagocytic cells? **(DEC 2019 ASSAM)**

(a) *Listeria monocytogenes*

(b) *Mycobacterium tuberculosis*

(c) *Leishmania donovani*

(d) *Streptococcus pneumonia*

2. Which of the following bacteria has subcellular localization in lysosomes? **(JUNE 2016)**

(a) *Salmonella typhii.*

(b) *Streptococcus pneumoniae*

(c) *Vibrio cholera*

(d) *Mycobacterium tuberculosis*

3. Intracellular pathogens like *Mycobacteria, Salmonella, Leishmania* and *Listeria* survive in macrophages by modulating host cellular machinery. In order to study the fate of these intracellular pathogens in macrophages, cells were labelled with lysotracker Red arid infected with GFP-Iabelled organisms.

After 2 hours at 37°C, cells were fixed, stained with anti-transferrin receptor antibody and probed with secondary antibody conjugated-blue dyes. Cells were viewed under confocal microscope. .

Observation: GFP-labelled *Mycobacteria, Salmonella* and *Listeria* were localized in the same compartment labelled with blue dyes; whereas GFP-*Leishmania* colocalize with red labelled compartment.

Which of the following statement is true based on these observations? **(JUNE 2011)**

(a) *Mycobacteria, Salmonella* and *Listeria* reside in the lysosomes.

(b) *Leishmania* reside in lysosome like compartment

(c) *Leishmania* reside in a compartment which bears characteristics of early endocytic compartment.

(d) *Mycobacteria, Salmonella* and *Listeria* lyse the phagosomal membrane and reside in cytosol.

4. *Mycobacterium tuberculosis* is an intra-cellular bacterium. It prefers to infect? **(JUNE 2012)**

(a) Macrophages (b) B-cells

(c) T-cells (d) neutrophils

5. A pathogen like *Mycobacterium*, which colonizes inside the cells of the host, is likely to be least affected by which one of the following host immune defense mechanisms?

(JUNE 2014)

(a) Cell-mediated immune response

(b) $CD4^+$ T lymphocytes

(c) Cytokines

(d) Humoral immune response

6. Survival of intracellular pathogens depends on the levels of pro-inflammatory and anti-inflammatory cytokines in macrophages. In an experimental condition, *Mycobacteria* infected macrophages were treated with IL-6 or IL-12 for 4 hours at 37° C. untreated cells were used as control. Cells were lysed and number of bacteria in each experimental set was counted by measuring colony forming unit (CFU). Which of the following observations is true?

(DEC 2011)

(a) IL-12 treated cells contain more intracellular bacteria than control

(b) IL-12 treated cells contain less intracellular bacteria than control

(c) IL-6 treated cells contain more intracellular bacteria than control

(d) IL-6 treated cells contain less intracellular bacteria than control

7. Opsonisation of a bacterium is a process by which specific antibody binds with the surface molecule of the bacteria. In an experimental condition, macrophage were infected with either WT: Myccbacteria or with opsonised: Mycobacteria for 2 hrs at 37°C. Subsequently, cells were washed and further incubated for 24 hrs at 37°C. Finally, bacterial load in macrophages were determined by colony forming unit (CFU). Which of the following observation is true? **(MODELPAPER)**

(a) WT:Mycobacteria inhibits its transport to the lysosomes and survive in macrophages.

(b) Opsonised: Mycobacteria inhibits its transport to the lysosomes and survive in macrophages.

(c) WT: Mycobacteria are targeted to the lysosomes and killed in macrophages.

(d) Opsonised: Mycobateria are targeted to the lysosomes and survive in macrophages.

8. Bacteria adopt different strategies to evade host defense mechanisms. From the lists of various different mechanisms and bacterial strategies against host defense given below, select the option representing all correct pairing. **(DEC 2018)**

Host Defence Mechanism		Bacterial strategies against host defence	
a	Phagocytosis	W	Change of bacterial surface charge, making it more positive
b	Release antibodies, like IgG	X	Capsular polysaccharides, such as that of *Klebsiella pneumoniae*
c	Antibody mediated agglutination	Y	Release of soluble proteins like protein A of *Staphyloccus aureus*
d	Anti microbial peptides	Z	Secretion of elastase to inactivate C3a and C5a

(a) a-W; b-X; c-Y; d-Z
(b) a-X; b-Y; c- Z; d-W
(c) a-Z; b-Y; c-X; d-W
(d) a-Y; b-W; c- Z; d-X

9. Which one of the following does NOT belong to human antimicrobial proteins and peptides at epithelial surface forming part of innate immunity? **(JUNE 2019)**

(a) Lactoferrin (b) Defensin
(c) Calprotectin (d) Vimentin

10. Bacterial infections are generally divided into two broad classes: intracellular and extracellular bacterial infections. Given below are some of the properties, which are applicable for bacterial infections.

A. Humoral immune response is the main protective response against extracellular bacteria.

B. Innate immunity is not effective against intracellular bacterial pathogens.

C. Bacterial endotoxins do not induce an innate immune response.

D. Intracellular bacterial infections generally induce a cell-mediated immune response resulting in secretion of cytokines which activate macrophages.

Which one of the following combination of statements is correct? **(DEC 2019)**

(a) A and B (b) B and C
(c) C and D (d) A and D

11. Given below are some of the major characteristics of inflammatory responses against bacterial infections in general:

A. TGF β levels are elevated

B. Phagocytic capacity of macrophages is increased

C. Complement system may get activated

D. Self-reactive $CD8^+$ T cells may get activated

Which one of the following combinations of above characteristics is most appropriate for inflammatory response against extracellular bacterial infections? **(DEC 2019 ASSAM)**

(a) A and B (b) B and C
(c) C and D (d) A and D

12. Given below are some statements related to bacterial toxins.

A. Exotoxins are usually heat stable proteins secreted by bacteria.

B. Exotoxins are usually heat labile proteins secreted by bacteria.

C. Endotoxins are heat stable lipopolysaccharide-protein complexes of the outer membrane of gram-negative bacteria.

D. Endotoxins are heat labile lipoproteins of the outer membrane of gram-negative bacteria.

Which one of the following options represents combination of correct statements? **(DEC 2019 ASSAM)**

(a) A and B (b) B and C
(c) B and D (d) A and D

Answer Key

1.(d)	2.(d)	3.(a)	4.(a)	5.(d)	6.(b)	7.(a)
8.(b)	9.(d)	10.(d)	11.(b)	12.(b)		

21. Plasmodium falciparum and Leishmania

1. Which of the following factors is **NOT** true for the low levels of immune response in *Plasmodium* infection? **(DEC 2012)**
   (a) Different types of antigens are expressed at various stages of *Plasmodium* life cycle.
   (b) Most of the phases in the life cycle of *Plasmodium* are intracellular.
   (c) Sporozoites are rapidly cleared from blood circulation.
   (d) *Plasmodium* infection primarily destroys macrophages and dendritic cells.
2. Following sets of *Plasmodium falciparum* sporozoites **(JUNE 2014)**
   (i) normal sporozoites
   (ii) sporozoites with mutation in the C-terminal of circumsporozoite (CS) antigen
   (iii) sporozoites with mutation in the N-terminal region of circumsporozoite (CS) antigen

   are injected into 2 groups of mice one normal (Group A) and one (Group B) where localized knock down of heparin sulfate (receptor for CS antigen in liver cells) is achieved by injecting specific shRNA expressing lentiviral particles in the liver prior to infection. 15 days post infection parasitemia is measured by counting infected RBC through Giemsa staining.

   Which of the following groups will show maximum level of parasitemia?
   (a) Group B with set (i)
   (b) Group A with set (ii)
   (c) Group A with set (iii)
   (d) Group B with set (ii)
3. A novel vaccine (against malaria liver stage specific) has been developed by an investigator. This happened to be the parasites unique cell surface protein molecule (M_p). This M_p when injected in mice elicits humoral antibody response (I_gG in nature) and can efficiently neutralize sporozoites by inhibiting their binding to the liver cells. Upon a pre-clinical trial this vaccine failed universally despite a high titre I_gG response. Which one out of the choices below is the correct answer of this failure? **(MODELPAPER)**
   (a) The liver cell surface receptor to which M_p binds is changed.
   (b) The malaria parasite is successful in changing the epitope in Mp to which I_gG binds
   (c) I_gG molecules change its binding sites for M_p.
   (d) The affinity of I_g to M_p epitopes is reduced.
4. Upon ligand binding, cell surface receptors move laterally to be capped and internalized. *Leishmania* a protozoan parasite, can use several receptors on macrophages to get internalized.one of them is Toll-like receptor 2 (TLR2) that binds lipophosphoglycan on *Leishmania.*

   Once internalized, the parasite is destroyed in the phagolysosome. Which of the following treatments of *Leishmania*-infected macrophages will result in lowest parasite number in macrophages? JUNE 2012
   (a) membrane cholesterol-depleting drug, β-methyl cyclodextrin (β-MCD)
   (b) ammonium chloride that increases lysosomal pH
   (c) both β-MCD and ammonium chloride.
   (d) anti-TLR2 antibody.
5. Clearance of phagocytosed intracellular parasite like *Leishmania* requires the involvement of reactive oxygen species (ROS) and reactive nitrogen species (RNS). Administration of IFN-γ to macrophages harbouring an intracellular pathogen leads to the production of ROS and RNS by JAK/STAT pathway.

A macrophage cell line J774 infected with *Leishmania* is given the following treatments.

A. IFN- γ

B. IFN- γ +AMT, a potent iNOS inhibitor.

C. IFN- γ + apocyanin, a NADPH oxidase inhibitor.

D. IFN- γ + NMMA (N-monomethyl arginine), an arginine analogue.

What will be the most appropriate graph showing the survival of parasites after these treatments? **(DEC 2013)**

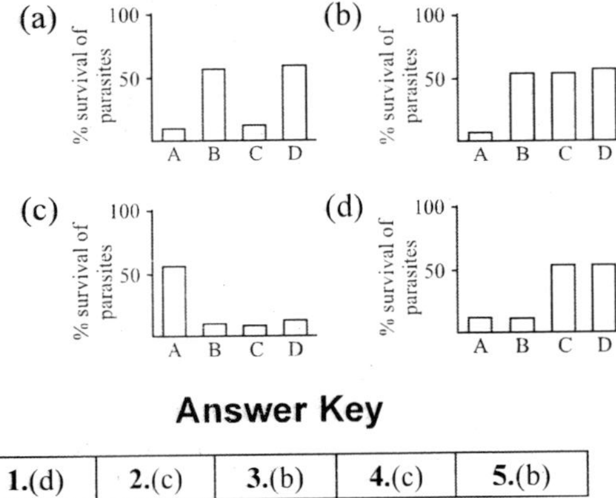

Answer Key

1.(d)	2.(c)	3.(b)	4.(c)	5.(b)

22. Viral Infections

1. Fruit bats are known to harbor and spread several viruses that can infect other animals and humans. Which one of the following viruses NOT reported to spread by fruit bats? **(JUNE 2019)**

   (a) Ebola (b) Nipah
   (c) SARS (d) HIV

2. Viruses adopt different strategies to suppress immune response of the host. Which one of the following statements is NOT correct? **(DEC 2018)**

   (a) Human Immunodeficiency Virus (HIV) destroys CD4$^+$T cells.

   (b) Epstein-Barr Virus (EBV) produces a homolog of human IL- 10.

   (c) Human influenza virus directly infects CD8$^+$ T cells.

   (d) Human Cytomegalo Virus (CMV) establishes latent· infection in bone marrow stem cells.

3. The relation between cellular immune response generated against hepatitis C virus is the critical determinant of the outcome of infection. Given below are the representative figures of cellular immune response in column I and various outcome of infection in column **(DEC 2015)**

Column I	Column II
(a) Response → Time →	(i) Acute
(b) Response → Time →	(ii) Resolution
(c) Response → Time →	(iii) Chronic

Choose the best possible combination

(a) A - (ii), B - (iii), C-(i)

(b) A-(i), B-(iii), C-(ii)

(c) A- (iii), B- (ii), C-(i)

(d) A-(i), B-(ii), C-(iii)

4. Certain chemokines are known to suppress HIV infection whereas proinflammatory cytokines are known to enhance infection. In order to explain these findings, control and chemokine receptor knock-out animals were treated with pro inflammatory cytokines followed by HIV administration and then infection was assessed periodically. Which one of the graphical

representation given below best explain the experimental results. **(JUNE 2015)**

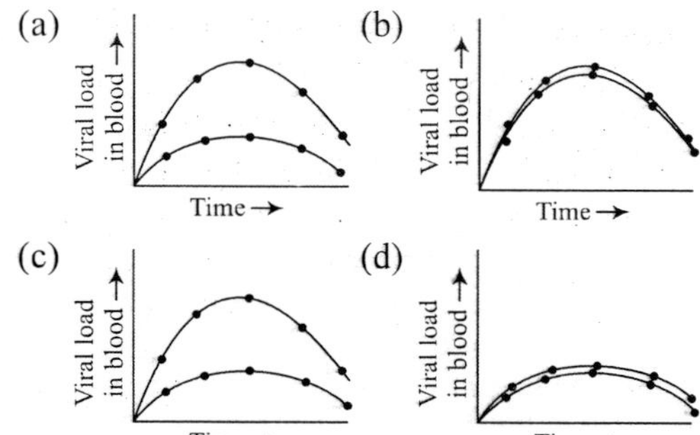

5. p24 is an important core protein of HIV. This protein is abundant during active replication of tile virus. The serum of an mv patient was examined for the presence of p24 and antibody against p24 for proper diagnosis of the infection stage. Match the clinical observations in column A with the inferences in column B.

Column A	Column B
A. p24 is present in the serum.	a. viral latency
B. Anti-p24 antibody is high in the serum.	b. progression of HIV from latency to lytic stage
C. Anti-p24 antibody begins to decline with corresponding increase in p24.	c. early stage of infection

Choose the correct matching **(DEC 2012)**

(a) A-a, B-b, C-c (b) A-b, B-a, C-c
(c) A-c, B-a, C-b (d) A-c, B-b, C-a

6. While attempting to create a disease model of poliomyelitis in mice, it was found that mice can not be infected with the said virus. Since human beings are susceptible to this viral infection, which kind of transgenic mice should be generated to have a transgenic mouse model that can be infected with polio virus? Select the right approach from below: **(JUNE 2013)**
   (a) A mouse expressing surface protein of polio virus.
   (b) A mouse expressing human receptor gene which makes cell surface protein for docking and internalization of polio virus.
   (c) A mouse expressing human MHC class II invariant chain.
   (d) A mouse expressing human receptor gene which makes cell surface protein for docking and internalization of polio virus along with a gene designed to express surface protein of this virus at puberty.

Answer Key

1.(d)	2.(c)	3.(b)	4.(c)	5.(c)	6.(b)

23. Vaccines

1. Which of the following food crops has recently been genetically engineered to obtain edible vaccine to develop immunity against hepatitis B? **(JUNE 2011)**
   (a) Banana (b) Maize
   (c) Potato (d) Tomato
2. An efficient vaccine candidate is **(DEC 2013 GU)**
   (a) a hapten-like small molecule.
   (b) an unconjugated peptide.
   (c) a polysaccharide.
   (d) a polypeptide
3. Among the following antigens specific to a pathogen, which one is most likely to be ineligible as a vaccine with long lasting host protective effect? **(JUNE 2014)**
   (a) A cell surface protein
   (b) An enzyme involved in pathogen metabolism.
   (c) A signaling intermediate, which is a kinase.
   (d) A long chain fatty acid.
4. The tetanus vaccine given to humans in the case of a deep cut is a **(DEC 2016)**
   (a) DNA vaccine
   (b) Recombinant vector vaccine
   (c) Subunit vaccine
   (d) Toxoid vaccine

5. DNA vaccines offer several advantages over other existing vaccine approaches. Which one of the following statements related to DNA vaccine is NOT correct? **(DEC 2018)**
   (a) The immune response is directed to the antigen encoded by the DNA and able to induce both humoral and cell-mediated immunity.
   (b) DNA vaccine can induce prolonged expression of the antigen, enhancing the induction of immunological memory
   (c) DNA vaccine could remain stable and potent for long time without refrigeration, eliminating the challenges of storage and transportation.
   (d) DNA vaccine construct can be engineered to carry several antigens to infect host and replicate in neuronal cells.

6. Susceptible individuals were infected with pathogen A and pathogen B separately. Pathogen A has a very short incubation period and disease symptoms are already underway by the time memory cells are activated. Pathogen B on the other hand has a long incubation period which allows the memory cells to be activated and respond. Which one of the following will be the most appropriate vaccination strategy against both pathogens A and B? **(DEC 2018)**
   1. Repeated vaccination against both A and B for maintaining high levels of neutrallizing antibodies.
   2. Repeated vaccination against A and a single injection of pathogen B vaccine for maintaining high levels of neutrallizing antibodies.
   3. Single injection of pathogen A vaccine and repeated vaccination against pathogen B for maintaining high levels of neutrallizing antibodies.
   4. Single injection of both pathogens A and B vaccine so that memory cells can respond by producing high levels of serum antibodies.

7. A researcher wanted to immunize individuals of a particular area with viral infections. The researcher developed two different vaccine types (A and B) with the following properties.
   (i) When vaccine type A specific for a viral strain is administered to individuals, they develop strong neutralizing antibody response with very poor immunological memory. Hence it has to be administered in repetitive doses.
   (ii) When vaccine type B specific for a viral strain is administered to individuals, they fail to develop circulating antibody response at the time of infection but they develop strong immunological memory.

   If two viral strains V1 (incubation period-2 days) and V2 (incubation period-15 days) are likely to infect the area, which of the following vaccine combination would provide maximum immunization? **(DEC 2014)**
   (a) V1 specific type A and V1 specific type B
   (b) V1 specific type A and V2 specific type B
   (c) V2specifictype A and V1 specific type B
   (d) V2 specific type A and V2 specific type B

8. In order to develop a vaccine against a regulatory T cell-promoting but Th 1 suppressing viral infection, four groups (A-D) of mice were primed with either killed virus (A) or a virus-derived immune dominant peptide (B), or the same peptide but with two substitutions (C) or were left unprimed (D). Upon infection challenge, the order of increasing severity of infection was observed to be B>A>D>C.

   To explain the contrasting effects of these two peptides (B) and (C), their MHC-binding affinities were assessed but no difference was found. Which of the following possibilities most likely to explain their contrasting effects? **(JUNE 2014)**
   (a) The wild-type peptide (B) elicits T cell expansion but the mutant peptide (C) fails.
   (b) The wild-type peptide induces T cell deletion but the mutant peptide does not

(c) The wild-type peptide-MHC complex binds T cell receptor with significantly higher affinity than the mutant peptide-MHC complex.

(d) The wild-type peptide induces deletion of T-Reg cells but increases IFN γ production by T cells, whereas the mutant peptide fails to induce these effects.

9. In order to prevent tetanus in neonates, one the following treatments can be adopted.

A. Treatment of the infant with anti-toxin and the toxoid

B. immunize the mother with the toxoid

In case of A, the treatment can be given

a. immediately after birth

b. after the onset of the condition

In case of B, the immunization has to be done

c. before pregnancy

d. late in the pregnancy **(DEC 2011)**

(a) A / a (b) A / b

(c) B / c (d) B / d

10. Given below are the types of vaccination (column A), the diseases or conditions against which these vaccination types are used (column 8) and the advantages or disadvantages for using these vaccination types (column C).

	A		B		C
a	Live attenuated	i	Rabies	x	Strong immune response; often life- long immunity with a few dose
b	Inactivated or killed	ii	Measles	y	Stable immune response; cold chain is not required
c	Inactivated exotoxin	iii	Diptheria	z	Chance of untoward immunological reactions are very low; less successful in inducing long-term immunity; need to be administered repeatedly

Which one of the following combinations is the most appropriate match? **(JUNE 2018)**

(a) a-i-z, b-ii-y, c-iii-x

(b) a-ii-x, b-i-y, c-iii-z

(c) a-iii-y, b-ii-x, c-i-z

(d) a-ii-z, b-iii-x, c-i-y

Answer Key

1.(c)	**2.**(d)	**3.**(d)	**4.**(d)	**5.**(d)	**6.**(b)	**7.**(b)
8.(d)	**9.**(d)	**10.**(b)				

SECTION-5

DEVELOPMENTAL BIOLOGY

1. Determination, Fate, Morphogen

1. The ability of cells to achieve their respective fates by interacting with other cells is known as **(JUNE 2017)**
   (a) autonomous specification
   (b) conditional specification
   (c) induction
   (d) competence
2. If you remove a set of cells from an early embryo, you observe that the adult organism lacks the structure that would have been produced from those cells. Therefore, the organism seems to have undergone **JUNE 2015**
   (a) autonomous specification.
   (b) conditional specification.
   (c) morphogenic specification.
   (d) syncytial specification.
3. With respect to development of any organism, "autonomous specification" would result in which type of development? **DEC 2012**
   (a) Regulative. (b) Syncytial.
   (c) Mosaic. (d) Definitive.
4. With respect to development of any organism, "conditional specification" would result in which type of development? DEC 2013 GU
   (a) Regulative (b) Mosaic
   (c) Syncytial (d) Definitive
5. Following statements are being made regarding specification determination during animal development:
   A. During the course of commitment, the cell may not appear different from its nearest or most distant neighbours in the embryo and show no visible signs of differentiation; but its developmental fate is restricted.
   B. At the stage of specification, cell commitment is not labile.
   C. A cell or tissue is determined when it is capable of differentiating autonomously even when placed into another region of the embryo, or a cluster of differently specified cells in a petri- dish
   D. Cytoskeletal arrangements maintain positioning of nuclei in the syncytium, which enables specification of these nuclei by opposing morphogen gradients namely Bicoid and Caudal in Drosophila
   E. Capacity for "mosaic" development allows cells to acquire different functions as a result of interactions with neighbouring cells.

   Which of the above statements are correct? **(DEC 2019)**
   (a) A B and C (b) B, C and D
   (c) C, D and E (d) A, C and D
6. Certain proteins or mRNAs that are regionally localized within the unfertilized egg and regulate development are called **(DEC 2016)**
   (a) gene regulators.
   (b) morphometric determinants .
   (c) cytoplasmic determinants.
   (d) mosaic forming factors.
7. Cell to cell communication is important in development of an organism. The ability of cells to respond to a specific inductive signal is called **(DEC 2016)**
   (a) Regional specificity of induction
   (b) Competence
   (c) Juxtracrine signalling
   (d) Instructive interaction
8. In chick, development of wing feather, thigh feather and claws depends on epithelial specificity conferred by induction from mesenchymal components from different sources of the dermins. This may be attributed to? **(DEC 2015)**
   (a) Autocrine interaction
   (b) Regional specificity of induction
   (c) Receptor activation by hormones
   (d) Inactivation of genetic interactions

9. During development, if a cell has committed to a particular fate, it is said to be

(JUNE 2016)

(a) pluripotent (b) totipotent
(c) determined (d) differentiated

10. Given are certain facts which define 'determination' of a developing embryo.

A. Cells have made a commitment to a differentiation program.
B. A phase where specific biochemical actions occur in embryonic cells.
C. The cell cannot respond to differentiation signals.
D. A phase where inductive signals trigger cell differentiation.

Which of the above statements best define determination? **(JUNE 2015)**

(a) B and D (b) A and C
(c) Only A (d) Only B

11. The fate of a cell or a tissue is specified when it is capable of differentiating autonomously on being placed in a neutral environment with respect to the developmental pathway. An embryo will show development pattern based on its type of specification: *Based on the above facts it can be said that potency of a cell is:*

(JUNE 2012)

A. Equal to its normal fate in regulative development
B. greater than its normal fate in regulative development
C. equal to its normal fate in mosaic development
D. greater than its normal fate in mosaic development

Which of the above statements are true?

(a) (B) and (C) (b) (A) and (D)
(c) (A) and (C) (d) (B) and (D)

12. During early cleavage of *Caenorabditis elegans* embryos, each asymmetrical division produces one founder cell which produces differentiated descendants and one stem cell. The very first cell division produces one anterior founder cell, namely AB and one posterior stem cell, namely P1. When these blastomeres are experimentally separated and allowed to proceed further with development, one could get the following possible outcomes:

(JUNE 2011)

(a) P1 cell would develop autonomously while the AB would show conditional development.
(b) P1 cells would show conditional development while AB would show autonomous development.
(c) Both would show autonomous specification and result in mosaic development.
(d) Both would show conditional specification and result in regulative development.

13. At the 2-celled stage of *Caenorabditis elegans* development the blastomeres were experimentally separated and allowed to proceed in development. One of the blastomere (P1) developed generating all types of cells it would normally make while the other blastomere (AB) made only a fraction of the cell types it would normally make.

The following conclusion could be drawn:

A. The determination of both P1 and AB was autonomous.
B. The determination of both P1 and AB was conditional.
C. The determination of P1 was autonomous and AB was conditional .
D. Both asymmetric division and cell-cell interactions specify cell fate in early development.

Which of the above combinations is correct?

JUNE 2014

(a) A and C (b) B and D
(c) A and D (d) C and D

14. The zygote of *C. elegans* exhibits rotational cleavage. When the first two blastomeres formed (P1 and AB) are experimentally separated, the following outcomes may be possible:

A. The P1 cell in isolation generates all the cells it would normally make, showing autonomous specification.

B. The P1 cell in isolation generates all the cells it would normally make, showing conditional specification.

C. The AB cell in isolation generates a small fraction of cell types it would normally make, showing conditional specification.

D. The AB cell in isolation generates a small fraction of cell types it would normally make, showing autonomous specification.

Which one of the above combination of statements is true? **(DEC 2018)**

(a) A and C (b) BandC

(c) B and D (d) A and D

15. A two-celled embryo is made of blastomeres A and B. If the two blastomeres are experimentally separated, the 'A' blastomere generates all the cells it would normally make. However, the 'B' blastomere in isolation makes only a small fraction of cells it would normally make.

Based on the above observations only, which one of the following conclusions is correct? **(DEC 2014)**

(a) 'A' blastomeres is autonomously specified while 'B' blastomere is conditionally specified

(b) 'A' blastomeres is conditionally specified while 'B' blastomere is autonomously specified

(c) Descendants of 'A' blastomeres are autonomously specified

(d) Descendants of 'B' blastomeres can either be autonomously specified or conditionally specified.

16. *C. elegans* embryo uses both autonomous and conditional modes of specification. Conditional specification at the 4-cell stage can be seen in the development of the endoderm cell lineage and also in the establishment of dorsal-ventral axis. Following are few statements regarding this:

A. If the P_2 cell is removed at the early 4-cell stage, the EMS cell will divide into two MS cells and no endoderm will be made.

B. In *pop-1* deficient embryos, both EMS daughter cells become E cells.

C. When the position of ABa and ABp was reversed, their fates get reversed and no normal embryo forms.

D. In embryos whose mother have mutant *glp-1*, ABp is transformed into ABa cell.

Which of the above statements are true? **(DEC 2017)**

(a) A, B and D (b) A, B and C

(c) B, C and D (d) A, C and D

17. When 8-cell embryo of tunicates is separated into 4 blastomere pairs and allowed to grow independently in culture medium, then each blastomere pair can form most the cell types; however, cells for nervous system are not developed. The following statements are formed from the above observations:

A. Nervous system development demonstrated autonomous specification.

B. The other tissue types are formed due to conditional specification.

C. All the tissue types, except nervous tissues that developed demonstrated autonomous specification.

D. Nervous system development demonstrated conditional specification.

The correct combination of statements that explains the above result is: **(JUNE 2019)**

(a) A and B (b) B and C

(c) C and D (d) A and D

18. When the prospective neurons from an early gastrula of a frog were transplanted into the prospective epidermis region, the donor cells differentiated into epidermis. However, when a similar experiment was done with the late gastrula of frog, the prospective neurons developed into neurons only. These observations could possibly be explained by the following phenomena.

A. The early gastrula show conditional development whereas the late gastrula shows autonomous development.

B. The early gastrula show autonomous development whereas the late gastrula shows conditional development.

C. The prospective neurons from the early gastrula are only specified whereas those from the late gastrula are determined.

D. The prospective neurons from the early gastrula are determined whereas those from the late gastrula are specified.

Which of the conclusions drawn above are correct? **(DEC 2011)**

(a) A and B (b) A and C

(c) A and D (d) B and C

19. Which of the inferences (A-D) given below would you draw from the following tissue transplantation experiments performed with the early and late gastrula stages of the newt?

	Host regions	**Donor regions**	**Differentiation of donor tissue**
	EARLY GASTRULA		
(i)	Prospective neurons	Prospective epidermis	Epidermis
(ii)	Prospective epidermis	Prospective neurons	Neuron
	LATE GASTRULA		
(i)	Prospective neurons	Prospective epidermis	Neuron
(ii)	Prospective epidermis	Prospective neurons	Epidermis

A. Cells of early newt gastrula exhibit conditional development.

B. Cells of early newt gastrula exhibit autonomous development

C. Cells of late newt gastrula exhibit conditional development.

D. Cells of late gastrula exhibit autonomous development.

The correct inferences are: JUNE 2013

(a) (A) and (D) (b) (B) and (C)

(c) (A) only (d) (D) only

20. When prospective neuroectoderm from an early amphibian gastrula is transplanted in the prospective epidermal region of a recipient (early gastrula) embryo, the donor tissue will give rise to **(MODELPAPER)**

(a) neural tube.

(b) epidermis.

(c) neural tube and notochord.

(d) neural tube and epidermis.

21. Which of the following cannot be used for determination of tissue lineage of a given progenitor cell population in an animal? **(MODEL PAPER)**

(a) Marking progenitor cells with vital dye.

(b) Transplanting equivalent progenitor cells from immunologically distinct but related organism.

(c) Marking progenitor cells by genomic recombination coupled with reporter gene expression.

(d) Marking progenitor cells by reporter gene expression under the control of a promoter – enhancer element specific for the given progenitor population.

22. In an experiment, the cells that would normally become the middle segment of a Drosophila leg were removed from the leg forming area of the larva and were placed in the tip of the fly's antenna. Based on the "French flag" analogy for the operation of a gradient of positional information, which of the following statements is true? **(JUNE 2013)**

(a) The transplanted cells retain their committed status as leg cells, but respond to the positional information of their environment by becoming leg tip cells-i.e., claws.

(b) The transplanted cells are determined as leg cells and therefore would form a complete limb instead of an antenna.

(c) The transplanted cells would intermingle with the cells present in the new environment and develop accordingly to give rise to an antenna.

(d) The transplanted cells retain their committed status as leg cells and would develop to form a chimeric structure having proximal region made of antenna and the distal region ending in a complete leg.

23. Following are certain statements regarding morphogen gradients and cell specification.

A. Morphogens are always transcription factors.

B. Morphogens can be paracrine factors that are produced in one group of cells and travel to another population of cells

C. When the concentration of a morphogen drops below a certain threshold cells stop differentiating and never get determined to another fate.

D. Morphogen gradients are involved in conditional specification.

Which combination of the above statements is true? **(DEC 2015)**

(a) A and B (b) B and D

(c) C and D (d) A and C

24. Instructive and permissive interactions are two major modes of inductive interaction during development. The following compares some properties of cell lines and cord blood stem cells. Cell lines which are stored in liquid nitrogen, can be retrieved for experiments, where they behave as per their original self. Cord blood can also be retrieved from liquid nitrogen for procuring stem cells. Unlike cell lines, the stem cells can be additionally induced to undergo differentiation into desired lineage, which are very different from their original self.

The behaviour of cell lines and stem cells is analogous to which of the interactions? **(DEC 2015)**

(a) Both cell lines and stem cells show instructive interaction

(b) Cell lines show instructive interaction whereas stem cells show permissive interaction

(c) Cell lines show permissive interaction whereas stem cells show instructive interaction

(d) Both types of cells show permissive instruction

Answer Key

1.(b)	**2.**(a)	**3.**(c)	**4.**(a)	**5.**(d)	**6.**(c)	**7.**(b)
8.(b)	**9.**(c)	**10.**(c)	**11.**(a)	**12.**(a)	**13.**(d)	**14.**(a)
15.(a)	**16.**(a)	**17.**(c)	**18.**(b)	**19.**(b)	**20.**(b)	**21.**(d)
22.(a)	**23.**(b)	**24.**(c)				

2. Gametogenesis and fertilization

1. In case of sea urchin, which of the following is the correct sequence of events taking place during the interaction of sperm and egg? **(JUNE 2011)**

(a) Chemoattraction of sperm to the egg by soluble molecules secreted by the egg→ exocytosis of the sperm acrosomal vesicle to release its enzymes →binding of the sperm to the extracellular matrix of the egg →passage of sperm through this extracellular matrix →fusion of egg and sperm cell membranes.

(b) Chemoattraction of sperm to the egg by soluble molecules secreted by the egg →binding of the sperm to the extracellular matrix of the egg → exocytosis of the sperm acrosomal vesicle to release its enzymes →passage of sperm through this extracellular matrix →fusion of egg and sperm cell membranes.

(c) Chemoattraction of sperm to the egg by soluble molecules secreted by the egg →binding of the sperm to the extracellular matrix of the egg →passage of sperm through this extracellular matrix →exocytosis of the sperm acrosomal vesicle to release its enzymes →fusion of egg and sperm cell membranes.

(d) Chemoattraction of sperm to the egg by soluble molecules secreted by the egg →passage of sperm through this extracellular matrix→binding of the sperm to the extracellular matrix of the egg→ exocytosis of the sperm acrosomal vesicle to release its enzymes →fusion of egg and sperm cell membranes.

2. Successful fertilization in sea urchin demands specific interaction between proteins and receptors of sperms and eggs. In view of the above, which one of the following combinations is correct? **(DEC 2015)**

(a) Bindin in acrosomes and bindin receptors on egg vitelline membrane

(b) Bindin in egg membrane and bindin receptors in acrosomes

(c) Resact on egg jelly and bindin on sperm membrane

(d) Proteasomes on egg membranes and complex sugars on sperm membranes

3. Consider the following events which occur during fertilization of sea urchin eggs.

A. Resact/Speract are peptides released from the egg jelly and help in sperm attraction.

B. Bindin, an acrosomal protein interacts in a species specific manner, with eggs.

C. A "respiratory burst" occurs during cross-linking of the fertilization envelope, where a calcium- dependent increase in oxygen level is observed.

D. IP_3, which is formed at site of sperm entry, releases sequestered calcium leading to cortical granule exocytosis.

Which of the above statement(s) is NOT true? **(DEC 2016)**

(a) Only C (b) A and C

(c) Only D (d) B and D

4. For successful fertilization in sea urchin, interaction between the surface of the egg and acrosomal proteins, specifically a 30.5 kDa protein called bindin, is necessary. The following factors could affect this interaction and prevent fertilization:

A. Removal of egg jelly polysaccharides.

B. Removal of bindin receptors on the egg vitelline membrane.

C. Removal of bindin receptors from the egg jelly.

D. Removal of bindin receptors from a single cluster on the vitelline membrane.

Which one or the combination of the above statements is correct? **(DEC 2014)**

(a) A and D (b) Only B.

(c) A and B (d) Only C

5. Polyspermy results when two or more sperms fertilize an egg. It is usually lethal since it results in blastomeres with different numbers and types of chromosomes. Many species therefore, have two blocks to polyspermy: the fast block and the slow block.

In the case of sea urchins:

A. the fast block is immediate and causes the egg membrane resting potential' to rise which does not allow the sperm to fuse with the egg and is mediated by an influx of sodium ions.

B. the fast block is immediate and causes the egg membrane resting potential to rise which does not allow the sperm to fuse with the egg and is mediated by an efflux of sodium ions.

C. the slow block or cortical granule reaction is mediated by calcium ions

D. the slow block or cortical granule reaction is mediated by potassium ions.

Which of the above statements are true?

(JUNE 2013)

(a) (A) and (C) (b) (A) and (D)

(c) (B) and (C) (d) (B) and (D)

6. Fertilization in sea urchin involves interaction of sperm Bindin with its receptor EBR1, a 350 kDa glycoprotein on the egg vitelline membrane. The plot given below shows the status of membrane potential and levels of EBR1, Na^+ and K^+ in an unfertilized egg.

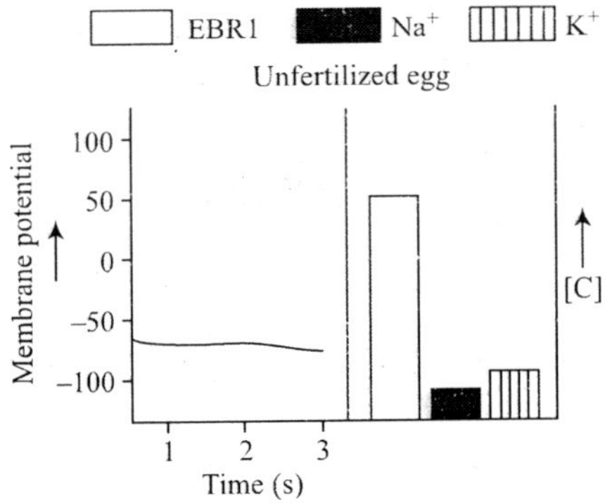

Which one of the following graphs best represents the condition within an egg 1-3 seconds after fertilization? **(DEC 2017)**

(a)

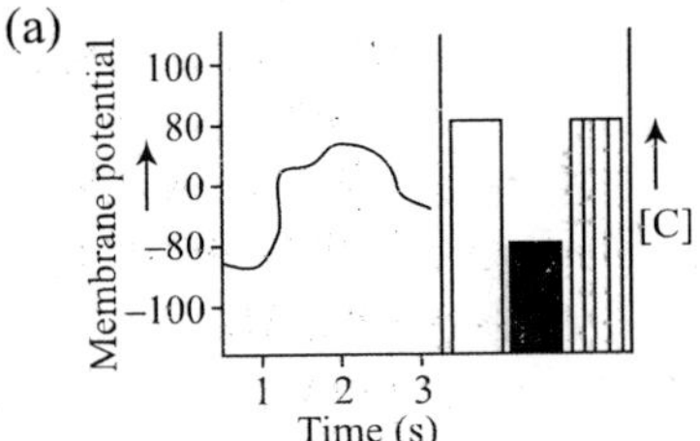

(b)

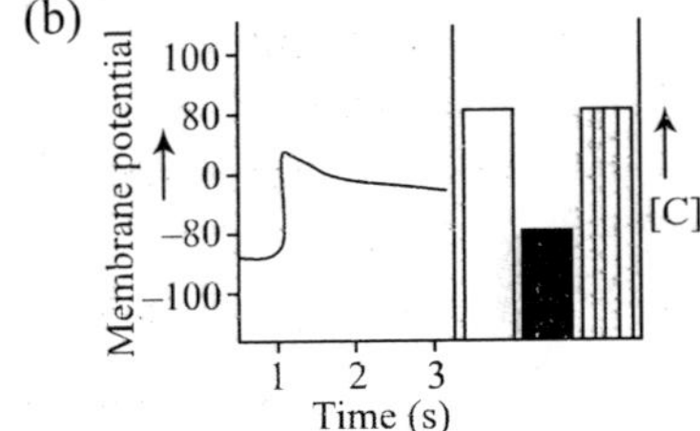

(c)

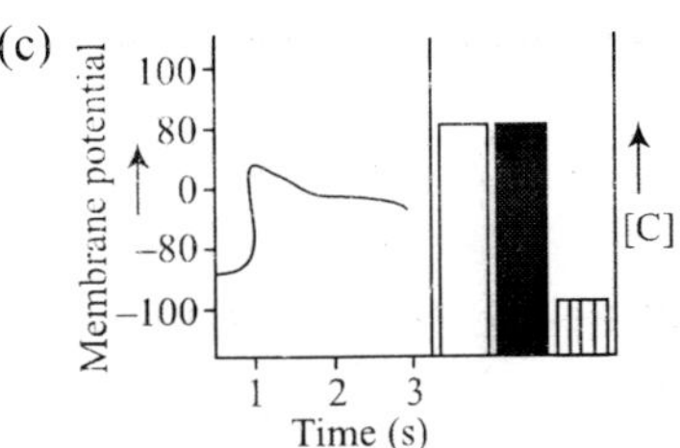

(d)

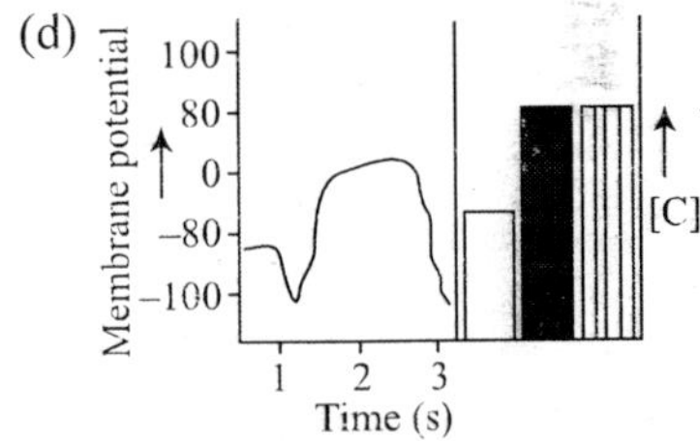

7. During fertilization in amphibians, the fusion of egg and sperm plasma is preceeded by

A. release of enzymatic contents from the acrosomal vesicle through exocytosis

B. binding and interaction of the sperm to vitelline membrane

C. chemoattraction of the sperm to the egg by soluble factors secreted by egg

D. passage of sperm through extracellular envelope

Which of the following is the correct sequence?

(MODEL PAPER)

(a) (A) → (B) → (C) → (D)

(b) (B) → (A) → (C) → (D)

(c) (C) → (A) → (B) → (D)

(d) (C) → (B) → (A) → (D)

8. Amphibian oocytes remain for years in the diplotene stage of meiotic prophase. Resumption of meiosis is initiated by

(MODEL PAPER)

(a) gonodatropic hormone.

(b) growth hormone.

(c) oestrogen.

(d) progesterone.

9. Capacitation of sperms in humans **(DEC 2012)**

(a) occurs during copulation.

(b) occurs after the acrosome reaction.

(c) takes place in the ampulla of the oviduct.

(d) takes place in the epididymis of testis.

10. In an experiments, sperm removed from epididymis of a male mouse was added in a dish containing appropriate media and oocyte. No fertilization was seen. However, when sperm from epididymis were directly placed in uterus of an ovulated female, she became pregnant. These observation suggest that **(JUNE 2012)**

(a) The sperm need to travel some distance to attain fertilizing ability.

(b) The oocyte secrets some biochemicals or factors which help sperm to fertilize.

(c) The hormones in body help sperm to attain fertilizing ability.

(d) The contents of female reproductive tract interact with sperm and activate it for fertilization

11. Capacitation of mammalian sperms allows them to be activated within the uterus and facilitate fertilization. The following statements were made regarding events occurring during capacitation:

A. removal of cholesterol from sperm head.

B. Removal of non-covalently bound glycoproteins.

C. Increased expression of fibronectin.

D. Decreased permeability of calcium ions.

Identify the correct statements: **(DEC 2013)**

(a) B, C and D (b) A, B and D

(c) A, B and C (d) A, C and D

12. During fertilization "in mammals, sperm-egg interaction is mediated by zona pellucida (ZP) membrane proteins and their receptors present in sperm membrane. ZP3 has been identified to be the principle ZP protein whose post-translational modification is important for sperm - egg interaction. In a competitive inhibition assay the sperm is saturated with either active ZP3 or its modified forms, before studying sperm-egg-interaction. Which of the following experiments will NOT inhibit sperm-egg-interaction **(JUNE 2013)**

(a) Saturate sperm with ZP3 protein prior to use.

(b) Deglycosylate the ZP3 protein and use it for saturation of sperm.

(c) Phosphorylate the ZP3 protein and use it for saturation of sperm.

(d) Dephosphorylate the ZP3 protein and use it for saturation of sperm.

13. Membrane-bound, Golgi-derived structures containing proteolytic enzymes in sperms of sea urchin are called **(DEC 2018)**

(a) cortical granules

(b) micromeres

(c) acrosomal vesicles

(d) macromeres

14. Human sperms are allowed to fertilize ova having non-functional ovastacin. The following possibilities may be of significance in the fusion of these gametes:

A. The sperms will not fertilize ova.

B. The sperms will bind and penetrate the zonapellucida but will not be able to fuse with ovum membrane.

C. ZP2 will not be clipped by cortical granule protease.

D. CD9 protein of egg membrane microvilli will not be able to interact with sperm membrane proteins in the absence of ovastacin.

E. Polyspermy may occur frequently.

Which combination of statements represent the outcome of the above event? **(DEC 2018)**

(a) A and B (b) C and E

(c) C and D (d) B and C

15. During fertilization in mammals proteins *Izumo* and *Juno* are required for recognition of sperm and egg. *Izumo* and *Juno* are found specifically in sperm and egg, respectively. Which one of the following in vivo experiments will demonstrate that *Izumo* and Juno interact with each other? **(DEC 2019)**

(a) If sperms from a male mouse where *Izumo* has been knocked out is used to fertilize eggs from a normal female and no fertilization occurs.

(b) Whole mount immunostaining for *Izumo* and Juno shows its presence on the sperm and egg respectively

(c) If a CFP fused *Izumo* protein is mixed with YFP fused Juno protein in a tube, FRET occurs, i.e., when CFP is excited, emission of YFP is observed.

(d) Two independent kidney cell lines are developed, one expressing *Izumo* and the other *Juno*. If the two cells are mixed, they tend to aggregate with each other.

16. Fertilization in sea urchin eggs involves Ca^{2+} release from the endoplasmic reticulum for cortical granule reactivation. The major molecule responsible for releasing Ca^{2+} from intracellular stores is **(JUNE 2018)**

(a) zonapellucida glycoproteins

(b) protamines

(c) inositol 1, 4, 5-trisphosphate

(d) N-acetylglucosaminidase

Answer Key

1.(a)	2.(a)	3.(a)	4.(c)	5.(a)	6.(c)	7.(c)
8.(d)	9.(c)	10.(d)	11.(c)	12.(b)	13.(c)	14.(b)
15.(d)	16.(c)					

3. Zygote formation, cleavage, blastula formation

1. If an embryo undergoes 13 cleavage divisions during embryogenesis, then the size of the embryo compared to zygote **(DEC 2012)**

(a) increases 13 times.

(b) increases only 6-7 times.

(c) increases in an exponential fashion.

(d) Remains almost the same.

2. Which kind of cleavage is shown in mammals? **(JUNE 2017)**

(a) Holoblastic rotational

(b) Meroblastic rotational

(c) Holobastic radial

(d) Meroblastic radial

3. Amphibian zygote will generally undergo **(DEC 2013 GU)**

(a) Holoblastic radial cleavage.

(b) Holoblastic rotational cleavage.

(c) meroblastic bilateral cleavage.

(d) meroblastic rotational cleavage.

4. Centrolecithal eggs show **(DEC 2019)**

(a) superficial cleavage

(b) displaced radial cleavage

(c) bilateral cleavage

(d) discoidal cleavage

5. The pattern of embryonic cleavage specific to a species is determined by two major parameters.

(A) The amount and distribution of yolk protein within the cytoplasm.

(B) The factors in the cytoplasm that influence the angle mitotic spindles and the timings of its formation.

Which of the following statements are true? **(JUNE 2012)**

(a) Species having teloecithal egg follow a holoblastic cleavage.

(b) Species having isolecithal egg follow a holoblastic cleavage.

(c) Species having centrolecithal egg follow a holoblastic cleavage.

(d) Species having isolecithal egg follow a meroblastic cleavage.

6. Match the following cleavage patterns with the species in which they occur. **(DEC 2017)**

	Species		Cleavage Pattern
A.	Flatworm	i.	Meroblastic discoidal
B.	Frog	ii.	Meroblastic superficial
C.	Birds	iii.	Holoblastic displaced radial
D.	Insect	iv	Holoblastic spiral

(a) A-iv, B-iii, C-i, D-ii
(b) A-iii, B-i, C-iv, D-ii
(c) A-ii, B-iii, C-i, D-iv
(d) A-ii, B-iv, C-iii, D-i

7. Given below are fate map of two organisms and the pattern by which embryos undergo cleavage. Which of the following is/are the right combination(s)? **(JUNE 2012)**

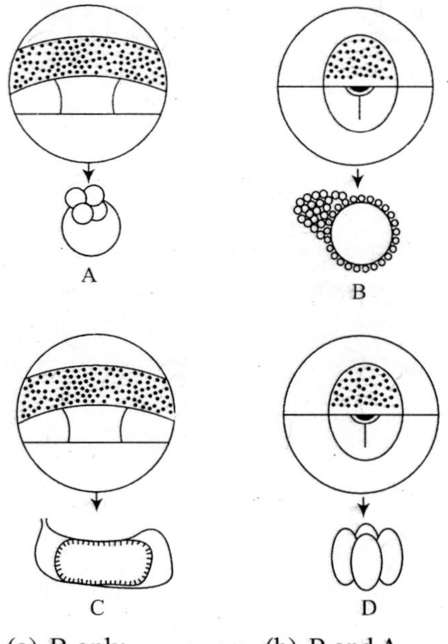

(a) B only (b) B and A
(c) A and C (d) B and D

8. Gradients of morphogens determine the future dorsal-ventral and anterior-posterior axes of the developing embryo in many organisms. How is the anterior-posterior axis developed in *C. elegans*? **(DEC 13 GU)**

(a) Sperm entry leads to reorgaization of the cytoskeleton and redistribution of naturally packed PAR proteins, which in turn determine the anterior-posterior axis.
(b) After fertilization β-catenin gets localized into the nucleus of the future anterior cells.
(c) The P-granules are localized in a way consistent with a role as a morphogenetic determinant and they act through translational regulation to initiate anterior-posterior axis.
(d) The localization of the maternally expressed polypeptide SKN-1 is responsible for the anterior-posterior axis.

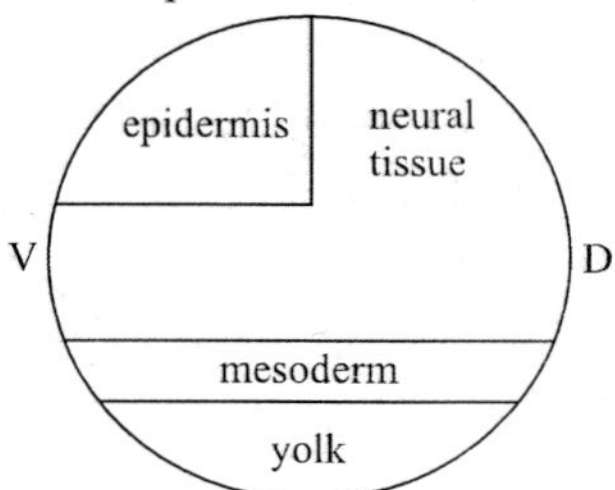

9. The figure above represents a late zebrafish gastrula. The following concepts may be proposed during further development of the embryo.

A. The concentration of FGF decreases from the yolk towards the epidermis, along with the increase of BMP activity from the dorsal to the ventral axis.
B. Increase in FGF activity in the epidermis with concomitant decrease in BMP activity towards the ventral axis.
C. Neural induction in zebrafish is independent of the organizer and depends on activation of BMP signalling.
D. In comparison, both Xenopus and chick embryos require activation of FGF for neural induction to occur in addition to BMP inhibition.

Which of the above statements are true? **(DEC 2011)**

(a) A and C (b) B and C
(c) A and D (d) C and D

10. Compaction during early embryonic development involves activation of actin filaments. Which one of the following inhibitors would prevent formation of the blastula from the morula? **(DEC 2019 ASSAM)**
   (a) An inhibitor that would block the action of Glin.
   (b) A specific inhibitor for stabilizing the function of Axin.
   (c) A specific inhibitor for GSK3B
   (d) An inhibitor which would block Smad4.

11. Human chorionic gonadotropin (hCG) is known to facilitate attachment of blastocyst to uterus. In women with mutation in hCG gene, biologically inactive hCG was formed but implantation occurred. When hcG was immune-neutralized in the uterus of normal woman, implantation failed. This suggests that for implantation in humans: **(JUNE 2013)**
   (a) biologically active circulating hCG is not required.
   (b) blastocyst can produce the required hCG, which helps locally in uterine attachment.
   (c) trophoblastic cells do not require hCG for the invasion of uterus.
   (d) extra-embryonic tissue is not responsible for the attachment of embryo to uterus

Answer Key

1.(d)	2.(a)	3.(a)	4.(a)	5.(b)	6.(a)	7.(a)
8.(a)	9.(a)	10.(c)	11.(b)			

4. Embryonic Stem cell and Potency

1. The cells of inner cell mass of a blastocyst stage mammalian embryo are **(DEC 2019)**
   (a) totipotent (b) pluripotent
   (c) multipotent (d) unipotent

2. Following statement were given regarding decisions taken during development of mammalian embryos
   A. Pluripotency of inner cell mass is maintain by a core of three transcription factors, Oct 4, Sox 2 and nanog.
   B. Prior to blastocyst formation each blastomere expresses both Cdx 2 and the Oct 4 transcription factors and appears to be capable of becoming either ICM or trophoblast .
   C. Both ICM and trophoblast cells synthesize transcription factors Cdx 2.
   D. Oct4 activates Cdx2 expression enabling some cells to become trophoblast and other cells to become ICM.

   Which of the above statement are true? **(DEC 2016)**
   (a) A and B (b) A and C
   (c) B and D (d) B and C

3. The decision to become either a trophoblast or inner cell mass blastomere is one of the first decisions taken by any mammalian embryo. Below is a diagrammatic representation of the different cells formed during development from the morula with the help of different molecules. Identify the molecules 1-4, sequentially. **(DEC 2011)**

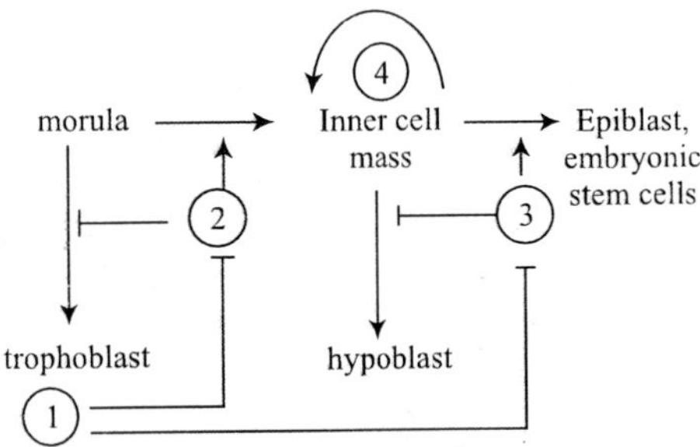

   (a) cdx 2, Oct 4, Nanog, Stat 3
   (b) cdx 2, Nanog, Stat 3, Oct 4
   (c) cdx 2, Nanog, Oct 4, Stat 3
   (d) cdx 2, Oct 4, Stat 3, Nanog

4. The following are certain statements regarding stem cells:
   A. All types of stem cells have the ability to give rise to a complete embryo.

B. Multipotent stem cells are those whose commitment is limited to a relatively small subset of all possible cell types.

C. Stem cell niches allow controlled self-renewal and also survival of the cells that leave the niche.

D. The pluripotency of the stem cells in an embryo is essentially maintained by Fgf8, Nanog and TGFβ.

E. Adult cells may be reprogrammed to gain pluripotency by modifying the following genes: Oct 3/4, Sox2, c-myc, Klf-4.

Which one of the following combinations of statements is correct? **(JUNE 2017)**

(a) A , B and D (b) B and E

(c) C and E (d) A , C and D

5. The following demonstrates proposed functions of different genes which determine the decision to become either trophoblast or inner cell mass (ICM) blastomere during early mammalian development:

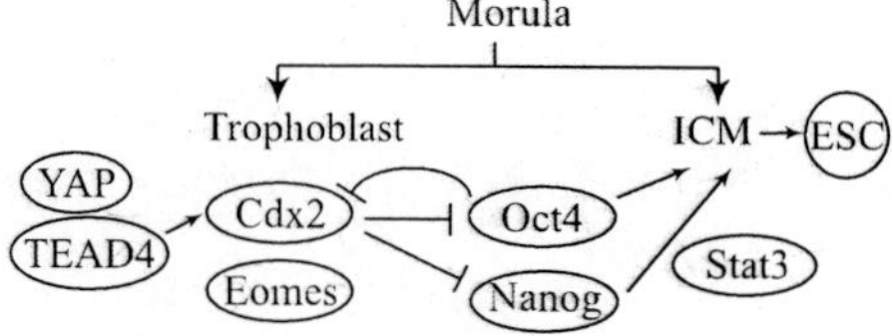

Based on the above figure, which one of the following assumptions is correct?

(JUNE 2019)

(a) The interplay between Cdx2 and Oct4 can influence the formation of ICM

(b) The ICM would form even if expression ofOct4 was inhibited.

(c) YAP and TEAD4 are upstream components of Cdx2 and can be inhibited by Nanog.

(d) The expression of Stat3 is optional for maintaining pluripotency of the ICM.

6. The pluripotency of the inner cell mass in mammals is maintained by a core of. three transcription factors namely, **(JUNE 2015)**

(a) Oct 4, Sox 2 and Nanog

(b) Oct 4, Sox 2 and Cdx2

(c) Sox 2, Nanog and Cdx2

(d) Oct 4, Cdx2 and Nanog

7. Hippo pathway is responsible for the trophoblast and inner cell mass (ICM) differentiation during mammalian development. Which one of the following is true for ICM formation?

(DEC 2019 ASSAM)

(a) The Tead4 transcription factor, when active, promotes transcription of Cdx2 geneleading to ICM formation.

(b) If LATS kinase phosphorylates the YAP transcriptional coactivator, thephosphorylated form of YAP does not enter the nucleus and gets degraded which leads to ICM formation.

(c) In the absence of functional LATS protein, the YAP transcriptional cofactor can bindto Tead4 to activate Cdx2 gene promoting ICM

(d) Synthesis of Cdx2 upregulates Oct4 and Nanog leading to ICM formation.

Answer key

1.(b)	2.(a)	3.(a)	4.(a)	5.(a)	6.(a)	7.(b)

5. Gastrulation and formation of germ layers

1. Alveolar cells of the lung arise from which one of the following layer(s)? **(DEC 2015)**

(a) Mesoderm (b) Endoderm

(c) Ectoderm

(d) Both ectoderm and endoderm

2. Bones of vertebrates embryonic are derived from **(JUNE 2016)**

(a) ectoderm (b) epiderm

(c) mesoderm (d) endoderm

3. Inward movement of an expanding outer layer spreading over the internal surface during gastrulation is termed as **(JUNE 2017)**

(a) invagination (b) Ingression

(c) involution (d) delamination

4. The splitting or migration or one sheet of cells into two sheets as seen during hypoblast formation in bird embryogenesis is termed as **(DEC 2014)**

(a) delamination (b) ingression
(c) involution (d) invagination

5. Migration of individual cells from the surface into the embryo's interior is termed as **(DEC 2015)**

(a) ingression (b) involution
(c) invagination (d) delamination

Column A	Column B	Column C
(i) Invagination	(i) Movement of epithelial cells as a unit to enclose deeper layers of the embryo	(i) Hypoblast in birds
(ii) Involution	(ii) Splitting of one cellular sheet into two parallel sheets	(ii) Ectoderm in amphibians
iii) Ingression	(iii) Infolding of epithelium	(iii) Mesoderm in amphibians
iv) Delamination	(iv) Migration of individual cells from surface into interior of the embryo	(iv) Endoderm in sea urchin
(v) Epiboly	(v) Inward movement of expanding outer layer so that it spreads over the internal surface of remaining external cells	(v) Mesoderm in sea urchin

6. Which one of the following is the correct combination? **(JUNE 2015)**

(a) A(i), B(iv), C(ii)
(b) A(iv), B(iii), C(i)
(c) A(iii), B(iv), C(v)
(d) A(v), B(ii), C(iii)

7. Which one of the following is NOT a typical movement pattern observed during gastrulation? **(DEC 2019 ASSAM)**

(a) Involution (b) Epiboly
(c) Furrowing (d) Delamination

8. Which one of the following developmental processes in animals is more dependent on cellular movements? **(JUNE 2014)**

(a) pattern formation
(b) morphogenesis
(c) cell differentiation
(d) growth

9. With respect to the extra embryonic structures formed in the mammals, the possible functional attributes have been designated:

A. Allantoin stores urinary waste and helps mediate gas exchange. It is derived from splanchnopleure at the caudal end of the primitive streak.

B. Amnion is a water sac and protects the embryo and its surrounding amniotic fluid. This epithelium is derived from somatopleure.

C. Chorion is essential for gas exchange in amniote embryos. It is generated from the splanchnopleure.

D. Yolk sac is the last embryonic membrane to form and is derived from somatopleure.

Which of the above statements are correct? **DEC 2011**

(a) A and B (b) A and C
(c) B and C (d) A and D

Answer Key

1.(b)	2.(c)	3.(c)	4.(a)	5.(a)	6.(c)	7.(c)
8.(a)	9.(a)					

6. Cell aggregation in Dictyostelium

1. cAMP signalling plays a very important role in the development and differentiation of *Dictyostelium discoideum*. This morphogen' is synthesized by different adenyl cyclases expressed at different stages of its life cycle. The following statements (AD) refer to the

effect of mutations in different adenyl cyclase genes:

A. *aca* deficient cells can be allowed to aggregate by exposing them to pulses of cAMP.

B. *acb* deficient cells would form normal fruiting bodies and the spores can germinate when exposed to favourable conditions.

C. *acg* deficient cells develop normally and the spores germinate in the spore head itself.

D. Spores formed from the *acg* deficient cells will germinate irrespective of the osmotic conditions.

Which of the above statements are correct?

(JUNE 2011)

(a) A and D (b) A only

(c) A and B (d) C and D

2. cAMP signalling plays a very important role in the development of *Dictyosteliumdiscoideum.* Below are few statements related to it.

A. Every amoeba at the time of aggregation has the potential to make, receive and relay cAMP.

B. acb$^-$ mutants develop normally but the spores formed appear glassy and are unable to germinate.

C. The spores formed by the acg$^-$ mutants germinate in the sorus itself.

D. RegA is an extracellular phosphodiesterase.

E. cAMP is continuously secreted in nanomolar amounts during aggregation.

Which combination of the above statements is correct? **(JUNE 2018)**

(a) A and D (b) A and B

(c) A and E (d) B and D

Answer Key

1.(a)	2.(b)

7. Development in Sea Urchin

1. Driesch performed famous "pressure plate" experiments involving intricate recombination with 8-celled Sea urchin embryo. This procedure reshuffled the nuclei that normally would have been in the region destined to form endoderm into the presumptive ectoderm region. If segregation of nuclear determinants had occurred, resulting embryo should have been disordered.

However, Driesch obtained normal larvae form these embryos possible interpretations regarding the 8-celled sea urchin embryo are:

A. The prospective potency of an isolated blastomere is greater than its actual prospective fate

B. The prospective potency and prospective fate of blastomere were identical

C. Sea-urchin embryo is a "harmoniously equipotential system" because all of its potentially independent parts interacted together to form single embryo.

D. Regulative development occurs where location of a cell in the embryo determines its fate.

Which of the interpretation(s) is/are true?

(DEC 2016)

(a) Only A (b) Only D

(c) Only A and B (d) A, C and D

2. Which one of the following about development of sea urchin embryos is TRUE?

(JUNE 2015)

(a) Each blastomere of a 4-cell stage possesses a portion of the original animal-vegetal axis and if isolated and allowed to develop will form a complete but smaller size larva.

(b) Each blastomere of a 8-cell stage has the capacity to form a complete embryo but by the 16-cell stage, blastomeres will develop according to their presumptive fate.

(c) Any blastomere isolated till the pluteus larva formation will regulate to go on and develop into a full sized embryo.

(d) After an intricate recombination at the 16 cell stage, the resulting embryo looses its ability to form a complete larva.

3. When the 4 blastomere pairs of the 8-cell stage tunicate embryo are dissociated, each forms most of the structures it would have formed had it remained in the embryo. However, the notochord and nervous system get specified only if different blastomeres get the chance to interact. Given below are certain interpretations formulated from the above results:

A. Each pair of blastomeres forming respective structures indicate autonomous specification

B. Each pair of blastomeres forming respective structures indicate conditional specification

C. The notochord and nervous system development indicate autonomous specification

D. The notochord and nervous system development indicate conditional specification.

Which combination of interpretations is most appropriate? **(JUNE 2017)**

(a) A and C (b) B and D

(c) A and D (d) B and C

4. In a given experiment, transplantation of micromeres from the vegetal pole of a 16-cell sea urchin embryo onto the animal pole of a host 16-cell sea urchin embryo would initiate: **(DEC 2017)**

(a) The transplanted micromeres to invaginate into the blastocoel to create a new set of skeletogenic mesenchyme cells

(b) The transplanted micromeres to ingress into the blastocoel to create a new set of skeletogenic mesenchymal cells

(c) The transplanted micromeres will mingle with the host micromeres to ingress into the blastocoel to create skeletogenic mesenchyme cells.

(d) The transplanted micromeres will form the secondary archenteron

5. Given below are few statements regarding the role of Disheveled (Dsh) and β-catenin (β-cat) in the development of sea urchin.

A. Dsh is localized in the vegetal cortex of the oocyte before fertilization and in the region of the 16-cell embryo about to become the micromeres.

B. Dsh is localized in the cytosol of the oocyte during oogenesis and in the micromere forming blastomeres of a 16- cell embryo.

C. β-cat accumulates predominantly in the micromeres and somewhat in the veg_2 tier cells.

D. Treatment of embryos with lithium chloride does not allow the accumulation of β-cat in the nuclei of all blastula cells, and the animal cells thus become specified as endoderm and mesoderm.

E. When β-cat is prevented from entering the nucleus, the embryo develops as a ciliated ectodermal ball.

Which one of the following options represents a combination of correct statements? **(JUNE 2018)**

(a) B, C and E (b) A, C and D

(c) A, C and E (d) B, D and E

6. In sea urchins, a group of cells at the vegetal pole become specified as the large micromere cells. These cells are determined to become skeletogenic mesenchyme cells that will leave the blastula epithelium to ingress into the blastocoel. This specification is controlled by the expression of Pmar1 which is a repressor of HesC. HesC represses the genes encoding transcription factors activating skeleton forming genes. The gene regulatory network is given below.

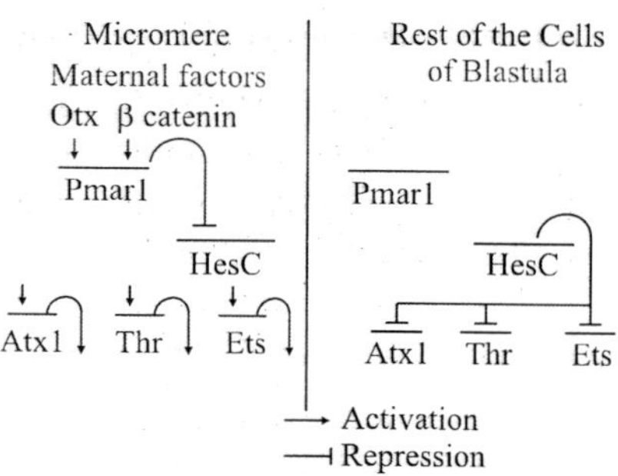

Below, column I lists the experiments carried with mRNA/antisense RNA of different genes injected into single-celled sea urchin embryo while column II lists the developmental outcomes: Match the following: DEC 2012

Column I (Injection of)	Column II (developmental outcomes)
A. mRNA of Pmar1	I. All cells will start ingressing into the blastocoel
B. mRNA of HesC	2. Skeleton mesenchyme will not be formed
C. Antisense of Pmar1	
D. Antisense of HesC	

Which of the following combinations is correct?

(a) A-2, B-1, C-1, D-2

(b) A-1, B-1, C-2, D-2

(c) A-1, B-2, C-2, D-1

(d) A-2, B-2, C-2, D-2

Answer Key

1.(d)	2.(a)	3.(c)	4.(a)	5.(c)	6.(c)

8. Axes and pattern formation in Drosophila

1. Which of the following maternal effect gene products regulate production of anterior structures in Drosophila embryo?**(DEC 2011)**

   (a) Bicoid and Nanos

   (b) Bicoid and Hunchback

   (c) Bicoid and Caudal

   (d) Nanos and Caudal

2. Cytoplasmic determinants coding for anterior structure of Drosophila embryo if injected elsewhere in the recipient embryo, would lead to **(JUNE 2011)**

   (a) normal development.

   (b) formation of additional ectopic head.

   (c) degeneration.

   (d) a phenotype with two heads and two tails.

3. Three embryos, X (wild type), Y (mutant for bicoid) and Z (mutant for nanos) were injected with bicoid mRNA in their posterior pole at early cleavage stage. What would be the phenotypes of the resulting embryos? **(EC 2012)**

   (a) Embryo X will develop head on both anterior and posterior side, while embryos Y and Z will develop head on posterior side only.

   (b) Embryos X and Z will develop head on both anterior and posterior side, while embryo Y will develop head on posterior end only.

   (c) Embryos X, Y and Z will develop head on both anterior as well as posterior side.

   (d) Embryo X will develop head on anterior side, embryo Y will develop no head, while embryo Z will develop head on anterior as well as posterior side.

4. The homologue of β-catenin in Drosophila is **JUNE 2013**

   (a) Fushi tarazu (b) Engrailed.

   (c) Armadillo (d) Cubitus interruptus.

5. Segmentation genes in Drossophila are divided into three groups (gap, pair rule and segment polarity) based on their mutant phenotype. Below are some of the major genes expressed in a sequential manner (with respect to the groups) affecting segmentation pattern.

   A. *hairy* → *paired*→ *tailless* → *patched*

   B. *hunchback* → *even-skipped* → *fushi tarazu* → *wingless*

C. *odd-skipped* → *giant* → *paired* → *wingless*

D. *tailless* → *hairy* → *fushi tarazu* → *gooseberry*

Which of the above sequence(s) of genes expressed from early to late embryo is/are correct? **(JUNE 2013)**

(a) (D) only (b) (A) and (B)

(c) (C) and (B) (d) (B) and (D)

6. Engrailed expression in *Drosophila melenogaster* defines **(JUNE 2014)**

(a) anterior margin of the segment.

(b) anterior compartment of each segment.

(c) posterior margin of each para-segment.

(d) posterior compartment of each segment

7. The following are statements regarding the development and maintenance of anterior and posterior compartments in each segment of *Drosophila*:

A. Expression of *wingless* and *engrailed* is activated by pair-rule genes

B. continued expression of *wingless* and *engrailed* is maintained by interaction between the cells expressing *engrailed* and *wingless* proteins

C. *Hedgehog* is expressed wingless expressing cells and forms short range gradient

D. *Hedgehog* is a transcription factor

E. *Engrailed* is a secretory factor and binds with the patched receptor of the wingless expressing cells.

Which one of the following combination of above statements is correct? **(DEC 2015)**

(a) C and E (b) C, D and E

(c) D and E (d) A and B

8. What will happen if wingless *RNAi* is expressed in wingless expressing cells from the stage when this gene initiates its expression in a developing Drosophila embryo?

A. The enhanced expression of wingless thus caused will broaden the area of engrailed expression.

B. Since wingless protein makes a long range gradient, its effect will not be seen in the same segment.

C. The posterior compartment of each future segment will get affected.

D. Since engrailed expression is initiated by pair rule genes, the posterior segment will not be affected.

Which one of the following will most appropriately answer the question? **(JUNE 2015)**

(a) A and C (b) Only C

(c) B and D (d) Only D

The anterior-posterior compartment of each segment of Drosophila is defined by wingless and engrailed genes. The following statements are given towards explaining their regulation:

A. Wingless is a secretory factor

B. Engrailed is a secretory factor and forms a long-range concentration gradient

C. Engrailed regulates Wingless through Hedgehog which forms a short-range concentration gradient

D. β-catenin homologue is the signalling molecule upstream of Engrailed, which gets cleaved by GSK3 homologue

E. Cubitus interruptus is an intracellular signalling molecule in the Engrailed expressing cells.

Which one of the following options has all the correct statements towards the regulation ofanterior-posterior compartment of segments? **(DEC 2019)**

(a) B only (b) C only

(c) B and E (d) A, C and D

9. The following statements regarding the generation of dorsal/ventral axis in Drosophila was made:

A. Gurken protein moves along with the oocyte nucleus and signals follicle cells to adopt the ventral fate.

B. Maternal deficiencies of either the gurken or torpedo gene cause ventralization of the embryo.

C. Gurken is active only in the oocytes, Torpedo is active only in the somatic follicle cells.

D. The Pipe protein is made in the dorsal follicle cells.

E. The highest concentration of Dorsal is in the dorsal cell nuclei, which becomes the mesoderm.

Which one of the following combination of the above statements is true? **(JUNE 2019)**

(a) A and E (b) C and D
(c) B and C (d) B and E

10. Following are the events that might take place during demo-ventral axis specification in early embryonic development of *Drosophila:*

A. 'Torpedo' receptor activation
B. 'Pipe' synthesis
C. A cascade of protease activity
D. 'Cactus' dephosphorylation
E. Entry of 'Dorsal' in the nuclei of syncytial blastoderm stage embryo.

Which combination of the above events will occur in the presumptive dorsal side of the embryo deficient in maternal *gurken?*

(DEC 2017)

(a) A only
(b) B and C only
(c) B, C and E only
(d) B, C, D and E only

11. Torpedo is a trans-membrane receptor on follicle cells that binds with Gurken protein located in the presumptive dorsal surface of the oocytes and inhibits a cascade leading to nuclear localization of the Dorsal Protein. In an experiment, *Drosophila* germ line chimeras were made by interchanging pole cells (germ line precursors) between wild type embryos and embryos from mother homozygous for a mutation of *torpedo* gene. These transplants produced:

(i) wild type females whose egg came from mutant mother, and
(ii) *torpedo* deficient females whose egg came from wild type mother.

The possible outcome of this experiment can be:

A. Torpedo deficient eggs developed in wild type ovary induced normal embryos.

B. Wild type eggs developed in Torpedo deficient ovary produced ventralized embryos.

C. Torpedo deficient eggs developed in wild type ovary produced ventralized embryos.

D. Dorsal protein enters in the nuclei of dorsal side of embryos which came from wild type eggs developed in Torpedo deficient ovary.

E. Dorsal protein remains cytoplasmic in the dorsal side of the embryos which came from wild type eggs developed in Torpedo deficient ovary.

Which of the above combination is correct?

JUNE 2014

(a) A, B and D (b) B, C and E
(c) B, D and E (d) A, C and E

12. Torpedo, is known to serve as a receptor for Gurken. Deficiencies of the *torpedo* gene in *Drosophila* cause ventralization of the embryo. In an experiment, the germ cell precursors from a wild type embryo were transplanted into embryos whose mother carried the *torpedo* mutation. Also, the reverse experiment, i.e., transplantation of germ cell precursors from torpedo mutants into wild type embryos was done. The *torpedo* deficient germ cells developed in a wild type female showed normal dorso-ventral axis, while the wild type germ cells developed in a torpedo deficient female showed ventralized egg.

Some of the following statements are drawn from the above experiments and some from known facts to understand the functioning of Torpedo.

A. Zygotic contribution of Torpedo is essential for the development of dorso- ventral axis.

B. Maternal contribution of Torpedo is essential for the development of dorso-ventral axis.

C. Since Torpedo is a receptor for Gurken and follicle cells surround the part of the oocyte where Gurken is expressed, it is likely that Torpedo is expressed in follicle cells.

D. Gurken signalling initially dorsalizes the follicle cells, which in turn send signal to organize the dorso-ventral polarity in oocyte.

E. Gurken signalling initially dorsalizes the nurse cells which help in generation of dorso-ventral polarity in oocyte.

Which one of the following combination of statements is most appropriate?**(JUNE 2018)**

(a) B, C and D (b) A, C and D
(c) B, C and E (d) A, D and E

13. During embryogenesis, a Drosophila embryo with a mutation in a homeotic gene shows DEC **(2013 GU)**

(a) changed polarity of each segment.
(b) defects in dorso-ventral polarity.
(c) deletion of several segments in a row.
(d) replacement of one body part by another body part

14. Which of the following statements is true about dorso-ventral patterning of drosophila embryo? **(MODEL PAPER)**

A. This is dictated by the location of the nurse cells

B. Dorsal is the default fate

C. The whole process is regulated by preventing the entry of a transcription factor to the nucleus of dorsal cells

D. Homeobox containing genes play a critical role in this process.

(a) (A), (B) and (C)
(b) (A),(B), (C) and (D)
(c) (A) and (D)
(d) (B) and (C)

15. A mutant was experimentally generated which had wings reduced to haltere like structure.

The following statements are put forward regarding this phenotype:

A. *ultrabithorax* gene ectopically expressed in second thoracic segment

B. *antennapedia* gene ectopically expressed in second thoracic segment

C. A homeotic mutation,

D. A mutation in gap gene

The following combination of statements will be most appropriate explaining the molecular basis of mutant phenotype: **(DEC 2014)**

(a) A and B (b) B and C
(c) C and D (d) A and C

16. Antennapedia complex in Drosophila contains five genes, *lab, pb, dfd, scr* and *Antp* and they express in parasegments 1 to 5, respectively in a non-overlapping manner. In the larva or in later stages of development, the region of *Antp* (Antennapedia) expression corresponds to a part of second thoracic segment.

A mutation in *Antp* is known to cause transformation of antenna to leg-like structures. Below are certain statements made in respect to the functions of Antennapedia:

A. In the above described *Antp* mutation, the gene ectopically expresses in the head region

B. One of the functions of *Antp* is torepress genes that induce antenna development

C. *Antp* expresses in thorax and forms a concentration gradient in the posterior-anterior direction, thus affecting head development

D. A homozygous recessive mutation of *Antp* is expected to transform the leg to antenna in the second thoracic segment.

Which combination of the above statements correctly describes the function of Antennapedia? **(JUNE 2017)**

(a) A, B and C (b) B and C

(c) C and D (d) A, B and D

17. What is the observed phenotype when the *ultrabithorax* gene is deleted in *Drosophila?*

(a) The third thoracic segment is transformed into another second thoracic segment resulting in a fly with four wings.

(b) Since it specifics the second thoracic segment, instead of antenna leg growsout of the head socket.

(c) Since it specifies the third thoracic segment, a fly with two pairs of haltersdevelop.

(d) Since this gene fails to be expressed in the second thoracic segment, theantennae sprout in the leg position.

18. The table given below provides categories and names of genes involved in Drosophila development **(DEC 2019 ASSAM)**

	Category		Gene Name
I	Pair rule gene	A	abdominal A
II	Homeotic gene	B	gooseberry
III	Gap gene	C	fushi tarazu
IV	Segment polarity gene	D	giant

Which one of the following options is a correct match between the categories and gene names?

(a) I-A, II–B, III- C, IV-D

(b) I-D, II-C, III–B, IV-A

(c) I-B, II - D, III - A, IV-C

(d) I-C, II - A, III-D, IV-B

19. The change in the state of specification of imaginal disc of Drosophila to that of a different disc type is known as **(DEC 2013)**

(a) transdetermination

(b) transdifferentiation

(c) transformation

(d) transduction

Answer Key

1.(a)	**2.**(b)	**3.**(b)	**4.**(c)	**5.**(d)	**6.**(d)	**7.**(d)
8.(b)	**9.**(c)	**10.**(c)	**11.**(a)	**12.**(a)	**13.**(d)	**14.**(d)
15.(d)	**16.**(d)	**17.**(a)	**18.**(d)	**19.**(a)		

9. Axes and pattern formation amphibia and chick

1. The initial dorsal-ventral axis in amphibian embryos is determined by **(JUNE 2016)**

(a) the point of sperm entry.

(b) gravity.

(c) the point of contact with the uterus.

(d) genetic differences in the cells.

2. The blastopore region of amphibian embryo that secretes BMP inhibitors and dorsalizes the surrounding tissue is known as **(DEC 2011)**

(a) Brachet's cleft

(b) Nieuwkoop center

(c) Hensen's node

(d) Spemann's organizer

3. In case of is *Xenopus levis* which cells make up the Knewkoop center and Spemann's organizer? **(JUNE 2012)**

(a) endodermal and mesodermal, respectively

(b) mesodermal and endodermal, respectively

(c) endodermal and ectodermal, respectively

(d) ectodermal and endodermal, respectively

4. Which protein secreted by the amphibian organizer induces neural tissue formation by inhibiting Bone Morphogenetic Protein? **(JUNE 2013)**

(a) β-catenin. (b) Noggin.

(c) Dickkopf. (d) Dishevelled.

5. The group of cells of amphibian blastula capable of inducing the organizer is called as **(DEC 2014)**

(a) Hensen's node

(b) Nieuwkoop centre

(c) Dorsal blastopore lip
(d) Hypoblast

6. The dorsal-most vegetal cells of the amphibian embryo that is capable of inducing the organizer is called as Nieuwkoop centre and is marked by the presence of **(JUNE 2017)**

(a) Chordin (b) β-catenin
(c) Goosecoid (d) Nanos

7. In amphibian oocyte, the germplasm which gets segregated during cleavage to give rise to primordial germ cells (PGC's) is normally **JUNE 2011**

(a) distributed evenly throughout the oocyte.
(b) localized at animal pole.
(c) localized at vegetal pole.
(d) aggregated in central part of oocyte.

8. The presence of β-catenin in the nuclei of blastomeres in the dorsal portion of the amphibian embryo is one of the determinants for laying down the dorso-ventral axis. What will be the outcome of expressing a dominant negative form of GSK3 in the ventral cells of early embryo? **(JUNE 2017)**

(a) The dorsal cells will be ventralized
(b) A second axis will be formed
(c) The primary organizer will not be formed
(d) The embryo will develop normally

9. In *Xenopus* embryos, β-catenin plays an important role in the Dorsal/Vental axis development. What would you expect if the endogenous glycogen synthase kinase 3 (GSK3) is knocked out by a dominant negative form of GSK3 in the ventral cells of the early embryos? **(JUNE 2019)**

(a) Blocking of GSK3 on the ventral side has no effect. A normal embryo will form.
(b) The resulting embryo will only have ventral sides
(c) A second axis will form
(d) The dorsal fate is suppressed.

10. In case amphibians, the dorsal cells and their derivatives are called as "Spemann – Mangold organizer". Following statements are related to the "organizer" were made:

A. It induces the host's ventral tissues to change their fates to form neural tube and dorsal mesodermal tissues.
B. It cannot organize the host and donor tissues into a secondary embryo.
C. It does not have the ability to self-differentiate into dorsal mesoderm
D. It has ability to initiate the movements of gastrulation.
E. Both β-catenin and Chordin are produced by the organizer

Which of the above statements are correct? **(DEC 2016)**

(a) A and D (b) D and E
(c) A and E (d) B and C

11. Which one of the following statements with respect to amphibian development is correct?

(a) The organizer is itself induced by the Nieuwkoop Centre located in the dorsal most mesodermal cells.
(b) The organizer functions by secreting proteins like Noggin, Chordin and Follistatin that blocks BMP signal that would otherwise dorsalize the mesoderm.
(c) In the presence of BMPactivators the ectodermal cells form neural tissue.
(d) Wnt signalling causes a gradient of β-catenin along the anterior-posterior axis of the neural tube that appears to specify the regionalization of the neural tube.

2. Injection of Noggin mRNA in cells that will become the future ventral side of a frog embryo mimics the effect of an organizer graft to the ventral side. This experiment demonstrates that

A. Noggin is a transcription factor
B. Noggin induces ventral fates
C. Noggin is involved in organizer fate
D. Noggin is required to induce a secondary axis

Which one of the following options represents correct combination of statement/s?

(JUNE 2017)

(a) A and C (b) C and D
(c) A and B (d) B and C

13. A "morphogen" can determine the fate of a cell by its concentration. Given below are some statements on the experiment performed to study the gradient-dependent effect of the morphogen, activin on cell fate by placing activin (4 nm)-secreting beads on unspecified cells from an early Xenopus embryo:

A. Beads without activin did not elicit expression of either *Xbra* or *goosecoid* genes.

B. Cells nearest to the beads getting highest concentration of activin induced goosecoid gene whose product is a transcription factor, specifies the frog's dorsal-most structures.

C. Cells nearest to the beads getting highest concentration of activin induced *Xbra* gene whose product is a transcription factor, specifies the frog's dorsal-most structures.

D. Cells farthest from the beads getting negligible activin activate *Xbra* gene and become blood vessels and heart

E. Cells farthest from the beads getting negligible activin, activated neither *Xbra* nor *goosecoid* and the default gene expression instructed the cells to become blood vessels and heart.

Which of the above observations and conclusions drawn are correct? **(DEC 2019)**

(a) A, B and C (b) B, C and D
(c) C, D and E (d) A, B and E

14. Cells from an early frog blastula were remove from the animal pole and used to replace cells from the vegetal pole of the blastula. The following events may be expected.

A. Transplanted cells would develop normally as part of the cells of the vegetal pole.

B. Transplanted cells would develop as cells of the animal pole of the adult on the vegetal pole.

C. Region of the animal pole from where the cells were removed would be missing in the adult.

D. Remaining cells in the animal pole would compensate for the cells that were removed.

Which of the following are true? **(DEC 2013)**

(a) B, C and D (b) A, B and D
(c) A, B and C (d) A, C and D

15. The part of the embryo from which the ectoderm, mesoderm and endoderm are formed in chick is known as **(JUNE 2013)**

(a) primitive streak. (b) hypoblast.
(c) epiblast. (d) cytotrophoblast.

16. Hensen's node is established as the avian equivalent of the amphibian dorsal blastopore lip. The following observations are presumed to be support of the same.

A. It is the region whose cells induce and pattern a second embryonic axis when transplanted into other locations of the gastrula.

B. It is equivalent in terms of tissue structure.

C. It expresses the same marker genes as the Spemann's organizer in Amphibians.

D. The same micro RNA can interfere with the formation of pre-chordal plate in both Hensen's node and Spemann's organizer.

Choose the correct set among the following:

(DEC 2013)

(a) A and D (b) A and C
(c) B and C (d) A and B

17. Following statements were made about the events occurring during chick development.

A. The fertilized chick egg undergoes discoidal meroblastic cleavage; however the cleavage does not extend into the yolk cytoplasm.

B. *Development* of primary hypoblast is mediated by localized migration of a group of highly specified and connected cluster of 30-40 cells.

C. By the stage XIII of chick embryogenesis and little prior to primitive streak formation, the formation of the hypoblast is just complete.

D. Hansen's node of the chick embryo signifies a region at the anterior end of the primitive streak with regional thickening of cells.

E. Inhibition of *Wnt* planar cell polarity pathway in the epiblast causes the mesoderm and endoderm to form centrally instead of peripherally.

Which one of the following combinations represents all correct statements?**(DEC 2019)**

(a) A, B and D (b) A, C and E

(c) A, B and C (d) A, C and D

Answer Key

1.(a)	2.(d)	3.(a)	4.(b)	5.(b)	6.(b)	7.(c)
8.(b)	9.(c)	10.(a)	11.(d)	12.(a)	13.(d)	14.(*)
15.(c)	16.(b)	17.(d)				

10. Vulva formation in Caenorhabditis elegans

1. A group of six cells called 'equivalence group cells' divide to form the vulval structure in *Caenorhabditis elegans*. They are called so because **(MODELPAPER)**

(a) they have similar fates during development of vulva.

(b) all the six cells are competent to form vulva and can replace each other under various experimental conditions.

(c) they are all under the influence of the anchor cell, signals from which initiate vulval development.

(d) they interact with each other to form the vulval structure.

2. In *C. elegans*, an anchor cell and a few hypodermal cells take part in the formation of vulva. The experiment performed to understand the role of these cells in vulva formation and the results obtained are as follows:

If the anchor cell is killed by laser beam, hypodermal cells do not participate in vulva formation and no vulva develops.

If six hypodermal cells closely located with anchor cell (called vulval precursor cells) are killed, no vulva develops

If the three central vulval precursor are destroyed, the three outer cells, which normally form hypodermis, take the fate of vulval cells instead.

Following are certain statements regarding vulva formation:

A. Anchor cells acts as an inducer

B. Six hypodermal cells with the potential to form vulva form an equivalence group.

C. Three, out of six, hypodermal cells participate in vulva formation

D. The central cell functions as the 1^0 cell and the two cells on both side act as the 2^0 cells

E. The 1^0 cell secretes a short range juxtacrine signal

Which combinations of the above statements have been derived from the above experimental results? **(DEC 2015)**

(a) A, B and C (b) A, B and D

(c) D and E (d) B, D and E

3. The following experiments related to vulva development in *C. elegans* were performed:

When the anchor cell was killed by laser beam at early embryonic development and embryo was allowed to develop, a vulva did not form.

When any three out of the SIX hypodermal cells were killed and embryo was allowed to develop, a normal vulva formed.

When any four out of six hypodermal cells were killed and embryo was allowed to develop a normal vulva did not form.

When anchor cell as well as anyone of the six hypodermal cells were killed, the vulva did not form.

Based on the above observations the following conclusions were made:

A. Out of six hypodermal cells, any three get involved in vulva formation.

B. The fourth hypodermal cell plays a major role in vulva formation.

C. Anchor cell as well as the middle three hypodermal cells only has the ability to participate in vulva formation.

D. Anchor cell is essential for vulva formation.

Which of the above statement or combination of statements is correct? **(DEC 13 GU)**

(a) A and B (b) only C

(c) A and D (d) A and C

4. In *C. elegans* during embryogenesis, an anchor cell and 6 hypodermal vulval precursor cells (VPCs) get involved in forming the vulva. If 3 of the hypodermal VPCs are killed by a laser beam, a normal vulva is still formed. This could be due to the following possible reasons.

A. Six hypodermal VPCs form equivalence group of cells, out of which only 3 participate in vulva formation and 3 cells remain as reserve cells.

B. When 3 hypodermal VPCs are killed, the 3 neighboring hypodermal non- VPCs get freshly recruited.

C. Anchor cell functions as an inducer which can induce epithelial cells of the gonad to get recruited to compensate for the loss.

D. Anchor cell acts as an inducer which can spatially induce only 3 hypodermal cells to form the vulva.

Which combination of the above statements is correct? **(DEC 2012)**

(a) A and B (b) B and C

(c) C and D (d) A and D

5. Development of vulva in *C. elegans* is initiated by the induction of a small number of cells by short range signals from a single inducing cell. With reference to this, following statements were put forward.

A. When the anchor cell was ablated early in development no vulva formed.

B. In a dominant negative mutant of let-23, a primary vulva formed but the secondary vulva formation did not take place. .

C. A cell adopting a primary fate inhibits adjacent cells from adopting the same fate by lateral inhibition involving LIN-39 and also induces the secondary fate in these cells.

D. A constitutive signal from the hypodermis inhibits the development of both the primary and secondary fates but it is overruled by the initial signal from the anchor cell.

Which of the above statements is true?

(JUNE 2015)

(a) A and B (b) A and C

(c) A and D (d) B and D

6. During vulva development in *C. elegans,* the anchor cell produces Lin-3 protein which interacts with the Let-23 protein present on the six vulval precursor cells (VPCs) that form an equivalence group.

The central lineage cell (P6.p) adopts the primary fate, the adjacent VPCs (P5.p and P7.p) adopt the secondary fate and the rest VPCs adopt the tertiary fate.

Few mutants (Column A) and phenotypes (Column B) are listed in the table given below.

(DEC 2017)

	Column A		Column B
A	Loss of function of *lin*-3	(i)	P5.p, P6.p and P7.p adopt primary fate, P4.p and P8.p adopt secondary fate

B	Loss of function of *lin-B* and gain of function of *let-23*	(ii)	Multivulva
C	Reduced function of *lin*-3	(iii)	P6.p adopt primary fate and the rest of the VPCs adopt tertiary fate
D	Overexpression of *lin*-3	(iv)	All VPCs adopt tertiary fate

Match the correct mutant with the observed phenotype.

(a) A-iv, B-ii, C-iii, D-i

(b) A-iv, B-iii, C-i, D-ii

(c) A-ii, B-iii, C-iv, D-i

(d) A-iii, B-i, C-ii, D-iv

7. Following statements were made regarding vulval development in *Caenorhabditis elegans*:

A. The six vulval precursor cells (VPCs) are influenced by the anchor cell to form an equivalence group.

B. In the loss of function lin-12 mutants, both cells become uterine whereas in gain of function mutants, both become anchor cell.

C. If the anchor cell is destroyed early in development, all the six VPCs divide once and contribute towards the formation of hypodermal cells.

D. The anchor cell/ventral uterine precursor decision is due to Notch-Delta mediated mechanism of restricting adjacent cell fates.

E. The paracrine factor secreted by the anchor cell directly activates the Notch-delta pathway.

Which one of the following options represents a combination of correct statements?

(JUNE 2018)

(a) A, C and D (b) A, B and D

(c) C, D and E (d) B, D and E

8. The two cell types, Anchor cell and vulval precursor cells are involved in vulva formation in *C. elegans*. The following statements are given towards understanding the roles of these two cell types and their signalling activities:

A. Anchor cell is a germ cell from the gonad, which sends induction signal to vulva for maintaining differentiation states.

B. The six vulval precursor cells, influenced by anchor cell, form an equivalence group

C. The cell directly beneath the anchor cell divides to form the central vulval cells, while the two flanking cells divide to become the lateral vulval cells

D. The three cells further away from anchor cell generate hypodermal cells

E. *Lin3* signal from anchor cell forms a gradient and activates vulva forming genes in central and lateral cells

F. The Notch-Delta mediated mechanism of restricting adjacent cell fates is called Lateral inhibition.

Which combination of the above statements is correct towards vulva formation in *C. elegans*? **(DEC 2019 ASSAM)**

(a) A, B and E only

(b) A, B, E and F

(c) A, B, C and D

(d) B, C, D, E and F

Answer Key

1.(b)	2.(a)	3.(c)	4.(d)	5.(c)	6.(a)	7.(a)
8.(d)						

11. Eye lens induction

1. During lens formation in the *Xenopus*, the following statements have been proposed:

A. Lens induction can be achieved in the absence of optic vesicle after priming of head ectoderm by the anterior neural plate.

B. The optic vesicle can induce the presumptive trunk ectoderm to form the lens.

C. Only the head ectoderm can respond to direct signals from the optic vesicle form the lens.

D. The anterior neural plate primes the head ectoderm via BMP4 and Fgf8 prior to signals from the optic vesicle.

Which of the above combinations is correct? **(JUNE 2014)**

(a) C and D (b) B and D
(c) A and D (d) A and C

2. The functionality of the pax6 gene in the formation of optic and nasal structure may be attributed to the following

A. Pax6 makes the optic vesicle competent and allows lens formation.

B. The optic vesicle can induce any part of the head ectoderm to form the nasal and optic structures, due to presence of Pax6.

C. Pax6 renders the head ectoderm competent to receive signals from optic vesicle.

D. Apart from the optic vesicle, the head ectoderm may also be induced by BMP and FGF, so pax6 is not exclusive for lens formation.

Which of the above attributes are true? **(JUNE 2012)**

(a) (A) and (D) (b) (C) and (D)
(c) (B) and(C) (d) (C) only

3. Following are some statements given for vertebrate eye lens induction:

A. Paired box6 (Pax6) is a transcription factor synthesized in specific region of head surface ectoderm.

B. Pax6 is secreted by optic vesicle.

C. Optic vesicle serves as an inducer for competent head surface ectoderm.

D. Studies on amphibians suggest that the first inducers of lens may be the foregut endoderm and heart forming mesoderm.

E. The genes for lens proteins get induced in the surface ectoderm of optic vesicle.

Which combination of the above statements is correct towards vertebrate eye lens induction? **(DEC 2019 ASSAM)**

(a) A, C, D (b) A, C, E
(c) B, E (d) B, C, D

4. Lens formation requires sequential events whereby the anterior neural plate signals the anterior ectoderm to promote secretion of Pax 6, which renders the anterior ectoderm more receptive to secretions from the optic vesicle. The above can be best explained by which of the following phenomenon? **(DEC 2014)**

(a) Instructive interactions only
(b) Epithetial- Mesenchymal interactions
(c) Permissive interactions
(d) Induction and competence

5. In Amphibians, when due to some injury, the eye lens is damaged, the fully differentiated iris cells can regenerate the lens. It is achieved through the possible processes:

A. Iris cells through some signaling undergo dedifferentiation and transdifferentiation into lens cells to regenerate the lens.

B. Iris cells transform into lens cells spontaneously.

C. Iris cells induce in a stepwise manner, specific genes responsible for their dedifferentiation and then conversion to lens cells.

D. Stem cells present in iris tissue differentiate into lens cells.

Which of the following is correct? **(DEC 2013)**

(a) A and B (b) A and C
(c) B and D (d) B and C

Answer Key

1.(a)	2.(b)	3.(a)	4.(d)	5.(b)

12. Limb development in vertebrates

1. Which one of the following statements regarding limb development in mice is true? **(DEC 2017)**
   (a) The gene encoding *Tbx5* is transcribed in the limb fields of the hindlimbs.
   (b) The gene encoding *Tbx4* is transcribed in the limb fields of the forelimbs.
   (c) Genes encoding *Islet 1*, *Tbx4* and *Pitx* are expressed in the presumptive hindlimb.
   (d) Genes encoding *Islet 1*, *Tbx4* and *Pitx* are expressed in the presumptive forelimb.
2. Over-expression of a dominant negative FGF receptor during amphibian development would prevent formation of **(DEC 2013)**
   (a) trunk and tail.
   (b) head and trunks.
   (c) trunk and fore limbs.
   (d) head and forelimbs.
3. Apical ectodermal ridge induction is essential tetrapod limb development. Which of the following is NOT essential for the formation of a functional limb? **(DEC 2016)**
   (a) Tbx genes and Wnt
   (b) Androsterone
   (c) Apoptotic gene
   (d) Fibroblast growth factor
4. What would happen as a result of a transplantation experiment in a chick embryo where the leg mesenchyme is placed directly beneath the wing apical ectodermal ridge (AER)? **(JUNE 2015)**
   (a) Distal hind limb structures develop at the end of the limb.
   (b) A complete hindlimb will form in the region where the forelimb should be.
   (c) The forelimb would form normally.
   (d) Neither a forelimb nor a hindlimb would form since the cells are already determined.
5. During vertebrate limb development, a specialised ectodermal structure, called Apical Ectodermal Ridge (AER), forms at the dorso-ventral ectodermal boundary at the distal tip of the developing limb bud.

   The following experimental facts about the AER is available:

   A. FGF 2, 4, and 8 are expressed in the AER
   B. Removal of the AER causes cessation of limb growth
   C. Removal of AER along with implantation of beads soaked in FGF 8 or
   D. FGF 4 or FGF 2 protein rescues the AER removal phenotype and gives rise to normal limb

   Which of the following statements cannot be made based on the above facts? **(MODEL PAPER)**
   (a) FGF 2, 4, and 8 are secreted proteins.
   (b) FGF 2, 4, and 8 are necessary and sufficient for AER function
   (c) FGF 2, 4, and 8 are sufficient for AER function
   (d) FGF 2, 4, and 8 have largely redundant functions
6. The proximal distal growth and differentiation of the tetrapod limb bud are made possible by a series of interactions between the apical ectodermal ridge (AER) and limb bud mesenchyme directly beneath it. Some of the interactions performed in chick demonstrated the following results:

   A. When the AER was removed at any time of development, further development of distal limb skeletal elements ceased.
   B. When leg mesenchyme was placed directly beneath the wing AER, distal hindlimb structures developed at the end of the wing.
   C. When limb mesenchyme was replaced by a non-limb mesenchyme beneath the AER, the limb still developed
   D. When an extra AER was grafted onto an existing limb bud, the development of the limb ceased.

Which of the above combinations is correct?

(JUNE 2014)

(a) A and B (b) A and C

(c) B and D (d) B and C

7. A set of experiments that were carried out to demonstrate the effect of Apical Ectodermal Ridge (AER) of the chick limb bud on the underlying mesenchyme are enlisted below, along with their expected outcomes:

A. Removal of the AER of forelimb leads to cessation of limb development.

B. If an extra AER is placed in the forelimb bud, duplication of the distal region of the wing takes place.

C. If an extra AER is placed in the forelimb bud, a leg develops instead of a wing.

D. If AER of forelimb bud is replaced with beads soaked in FGF2, a normal wing develops.

E. If a non limb mesenchyme is placed below an AER, the AER directs the mesenchyme to form a normal wing.

Which of the above statements are correct?

(DEC 2013)

(a) A, C and E (b) C, D and E

(c) B, D and E (d) A, B and D

8. In the context of the proximal-distal growth and differentiation of a tetrapod limb following experiments were visualized

(A) If the apical ectodermal ridge (AER) is removed at any time during the limb development, further development of distal limb skeletal elements ceases.

(B) If leg mesenchyme is placed directly beneath the wing AER, proximal hind limb structures develop at the end of the limb

(C) If an extra AER is grafted onto an existing limb bud, supernumerary structures are formed usually at the distal end of the limb.

(D) If leg mesenchyme is placed directly beneath the wing AER, proximal hind limb structures develop at the end of the limb

Which of the above experiments would show the possible interactions between the AER and the limb mesenchyme directly beneath it during limb development? **(JUNE 2012)**

(a) (A) and (B) only

(b) (B) and (C) only

(c) (C) and (D) only

(d) (A) and (C) only

9. During wing development in chick, if Apical Ectodermal Ridge (AER) is removed, the limb development ceases, on the other hand placing leg mesenchyme directly beneath the wing AER, distal hindlimb structures deveolpa at the end of the wing, and if replaced by non-limb mesnchyme sheath AER, the AER regresses. This may demonstrate that:

A. the limb mesenchyme cells induce and sustain AER.

B. the mesenchyme cells specify the type: wing or limb.

C. the AER is responsible for specifying the type: wing or limb.

D. the AER is responsible for sustained outgrowth and development of thelimb

E. the AER does not specify the type: wing or limb.

Which combination of above statements is demonstrated by the experiment?

(JUNE 2019)

(a) A, B, C and D only

(b) A, B, D and E only

(c) C, D and E only

(d) A and E only

10. The mechanisms specifying the three axes of a tetrapod limb are interrelated and coordinated. In case of chick limb development the following statements were made:

A. In the positive feedback loop, the fibroblast growth factor (FGF) from the apical ectodermal ridge (AER) activates *sonic hedgehog* (Shh) transcription thereby stabilizing the zone of polarizing activity (ZPA).

B. In the inhibitory loop, *Shh* from the ZPA inhibits Gremlin which blocks the bone morphogenetic proteins (BMPs) thus preventing the BMP mediated inactivation of the FGFs in the AER.

C. The dorsal-ventral axis is formed by the expression of *Wnt7a* in the ventral portion of the limb ectoderm.

D. The BMPs are involved both in inducing apoptosis and in differentiating the mesenchymal cells into cartilage.

Which of the above combination of statements is true? **(DEC 13 GU)**

(a) A and B (b) A and D

(c) B and C (d) C and D

11. Which one of the following signalling pathways will best support the formation of the zone of polarizing activity (ZPA) during limb development in mice **(DEC 2019 ASSAM)**

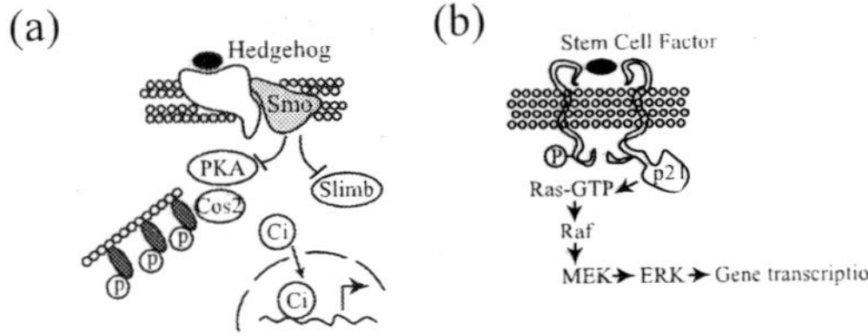

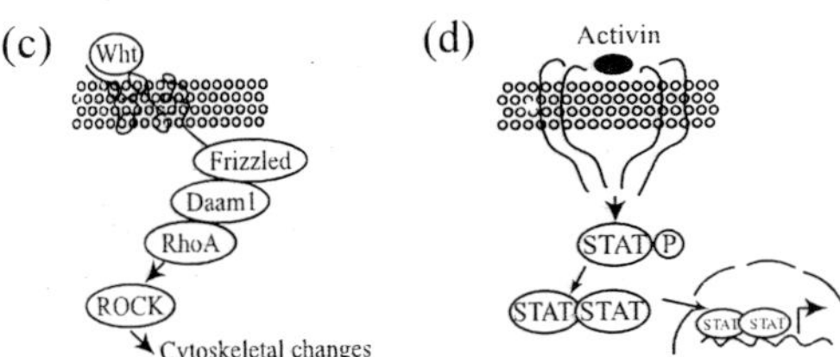

12. In tadpoles, if the tail is amputated it can regenerate. However, if the tail is amputated and then exposed to retinoic acid, it develops limbs instead of regenerating the tail. This could be due to the following reasons:

A. Retinoic acid is a morphogen and induces genes responsible for limb formation.

B. Retinoic acid raises the positional values in that region for limb development to take place.

C. This is a random phenomenon and is not well understood.

D. Retinoic acid possibly acts as a mutagen and the phenotype observed is a result of several mutations.

Which combination of the above statements is true? ***(DEC. 12)***

(a) A and B (b) C and D

(c) B and D (d) B and C

13. Dose-dependence of retinoic acid treatment supports the notion that a gradient of retinoic acid can act as a morphogen along the proximo-distal axis in a developing limb. Following are certain facts related to the above notion.

A. Treatment with high level of retinoic acid causes a proximal blastema to be re-specified as a distal blastema and only distal structures are regenerated.

B. Treatment with high level of retinoic acid causes a distal blastema to be re-specified as a proximal blastema and regeneration of a full limb may be initiated.

C. Treatment with retinoic acid affects only distal blastemas and causes them to form only proximal structures.

D. Treatment with high level of retinoic acid causes any blastema to form only distal structures.

Which one of the following is correct? **(JUNE 2015)**

(a) B and D (b) Only C

(c) A and C (d) Only B

14. Which of the following is true about amphibian limb regeneration? **(MODELPAPER)**

(A) It requires a minimum number of functional nerves.

(B) The blastema of an amputated limb, if transplanted in the trunk region between two existing limbs in a host, will still give rise to a limb.

(C) The size of the regenerated limb is often grossly different from the original limb.

(a) (A) (b) (B) and (C)
(c) (A) and (C) (d) (A), (B) and (C)

15. What would be the effect on newt limb regeneration, if more than 90% of the nerve supply is severed before amputation? **(DEC 2017)**

(a) The apical ectodermal cap stimulates growth of the blastema by secreting FGF8 but regeneration does not take place.
(b) Limb regeneration will take place and form a limb with no nerve supply.
(c) Outgrowth will occur but the identity of the limb formed will be lost with no clear anterior-posterior polarity.
(d) Limb regeneration with nerve supply will take place.

16. How will the regeneration of a newt limb be affected if the nerve supply is severed before amputation? **(DEC 13 GU)**

(a) Partial regeneration will occur where only the proximal structures will be formed.
(b) Regeneration of most of tissues of the limb occurs normally but the regeneration of the nerve will fail.
(c) Regeneration will occur but the identity of the limb is lost and normal proximal-distal patterning will not occur.
(d) blastema will form but not grow and regeneration will fail.

17. Which one of the following statements regarding limb regeneration in Salamander is correct? **(JUNE 2018)**

(a) A normal limb is regenerated after amputation, irrespective of whether the cut was made below the elbow or through the humerus.
(b) It occurs by compensatory regeneration and does not include formation of an apical ectodermal cap.
(c) Regeneration occurs through formation of a blastema, which essentially consists of unspecified multipotential progenitor cells.
(d) Proliferation of the blastema cells does not require nerves or factors secreted by the nerves.

18. Formation of digits and sculpting the tetrapod limb requires death of specific cells in the limb in a programmed manner. Which one of the following interactions could explain proper limb formation? **(JUNE 2015)**

(a)
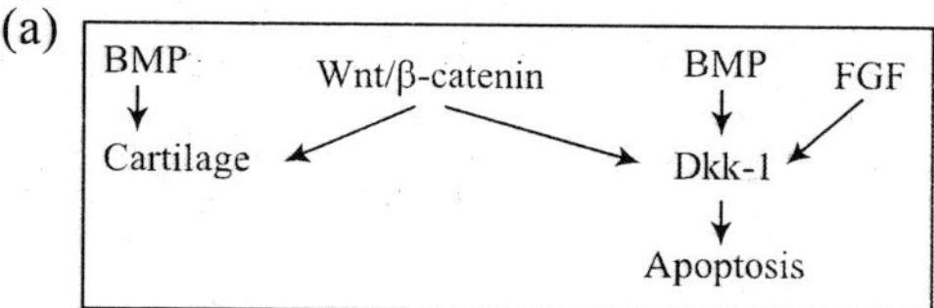

(b)
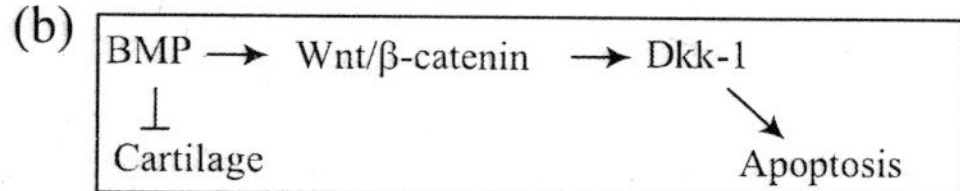

(c)
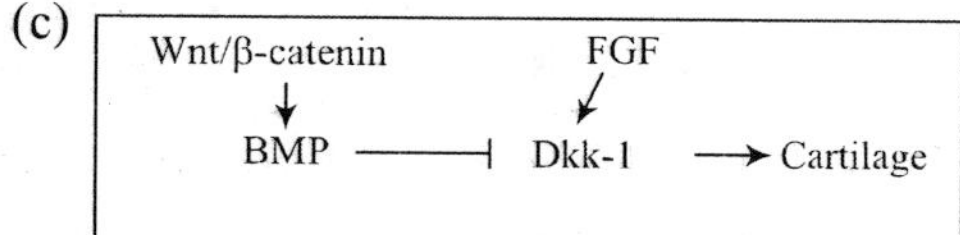

(d)
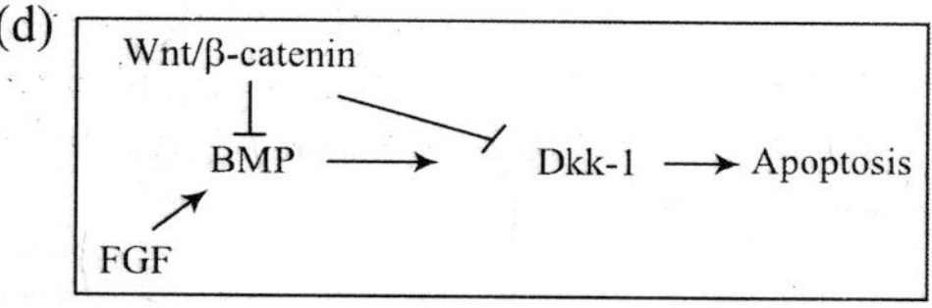

19. When a wrist blastema from a recently cut *Axolotl* forelimb is placed on a host hindlimb cut at the mid thigh level, it will generate only a wrist. The host (whose own hind limb was removed) will fill the gap and regenerate the limb upto the wrist. However if the donor blastema is treated with retinoic acid on grafting, the wrist blastema will regenerate a complete limb and will not allow the host to fill the gap. This happens because retinoic acid. **(JUNE 2011)**

(a) helps in proximalization of the blastema and activates the Hox genes differentially across the blastema.
(b) helps in the distalization of the blastema and activates the Hox genes differentially across the blastema.

(c) helps block the receptors of FGF essential for limb development

(d) helps vigorous proliferation of the cells at the cut surface.

20. Homeobox transcription factor (Hox proteins), play important roles in specifiying wheather a particular mesenchymal cell will become stylopod, zeugopod or autopod. Based on the expression patterns of these genes, a model was proposed wherein these HOX genes specify the identity of a limb region. What would be the observed phenotype for human homozygous for a *HOXD13* mutation? **(JUNE 2019)**

(a) No zeugopod formation

(b) Abnormalities of the hands and feet wherein the digits fuse

(c) Deformities in stylopods

(d) No femur of patella formation.

Answer Key

1.(c)	2.(a)	3.(b)	4.(a)	5.(b)	6.(a)	7.(d)
8.(d)	9.(b)	10.(b)	11.(a)	12.(a)	13.(d)	14.(c)
15.(a)	16.(d)	17.(a)	18.(a)	19.(a)	20.(b)	

13. Regeneration and Differentiation of neurons

1. In case of morphallactic regeneration: **(DEC 2011)**

(a) there is repatteming of the existing tissues with little new growth

(b) there is repatteming of the existing tissues after the stem cell division has taken place

(c) there is cell division of the differentiated cells which maintain their differentiated state to finally form a complete organism.

(d) there is dedifferentiation of the cells at the cut surface which become undifferentiated. These undifferentiated cells then divide to redifferentiate to form the complete structure

2. Hydra shows morphallactic regeneration and involves which one of the following signal transduction pathway in its axis formation? **(JUNE 2015)**

(a) Wnt/β-catenin pathway

(b) Retinoic acid pathway

(c) FGF pathway

(d) Delta-Notch pathway

3. In case of Hydra, the major head inducer of the hypostome organizer is a set of Wnt proteins acting through the canonical β-catenin pathway. What would be the result, if a transgenic Hydra is made to globally mis-express the downstream Wnt effector β-catenin? **(DEC 2018)**

(a) Ectopic buds will be formed all along the body axis and even on the top of the newly formed buds.

(b) Ectopic tentacles form at all levels.

(c) Both ectopic tentacles and buds would be formed along the body axis.

(d) There would be no change observe

4. Given below are some of the statements regarding regeneration:

A. The type of regeneration characteristic of mammalian liver is considered as compensatory regeneration.

B. Regrowth of hair shaft from follicular cells exemplifies stem cell mediated regeneration.

C. Regeneration occurring through the re-patterning of existing tissues with little new growth is known as morphallaxis.

D. Adult structures undergoing dedifferentiation forming a blastema, that then re-differentiates to form the lost structure is called epimorphosis.

Choose the most appropriate combination of correct statements: **(DEC 2018)**

(a) D only (b) C and D only

(c) A, Band C only (d) A, B, C and D

5. In two major ways the neural plate is converted into a neural tube, primary neurulation and secondary neurulation.
   A. In primary neurulation, a hollow neural tube is formed.
   B. In secondary neurulation, the neural tube arises from the coalescence of mesenchymal cells into a solid cord.
   C. The solid cord formed in secondary neurulation subsequently gets hollowed.

   Which of the above processes is/are involved in brain development in vertebrates?

   (DEC 13 GU)

   (a) A and B (b) B and C
   (c) only A (d) only B

6. In many different contexts of cell differentiation, two distinct cell populations emerge from a uniform cell population. This process is referred to as lateral inhibition. Which one of the following must not be true about lateral inhibition? **(MODELPAPER)**
   (A) lateral inhibition results from morphogen action
   (B) lateral inhibition requires direct cell cell contacts
   (C) lateral inhibition requires reciprocal signalling between two neighbouring cells
   (D) lateral inhibition is preceded by stochastic changes in gene expression in two neighbouring cells

   (a) (D) (b) (A) and (D)
   (c) (B) and (C) (d) (A)

7. Both TGF β and Sonic hedgehog signals play important roles in both neurulation and cell-fate patterning of the neural tube. Which one of the following statements is true? **(DEC 2018)**
   (a) High levels of BMP specify the cells to become epidermis.
   (b) Very low levels of BMP specify the cells to become epidermis.
   (c) High levels of BMP specify the cells to become neural plate.
   (d) Intermediate levels of BMP do not effect the formation of neural crest cells.

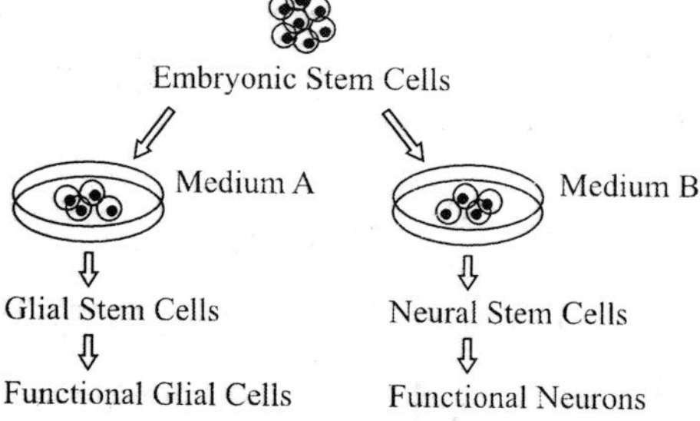

8. The following assumptions were derived from the above experiment:
   A. Medium A contained FGF and PDGF.
   B. Medium B contained retinoic acid.
   C. Cells cultured in Medium B were determined to become functional neurons prior to addition of the medium.

   Which one of the following combinations represents correct statements? **(DEC 2018)**

   (a) A and B only (b) A, B and C
   (c) Band Conly (d) A and C only

Answer Key

1.(a)	2.(a)	3.(a)	4.(d)	5.(a)	6.(c)	7.(a)
8.(a)						

14. Sex determination

1. A mutant embryo of Drosophila in which one of the major sex determining gene, sex lethal, can only undergo default splicing, was allowed to develop. The following statements are towards explaining the determination of sex of the embryo:
   A. The embryo will develop into a male fly
   B. The embryo will develop into a female fly
   C. sex lethal gene product directly regulates sex specific alternate splicing of double sex RNA
   D. sex lethal gene product regulates sex specific splicing of transformer RNA which in turn regulates splicing of double sex RNA

The correct combination of above statements to explain sex determination of the given embryo is: **(DEC 2014)**

(a) A and C (b) A and D

(c) B and D (d) B and C

Answer Key

1.(b)

15. Fertilization and development in Plants

1. During double fertilization in plants, one sperm fuses with the egg cell and the other sperm fuses with **(DEC 2013)**

   (a) synergid cell. (b) central cell.

   (c) antipodal cell (d) nucellar cell.

2. During reproductive development in plants:

   (A) male and female gametes are produced as a result of two mitotic divisions after meiosis

   (B) vegetative cells in pollen contribute to pollen development

   (C) antipodal provide nourishment to developing embryo

   (D) Pollen tube ruptures and releases both the male gametes in one of the degenerated synergids

   Which of the above statement are true? **(JUNE 2013)**

   (a) A and B (b) B and D

   (c) B and C (d) A and D

3. Sperm cell behaviour during double fertilization in *Arabidopsis* can be stated as follows. Identify the INCORRECT statement: **(JUNE 2016)**

   (a) Pollen tube bursts and discharges sperm cells.

   (b) Pollen cells produce pollen tubes and enter into female gametophyte.

   (c) The receptive antipodal cells break down when pollen tube enters the female gametophyte.

   (d) One sperm nucleus fuses with the egg cell and the other fuses with the central cells.

4. Which of the following statement is WRONG? **(DEC 2016)**

   (a) Megasporocyte develops within the megasporangium of the ovule

   (b) Megasporocyte undergoes meiosis to produce four haploid megaspores

   (c) All the four megaspore undergo several mitotic division to form female gametophyte in most angiosperms

   (d) Female gametophyte is haploid

5. A few statements on early developmental stages in plants are given below:

   A. The cells of flower are diploid in nature.

   B. Only some specialized cells in reproductive organs undergo meiosis produce haploid cells.

   C. The haploid cells produced in (B) above, undergo a few normal mitotic cell divisions.

   D. All the progeny cells produced in (B) above, differentiate either into haploid egg cells or into haploid sperm cells.

   Which one of the following combinations of above statements is correct? **(DEC 2013)**

   (a) A, B and C (b) B, C and D

   (c) A, C and D (d) A, B and D

6. During embryo germination in a grass family an absorptive organ that forms interface between the embryo and the starchy endosperm tissue is called **(JUNE 2017)**

   (a) Coleorhiza (b) Coleoptile

   (c) Scutellum (d) Mesocotyl

7. Match the two columns following asexual reproduction of plants and apomixes: **(JUNE 2015)**

A. Agamospermy	(i) No seed formation
B. Clonal propagation	(ii) Seed formation

C. Embryo sac formed from nucellus or integument of the ovule	(iii) Diplospory
D. Gametophyte develops without fertilization from unreduced megaspore	(iv) Apospory

(a) A - (i); B - (ii); C - (iii); D - (iv)
(b) A - (ii); B - (iii); C - (iv); D - (i)
(c) A - (ii); B - (i); C - (iii); D - (iv)
(d) A - (ii); B - (i); C - (iv); D- (iii)

8. Following are certain statements regarding apomixis in plants:

A. Apomixis cannot be used to maintain hybrid vigor over many generations in plants.

B. In sporophytic apomixis maternal genotype is maintained.

C. There is an event of meiosis during gametophytic apomixis and is also referred as apomeiosis.

D. In diplospory, meiosis of the megaspore mother cell is aborted, resulting in two unreduced spores, out of which one forms the female gametophyte.

Which one of the following combinations is correct? **(DEC 2018)**

(a) Aand B (b) A and C
(c) B and C (d) B and D

9. In certain plants, the mechanism where timing of anther dehiscence and stigma receptivity do not coincide to avoid self-pollination is called **(JUNE 2019)**

(a) dichogamy (b) herkogamy
(c) monoecy (d) Dioecy

10. Distyly is **(DEC 2019 ASSAM)**

A. presence of two styles in a flower

B. presence of two lengths of styles in a species

C. a form of dichogamy.

D. a form of herkogamy.

Which one of the following options represents all correct statements?

(a) B only (b) A and C
(c) B and C (d) B and D

11. In mature Arabidopsis embryo, root apical meristem consists of cells derived from **(JUNE 2011)**

(a) embryo and apical suspensor cells
(b) embryo only
(c) suspensor only.
(d) hypophysis only.

12. During development of embryos in plants, PIN proteins are involved in **(DEC 2011)**

(a) establishment of auxin gradients
(b) regulation of gene expression
(c) induction of programmed cell death
(d) induction of cell division

13. During which one of the following stages of *Arabidopsis* embryogenesis, cell elongation throughout the embryonic axis and further development of the cotyledons occur? **(JUNE 2014)**

(a) Globular stage. (b) Torpedo stage.
(c) Heart stage. (d) Mature stage.

14. In which stage of *Arabidopsis* embryogenesis is hypophysis first observed? **(DEC 2018)**

(a) Octant (b) Dermatogen
(c) Globular (d) Transition

15. Following statements are made in relation to the five widely recognized stages of *Arabidopsis* embryogenesis:

A. The fusion of haploid egg and sperm takes place in globular stage

B. Rapid cell division in two regions on either side of the future shoot apex forms heart stage

C. The cell elongation throughout the embryo axis and further development result in torpedo stage

D. The embryo loses water and becomes metabolically inactive in the zygotic stage

Which combination of the above statements is correct? **(DEC 2015)**

(a) A and B (b) B and C

(c) C and D (d) D and A

16. Following are certain statements regarding seed development in plants:

A. During final phase of development embryo's of "orthodox" seeds became tolerant to desiccation, dehydrate losing up to 90% of water

B. Dormant seeds will germinate upon rehydration while quiescent seeds require additional treatments or signals for the germination

C. Precocious germination is germination of seeds without passing through the normal quiescent and/or dormant stage of development

D. Abscisic acid is known to inhibit precocious germination

Which one of the following combinations is correct? **(JUNE 2015)**

(a) A, B and C (b) A, B and D

(c) B, C and D (d) A, C and D

17. Due to the presence of cellulose in the cell wall of plants, leaf shape is determined in the leaf primorida by **(MODELPAPER)**

(a) rates of cell division.

(b) planes of cell division.

(c) cell migration.

(d) cell-cell interactions.

18. The following statements have been proposed for plant vegetative development:

(A) Lateral roots develop from epidermal cells.

(B) Shoots axillary meristem develops from shoot apical meristem during differentiation of leaf primordia.

(C) Root cap is made of dead cells.

(D) Lateral meristem and cylindrical meristem found in roots and shoots results in secondary growth.

Which of the above statements are true? **(JUNE 2012)**

(a) (A) and (B) (b) (B) and (D)

(c) (A), (B) and (D) (d) (C) and (D)

19. The group of cells which generates the vascular tissues including the pericycle in roots of higher plants are called **(DEC 2012)**

(a) procambium.

(b) protoderm.

(c) ground meristem.

(d) apical meristem.

20. Mutation in a gene x in *Arabidopsis thaliana* results in more number of lateral root formation. Which one of the following is the correct statement? **(JUNE 2014)**

(a) The gene product acts as a positive factor of lateral root formation.

(b) The gene product acts as a negative regulator of lateral root formation.

(c) The gene product is not likely to be involved in lateral root formation.

(d) The gene product promotes replication for lateral root development

21. Identification of genes that are associated with the development of male and/or female gametophyte and embryogenesis in plants is facilitated by T-DNA mediated insertional mutagenesis.

In an experiment, a transgenic. plant was generated by insertion of T-DNA (containing a Kanamycin-resistance gene) into a gene "A". Self-pollination of the T_0 plant generated F_1 progeny that segregated in a 2 : 1 ratio for resistance : sensitivity to Kanamycin. These observations indicate that **(DEC 2017)**

(a) the mutant allele did not segregate from the wild type allele.

(b) mutation in gene "A" induces lethality in the male gametophyte.

(c) mutation in gene "A" induces lethality in the female gametophyte.

(d) mutation in gene "A" induces zygotic lethality.

22. Extensive molecular genetic studies on miR156, miR172, SPL genes and AP2- like genes have yielded the following functional model on the juvenile → adult → reproductive transition in Arabidopsis:

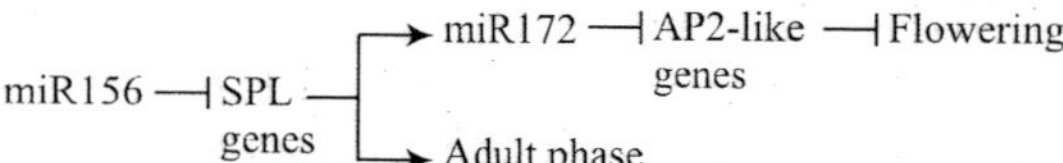

Based on these results, the following schematic diagram has been proposed to predict the expression kinetics of these genetic factors:

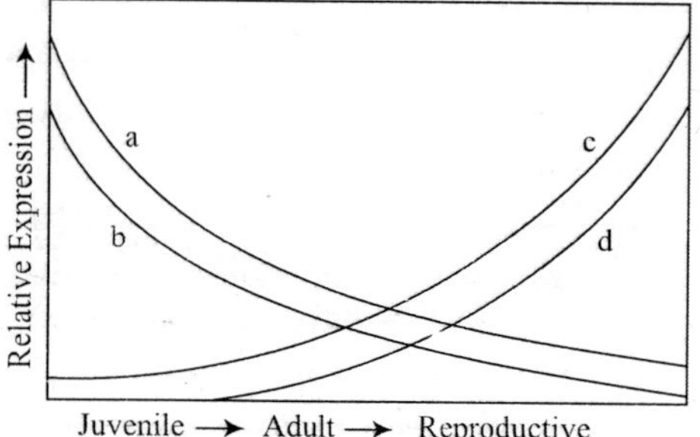

Which of the following combinations is most likely to be correct? **(JUNE 2017)**

(a) a-miR156; b-SPL genes; c -miR172; d - AP2 like genes

(b) a-miR156; b-miR172; c-:-SPL genes; d - AP2 like genes

(c) a-miR 172; b-SPL genes; c-AP2 like genes; d-miR156

(d) a-miR156; b-AP2 like genes; c. miR172; d-SPL genes

23. Consider the following facts regarding the control of shoot apical meristem (SAM) size in Arabidopsis

(a) Loss of the CLAVATA1 (CLV1) gene leads to bigger SAM

(b) Loss of the CLAVATA3 (CLV3) gene leads to bigger SAM

(c) Loss of the WUSCHEL (WUS) gene leads to smaller SAM

(d) Loss of both CLV1 and WUS leads to smaller SAM

(e) Loss of both CLV3 and WUS leads to smaller SAM

(f) Loss of both CLV1 and CLV3 leads to bigger SAM

(g) Over expression of CLV3 leads to smaller SAM

(h) Over expression of CLV3 in the loss of function mutant of CLV1 leads to bigger SAM.

Based on the above information, which of the following genetic pathways describes the relationship among CLV1, CLV3 and WUS most appropriately? **(JUNE 2017)**

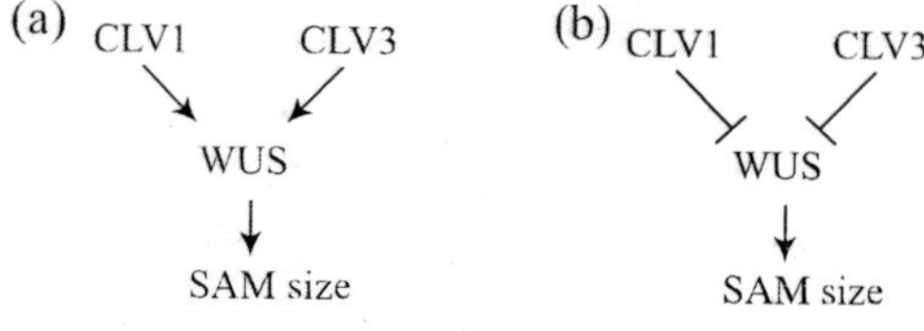

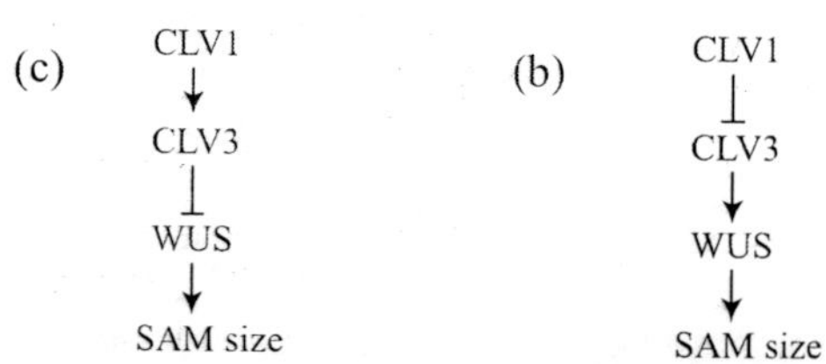

24. The following diagram represents a longitudinal section through an *Arabidopsis* shoot apical meristem (SAM) and leaf primordium at its flank. The dorsal (*D*) and ventral (*V*) domains are marked. The *D* and *V* genes are expressed in the dorsal and ventral domains, respectively.

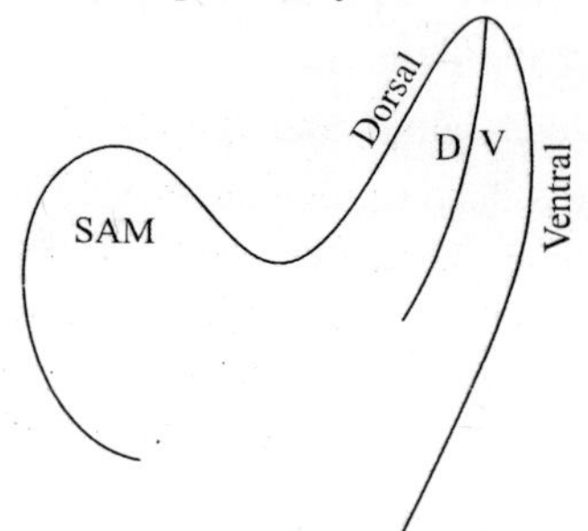

Consider the following statements describing the phenotypes of leaf polarity.

A. Loss of *D* function makes the leaf ventralized whereas its overexpression dorsalizes the leaf.

B. Loss of *V* function makes the leaf dorsalized whereas its overexpression ventralizes the leaf.

C. Loss of m icroRNA *miR166* dorsalizes the leaf whereas its overexpression ventralizes the leaf.

D. *miR166* functions by inhibiting its target mRNA.

Which one of the following functional models best describes the above results?

(DEC 2019) (c)

(a)
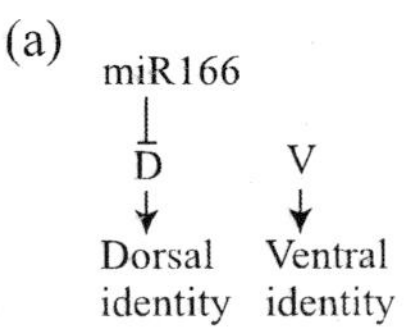

(b)
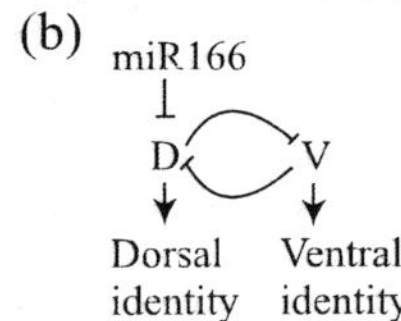

(c)
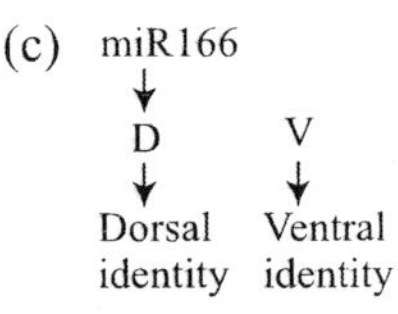

(b)
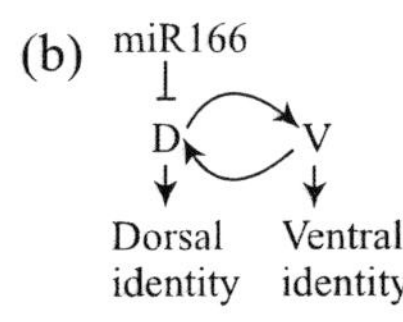

25. Given below are factors/terms associated with various aspects of plant growth and development:

	Column A		Column B
A.	Vacuolar-type	i.	Self-incompatibility
B.	S-locus	ii.	Adaxial-Abaxial patterining of leaf
C.	CONSTANS	iii.	Megasporogenesis
D.	Knox	iv.	Flowering in long photoperiods

Which one of the following options represents the most appropriate match between allterms of Column A and Column B?

(DEC 2019 ASSAM)

(a) A-ii, B-iii, C-iv, D-i
(b) A-iii, B-, C-iv, D-ii
(c) A-ii, B-iv, C-, D- iii
(d) A-iv, B-i, C-ii, D-iii

26. Change in leaf morphology is observed during transition from vegetative to reproductive phase in several plants. The following statements are proposed to explain the above observation:

A. Alteration in the gene content of leaves of reproductive phase from those of vegetative phase.

B. Differential methylation pattern of genes influencing leaf development and morphology.

C. Mutation in transcription factor that prevents its association with promoter elements of genes regulating leaf development.

D. Small RNA mediated inhibition of gene expression of a homeotic gene.

Which one of the following options represents a correct combination of statements that could explain the observed changes? **(JUNE 2018)**

(a) B and C (b) A and D
(c) B and D (d) A and C

27. Given below are statements related to different aspects of plant growth and development.

A. Leaf longevity is increased in ethylene insensitive mutants eir1-1 and ein2 of Arabidopsis.

B. Programmed cell death (PCD) is responsible for the formation of prickles, thorns and spines in plants.

C. Senescence and PCD occur only in the development of vegetative tissues and does not occur in reproductive tissues

D. Re-differentiation of organelles is an integral component during initial stages of senescence in plants.

Which one of the following represents the combination of all correct statements?

(JUNE 2019)

(a) A, C and D (b) B and C
(c) A, B and D (d) C and A

Answer Key

1.(b)	**2.**(c)	**3.**(c)	**4.**(c)	**5.**(a)	**6.**(c)	**7.**(d)
8.(d)	**9.**(a)	**10.**(d)	**11.**(a)	**12.**(a)	**13.**(b)	**14.**(b/c)
15.(b)	**16.**(d)	**17.**(b)	**18.**(b)	**19.**(a)	**20.**(b)	**21.**(d)
22.(d)	**23.**(c)	**24.**(c)	**25.**(b)	**26.**(c)	**27.**(c)	

16. Floral development in Arabidopsis

1. Floral organ development is controlled by overlapping expression of 'A' class, 'B' class and 'C' class genes in different whorls. In an *Arabidopsis* mutant, the flowers had sepals, sepals, carpels and carpels in the four whorls. Mutation in which one of the following is the cause for the mutant phenotype? **(DEC 2015)**
   (a) 'A' class gene alone
   (b) 'B' class gene alone
   (c) 'A' and 'B' class genes
   (d) 'C' class gene alone
2. Which of the floral whorls is affected in agamous (ag) mutants **(DEC 2011)**
   (a) Sepals and petals
   (b) Petals and stamens
   (c) Stamens and carpels
   (d) Sepals and carpels
3. Which of the floral whorls is affected in apetala 3/pistillata (*ap3/pi*) mutants? **(JUNE 2013)**
   (a) Sepals and petals.
   (b) Petals and stamens.
   (c) Stamens and carpels.
   (d) Sepals and stamens.
4. Plants with *apetala2* mutation show a phenotype that lacks **(DEC 2013 GU)**
   (a) sepals only.
   (b) both sepals and petals.
   (c) petals only.
   (d) stamens only.
5. In agamous mutant (flower within flower phenotype) which of the following statements is valid? **(MODELPAPER)**
   (a) Class A genes are expressed in the first two whorls, Class B genes are expressed in the second and third whorls and Class C genes are expressed in the third and fourth whorls.
   (b) Class A genes are not expressed. Class B and C genes are expressed in all the whorls.
   (c) Class A genes are not expressed. Class B genes are expressed in the second and the third whorls and Class C genes are expressed in all the whorls.
   (d) Class A genes are expressed in all the whorls. Class B genes are expressed in the second and the third whorls.
6. Based on ABC model during flower development, loss of class A activity results in the formation of only stamen and carpel. Which of the following floral organ identity genes controls the class A activity? **(DEC 2017)**
   (a) APETALA 1 and APETALA 2
   (b) APETALA 3 and PISTILLATA
   (c) Only PISTILLATA
   (d) Only AGAMOUS
7. Following are certain statements regarding the activities of homeotic genes of classes A, B and C involved in floral organ identity:
   A. Activity of A alone specifies sepals
   B. Activity of B alone specifies petals
   C. Activities of B and C form stamens
   D. Activity of C alone specifies carpels
   Which one of the following combinations of above statements is correct? **(DEC 2014)**
   (a) A, B and C (b) A, B and D
   (c) B, C and D (d) A, C and D
8. Individual and overlapping expression of homoeotic genes in adjacent whorls of a flower determine the pattern of floral organ development. In an *Arabidopsis* mutant, floral organs are distributed as follows:
   Whorl 1 (outer most) – carpel
   Whorl 2 – stamens
   Whorl 3 - stamens
   Whorl 4 (inner most) – carpel
   Loss of function mutation in which one of the following genes would have caused the above pattern of floral organ development? **(DEC 2016)**

(a) APETALA 2 (b) APETALA 3
(c) PISTILLATA (d) AGAMOUS

9. According to the ABC model of floral development in *Arabidopsis* as shown below

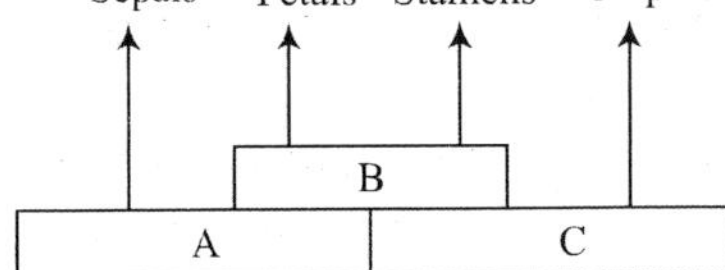

Several genes/transcription factors e.g. AP 1, AP2, AP3, AG etc., are involved. Which one of the following statements is correct?

(JUNE 2015)

(a) Apetala 2 (AP2) transcripts expressed during sepal and petal development.
(b) Agamous AG is considered as class A gene.
(c) AP1 expressed during carpel development.
(d) AP3 expressed during sepal development.

10. The control of flowering is a complex process involving several key regulatory genes. Some statements on flower development are given below:

A. Two major types of genes regulate floral development: meristem identity genes and floral organ identity genes.
B. The important genes in Arabidopsis that play key regulatory roles in meristem identity are: APETALA1, LEAFY and SUPPRESSOR OF CONSTANS1.
C. The genes that determine floral organ identity were discovered as floral homeotic mutants.
D. Most plant homeotic genes belong to a class of related sequences known as FAD box genes.

Which one of the following combinations of the above statements is correct? **(DEC 2013)**

(a) A, B and C (b) B, C and D
(c) A, C and D (d) A, B and D

11. Which one of the following statements about LEAFY (LFY), a regulatory gene in *Arabidopsis thaliana*, is correct? JUNE 2015

(a) LEAFY (LFY) is involved in floral meristem identity
(b) LEAFY (LFY) is involved in leaf expansion.
(c) LEAFY (LFY) 'is involved in root meristem identity.
(d) LEAFY (LFY) is responsible far far-red light mediated seedling grawth.

12. Flowers represent a complex array of functionally specialized structures that differ substantially from the vegetative plant body in form and cell types.

Following are statements made regarding floral meristems.

A. Floral meristems can usually be distinguished from vegetative meristems by their larger size.
B. The increase in the size of the meristem is largely a result of increased rate of cell division in central cells.
C. The increase in the size of the meristem is due to larger size of the cells, which in turn results from rapid cell expansion only.
D. A network of genes control floral morphogenesis in plants.

Which combination of the above statements is true? **(DEC 2012)**

(a) A, B and D (b) A, Band C
(c) B, C and D (d) A, C and D

13. *AP1* (APETLA 1) is one of the floral meristem identifying genes. In wild type *Arabidopsis thaliana* plants transformed with *AP1*:GUS, β glucuronidase (GUS) activity is seen in floral meristem, only after the commitment to flowering. Ectopic expression of *AP1*::GUS in the EMBRYONIC FLOWER (*emf*) mutant background results in GUS activity throughout the shoots in four day old seedlings. These observations suggest that *AP1* is:

(DEC 2011)

(a) not involved in flowering
(b) involved in repression of flowering.

(c) involved in promoting flowering .

(d) stimulation of flowering in the *emf* background

14. Which one of the following options related to plant growth and development is matched correctly? **(DEC 2019 ASSAM)**

(a) LEAFY: regulator of genomic imprinting

(b) Xylogenesis: wood formation

(c) MET1: Inflorescence architecture

(d) Caspases: inhibition of apoptosis

15. In most angiosperms, flowers have a perianth that consists of whorls of organs. Select the statement that is INCORRECT about the floral organs in angiosperms. **(DEC 2019 ASSAM)**

(a) All petals are not homologous

(b) If there is only one whorl of organs, these structures are referred to as tepals

(c) Sepals tend to be initiated almost simultaneously and never in a spiral sequence.

(d) Both sepals and petals can act as protective organs for inner floral organs and also as organs to attract pollinators

Answer Key

1.(b)	**2.**(c)	**3.**(b)	**4.**(b)	**5.**(d)	**6.**(a)	**7.**(d)
8.(a)	**9.**(a)	**10.**(a)	**11.**(a)	**12.**(a)	**13.**(b)	**14.**(b)
15.(c)						

SECTION-6

PLANT PHYSIOLOGY

1. Photosynthesis - Light Reaction

1. The quantum yield of oxygen evolution during photosynthesis drastically drops in far-red light. This effect is known as: **(JUNE 2015)**
   (a) Far red drop.
   (b) Red drop.
   (c) Blue drop.
   (d) Visible spectrum drop.
2. Photosystem II functions as a light dependent water plastoquinone oxidoreductase. What are the names of two reaction center protein that bind electron transfer prosthetic group, such as P680, pheophytin and plastoquinone? **(JUNE 2012)**
   (a) CP43 and CP47
   (b) D1 and D2
   (c) 33 kDa and 23 kDa
   (d) F_A and F_B
3. The photon intensity captured by green plants is used in the following processes: **(DEC 2017)**
   (i) Photosynthesis
   (ii) Generation of heat
   (iii) Production of toxic products such as superoxide, singlet oxygen etc.
   (iv) Damage to D1 protein of PSII
   Based on the above facts, photoinhibiton will happen when the:
   (a) entire photon intensity is used for photosynthesis
   (b) excess photon intensity is completely used for heat generation
   (c) excess photon intensity is used for heat generation and formation of toxic products
   (d) excess photon intensity leads to the damage of D1 protein
4. Which of the following mechanisms is NOT involved in providing photo protection to plants? **(DEC 2013)**
   (a) Degradation of D1 protein.
   (b) Zeaxanthin formation.
   (c) Photolysis of water.
   (d) Thermal dissipation
5. In chloroplast, the site of coupled oxidation-reduction reactions is the **(DEC 2014)**
   (a) outer membrane
   (b) inner membrane
   (c) thylakoid space
   (d) stromal space
6. Which one of the following is the function of cytochrome b_6f complex? **(DEC 2013 GU)**
   (a) Oxidation of water to O_2
   (b) Oxidation of plastohydroquinone (PQH_2)
   (c) Reduction of $NADP^+$ to NADPH.
   (d) Production of ATP.
7. Which one of the following is NOT a characteristic property of carotenoids? **(DEC 2014)**
   (a) They possess complex porphyrin ring.
   (b) They are integral constituent of thylakoid membrane.
   (c) They are also called accessory pigments.
   (d) They protect plants from damages caused by light.
8. If an *Arabidopsis* plant, mutated in lycopene biosynthetic pathway is grown in sunny tropical climate in the presence of oxygen: **(JUNE 2013)**
   (a) it would accumulate higher biomass due to higher rate of photosynthesis.
   (b) there will not be any influence of this mutation on the rate of photosynthesis and plant growth.
   (c) it would show reduced biomass due to photo oxidative damage.
   (d) the leaves would be bluish purple in color because of higher accumulation of xanthophylls.
9. Upon absorption of a photon, a chlorophyll molecule get converted to its excited state when the energy of photon is **(JUNE 2012)**

(a) more than that of ground state of pigment molecule

(b) equal to that of pigment molecule's excited state

(c) more than that of ground state but lesser then excited state of the pigment molecule

(d) equal to energy gap between ground state and excited state energy

10. After absorbing light, chlorophyll molecules in green plants exist in singlet and triplet states. Following are certain statement on singlet and triplet states of chlorophyll molecules:

A. Singlet state is short lived compared to triplet state.

B. Singlet state is long lived compared to triplet state.

C. Singlet state contains electrons with anti parallel. spins while triplet state has electrons with parallel spins.

D. Singlet state contains electrons with parallel spins while triplet state has electrons with anti-parallel spins.

Which one of the following combinations is correct? **(JUNE 2019)**

(a) A and B (b) B and C

(c) A and C (d) B and D

11. In plants, the energy of sunlight is first absorbed by the pigments present in their leaf cells followed by the fixation of carbon through photosynthesis. Consider the following statements.

(A) Chlorophylls a and b are abundant in green plants.

(B) Chlorophylls c and d are found in some protists and cyanobacteria.

(C) Out of different types of bacteriochlorophyll, type a is the most widely distributed.

(D) Out of different types of bacteriochlorophyll, type b is the most widely distributed.

Which one of the following combination of above statements is correct? **(DEC 2012)**

(a) A, B and C (b) A, C and D

(c) B, C and D (d) A, B and D

12. In photosynthetic electron transport, electrons travel through carriers organized in the "Z-scheme". The following are indicated as directions of electron flow:

A. P680 $\rightarrow$ PQ_A $\rightarrow$ PQ_B $\rightarrow$ Cytb_6f $\rightarrow$ Pheo $\rightarrow$ PC $\rightarrow$ P700

B. P700 $\rightarrow$ A_0 $\rightarrow$ A_1 $\rightarrow$ FeS_x $\rightarrow$ FeS_A $\rightarrow$ FeS_B $\rightarrow$ Fd

C. P680 $\rightarrow$ Pheo $\rightarrow$ PQ_A $\rightarrow$ PQ_B $\rightarrow$ Cytb_6f $\rightarrow$ PC $\rightarrow$ P700

D. P700 $\rightarrow$ A_1 $\rightarrow$ A_0 $\rightarrow$ FeS_B $\rightarrow$ FeS_A $\rightarrow$ FeS_X $\rightarrow$ Fd

Which of the following combinations is correct? **(DEC 2016)**

(a) A and B (b) B and C

(c) C and D (d) A and D

13. Which one of the following is the correct order of electron transport during light reaction in the thylakoid membrane of chloroplast? **(JUNE 2016)**

(a) P680 $\rightarrow$ Cytochrome b_6f $\rightarrow$ PC $\rightarrow$ PQ

(b) P680 $\rightarrow$ PC $\rightarrow$ Cytochrome b_6f $\rightarrow$ PQ

(c) P680 $\rightarrow$ PQ $\rightarrow$ PC $\rightarrow$ Cytochrome b_6f

(d) P680 $\rightarrow$ PQ $\rightarrow$ Cytochrome b_6f $\rightarrow$ PC

14. A 'Z' scheme describes electron transport in O_2-evolvingphotosynthetic organisms. The direction of electron flow is presented in the following sequences:

A. P680* $\rightarrow$ Pheophytin $\rightarrow$ Q_A $\rightarrow$ Q_B $\rightarrow$ PC $\rightarrow$ Cytochrome b_6f $\rightarrow$ P700

B. P700* $\rightarrow$ Phylloquinone $\rightarrow$ FeS_A $\rightarrow$ FeS_B $\rightarrow$ FeS_x $\rightarrow$ Fd

C. P680* $\rightarrow$ Pheophytin $\rightarrow$ Q_A $\rightarrow$ Q_B $\rightarrow$ Cytochrome b_6f $\rightarrow$ PC $\rightarrow$ P700

D. P700* $\rightarrow$ Phylloquinone $\rightarrow$ FeS_x $\rightarrow$ FeS_A $\rightarrow$ FeS_B $\rightarrow$ Fd

Which one of the following combinations is correct? **(JUNE 2015)**

(a) A and B (b) B and C

(c) C and D (d) D and A.

15. Given below are a few statements regarding light reaction of photosynthesis

A. Photo system II oxidizes water to O_2 and in the process releases protons into lumen.

B. Photosystem II oxidizes water to O_2 and in the process releases protons into stroma.

C. ATP synthase . produces ATP while protons go back through it from lumen to stroma.

D. ATP synthase produces ATP while protons go back through it from stroma to lumen.

Which one of the following combinations of above statements is correct? **(DEC 2013 GU)**

(a) B and D (b) A and D

(c) Band C (d) A and C

16. Light reactions of photosynthesis are carried out by four major protein complexes: Photosystem I (PSI), photosystem II (PSII), the cytochrome b_6f complex and ATP synthase. The following are certain statements on PSI:

A. PSI reaction centre and PSII reaction centre are uniformly distributed in the granal lamellae and stromal lamellae.

B. The electron donor for the P700 of PSI is plastocyanin and electron acceptor of P700* is a chlorophyll known as A_0.

C. The core antenna and P700 are bound to two key proteins *PsaA* and *PsaB*.

D. Cyclic electron flow occurs from the reducing side of PS I via plastohydroquinone and b_6f complex. This supports ATP synthesis but does not reduce $NADP^+$.

Which one of the following combinations of the above statements is correct?**(JUNE 2016)**

(a) A, B and C (b) A, C and D

(c) A, B and D (d) B, C and D

17. The names of some genes in Column X and their involvement in photosystem I (PSI) or photosystem II (PS II) is provided in Column Y.

Column X		Column Y	
A	*psb* A	i.	PS II
B	*psa* A	ii.	PS I
C	*lhcbl*		
D	*lhcal*		

Which one of the following combinations represents correct matches between columns X and Y? **(DEC 2019 ASSAM)**

(a) A-i, B-ii, C-i, D-ii

(b) A-ii, B-, C-ii, D-i

(c) A-i, B-ii, C-ii, D-I

(d) A-ii, B-, C-i, D-ii

18. Following are statements related to the organization of the four major protein complexes of thylakoid membrane,

A. Photosystem II is located predominantly in the stacked regions of the thylakoid membrane.

B. Photosystem I is found in the unstacked regions protruding into stroma.

C. Cytochrome B_6f complex is confined to stroma only.

D. ATP synthase is located in the unstacked regions protruding into stroma.

Which one of the following combinations of above statements is correct? **(DEC 2012)**

(a) A, B and C (b) A, B and D

(c) B, C and D (d) C, D and A

19. Following are some facts regarding localization of photosynthetic supramolecular complexes on plastid lamellae:

A. PSII is preferentially localized on granal lamellae

B. ATP synthase and PSI is preferentially localized in stromal lamellae

C. PSI and PSII are located adjacent to each other in stromal lamellae

D. Cytchrome b_6/f complex is not a membrane bound complex

Which of the following combinations of the above statements is true? **(JUNE 2012)**

(a) A and B (b) C and D
(c) B and D (d) B and C

20. DCMU inhibits electron transport in chloroplast by preventing the reduction of **(MODEL PAPER)**

(a) P680. (b) Q_A.
(c) PQ. (d) Q_B.

21. The herbicide, dichlorophenyldimethylurea (DCMU), is an inhibitor of **(DEC 2016)**

(a) Shikimate pathway for biosynthesis of aromatic amino acids.
(b) Electron transport from P680 to P700.
(c) Branch in amino acid pathway.
(d) Electron transport from P700 to ferrodoxin.

22. A researcher wanted to study light reaction during photosynthesis by blocking photosynthetic electron flow using the herbicide, di chloro phenyl dimethyl urea (DCMU) and paraquat. The researcher listed the following observations:

A. Both DCMU and paraquat block the electron flow in Photosystem II
B. Both DCMU and paraquat block the electron flow in Photosystem I
C. DCMU blocks electron flow in Photsystem I while paraquat blocks in Photosystem II
D. DCMU blocks electron flow in Photosystem II while paraquat blocks in Photosystem I

Which of the following combinations of the above statements is ***INCORRECT****?* **(JUNE 2017)**

(a) A, B and C (b) A, B and D
(c) A, C and D (d) B, C and D

23. In an experiment, spinach chloroplasts are illuminated in the absence of ADP and Pi, then the light is turned off and ADP and Pi are added. Identify the correct statement regarding the status of ATP synthesis **(DEC 2013 GU)**

(a) The rate of hydrolysis of ATP will be enhanced in dark
(b) No synthesis of the ATP at all in the dark
(c) ATP cannot be synthesized from ADP and Pi without the addition of appropriate enzyme
(d) ATP is synthesized for a short time in the dark

24. 'A' is an inhibitor of chloroplast function. The production of O_2 and synthesis of ATP are measured in illuminated chloroplast before and after addition of 'A' as shown below. **(DEC 2016)**

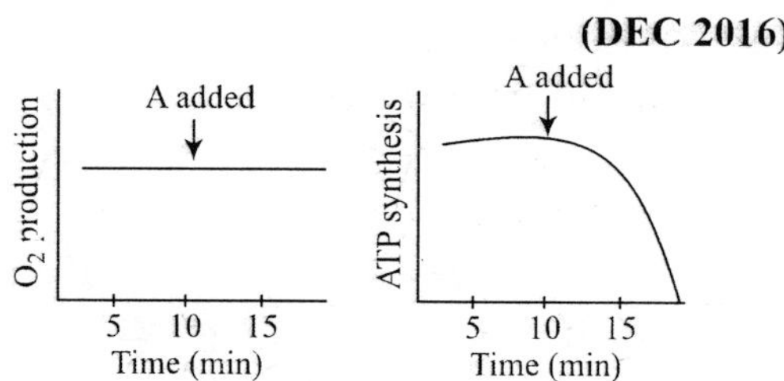

Which statement is correct?

(a) 'A' inhibits the reduction of $NADP^+$
(b) 'A' inhibits the proton gradient and the reduction of $NADP^+$
(c) 'A' inhibits the proton gradient but not the reduction of $NADP^+$
(d) 'A' inhibits neither the proton gradient nor the reduction of $NADP^+$

Answer Key

1.(b)	2.(b)	3.(d)	4.(c)	5.(*)	6.(b)	7.(a)
8.(c)	9.(d)	10.(c)	11.(a)	12.(b)	13.(d)	14.(c)
15.(d)	16.(d)	17.(a)	18.(b)	19.(a)	20.(d)	21.(b)
22.(a)	23.(d)	24.(c)				

2. CO_2 fixation-C3, C4 and CAM pathways.

1. During photosynthetic carbon reduction cycle in green leaves, net production of one molecule of glyceraldehyde 3-phosphate requires one of the following combinations of energy equivalents: **(JUNE 2013)**

(a) 9 NADPH and 6 ATP.
(b) 3 NADPH and 9 ATP.

(c) 2 NADPH and 3 ATP.

(d) 6 NADPH and 9 ATP.

Which one of the following reactions takes place during the reduction phase of the Calvin- Benson cycle? **(DEC 2019)**

(a) Ribulose 1,5-bisphosphate to 3-phosphoglycerate

(b) 1,3-bisphosphoglycerate to glyceraldehyde-3 phosphate

(c) Dihydroxyacetone phosphate to fructose 1,6-bisphosphate

(d) Ribulose 5-phosphate to ribulose 1,5 bisphosphate

2. In which molecule would the radiolabel appear the earliest when wheat and sugar cane leaves are fed with $^{14}CO_2$? **(MODELPAPER)**

(a) Wheat : malate, sugarcane : 3 phosphoglycerate.

(b) Wheat : aspartate, sugarcane : malate.

(c) Wheat: 3 phosphoglycerate, sugarcane : 3 phospho glycerate.

(d) Wheat : 3-phosphyoglycerate, sugarcane : malate.

3. Atmospheric CO_2 contains the naturally occurring stable carbon isotopes ^{12}C and ^{13}C in the proportion of 98.9% and 1.1%, respectively. Following are some of the statements regarding CO_2 assimilation:

A. Both C_3 and C_4 plants assimilate less $^{13}CO_2$ than $^{12}CO_2$.

B. Both C_3 and C_4 plants assimilate less $^{12}CO_2$ than $^{13}CO_2$.

C. C_3 plants assimilate lesser $^{13}CO_2$ than $^{12}CO_2$ as compared to C_4 plants.

D. C_4 plants assimilate lesser $^{13}CO_2$ than $^{12}CO_2$ as compared to C_3 plants.

Which one of the following combinations of above statements is true? **(JUNE 2011)**

(a) A and B (b) A and C

(c) C and D (d) A and D

4. For which one of the following physiological studies $^{12}CO_2$ and $^{13}CO_2$ are used? **(DEC 2015)**

(a) Estimate the rate of photosynthesis

(b) Determine rate of photorespiration

(c) The ratio of C_3 and CAM pathways of CO_2 fixation.

(d) The ratio of C_3 and C_4 pathways of CO_2 fixation

5. The C_4 carbon cycle is a CO_2 concentrating mechanism evolved to reduce photorespiration. The following are stated as important features of the C_4 pathway:

A. The leaves of C_4 plants have Kranz anatomy that distinguishes mesophyll and bundle sheath cells.

B. In the peripheral mesophyll cells, atmospheric CO_2 is fixed by phosphoenol pyruvate carboxylase yielding a four-carbon acid.

C. In the inner layer of mesophyll, NAD-malic enzyme decarboxylates four-carbon acid and releases CO_2.

D. CO_2 is again re-fixed through Calvin cycle in the bundle sheath cells.

Which one of the following combinations is correct? **(DEC 2016)**

(a) B, C and D (b) A, B and C

(c) A, B and D (d) A, C and D

6. Following are some of the statements regarding the effect of CO_2 concentration on photosynthesis in plants.

A. With elevated CO_2 levels, C_3 plants are much more responsive than C_4 plants under well watered conditions.

B. In C_3 plants, increasing intracellular CO_2 partial pressure can stimulate photosynthesis only over a narrow range.

C. In C_4 plants, CO_2 compensation point is nearly zero.

Which one of the following combination of above statements is correct? **(DEC 2011)**

(a) A and B (b) B and C

(c) A and C (d) Only C

7. The CO_2 compensation point for C_3 plants is greater than C_4 plants because in C_3 plants **(DEC 2018)**

(a) dark respiration is higher
(b) dark respiration is lower
(c) photorespiration is present
(d) photorespiration is absent

8. Following are certain statements regarding C_3, C_4 and CAM plants? **(DEC 2014)**

A. The ratio of water loss to CO_2 uptake is higher in CAM plants than it is in either C_3 and C_4 plants.
B. The rate of photosynthesis attains maximum rate at lower intracellular CO_2 partial pressure in C_4 plants than in C_3 plants.
C. The compensation point in C_3 plants are always lower than C_4 plants.
D. Plants with C_4 metabolism need less rubisco than C_3 plants to achieve a given rate of photosynthesis..

Which one of the following combinations of above statements is correct?

(a) A and B (b) A and C
(c) C and D (d) B and D

9. The quantum yield of photosynthetic carbon fixation in a C_3 plant and C_4 plant is studied as a function of leaf temperature. Following are some statements based on this study.

A. At lower temperature the quantum yield of C_3 plant is lower than C_4 plant.
B. In C_4 plant quantum yield does not show a temperature dependence.
C. Since the photorespiration is low in C_4 plants because of CO_2 concentrating mechanism, quantum yield is not affected.
D. At higher temperature the quantum yield of C_3 plant is lower than C_4 plant.

Which one of the following combination of above statements is correct? **(DEC 2011)**

(a) A, B and D (b) B, C and D
(c) A, B and C (d) A, C and D

10. Which one of the following graphs best represents the net CO_2 fixation of typical C_3 and C_4 plants under increasing CO_2 concentration and saturating light? **(DEC 2019)**

(a)
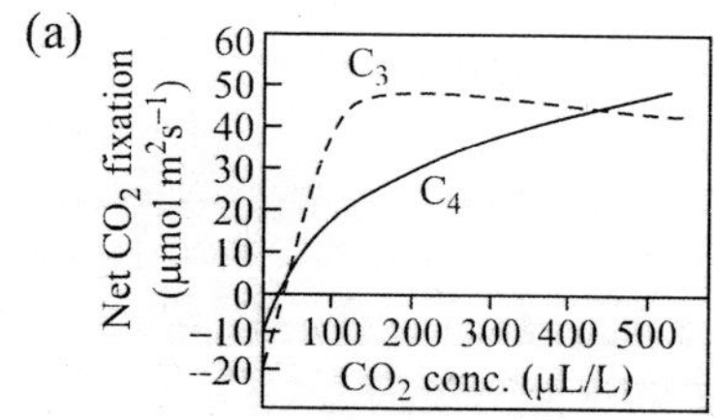

(b)
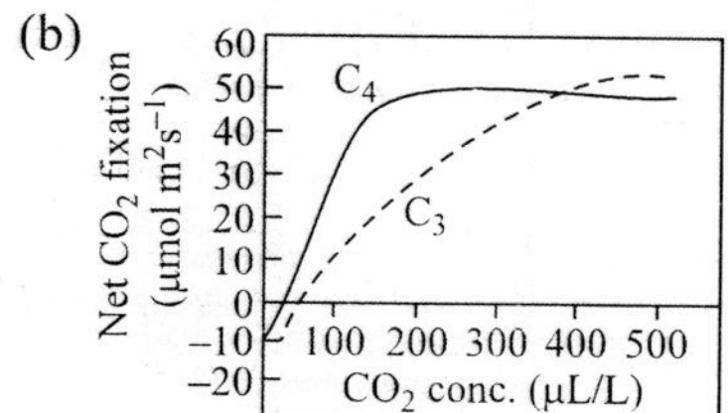

(c)
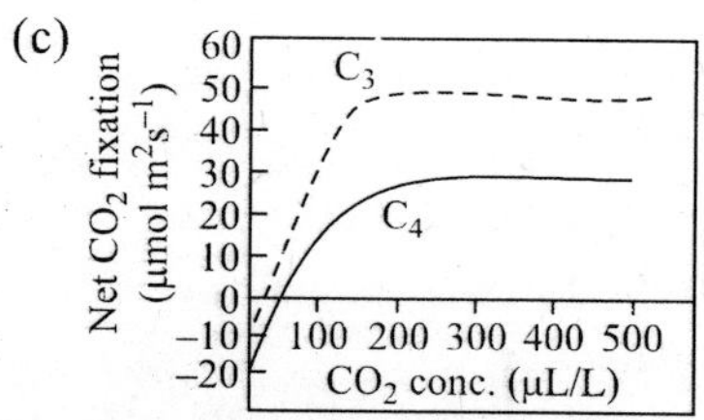

(d)
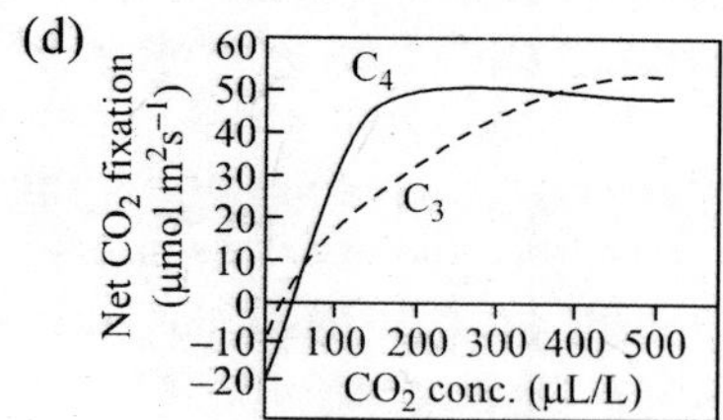

11. Following are certain statements regarding CO_2 assimilation in higher plants:

A. The action of aldolase enzyme during Calvin-Benson cycle produces fructose 1,6-bisphosphate.
B. The conversion of glycine to serine takes place in mitochondria during C_2 oxidative photosynthetic carbon cycle.
C. During C_4, carbon cycle, NAD-malic enzyme releases the CO_2 from the 4-carbon acid, malate yielding a 3 carbon acid, pyruvate.

D. Malic acid during crassulacean acid metabolism (CAM) is stored in mitochondria during dark and released back to cytosol during day.

Which one of the following combinations of above statements is correct? **(DEC 2015)**

(a) A, B and C (b) A, C and D
(c) B, C and D (d) A, B and D

12. Following are certain statements regarding Rubisco, the predominant protein in plant leaves that catalyzes the initiai reaction of the Calvin-Benson cycle.

A. During the oxygenase activity of Rubisco, O_2 is used as substrate to produce three carbon molecule, 3-phosphoglyceratc and two-carbon molecule, 2-phosphoglycolate.

B. In red and brown algae, the large subunit of Rubisco is localized in the chloroplast while small subunit is localized in the nucleus.

C. The bound sugar phosphates in Rubisco are specifically removed by an ATP dependent enzyme, Rubiscoactivase.

D. The active form of Rubisco catalyzes carboxylation or oxygenation reactions in five steps.

Which one of the following combinations of above statements is correct? **(JUNE 2018)**

(a) A, B and C (b) A. B and D
(c) B, C and D (d) A, C and D

13. Following observations were recorded while studying physiological parameters of sorghum and wheat under similar conditions:

A. Sorghum RUBISCO exhibits relatively higher affinity for CO_2 compared to that of wheat.

B. Light saturation of net photosynthetic flux is relatively lower for sorghum compared to that of wheat.

C. Warburg effect is difficult to record for sorghum and could be said as "not measurable" whereas it could be easily recorded for wheat.

D. Temperature optimum for net photosynthesis is lower for sorghum compared to that of Wheat.

E. $^{13}C/^{12}C$ ratio of assimilate is relatively higher for sorghum compared to that of wheat.

Which one of the following combination of the above observation is correct? **(DEC 2018)**

(a) Only A, B and C (b) Only B, C and E
(c) Only A, B and D (d) Only A, C and E

14. The table below shows photosynthetic type, temperature and sunlight intensity levels.

Photosynthetic type	Temperature	Sunlight Intensity
A. C_3 Plant	i. High	P. High
B. C_4 Plant	ii. Moderate	Q. Moderate
	iii. Low	R. Low

Which of the following correctly matches the plant photosynthetic type with the temperature and sunlight conditions in which photosynthetic rate per unit leaf area is maximum for that plant? **(DEC 2019)**

(a) A-i-P; B-iii-R (b) A-iii-P; B-i-Q
(c) A-i-R; B-ii-Q (d) A-ii-Q; B-i-P

Answer Key

1.(d)	**2.**(d)	**3.**(b)	**4.**(d)	**5.**(c)	**6.**(c)	**7.**(c)
8.(d)	**9.**(b)	**10.**(c)	11.(a)	**12.**(d)	**13.**(d)	**14.**(d)

3. Sucrose-Starch Synthesis

1. The photosynthetic assimilation of atmospheric CO_2 by leaves yield sucrose and starch as end products of two gluconeogenic pathways that are physically separated. Which one of the following combination of cell organelles are involved in such physical separation of the process? **(DEC 2015)**

(a) Sucrose in cytosol and starch in mitochondria.

(b) Sucrose in chloroplasts and starch in cytosol.

(c) Sucrose in mitochondria and starch in cytosol.

(d) Sucrose in cytosol and starch in chloroplasts.

2. Following are some enzyme-substrate combinations involved in the conversion of photosynthetically formed triose-phosphate to sucrose.

A. Fructose-1, 6-bisphosphate aldolase: Dihydroxyacetone phosphate.

B. UDP-glucose pyrophosporylase: UDP-Glucose.

C. Sucrose 6-phosphate synthase: Fructose-6-phosphate.

D. Phosphofructo kinase: Fructose 6 phosphate

Which one of the following combinations of above enzyme-substrate combinations is correct? **(DEC 2013 GU)**

(a) A, B and D (b) A, C and D

(c) B, C and D (d) A, B and C

3. From the following statements:

A. Triose phosphate is utilized for the synthesis of both starch and sucrose.

B. Triose phosphate is translocated to cytosol from chloroplast.

C. Triose phosphate is confined to chloroplast and is utilized for synthesis of starch only.

D. Triose phosphate is translocated from cytosol to chloroplast.

Which one of the following combinations is correct regarding starch and sucrose synthesis during day time? **(DEC 2012)**

(a) A and B (b) B and C

(c) C and D (d) D and A

4. The following statements were made with the assumption that the concentration of 3 phosphoglycerate is high inside chloroplasts of an actively photosynthesizing leaf.

A. There will be high concentration of triose phosphate in the chloroplast.

B. The activity of ADP-glucose pyrophosphorylase will be inhibited.

C. The carbon flow will be diverted from sucrose to starch.

D. Starch synthesis will be inhibited and carbon flow will be more towards sucrose synthesis.

Which one of the following combinations of above statements is correct? **(DEC 2019)**

(a) A and B (b) B and D

(c) C and D (d) A and C

5. The following statements were made regarding photosynthesis:

A. Conversion of 3-phosphoglycerate into glyceraldehyde 3- phosphate is a reduction process which utilizes ATP as well as NADH in Calvin cycle

B. Conversion of ribulose 5-phosphate to ribulose 1,5-bisphosphate utilizes one third of the total ATP requirements of carbon fixation in Calvin cycle.

C. Sucrose is synthesized outside the chloroplast and uses uridine triphosphate to activate glucose.

D. Starch is a glucose polymer which is synthesized in chloroplast stroma and uses ATP

Which one of the following options represents a combination of correct statements?

(DEC 2019 ASSAM)

(a) C and D (b) B and D

(c) A and C (d) A and D

6. The following statements are made regarding a plant cell:

A. Polysaccharides of various sugars with methyl esters of galacturonic acid are generically referred as chitin.

B. Suberin and cutin are polymers and are mixtures of polyesters of hydroxy fatty acids and glycerol

C. Cellulose is a polymer of glucose and fructose which can be stained by light green and haematoxylin

D. Plasmodesmata maintain continuity between the cytoplasm of adjacent cells and allow large molecules to pass between them without crossing membrane.

Which one of the following options represents a combination of correct statements?

(DEC 2019 ASSAM)

(a) A and C (b) B and C

(c) B and D (d) A and B

Answer Key

1.(d)	2.(c)	3.(a)	4.(d)	5.(b)	6.(c)

4. Photorespiration, Respiration, ETC, PPP

1. Which one of the following metabolites moves from mitochondria to peroxisome during the operation of the C_2 oxidative photosynthetic cycle? **(JUNE 2018)**

   (a) Glycerate (b) Glycolate

   (c) Glycine (d) Serine

2. During the operation of C_2 oxidative photosynthetic cycle, which of the following metabolites is transported from chloroplast to peroxisome? **(JUNE 2014)**

   (a) Glycerate (b) Serine

   (c) Glycine (d) Glycolate

3. Which of the following statements are associated with the process of photorespiration in plants? **(MODELPAPER)**

   (A) Photorespiration takes place in only C_3 plants.

   (B) Photorespiration takes place in only C_4 plants.

   (C) Photorespiration takes place in both C_3 and C_4 plants.

   (D) Glycolate is oxidized to glyoxylate in the peroxisome.

   (E) Glycolate is oxidized to glyoxylate in the mitochondira.

   (a) (A) and (D) (b) (C) and (D)

   (c) (B) and (E) (d) (C) and (E)

4. Ribulose bisphosphate carboxylase (Rubisco) catalyzes both carboxylation and oxygenation of ribulose-1, 5-bisphbsphate. The latter reaction initiates a physiological process known as 'photorespiration'. The following are certain statements on photorespiration:

   A. The active sites on Rubisco for carboxylation and oxygenation are different.

   B. One of the steps in photorespiration is conversion of glycine to serine.

   C. 50% of carbon lost in chloroplast due to oxygenation is recovered through photorespiration.

   D. The pathway of photo respiration involves chloroplast, peroxisome and mitochondria.

   Which one of the following combinations of above statements is correct? **(JUNE 2016)**

   (a) A and C (b) A and D

   (c) B and D (d) C and D

5. Pyrurate dehydrogenase is subject to feed back inhibition by its products in glycolysis. Some of the chemical compounds which might be involved in the process, are listed below:

   A. NADH B. FADH

   C. Acetyl-CoA D. Acetaldehyde

   Which one of the following combinations of above chemical compounds is involved in feedback inhibition of pyruvate dehydrogenase?

   (JUNE 2015)

   (a) A and B (b) B and C

   (c) C and D (d) A and C

6. During episodes of anoxia in plants, pyruvate produced in glycolysis is initially fermented to lactate. During later stage, there is an increase in the fermentation to ethanol and decrease in the fermentation to lactate, a phenomena which helps plants survive anoxia. Which of the following statements is correct about this change of fermentation flux from lactate towards ethanol? **(JUNE 2011)**

   (a) The cytosolic pH increases, thus activating both lactate dehydrogenase and pyruvate decarboxylase activity.

(b) The cytosolic pH increases, thus inhibiting lactate dehydrogenase activity and activating pyruvate decarboxylase activity.

(c) The cytosolic pH decreases, thus activating both lactate dehydrogenase and pyruvate decarboxylase activity.

(d) The cytosolic pH decreases, thus inhibiting lactate dehydrogenase and activating pyruvate decarboxylase activity

7. Pyruvate dehydrogenase complex has sophisticated regulatory mechanisms including reversible phosphorylation. Various metabolites are also able to modulate the dehydrogenase complex. Which one of the following statements is NOT correct? **(DEC 2013 GU)**

(a) Pyruvate stimulates the kinase.

(b) Pyruvate inhibits the kinase.

(c) Ammonia stimulates the kinase.

(d) Ammonia inhibits pyruvate oxidation.

8. Which one of the following statements is INCORRECT about the role of oxidative pentose phosphate pathway in plant metabolism? **(JUNE 2013)**

(a) Generation of NADPH required to drive biosynthetic reactions.

(b) Production of pentose phosphate for the synthesis of nucleic acids.

(c) Formation of erythrose 4-phosphate for biosynthesis of aromatic amino acids.

(d) Production of NADH to generate ATP

9. The oxidative pentose phosphate pathway provides the reducing equivalents for nitrite reduction in plastids (Ieucoplasts) of non-green tissues. Which one of the following statements would be correct for the above mentioned pathway? **(JUNE 2013)**

(a) Glutamate synthesized from NH_4^+ is translocated from cytosol to leucoplast

(b) α-ketoglutarate is translocated from cytosol to leucoplast.

(c) Glucose-6-phosphate is translocated and moves from leucoplast to cytosol.

(d) Triose phosphate is translocated from cytosol to leucoplast.

10. Following statements are related to oxidative phoshorylation.

A. Redox reactions of electron transport chain coupled with ATP synthesis are collectively called oxidative phosphorylation.

B. Three major processes: glycolysis, oxidative pentose phosphate pathway and citric acid cycle are related to oxidative phosphorylation.

C. Electron transport proteins are bound to outer of the two mitochondrial membranes.

D. In the electron transport chain electrons are transferred to oxygen from NADH.

Which one of the following combinations of above statements is correct? **(JUNE 2014)**

(a) A and D (b) B and C

(c) C and D (d) A and C

11. Following are certain statements with regard to plant respiration:

A. Metabolism of glucose into pyruvate through glycolysis generates NADH and not NADPH.

B. Metabolism of glucose through oxidative pentose phosphate cycle does not produce NADPH.

C. Cyanide forms a complex with haem iron of cytochrome oxidase leading to prevention of change in valency, which in turn stops electron transport in therespiratory chain.

D. Alternative oxidase is insensitive to cyanide and has higher Km than that of cytochrome oxidase.

Which one of the following combinations is correct? **(JUNE 2019)**

(a) A, B and C (b) B, C and D

(c) B and D (d) A, C and D

12. According to the current model of alternative oxidase regulation, the following factors cause induction of alternative oxidase:

(A) significant increase in the ubiquitin pool in the cytosol.

(B) presence of α-keto acids (like pyruvate and glyoxylate).

(C) cold stress.

(D) increase in cytosolic ATP concentration.

Which one of the following combinations of above statements is true? (JUNE 2013)

(a) (A) and (D) (b) (B) and (C)
(c) (A) and (B) (d) (A) and (C)

13. Which of the following statements with respect to alternate oxidase activity in cyanide-resistant respiration in plants, is not correct? **(JUNE 2011)**

(a) Alternate oxidase accepts electrons directly from cytochrome C.

(b) Some plants exhibit thermogenesis during inflorescence development.

(c) Transcription of alternate oxidase gene is often induced by various abiotic stresses.

(d) When electrons pass to alternate oxidase, two sites of proton pumping are bypassed.

14. Following are certain statements regarding respiratory metabolism in plants:

A. Respiratory quotient during partial breakdown of carbohydrate (alcoholic fermentation) will be infinity.

B. Respiratory quotient indirectly provides information about (i) nature of the substrate used for respiration and (ii) the relative rate of competing respiratory processes.

C. Breakdown of organic acids in mature fruit will exhibit a respiratory quotient value of more than one since organic acids are relatively oxygen-rich compared to other common substrates.

D. Anabolic metabolism can influence respiratory quotient by removing reduction equivalents for respiration leading to decrease in oxygen uptake.

Which one of the following combination of the above statement is correct? **(DEC 2018)**

(a) Only A (b) Only Band C
(c) Only D (d) A, B, C and D

15. First committed precursor in the biosynthesis of chlorophyll is: **(DEC 2019 ASSAM)**

(a) δ-aminolevulinic acid
(b) Porphobilinogen
(c) Protoporphyrin IX
(d) Coproporphyrinogen III

Answer Key

1.(d)	**2.**(d)	**3.**(a)	**4.**(c)	**5.**(d)	**6.**(d)	**7.**(a)
8.(d)	**9.**(b)	**10.**(a)	**11.**(d)	**12.**(b)	**13.**(a)	**14.**(d)
15.(a)						

5. Microbial Nitrogen Fixation

1. Which of the following organism do not possess the ability to fix nitrogen? **(DEC 2012)**

(a) Organisms specialized for high altitude
(b) Marine plankton
(c) Eukaryotic organisms
(d) Acidophilic organisms

2. In the process of nitrification by organisms, the respective bacteria A and B in the following reaction are: **(MODELPAPER)**

$$NH_4 + 1/2O_2 \xrightarrow{A} NO_2 + 2H^+ + H_2O$$

$$NO_2 + 1/2O_2 \xrightarrow{B} NO_3$$

(a) *Azotobacter, Nitrobacter*
(b) *Nitrobacter, Azotobacter*
(c) *Nitrosomonas, Nitrobacter*
(d) *Nitrobacter, Nitrosomonas*

3. Free-living nitrogen fixers can survive in different ecological niches. Identify the incorrect combination from the following list: **(JUNE 2013)**

(a) *Azotobacter* - acidic soil.
(b) *Deraxia* - alkaline soil.
(c) *Beijernckia* - acid soil.
(d) *Frankia* - neutral soil.

4. Important chemical reactions involved in nutrient cycling in ecosystems are given below:

a. $NO_2^- \rightarrow NO_3^-$ b. $N_2 \rightarrow NH_3$

c. $NH_4^+ \rightarrow NO_2^-$ d. $NO_3^- \rightarrow N_2$

The organisms associated with these chemical reactions are **(DEC 2015)**

(a) *a - Nitrosomonas b-Pseudomonas c - Nostoc d - Nitrobacter*

(b) *a - Pseudomonas b-Nitrobacter c - Nostoc d -Nitrosomonas*

(c) a - *Nitrobacter b-Nostoc c-Nitrosomonas d - Pseudomonas*

(d) *a-Nostoc b-Nitrosomonas, c - Nitrobacter d - Pseudomonas*

5. In the global nitrogen cycle, the following microbial organisms are involved in three important process- denitrification, nitrification and nitrogen fixation.

(a) *Rhyzobium* *(b) Nitrosomonas*

(c) Nitrobacter *(d) Pseudonomas*

(e) Azotobacter

Which of the following is correctly matched pair of process and its causative species?

(DEC 2011)

(a) Denitrification – (b); nitrogen fixation – (c) and (e); nitrification – (d)

(b) Denitrification – (d); nitrogen fixation – (a) and (e); nitrification – (c)

(c) Denitrification – (e); nitrogen fixation – (a) and (d); nitrification – (d)

(d) Denitrification – (b); nitrogen fixation – (a) and (d); nitrification – (c)

6. Match the following associations involved in dinitrogen fixation with their representative genera **(JUNE 2016)**

	Associations		Genera
A.	Heterotrophic nodulate	(i)	Azotobacter
B.	Heterotrophic non-nodulate	(ii)	Frankia
C.	Phototrophic associative	(iii)	Nostoc
D.	Phototrophic free living	(iv)	Rhodospirillum

(a) A - (ii); B -(i); C - (iv); D - (iii)

(b) A - (iii); B - (i); C - (ii); D - (iv)

(c) A - (i); B -(ii); C - (iii); D - (iv)

(d) A - (ii); B - (i); C - (iii); D - (iv)

7. Nitrogen gas is reduced to ammonia by nitrogen fixation' method. In order to -execute the process, which one of the following compounds is usually required? **(JUNE 2015)**

(a) ATP (b) GTP

(c) UDP (d) ADP

8. Symbiotic biological nitrogen fixation takes place with the association between a plant and a nitrogen fixing prokaryote as shown in the following table: **(JUNE 2012)**

List of plants	**Nitrogen fixing**
A. Soyabean	(a) *Frankia*
B. *Casurina*	(b) *Bradyrhyzobium*
C. *Guneria*	(c) *Anabena*
D. *Azolla*	(d) *Nostoc*

The correct combination is:

(a) A-1, B-2, C-3, D-4

(b) A-2, B-1, C-4, D-3

(c) A-3, B-2, C-1, D-4

(d) A-4, B-3, C-2, D-1

9. Nitrogenase, a complex metal containing enzyme is involved in conversion of N_2 to $NH_{(c)}$ Which one of the following metal is NOT involved in the activity of nitrogenase?

(JUNE 2019)

(a) Molybdenum (Mo)

(b) Iron (Fe)

(c) Vanadium (V)

(d) Cobalt (Co)

10. Nitrogen fixation is basically a process of converting nitrogen gas into ammonia (NH_3). One of the key enzymes in the process is "nitrogenase". The production and activity of nitrogenase is very highly regulated as highlighted below:

A. Nitrogen fixation through nitrogenase is an energetically expensive process.

B. Nitrogenase encoding gene is under a constitutive promoter.

C. Nitrogenase is highly sensitive to oxygen.

D. Endogenous availability of the cofactor of nitrogenase enzyme is very low.

Which one of the following combinations of above statements is correct? **(JUNE 2014)**

(a) A and B (b) A and C

(c) B and C (d) B and D

11. The following are certain facts regarding biological nitrogen fixation in plants:

A. Oxygen irreversiblyinactivates nitrogenase enzyme involved in nitrogen fixation.

B. The nod genes that code for nodulation proteins are activated by NodD

C. The two components of nitrogenase enzyme complex, the Fe protein and MoFeprotein, can show catalytic activity independently.

D. During the reaction catalyzed by nitrogenase enzyme, the Fe protein reduces the MoFe protein while the MoFe protein reduces N_2.

Which one of the following combination of the above statements is correct? **(DEC 2013)**

(a) A, B and C (b) B, C and D

(c) A, C and D (d) A, B and D

12. Rhizobial genes that participate in legume nodule formation are called nodulation (nod) genes. The nodD-encoded protein. **(JUNE 2016)**

(a) is an acetyl transferase that adds a fatty acyl chain to the Nod factor.

(b) binds to the nod box and induces transcription of all nod genes.

(c) catalyzes the linkage of N-acetyl glucosamine residues.

(d) influences the host specificity of Rhizobium.

13. Which one of the following combinations of nodulation (*nod*) genes are found in all rhizobial strains? **(DEC 2019 ASSAM)**

(a) *nod A, nod B and nod C*

(b) *nod P, nod Q and nod H*

(c) *nod F, nod E and nod L*

(d) *nod A, nod E and nod P*

14. The nodulation (*nod*) genes are classified as common nod genes or host specific nod genes. Some statements related to such classification are given below:

A. *nodA* is a common *nod* gene and *nodC* is a host specific gene.

B. *nodB* is a common *nod* gene and *nodP* is a host specific gene.

C. *nodQ* is a common *nod* gene and *nodA* is a host specific gene.

D. *nodH* is a common *nod* gene and *nodQ* is a host specific gene.

Choose the correct answer from the above statements: **(DEC 2015)**

(a) A and B (b) C and D

(c) A only (d) B only

15. Following statements were made with respect to symbiotic association of rhizobia with legumes:

A. *nodD* is a regulatory gene.

B. Nod factors are lipochitin oligosaccharides.

C. Nod factors predominantly have α 1- 4 linked N-acetyl-D-glucosamine backbone.

D. Receptors (or Nod factors are protein kinases with extracellular sugar-binding Lys M domain.

Which one of the following combinations represents all correct statements? **(DEC 2018)**

(a) A, B and C (b) A, C and D

(c) B, C and D (d) A, B and D

16. Symbiotic nitrogen fixation. in legume nodules involves complex interaction between Rhizobium and legume roots. This complex interaction is governed by

A. Integration of *sym* plasmid of *Rhizobium* in the root nuclear genome.

B. Sensing of plant flavonoids by rhizobia.

C. Activation of nod genes in rhizobia.

D. Activation of NODULIN genes in legume roots.

Which one of the following combinations is correct? **(JUNE 2015)**

(a) A, B and C (b) A, C and D

(c) B, C and D (b) A, B and D

17. In an experiment to show that biogeochemical cycles interact, nitrogen fixing vines *(Galactia* sp.) were grown in plots under normal levels of CO_2 (Control) and under artificially elevated atmospheric CO_2 (Experimental). Effect of elevated CO_2 levels on nitrogen fixation was measured over a period of 7 years (Plot A) and the concentrations of iron and molybdenum in the leaves of these plants were quantified at the end of the study (Plot B). Which one of the following inferences CANNOT be made from the above experiment? **(JUNE 2019)**

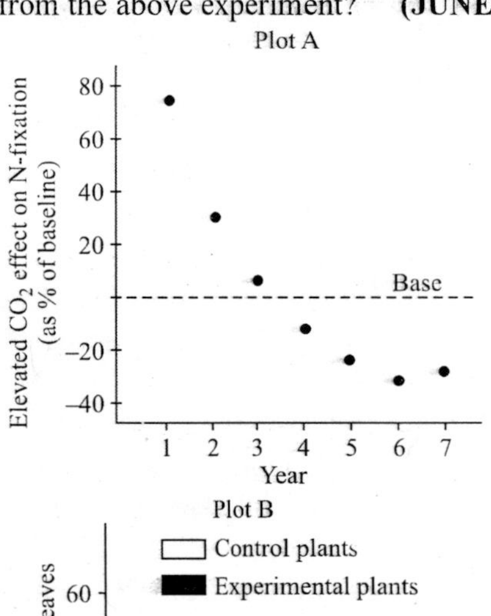

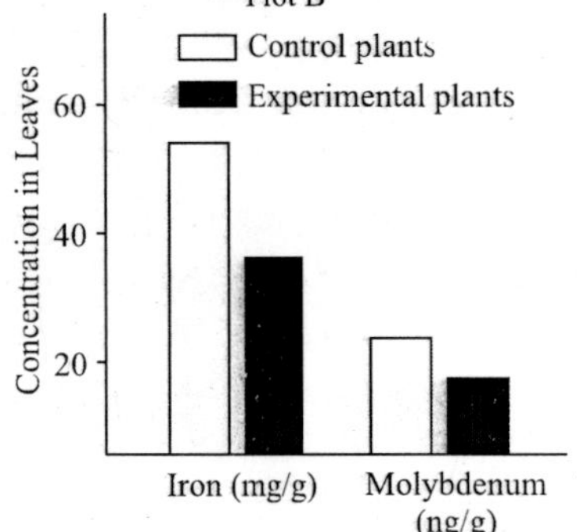

(a) Decreasing rate of N-fixation correlates with decreased levels of leaf iron and molybdenum, two micronutrients essential for N-fixation

(b) An initial exposure to elevated CO_2 increased N-fixation by these plants

(c) There is a continuous decrease in N-fixation due to elevated CO_2 treatment.

(d) Plants exposed to continuous elevated levels of CO_2 had lower levels of iron and molybdenum in their leaves.

Answer Key

1.(c)	**2.**(c)	**3.**(a)	**4.**(c)	**5.**(b)	**6.**(a)	**7.**(a)
8.(b)	**9.**(d)	**10.**(b)	**11.**(d)	**12.**(b)	**13.**(a)	**14.**(d)
15.(d)	**16.**(c)	**17.**(c)				

6. Nitrate and ammonium assimilation

1. In higher plant leaves, the reduction of nitrate to ammonium takes place by the combined action of nitrate reductase localized in cytosol and nitrite reductase localized in **(MODEL PAPER)**

(a) peroxisomes. (b) mitochondria.

(c) chloroplasts. (d) cytosol.

2. Which of the following is NOT a prosthetic group of nitrate reductase **(DEC 2011)**

(a) FAD (b) Heme

(c) Mo (d) pterin

3. Nitrate reductase is a key enzyme of nitrate assimilation process. Which of the following statements is INCORRECT about the nitrate reductase? **(DEC 2013 GU)**

(a) Nitrate reductase-mediated reaction involves the transfer of one electron.

(b) Nitrate reductase is influenced by nitrate and carbohydrate.

(c) Nitrate reductase converts nitrate to nitrate in the cytosol.

(d) Nitrate reductase is influenced by light.

4. Only members of the plant kingdom and many bacteria have capability of biological nitrogen reduction. In this regard following statements are given:

A. Nitrogen is normally taken by the plant in their fully oxidized form but needs to be reduced before incorporation in organic molecules.

B. Conversion of oxidized nitrogen into reduced nitrogen needs energy in the form of NAD(P).

C. The metal associated with the enzyme nitrate reductase is Magnesium.

D. Nitrate reduction takes place in the cytoplasm, whereas nitrite reduction takes place in chloroplast.

Which one of the following combinations of the above statements is correct? **(JUNE 2019)**

(a) A and C (b) A, B and C

(c) B and D (d) A and D

5. Nitrate reductase is an important enzyme for nitrate assimilation. Given below are some statements on nitrate reductase enzyme:

(A) Nitrate reductase of higher plants is composed of two identical subunits.

(B) One subunit of nitrate reductase contains three prosthetic groups.

(C) One of the prosthetic groups attached to both subunits is heme.

(D) One of the prosthetic groups complexed with pterin is magnesium.

Which one of the following combination of statements on nitrate reductase mentioned above is correct? **(DEC 2012)**

(a) A, B and C (b) A, C and D

(c) B, C and D (d) A, B and D

6. Which one of the following is a precursor for the biosynthesis of tyrosine, phenylalanine and tryptophan? **(DEC 2013 GU)**

(a) 3-Phosphoglycerate.

(b) phosphoenolpyruvate.

(c) Pyruvate.

(d) Oxaloacetate.

7. Aspartate kinase is a key enzyme in the lysine amino-acid biosynthesis in plants. With an objective of increasing the lysine content in maize seeds, maize plants were transformed with *E. coli* aspartate kinase with a strong seed specific plant promoter. Resulting transgenic plants were found to express the transgene; however, the content of lysine did not increase. Which of the following option best explain the possible reason? **(MODELPAPER)**

(a) Bacterial proteins are not stable in plants.

(b) Bacterial proteins are not properly folded in plants.

(c) Proper post-translational modification did not take place in plants.

(d) Lysine causes feed back inhibition of aspartate kinase

Answer Key

1.(c)	2.(d)	3.(a)	4.(d)	5.(a)	6.(b)	7.(d)

7. Plant hormones

Which one of the following parts of root is involved in perceiving gravity? **(DEC 2019)**

(a) Quiescent center

(b) Endodermis

(c) Root cap

(d) Elongation zone

1. The plant hormone indole-3-acetic acid (IAA) is present in most plants. The structure of this hormone is related to which one of the following amino acids? **(DEC 2014)**

(a) Glutamic acid (b) Aspartic acid

(c) Lysine (d) Tryptophan

2. A young dicot seedling (e.g soyabean) is subjected to gravity stimulus by laying it horizontally on a surface the shoot bends upwards and root bends downward. Indicate the reason. **(MODELPAPER)**

(a) Redistribution of auxin throughout the seedlings is responsible for stimulatory unequal growth in shoots and roots.

(b) Redistribution of auxin in shoots while cytokinine in roots is responsible for stimulatory unequal growth.

(c) Redistribution of auxin in roots while cytokinine in shoots is responsible for stimulatory unequal growth.

(d) Redistribution of cytokinine throughout the seedlings is responsible for stimulatory unequal growth in shoots and roots.

3. The plant hormones, auxins and cytokinins, and their interactions play an important role in regulating apical dominance. The following figure represents an experiment related to the study of gene interactions that influence axillary bud outgrowth or dormancy. Q, Z and M represent genes involved in phytohormone pathway.

Based on the above figure, the following statements were made:

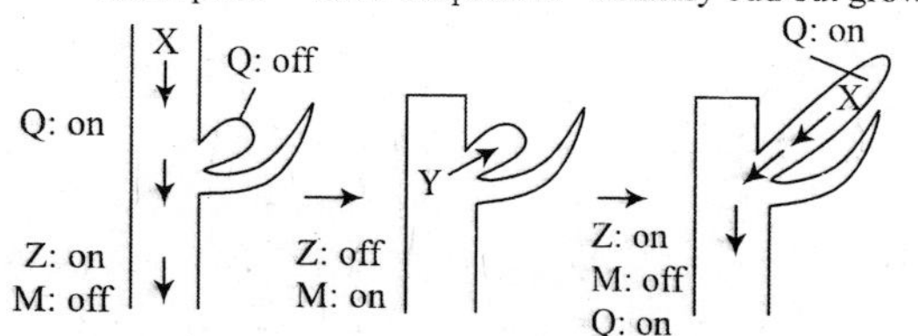

A. 'X' is an auxin that maintains expression of 'Q' and 'Z' and represses 'M'.

B. 'Y' is a cytokinin that promotes axillary bud growth and is induced by 'M'.

C. Decapitation (removal of apex) activates

D. 'X' is a cytokinin that represses 'M'. 'X'

Which one of the following options represents correct statement(s)? **(DEC 2018)**

1. A and C only 2. B and D only
3. A and B only 4. C only

4. Which one of the following statements is correct for the transport of cytokinin in plants? **(DEC 2013 GU)**

(a) It takes place only from root to shoot.

(b) It takes place only from shoot to root

(c) Root and shoot make their own cytokinin therefore, no transport is required.

(d) transport can occur from root to shoot and vice versa.

5. A transgenic lettuce plant was generated by over-expressing isopentenyl transferase (IPT) gene under the control of the promoter of senescence activator gene (SAG12). Following are some statements regarding this transgenic plant. The transgenic plants

A. exhibit delayed senescence.

B. exhibit fast senescence.

C. have higher amount of cytokinin during senescence.

D. have higher amount of gibberellins during senescence.

Which one of the following combinations of above statements is correct? **(DEC 2014)**

(a) A and B (b) A and C
(c) B and D (d) C and D

6. Transgenic tobacco plants over-expressing isopentenyl transferase *(IPT)* under the control of promoter region of Senescence Associated Receptor kinase (P_{SARK}) were exposed to drought for 15 days followed by re-watering for 7 days. The following hypotheses were proposed regarding ch^nges in the transgenic plants at the end of 7 days of re-watering:

A. The plants would be wilted and fail to survive.

B. The plants would be healthy and survive.

C. The plants would show higher production of cytokinin compared to wild type plants.

D. The plants would show higher production of absicic acid compared to wild type plants.

Which one of the following combinations of the above hypotheses is correct? (DEC 2017)

(a) A and C (b) A and D
(c) B and C (d) B and D

7. Which one of the following statements related to components/features of senescence in plants is **INCORRECT?** **(JUNE 2018)**
   (a) Programmed cell death in plants may generate functional cells or tissues.
   (b) Senescence can be induced by application of cytokinins and delayed by overexpression of salicylic acid.
   (c) Plants defective in autophagy demonstrate accelerated plant senescence.
   (d) Leaf senescence is regulated by NAC and WRKY genes families.

8. The dwarf pea mutant (*le*) used by Mendel was defective in which of the following enzyme involved in gibberellin biosynthesis? **(JUNE 2011)**
   (a) ent-Kaurene synthase.
   (b) GA 3 β-hydroxylase.
   (c) GA 20-oxidase.
   (d) ent-Kaurenoic acid hydroxylase.

9. Which one of the following statements regarding seed germination of a wild type plant is NOT correct? **(DEC 2014)**
   (a) Low ABA and high bioactive GA can break seed dormancy.
   (b) Light accompanied with high temperature can break seed dormancy.
   (c) GA induces synthesis of hydrolytic enzymes in cereal grains
   (d) Degradation of carbohydrates and storage proteins provide nourishment and energy to support seedling growth.

10. Gibberellic acid (GA) controls seed germination by directing breakdown of the stored starch. In which one of the following tissues of the barley seed, α-amylase gene is induced in response to GA? **(DEC 2015)**
   (a) Endosperm (b) Coleoptile
   (c) Aleurone layer (d) Embryo

11. A farmer growing a particular variety of grape plants in vineyard, observes the following:
   (A) Fruit size normally remained small.
   (B) Natural seed abortion.
   (C) Development of fungal infection as the pedicels are small in size due to which moisture is retained in the bunches of grapes.

   Experts suggested spraying gibberellic acid during the fruit development. This treatment would help in getting rid of **(DEC 2012)**
   (a) A, B and C (b) only A and B
   (c) only A and C (d) only B and C

12. The plant hormone, Gibberellic Acid is generally **NOT** associated with **(DEC 2017)**
   (a) stem elongation. (b) Parthenocarpy
   (c) parthenogenesis (d) malt production

13. Out of several gibberllins identified in plants, which one of the following is NOT bioactive **(JUNE 2019)**
   (a) GA_1 (b) GA_3
   (c) GA_4 (d) GA_5

14. For which one of the plant hormone biosynthetic pathways, 1-aminocyclo- propane- 1 -carboxylic acid is an intermediate? **(JUNE 2017)**
   (a) Abscisic acid (b) Brassinosteroid
   (c) Ethylene (d) Gibberellic acid

15. Plants make several hormones that are important for growth and development. Some statements on plant hormones are given below:
   A. Auxin is produced primarily in the root apices
   B. Cytokinins are a smaller group of related compounds.
   C. Gibberellins are a large group of related compounds defined not by their biological functions but by their structures.
   D. Brassinosteroids are an important class of plant hormones, which control a broad spectrum of developmental reposes including pollen tube growth.

Which one of the following combination of the above statements is correct? **(DEC 2013)**

(a) A, B and C (b) B, C and D
(c) A, C and D (d) A, B and D

16. A plant hormone that promotes the acquisition of desiccation tolerance in developing seed is **(DEC 2012)**

(a) ABA (b) Ethylene
(c) IAA (d) GA3.

17. Following are certain facts about the effect of abscisic acid (ABA) on the development and physiological effect of plants:

A. ABA promotes leaf senescence independent of ethylene.

B. ABA promotes shoot growth and inhibits root growth at low water potential.

C. ABA inhibits gibberellin induced enzyme production

D. Seed dormancy is controlled by ratio of ABA and gibberllin

Which one of the following combination of above statements is true? **(JUNE 2012)**

(a) A, B and C (b) B, C and D
(c) A, B and D (d) A, C and D

18. Which of the following factors is known to be involved in postponing programmed cell death in cereal aleurone until endosperm mobilization is complete? **(DEC 2019)**

(a) Gibberellic acid
(b) Abscisic acid
(c) Acidic pH of the vacuoles
(d) c-GMP mediated signal transduction pathway

19. Which one of the following statements on plant hormones is **INCORRECT**? **(DEC 2013 GU)**

(a) Salicyclic acid is intimately involved in shoot and root elongation.
(b) Ethylene is involved in senescence and fruit ripening.
(c) Abscisic acid plays a key role in seed germination and drought responses.
(d) Jasmonic acid inhibits root growth and is involved in pathogenesis.

20. Jasmonic acids (JA) that are known to activate many defense responses in plants are biosynthesized in chloroplast and peroxisome. Which one of the following reactions is confined to perosoxisome only during biosynthesis of JA? **(DEC 2013 GU)**

(a) Conversion of 12, 13-epoxy linolenic acid to 12-oxo-phytodienoic acid by the action of allene oxide cyclase.
(b) Conversion of 12-oxo-phytodienoic acid to 3-oxo-2(2'[Z]-pentenyl) cylopentane-1-octanoate by the action of oxo-phytodienoate reductase.
(c) Conversion of linolenic acid to β-hydroperoxy linolenic acid by lipoxygenase.
(d) Conversion of 13-hydroperoxy linolenic acid to 12, 13-Epoxy linolenic acid by allene oxide synthase

21. Which of the following set of cell organelles are involved in the biosynthesis of jasmonic acid through octadecanoid signalling pathway? **(DEC 2011)**

(a) Chloroplast and peroxisomes
(b) Chloroplast and mitochondria
(c) Mitochondria and peroxisomes
(d) Golgi bodies and mitochondria

22. Following are some statements for synthesis of jasmonic acid in plants

A. 12-oxo-phytodienoic acid is produced in chloroplast and transported to peroxisome.

B. Action of lipoxygenase, allene oxide synthase and allene oxide cyclase takes place in peroxisome.

C. 12-oxo-phytodicnoic acid is first reduced and then converted to jasmonic acid by β-oxidation.

D. Final production of jasmonic acid takes place in chloroplast.

E. Action of allene oxide synthase and allene oxide cyclase takes place in chloroplast.

Which one of the following combination of above statements is correct? **(DEC 2011)**

(a) A, B and C (b) B, D and E

(c) C, D and E (d) A, C and E

23. Which one of the following plant derived signalling molecule induces hyphal branching of arbuscular mycorrhizal fungi, a phenomenon that is observed at the initial stages of colonisation of these fungi? **(DEC 2016)**

(a) Salicylic acid (b) Abscisic acid

(c) Strigolactones (d) Systemin

24. Given below are names of phytohormones in column I and their associated features/effects/ function in column II.

I	II
A. Auxin	i. Delayed leaf senescence
B. Gibberellins	ii. Epinastic bending of leaves
C. Cytokinin	iii. Polar transport
D. Ethylene	iv. Removal of seed dormancy.

Select the correct set of combinations from the options given below; **(DEC 2016)**

(a) A-iii, B-ii, C-iv, D-I

(b) A-iv, B-iii, C-i, D-ii

(c) A-iii, B-iv, C-i, D-ii

(d) A-i, B-iv, C-iii, D-ii

25. Following are some statements regarding plant growth hormones.

A. Ethylene regulates abscission.

B. Gibberlins do not play any role in flowering.

C. Auxin and cytokinin promote cell division.

D. Over expression of cytokinin oxidase would promote root growth.

E. ABA inhibits root growth and promotes shoot growth at low water potential.

F. ABA promotes leaf senescence independent of ethylene.

Which one of the following combination of above statements is correct? **(DEC 2011)**

(a) A, C and F (b) B, C and D

(c) D, E and F (d) B, D and E

Answer Key

1.(d)	**2.**(a)	**3.**(c)	**4.**(d)	**5.**(b)	**6.**(c)	**7.**(b)
8.(b)	**9.**(b)	**10.**(c)	**11.**(a)	**12.**(c)	**13.**(d)	**14.**(c)
15.(b)	**16.**(a)	**17.**(d)	**18.**(b)	**19.**(a)	**20.**(b)	**21.**(a)
22.(d)	**23.**(c)	**24.**(c)	**25.**(a)			

8. Plant Hormone –Signaling

Column I	Column II	Column III
(A) Auxin and Gibberellins	(a) Trans-membrane	(i) Response mediated by receptor phosphorylation / dephosphorylation
(B) Cytokinin and Brassino-steroid	(b) Soluble receptor	(ii) Response mediated by proteasome mediated protein degradation

1. Match the above columns involving plant hormones and their signalling pathways: **(DEC 2019)**

(a) A)-(a)-(i) and (B)-(b)-(i)

(b) (A)-(b)-(ii) and (B)-(a)-(i)

(c) (A)-(b)-(i) and (B)-(a)-(i)

(d) (A)-(a)-(ii) and (B)-(b)-(i)

2. The following statements are made to describe auxin signal transduction pathway, from receptor binding to the physiological response:

A. Auxin response factors (ARFs) are nuclear proteins that bind to auxin response elements (Aux REs) to activate or repress gene transcription.

B. AUX/IAA proteins are secondary regulators of auxin-induced gene expression. Binding of AUX/IAA proteins to the ARF protein blocks its transcription regulation. .

C. Auxin binding to TIRI/AFB promotes ubiquitin-mediated degradation and removal of AUX/IAA proteins.

D. Auxin binding to auxin response factors (ARFs) causes their destruction by the 26S proteasome pathway.

Which one of the following combinations of above statements is correct? **(JUNE 2016)**

(a) A, B and C (b) A, C and D
(c) B, C and D (d) A, B and D

3. Which one of the following plant hormones use the two-component histidine kinase receptor system for signal transduction? **(JUNE 2016)**
(a) Auxin (b) Gibberellin
(c) Cytokinin (d) Abscisic acid

4. Which of the following plant hormones can mimic the *det1* mutation, causing de-etiolation and chloroplast development in dark?
(DEC 2013)
(a) Cytokinin (b) Gibberellin
(c) Auxin (d) Ethylene

5. Examples of many factors that regulate plant height in response to gibberellic acid (GA) are listed below:
A. Binding of a GA bound repressor to the promoter of the DELLA domain containing GRAS protein gene and blocking its expression.
B. Binding of the GA receptor complex to GRAS.
C. Directing GRAS for ubiquitination and degradation by the 26S proteasome.
D. Micro RNA directed down regulation of the GRAS protein expression.
Which one of the following combinations is correct? **(DEC 2015)**
(a) A and B (b) B and C
(c) C and D (d) A and D

6. While screening an EMS-mutagenized population of a plant, a researcher identified a mutant with reduced gibberellic acid sensitivity. Which one of the following proteins is most likely to be defective in this mutant?
(DEC 2019)
(a) Sucrose non-fermenting related kinase 2 (SnRK2)
(b) Constitutive triple response 1 (CTR1)
(c) Phytochrome interacting factor (PIF)
(d) Coronative-insensitive 1 (COI1)

7. Ethylene binding to its receptor does NOT lead to **(DEC 2011)**
(a) dimerization of the receptor
(b) phosphorylation of the receptor
(c) activation of CTR Raf kinase
(d) endocytosis of ethylene-receptor complex

8. Ethylene signaling pathway is important for a fruit ripening. Which one of the following responses is routinely used to identity ethylene signaling pathway components? **(JUNE 2014)**
(a) Cotyledon expansion response
(b) Lateral root formation response
(c) Triple response
(d) Flowering time response

9. Dark-grown seedlings display 'triple response' when exposed to ethylene. Which one of the fallowing is NOT a part of 'triple response'?
(JUNE 2015)
(a) Decrease in epicotyl elongation.
(b) Rapid unfolding and expansion of leaves.
(c) Thickening of shoot.
(d) Horizontal growth of epicotyl.

10. Dark grown Arabidopsis seedlings show 'triple response' when exposed to ethylenehormone. Which one of the following options is characteristic of 'triple response'?
(JUNE 2018)
(a) Reduced shoot elongation, increased shoot thickness and tightening of apical hook.
(b) Reduced shoot elongation, reduced shoot thickness and loosening of apical hook.
(c) Increased shoot elongation, increased shoot thickness and loosening of apical hook.
(d) Increased shoot elongation, reduced shoot thickness and tightening of apical hook.

11. Typical morphological defects are routinely used in genetic screens to identify novel genes in signal transduction pathways. Which one of the following morphology has been used to decipher the ethylene signaling pathways?
(JUNE 2013)

(a) Light grown morphology of seedling.
(b) Triple response morphology of seedling.
(c) Dark grown morphology of seedling.
(d) Morphology of true leaves.

12. Ethylene is an important plant hormone that regulates several aspects of plant growth and development. Some statements are given below in relation to ethylene signalling pathways:

A. Unbound ethylene receptors work as positive regulators of the response pathway.
B. There are more than two ethylene receptors known to date.
C. The carboxy terminal half of the ethylene receptor, *ETR1* (Ethylene response 1), contains a domain homologous to histidine kinase catalytic domain.
D. *EIN2* (Ethylene insensitive 2) encodes a transmembrane protein. The ein2 mutation promotes ethylene responses in both seedlings and adult *Arabidopsis* plants.

Which combination of the above statements is correct? **(DEC 2015)**

(a) A and B (b) B and C
(c) C and D (d) D and A

13. Which one of the following is the correct function of JAZ (JASMONATE ZIM-DOMAIN) protein family, a key regulator of Jasmonic Acid (JA) signalling response? **(DEC 2017)**

(a) Binds to MYC2 and represses the JA dependent genes.
(b) Binds to MYC2 and transcribes JA dependent genes.
(c) Acts as receptor of JA signal.
(d) Involved in directly inducing JA dependent genes.

14. Many factors related to the role of abscisic acid (ABA) in contributing to drought, cold and salt resistance in plants are listed below:

A. The transcription factors *DREB 1* and *DREB2* bind to the cis-acting elements of the promoter of ABA responsive genes in an ABA dependent manner.
B. ABA induces many genes such as *LEA* and *RD29*.
C. ABA-responsive genes contain six nucleotide ABRE elements in the promoter.
D. Nine-nucleotide dehydration- responsive elements (ORE) are present in ABA responsive genes.

Which one of the following combinations of the above statements is correct with respect to ABA? **(DEC 2015)**

(a) A, B and C (b) A, C and D
(c) B, C and D (d) A only

15. Brassinosteroids are a group of steroid hormones that function in a variety ofcellular and developmental contexts in plants. Which one of the following acts asan inhibitor of the brassinosteroid receptor? **(JUNE 2018)**

(a) BRI 1 (b) BKI 1
(c) BAK 1 (d) BSK1

16. A specific Mitogen Activated Protein Kinase (MAPK) cascade comprising MEKK1(a MAP3K), MKK1 (a MAP2K) and MPK6 (a MAPK) is activated sequentially in that order in Arabidopsis plants upon perceiving certain abiotic stress stimuli. The activated MPK6 phosphorylates and activates a transcription factor 'X', thereby making plant tolerant to the abiotic stress. Two different variants of MKK 1 protein, a kinase inactive (KI) and a constitutively active (CA) forms were expressed independently in *mkk1* mutant Arabidopsis plant. Considering above facts, which one of the following statements is **CORRECT**? **(DEC 2017)**

(a) 'X' will be activated even in the absence of stimuli in CA plants
(b) 'X' will not be activated in the absence of stimuli in CA plants
(c) 'X' will be activated in the absence of stimuli in KI plants
(d) The KI plants will be tolerant to the abiotic stress.

17. Jasmonate is known to inhibit root growth while auxin facilitates root growth. Upon infection with pathogenic bacteria that produce coronatine, we may expect the following in plants:

A. Upregulation of *COI-1* gene and inhibition of root growth.

B. Upregulation of *Auxl-1* gene and inhibition of root growth.

C. Inhibition of *Auxl-1* gene and promotion of root growth.

D. Inhibition of *COI-1* gene and promotion of root growth.

Which one of the following is correct?

(JUNE 2019)

(a) A, B and C (b) Only A
(c) Only B (d) Only C

Answer Key

1.(b)	**2.**(a)	**3.**(c)	**4.**(a)	**5.**(b)	**6.**(c)	**7.**(d)
8.(c)	**9.**(b)	**10.**(a)	**11.**(b)	**12.**(b)	**13.**(a)	**14.**(a)
15.(b)	**16.**(a)	**17.**(b)				

9. Phytochromes

1. Which one of the following amino acids is induced drastically in dark adapted plants?

(DEC 2019 ASSAM)

(a) Glutamine (b) Glycine
(c) Asparagine (d) Serine

2. Which one of the following is **NOT** a phenotype of dark-grown seedlings that are etiolated? **(DEC 2017)**

(a) Short hypocotyls
(b) An apical hook
(c) Closed cotyledons
(d) Non-photosynthetic proplastids

3. A researcher developed quadruple mutant that disrupted the function of all phytochrome interacting factor (PIF) family members. The following hypotheses were proposed regarding the phenotype of the mutant plants when grown in dark:

A. Plants would show short hypocotyls

B. Plants would be etiolated

C. Light induced genes would be activated

D. The cotyledons would be open

Which one of the following combinations of the above hypotheses is correct? **(DEC 2017)**

(a) A, B and C (b) A, B and D
(c) A, C and D (d) B, C and D

4. In a photoresponse experiment, imbibed seeds were kept under the following light regime's and their germination status was noted as follows:

Light regime				Status
D				Not Germinated
R	D			Germinated
R	FR	D		Not Germinated
R	FR	R	D	Germinated
R	FR	R	FR	Not Germinated

D: Darkness; R: Red light; FR: Far-red light

In an independent biochemical experiment, it was demonstrated that the red light photoreceptor phytochrome is interconverted between two forms, P and P', by red or far-red light. Keeping these information in mind, which of the following combination of conclusions is correct? **(JUNE 2017)**

(a) Red light converts P to P'; P' promotes seed germination

(b) Far-red light converts P to P'; P' promotes seed germination

(c) Red light converts P' to P ; P' promotes seed germination

(d) Far-red light converts P' to P; P promotes seed germination

5. Consider the following statements:

A. Chlorophyll absorbs more in the red region of the visible spectrum than in far-red.

B. The phytochrome photoreceptor (P) of plants occurs in two inter convertible forms, P_r and P_{fr} where red light converts P_r to P_{fr} and far-red light converts P_{fr} to P_r.

C. Growing a sun plant under the canopy shed causes increased stem elongation

Which one of the following combination of statements is correct for the plants growing under the canopy as compared to those growing above the canopy? **(DEC 2019)**

(a) Red far-red ratio is lower; P_r: P_{fr}: ratio is higher, P_{fr} inhibits stem elongation.

(b) Red: far-red ratio is higher, P_r: P_{fr} ratio is higher, P_r, inhibits stem elongation.

(c) Red: far-red ratio is lower; P_r: P_{fr} ratio is lower; P_{fr} promotes stem elongation.

(d) Red: far-red ratio is higher; P_r: P_{fr} ratio is lower; P_r promotes stem elongation.

6. Red and far-red lights are perceived by plants through various photoreceptors including phytochromes. The activation of phytochromes is caused by? **(JUNE 2012)**

(a) Conversion of Pr to Pfr form through the effect of red light

(b) repression of Pr form through the effect of far-red light

(c) equal proportion of red and far red light at same fluence rate

(d) presence of red and far-red light at different fluence rate

7. In higher plants, the red/far-red sensory photoreceptor, phytochrome, is a light-regulated kinase. Which of the following classes of kinases does it represent? **(MODEL PAPER)**

(a) Two-component sensor regulator (histidine kinase).

(b) Two-component sensor regulator (serine/threonine kinase).

(c) Leucine rich repeat (LRR) receptor kinase.

(d) Calcium-dependent protein kinase.

8. The amount of each enzyme present in the chloroplast stroma is regulated by mechanisms that control the concerted expression of nuclear and chloroplast genomes. Following are certain statements regarding the regulation of chloroplast enzymes:

A. Nucleus-encoded enzymes are translated on 70S ribosomes in the cytosol and subsequently transported into the plastid.

B. Plastid encoded enzymes are translocated in the stroma on prokaroyte like 70S ribosomes.

C. Light modulates the expression of stromal enzymes encoded by the nuclear genome via specific photoreceptors.

D. The eight small subunits of rubisco is encoded in plastid.

Which one of the following combinations of above statements is correct? **(JUNE 2014)**

(a) A and B (b) A and C

(c) B and C (d) C and D

9. Which one of the following proteins regulates photo morphogenesis in plants by adding ubiquitin tags to a subset of nuclear proteins? **(DEC 2019 ASSAM)**

(a) PhyA (b) Hy5

(c) LAF1 (d) COP1

10. Phytochrome-mediated control of photomorphogenesis is linked to many other gene functions. The following statements are made on the mechanism of phytochrome action:

A. Phytochrome function requires COP1, an E3 ubiquitin ligase that brings about protein degradation.

B. COP1 is slowly exported from the nucleus to the cytoplasm in the presence of light.

C. HY5 is targeted by COP1 for degradation in the presence of light.

D. HY5 is a transcription factor involved in photomorphogenetic response.

Which of the following combinations is correct? **(DEC 2016)**

(a) A, B and C (b) B, C and D

(c) A, B and D (d) A, C and D

11. Constitutive photomorphogenesis (COP 1) protein, an E3 ubiquitin ligase, regulates the turnover of proteins required for photomorphogenic development. Following are certain independent statements related to the function of COP 1 protein:

A. In light, COP1 along with SPA1 adds ubiquitin tags to a subset of nuclear proteins.

B. The proteins ubiquinated by COP1 and SPA1 are targeted for degradation by the 26S proteasome.

C. In dark COPI is slowly exported to the cytosol from nucleus.

D. The absence of COP 1 in the nucleus permits the accumulation of transcriptional activators necessary for photomorphogenic development.

Which one of the following combinations is correct? **(JUNE 2016)**

(a) A and C (b) A and D

(c) B and C (d) B and D

Answer Key

1.(c)	2.(a)	3.(c)	4.(a)	5.(a)	6.(a)	7.(b)
8.(c)	9.(d)	10.(c)	11.(d)			

10. Cryptochromes, phototropins, stomatal movement

1. Directional growth of plants induced by light is called phototropism. Some statements on phototropism are given below:

A. Phototropism is a photomorphogenetic response.

B. PHOT1 and PHOT2 genes mediate phototropism.

C. CRY1 and CRY2 genes although help to perceive blue light are not involved in phototropism.

D. Perception of blue light by phyA photoreceptor initiates phototropism.

Which one of the following combination of the above statements is correct? **(DEC 2013)**

(a) A, B, and C (b) B, C and D

(c) A, C and D (d) A, B and D

2. Blue light receptor cry1 binds to COP1 and SPA1 complex by interacting with C-terminal region of cry1 (CCT) in a light dependent manner and regulates photomorphogenesis via transcription factor HY5.

Read the following statements:

A. cry 1 binds to COP 1 and SPA1 complex leading to degradation of HY5.

B. cry1 binds to CON and SPA1 complex and prevents degradation of HY5.

C. CCT is overexpressed and the plant are kept in dark.

D. CCT is overexpressed and the plants are kept in light.

Which of the following combination of above statements will result in photo-morphogenesis? **(JUNE 2019)**

(a) Only A (b) Only B

(c) A, B and C (d) B, C and D

3. Which one of the following statements is NOT correct? **(DEC 2013 GU)**

(a) Phototropins are autophosphorylating protein kinases.

(b) Phototropin 1 and 2 are flavoproteins.

(c) Phototropin 1 and 2 are the photoreceptors of blue light.

(d) Phototropin 1 and 2 are not directly involved in phototropic movements.

4. Plants are able to perceive light through various photoreceptors and downstream genes. Which one of the following genes is NOT involved in light perception? **(JUNE 2013)**

(a) *PIF3* (b) *NPR1*

(c) *PHYE* (d) *CRY3*

5. Chloroplastdistribution in a photosynthesizing cell is governed by blue light sensing phototropin 2 (PHOT2). When the cells are irradiated with high intensity blue light the chloroplasts **(DEC 2011)**

(a) move to the side walls

(b) aggregate in the middle of the cell

(c) are sparsely distributed

(d) aggregate in small clusters

6. Plants have evolved with multiple photoreceptors, which can perceive specific wavelength of light. Which of the following statement is correct about photoreceptors? **(JUNE 2012)**

(a) Phytochrome A can receive both far-red and blue light

(b) Phytochrome C can receive far-red light

(c) Cryptochrome I and phytochrome B are for perceiving blue light

(d) Phytochrome B can predominantly perceive far red light

7. Light is crucial for plant growth and development. Following are certain statements related to photoreceptors in model plant *Arabidopsis thaliana.*

A. Among the five phytochrome genes, representing a gene family, PHYB plays a predominant role in red-light perception.

B. Cryptochromes are involved in the regulation of flowering time and hypocotyl length.

C. phyA photoreceptor is predominantly involved in far-red light perception.

D. The LOV domain of phytochrome C (pHYC) is an important domain for signal transmission.

Which one of the following combinations of above statements is correct? **(DEC 2014)**

(a) A, B and C (b) A, C and D

(c) B, C and D (d) A, B and D

8. Light is perceived by various photoreceptors in plants. The photoreceptors predominantly work at specific wavelengths of light. Some of the following statements are related to the functions of plant photoreceptors.

(A) Phytochrome A predominantly perceives the red and far-red light.

(B) Phytochrome B predominantly perceives red light.

(C) Cryptochromes regulate plant development.

(D) Phototropins are involved in blue light perception and chloroplast movements.

Which one of the following combinations based on above statements is correct? **(DEC 2012)**

(a) A, B and C (b) B, C and D

(c) C, D and A (d) A, B and D

9. Light is an important factor for plant growth and development. There are several photoreceptors in higher plants such as *Arabidopsis thaliana* involved in perception of various wavelengths of light. Some statements are given below related to photoreceptors:

A. Red light photoreceptors are represented by a gene family.

B. Phytochrome C is the most prominent photoreceptor to perceive red light.

C. Cryptochrome 1 and cryptochrome 2 have evolved from bacterial DNA photolyases.

D. Far-red light is perceived by phytochrome D.

Which one of the following combinations of above statements is correct? **(JUNE 2015)**

(a) A and B (b) B and C

(c) C and D (d) A and C

10. Light is the dominant environmental signal that controls stomatal movement in leaves of well-watered plants grown in natural environment. Which one of the following wavelengths of light is responsible for such regulation? **(DEC 2014)**

(a) Red light (b) Blue light

(c) Green light (d) Far-red light

11. Perception of blue light in plants causes **(JUNE 2013)**

(a) inhibition of cell elongation and stimulation of stomatal opening.

(b) stimulation of cell elongation and inhibition of stomatal opening.

(c) inhibition of stomatal opening.

(d) inhibition of cell elongation.

12. Stomata from detached epidermis of common dayflower (*Commelina communis*) were treated with saturating photon fluxes of red light. In a parallel treatment, stomata treated with red light were also illuminated with blue light (indicated by arrow). From the graphs shown below, select the correct pattern of stomata opening (solid lines and dotted lines represent stomatal aperture under red and blue lights, respectively). **(JUNE 2018)**

(a)

Stomatal aperture →

Time (h) →

(b)

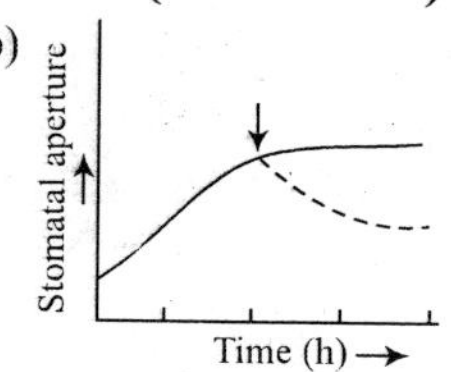

(c)

Stomatal aperture →

Time (h) →

(b)

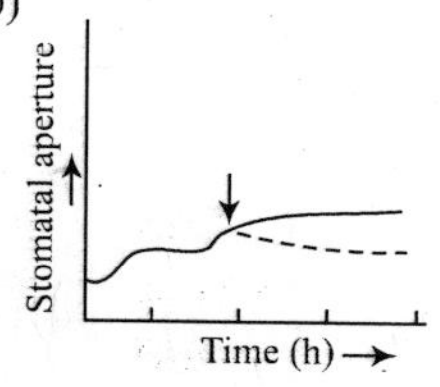

Answer Key

1.(a)	2.(d)	3.(d)	4.(b)	5.(a)	6.(a)	7.(a)
8.(b)	9.(d)	10.(b)	11.(a)	12.(a)		

11. Photoperiodism and biological clocks

1. The following scheme shows the flowering status of a plant species and the photoperiod regimes in which it is grown (L denotes light period; D denotes dark period). **(JUNE 2017)**

Photoperiod	Status
L | D	Flowered
L | D	Not flowered
L | D | D | D	Not flowered
L | D | L | D	Not flowered
L | D	Not flowered
L | D	Flowered

Which of the following conclusions is most appropriate?

(a) The species is a short day plant; length of the dark phase determines flowering status.

(b) The species is a long day plant; length of the dark phase determines flowering status.

(c) The species is a short day plant; length of the light phase determines flowering status.

(d) The species is a long day plant; length of the light phase determines flowering status.

2. Fill in the blanks (a, b, c and d) in the following statements with a proper combination of m, n, 0 and p.

Where

m represents - longer

n represents - shorter

o represents -prevents

p represents - induces

Short day (SD) plants flower when night lengths are ______a_____ than a critical dark period. Interruption of the dark period by a brief light treatment _______b_____ flowering in SD plants. Long day (LD) plants flower when night length is _______c_____ than a critical period. Shortening of the night with a brief light treatment ________d_____ flowering in LD plants. **(JUNE 2011)**

a b c d

(a) m o n p

(b) n p m o

(c) n o m p

(d) m p n o

3. Which one of the following photoreceptors plays a role in day length perception and circadian rhythms? **(JUNE 2016)**

(a) Zeitlupe family

(b) Cryptochromes

(c) Phototropins

(d) UV Resistance locus 8

4. The transition to flowering in plants requires **(JUNE 2011)**
   (a) growth of plants under long- day conditions.
   (b) growth of plants under short- day conditions.
   (c) reprogramming of the shoot apical meristem.
   (d) synthesis of the flowering hormone florigen.

5. Mutations in *CONSTANS* (CO) of *Arabidopsis thaliana* results in late flowering phenotype. Transcript levels of CO were determined in long day and short day seedlings. Which of the following would likely represent the transcript profile of CO? **(JUNE 2011)**

	Long-day seedlings	**Short-day seedlings**
(a)	▬▬▬▬▬	________
(b)	________	________
(c)	________	▬▬▬▬▬
(d)	▬▬▬▬▬	▬▬▬▬▬

6. Flower development is controlled by multiple developmental pathways in *Arabidopsis thaliana*. Which of the following statement on flower development is NOT correct? **(DEC 13 GU)**
   (a) Gibberellins can affect flowering.
   (b) Photoperiod, such as short day cycle (8 hour light and 16 hour darkness), promotes flowering.
   (c) Vernalization or low temperatures control flowering.
   (d) FT protein plays a key role in flowering.

Answer Key

1.(a)	2.(a)	3.(a)	4.(c)	5.(a)	6.(a)

12. Solute/Water transport

1. Which one of the following mineral deficiency will first be visible in younger leaves? **(DEC 2018)**
   (a) Calcium (b) Nitrogen
   (c) Zinc (d) Molybdenum

2. Under which conditions do members of the family Gramineae synthesize and release phytosiderophores? **(DEC 2013)**
   (a) Iron deficiency.
   (b) Phosphorus deficiency.
   (c) Availability of iron complexes in rhizosphere.
   (d) Availability of phosphorus complexes in rhizosphere.

3. Which one of the following essential micronutrients is associated with urease enzyme found in higher plants? **(JUNE 2013)**
   (a) Nickel. (b) Molybdenum.
   (c) Zinc. (d) Copper.

4. The chlorosis (yellowing) symptom of iron deficiency is influenced by **(DEC 2012)**
   (a) Sodium and Potassium
   (b) Sodium and Phosphorus.
   (c) Calcium and Nitrogen
   (d) Potassium and Phosphorus.

5. Which one of the following describes the function of silicon in plants? **(JUNE 2019)**
   (a) Constituent of amino acids
   (b) Contributes to cell wall rigidity and elasticity
   (c) Constituent of the photosynthesis reaction centre
   (d) Maintenance of cell turgor and electroneutrality

6. Vesicular-arbuscular mycorrhiza (VAM) represents a beneficial association between plant roots and fungus, where fungus assists plants in obtaining from the soil **(MODEL PAPER)**
   (a) iron. (b) zinc.
   (c) sulphate. (d) phosphate.

7. Which one of the following statement is correct? **(JUNE 2019)**
   (a) Ecto-mycorrhizal associations predominantly reduce phosphorous limitation, and endo-mycorrhizal associations reduce both nitrogen and phosphrous limitations.
   (b) Endo-mycorrhizal associations predominantly reduce phosphorus limitation, and ecto-mycorrhizal associations reduce both nitrogen and phosphorous limitations.
   (c) Ecto and endo-mycorrhizal associations do not reduce nitrogen and phosphorous limitation.
   (d) Ecto and endo-mycorrhizal associations are able to reduce only phosphorous limitation.
8. Change in Ca^{2+} concentration can initiate various responses in plants. Which one of the following responses is NOT known to be initiated by change in Ca^{2+} concentration? **(DEC 2012)**
   (a) Closure of stomata.
   (b) Reorientation of growth in pollen tubes.
   (c) Thickening of cell walls in young tobacco seedlings in response to wind.
   (d) Lateral root formation.
9. Both halophytes and glycophytes compartmentalize cytotoxic ions into the intracellular compartment or actively pump them out of the cell to the apoplasts with the help of membrane transport proteins. Among these, the Na^+-H^+ antiporter, NHX1, is localized in **(JUNE 2011)**
   (a) the plasma membrane.
   (b) chloroplast (inner envelope).
   (c) mitochondria (outer membrane).
   (d) Tonoplast.
10. The dependence of the rate of sucrose uptake with respect to sucrose concentration in plant cell was studied and data are shown in the following graph. **(JUNE 2011)**

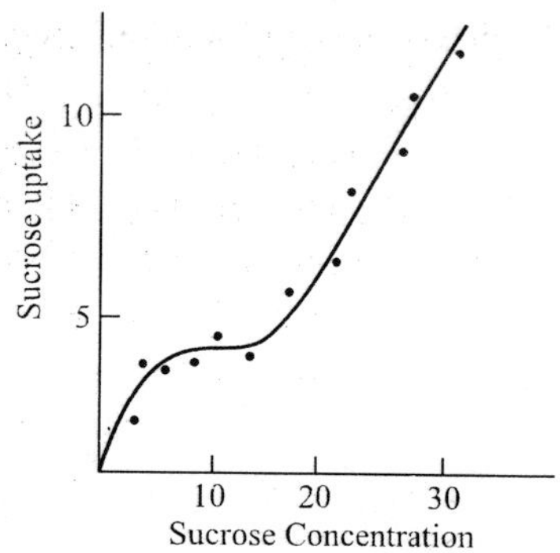

From the above data it can be inferred that
   (a) the sucrose uptake is energy independent and no special carrier is involved.
   (b) the sucrose uptake is energy dependent and a special carrier is involved.
   (c) at lower concentration of sucrose the uptake of sucrose is energy dependent and carrier mediated.
   (d) at higher concentration of sucrose the uptake is energy dependent and carrier mediated.
11. In a study, it was found that K^+ ion concentration in root cells of a pea plant was ~75 times greater than that of the nutrient medium in which the plant was grown. This indicated that K^+ ions were absorbed from the medium **(JUNE 2017)**
   (a) Because plant was grown continuously in the dark
   (b) by an active, energy-dependent process
   (c) by simple diffusion
   (d) through plasmodesmatal connections between the epidermis and the medium
12. In the endodermis of higher plants, the role of Casperian strip is to control the water movement so that it flows **(MODEL PAPER)**
   (a) between the cells.
   (b) through the plasma membrane.
   (c) through the cell wall.
   (d) through the transfusion tissue.

13. Which one of the following statements is **NOT** correct for Casparian strip? **(DEC 2017)**

(a) It breaks the continuity of water movement through the apoplast pathway

(b) It is formed in the growing part of the root

(c) It is formed several millimetres or several centimetres behind the root tip

(d) It is a band within the radial cell walls of the endodermis that is impregnated with lignin

14. Which one of the following best describes the function of Casparian bands during the translocation of nutrients and water across the root? **(DEC 2018)**

(a) Block apoplastic nutrient transport

(b) Block symplastic nutrient transport

(c) Act as a nutrient carrier

(d) Help in creating passage cells

15. Which one of the following best describe the symplast pathway of water flow from the epidermis to endodermis in plant root? **(DEC 2016)**

(a) Water moves through cell walls and extracellular spaces without crossing any membrane

(b) Water travel across the root cortex via the plasmodesmata

(c) Water crosses the plasma membrane of each cell in its path twice, once on entering and once on exiting

(d) Transport across the tonoplast

16. Vascular wilts are wide spread and destructive plant diseases. The symptoms of this disease are primarily caused by the clogging of **(DEC 2019)**

(a) xylem vessels (b) phloem vessels

(c) stomata (d) hydathodes

17. Which one of the following statements is true about xylem in plants? **(DEC 2017)**

(a) It is characterized by the presence of tracheary elements responsible for the conduction of water in vascular plants

(b) It is responsible for the transport of water and characterized by sieve elements like parenchyma cells and sclereids

(c) It is responsible for the transport of assimilates in the vascular plants and characterized by tracheary elements.

(d) It is responsible for the transport of assimilates in both vascular and non-vascular plants

18. Which of the following statement regarding aquaporin or water channels is NOT correct? **(JUNE 2012)**

(a) Aquaporins are found in both plants and animals membranes

(b) Aquaporins cannot transport uncharged molecules like ammonia

(c) Phosphorylation and calcium concentration regulates aquaporin activity

(d) Activity of aquaporin is regulated by pH and reactive oxygen species

19. Aquaporins are a class of proteins that are relatively abundant in plant membranes. Following are certain statements regarding the properties of aquaporins:

A. Aquaporins form water channels in membrane.

B. Some aquaporins also transport uncharged molecules such as NH_3

C. The activity of aquaporins is not regulated by phosphorylation.

D. The activity of aquaporin is regulated by calcium concentration reactive oxygen species.

Which one of the following combinations of above statements is correct? **(JUNE 2014)**

(a) A, B and D (b) B, C and D

(c) A, C and D (d) A, B and C

20. Transport of water across aquaporins is regulated by the presence of which of the following sequence of three highly conserved amino acids? **(JUNE 2011)**

(a) Ala-Asn-Pro. (b) Pro-Asn-Ala.

(c) Asn-Pro-Ala. (d) Pro-Ala-Asn

20. Which one of the following is NOT the main factor that controls to water potential among plant growth under normal conditions? **(DEC 2014)**

(a) Solute potential
(b) Hydrostatic pressure
(c) Gravity
(d) Temperature

22. Following are a few statements regarding water potential in plants:

A. Solute concentration and pressure potential contribute to water potential of a plant cell in a given state.
B. When a flaccid cell is placed in a solution that has a water potential less negative than the intracellular water potential, water will move from solution into the cell.
C. When a flaccid cell is placed in a solution that has a water potential less negative than the intracellular water potential, water will move out from cell into the solution.
D. Water potential of a plant cell under severe water stress is always less negative as compared to that of unstressed cells.

Which combination of the above statements is correct? **(JUNE 2017)**

(a) A and B (b) B and C
(c) A and C (d) C and D

23. Following are few statements regarding water potential of soil.

(A) The osmotic potential (ψ_s) of soil water is generally negligible, except in saline soils.
(B) The osmotic potential (ψ_s) of saline soil is always more than zero.
(C) In dry soils the hydrostatic pressure (ψ_p) of soil water potential is always positive.
(D) Gravitational potential (ψ_g) of soil water is always proportional to height of the tree.

Which one of the following combinations of above statements is true? **(JUNE 2013)**

(a) (A) and (C) (b) (B) and (D)
(c) (C) and (B) (d) (D) and (A)

24. A plant cell with turgor pressure of 0.1 MPa and osmotic pressure of 0.7 MPa is placed in a salt solution of 0.6 MPa osmotic pressure. Considering that all other pressures and coefficients have no effect on water potential, the following statements were recorded:

A. Water potential of the salt solution will be higher as compared to that of plant cell.
B. Higher osmotic pressure will lower water potential, but higher turgor pressure will raise water potential.
Therefore, both of these components are important to calculate water potential.
C. There will be no net movement of water in the cell.
D. Higher turgor pressure will force the water to move out of the cell.

Which one of the following options represents a combination of correct statements? **(DEC 2019 ASSAM)**

(a) A and D (b) A and B
(c) C and D (d) B and C

25. Water can move through the soil-plant-atmosphere continuum, only if water potential (ψ_w) along that path **(DEC 2012)**

(a) Decreases
(b) Increases
(c) Remains unchanged.
(d) Fluctuates rapidly in either direction.

26. The table given below represents the types of intercellular transport in "Column I" in land plants and their transport pathways in "Column II".

	I	II
A	Apoplastic	i. Via interconnecting plasmodesmata
B	Symplastic	ii. Via the water filled spaces of the cell wall matrices and lumen of xylem trachearv elements
C	Transcellular Transport	iii. Via the vacuole transport across the tonoplast followed by exit across the plasma membrane before regaining entry to the adjacent cell through the plasma membrane.

Which one of the following combinations matches column I correctly with column II. **(JUNE 2018)**

(a) A-i, B-ii, C-iii (b) A-ii, B-i, C-iii
(c) A-iii, B-ii, C-i (d) A-i, B-iii, c-ii

27. One of the important functions of program cell death (PCD) in plants is protection against pathogens. PCD also appears to occur during the differentiation of xylem tracheary elements that leads to nuclei and chromatin degradation. These changes result from the activation of certain genes. Following are certain genes encoding

A. Topoisomerase B. Nuclease
C. RNA polymerase D. Protease

Which one of the following combinations of the above is involved in differentiation of xylem tracheary elements? **(DEC 2014)**

(a) A and B (b) B and C
(c) C and D (d) D and A

Answer Key

1.(a)	**2.**(a)	**3.**(a)	**4.**(d)	**5.**(b)	**6.**(d)	**7.**(b)
8.(d)	**9.**(d)	**10.**(c)	**11.**(b)	**12.**(d)	**13.**(b)	**14.**(a)
15.(b)	**16.**(a)	**17.**(a)	**18.**(b)	**19.**(a)	**20.**(c)	**21.**(d)
22.(a)	**23.**(d)	**24.**(d)	**25.**(a)	**26.**(a)	**27.**(d)	

13. Phloem translocation

1. Which of the following statement describes the process of phloem loading? **(JUNE 2012)**
   (a) Triose phosphate is transported from chloroplast to cytosol
   (b) Sugars are transported into sieve elements and companion cells.
   (c) Sugars are transported from producing cells in mesophyll in the vicinity of sieve element
   (d) Solutes are transported from roots to the shoots

2. Which one of the following components is expected to be most abundant in the phloem sap of a plant? **(DEC 2018)**
   (a) Proteins (b) Organic acids
   (c) Sugars (d) Phosphates

3. Given below are statements describing various features of solute transport and photoassimilate translocation in plants:
   A. Apoplastic phloem loading of sucrose happens between cells with no plasmodesmatal connections.
   B. Growing vegetative sinks (e.g. young leaves and roots) usually undergo symplastic phloem unloading
   C. Movement of water between the phloem and xylem occurs only at the source and sink regions
   D. Symplastic loading of sugars into the phloem occurs in the absences of plasmodesmatal connections

   Select the option that gives combination of correct statements: **(DEC 2016)**
   (a) Only A and C (b) Only B and C
   (c) Only B and D (d) Only A and B

4. Several transport steps are involved in the movement of photosynthate from the chloroplasts. Following are certain statements regarding the transport of photosynthate:
   A. Pentose phosphate formed by photosynthesis during the day is transported from the chloroplast to the cytosol, where it is converted to sucrose. .
   B. Carbon stored as starch exits the chloroplast at night primarily in the form of maltose and is converted to sucrose in cytosol.
   C. During short distance transport, sucrose moves from producing cells in the mesophyll to cells in the vicinity of the sieve elements in the smallest veins of the leaf.
   D. In the process of phloem loading, sugars are transported into phloem parenchyma cells.

Which one of following combinations of above statements is correct? **(JUNE 2016)**

(a) A and B (b) B and C
(c) C and D (d) A and D

5. Which one of the following compounds is generally translocated in the phloem? **(JUNE 2015)**

(a) Sucrose (b) D-Glucose
(c) D-Mannose (d) D-Fructose

6. Carbohydrates synthesized by photosynthesis are converted into sucrose and transported via phloem to other parts of the plant. The following aspects are associated with sucrose uploading in phloem and its transport:

A. Both reducing and non-reducing sugars are transported efficiently through phloem.
B. Sucrose uploading can be both symplastic and apoplastic.
C. The route of phloem uploading is- mesophyll cells$\rightarrow$ phloem parenchyma $\rightarrow$ companion cells $\rightarrow$ sieve tubes.
D. Transport in sieve tubes is as per the 'pressure-flow model'.

Which one of the following combinations is correct? **(JUNE 2015)**

(a) A, B and C (b) B, C and D
(c) C, D and A (d) D, B and A

7. The following are some statements about long distance translocation of photoassimilates in higher plants:

A. Sugars are translocated in the phloem by mass transfer along a hydrostatic pressure.
B. Gibberellic acid stimulates the unloading of sugars from phloem tissue into apoplasts.
C. Munch pressure-flow hypothesis is crucial to drive translocation in the phloem.
D. Allocation and partition of carbon within a source leaf determine the phloem loading phenomenon.

Which one of the following combinations of the above statements is true? **(DEC 2013)**

(a) A, B and D (b) A, B and C
(c) A, C and D (d) B, C and D

8. Sieve elements of phloem conduct sugars and other organic materials throughout the plant. The following statements were made about characteristics of sieve elements in seed plants:

A. Angiosperms contain sieve plate pores.
B. There are no sieve plates in gymnosperms.
C. P-protein is present in all eudicots and many monocots.
D. There is no P-protein in angiosperms.

Which of the following combination is correct? **(DEC 2018)**

(a) B, C and D (b) A, B and C
(c) A, B and D (d) A, C and D

Answer Key

1.(b)	2.(c)	3.(d)	4.(b)	5.(a)	6.(b)	7.(c)
8.(b)						

14. Secondary metabolites - Biosynthesis of terpenes

1. The following statements are made regarding secondary metabolites of plants:

A. All secondary metabolites are constitutively produced in all cells of a plant during its entire life
B. They serve as signals to help the plant survive adverse conditions
C. They may be volatile compounds
D. They contribute to flower colour

Which one of the following options represents a combination of correct statements? **(JUNE 2017)**

(a) A, B and C (b) B, C and D
(c) A, C and D (d) A, B and D

2. Which one of the following combinations of secondary metabolite biosynthetic pathway result in the biosynthesis of terpenes? **(JUNE 2012)**

(a) Mevalonic acid and MEP pathways
(b) Malonic acid and MEP pathways
(c) Shikimik acid and Malonic acid pathway
(d) Shikimik acid and Mevalonic acid pathways

3. Which of the following acts as a branch point for the bioosynthesis of sesquiterpene and triterpenes? **(DEC 2011)**

(a) Farnesyl pyrophosphate
(b) Geranyl pyrophosphate
(c) Isopentyl pyrophosphate
(d) Hydroxymethylglularyl -CoA

4. Following are a few statements regarding the structure of terpenes:

A. Isopentenyl diphosphate and farnesyl diphosphate are monoterpene and sesquiterpene, respectively.
B. Squalene and geranyl diphosphate are triterpene and monoterpene, respectively.
C. Dimethylally diphosphate have 10 and 20 carbons, respectively.
D. Diterpenes have 20 carbons, whereas sesquiterpenes have 15 carbons

Which combination of the above statements is correct? **(JUNE 2017)**

(a) A and B (b) B and D
(c) A and C (d) C and D

5. Following are certain statements regarding terpene class of secondary metabolites in plants:

A. Isopentenyl diphosphate and its isomer combine to form larger terpenes.
B. Diterpenes are 20 carbon compounds.
C. All terpenes are derived from the union of 4-carbon elements.
D. Pyrethroids are monoterpene esters.

Which one of the following combination of above statements is correct? **(DEC 2015)**

(a) A, B and C (b) A, B and D
(c) B, C and D (d) A, C and D

6. Following are certain statements regarding secondary metabolites found in plants:

A. All terpenes are derived from a six carbon element.
B. Alkaloids are nitrogen containing compounds.
C. Pyrethroids, a monoterpene ester found in the leaves and flower of Chrysanthemum species, show insecticidal activity.
D. Limonoids are groups of alkaloids and have anti-herbivoral activity.

Which one of the following combinations of above statements is correct? **(DEC 2014)**

(a) A and B (b) A and D
(c) B and C (d) C and D

7. In terpene biosynthesis pathways, three acetyl-CoA are joined together stepwise to form mevalonic acid. Which one of the following three- steps is required by mevalonic acid to form isopentenyl diphosphate or isopentenyl pyrophosphate (IPP)? **(JUNE 2014)**

(a) Pyrophosphorylation, decarboxylation, and dehydration.
(b) Alkylation, pyrophosphorylation and decarboxylation.
(c) Methylation, dehydration and alkylation.
(d) Phosphorylation, carboxylation and methylation.

8. Secondary metabolites are diverse array of organic compounds in plants. The following are certain statements about secondary metabolites:

A. They protect plants against being eaten by herbivores and against being infected by microbial pathogens.
B. Terpenes, the largest class of secondary metabolites are synthesized by methyl erythritol phosphate(MEP) pathway and shikimic acid pathway.
C. The most abundant classes of phenolic compounds in plants are derived from phenylalanine.

D. Alkaloids are nitrogen containing secondary metabolites in plants.

Which one of the following combinations of the above statements is correct? **(DEC 2013)**

(a) A, B and C (b) B, C and D

(c) A, C and D (d) A, B and D

9. Following are some statements for synthesis of secondary metabolites in plants.

A. Terpenes are synthesized by shikimic acid pathway and mevalonic acid pathway.

B. Alkaloids are nitrogen containing compounds and are synthesized by shikimic acid pathway.

C. Phenolic compounds are synthesized by shikimic acid pathway and mevalonic acid pathway.

D. Both alkaloids and terpenes are synthesized by mevalonic acid pathway and MEP pathway.

Which one of the following combinations of the above statement is true? **(JUNE 2011)**

(a) A and D (b) A and C

(c) B and C (d) B and D

10. Certain plant species produce cyanogenic glycosides to protect them from pathogens. A researcher has identified a variant of such a plant that has higher level of cyanogenic glycoside yet it is highly susceptible to a specific fungal pathogen. To interpret this counter-intuitive observation, the researcher hypothesizes that the fungal pathogen has higher level of

A. B-glucosidase activity

B. formamide hydrolyase activity

C. cytochrome P-450 enzyme

D. cyanide-resistant, alterative oxidase activity

Which one of the following combinations of the above hypotheses is correct? (DEC 2019)

(a) A and B (b) B and C

(c) C and D (d) B and D

Answer Key

1.(b)	2.(a)	3.(a)	4.(b)	5.(b)	6.(c)	7.(a)
8.(c)	9.(c)					

15. Phenols, Alkaloids and Glucosinolates

1. Phenylalanine, a precursor of most of the phenolics in higher plants is a product of which one of the following pathways? **(DEC 2015)**

(a) Shikimic acid pathway

(b) Malonic acid pathway

(c) Mevalonic acid pathway

(d) Methylerythritol pathway

2. Which One of the following pairs of precursor amino acid and alkaloid is correct?

(JUNE 2013)

(a) 'Ornithine aspartate-nicotine' and 'tryptophan –quinine'.

(b) 'Ornithine-nicotine' and 'tyrosine-orphine'

(c) 'Tyrosine-quinine' and 'tryptophan-orphine'

(d) 'Ornithine-quinine' and 'ornithine aspartate - nicotine'

3. Which of the following is a mismatch between the plant drug and its source? **(JUNE 2013)**

(a) Codeine - *Papaver somniferum*

(b) Vinblastine - *Catharanthus roseus*

(c) Quinine - *Cinchona ledgeriana*

(d) Digitalin - *Artemisia annua*

4. Several plants produce metabolites with important medicinal properties and have been extensively used in traditional medicine across the world. Many of these compounds can now be purified or synthesized and are used in modern medicine.

Given below is a list of metabolites, their plant source and medicinal use:

Metabolite	Plant Source	Medicine Use
A. Digitoxin	*i. Artemisia annua*	Q. Aspirin
B. Salicin	*ii. Papaver somniferum*	R. Anti-malarial
C. Morphine	*iii. Digitalis purpurea*	S. Cardiac ailment
D. Artemisin	*iv. Willow tree*	T. Narcotic analgesic

Which one of the following options is the most appropriate match of the compound with its plant source and use? **(DEC 2019)**

(a) A-iii-R; B-i-T; C-iv-Q; D-ii-S

(b) A-iv-Q; B-iii-R; C-ii-S; D-i-T

(c) A-ii-T; B-iii-S, C-i-R; D-iv-Q

(d) A-iii-S; B-iv-Q; C-ii-T; D-i-R

5. Phenyl ammonia lyase (PAL) and chalcone synthase (CHS) involved in biosynthesis of phenolic compound in plants. Following are some statements regarding the action of PAL and CHS:

A. Substrate for PAL is phenyl alanine and for CS is chalcone

B. PAL catalyze conversion of phenyl alanine to trans-cinnamic acid

C. PAL catalyze conversion of phenyl alanine to p-coumaric acid

D. Coumaryl CoA is converted to chalcone by chalcone synthase

Which of the following combinations of above statement is true? **(JUNE 2012)**

(a) A and B (b) A and C

(c) B and C (d) B and D

6. Given below are some statements on secondary metabolites in plants.

A. Glucosinolates are synthesized by elongation in the length of side chains of their precursor amino acids. B. All terpenes produced in herbs and spices are sticky, oily liquids which reduces the palatability (edible value) of these plants.

C. Cyanogenic glycosides are produced and stored in a toxic form in plants and are therefore more effective in defense against invading pathogens and herbivores.

D. The defense molecules, alkaloids (or their precursors), are gathered from plants and used by some insects for their own protection against predators.

Which one of the following represents correct statements? **(DEC 2017)**

(a) B and C only (b) A and B only

(c) C only (d) A and D only

7. Which one of the following compounds is NOT a part of alkaloid class of secondary metabolites? **(DEC 2016)**

(a) Lignin (b) Indole

(c) Tropane (d) Pyrroidine

8. Which one of the following secondary metabolites is characterized by the presence of a central carbon atom that is bound by a sulphur to a glycone group, and N by a nitrogen to a sulfonatedoxime group? **(JUNE 2018)**

(a) Alkaloids (b) Terpenes

(c) Phenolics (d) Glucosinolates

Answer Key

1.(a)	2.(a)	3.(d)	4.(d)	5.(d)	6.(d)	7.(a)
8.(d)						

16. Responses of plants to abiotic stress

1. While studying the primary effect of different abiotic stresses on plants, a researcher observed water potential (ψ_p) reduction and cellular dehydration. Which of the following combination of abiotic stress may cause observed effect? **(JUNE 2012)**

(a) Water deficit, salinity and chilling

(b) Salinity, high temperature and Flooding

(c) Freezing, salinity and water deficit

(d) Freezing, chilling and flooding

2. The membranes of chilling-sensitive plants is characterized by **(DEC 2013)**

(a) high proportion of saturated fatty acids
(b) lower transition temperature
(c) lower proportion of saturated fatty acids
(d) lower transition temperature and higher proportion of unsaturated fatty acids

3. Following are certain statements related to plants exposed to dehydration stress:

A. When the water potential of the rhizosphere decreases due to water deficit, plants continue to absorb water as long as plant water potential is lower than that of soil water.

B. The ratio of root to shoot growth, increases in response to water deficit

C. Plant cells tend to release solutes to lower water potential during periods of osmotic stress.

D. Abscisic acid is synthesized at higher rate when leaves are dehydrated, and more ABA accumulates in the leaf apoplast.

Which one of the following combinations of above statements is correct? **(JUNE 2015)**

(a) A, B and C (b) B, C and D
(c) A, B and D (d) A, C and D

4. Following are some statements to osmotic stress in plants.

A. The accumulation of ions during osmotic adjustment is predominantly restricted to the vacuoles.

B. In order to maintain the water potential equilibrium within the cell, other solutes or compatible osmolytes accumulate in the cytoplasm.

C. Galactose is one of the compatible osmolytes involved in osmotic stress in plants.

D. There are mainly four groups of molecules that frequently serve as compatible solutes.

Which one of the following combinations of above statements is correct? **(DEC 2014)**

(a) A, B and C (b) B, C and D
(c) A, B and D (d) A, C and D

5. Asada-Halliwell pathway protects plants against oxidative stress during unfavorable environmental growth regimes. The following are some statements related to the stress-tolerance mechanism through this pathway in plants.

A. Oxygen accepts electrons as an alternative electron acceptor.

B. Hydrogen peroxide is reduced by catalase to form water.

C. Ascorbate is oxidized and regenerated.

D. Glutathione is oxidized and reduced.

Which one of the following combinations of the above statements is true? **(DEC 2013)**

(a) B, C and D (b) A, B and C
(c) A, B and D (d) A, C and D

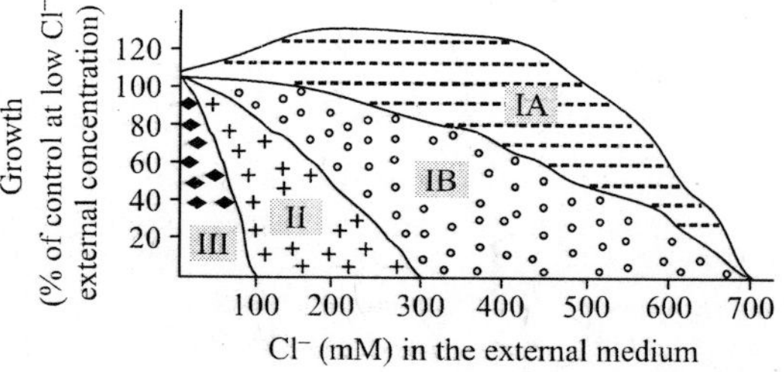

6. Shown above, is a graph representing the growth of different plant species subjected to salinity relative to that of unsalinized control. Which of the following statements is NOT true? **(DEC 2012)**

(a) Plants In group IA are extreme halophytes while very salt sensitive species will be part of group III.

(b) Plants in group IA are very salt sensitive and extreme halophytes will be part of group III.

(c) Halophytes, which can tolerate salt but their growth is retarded will be of part of group IB.

(d) Non-halophytes, which are salt tolerant but lacks salt glands will be a part of group II.

7. Following are some statements about low temperature stress in plants.
   A. Fatty acid composition of mitochondria isolated from chilling resistant and chilling sensitive plants differs significantly.
   B. Ratio of unsaturated fatty acids to saturated fatty acids is lower in chilling resistant species.
   C. The cellular water does not freeze even at -40°C, because of the presence of solutes and other antifreeze proteins.
   D. Heat shock proteins do not play any role during low temperature stress.

   Which one of the following combination of above statement is correct? **(DEC 2011)**
   (a) A and B (b) A and C
   (c) B and C (d) B and D
8. Plants grown in greenhouse at 25°C when exposed first to 35°C for 6 hours and subsequently to 42°C for 12 hours adapt better to the high temperature (42°C) in comparison to those directly transferred to 42°C for the same duration. What is the phenomenon called and what is its main physiological basis? **(MODEL PAPER)**
   (a) Acquired thermo-tolerance because of the induction of mutagens resulting into improved stability of all the proteins.
   (b) Induced thermo-tolerance because of the induction of heat shock proteins.
   (c) Induced thermo-tolerance because changes in RNA polymerase II resulting in efficient and improved transcription.
   (d) Acquired thermo-tolerance because of efficient post translational modification of proteins.
9. To characterize the mechanism/s by which heat-stress is perceived in Arabidopsis, a team of researchers fused a Heat Shock promoter with luciferase gene. Transgenic plants having promoter: luciferase fusion were raised. Such plants revealed strong luciferase expression upon heat-stress but they showed no expression under unstressed control condition. Subsequently, these transgenic plants were mutagenized by EMS and seeds from F_2 generation were obtained. To analyze the downstream positive regulators of heat-stress, the researchers should analyze seedlings that are **(JUNE 2018)**
   (a) expressing luciferase in the presence of heat-stress
   (b) not expressing luciferase in the presence of heat-stress
   (c) expressingluciferase in the absence of heat-stress
   (d) not expressing luciferase in the absence of heat-stress

Answer Key

1.(c)	2.(a)	3.(b)	4.(c)	5.(d)	6.(b)	7.(b)
8.(b)	9.(b)					

17. Responses of plants to biotic stress

1. Which of the following phenomena is observed in compatible plant-pathogen interactions? **(DEC 2013)**
   (a) Virulence in pathogen.
   (b) Hypersensitive response in host.
   (c) Resistance in host.
   (d) Avirulence in pathogen.
2. Satellite RNAs (sat-RNAs) are species of RNA associated with specific strains of some plant RNA viruses, although it is not necessary for their replication. Few statements are given below on sat-RNA **(JUNE 2014)**
   A. Presence of sat-RNA leads to reduction in severity of disease symptoms.
   B. Presence of sat-RNA leads to increase in severity of disease symptoms.
   C. sat-RNA is constitutively expressed like coat proteins and is independent of virus infection.

D. sat-RNA is not constitutively expressed like coat proteins but is expressed only after virus infection.

Which one of the following combinations of above statements regarding sat-RNA is correct?

(a) A and B (b) A and C
(c) C and D (d) A and D

3. Most of the plant disease resistance (R) gene products contain: **(JUNE 2019)**

(a) G-Box domains
(b) Transcription repression domains
(c) Leucine-rich repeats
(d) Enzymatic activities

4. Plants expressing bacterial NahG (Salicylate hydroxylase) are known to be defective in inducing systemic acquired resistance (SAR). A researcher applied the following synthetic chemicals on the NahG expressing plants.

A. 2,6-dichloro iso-nicotinic acid (INA)
B. Benzo (1,2,3) thiodiazole-7-carbothionic acid 5-methyl ester (BTH)

Which one of the following statements is correct? **(DEC 2019 ASSAM)**

(a) INA induces SAR but BTH does not induce SAR
(b) INA does not induce SAR but BTH induces SAR
(c) Both INA and BTH induce SAR
(d) Neither INA nor BTH induce SAR

5. Read the following statements related to plant pathogen interaction

A. Systemic acquired resistance is observed following infection by compatible pathogen
B. Induce systemic resistance is activated following infection by compatible pathogen
C. A bacterial infection can induce effector triggered immunity (ETI) leading to hypersensitive response locally
D. NPR1 monomers that are released in cytosol due to salicylic acid accumulation is rapidly translocated to nucleus

Which combination of above statements is correct? **(DEC 2016)**

(a) A, B and C (b) A, C and D
(c) A, B and D (d) B, C and D

6. Following are certain statements that describe plant-pathogen interactions:

A. Hemibiotrophic pathogens are characterized by initially keeping host cells alive followed by extensive tissue damage during the later part of the infection.
B. Effectors are molecules present in host plants that act against the pathogen attack.
C. Plants possess pattern recognition receptors (PRRs) that perceive microbe-associated molecular patterns (MAMPs) present in specific class of microorganisms but are absent in the hosts.
D. Phytoalexin production is a common mechanism of resistance to pathogenic microbes in a wide range of plants.

Which one of the following combinations is correct? **(JUNE 2016)**

(a) A, B and C (b) A, C and D
(c) B, C and D (d) A, B and D

7. During interaction with host, phytopathogens are known to deliver effector is directly into the host cells. The following statements were made regarding the role of these effector proteins.

A. May promote pathogen virulence.
B. May elicit avirulence response.
C. May suppress defense response.
D. May promote plant growth.

Which one of the following combinations of the above statements is correct? **(JUNE 2019)**

(a) A, B and D (b) A. C and D
(c) A, B and C (d) B, C and D

8. With reference to plant biotic interactions, match the terms of Column I with the most appropriate term of Column II **(DEC 2017)**

	Column I		Column II
A	Phytoliths of Poaceae	(i)	Phloem feeders
B	Salicyclic acid signalling pathway	(ii)	Decrease in stomatal aperture
C	MAMP	(iii)	R gene/NBS-LRR receptor
D	Effector-triggered immunity	(iv)	Mechanical barrier to herbivory

(a) A - (ii), B - (iv), C - (i), D - (iii)
(b) A - (iii), B - (i), C - (ii), D - (iv)
(c) A- (iv), B - (ii), C - (iii), D - (i)
(d) A- (iv), B -(i), C - (ii), D - (iii)

9. Given below are certain statements regarding plant-pathogen Interactions:

A. The pattern recognition receptor (PRR), upon perceiving pathogen or microbe associated patterns (PAPMs/MAMPs), activates plant defenses resulting in pattern triggered immunity (PTI).

B. AvrPto is a resistance gene in tomato that acts against pathogenic attack by the bacterium *Pseudomonas syringae*pv. tomato.

C. The effector molecules produced by pathogen is recognized by resistance (R) gene present in plants resulting into a defense strategy known as effector triggered immunity (ETI).

D. Defense mechanisms triggered in plants during PTI are usually stronger than those during ETI.

Which one of the following combinations of above statements is correct? **(JUNE 2018)**

(a) A and B (b) C and D
(c) A and C (d) B and D

10. A quadratic check of gene combinations and disease reaction types in a host-pathogen system where the gene-for-gene concept operates is represented below:

Virulence or avirulence genes in the pathogen	Resistance or susceptibility genes in the plant	
	R (resistant) dominant	r (susceptible) recessive
A (avirulent) dominant	AR	Ar
a (virulent) recessive	aR	ar

The following statements were made about. the above genotypes:

A. AR genotype had incompatible (resistant) reactions.

B. Ar genotype had compatible (susceptible) reactions.

C. ar genotype had compatible (susceptible) reactions.

D. aR genotype had incompatible (resistant) reactions.

Choose the combination with all correct statements. **(DEC 2018)**

(a) A, B and D (b) A, B and C
(c) B, C and D (d) A, C and D

Answer Key

1.(a)	2.(1/4)	3.(c)	4.(c)	5.(b)	6.(b)	7.(c)
8.(d)	9.(c)	10.(b)				

SECTION-7

ANIMAL PHYSIOLOGY

1. Blood, plasma, groups and hemostasis

1. The maturation of red blood cells does not depend on **(DEC 2018)**
   (a) folic acid (b) vitamin B12
   (c) pyridoxine (d) tocopherol
2. In order to separate red and white blood cells, which of the following methods can be used? **(JUNE 2017)**
   (a) Ion-exchange chromatography and FACS
   (b) Hydrophobic chromatography and density gradient centrifugation
   (c) Density gradient centrifugation and FACS
   (d) Hydrophobic chromatography and FACS
3. The only bone marrow cell that never appears in peripheral blood is **[DEC 2013]**
   (a) myeloblast (b) myelocyte.
   (c) lymphoblast. (d) megaloblast.
4. Menstrual blood does not contain clots due to higher level of **[DEC 2013 GU]**
   (a) Plasmin. (b) Fibrinogen.
   (c) Heparin. (d) Prothrombin.
5. In high altitude, hypoxia induces increased number of circulating red blood cells, which can be explained by the following changes:
   A. The transcription factors, HIPs are produced.
   B. Erythropoietin secretion is increased
   C. Myoglobin content is decreased
   D. Cytochrome oxidase is decreased
   Which one of following is NOT true? **[JUNE 2015]**
   (a) Only A (b) A and B
   (c) B and C (d) C and D
6. During the spanish conquest of the Inca Empire at the high altitude in Peru, many soldiers fell sick. It was found that the sickness was due to low partial pressure of O_2 in the atmosphere at that altitude. To determine the reason, blood was collected from those patients. The circulating erythropoietin (EPO) levels were estimated and the O_2-dissociation curve of haemoglobin were drawn and compared with the same in native people as depicted below.

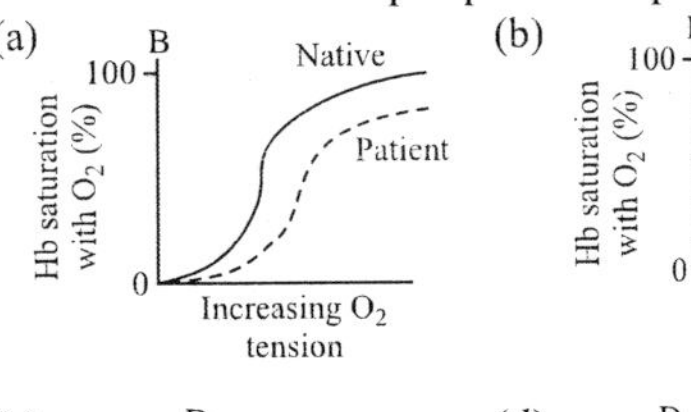

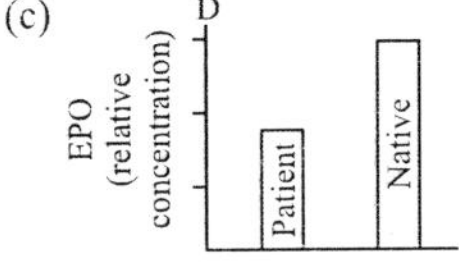

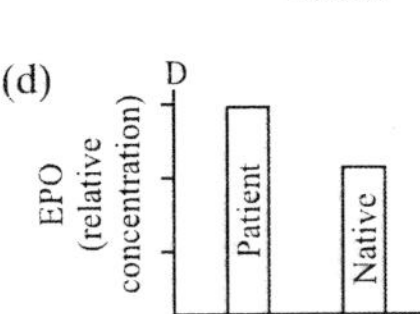

Which one of the following combinations is logically correct? **[DEC 2011]**
   (a) A and C (b) A and D
   (c) B and C (d) B and D
7. Which one of the following is NOT true in the process of acclimatization to high altitude? **(DEC 2019)**
   (a) Respiratory alkalosis
   (b) Increased 2,3-DPG in RBC
   (c) Rise in pH of cerebrospinal fluid
   (d) Increased cytochrome oxidase in tissues
8. Relative rates of red blood cells production in bone marrow of different bones of different ages are shown below: Identify the correct figure. **(DEC 2017)**

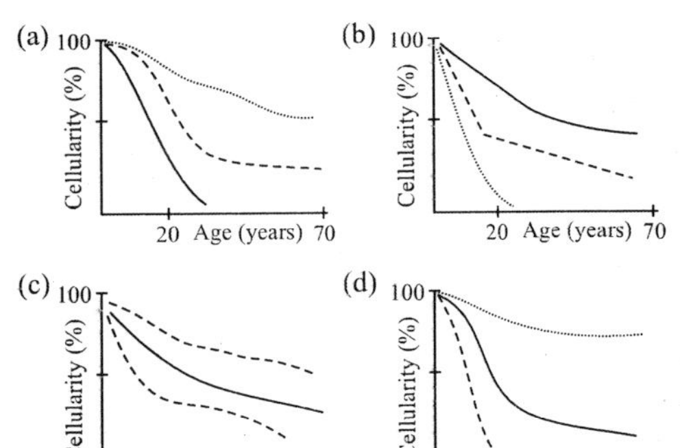

9. Which one of the following functions is NOT served by the plasma proteins? **[DEC 2014]**
(a) Blood clotting
(b) O_2 transport
(c) Hormone binding and transport
(d) Buffering capacity of blood

10. Serum has essentially the same composition as plasma EXCEPT that it lacks **[DEC 2016]**
(a) Albumin
(b) Stuart-Prower factor
(c) Antihemophilic factor
(d) Hageman factor

11. If in a blood transfusion, type A donor blood is given to recipient having type B blood, the red blood cells (RBC) of donor blood would agglutinate but the recipients RBCs would be least affected. These observations can be explained in the following statements.
A. Agglutininins in recipient's plasma caused agglutination by binding with type A agglutinogens.
B. The agglutinins of donor blood was diluted in recipient's plasma resulting in low agglutination.
C. Low titre of anti-A agglutinins is the cause of low agglutinations of recipients RBC's.
D. High agglutination of donor RBC's is the outcome high titre of anti-B agglutinins
Which of the above statement(s) is/are INCORRECT? **[DEC 2016]**
(a) Only A (b) A and B
(c) Only B (d) C and D

12. Individuals in a population are divided into various blood groups (in column 'X') based on the set of enzymes they have in column 'Y':

COLUMN X	COLUMN Y
A	(i) Fucose transferase
B	(ii) GalNAc transferase
AB	(iii) Gal transferase
O	

Which one of the following combinations is NOT correct for the blood group type and the set of enzymes a person has? **(DEC 2019)**
(a) A - (i) and (ii)
(b) B - (i) and (iii)
(c) AB - (i) (ii) and (ii)
(d) O - (i), (ii) and (iii)

13. Given below are few statements with reference to blood clot formation which results from triggered chain of reactions:
A. Conversion of fibrinogen to fibrin.
B. Activation of factor XIII, which stabilizes fibrin mesh work.
C. Activation of factor XII, which promotes plasmin activation.
D. Enhancement of platelet aggregation.
Which one of the following of statements is correct with reference to roles of thrombin in hemostasis? **[DEC 2014]**
(a) B, C and D (b) A, B and D
(c) A, C and D (d) A, B and C

14. Which one of the following is NOT secreted by capillary endothelium? **(DEC 2018)**
(a) Prostacyclin (b) Guanosine
(c) Endothelin (d) Nitric oxide

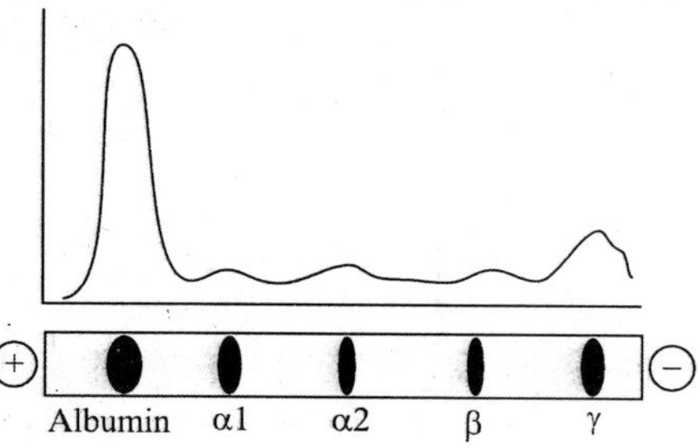

15. The blood plasma proteins (albumin and globulins) from a healthy person were separated by electrophoresis as shown above. The diagnosis of acute inflammation can be done based on one of the following observations. **(JUNE 2019)**
(a) Increase in both α1 and α2; decrease in Albumin

(b) Increase in albumin; decrease in α1, α2 and β.

(c) Increase in albumin and decrease in □ globulin

(d) Only decrease in albumin.

Answer Key

1.(d)	2.(c)	3.(a)	4.(a)	5.(d)	6.(a)	7.(c)
8.(a)	9.(b)	10.(c)	11.(d)	12.(d)	13.(b)	14.(b)
15.(a)						

2. Cardiovascular System

1. Which one of the following is NOT involved with the pacemaker potential of heart? **[DEC 2015]**

(a) "h"- channel

(b) Transient calcium channel

(c) Long-lasting calcium channel

(d) "f"- channel

2. Action potentials were recorded intracellularly from different parts of mammalian heart and these are shown below. Which one-of these has been recorded from sinoatrial node? **[JUNE 2016]**

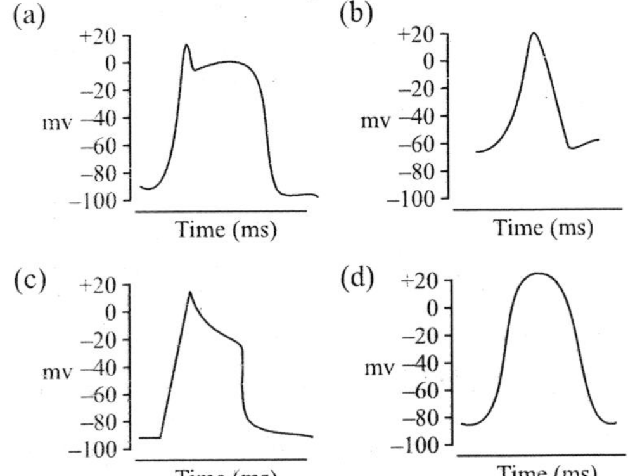

3. For a normal heart, the time taken for atrial systole and diastole are A_s and A_d seconds, respectively, while the same for ventricular systole and diastole are V_s and V_d. Which one of the following equations is correct? **[DEC 2011]**

(a) $A_s + A_d = V_s + V_d$

(b) $A_s + A_d < V_s + V_d$

(c) $A_s + A_d - V_s + V_d = 0$

(d) $A_s + A_d > V_s + V_d$

4. The time taken for atrial systole and diastole in a normal heart are t_{as} and t_{ad} seconds, respectively. If ventricular systole takes t_{vs} seconds, calculate the ventricular diastolic time (seconds) **[JUNE 2011]**

(a) $(t_{as} + t_{ad}) - t_{vs}$ (b) $(t_{as} - t_{ad}) + t_{vs}$

(c) $(t_{ad} - t_{as}) - t_{vs}$ (d) $(t_{as} + t_{ad}) \times t_{vs}$

5. The changes in left atrial, left ventricular and aortic pressure in a cardiac cycle are shown below in the figure

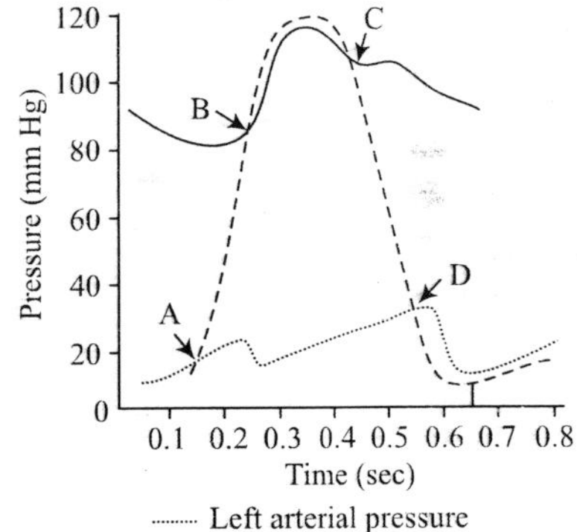

Given below are the events of cardiac cycle (column A) associated with marked points (A, B, C, D) in the figure (column B).

	Column A	Column B	
a.	Aortic valve opens	(i)	D
b.	Mitral valve closes	(ii)	B
c.	Mitral valve opens	(iii)	A
d.	Aortic valve closes	(iv)	C

Choose the option that matches the events with marked points in the figure. **(DEC 2018)**

1. a- (ii), b - (iii), c - (i), d - (iv)
2. a- (i), b- (iv), c - (ii), d - (iii)
3. a - (iv), b - (i), c- (iii), d-(ii)
4. a - (iii), b - (ii), c - (iv), d - (i)

6. The blood forced into aorta during systole sets up a pressure wave that travels along arteries (arterial pulse wave). The following figure depicts the pressure wave recorded from an artery during a cardiac cycle and the table below represents points in the graph (column A) and associated features (column B).

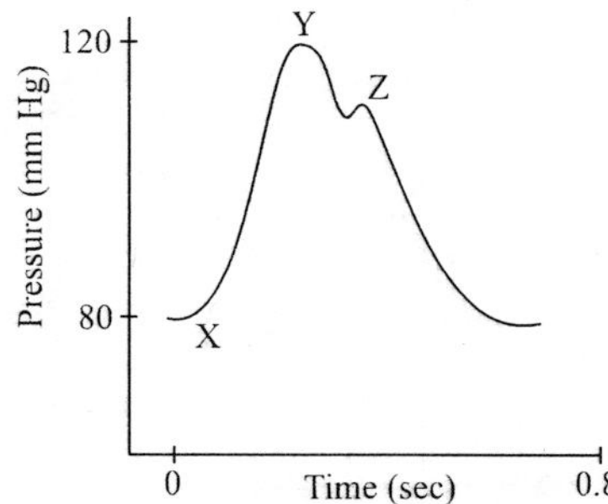

Column A		Column B	
a.	X	(i)	Dichrotic notch
b.	Y	(ii)	Systolic blood pressure
c.	(Y – X)	(iii)	Diastolic blood pressure
d.	Z	(iv)	Pulse pressure
e.	$\left[X + \frac{Y-X}{3}\right]$	(v)	Mean arterial pressure

Which one of the following options represents a correct match between all terms of columns A and B? **(DEC 2019 ASSAM)**

(a) a-i, b - iii, c-ii, d - v, e-iv

(b) a - ii, b - iv, c-iii, d-v, e-i

(c) a - iii, b - ii, c-iv, d-i, e-V

(d) a - iv, b-V, C-i, d - iii, e – ii

7. The changes in left ventricular stroke work (LVSW) according to the different left ventricular end-diastolic pressures (LVEDP, which indicates the initial myocardial fiber length) in a dog, under control conditions, were recorded, which follows Starling's law of the heart. This LVSW-LVEDP relationship was investigated in the same dog after constant infusion of norepinephrine, and these two data sets were plotted. Which one of the following graphs correctly represents the results obtained? **[JUNE 2015]**

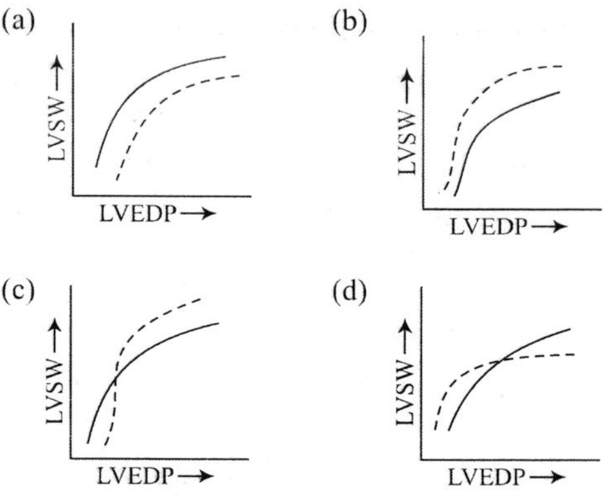

8. A cardiac patient was admitted to the hospital after a heart block was detected, specifically 'infra-Hisian block'. The doctors identified the problem in the bundle of His. Which one of the following statements is true regarding the bundle of His? **[DEC 2013 GU]**

(a) The bundle of His is a collection of specialized nerve fibres which help in electrical conduction.

(b) The intrinsic rate of the bundle of His is 20 or less beats per minute.

(c) The bundle is located only in the ventricles to allow for contraction.

(d) The bundle arises in blood vessels that are connected to the lung and bring O_2 to the heart.

9. Which one of the following does NOT occur due to stimulation of baroreceptors? **[DEC 2016]**

(a) Bradycardia (b) Hypotension

(c) Venodialation (d) Vasoconstriction

10. The "Mayer waves" in the blood pressure originate due to **(DEC 2018)**

(a) systole and diastole of ventricle

(b) inspiration and expiration

(c) reflex oscillation of neural pressure control mechanisms

(d) Bainbridge reflex

11. The pacemaker cells of sinoatrial node (SA node) are inhibited by the stimulation of vagus nerve. The probable mechanisms of this inhibition are stated as follows:

A. The acetylcholine-regulated K^+ channels are activated

B. The outward K^+ causes hyperpolarization of pacemaker cells

C. The inward "funny current" of pacemaker potential is increased

D. The increased intracellular cAMP, induced by activation of M2 muscarinic receptors, slows the opening of Ca^{++} channels

Choose the answer with correct statements. **(DEC 2017)**

(a) A and C (b) B and C

(c) A and B (d) C and D

12. In an experiment on healthy young men, the muscarinic receptor antagonist, atropine was administered to one group (Group A) while the P-adrenergic receptor antagonist, propranolol was administered to another group (Group B) in four increasing doses of equal concentration for both the drugs. The effects of these two drugs on the heart rate are shown below:

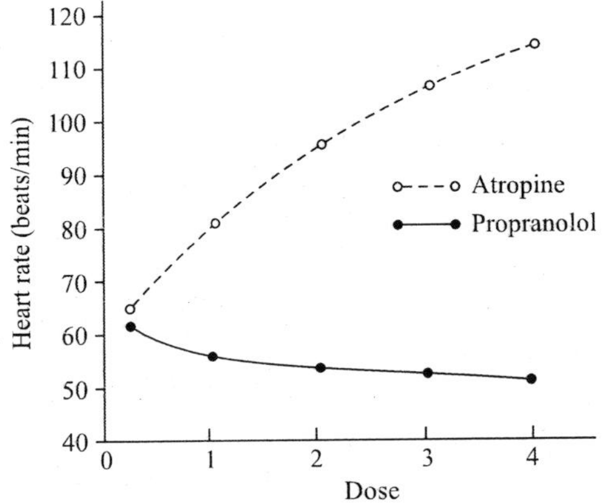

On the basis of these observations, an investigator proposed the following statements:

A. Atropine and propranolol block sympathetic and parasympathetic effects on the heart, respectively

B. As the change of heart rate is more in Group A than in Group B, the sympathetic tone usually predominates in healthy resting individuals.

C. Atropine and propranolol block parasympathetic and sympathetic effects on the heart, respectively D. As substantial changes occur in the heart rate with atropine, the parasympathetic tone is predominant in healthy resting individuals.

Select the option with **INCORRECT** statement(s) **(JUNE 2017)**

(a) Only A (b) A and B

(c) Only C (d) A and D

13. The arterial pressure usually raises and falls 4 to 6 mm Hg in a wave like manner causing "respiratory waves". The probable mechanism of these waves has been proposed in the following statements:

A. The more negative intrathoracic pressure during inspiration reduces the quantity of blood returning to the left side of the heart causing decreased cardiac output.

B. The changes of intrathoracic pressure during respiration can excite vascular and atrial stretch receptors which affect heart and blood vessels.

C. The activity of medullary respiratory centers can influence the vasomotor center.

D. The "respiratory waves" are outcome of the oscillation of the central nervous system ischemic pressure control mechanism.

Which of the above statement(s) is/are NOT appropriate? **[DEC 2016]**

(a) Only A (b) A and B

(c) B and C (d) Only D

14. The heart rate shows variation during respiratory rhythm in most human subjects. Which one of the following statements describing the changes of heart rate during respiratory phases is true? **[DEC 2014]**

(a) The heart rate is accelerated during expiration, but no change occurs during inspiration.

(b) The heart rate is accelerated during inspiration and decelerated during expiration

(c) The heart rate is accelerated during expiration and decelerated during inspiration.

(d) The heart rate is accelerated during inspiration and no change occurs during expiration.

15. Which one of the following changes will occur in the cell membrane of nodal tissue of heart, which results in an increased heart rate due to stimulation of sympathetic nerves?

[JUNE 2012]

(a) Opening of sodium channels is facilitated

(b) Potassium conductance is increased

(c) Opening of L-calcium are facilitated

(d) 'h' channels are inhibited

16. (A) In an experiment, 2mg of a substance 'A' dissolved in 4.5 ml sterile solvent was injected as a bolus into the femoral vein of an intact frog. It was observed that the frog's heart rate increased significantly.

(B) The same solution as in (a) when applied directly on the heart of the same frog after exposing the heart, the heart rate did not change.

(C) The same solution as in (a) when injected to an intact cat femoral vein, the heart rate did not change significantly.

From the above observations, which one of the following statements is most likely to be correct? **(MODELPAPER)**

(a) Substance 'A' is stimulatory on heart and the effect was inotropic.

(b) substance 'A' acted in the brain of frog and also might have released other hormones to increase the heart rate.

(c) substance 'A' could not have acted on the brain but must have induced other substances in the blood to increase the heart rate.

(d) the increased heart rate was merely due to increased volume of the heart muscles.

17. Thyroid hormone (T_3) increases the heart rate and the force of contraction of cardiac muscles. The mechanisms of these effects of T_3 have been explained by a researcher in the following statements:

A. T_3 inhibits the expression of gene for α-myosin heavy chain and enhances the expression of gene for β-myosin heavy chain

B. The expression of gene for Na^+-Ca^{++} antiporter is enhanced by T_3

C. The sarcoplasmic reticulum Ca^{++} ATPase is increased by T_3

D. T_3 increases ryanodine Ca^{++} channels in the sarcoplasmic reticulum

E. The number of β-adrenergic receptors in heart muscles is inhibited by T_3

Which one of the following combinations contains both correct explanations?

(DEC 2019)

(a) A and B (b) B and C

(c) C and D (d) D and E

Answer Key

1.(c)	**2.**(b)	**3.**(a)	**4.**(a)	**5.**(a)	**6.**(c)	**7.**(b)
8.(b)	**9.**(d)	**10.**(c)	**11.**(c)	**12.**(b)	**13.**(d)	**14.**(b)
15.(c)	**16.**(b)	**17.**(c)				

3. ECG – its principle and significance

1. Which of the following waves is likely to be absent in a normal frog ECG? **[DEC 2011]**

(a) P (b) Q

(c) T (d) R

2. The T-waves of ECG indicates **[DEC 2012]**

(a) Atrial depolarization.

(b) Ventricular depolarization

(c) Ventricular repolarization

(d) Atrial depolarization

3. The S wave of normal human ECG originates due to **[JUNE 2015]**
   (a) septal and left ventricular depolarization.
   (b) late depolarization of the ventricular walls moving back toward the AV junction.
   (c) left to right septal depolarization.
   (d) repolarization of atrium

4. Different leads are used to record ECG of humans. Which one of the following is NOT unipolar leads? **(DEC 2016)**
   (a) Augmented limb leads
   (b) V1 and V2 leads
   (c) Standard limb leads
   (d) VR and VL leads

5. The different waves of normal electrocardiogram (ECG) of a human subject are shown below:

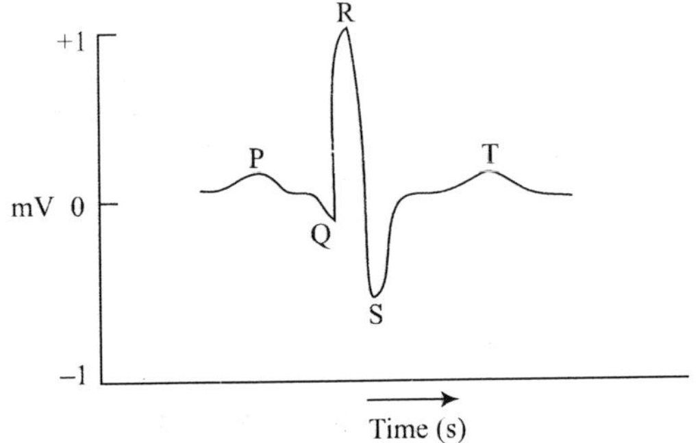

The relationship of the events of cardiac cycle to these ECG waves are proposed in the following statements:

A. The P wave occurs due to the depolarization of atria
B. The atrial repolarization is responsible for the T wave
C. The QRS complex occurs during ventricular depolarization
D. Q - T interval indicates plateau portion of auricular action potential

Select the combination with **INCORRECT** statements from the following options: **(JUNE 2018)**

(a) A and B
(b) B and C
(c) C and D
(d) B and D

6. Given below are the different intervals/ durations of electrocardiogram of a humansubject (column A) and the events in heart during the process (column B).

	COLUMN A		COLUMN B
a	PR interval	(i)	Ventricular action potential
b	QRS duration	(ii)	Atrio ventricular conduction
c	QT interval	(iii)	Ventricular depolarization
d	ST interval	(iv)	Plateau portion of the ventricular action potential

Which one of the following options is a correct match of entries in columns A and B? **(JUNE 2019)**

(a) A – (i); B – (iv); C – (ii); D – (iii)
(b) A – (ii); B – (iii); C – (i); D – (iv)
(c) A – (iv); B – (ii); C – (iii); D – (i)
(d) A – (iii); B – (i); C – (iv); D – (ii)

7. The ECG recorded by different leads is analysed on the basis of variation of electrical potential at various loci on the surface of the body, and the time scale relation of different waves. After analysing the ECG, following particulars of heart are proposed to be obtained

A. Stoke volume and cardiac output
B. Volume and pressure changes during cardiac cycle
C. Anatomical orientation of heart
D. Various disturbances in the rhythm and conduction of cardiac excitation
E. The extent, location and progress of ischemic damage to myocardium

Which one of the following combinations represents both INCORRECT particulars of heart? **(DEC 2019)**

(a) A and B
(b) B and C
(c) C and D
(d) D and E

Answer Key

1.(b)	2.(c)	3.(b)	4.(c)	5.(d)	6.(b)	7.(a)

4. Respiratory system

1. Which one of the following is the most powerful buffer system of blood? **(JUNE 2017)**

   (a) Bicarbonate (b) Phosphate
   (c) Proteins (d) Haemoglobin

2. The P_{50} value of haemoglobin for oxygen is increased during exercise. The mechanism of this change is described in the following proposed statements.

   A. Increased CO_2 production by muscles elevated pCO_2 of blood which affects P_{50} value
   B. The affinity of haemoglobin for oxygen increases as 2,3-bisphosphoglycerate (2,3 - BPG) level is elevated
   C. Increased body temperature shifts the oxyhaemoglobin dissociation curve of the left
   D. The decreased pH of blood reduces the affinity of haemoglobin for oxygen.

   Which of the above statement(s) is (are) **INCORRECT**? **(JUNE 2017)**

   (a) Only A (b) B and C
   (c) Only C (d) A and D

3. The size of red blood cells (RBC) in venous blood is greater than that of arterial blood. This increased size of red blood cell in the venous blood is due to **[JUNE 2013]**

   (a) the increased permeability of red blood cell (RBC) membrane.
   (b) the decreased osmotic pressure in plasma.
   (c) the increased osmotic pressure in RBC.
   (d) the dissociation of cytoskeletal proteins in RBC.

4. Myoglobin (Mb) in muscles, Hemoglobin A (HbA) in adult RBC, Hemoglobin C (HbC) in patients with thinner RBC and Hemoglobin S (HbS) in sickle cell disease are four different hemoproteins. Oxygen saturation kinetics of these four proteins is different. Which of the following profile is most plausible? **[DEC 2013]**

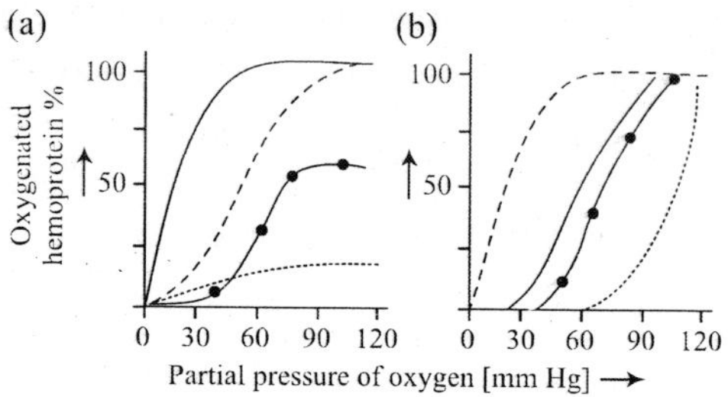

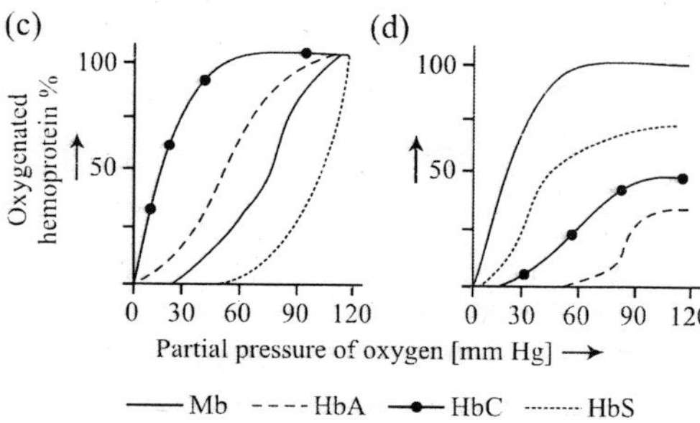

5. One hemoglobin molecule containing four heme groups can bind with four O_2. The reactions of Hb and O_2 are shown below. Which one of the following reactions is fastest? **[JUNE 2014]**

   (a) $Hb_4 + O_2 \rightarrow Hb_4O_2$
   (b) $Hb_4O_2 + O_2 \rightarrow Hb_4O_4$
   (c) $Hb_4O_4 + O_2 \rightarrow Hb_4O_6$
   (d) $Hb_4O_6 + O_2 \rightarrow Hb_4O_8$

6. The oxygen-haemoglobin dissociation curve illustrates the relationship between pO_2 in blood and the number of O_2 molecules bound to haemoglobin. The 'S' shape of the curve has been explained in the following proposed statements:

   A. The quaternary structure of haemoglobin determines its affinity to O_2.
   B. In deoxyhaemoglobin, the globin units are tightly bound in a T-configuration.
   C. The interactions between globin subunits are altered when O_2 binds with deoxyhaemoglobin.
   D. The affinity to O_2 in T-configuration of haemoglobin is increased.
   E. In the relaxed configuration of haemoglobin, the affinity to O_2 is reduced.

Choose one of the following combinations with both INCORRECT statements. **(DEC 2018)**

(a) A and B (b) B and C

(c) C and D (d) D and E

7. The CO_2 dissociation curves of oxygenated and deoxygenated blood are given along with dissolved CO_2 below

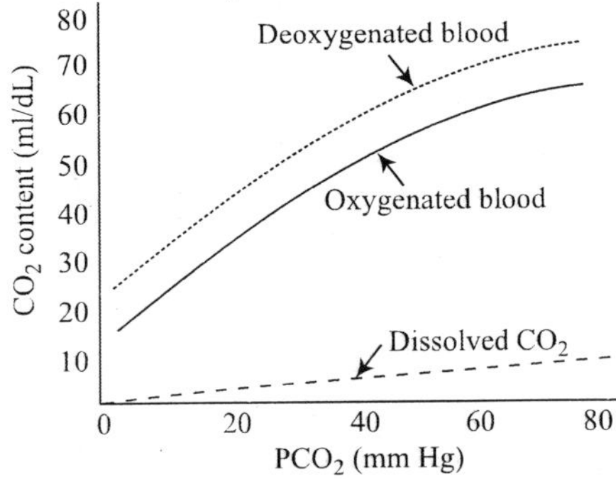

Following are the statements deduced from the curves above and or based on the knowledge about CO_2 transport, which may or may not be correct:

A. The deoxygenated haemoglobin has greater affinity for CO_2, than oxygenated haemoglobin.

B. The deoxygenated haemoglobin does not bind with free H^+ ions released during theformation of HCO_3^- from CO_2.

C. The haemoglobin saturation with O_2 has no effect on CO_2, dissociation curve.

D. O_2 and CO_2 bind to haemoglobin at different sites.

Which one of the following options represents a combination of all correct statements?

(JUNE 2019)

(a) A and B (b) B and C

(c) C and D (d) A and D

8. Given below are some statements made on the buffering capacity of hemoglobin (Hb) in contributing to acid-base balance in the body.

A. Hb buffering capacity is due to large number of histidine residues

B. Imidazole groups of Hb dissociate less than those of oxyhemoglobin (HbO_2)

C. Hb buffering capacity is due to large number of threonine residues

D. Hb is a strong acid

Which one of the following options represents a combination of correct statements?

(DEC 2019 ASSAM)

(a) A and B (b) B and C

(c) C and D (d) A and D

9. The volume of RBCs in venous blood is greater than that of arterial blood, as water accumulates in these cells. Which one of the following is true for the accumulation of water in venous RBCs? **[DEC 2013 GU]**

(a) Proton pump.

(b) Sodium-potassium pump.

(c) Chloride/bicarbonate exchanger.

(d) Calcium/magnesium exchanger.

10. The Cl content of red blood cells (RBCs) in the venous blood was found to be higher than that in arterial blood in a human subject. Following proposals were made to explain these observations:

A. The high pCO_2 in venous plasma leads to increased diffusion of CO_2 into RBC and the formation of $H_2CO_{(c)}$

B. HCO_3^- content in the RBC of venous blood becomes much greater than that in plasma.

C. The excess HCO_3^- leaves the RBC of venous blood along with Na^+ to plasma by a Na^+ - HCO_3^- symporter,

D. The increased Na^+ in the venous plasma is transported to the RBC along with Cl^-

Select the combination with **INCORRECT** statements from the following options.

(JUNE 2018)

(a) A and B (b) B and C

(c) A and D (d) C and D

11. The uptake of nitrous oxide (N_2O) and carbon monoxide (CO) in the blood of lung alveolar capillary relative to their partial pressure and the transit time of red blood cell in capillary is shown in the figure below:

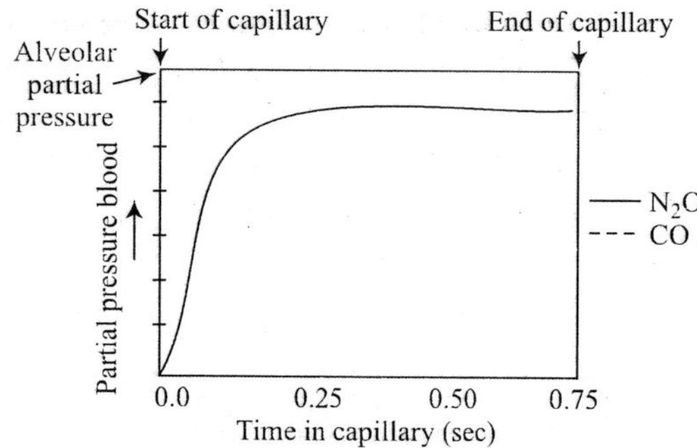

The reason for difference in the pattern of alveolar gas exchange of N_2O and CO have been proposed in the following statements:

A. N_2O does not chemically combine with proteins in blood but equilibrate rapidly between alveolar gas and blood.

B. CO has high solubility in the blood

C. CO has high solubility in the alveolar capillary membrane.

D. The dispersion of N_2O between alveolar gas and blood is considered as diffusion limited.

Which of the above statement(s) is/are INCORRECT? **[DEC 2016]**

(a) Only A (b) A and B

(c) Only C (d) C and D

12. During physical exercise, the oxygen supply to the active muscles is increased which has been explained by the following statements:

A. PO_2 declines and PCO_2 rises in the active muscles

B. The temperature is increased and pH is decreased in active muscles

C. 2, 3-biphosphoglycerate is decreased in RBC and PO_2 rises

D. Metabolites accumulating in the active muscles increase the affinity of hemoglobin to oxygen

Which one of the following is NOT correct? **[JUNE 2015]**

(a) A only (b) A and B

(c) B and C (d) C and D

13. In the process of respiration the oxygen inspired reaches the cells through haemoglobin of blood. **[DEC 2013 GU]**

A. The oxygen is used to oxidize various toxic substances in the peroxisomes.

B. The molecular oxygen enters in Krebs cycle occurring in the mitochondrial matrix and forms CO_2.

C. The oxygen combines with H^+ ions entering through ATP synthase in the matrix and forms water.

D. H^+ ions combine with electrons donated by electron transport chain forming molecular hydrogen.

Which of the above four statement(s) is/are correct for utilization of oxygen in respiration?

(a) B and D (b) Only B

(c) A and B (d) A and C

14. Pulmonary surfactant is synthesized primarily by **(DEC 2019 ASSAM)**

(a) Type I alveolar epithelial cells

(b) Alveolar macrophages

(c) Type II alveolar epithelial cells

(d) Alveolar mast cells

15. An organism having heart for circulation, excretes through green glands. It has several ganglia and tactile organs on its body and its larval form is very different than its adult form. This organismis most likely to respire by:

A. exchanging oxygen and carbon dioxide through an extensive tracheal system.

B. gaseous exchange over thinner areas of cuticle or by gills

C. an efficient tracheal system that delivers oxygen directly to the tissues

D. a double transport system, where the circulating fluid contains a dissolved respiratory pigment.

Choose the correct option. **[JUNE 2011]**

(a) A and C (b) Only D

(c) Only B (d) B and D

16. The following statements may be proposed for respiration in humans:

A. Capacity of lungs depends on the height, age, sex and weight of a person.

B. People born and living at sea level have higher lung capacity than people living at high altitudes.

C. Human lungs are 'overbuilt' and have tremendous reserve volume as compared to oxygen exchange requirements when at rest.

D. Apnea prevents absorption of oxygen and releases CO_2 from the lungs.

Which combination of the above statements is INCORRECT? **[DEC 2013 GU]**

(a) C and D (b) A and B

(c) B and D (d) B and C

17. The atmosphere in sealed space craft contains **[DEC 2011]**

(a) pure oxygen

(b) a mix of oxygen and nitrogen

(c) mix of oxygen and carbon dioxide

(d) pressurized atmospheric air available normally on earth

18. Respiration can be inhibited voluntarily for some time. The point at which respiration cannot be voluntarily inhibited is known as breaking point. Following explanations are offered for the breaking point:

A. J-receptors stimulate respiratory centers

B. Hering-Breuer reflex operates

C. The rise of arterial pCO_2 stimulates the respiratory centre

D. The fall of arterial pO_2 stimulates the respiratory centre

Which of the above combination is correct? **[JUNE 2014]**

(a) A and B (b) B and C

(c) C and D (d) A and D

19. Spinal cord of an animal was transected at C1/C2 level. The respiration of the animal stopped and it needed artificial respiration. However, the heart continued to beat although at a slower rate.

Some of the explanations given were:

A. respiration regulatory centre is located in the medulla.

B. respiration regulatory centre is located above the C1/C2 cut.

C. heart regulatory centre is above the C1/C2 cut.

D. heart has autoregulation.

Which one of the following is most appropriate? **[JUNE 2011]**

(a) A only. (b) B and C only.

(c) A, B and D only (d) B, C and D only.

20. The type I glomus cells present in the carotid bodies contain granules which release some substances during hypoxia. Which one of the following is released in hypoxia? **[DEC 2014]**

(a) Serotonin (b) GABA

(c) Dopamine (d) IL 8

21. A diabetic patient developed metabolic acidosis resulting in deep and rapid breathing which is called **[DEC 2015]**

(a) Kussmaul breathing

(b) Cheyne-Stokes respiratory pattern

(c) Apneustic breathing

(d) Periodic breathing

Answer Key

1.(a)	**2.**(b)	**3.**(c)	**4.**(a)	**5.**(d)	**6.**(d)	**7.**(d)
8.(a)	**9.**(c)	**10.**(d)	**11.**(d)	**12.**(d)	**13.**(d)	**14.**(c)
15.(c)	**16.**(c)	**17.**(b)	**18.**(c)	**19.**(b)	**20.**(c)	**21.**(a)

5. Action potential, Nerve Impulse Transmission

1. SNARE complexes are important for guiding synaptic transmission and Ca^{2+} binding is essential for activation of neurotransmitter release from the pre-synaptic terminal. In lieu of the above, identify the components of the SNARE complex that are activated in a sequential manner in order to facilitate neurotransmitter uptake, docking and release. **[DEC 2013 GU]**
   (a) Syntaxin→synaptotagmin→synaptobrevin
   (b) Synaptotagmin → synaptobrevin →SNAP-25.
   (c) Synaptobrevin → syntaxin → SNAP-25
   (d) SNAP-25→synaptotagmin → syntaxin
2. Which one of the following is required to anchor the synaptic vesicles to the cytoskeletal proteins in the pre-synaptic nerve terminals? **[JUNE 2014]**
   (a) Syntaxin (b) Synaptobrevin
   (c) Synaptotagmin (d) Synapsin
3. Which one of the following synaptic vesicle proteins is involved in tethering of the vesicle to the cytoskeletal system in the nerve terminus? **(DEC 2019)**
   (a) Synaptotagmin (b) Synapsin
   (c) Synaptophysin (d) Synaptobrevin
4. A neuron that fires when an individual is eating by hand, also fires when he sees someone else eating with hand. Such neurons are called **[DEC 2013]**
   (a) mirror neurons. (b) mimicry neurons.
   (c) motor neurons. (d) reward neurons
5. A nerve fibre cannot be stimulated during the absolute refractory period of a previous stimulation because **[DEC 2012]**
   (a) Sodium permeability remains high.
   (b) Sodium-potassium pump does not operate.
   (c) Voltage-gated calcium channels remain closed
   (d) Potassium conductance remains low.
6. During generation of an action potential, depolarization is due to **[JUNE 2011]**
   (a) K^+ efflux. (b) Na^+ efflux.
   (c) Na^+ influx. (d) K^+ influx.
7. Unidirectional propagation of electrical signal in nervous system is **(MODELPAPER)**
   (a) proportional to the length of axon.
   (b) due to chemical synapse.
   (c) due to electrical synapse.
   (d) proportional to myelination.
8. In the nervous system, the action potential is generated at the axon hillock in physiological conditions and it is conducted to the terminal end of axon. The location specific origin of action potential and its direction-specific conduction are explained by a researcher in the following proposed statements·
   A. The membrane of axon hillock has highest threshold for the generation of action potential
   B. The membrane of axon hillock contains large numbers of voltage-gated Na^+ channels and that makes it more excitable
   C. The propagating action potential in the middle of the axon cannot generate another action potential in the direction of cell body since a large fraction of voltage-gated Na^+ channels in the preceding portion is voltage inactivated
   D. As the number of voltage-gated Na^+ channels is less in the preceding portion of axonal membrane, the propagating action potential in the middle of the axon cannot generate another action potential in the direction of cell body.

   Which one of the following combinations represents both correct explanations? **(DEC 2019)**
   (a) A and B (b) B and C
   (c) C and D (d) A and D

10. An experimentalist stimulates a nerve fibre in the middle of an axon and records the following observations. Which one of the observation is correct? **[DEC 2012]**

 (a) Nerve impulse is travelling in a direction towards cell body.

 (b) Nerve impulse is travelling in a direction towards telodendrons.

 (c) Nerve impulses are travelling in both the directions opposite to each other.

 (d) Nerve impulse is not moving in either direction

11. An action potential of a nerve fibre is described by different components including after-hyperpolarization. The mechanism of generation of this after-hyperpolarization has been proposed in the following statements:

 A. The increased conductance of Na^+ has returned to the base line level but the conductance of K^+ remains elevated during after-hyperpolarization phase

 B. The membrane potential is pulled even closer to the K^+ equilibrium potential at the after - hyperpolarization phase

 C. The conductance of Na^+ is increased before any change of K^+ conductance during after-hyperpolarization phase

 D. At the after- hyperpolarization phase, the membrane potential is driven closer to Na^+ equilibrium potential

 Choose the option with both correct Statements: **(JUNE 2017)**

 (a) A and B (b) B and C

 (c) C and D (d) A and D

12. The membrane potential in a giant squid axon recorded intracellularly at the resting condition (-70 mV) was reversed at the peak of action potential (+35 mV) after stimulation of the nerve fibre with a threshold electrical stimulus. This overshoot of the membrane potential has been explained in the following proposed statements:

 A. The rapid increase in Na^+-conductance during early phase of action potential uses membrane potential to move toward the equilibrium potential of Na^+ (+45 m V).

 B. The Na^+-conductance quickly decreases toward resting level after peak in the early phase and Na^+-ions are not able to attain its equilibrium potential within this short time.

 C. The conductance of K^+ at the early phase of action potential is increased and that leads to the reversal of membrane potential.

 D. The increase of K^+-conductance due to stimulation of nerve occurs before the changes of Na^+ -conductance is initiated and thus causes overshoot at the peak of action potential.

 Which one of the following is correct? **[JUNE 2016]**

 (a) A only (b) A and B

 (c) C only (d) C and D

13. When a nerve fiber is stimulated with increasing strength of stimulus, the action potential fails to generate even though the threshold level may be passed. The following statements may explain this accommodation of nerve fiber:

 A. The critical number of open sodium channels required to trigger the action potential may never be attained due to slow depolarization.

 B. Potassium channels open in response to slow depolarization, which makes the nerve fiber refractory to depolarization.

 C. The low threshold sodium channels remain open, which increases the threshold of firing of action potential.

 D. The efflux of sodium and influx of potassium due to operation of Na^+, K^+ ATPase oppose the depolarization.

 Which one of the following is correct? ***[DEC 2014]***

 (a) A only (b) A and B

 (c) C only (d) C and D

14. An action potential was generated on a nerve fibre by a threshold electrical stimulus. When a second stimulus was applied, no matter how strong it was, during the absolute refractory period of the action potential, the nerve fibre was unable to generate second action potential. This observation was explained in the following statements:

A. A large fraction of potassium channel was voltage inactivated

B. The critical number of sodium channels required to produce an action potential could not be recruited.

C. A large fraction of sodium channel was voltage inactivated.

D. The critical number of potassium channels required to produce an action potential could not be recruited.

Which one of the following is true?

[DEC 2013]

(a) Only A (b) A and B

(c) Only C (d) C and D

15. The conduction velocity of action potential in a myelinated nerve fibre was much greater than that of an unmyelinated fibre of the same diameter. The following statements were proposed to explain this observation:

A. The speed of conduction in a nerve fibre is determined by the plasma membrane resistance and axial resistance of axonal cytoplasm.

B. The electrical properties of myelinated and unmyelinated nerve fibres are not similar.

C. The myelin sheath decreases theeffective membrane resistance.

D. The magnitude of an electrotonic potential decreases more with distance along the axon in myelinated nerve fibres than that of unmyelinated fibres.

E. The voltage-gated Na^+ channels are highly concentrated at the nodes of Ranvier.

Choose one of the following combinations with both INCORRECT statements. **(DEC 2018)**

(a) A and B (b) B and C

(c) C and D (c) D and E

16. The action potential was recorded intracellularly from a squid giant axon bathed in two types of fluid such as sea water and artificial sea water having lower concentration of sodium ions while maintaining the same osmotic pressure with choline chloride. The nature of action potential was different in the two bathing fluids. Which of the following results is most likely? **[JUNE 2013]**

(a) The resting transmembrane potential was not changed but the amplitude of action potential was increased with lower sodium concentration in the bathing fluid.

(b) The amplitude of action potential was gradually decreased with reduction of sodium concentration in bathing fluid but the duration of action potential was prolonged.

(c) The resting transmembrane potential was decreased and the amplitude of action potential was also decreased with lower sodium concentration in the bathing fluid.

(d) The amplitude of action potential was not changed with reduction of sodium concentration in the bathing fluid but the duration of action potential was prolonged.

17. Electrical stimulation of a nerve bundle maintained at 37±1°C showed A, B, C and D peaks (as shown in the diagram) when recorded at a distance of 15 cms from the stimulating site on the same bundle.

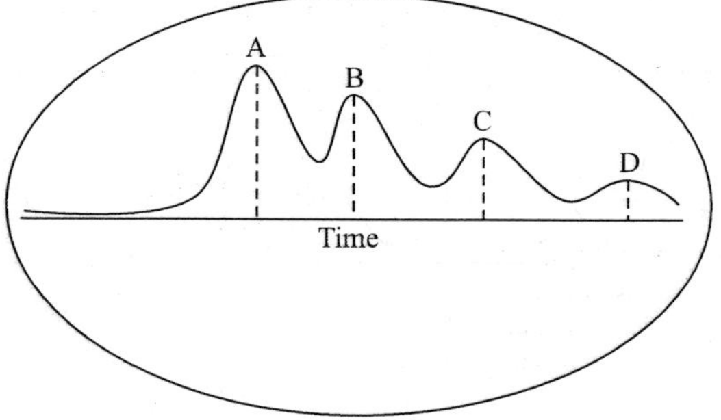

The same experiment was conducted at significantly lower temperature 15±1°C. Which of the following statements is correct?

(MODEL PAPER)

Lower temperature would

(a) not affect the record.

(b) not show A, but would show B, C and D peaks.

(c) not show peak D and may not show C, but would show A and B.

(d) show all the peaks along with a new peak

18. The peaks of the compound action potential (i.e., A, B and C) recorded from a mammalian mixed nerve were affected after application of increasing pressure on the nerve. Some probable changes of compound action are stated below:

A. 'A' peak was inhibited by lower intensity of pressure

B. 'C' peak was inhibited by higher intensity of pressure

C. 'B' peak was inhibited by lower intensity of pressure

D. 'C' peak was inhibited by lower intensity of pressure

E. 'A' peak was inhibited by higher intensity of pressure

Select the option with the combination of CORRECT statements. **(JUNE 2018)**

(a) A and B (b) B and C

(c) C and D (d) D and E

Answer Key

1.(b)	**2.**(d)	**3.**(b)	**4.**(a)	**5.**(a)	**6.**(c)	**7.**(b)
8.(b)	**9.**(b)	**10.**(c)	**11.**(a)	**12.**(b)	**13.**(b)	**14.**(*)
15.(c)	**16.**(b)	**17.**(c)				

6. Brain, Central and peripheral nervous system

1. Mammillary bodies are present in **(DEC 2017)**

(a) thalamus (b) epithalamus

(c) hypothalamus (d) midbrain

2. Memory of facts and events is not consolidated if there is an injury to **[DEC 2013 GU]**

(a) striatum. (c) amygdala.

(b) cerebellum. (d) Hippocampus.

3. Which one of the following functions is correct for the microglia cells in the brain?

(DEC 2019 ASSAM)

(a) Myelin formation

(b) Making blood-brain barrier

(c) Scavenging injured cell debris

(d) Enveloping synapses

4. A monkey undergoes cerebellectomy. After the post-operative recovery, the monkey was given a task to press a bar. The possible observations are:

A. Its hand would overshoot the target while reaching the bar.

B. It would be unable to move forelimbs.

C. It would show intention tremor while trying to press the bar.

D. It would press the bar with mouth instead of hand.

Which one of the following is correct?

[DEC 2011]

(a) A and C (b) B only

(c) D only (d) B and D

5. A response was observed when a specific site in a rat brain was stimulated by passing electrical pulses through indwelling electrode implanted surgically. In another experiment in another rat, a cannula was surgically implanted instead of the electrode and stimulated the area by injecting excitatory neurotransmitter. However, the result of the two experiments did not match. The possibilities of variations in the results could be due to **(MODELPAPER)**

(a) animal variations only.

(b) stimulation of cell bodies or nerve fibers only.

(c) difference in anatomical brain areas only.

(d) variations in all the reasons mentioned in 1, 2 and 3.

6. Which one of the following neurotransmitters is secreted by the pre-ganglionic neurons of sympathetic nervous system? **[DEC 2015]**
   (a) Epinephrine (b) Acetylcholine
   (c) Dopamine (d) Norepinephrine
7. The cell bodies of sympathetic-preganglionic neurons are located in **[JUNE 2016]**
   (a) Intermediolateral cell column of spinal cord
   (b) Posterior cell column of spinal cord
   (c) Celiac ganglion
   (d) Paravertebral ganglion
8. Sympathetic post-ganglionic neurons that are cholinergic, innervate **(JUNE 2018)**
   (a) sweat glands (b) parotid glands
   (c) hair follicles (d) Pancreas
9. A football player injured in the heat of play did not feel pain until the game was over.
   This stress-induced analgesia was explained by the following proposed statements:
   A. The release of norepinephrine from brain stem catechol aminergic neurons in the amygdala reduced during stress
   B. The reduced release of 2-arachidonoglycerol (2AG) in brain contributed to stress-induced analgesia
   C. The inhibition of CB, receptors in many brain regions accounted for reduced analgesia in stress
   D. The activation of CB2 receptors on microglia in stress caused analgesic effect
   Which one of the following options represents both correct statements? **(DEC 2019 ASSAM)**
   (a) A and B (b) B and C
   (c) C and D (d) A and D

Answer Key

1.(c)	2.(d)	3.(c)	4.(a)	5.(b)	6.(b)	7.(a)
8.(a)	9.(d)					

7. Neural control of muscle tone and posture

1. When ATP remains attached with the myosin head, which one of the following event happens in the skeletal muscle? **[DEC 2013 GU]**
   (a) Muscle contraction.
   (b) Muscle relaxation.
   (c) Muscle rigor.
   (d) Muscle spasm.
2. Which one of the following skeletal muscles of human body contains highest number of muscle fibre in a motor unit? [JUNE 2015]
   (a) Muscles of hand
   (b) Extraocular muscles
   (c) Muscles of leg
   (d) Muscles of face.
3. A patient has episodes of painful spontaneous muscles contraction, followed by periods of paralysis of the affected muscles. It was identified as primary hyperkalemic paralysis, an inherited disorder. The possible causes of the paralysis are
   A. The elevation of extracellular K^+ causes hyperpolarization of skeletal muscle cells.
   B. The hyperpolarization of the muscle cell membrane inactivates sodium channels.
   C. The elevation of extracellular K^+ causes depolarization of skeletal muscle cells.
   D. The sodium channels are voltage inactivated in depolarized state
   Which one of the following is true?
   [DEC 2013]
   (a) Only A (b) A and B
   (c) Only C (d) C and D
4. A myasthenia gravis patient develops muscle paralysis because **(MODELPAPER)**
   (a) the nerve terminal at the neuromuscular junction fails to release acetylcholine.
   (b) although enough acetylcholine is released at the neuromuscular junction, it is destroyed by acetylcholinesterase.

(c) the patient develops antibody against his own acetylcholine receptor.

(d) the patient develops antibody against his own acetylcholine.

5. When a skeletal muscle with an intact nerve supply is stretched, the muscle contracts and the tone increases. The following explanations are offered for this observation:

A. Golgi tendon organ was stimulated by the stretching of muscle

B. γ-motor neurons were excited by the stimulated afferent nerve fibres from the stretched muscle

C. Muscle spindle was stimulated by stretching of muscle

D. α-motor neurons were excited by the stimulated afferent nerve fibres from the stretched musscle

Which of the combination is correct?

[JUNE 2014]

(a) A and B (b) B and C

(c) C and D (d) A and D

6. When a skeletal muscle was passively stretched, it contracted reflexly. However, when the muscle was over stretched in dos way it showed sudden cessation of contraction followed by relaxation, The following statements provided the possible explanation of these observations:

A. The passive stretching of muscle caused stretching of muscle spindle

B. The over stretching of muscle stimulated Golgi tendon organ

C. Group lb sensory nerve fibers were stimulated by stretching of muscle spindle

D. Group lb sensory nerve fibers stimulated α-motor neurons that supplied the muscle from which these fibers arose.

E. Group la sensory nerve fibers were connected to Golgi tendon organ.

F. The stimulation of group la sensory fibers led to the production of IPSP on the $\alpha\alpha$ - motor neurons that supplied the muscle from which these fibers arose.

Which of the above statement(s) is/are correct?

(DEC 2017)

(a) A and B (b) C and D

(c) E and F (d) C and F

Answer Key

1.(b)	2.(c)	3.(d)	4.(c)	5.(c)	6.(a)

8. Sense organs – Vision

1. A 1 meter tall object was placed 10 meter in front of a normal eye. The size of the image on the retina will be (consider distance between lens and retina = 1.7 cms) **[DEC 2011]**

(a) 0.17 mm (b) 1.7 mm

(c) 3.4 mm (d) 170 μm

2. Hawk's retina possesses a large number of **(MODEL PAPER)**

(a) rods. (b) melanocytes.

(c) cones. (d) kuffer cells.

3. Melanopsin is found in which cell of the retina? **(JUNE 2018)**

(a) Cones (b) Rods

(c) Ganglion cells (d) Bipolar cells

4. In a normal human eye, for sharp image formation on the retina, maximum dioptric power is provided by the **[DEC 2011]**

(a) retina

(b) cornea

(c) anterior surface of the lens

(d) posterior surface of the lens

5. In the dark, rods show a large inward "dark" current which is suppressed by a flash of light. Which one of the following statements, explaining the effect of light, is true?

[JUNE 2013]

(a) Sodium channels in the outer segment of rods are closed.

(b) Cytoplasmic cGMP concentration increases.

(c) Sodium channels in the inner segment of rods are closed.

(d) Transducin dissociates from beta-arrestin.

6. The probable effects of lesion of left optic tract on the vision of a human subject are given below. Identify the correct statement.

[DEC 2016]

(a) Blindness in the left eye but the visual field of right remains intact.

(b) Blindness in the right half of the visual fields of both the eyes.

(c) Blindness in the left half of the visual field of left eye and blindness in the right half of the visual field of the right eye.

(d) Blindness in the left half of the visual field of both the eyes.

7. A majority of humans with normal colour vision was found to be more sensitive to red light in Rayleigh match, where the subject mixed variable amount of red and green light to match monochromatic orange. Which one of the following statements is NOT true to explain the observation? **[JUNE 2016]**

(a) There are variations in the sensitivity of long-wave cone pigments.

(b) The short-wave cone opsin in red- sensitive subjects is different from others.

(c) The absorption curve of long-wave cone pigment peaks at 556 nm in red-sensitive subjects while it peaks at 552 nm in others.

(d) The long-wave cone opsin in red-sensitive subjects is different in primary structure from that of other.

8. Which one is the correct sequence of events that takes place during phototransduction when light falls onto the retina? **(JUNE 2019)**

(a) Closure of Na^+ channels → activation of transducing decreased release of glutamate → decrease in intracellular cGMP → structural changes in rhodopsin

(b) Decreased release of glutamate → structural changes in rhodopsin → activation of transducin → decrease in intracellular cGMP Closure of Na^+ channels

(c) Structural changes in rhodopsin → activation of transducin → decrease in intracellular cGMP → Closure of Na^+ channels → Decreased release of glutamate

(d) Decrease in intracellular cGMP → activation of transducin → decreased release of glutamate → structural changes in rhodopsin → Closure of Na^+ channels

9. When rods of retina kept in dark, were exposed to light, photo transduction occurred. Following are some explanations given by a researcher regarding phototransduction:

A. Activation of transducin

B. Inhibition of cGMP phosphodiesterase

C. Closure of Na channels

D. Hyperpolarization of rods

Which one did NOT occur in phototransduction?

[DEC 2015]

(a) only A (b) only B
(c) A and C (d) C and D

10. Upon prolonged illumination, activated rhodopsin does not activate transducing, hence the vision is impaired. This could be because of the following explanations:

A. Most of the activated rhodopsin gets phosphorylated and is unable to activate transducin.

B. Most of the activated rhodopsin gets dephosphorylated and is unable to activate transducin.

C. Arrestin further interacts with phosphorylated rhodopsin.

D. Arrestin further interacts with dephosphorylated rhodopsin.

Which of the above combinations is correct?

[JUNE 2014]

(a) A and C (b) B and D
(c) A and D (d) B and C

11. During prolonged illumination,rhodopsin is desentitized which leads to the termination of visual response.The associated proteins (Colum A) and their effects (Column B) are given below:

	Column-I		Column-II
a	Phosphorylated opsin	i	Phosphorylates opsin
b	Rhodopsin kinase	ii	Binds to phosphorylated opsin
c	Arrestin	iii	Decreases activation of transducing
d	Phosphatse	iv	Reverses the termination process.

Which one of the matched combinations is correct? **(JUNE 2019)**

(a) a-iv, b-ii, c-I, d-iii

(b) a-ii, b-iii, c-iv,d-i

(c) a-iii, b-i, c-ii, d-iv

(d) a-ii, b-iv, c-iii,d-i

Answer Key

1.(b)	2.(c)	3.(c)	4.(b)	5.(a)	6.(a)	7.(b)
8.(c)	9.(b)	10.(a)	11.(c)			

9. Hearing and Tactile response.

1. Prestin, a membrane protein, is found in which one of the following cells of theorgan of Corti? **(JUNE 2018)**

(a) Inner hair cells

(b) Inner phalangeal cells

(c) Outer hair cells

(d) Outer phalangeal cells

2. The mechanism of sound localization in a 'horizontal plane' by the human" auditory system can be explained by

A. the difference in time between the arrival of stimulus in two ears

B. the difference in phase of the sound waves on two ears

C. the difference in tuning curves of two auditory nerves

D. the activity of neurons in superior olivary nucleus, but not by the neuronal activity of auditory cortex

Which one of the following is NOT correct? **[JUNE 2015]**

(a) Only A (b) A and B

(c) B and C (d) C and D

3. Different frequencies of sound were presented on the ear and the movement of basilar membrane was experimentally determined. The characteristics of movement of basilar membrane after presentation of' 100 Hz sound are described in the following statements:

A. The base of basilar membrane showed resonance.

B. The apex of basilar membrane showed resonance.

C. A wave travelled from the base to apex of basilar membrane but the maximum displacement was noted near the apex.

D. A wave travelled from the base to apex of' basilar membrane but the maximum displacement was noted near the base.

Which one of the following is correct? **[DEC 2014]**

(a) A and C (b) B and D

(c) C only (d) D only

4. The stereocilia of auditory hair cells are arranged in rows but the height of stereocilia are not the same in all the rows. Though the height of stereocilia is the same within a particular row, the heights increase in subsequent rows. When the stereocilia of shorter rows are mechanically pushed toward the taller rows, the hair cells are depolarized but a push on opposite direction hyperpolarize them. The significance of this graded height of stereocilia is:

(A) Each row of stereocilia may be displaced independent of other rows in physiological conditions

(B) The tip of the taller stereocilia will show greater displacement as compared to shorter ones when all the rows are moving in same axis

(C) The hair cells will be depolarized or hyperpolarized in different grades when the axis of displacement is changed

(D) The taller stereocilia are involved with depolarization and shorter ones are responsible for hyperpolarization

Which of the following is correct?

[JUNE 2012]

(a) A only (b) B only
(c) B and C (d) A and D

5. The excitation of auditory hair cells by the displacement of stereocilia has been explained in the following proposed statements:

A. The gradual increased height of stereocilia is required for the transduction process

B. The changes of membrane potential of auditory hair cells are proportional to the direction and magnitude of the displacement of stereocilia

C. The higher concentration of K^+ in endolymph and higher concentration of Na^+ in perilymph are not required for the excitation of hair cells.

D. The mechanically sensitive cation channels on the top of stereocilia are not adapted to maintain displacement of stereocilia

Select the combination with INCORRECT statements from the following options:

(JUNE 2018)

(a) A and B (b) B and C
(c) C and D (d) A and C

6. The following statements describe the possible functions of the outer hair cells (OHC) of organ of Corti:

A. The outer hair cells (OHC) are depolarized or hyperpolarized depending on the direction of movement of stereocilia.

B. The OHC are lengthened in depolarization and shortened in hyperpolarization

C. The OHC decrease the amplitude and clarity of sound by shortening and lengthening.

D. The efferent nerve fibers of olivocochlear bundle modulate the sensitivity of the OHC

E. The effect of stimulation of olivocochlear bundle on the OHC is excitatory.

Which one of the options given below represents the correct statements? **(DEC 2017)**

(a) A and B (b) C and D
(c) A and D (d) D and E

7. Following are some statements about the mechanism of stimulation of receptors for touch, pain, vision and warmth that may or may not be correct.

A. The touch receptor does not require any voltage gated cation channel for its activation.

B. Light causes closing of Na^+ channels in the outer segments of rods and cones

C. Pain sensation is caused by opening of Na^+ or Na^+/Ca^{++} channels in free sensory nerve endings.

D. The warmth receptor is activated by non-selective anion channels.

Choose the option with both statements as correct. **(JUNE 2017)**

(a) A and C (b) B and C
(c) C and D (d) A and D

8. External pressure given on a mixed nerve causes loss of touch sensation while pain sensation remains relatively intact. On the other hand, application of local anesthetics on the same nerve, induces loss of pain sensation keeping touch sensation least affected. These observations can be explained by the following statements:

A. External pressure causes loss of conduction of impulses in small diameter sensory nerve fibres.

B. Local anesthetics depress the conduction of impulses in large diameter sensory nerve fibres.

C. Touch-induced impulses are carried by fibre Type A

D. Fibre type C is responsible for pain sensation

Which of the above statement(s) is/are INCORRECT? **[DEC 2016]**

(a) A and B (b) C and D

(c) Only C (d) Only D

9. A mechanical pressure was exerted on a specific location of a peripheral nerve of a mammal. The touch or pain receptors were stimulated from the skin surface innervated by the same nerve. The action potential generated by touch receptor stimulation was blocked beyond the point of mechanical pressure. But the action potential generated by pain receptor stimulation passes through the point of mechanical pressure.

Following explanations were offered for these observations:

A. The large diameter 'A' fibres were affected by mechanical pressure

B. The small diameter 'C' fibres were not affected by mechanical pressure

C. The intermediate diameter 'B' fibres were affected by mechanical pressure

D. The large diameter 'A' fibres were not affected by mechanical pressure

Which of the above combinations is correct? **[JUNE 2014]**

(a) A and B (b) B and C

(c) C and D (d) A and D

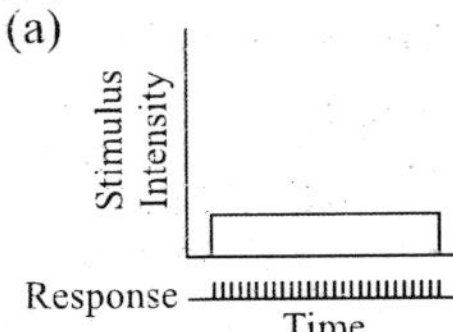

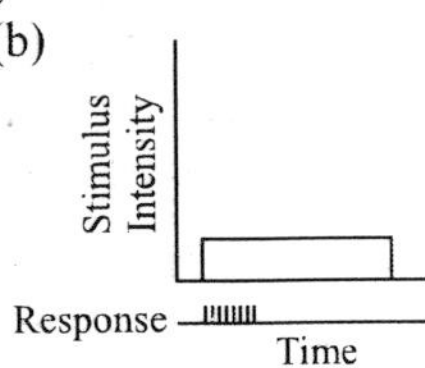

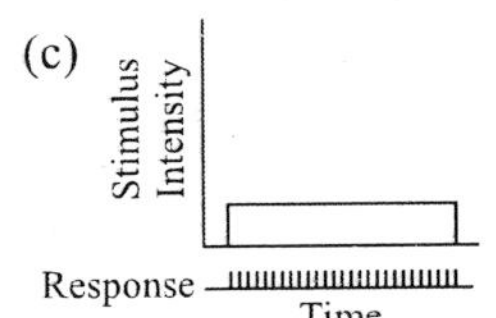

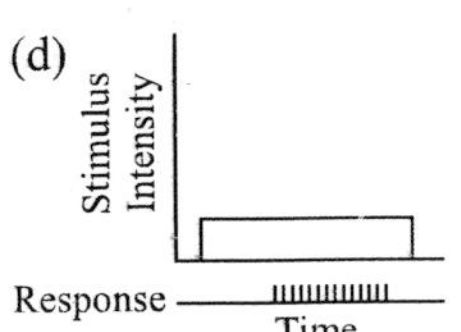

10. Various types of excitable tissues when stimulated showed response as shown in the above figures. Which one of them is an example of fast adapting tissue? **[DEC 2013]**

(a) A (b) B

(c) C (d) D

Answer Key

1.(c)	2.(d)	3.(c)	4.(c)	5.(c)	6.(c)	7.(b)
8.(a)	9.(a)	10.(b)				

10. Excretory system

1. The following statements are related to excretion in invertebrates:

A. Flame cells are found in molluscs and jelly fish.

B. Nephridia and Malpighian tubules convert ammonia to urea for water conservation.

C. Green glands are found in flatworm and help in the excreta elimination.

D. Excretory canals in nematodes carry waste materials to excretory pores in the body wall.

Choose the correct answer: **[JUNE 2014]**

(a) Only C (b) A and C

(c) Only B (d) B and D

2. On investigating the excretory organs of an organism that was found in an office backyard, the following structures were identified. A funnel-like structure to collect the waste, duct-like structures to carry the waste to nephrostones and finally nephridipores through which waste leaves the body. What species could the organism be? **[DEC 2013 GU]**

(a) Earthworm (b) Ascaris

(c) Snail (d) Spider

3. Filtration slits are formed by **(JUNE 2017)**

(a) podocytes

(b) endothelial cells of capillary

(c) mesangial cells

(d) Lacis cells

4. In kidney, Na^+ is reabsorbed across the second half of proximal tubule due to positive transepithelial voltage (i.e., tubular fluid becomes positive relative to blood) and by other mechanisms. The following proposed statements could explain the development of this positive transepithelial voltage.

Cl^- concentration gradient in the second half of the proximal tubule favours diffusion of Cl^- from tubular lumen to intercellular space via a paracellular route, which generates the positive transepithelial voltage.

The Na^+ - H^+ antiporters in the second half of proximal tubules create the positive transepithelial voltage.

The Na^+ - glucose symporters operating in the proximal part of renal tubules are responsible for this positive transepithelial voltage

The positive transepithelial voltage is created by the operation of $1Na^+$ - $1K^+$ - $2Cl^-$ symporter in the proximal tubules

Select the option with correct statement(s).

(JUNE 2017)

(a) only A (b) B and C
(c) C and D (d) only D

5. During urine formation the filtration of blood at the glomerulus is **[JUNE 2011]**

(a) an active process.
(b) an osmotic process.
(c) is a pressure-dependent physical process.
(d) a non energy- mediated transport process.

6. Which one of the following cells in the renal corpuscle can influence glomerular filtration by its contraction? **[DEC 2014]**

(a) Podocytes.
(b) Endothelial cells of glomerular capillaries
(c) Parietal epithelial cells of Bowman's capsule
(d) Mesanglial cells

7. In a healthy kidney given the following information:

(A) Glomerular hydrostatic pressure – 75 mm Hg
(B) Glomerular capillary colloid osmotic pressure – 40 mm Hg
(C) Hydrostatic pressure in the Bowman's capsule – 20mn Hg

The net filtration pressure will be

(MODEL PAPER)

(a) 55mm Hg. (b) 15 mm Hg.
(c) 35 mm Hg. (d) 135 mm Hg.

8. The difference in circulation between glomerular capillaries (GC) and true capillaries (TC) are described by a researcher in the following statements:

A. The hydrostatic pressure in GC is higher than that in TC
B. The endothelial cells are fenestrated in GC but not in TC
C. Both filtration and fluid movement into capillary takes place in TC but only filtration occurs in GC.
D. The plasma colloid osmotic pressures in both the ends of GC or TC are similar.

Which one of the following is NOT correct?

[DEC 2015]

(a) Only A (b) A and B
(c) Band C (d) Only D

9. In the glomerular capillary (GC), fluid moves into Bowman's capsule through its almost entire length. But in the muscle capillary (MC), fluid moves into interstitial space at its arteriolar end. The difference between these two capillaries is explained in the following proposed statements:

A. Afferent and efferent arterioles are present on the two ends of GC, but in MC, arteriole and venule are present on two ends.
B. The hydrostatic pressure in GC is higher than that in MC.
C. The efferent arteriole in GC has a relatively low resistance, but venules in MC has a high resistance.

D. The difference of hydrostatic pressure between two ends of GC is relatively more but it is negligible in MC.

E. The difference of oncotic pressure between two ends of MC is negligible but it is relatively more in GC.

F. The net filtration pressure falls to zero at the efferent end of GC but it is 9 mm Hg inward at the venular end of MC.

Which of the above statements are **INCORRECT**? **(DEC 2017)**

(a) A and B (b) C and D

(c) E and F (d) B and F

10. Maintaining the salt concentration and volume plasma are two key parameters for physiological processes achieved by kidney. Which of following structural and functional combines the most efficient renal regulatory system mammals? **[DEC 2013]**

	Structural	Functional
1	Large glomerulus, long proximal and distal tubules, long Henle's loop	Trans-epithelial in proximal counter current multiplier, ADH responsiveness of distal tubule
2	Small glomerulus, short proximal and distal tubule short Henle's loop	Transepithelial potential in distal tubule, very high ADH concentration in circulation
3	Very large glomerulus, short proximal tubule, very ong distal tubule, long Henle's loop	Very efficient glomerular filtration, prevention of solute loss
4	Small glomerulus, long proximal and distal tubules, Henle's loop	Preventing water and solute filtration, excreting solute, lowering ADH responsiveness

11. Desert animals have longer loop of Henle compared to that of humans. It may be due to the following reasons:

A. Long loop of Henle is associated with greater amount of vasopressin secretion.

B. In long loop of Henle, the counter-current exchanger is more effective.

C. Long loop of Henle conserves more water.

D. Long loop of Henle stimulates production of angiotensin II.

Which of the above reason(s) is/are correct? **[DEC 2012]**

(a) A and B (b) B and C

(c) C and D (d) only D

12. The different segments of renal tubule (column A) and the mechanism of Na^+ transport in the apical membrane of tubular cells (column B) are tabulated below:

	Column A		Column B
(a)	Proximaltubule	(i)	$Na^+ - Cl^-$ symporter
(b)	Thick ascending loop of Henle	(ii)	Diffusion through Na^+ Selective channel (ENaC)
(c)	Early distal tubule	(iii)	Na – glucose symporter
(d)	Late distal tubule and collecting duct	(iv)	1 Na^+ -1K^+-2 Cl^- Symporter

Select the option with the correct matches: **(JUNE 2018)**

(a) a - (iii), b - (iv), c - (i), d - (ii)

(b) a - (iv), b - (iii), c - (ii), d - (i)

(c) a - (i), b - (ii), c - (iii), d - (iv)

(d) a- (ii), b- (i), c - (iv), d - (iii)

13. Aldosterone increases the reabsorption of Na^+ from the tubular fluid in the thick ascending limb of loop of Henle and in the distal tubule.

These effects are explained in the following proposed statements:

A. Aldosterone increases the number of Na^+-Cl^- symporter in the apical membrane of principal cells in the early portion of distal tubule

B. The number of Na^+ channels (ENaC) is increased in the apical membrane of principal cells in the late portion of distal tubule by aldosterone

C. The synthesis of Na^+, K^+-ATPase in the basolateral portion of principal cells in distal tubule is decreased by the action of aldosterone

D. Aldosterone increases the reabsorption of Na^+ across the apical cell membrane in the thick ascending limb of loop of Henle by decreasing Na^+, K^+-ATPase in it

Which one of the following combinations represents both correct statements?

(DEC 2019)

(a) A and B (b) B and C
(c) C and D (d) A and D

14. Certain statements made on the mechanisms of action of a neuro-hormone, vasopressin (VP) are given below:

A. VP acts on collecting ducts of kidney to concentrate urine by binding to V_1 receptors and activating cAMP cascade
B. VP acts on collecting ducts of kidney to concentrate urine by binding to V_2 receptors and activating cAMP cascade
C. VP acts on blood vessels to produce constriction by binding to V_1 receptors and activating cAMP cascade
D. VP acts on blood vessels to produce constriction by binding to V_1 receptors and activating phosphatidylinositol cascade

Which one of the following options represents a combination of correct statements?

(DEC 2019 ASSAM)

(a) A and B (b) C and D
(c) B and D (d) A and D

15. Three forms of dextrans namely neutral, polyanionic and polycationic having different molecular radii were injected separately in three groups of rats. The concentrations of dextrans in glomerular filtrate were measured to determine the filterability of the dextrans.

The possible outcomes could be as follows:

(A) The dextrans having smaller diameter have greater filterability than larger dextrans.
(B) Neutral dextrans were filtered more than polycationic and polyanionic dextrans.
(C) Polycationic dextrans were filtered more than neutral and polyanionic dextrans.
(D) Polyanionic dextrans were filtered more than neutral and polycationic dextrans.

Which one of the following combinations is correct? **[JUNE 2013]**

(a) (A) only (b) (B) only
(c) (A) and (C) (d) (B) and (D)

16. The fractional clearance of neutral and cationic dextran molecules of various sizes through kidneys of a rat is shown in the figures below. Which one of the following is correct?

[JUNE 2015]

— Neutral dextran - - - Cationic dextran

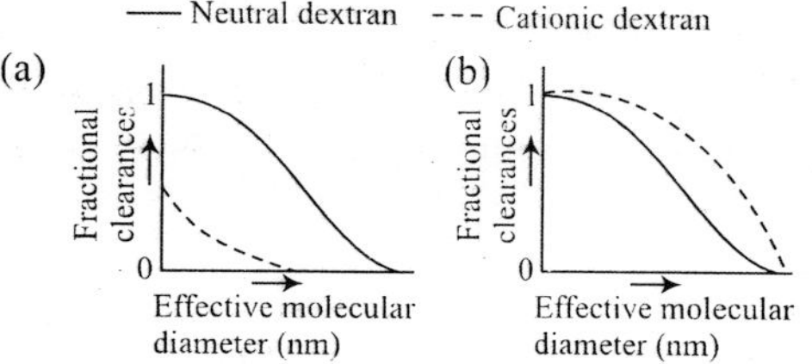

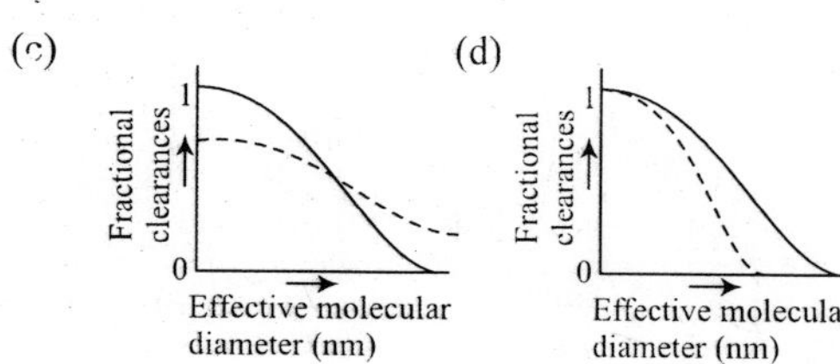

Answer Key

1.(d)	2.(a)	3.(a)	4.(a)	5.(c)	6.(d)	7.(b)
8.(d)	9.(b)	10.(a)	11.(b)	12.(a)	13.(a)	14.(c)
15.(c)	16.(b)					

11. Blood volume, electrolyte and, acid-base balance.

1. In which one of the body fluids is K^+ concentration higher than that of Na^+?

(DEC 2017)

(a) Plasma
(b) Perilymph
(c) Endolymph
(d) Cerebrospinal fluid

2. A boy eats a large serving of cheese having high amount of sodium. He hardly drinks any fluid. Inspite of this, the water and electrolyte balance was maintained. Which one of the following explanation is correct? **[DEC 2012]**
   (a) His aldosterone was decreased and alcohol dehydrogenase (ADH) was increased.
   (b) His aldosterone was increased and ADH was decreased.
   (c) There was no change in either of the hormones.
   (d) His sympathoadrenal system was stimulated.

3. Two individuals A and B, each of 75 kg body weight, have similar volume of body water. Both of them had high salt snack. However, individual A also had a glass of alcoholic drink. Based on this information, which one of the following statements is true? **(JUNE 2018)**
   (a) A will have lower circulating level of antidiuretic hormone (ADH) than B
   (b) B will have lower circulating level of ADH than A
   (c) The level of ADH will not change in these two individuals
   (d) The reabsorption of water in kidney will be more in A than B

4. After hemorrhage, a subject develops hypovolemia and hypotension. Following are some of the statements regarding homeostatic measure taken by the body after hemorrhage.
   A. Increased release of vasopressin
   B. Increased water retention and reduced plasma osmolality
   C. Increased rate of afferent discharge from low pressure receptors of vascular system
   D. Decreased rate of afferent discharge from high pressure receptors of vascular system

   Which one of the following is NOT correct in this condition? **[DEC 2015]**
   (a) Only A (b) A and B
   (c) Only C (d) B and D

5. A person recovered from a moderate degree of haemorrhagic shock. The participating physiological mechanisms in this recovery process are proposed in the following statements.
   A. The decrease in arterial pressure after haemorrhage causes inhibition of sympathetic-vasoconstrictor system.
   B. After haemorrhage, the angiotensin II level in blood is increased which causes increased re-absorption of Na^+ in renal tubules.
   C. The increased secretion of vasopressin after haemorrhage increases water retention by the kidneys.
   D. After haemorrhage, the reduced secretion of epinephrine and nor-epinephrine from adrenal medulla induces decreased peripheral resistance.
   E. In haemorrhage, the central nervous system ischemic response elicits sympathetic inhibition.

   Choose one of the following combinations with both the correct statements. **(DEC 2018)**
   (a) A and B (b) B and C
   (c) C and D (d) D and E

6. An isolated carotid sinus was prepared so that the pressure may be regulated by a pump and the resulting discharge in single carotid sinus nerve fibre could be recorded. The following are the possible observations.
   A. No discharge when carotid sinus perfusion pressure was below 30 mm Hg.
   B. Linear increase in discharge frequency when carotid sinus perfusion pressure was gradually increased from 70 to 110 mm Hg.
   C. Increase in discharge frequency was more prominent in greater pulsatile changes of carotid sinus pressure keeping the mean pressure identical in all cases.
   D. Increase in discharge was more prominent in the falling phase of pulsatile change of carotid sinus pressure than in the rising phase.

Which one of the following is correct? **[DEC 2011]**

(a) A, B and C (b) A and C

(c) B and D (d) D only

7. The blood volume decreased when a mammal was bled rapidly. However. the cardiovascular changes resulting from hemorrhage could be minimized by the following compensatory mechanisms:

A. Increased cerebral blood flow,

B. Reduction of baroreceptor activity and stimulation of chemoreceptors.

C. Reabsorption of tissue fluid in blood.

D. Increased release of enkephalins and beta-endorphins.

Which of the above is/are correct? **[DEC 2012]**

(a) A and B (b) B and C

(c) C and D (d) only D

8. In this flow diagram name the chemicals A, B, C and D in proper sequence. **[DEC 2011]**

Decreased arterial pressure

↓

(A) From kidney

↓

(B) Substrate for A

↓

(C)

↓ Converting enzyme from lung

(D)

↙ ↘

Renal retention of salt and water | Vasoconstriction

↘ ↙

Increased arterial pressure

(a) Renin, Angiotensin II, Angiotensin I, Angiotensinogen

(b) Angiotensin I, Angiotensinogen, Angiotensin II, Renin

(c) Renin, Angiotensin I, Angiotensin II, Angiotensinogen

(d) Renin, Angiotensinogen, Angiotensin I, Angiotensin II

9. Which one of the following is NOT a function of angiotensin II? **(DEC 2018)**

(a) Facilitates the release of norepinephrine from post-ganglionic sympathetic neurons

(b) Increases the sensitivity of baroreflex by acting on brain

(c) Produces arteriolar contraction

(d) Increases the secretion of vasopressin

10. Which one of the following agents cause relaxation of mesenglial cells? **(JUNE 2019)**

(a) Histamine (b) Thrombaxane A_2

(c) Norepinephrine (d) Dopamine

11. Desert rattlesnakes live in extremely dry environment. In order to conserve body water, they are adopted to convert ammonia to

(a) urea. (b) purines.

(c) uric acid (d) amino acids.

12. A healthy individual was immersed in water up to neck in an upright posture for 3h. The plasma concentration of atrial natriuretic peptide (ANP), renin and aldosterone were measured for 5 h at 1 h intervals including the immersion period. The results are graphically presented below.

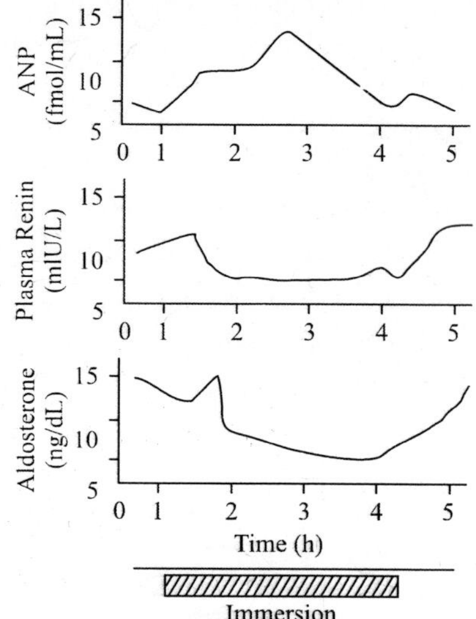

The results of this experimental condition (EC) are explained in the following proposed statements which may be correct or incorrect.

A. ANP secretion is proportional to the degree of stretch of atria.

B. The decreased plasma renin concen- tration in EC is due to increase in sympathetic activity.

C. The decreased aldosterone level in EC is the effect of plasma renin level.

D. The effect of gravity mi the circulation is counteracted in EC.

E. The central venous pressure is decreased in EC.

Choose one of the following combinations with all correct statements. **(DEC 2018)**

(a) A,B,C (b) A,C,D

(c) C,D,E (d) B,C,D

Answer Key

1.(c)	2.(a)	3.(a)	4.(c)	5.(b)	6.(a)	7.(b)
8.(d)	9.(b)	10.(d)	11.(c)	12.(b)		

12. Stresss and Thermoregulation

1. Polar bears maintain their body temperature because they have more of **[JUNE 2011]**

(a) transducin protein.

(b) uncoupling protein.

(c) myoglobin protein.

(d) F_oF_1 ATPase.

2. Name the ectothermic animal that can thermoregulate by behavioural means rather than by physiological means. **[JUNE 2011]**

(a) Bumble bee in an orchard.

(b) Tuna fish in the ocean.

(c) Lizard in a desert.

(d) Flatworm in a pond.

3. If the core body temperature of a human rises above normal, which of the following processes would be initiated sequentially for Thermo-regulation? **[JUNE 2011]**

(a) Peripheral vasodilation, increased rate of respiration, tachycardia.

(b) Peripheral vasoconstriction, increased rate of respiration, bradycardia.

(c) Peripheral vasodilation, decreased rate of respiration, tachycardia.

(d) Peripheral vasodilation, decreased rate of respiration, bradycardia.

4. There is evidence that following pyrogenic stimuli, cytokines produced by the CNS cause fever, possibly by local release of prostaglandins. Accordingly, the following statements have been proposed:

A. Cytokines act independently and directly on thermoregulatory centres

B. Intrahypothalamic injection of prostaglandin receptor agonists will prevent fever

C. Antipyretic effect of aspirin is exerted on the hypothalamus to prevent prostaglandin synthesis

D. Aspirin blocks infections and eventually prevents fever.

Which one of the following combination of above statements is correct? **(JUNE 2017)**

(a) A and D (b) B and C

(c) B and D (d) A and C

5. A visitor to a region of hot climate is more distressed by the heat than the regular resident. Within a few weeks, the visitor is more comfortable with the heat and capacity for work is increased. Following are some of the explanations given by a researcher regarding acclimatization to heat.

A. Sweating begins at a lower body temperature

B. Blood flow through skin is high for any body temperature

C. There is rise in resting body temperature.

D. Vasoconstriction starts at a lower body temperature

Which one of the following is NOT true? **[DEC 2015]**

(a) Only A (b) A and B

(c) Only C (d) C and D

6. Humans exposed to cold climate are able to maintain their body temperature. The physiological mechanisms operating in the condition were suggested in the following statements:
   A. The increased thermogenesis in brown adipose tissue occurred in cold
   B. Uncoupling protein 1 in brown adipose tissue participated in the increased heat production
   C. The decreased secretion of catecholamine in cold elevated heat production
   D. The reflex responses activated by cold were controlled from the anterior hypothalamus
   E. The counter-current exchanges of temperature between warm arterial blood flowing towards limbs and the cold venous blood coming from extremities conserved body heat.

   Which one of the following options represents a combination of INCORRECT statements?

   (DEC 2019 ASSAM)

   (a) A and B (b) B and C
   (c) C and D (d) D and E

Answer Key

1.(b)	2.(c)	3.(a)	4.(d)	5.(d)	6.(c)

13. Digestive system

1. A person showed the symptoms of diarrhea, gas and pain whenever milk was consumed. The doctor advised the person to take curd instead of milk and subsequently the symptoms mostly disappeared due to this change of dairy product. The following statements are proposed to explain this observation:
   A. The person has deficiency in the intestinal sucrase-maltase
   B. Curd is not deficient in sucrose and maltose
   C. The person has deficiency in the intestinal lactase
   D. The bacteria in curd contain lactase

   Which one of the following is true?

   [JUNE 2016]

   (a) A only (b) A and B
   (c) C only (d) C and D
2. Which type of cells located in gastric glands is responsible for the release of histamine?

   [DEC 2016]

   (a) Mucus neck cells
   (b) Enterochromaffin like cells
   (c) Cheif cells
   (d) Parietal cells
3. The transport of fructose into the enterocytes is mediated by: **[JUNE 2016]**
   (a) sodium-dependent glucose transporter (SGLT 1).
   (b) glucose transporter 5 (GLUT5).
   (c) SGLT 2.
   (d) GLUT 4.
4. The di- and tripeptides are transported in the enterocytes by peptide transporter I that

   [JUNE 2016]

   (a) Na^+ (b) Ca^{++}
   (c) H^+ (d) Cl^-
5. Which one of the following vitamins is **NOT** absorbed in the small intestine by Na^+ - co-transporters? **(JUNE 2017)**
   (a) Thiamine (b) Riboflavin
   (c) Folic acid (d) Ascorbic acid.
6. Which one of the following is NOT involved with the absorption of iron in the intestine?

   (DEC 2017)

   (a) Ferroportin 1
   (b) Divalent metal transporter
   (c) Hephaestin
   (d) Transferrin
7. Which one is required for vitamin B_{12} absorption in small intestine? **(JUNE 2019)**
   (a) Cobalophilin (b) Hephaestin
   (c) Hepcidin (d) Na^+-co-transporter

8. The sodium-independent iodide / chloride transporter is named as **(DEC 2019)**

(a) megalin (b) pendrin

(c) transthyretin (d) prestin

9. In which of the following conditions is Basal Metabolic Rate (BMR) the lowest? **[JUNE 2015]**

(a) Awake and resting

(b) Prolonged starvation

(c) Sleep

(d) Higher environmental temperature

10. An elderly person suffering from calcium deficiency was advised to take calcium rich food and to supplement the diet with vitamin D. The absorption of calcium in the intestine was increased with the supplementation of vitamin-D. Following explanations were offered for this increased calcium absorption by vitamin D:

A. The synthesis of Calbindin-D_{9K} and Calbindin-D_{28K} in enterocytes was stimulated.

B. The number of Ca^{2+} -ATPase molecules in enterocytes was increased.

C. The synthesis of divalent metal transporters 1 (DMT1) in the enterocytes was stimulated.

D. The number of hephaestin in the enterocytes was increased.

Which of the above combinations is correct? **[JUNE 2014]**

(a) A and B (b) B and C

(c) C and D (d) A and D

11. Water and electrolytes like Na^+ and cl^- are lost from the body in diarrhoea. Oral administration of NaCl solution in this condition does not improve the situation. When glucose is administered with normal NaCl solution through oral route, the absorption of electrolytes along with water occurs and the patient recovers.

A. Glucose enhances ATP production in the mucosal cells of small intestine and thus facilitates sodium absorption.

B. Glucose inhibits the diarrheal toxin induced cAMP production in the mucosal cells of small intestine.

C. Na^+ is co-transported with glucose on the apical surface of the mucosal cells of small intestine.

D. The epithelial sodium channels (ENaC) are activated by glucose in colon.

Which one of the following is true? **[DEC 2013]**

(a) Only A (b) A and B

(c) Only C (d) C and D

12. A patient comes to the hospital complaining of vomiting and diarrhea. The doctor suggested that the patient take glucose and electrolyte solution orally. Which one of following membrane proteins is likely to be involved in rehydrating the patient? **(JUNE 2019)**

(a) Cystic fibrosis transmembrane regulator (CFTR)

(b) Sodium glucose transporter protein 1(SGLT1)

(c) Insulin receptor protein (IRP)

(d) Sucrase –isomaltase protein (SIP)

13. Drawn below is an intestinal epithelial cell (IEC) performing the absorption of digested monosaccharides from the dietary carbohydrates ingested?

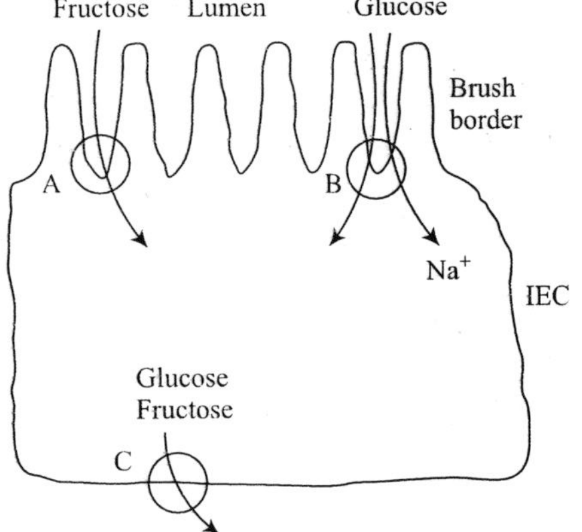

Which one of the following combinations of the transporter (A, B and C in the figure above) and the transported monosaccharide is correct? **(DEC 2019)**

(a) A-GLUT5; B-SGLT1; C-GLUT2

(b) A-GLUT2, B-GLUT5, C-SGLT1

(c) A-SGLT1; B-GLUT2; C-GLUT5

(d) A-SGLT1; B-GLUT5; C-GLUT2

14. A diabetic patient has a high blood glucose level due to reduced entry of glucose into various peripheral tissues in addition to other causes. There is no problem of glucose absorption, however, in the small intestine of these patients. The following statements are put forward to explain this observation:

A. Glucose is transported into the cells of muscles by glucose transporters (GLUTs) which are influenced by insulin receptor activation.

B. Glucose transport into the enterocytes is mediated by sodium-dependent glucose transporters (SGLTs) which are not dependent on insulin.

C. Glucose molecules are transported in the small intestine by facilitated diffusion.

D. The secondary active transport of glucose occurs in muscles.

Which one of the above statement(s) is INCORRECT? **[JUNE 2016]**

(a) Only A (b) A and B

(c) Only C (d) C and D

15. An individual was suffering from digestive complications. It was observed that the individual had dehydrated gastrointestinal tract. When an advanced investigation was done, the person was found to have defects in the following:

(A) cystic fibrosis transmembrane conductance regulator protein.

(B) glucose transporter protein.

(C) Na^+/K^+ ATPase

(D) Ca^{2+}ATPase

Which of the above could be the cause for such a digestive disorder? **[JUNE 2013]**

(a) (A) only (b) (B) and (C)

(c) (C) and (D) (d) (D) only

16. The stomach of a person was partially removed during surgery of a gastric tumour. Despite taking a balanced diet, the person developed anemia. Following possible explanations were offered:

A. Lower gastric secretion inhibits folic acid absorption

B. Protein digestion was disturbed in partial gastrectomy

C. Lower HCl secretion from stomach reduced iron absorption

D. Lower secretion of intrinsic protein factor from stomach reduced VitB absorption

Which of the above explanations were correct? **[DEC 2012]**

(a) A and B (b) B and C

(c) C and D (d) A and D

17. The intestinal absorption of glucose is impaired by use of ovabain, an inhibitor of Na^+-K^+ ATPase. Indicate the correct explanation **[JUNE 2012]**

(a) The inhibitor has blocked the transport of Na^+ from intestinal lumen to epithelial cells

(b) The inhibitor has blocked the transport of Na^+ from epithelial cells to intestinal lumen

(c) The inhibitor has blocked the transport of Na^+ from epithelial cells to interstitial space

(d) The inhibitor has blocked the transport of Na^+ from interstitial space to epithelial cells

18. A person has been suffering from night blindness. On consultation, the doctor advised person to eat carrots and/or cod fish oil. After some time having seen no improvement, doctor gave person vitamin A injection. Still no marked improvement was seen. The doctor mooted several suggestions indicating lack of the following enzymes for failure of treatment:

A. Retinol dehydrogenase

B. Retinal reducatase

C. Retinal isomerase

D. Retinal synthase

According to your opinion which is correct reason for night blindness in above case? [JUNE 2012]

(a) A only (b) B only
(c) B and C both (d) C and D both

19. In one study, a group of 5 day rat pups were fed for 3 weeks a diet A and the pups gained weight by 300%. In a second study, when the same diet fed for 3 weeks to rats of 350 gms, they did not gain weight significantly. In a third study, a diet B was fed to 250-350 gms rats and it was observed that they delivered normal pups after five weeks. Based on these observations which of the following statements is correct? [JUNE 2011]

(a) Diet A facilitates weight gain than diet B.
(b) Diet B facilitates pregnancy and child-bearing.
(c) More control experiments are to be conducted for definitive conclusion.
(d) Diet A is more energy containing that diet 'B'. Hence, its quantity should be reduced.

20. A four year old boy was brought to hospital for weak bones in spite of sufficient intake of calcium in his diet.

The attending doctor examined the functioning of the following organs:

A. Liver B. Kidney
C. Lung D. Pancreas

Which one of the following options represents a combination of probable malfunctioning organs? (DEC 2018)

(a) A and B (b) B and C
(c) C and D (d) A and D

21. Column X lists two diseases and column Y lists name of proteins which are commonly used for routine clinical diagnosis of these diseases

	Column-X		Column-Y
A	Myocardial infarction	i	Amylase
B	Pancreatitis	ii	Creatine kinase
		iii	Lipase
		iv	Troponin

Find out correct combination. (JUNE 2019)

(a) A-ii; B-i; A-iv; B-iii
(b) A-iii; B-iv; A-ii; B-i
(c) A-i; B-ii; A-iii; B-iv
(d) A-iv; B-iii ; A-i; B-ii

Answer Key

1.(d)	**2.**(b)	**3.**(b)	**4.**(c)	**5.**(c)	**6.**(d)	**7.**(a)
8.(b)	**9.**(b)	**10.**(a)	**11.**(c)	**12.**(b)	**13.**(a)	**14.**(d)
15.(a)	**16.**(c)	**17.**(c)	**18.**(a)	**19.**(c)	**20.**(a)	**21.**(a)

14. Endocrinology

1. Which one of the following options correctly relates the source gland/organ with its respective hormone as well as function? [JUNE 2016]

	Source gland	Hormone	Function
1	Thyroid	Thyroxine	Regulates blood calcium level
2	Anterior pituitary	Oxytocin	Contraction of uterine muscles
3	Posterior pituitary	Vasopressin	Resorption of water in distal tubules of nephron
4	Corpus luteum	Estrogen	Supports pregnancy

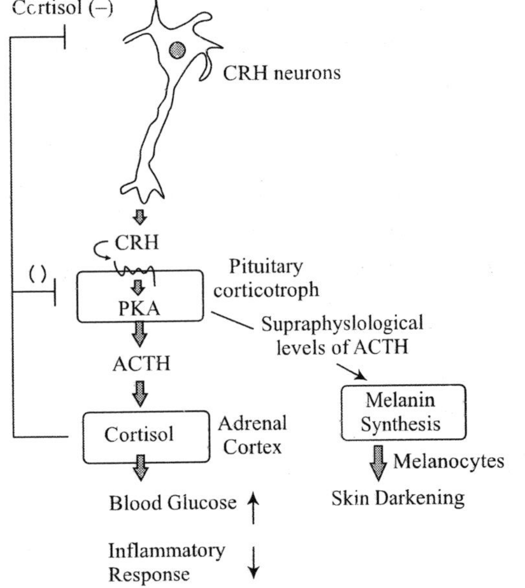

2. The above figure depicts the regulation of hypothalamo-pituitary-adrenal (HPA) axis. Changes in cortisol level in Addison's disease can lead to **(DEC 2017)**

(a) suppressed immune system and increased blood glucose level

(b) gain of body weight and lightening of skin

(c) loss of body weight, reduced blood glucose level and hyperpigmentation

(d) increased blood glucose and activated immune system

3. Hypophysiotropic hormones are secreted into blood at **(DEC 2019 ASSAM)**

(a) Posterior lobe of pituitary

(b) Median eminence

(c) Mammillary bodies

(d) Anterior lobe of pituitary

4. Given below is a figure of pro-opio melanocortin (POMC) polypeptide and its cleavage products (marked as A, B, C, D) which have different hormonal activities. The names of the cleaved products obtained from POMC are shown in the table below the diagram. **(JUNE 2019)**

POMC (1-134)

A (1-39) B (42-134)

C (1-13) D (104-134)

i.	Adrenocorticotropic hormone
ii.	β-lipotropin
iii.	α-melanocyte-stimulating hormone
iv.	β-endotropin

Which one of due following options represents A. B, C and D correctly?

(a) A – (i); B – (ii); C – (iii); D – (iv)

(b) A – (ii); B – (iii); C – (i); D – (iv)

(c) A – (i); B – (iv); C – (iii); D – (ii)

(d) A – (iii); B – (ii); C – (iv); D – (i)

5. In a stressful condition, ACTH secretion was increased and as a result glucocorticoid concentration was elevated in blood. One or a combination of the following changes most likely taking place in this condition:

A. Decreased circulating eosinophils and basophils.

B. Reduced IL2 release.

C. Potentiated inflammatory response to tissue injury.

D. Increased mitotic activity of lymphocytes in lymph nodes.

The correct answer is **[JUNE 2011]**

(a) B and C. (b) A and B.

(c) B and D. (d) C and D.

6. A steroid hormone receptor contains a functional nuclear localization signal in the ligand-binding domain. A specific hormone-based drug for that receptor, which was seen to enter the cytosol, could not efficiently carry out its function in tumors. This could be due to the following reasons:

A. Mutations in the ubiquitous bipartite signal clusters.

B. Lack of communication between importin-α and importin-β

C. Pairing of the NLS with proline-tyrosine.

D. Presence of Ran-GTP prior to entry of the complex through the nuclear pore.

Which one of the following combinations stated above is correct? **[DEC 2013 GU]**

(a) A, C and D (b) A and C

(c) B and C (d) A, B and D

7. Thyroxin releasing hormone (TRH) receptor belongs to **[DEC 2013]**

(a) nuclear receptor family.

(b) receptor tyrosin kinase family

(c) G-protein - coupled receptor family.

(d) guanylate cyclase receptor family.

8. Vasopressin acts on blood vessels leading to their constriction. Which one of the following signaling cascades will apply to the effect of vasopressin? **[JUNE 2014]**

(a) $V_1R \rightarrow$ Phosphatidylinositol hydrolysis $\rightarrow$ Ca^{2+} increase

(b) $V_1R \rightarrow$ Adenylate cyclase $\rightarrow$ cAMP

(c) $V_2R \rightarrow$ Phosphatidylinositol hydrolysis $\rightarrow$ Ca^{2+} increase

(d) $V_2R \rightarrow$ Adenylate cyclase $\rightarrow$ cAMP

9. Insulin increases facilitated diffusion of glucose in muscle cells by: **[JUNE 2016]**

(a) phosphorylation of glucose transporters.

(b) translocation of glucose transporter containing endosomes into the cell membrane.

(c) inhibition of the synthesis of mRNA for glucose transporters.

(d) dephosphorylation of glucose transporters.

10. Which of the following is NOT associated with insulin action? **(JUNE 2013)**

(a) Increased glucose transport.

(b) Increased glycogen formation.

(c) Enhanced lipolysis in adipose tissue.

(d) Decreased rate of gluconeogenesis.

11. A patient with primary hyper-aldosteronism develops diabetic glucose tolerance. Following statements are proposed on the patho-physiological condition of the patient:

A. the increase of serum potassium inhibits insulin secretion

B. the depletion of serum potassium decreases insulin secretion

C. the depletion of sodium decreases insulin secretion

D. thiazide diuretics worsen the condition of this patient

Which one of the following options represents a combination of correct statements?

(DEC 2019 ASSAM)

(a) A and B (c) B and C

(b) C and D (d) B and D

12. Vasopressin secretion does NOT increase with **[DEC 2016]**

(a) Exercise

(b) An increase in extracellular fluid volume

(c) Standing

(d) Vonmiting

13. At 17 years, a 7 feet tall human was diagnosed with gigantism caused by pituitary tumor. The condition was surgically corrected by removal of the person's pituitary gland. Doctors advised hormonal therapy. The possible hormonal therapies that would be required for survival are

A. Thyroid hormone

B. Glucocarticoids

C. Glucagon

D. Growth hormone

E. Insulin

Which one of the following combination can be used? **[DEC 2014]**

(a) A and B only (b) B and D

(c) A, B and D (d) A, C and E

14. Which one of the following is **NOT** formed after post-translational processing of pre-proglucagon? **(JUNE 2017)**

(a) Glicentin

(b) β-lipotropin

(c) Major proglucagon fragment

(d) Oxyntomodulin

15. Production of excessive amount of corticotropin (ACTH) occurs in which one of the following: **[DEC 2014]**

(a) Graves' disease

(b) Cushing's syndrome

(c) Grieg's syndrome

(d) Alport's syndrome

16. Graves disease is associated with **(JUNE 2011)**

(a) insufficiency of thyroid hormones.

(b) excess of thyroid hormones.

(c) insufficiency of corticosteroids.

(d) excess of growth hormones.

17. A person suffering from thyrotoxocosis has extremely high level of thyroid hormone in blood. There is failure of feedback regulation in hypothalamic-pituitary-thyroid axis. The detailed blood investigation exhibited high level of the following:

A. Thyroid stimulating hormone (TSH)

B. Thyroid stimulating Immunoglobulin (TSI)

C. Thyroid releasing hormone

D. Parathyroid hormone

In your opinion, which one of the following is the reason for such thyrotoxicosis

[JUNE 2012]

(a) A only (b) B only

(c) A and C (d) C and D

18. Which one of the following hormones is responsible for mobilizing calcium from the bone and increasing urinary excretion of phosphate? **(DEC 2019)**

(a) Calcitonin (b) Angiotensin II

(c) Parathormone (d) Vasopressin

19. A person takes 1.0 ml of insulin injection daily at 8:00 AM. His son gave 1.5 ml of insulin at 8:00 AM considering the father will go to party and eat more during lunch. The father also avoided breakfast, as he planned to eat more during lunch. Which one of the following events will occur? **[JUNE 2012]**

(a) father will be normo-glycemic

(b) father will be in hypoglycemic condition before lunch

(c) father will be in hyperglycemic condition before lunch

(d) blood glucose of father will be low after taking lunch

20. The following diagram represents steroidogenic pathway in the Zona Glomerulosa of the adrenal cortex:

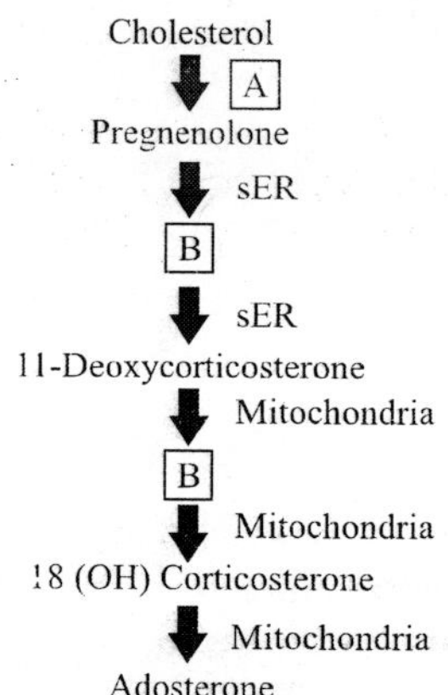

What do A, B and C represents, respectively?

[DEC 2016]

(a) sER , Progesterone , 11(OH) cortisol

(b) Mitochondria, Progesterone, Corticosterone

(c) Mitochondria, 3β-pregnenolone, 11(OH) cortisol

(d) sER, Progesterone, Corticosterone

21. The pathway of synthesis of aldosterone in zona glomerulosa along with the intracellular locations is shown below: **(JUNE 2019)**

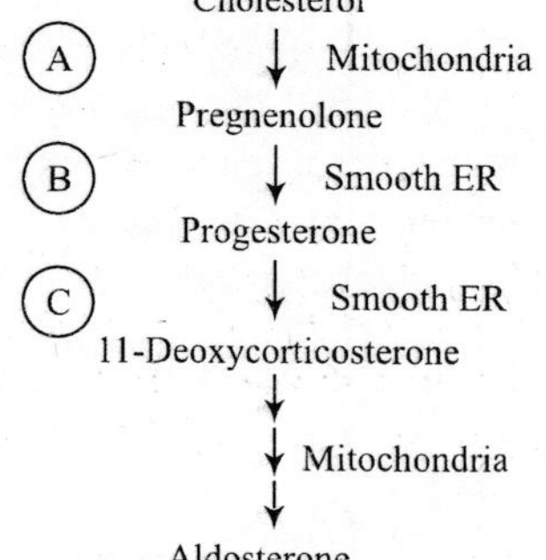

The enzyme below are required for different steps of synthesis of aldosterone:

i	21-Hydroxylase
ii	P450 side chain cleavage enzyme
iii	3β-Hydroxy steroid dehydrogenase

Which one of due following options represents correct matches for A, B and C?**(JUNE 2019)**

(a) A – (i); B – (ii); C – (iii)

(b) A – (iii); B – (i); C – (ii)

(c) A – (ii); B – (iii); C – (i)

(d) A – (ii); B – (i); C – (iii)

22. A specialized area of teratogenesis involves the misregulation of the endocrine system. Which one of the following statements regarding endocrine disruptors is true? **(JUNE 2019)**

(a) They can act as antagonist and inhibit the binding of a hormone to its receptors or block the synthesis of a hormone.

(b) They do not affect the synthesis, elimination or transportation of a hormone in the body.

(c) They do not mimic the effect of natural hormones.

(d) Low dose exposure to endocrine disruptors is not sufficient to produce significant disabilities later in life.

Answer Key

1.(c)	**2.**(c)	**3.**(b)	**4.**(a)	**5.**(b)	**6.**(d)	**7.**(c)
8.(a)	**9.**(b)	**10.**(c)	**11.**(d)	**12.**(b)	**13.**(d)	**14.**(a)
15.(b)	**16.**(b)	**17.**(b)	**18.**(c)	**19.**(b)	**20.**(b)	**21.**(c)
22.(a)						

15. Reproduction and hormones

1. GnRH is secreted during infancy (0-6 month) and puberty onwards (4 years and above) in monkeys. However, *i. v.* injection of GnRH during pre-pubertal period (about 2 years of age) led to elevated LH and FSH in blood compared to untreated 2 years old monkey. This suggest that

(A) Hypothalamus is active during pre-pubertal period

(B) GnRH action on pituitary is age dependent

(C) Pituitary matures during adulthood

(D) Pituitary is active in all stages of development in monkeys

Which one of the following is true?

[JUNE 2012]

(a) A and B (b) B and C

(c) C only (d) D only

2. What would be the outcome if the theca intema cells were destroyed in a Graafian follicle? **[JUNE 2013]**

(a) Immediate formation of corpus albicans.

(b) Increased progesterone synthesis in the granulosa cells.

(c) Decreased estrogen synthesis in the granulosa cells.

(d) Formation of corpus hemorrhagicum.

3. The figure below represents normal sex determination, differentiation and development in humans

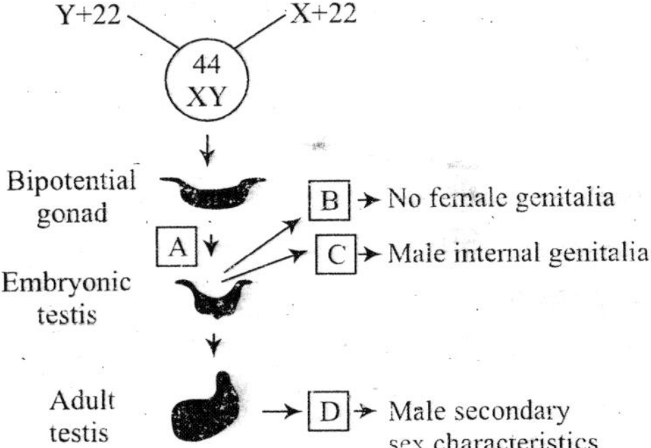

Identify A, B, C and D. (JUNE 2018)

(a) A= WT1 (Wilm's Tumor 1), B = MIS (Mullerian Inhibitory Substance), C = SRY, D = Testosterone

(b) A = GnRH, B = FSH, C = Testosterone, D = 5α Reductase

(c) A= SRY, B = MIS, C = Testosterone, D = DHT (Dihydro testosterone)

(d) A = WT1, B = LH, C = ABP (Androgen Binding Protein), D = Inhibin

4. Inhibin from sertoli cells of testes selectively inhibits **(MODELPAPER)**

(a) luteinizing hormone.

(b) follicle stimulating hormone.

(c) thyroid stimulating hormone.

(d) growth hormone.

5. Estradiol synthesis follows a 2 cell-2-gonadotropin theory, where partial synthesis occurs in the granulosa cells and the rest in the theca interna cells of the Graafian follicle. Which of the following correctly represents estradiol synthesis and secretion?

[JUNE 2015]

Oxidation — *Thace intena cell* — *Grandlose cell*

1. Chotestorol → LHR: Cholesterol → Androstenedlone → FSHR: Androstenedione → Testosterone → Estradiol → Estrodiol
2. Cholesterol → FSHR: Cholesterol → Estrone → LHR: Estrone → Estradiol → Estrodiol
3. Cholesterol → LHR: Cholesterol → Androstenedlone → FSHR: Androstenedlone → Estriol → Estrodiol
4. Cholesterol → LHR: Cholesterol → Testosterone → FSHR: Testosterone → Estrene → Estrodiol

6. Spermatogonial stem cell undergoes extensive metamorphosis to become a spermatozoan. Meiosis leads to the formation of spermatid containing 22 autosomes and one sex chromosome. A male mouse was found in a colony which always produced only female puff upon matting. Which one of following is a possible reason **[JUNE 2012]**

(a) Spermiogenesis was defective

(b) All spermgonial stem cells contained only X and no Y chromosomes

(c) Activation of Y-chromosome linked post meiotic death related gene may lead to such a situation

(d) Activation of X-chromosome linked post meiotic death related gene may lead to such a situation

7. A convenient and reasonably reliable indicator of the time of ovulation is usually a rise in the basal body temperature, possibly because progesterone is thermogenic. Of the four situations given below, which one is ideal for ensuring pregnancy after intercourse?

[DEC 2015]

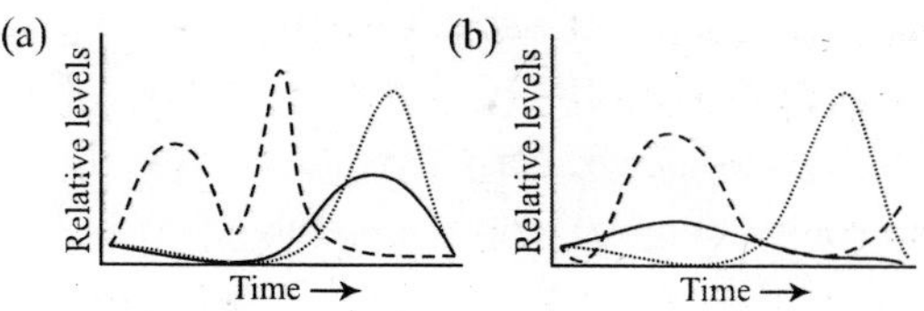

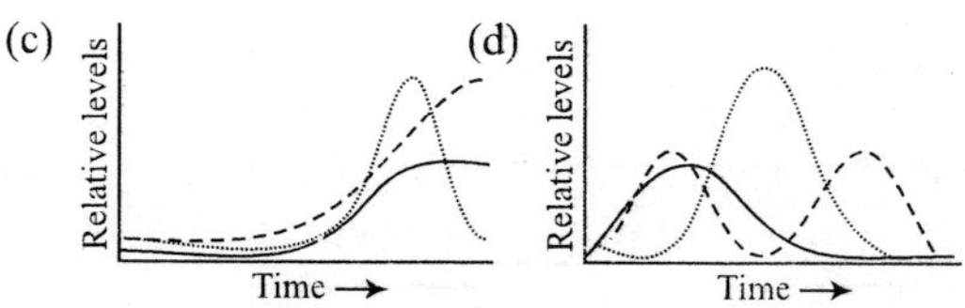

8. Which one of the following graphs best represents the hormone profile in a rat right after matting? **[DEC 2013]**

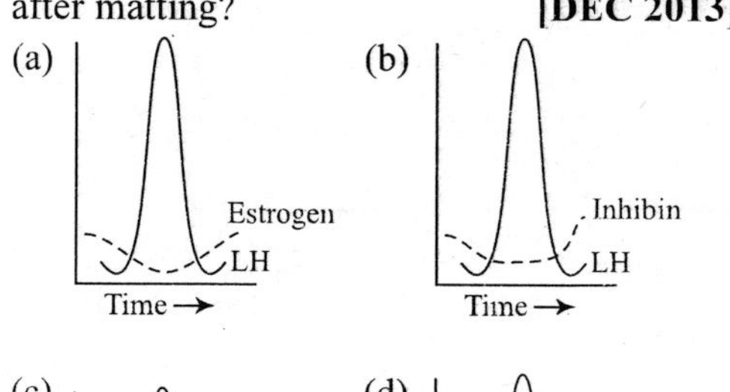

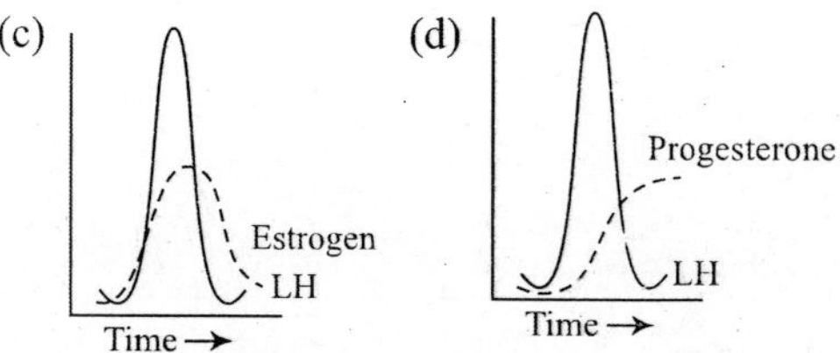

9. The graph represents relative plasma concentration of hormones (A and B) during reproductive cycle in a normal female. Which one of the following combinations is correct?

[DEC 2011]

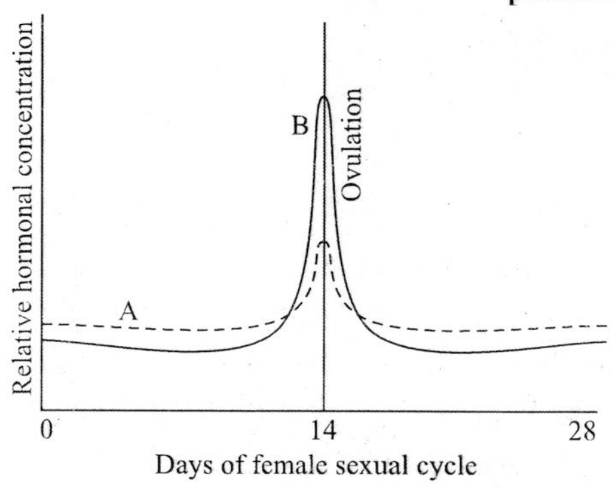

(a) (A) is FSH and (B) is estrogen
(b) (A) is estrogen and (B) is LH
(c) (A) is FSH and (B) is LH
(d) (A) is LH and (B) is FSH

10. Level of follicle stimulating hormone (FSH) during infancy and adulthood is the same but spermatogenesis is seen only during adulthood. mRNA levels coding for FSH receptor are also found to be the same in testis of both age groups.

Which of the following investigations will clarify this paradox a little more? **(JUNE 2012)**

(a) Culture testicular cells and add LH to see testosterone production.
(b) Culture testicular cells and add testosterone to see comparative rise in FSH mRNA from both age groups.
(c) Culture testicular cells and add FSH to see comparative rise in cAMP production by both age groups.
(d) Add both LH and FSH to testicular cells and evaluate cAMP production

11. Which of the following graph represents normal sexual cycle in a normal human female? **[JUNE 2011]**

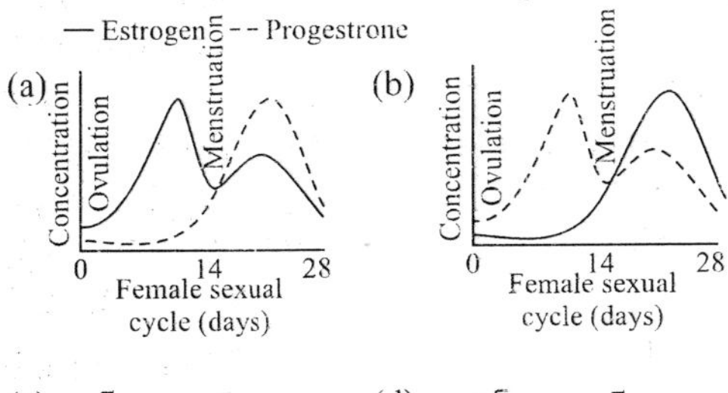

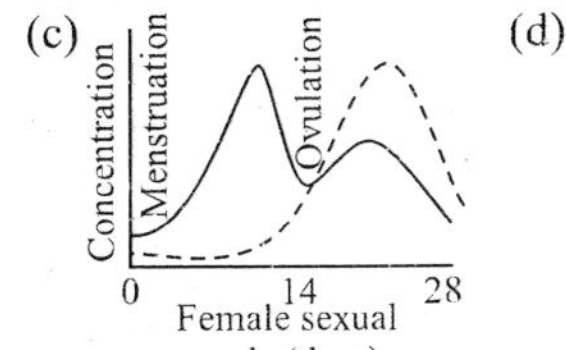

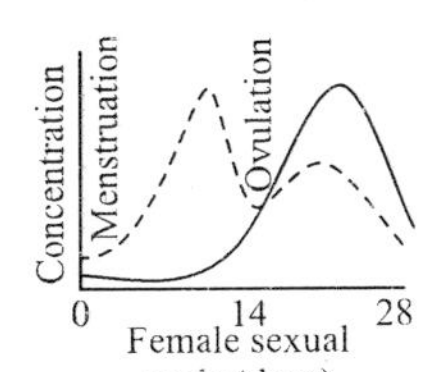

12. Which of the following is responsible for initiation of maternal behaviour in first time pregnant rats after parturition? **[JUNE 2012]**

(a) high prolactin levels in blood
(b) Stimulation of sensory receptors during delivery
(c) changes in uterine volume
(d) Presence of male rats

13. Which one of the following is responsible for the ejection of milk from mammary glands in mammals? **[DEC 2012]**

(a) Oxytocin (b) Prolactin
(c) Serotonin (d) Melatonin

14. You are asked to identify the stage of estrus cycle in vaginal smear of a mouse containing large number of leukocytes and very few nucleated epithelial cells. Which one of the following will be the correct stage of estrous cycle? **[DEC 2015]**

(a) Early estrus, late proestrus
(b) Late estrus, early metestrus
(c) Late metestrus, early diestrus
(d) Late diestrus, early proestrus

15. Estrus cycle in rats is controlled by pituitary and gonadal hormones. While treating a set of rat with vitamin D, a student accidentally injected the rat with an inhibitor of 17 α hydroxyl-pregnenolone and checked vaginal smear for 10 consecutive days. Which one of the following observation is correct?

(JUNE 2019)

(a) The Smears showed well formed nucleated epithelial cell through the period.
(b) The smears initially showed normal estrus stage but eventually entered a prolonged diestrus stage.
(c) The smear showed leukocytes and few epithelial cells.
(d) The cells showed metestrus for 3 days and then returned to the proestrus stage.

16. Foster pups were presented to a primiparous rat at the mid gestation period. Which of the following behaviour will be found in the pregnant rat? **(DEC 2013)**

(a) The rat shows maternal behaviour after few days of presentation of pups.
(b) The rat attacks and kills the pups time they are presented.
(c) The rat rejects the pups after repeated presentation.
(d) The rat shows fear response for a days.

17. Reflex ovulation does **NOT** occur in **(DEC 2017)**

(a) cats (b) rabbits
(c) mink (d) rats

Answer Key

1.(d)	2.(c)	3.(c)	4.(b)	5.(a)	6.(c)	7.(a)
8.(c)	9.(c)	10.(c)	11.(c)	12.(a)	13.(a)	14.(d)
15.(b)	16.(a)	17.(d)				

SECTION-8

GENETICS

1. Dominance, segregation, independent assortment

1. A cross is made between a pure breeding plant having red coloured flowers with a pure breeding plant having white coloured flowers. Such a cross is called as **(DEC 2013)**
   (a) test cross. (b) monohybrid cross.
   (c) dihybrid cross. (d) back cross.

2. In *Drosophila melanogaster* males, homologous chromosomes pair and segregate during meiosis but crossing over does not occur. At which stage of meiosis does segregation of 2 alleles of a gene take place in these individuals? **(JUNE 2017)**
   (a) Zygotene (b) Diakinesis
   (c) Anaphase I (d) Anaphase II

3. In *Drosophila* males, where no recombination occurs, segregation of the two alleles of a gene occurs at which stage of cell division? **(DEC 2019 ASSAM)**
   (a) Diplotene (b) Anaphase I
   (c) Anaphase II (d) Mitotic Telophase

4. In a heterozygous individual for a given gene, if a crossing over has occurred between the gene locus and the centromere of the chromosome, the segregation of the two alleles of the given gene will occur during meiosis at **(DEC 2015)**
   (a) either anaphase I or anaphase II
   (b) anaphase I only
   (c) anaphase II only
   (d) both anaphase I and II

5. Which of the following representations of chromosomal arrangement in meiotic metaphase I best explains the Law of Independent Assortment? **(DEC 2013)**

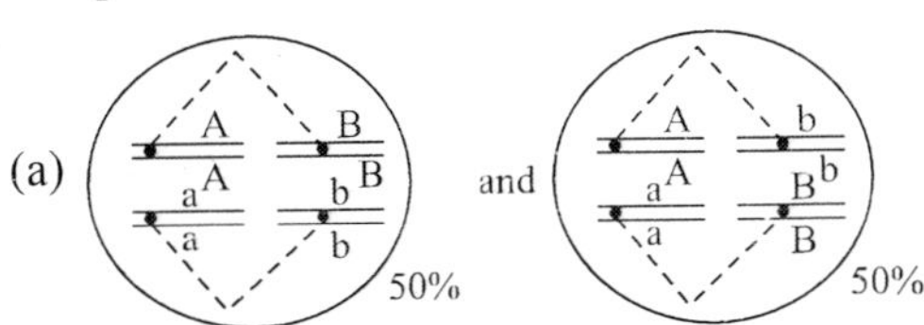

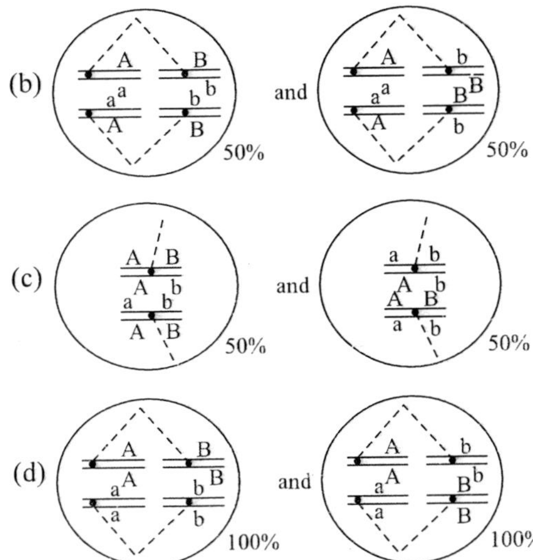

6. Two siblings who inherit 50% of the genome from the mother and 50% from the father show lot of phenotypic differences. Which one of the following events during gametogenesis of the parents will maximally contribute to this difference? **(DEC 2015)**
   (a) Mutation
   (b) Recombination
   (c) Independent assortment
   (d) Environment

7. Of the following meiotic events
   (A) formation of synaptonemal complex
   (B) crossing over
   (C) arrangements of chromosomes at the equatorial plate at Metaphase I
   (D) separation of chromatids at Anaphase II
   Which one(s) lead to variation? **(DEC 2013 GU)**
   (a) (B) only (b) (B) and (C)
   (c) (A) and (D) (d) (B) and (D)

8. Two strains of mice which are genetically identical except for a single genetic locus or region are said to be: **(DEC 2019)**
   (a) Syngenic (b) Allogenic
   (c) Congenic (d) Heterogenic

9. What is the probability of getting a sum of 9 from simultaneously throwing two dice? **(DEC 2019)**

(a) 1/6 (b) 1/8
(c) 1/9 (d) 1/12

10. Genes A, B and C. control three phenotypes which assort independently; A plant with the genotype Aa Bb Cc is selfed. What is the probability for progeny which show the dominant phenotype for AT LEAST ONE of the phenotypes controlled by genes A, B and C? **(JUNE 2015)**

(a) 1/64 (b) 27/64
(c) 63/64 (d) Cannot be predicted

11. In the following example, 3 independently assorting genes are known to govern coat color in mice. The genotype of few of the coat colors is given below:

Agouti: *A-B-C-*Black: *aa B-C -*

Albino: -- -- *cc*

What will be the expected frequency of abino, in the F2 progeny from crosses of Pure black with albino of the genotype *AAbbcc*? **(JUNE 2014)**

(a) 1/4 (b) 1/16
(c) 1/64 (d) 9/64

12. Assuming that the A, B, C and D genes are not linked. The probability of a progeny being AaBBccDd from a corss between AABbccDd and aaBBccDD parents will be **(JUNE 2019)**

(a) 4/32 (b) 3/16
(c) 1/4 (d) 3/32

13. How many genetically different gametes can be made by an individual of genotype AaBbccDDEe? **(JUNE 2012)**

(a) 3 (b) 5
(c) 8 (d) 32

14. A plant of the genotype AaBb is selfed. The two genes are linked and are 50 map units apart. What proportion of the progeny will haw the genotype aabb? **(DEC 2011)**

(a) 1/2 (b) 1/4
(c) 1/8 (d) 1/16

15. In a plant, male sterility is caused by the presence of a dominant allele for a nuclear gene (Ms). There are also lines which carry a gene whose dominant allele (F) restores male fertility. These are called restorer lines. A cross is made between a male sterile line and a homozygous restorer line which does not contain the Ms allele. Of the F_1 progeny, only those that carry the Ms allele are allowed to self-pollinate. What is the probability of obtaining a male sterile line in the resulting F_2 progeny? (The Ms and F genes assort independently). **(DEC 2019 ASSAM)**

(a) 1/4 (b) 3/16
(c) 3/4 (d) 1/16

16. Assuming a 1 : 1 sex ratio, what is the probability that three children from the same parents will consist of two daughters and one son? **(DEC 2011)**

(a) 0.375 (b) 0.125
(c) 0.675 (d) 0.75

17. Mendel crossed tall pea plants with dwarf ones. The F_1 plants were all tall. When these F_1 plants were selfed to produce F_2 generation, he got a 3: 1 tall to dwarf ratio in the offspring. What is the probability that out of three plants (of F_2 generation) picked up at random two would be dwarf and one would be tall? **(JUNE 2011)**

(a) 3/4 (b) 3/8
(c) 9/64 (d) 9/32

18. A phenotypically normal fruit fly was crossed to another fly whose phenotype was not recorded. Of the progeny, 3/8 were wild type, 3/8 had ebony body color, 1/8 had vestigeal wings and 1/8 had ebony body color and vestigeal wings. Ebony body color and vestigeal wings are recessive characters and their genes are located on two different autosomes.

Based on this information which one of the following is the likely genotype of the parents? **(JUNE 2017)**

(a) $ee\ vg\ vg$ and $e^+e^+\ vg^+vg$

(b) $ee\ vg^+\ vg$ and $e^+e\ vg^+vg$

(c) $e^+e\ vg\ vg$ and $e^+e^+\ vg^+vg$

(d) $e^+e\ vg^+\ vg$ and $ee\ vg^+vg$

19. In summer squash, white colour fruit (W) is dominant over yellow colour (w) and disc-shaped phenotype (D) is dominant over sphere-shaped phenotype (d).

Determine the genotype of the parents if the cross between white, sphere crossed with white, sphere gives ¼ white, sphere and 14 yellow, sphere. **(DEC 2019)**

(a) $WWDD \times wwdd$ (b) $Wwdd \times Wwdd$

(c) $WwDd \times wwdd$ (d) $WWDD \times WWdd$

Answer Key

1.(b)	**2.**(c)	**3.**(b)	**4.**(c)	**5.**(a)	**6.**(c)	**7.**(b)
8.(c)	**9.**(c)	**10.**(c)	**11.**(a)	**12.**(c)	**13.**(c)	**14.**(d)
15.(b)	**16.**(a)	**17.**(c)	**18.**(d)	**19.**(b)		

2. Multiple alleles

1. Blood group type A antigen is a complex oligosaccharide which differs from H antigen present in type O individual by the presence of terminal **(DEC 2012)**

(a) Glucose

(b) Galactose

(c) Fucose

(d) N-acetyl galactosamine

2. A mother of blood group O has a group A child. The father could be of blood type **(JUNE 2011)**

(a) A or B or O. (b) A only.

(c) A or AB. (d) AB only.

3. The ABO blood type in human is under the control of autosomal multiple alleles. Colour blindness is recessive X-linked trait. A male with a blood type A and normal vision marries a female who also has blood type A and normal vision. The couple's first child is a male who is colour blind and has a blood group O. What is the probability that their next female child has normal vision and a blood group O? **(DEC 2011)**

(a) ¼ (b) 3/4

(c) 1/8 (d) 1

4. In a hospital three babies were mixed up. The blood group of the babies were A, B and AB. In order to identify the parents of the babies, the blood groups of the parents were determined. The results obtained were:

Parent set 1 – A and AB

Parent set 2 – AB and O

Parent set 3 – B and AB

Which of the following conclusions can be definitively made? **(JUNE 2013)**

(a) The baby with blood group A is the child of the parent set 2.

(b) The baby with blood group AB is the child of the parent set 1.

(c) The baby with blood group B is the child of the parent set 3.

(d) The parentage of none of the babies can be determined from the given information.

5. The newborn baby of a mother having blood group AB, Rh^+ and father having blood group O, Rh^-, got mixed with other babies in the hospital. The baby with which of the following blood groups is expected to be of the said couple? **(JUNE 2019)**

(a) O, Rh^+ (b) O, Rh^-

(c) AB, Rh^- (d) B, Rh^+

6. A recessive inherited disease is expressed only in individuals of blood group O and not expressed in blood groups A, B or AB. Alleles controlling the disease and blood group are independently inherited. A normal woman with blood group A and her normal husband with blood group B already had one child with the disease. The woman is pregnant for second time. What is the probability that the second child will also have the disease? **(JUNE 2017)**

(a) 1/2 (b) 1/4

(c) 1/16 (d) 1/64

7. Mutation at two different loci of the same gene X results in altered functions. These two mutated versions of gene X are called **(JUNE 2012)**
   (a) alleles
   (b) complementation group
   (c) interrupted genes
   (d) linkage group

Answer Key

1.(d)	2.(c)	3.(a)	4.(d)	5.(d)	6.(c)	7.(a)

3. Complementation test and recon

1. A cis-trans complementation test is carried out to identify **(DEC 2013)**
   (a) if two mutations are allelic in nature.
   (b) if two genes interact with one another.
   (c) the number of genes influencing phenotype.
   (d) to understand the dominance/recessive relationships between alleles.
2. Two mutants of Drosophila melanogaster, one showing light red eye colour and the other showing dark brown eye colour, were crossed. In F_1 all flies showed normal red eye colour. This indicated that the two mutations are **(DEC 2013 GU)**
   (a) allelic. (b) non allelic.
   (c) linked. (d) unlinked.
3. When two mutants having the same phenotype were crossed, the progeny obtained showed a wild-type phenotype. Thus the mutations are **(MODEL PAPER)**
   (a) non-allelic.
   (b) allelic.
   (c) segregating from each other.
   (d) independently assorting
4. Four different mutant lines showing similar phenotype were identified from a genetic screen. When genetic crosses amont these mutants were carried out, the first mutant was found to complement the second, third and fourth mutant lines. However, no other complementation groups do the four mutant lines belongs to? **(JUNE 2012)**
   (a) 1 (b) 2
   (c) 3 (d) 4
5. The results of a complementation test for five independent mutants (1 to 5) are summarized below:

1	2	3	4	5	
0	0	0	+	+	**1**
	0	0	+	+	**2**
		0	+	+	**3**
			0	0	**4**
				0	**5**

'+' represents complementation;
'0' represents non-complementation.
Based on the above, which one of the following conclusion is correct? **(DEC 2013 GU)**
   (a) There are two cistrons. Mutations 1, 2 and 3 belong to one cistron; while 4 and 5 belong to a second cistron.
   (b) There is a single cistron. Mutations 1, 2 and 3 can recombine out from 4 and 5.
   (c) Each mutation represents a cistron.
   (d) There are two linkage groups. 1, 2, 3 comprise one group while 4 and 5 comprise the second group.
6. The results of a complementation test for five independent mutants (1 to 6) are summarized below

1	**2**	**3**	**4**	**5**	**6**	
0	+	0	+	+	0	**1**
	0	+	+	+	+	**2**
		0	+	+	0	**3**
			0	0	+	4
				0	+	**5**
					0	**6**

'+' represents complementation; '0' represents non-complementation

Based on the above, which one of the following conclusions is correct? **(JUNE 2013)**

(a) The mutations can be ordered in a single cistron as 1-3-5-2-4-6.

(b) All mutations belong to a single cistron, but their order cannot be determined.

(c) There are three cistrons, mutations 1, 3 and 6 represent one cistron, 4 and 5 represent the second cistron and 2 represents the third cistron.

(d) There are three linkage groups, mutations 1, 3 and 6 represent linkage group A, 4 and 5 represent linkage group B, and 6 represents linkage group C.

7. The concept of recon was proposed by Seymour Benzer by studying recombination between **(JUNE 2013)**

(a) lysis mutants of bacteriophage T4

(b) white eye mutants of *Drosophila melanogaster.*

(c) biochemical mutants of *Neurospora crassa.*

(d) auxotrophic mutants of *Escherichia coli.*

8. Upon studying a considerable number of different crosses in Drosophila, Morgan reached the conclusion that all genes of this fly were clustered into four linked groups corresponding to the four pairs of chromosomes. Further studies revealed that linkage is not absolute and it is broken frequently. It is broken in prophase by a process called **(MODELPAPER)**

(a) Recombination. (b) Jumping of genes.

(c) Integration. (d) Mutation.

9. Wild type T_4 bacteriophage can grow on Band K strains of *E. coli* forming small plaques. rll mutants of T_4 bacteriophage cannot grow on *E. coli* strain K (non-permissive host), but form large plaques on E. coli strain B (permissive host). The following two experiments were carried out:

Experiment I: E. coli K cells were simultaneously infected with two rII mutants (a^- and b^-). Several plaques with wild type morphology were formed.

Experiment II: E. coli B cells were simultaneously infected with the same mutants as above. T4 phages were isolated from the resulting plaques and used to infect E. coli K cells. Few plaques with wild type morphology were formed.

Which one is the correct conclusion made regarding the rII mutants, a^- and b^- from the above experiments? **(DEC 2012)**

(a) The mutations a^- and b^- belong to two different cistrons (experiment I) and there is no recombination between them (experiment II).

(b) The mutations a^- and b^- belong to two different cistrons (experiment I) and they recombined (experiment II).

(c) The mutations a^- and b^- belong to two different cistrons (experiment II) and they recombined (experiment I).

(d) The mutants a^- and b^- belong to the same cistron (experiment I) and they did not recombine (experiment II).

Answer Key

1.(a)	2.(b)	3.(a)	4.(b)	5.(a)	6.(c)	7.(a)
8.(a)	9.(b)					

4. Gene-Gene interactions

1. There are two mutant plants. One shows taller phenotype than wild type, whereas the other has the same height as the wild type. When these two mutations were brought in together by genetic crosses, the double mutant displayed even taller phenotype than the tall mutant plants. This genetic interaction is called **(JUNE 2013)**

(a) additive interaction.
(b) antagonistic interaction.
(c) synergistic interaction.
(d) suppressive interaction.

2. A cross is made between two plants with white flowers. All the F1 progeny had red coloured flower. This is because of **(JUNE 2013)**
(a) complementation (b) recombination
(c) translocation (d) reversion

3. Two mutant plants, both bearing white flowers, were crossed. All F_1 plants had red coloured flowers. When an F_1 plant was selfed it produced progeny with either red or white coloured flowers in 9:7 ratio. Based on this information, which one of the following conclusions is correct? **(DEC 2017)**
(a) The mutations in the parents do not complement each other
(b) The mutations in the parents are allelic
(c) The mutations in the parents are non-allelic
(d) The mutations in the parents are linked

4. When two independent pure lines of pea with white flowers are crossed, the F_1 progeny has purple flowers. The F_2 progeny obtained on selfing shows both pruple and white flower in a ratio of 9:7. The following conclusions were made
A. Two different genes are involved, mutation in which lead to formation of white flower.
B. These two genes show independent assortment
C. This is an example of complementary gene action
D. The is an example of duplicate genes
Which of the following conclusions are correct? **(JUNE 2012)**
(a) A and C only (b) A and D only
(c) A, B and D (d) A, B and C

5. The following is the biochemical pathway for purple pigment production in flowers of sweet pea:

Colourless Precursor 1 —*Allele A*→ Colourless Precursor 2 —*Allele B*→ Purple Pigment

Recessive mutation of either gene A or B leads to the formation of white flowers. A cross is made between two parents with the genotype: AaBb x aabb. Considering that the two genes are not linked, the phenotypes of the expected progenies are **(JUNE 2011)**
(a) 9 purple: 7 white.
(b) 3 white: 1 purple.
(c) 1 purple: 1 white.
(d) 9 purple: 6 light purple: 1 white

6. Two interacting genes (independently assorting) were involved in the same pathway. Absence of either genes function leads to absence of the end product of the pathway. A dihybrid cross involving the two genes is carried out. What fraction of the F 2 progeny will show the presence of the end product? **(JUNE 2016)**
(a) 1/4 (b) 3/4
(c) 9/16 (d) 15/16

7. A plant with red fruit is crossed to a plant with white fruit. The F1 progeny had red fruits. On selfing the F1 two kinds of progeny were observed, plants with red fruits and those with white fruits. To test whether it was a case of recessive epistatic interactions a chi-square test was performed. A value of 1.062 was obtained (chi-square value of $P_{0.05}$=3.841 for Degree of freedom=1). The following statements were made:
A. The null hypothesis was that plant with red and white fruits will occur in a 9:7 ratio
B. The null hypothesis was that plant with red and white fruits will occur in a 1:1 ratio
C. Based on the chi square value, it is a case of recessive epistatic interactions
D. Based on the chi square value, it is not a case of recessive epistatic interactions
Which of the combination of above statements is correct? **(JUNE 2015)**
(a) A and C (b) A and D
(c) Band C (d) B and D

8. Two plants with white flowers are crossed. White flowers arise due to recessive mutation. All F_1 progeny have red flowers. When the F1plants are selfed, both red and white flowered progeny are observed. In what ratio will red-flowered plants and white-flowered plants occur? **(DEC 2014)**

(a) 1:1 (b) 3:1
(c) 9:7 (d) 15:1

9. A black Labrador homozygous for the dominant alleles (*BBEE*) is crossed with a yellow Labrador homozygous for the recessive alleles (*bbee*). On intercrossing the F1, the F2 progeny was obtained in the following ratio: 9 black: 3 brown: 4 yellow. This is an example of **(JUNE 2014)**

(a) recessive epistasis where allele *e* is epistatic to *B* and *b*.
(b) dominant epistasis where allele *E* is epistatic to *B* and *b*.
(c) recessive epistasis where allele *e* is epistatic to *E*.
(d) complementary epistasis where allele *b* is epistatic

10. Following is a hypothetical biochemical pathway responsible for pigmentation of leaves. The pathway is controlled by two independently assorting genes 'A' and 'B' encoding enzymes as shown below.

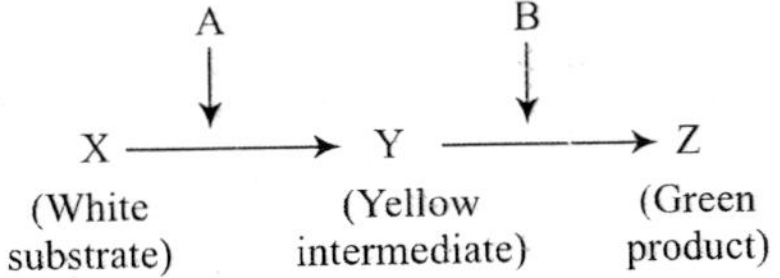

Mutant alleles 'a' and 'b' code for non-functional proteins.

What is the expected progeny after selfing a plant with the genotype AaBb? **(DEC 2015)**

(a) Green (9): White (d): Yellow (c)
(b) Green (9): Yellow (d): White (c)
(c) Green (9): Yellow (6): White (a)
(d) Green (9): White (7)

11. Consider the following hypothetical pathway:

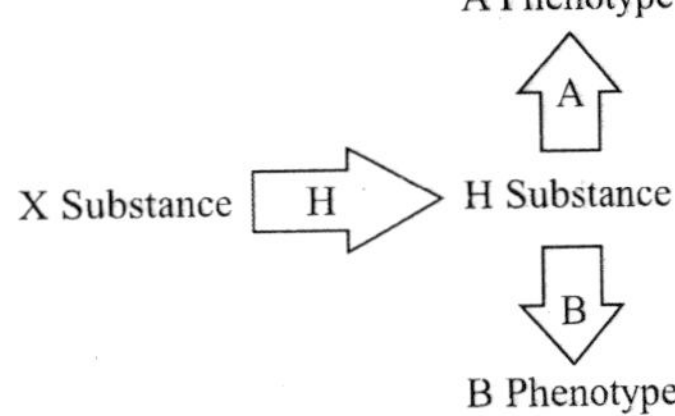

H allele converts X substance to H substance

h allele cannot convert X to H substance and leads to phenotype 'O'

A allele converts H substance leading to A phenotype

a allele cannot convert H substance

B allele converts H substance leading to B phenotype

b allele cannot convert H substance

An individual with 'A' phenotype when crossed with that of 'B' phenotype has a progeny with 'O' phenotype. Which one of the following crosses can lead to the above observation? **(DEC 2015)**

(a) Aahh X BbHH (b) AaHh X BBHh
(c) AaHh X BBHH (d) AAHH X BbHh

12. A hypothetical biochemical pathway for the formation of eye color in insect is given below.

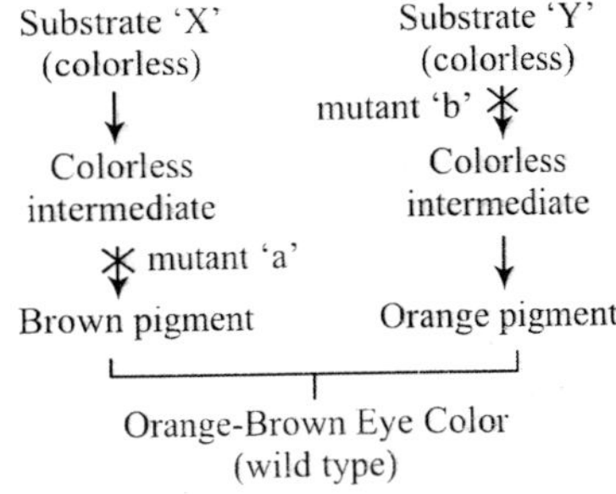

Two autosomal recessive mutants 'a' and 'b' are identified which block the pathway as shown above. Considering that the mutants are not linked, what will be the phenotype of the F2 progeny if crosses were made between parents of the genotype aaBB x AAbb, and the F_1 progeny are intercrossed? **(DEC 2014)**

(a) 9 orange-brown: 3 orange; 3: brown: 1 colorless
(b) 9 orange-brown: 7 colorless
(c) 1 orange: 2 colorless
(d) 15 orange-brown: 1 colorless

13. The following is a schematic representation of a hypothetical pathway involved in formation of eye color in an insect species. **(JUNE 2017)**

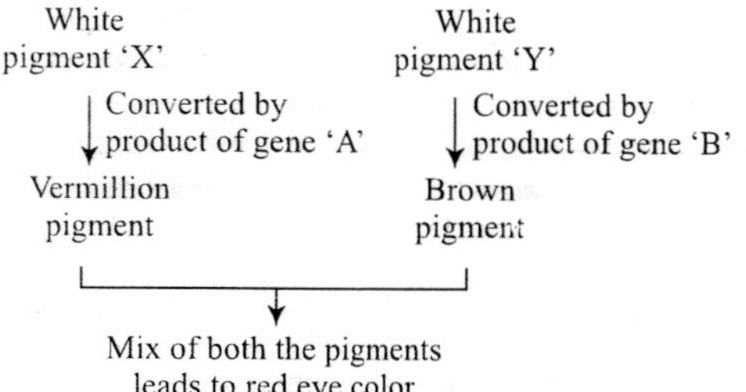

Genes A and B are linked and have a map distance of 10cM. Females with genotypes a^+ ab^+b are test crossed. Further, in these females, the two genes are linked in cis. a^+ and b^+ represent wild type alleles, while a and b are null alleles. The progeny of the test cross have individuals with four different eye colours.

What is the expected ratio of individuals with eye color Red: Vermillion: Brown: White in the progeny? **(JUNE 2017)**

(a) 9 : 3 : 3 : 1 (b) 1: 1: 1 : 1
(c) 9 : 1 : 1 : 9 (d) 1 : 9 : 9 : 1

14. The following is a hypothetical pathway for the development of wild type (red) eye colour in an insect:

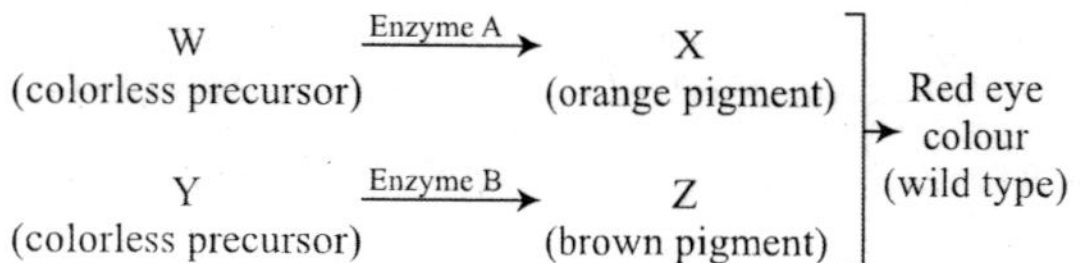

Enzymes A and B are encoded by the genes a^+ and b^+, respectively. The following statements are made regarding inheritance of the genes involved in the development of eye colour:

A. When two heterozygous individuals of the genotype a^+ab^+b are mated, progenies with red, orange, brown and white eye colour will be observed irrespective of whether the genes are independently assorting or showing incomplete linkage. .

B. When two heterozygous individuals of the genotype a^+ab^+b are mated, progenies with red, orange, brown and white eye colour will be observed in a ratio of 9:3:3:1, when the genes are independently assorting.

C. When an heterozygous individual of the genotype $a^+b/\cdot a\ b^+$ is test crossed, progenies with red and white eye colour will be more in number if genes are linked.

D. When an heterozygous individual of the genotype $a^+b/\ a\ b^+$ is test crossed, progenies with orange and brown eye colour will be more in number if genes are linked.

Which of the above statements is TRUE?

(JUNE 2011)

(a) A and C (b) B and C
(c) A, B and C (d) A, B and D

15. In a flower species, light pink locus is hypostatic to pigment development locus. R is for pigment development and W for light pink. The recessive allele (w) in light pink locus gives red colour and recessive allele in pigment locus (r) gives white colour.

What will be the phenotype of W/-, r/r and w/w, r/r? **(DEC 2019 ASSAM)**

(a) Red, Pink (b) Pink, White
(c) Red, Red (d) White, White

16. Two Yellow mice with straight hair were crossed and the following progeny was obtained:

1/2 Yellow, straight hair
1/6 Yellow, fuzzy hair
1/4 gray, straight hair
1/12 gray, fuzzy hair

In order to provide genetic explanation for the results and assign genotypes to the parents and progeny of this cross the following statements were given: **(JUNE 2019)**

(A) The 6:2:3:1 ratio obtained here indicates recessive epistasis.

(B) This cross concerns two independent characteristics – body colour and type of hair.

(C) The deviation of dihybrid ratio from 9:3:3:1 to 6:2:3:1 may be due to one of the genes being a recessive lethal

(D) The lethal allele is associated with straight hair

The most appropriate combination of statements to provide genetic explanation for the result is: **(JUNE 2019)**

(a) B and C (b) A Only

(c) B, C and D (d) A, C and D

Answer Key

1.(c)	**2.**(a)	**3.**(c)	**4.**(d)	**5.**(b)	**6.**(c)	**7.**(a)
8.(c)	**9.**(a)	**10.**(a)	**11.**(b)	**12.**(a)	**13.**(c)	**14.**(d)
15.(d)	**16.**(a)					

5. Genomic imprinting, dosage compensation, phenocopy, Sex linked characters

1. A mouse carrying two alleles of insulin-like growth factor II (*Igf2*) is normal in size; whereas a mouse that carries two mutant alleles lacking the growth factor is dwarf. The size of a heterozygous mouse carrying one normal and one mutant allele depends on the parental origin of the wild type allele. Such pattern of inheritance is known as **(JUNE 2016)**

   (a) Sex-linked inheritance

   (b) Genomic imprinting

   (c) Gene-envr interaction

   (d) Cytoplasm inheritance

2. Which one of the following statements is correct? **(DEC 2019 ASSAM)**

   (a) Epigenetic memory depends on DNA acetylation by trithorax group of proteins

   (b) An epigenetic change could be inherited from a cell to a daughter cell

   (c) Parental origin does not influence the expression level of imprinted loci

   (d) Epigenetic changes do not alter the chromatin landscape

3. In context of DNA methylation, which one of the following statements is FALSE? **(DEC 2019)**

   (a) Generally, methylation occurs at the 3rd carbon position of cytosine and converts it to 3-methyl cytosine

   (b) Maintenance methyl transferase acts constitutively on hemi-methylated sites and converts them to fully methylated sites

   (c) During mammalian gametogenesis, the genomic methylation patterns are erased in primordial germ cells

   (d) Replication converts a fully methylated site to hemi-methylated site

4. A male mouse cell line has a large translocation from X chromosome into chromosome 1. When a GFP containing transgene is inserted in this chromosome 1 with translocation, it is often silenced. However when inserted in the other homologue of chromosome 1 that does not contain the translocation, it is almost always expressed. Which of the following phenomenon best describes this effect? **(JUNE 2016)**

   (a) Genome imprinting

   (b) Gene balance

   (c) Sex-specific expression

   (d) Dosage compensation

5. The individuals considered in this question are having two haploid sets of autosomes and no Y. chromosome. The X:A ratio of the individuals, the type of organisms chosen, their primary sex and number of Barr bodies expected in their cells are shown in the table below:

   The individuals considered in this question are having two haploid sets of autosomes and no Y-chromosome. The X:A ratio of the individuals, the type of organisms chosen, their primary sex and number of Barr bodies expected in their cells are shown in the table below:

X:A ratio	Organism	Primary Sex	Number of Barr bodies
i. 0.5 ii. 2	A. Human B. *Drosophila*	I. Male II. Female III. Metafemale	a. Zero b. One c. Two d. Three

Select the option below with all correct matches:

(a) i-A-II-a; ii-A-II-d; i-B-I-a; ii-B-III-a

(b) i-A-I-a; ii-A-II-c; i-B-II-a; ii-B-I-a

(c) i-A-II-c; ii-A-I-d; i-B-I-c; ii-B-II-b

(d) i-A-II-a; ii-A-II-d; i-B-III-a; ii-B-I-a

Select the option below with all correct matches: **(DEC 2019)**

(a) i-A-II-a; ii-A-II-d; i-B-I-a; ii-B-III-a

(b) i-A-I-a; ii-A-II-C; i-B-II-a; ii-B-I-a

(c) i-A-II-c; ii-A-I-d; i-B-I-c; ii-B-II-b

(d) i-A-II-a; ii-A-II-d; :-B-III-a; ii-B-I-a

6. A researcher exposed Drosophila larvae to 37°C during their growth. One of the adult flies that emerged had a crossveinless phenotype. Crossveinless is a known mutant in Drosophila. When this crossveinless fly was crossed to a known crossveinless mutant fly all the progeny had normal phenotype. The observed phenotype can be best explained as an example of **(DEC 2017)**

(a) Conditional mutant

(b) Phenocopy

(c) Penetrance

(d) Pleiotropy

7. *Drosophila melonogaster* males with normal wing size and grey body colour is crossed with females having vestigial wings and yellow body colour. Vestigeal wing and yellow body colour are recessive characters. The F_1 progeny is sib-mated. In the F_2 progeny the following phenotypes are observed in the ratios given below:

S. No.	Phenotype		Sex	Ratio
	Wing Shape	Body Color		
1.	Normal	Grey	Male	3
2.	Normal	Yellow	Male	3
3.	Normal	Grey	Female	3
4.	Normal	Yellow	Female	3
5.	Vesitigeal	Grey	Male	1
6.	Vesitigeal	Yellow	Male	1
7.	Vesitigeal	Grey	Female	1
8.	Vesitigeal	Yellow	Female	1

The above observation suggests: **(DEC 2013 GU)**

(a) Yellow body colour is X-linked, while vestigeal wing is an autosomal character.

(b) Yellow body colour is autosomal, while vestigial wing is an X-linked character

(c) Both yellow body and vestigial wing are X-linked character.

(d) Both yellow body and vestigial wing are autosomal and un-linked.

8. Consider the following crosses involving grey (wild-type) and yellow body colour true breeding *Drosophila*:

	Cross	F_1 progeny	F_2 progeny
Cross 1	Grey female X yellow male	All males: grey All females: grey	Grey females:98 Yellow males : 45 Grey males : 49
Cross 2	Yellow females X grey males	All males: yellow All females: grey	?

Assuming 200 F_2 offspring are produced in cross 2, which one of the following outcome is expected? **(DEC 2011)**

(a) 97 grey males, 54 yellow females, 49 grey males

(b) 102 yellow males, 46 yellow females, 52 grey females

(c) 52 grey males, 49 yellow males, 48 yellow females, 51 grey females

(d) 98 grey males, 94 yellow females, 2 yellow males, 6 grey females

9. The allele l in Drosophila is recessive, sex-linked and lethal' when homozygous or hemizygous. If a female of the genotype Ll is crossed with a male, what is the ratio of females : males in the progeny? **(DEC 2018)**

(a) 3 female: 1 male (b) 2 female: 1 male

(c) 1 female: 1 male (d) 1 female: 2 male

10. In an organism, allele for red eye colour is dominant over the allele for white eye colour. A cross is made between a white eyed male and a red eyed female. In the progeny all males are red eyed while the females are white eyed. The reciprocal cross leads to all red eyed progeny. Based on the above information which one of the following conclusions is correct? **(JUNE 2018)**

(a) This is a sex-limited trait, and the male is the homomorphic sex

(b) This is a sex-linked trait, with male being the homomorphic sex

(c) This is a sex-linked trait, with female being the homomorphic sex

(d) This is a case of autosomal inheritance, with incomplete penetrance

Answer Key

1.(b)	2.(b)	3.(a)	4.(d)	5.(a)	6.(b)	7.(a)
8.(c)	9.(b)	10.(b)				

6. Recombination and Gene mapping methods: Linkage

1. In an organism, a^+ allele governs grey body colour, while its mutant allele a gives yellow body colour. Further, presence of b^+ allele gives long and thin hairs while b allele gives rise to short and thick hairs. The alleles a^+ and b^+ are dominant over a and b, respectively. An individual with the genotype

$$\frac{a^+ \quad b^+}{a \quad b}$$

Has a patch of yellow cells with short and thick hairs. Which one of the following events is most likely to lead to the above? **(DEC 2019)**

(a) Non disjunction of the homologous chromosomes during mitosis

(b) Somatic recombination involving a and b

(c) Translocation occurring in a few somatic cells

(d) Mutation of both a^+ and b^+ alleles in the somatic cells

2. In a linkage map, two genes A and B, are 70 cM apart. If individuals heterozygous for both the genes are test crossed number of progeny with parental phenotype will be: **(DEC 2017)**

(a) equal to the number of progeny with recombinant phenotype

(b) more than the number of progeny with recombinant phenotype

(c) less than the number of progeny with recombinant phenotype

(d) could be more or less than the number of progeny with recombinant phenotype depending on whether the genes are linked in cis or trans, respectively

3. A cross was made between pure wild type males and brown eyed curled winged females of D. *melanogaster.* The F_1 females were test crossed. The F_2 progeny obtained was as follows:

Wild type	200
Brown eyes, curled wings	150
Brown eyes, normal wings	30
Normal eyes, curled wings	20
Total	400

The genetic distance (cM) between brown eye and curled wing loci is: **(DEC 2012)**

(a) 12.5 (b) 50
(c) 150 (d) 25

4. A test cross was made with phenotypically wild type Drosophila flies having genes for sepia eye and curled wing in heterozygous condition. The following results were obtained: **(DEC 2019 ASSAM)**

Wild type	400 individuals
Sepia eyed	150 individuals
Curled winged	100 individuals
Sepia eyed curled winged	350 individuals

The result indicates

(a) Independent assortment, as wild and the double mutant types are in almost equal proportion
(b) Unequal segregation, as it is showing departure from 1:1:1:1 ratio
(c) Complete linkage, as the two genes are very closely placed
(d) Linkage, the two genes are separated by 25 cM distance

5. In a haploid organism, the loci *A/a* and *D/d* are 8 map units apart. In a cross *Ad* X *aD*, what will be the proportion of each of the following progeny classes:
(a) *Ad* (b) Recombinants. **(MODEL PAPER)**

(a) 92%, 8% (b) 46%, 8%
(c) 92%, 4% (d) 46%, 4%

6. In an experiment involving mapping of 3 genes (a, b and c) in Drosophila, a three point test cross is carried out. The parental cross was *AAbbCC* x *aaBBcc*. The genotypes of the double crossovers are: *Aabbcc* and *aaBbCc*. Based on this, determine the order of the genes. **(JUNE 2014)**

(a) *acb* (b) *cab*
(c) *abc* (d) *bac*

7. A three point test cross was carried out in Drosophila melanogaster involving three adjacent genes X, Y and Z. arranged in the same order. The distance between X to Y is 32.5 map unit (mu) and that between Y to Z is 20.5 map. The coefficient of coincidence = 0.886.

What is the percentage of double recombinants in the progeny obtained from the testcross? **(JUNE 2016)**

(a) 6% (b) 8%
(c) 12% (d) 16%

8. In *Drosophila,* Bar eye (B) is a dominant mutation while miniature wing (m) and yellow body colour (y) are recessive mutations. eterozygous females for these mutations were crossed to normal eyed miniature winged and yellow body coloured males. Assume the following progeny was obtained:

Phenotypes	**Numbers**
$B^+ m^+ y^+$	30
$B\ m\ y^+$	25
$B\ m^+ y^+$	165
$B^+ m\ y^+$	120
$B\ m\ y$	20
$B^+ m\ y$	185
$B\ m^+ y$	110
$B^+ m^+ y$	45

Based on the results obtained, the order of genes will be:

A. B m y B. m B y

The genetic distance between B and y will be:

C. 40cM D. 17.1cM

The correct combination of answer is **(DEC 2017)**

(a) A and C (b) B and C
(c) A and D (d) B and D

9. Poplar is a dioecious plant. A wild plant with genes AABBCC was crossed with a triple recessive mutant aabbcc. The F1 male hybrid (AaBbCc) was then back crossed with the triple mutant and the phenotypes recorded are as follows:

AaBbCc 300
aaBbCc 100

aaBbcc	16
AabbCc	14
AaBbcc	65
aabbCc	75
aabbcc	310
Aabbcc	120

The distance in map unit (mu) between A to B and B to C is **(JUNE 2016)**

(a) 25 and 17 mu, respectively
(b) 33 and14 mu, respectively
(c) 25 and 14 mu, respectively
(d) 33 and 11 mu, respectively

10. The genetic map of three genes in *Drosophila melanogaster* is given below:

a ——— b ——— c
(a to b: 10 cm; b to c: 5 cm)

A cross, as given below individuals of the genotype:

$$\frac{a^+ \quad b^+ \quad c}{a^+ \quad b^+ \quad c} \times \frac{a \quad b \quad c^+}{a \quad b \quad c^+}$$

The female F1 progeny are test-crossed and 1000 progeny are obtained. Assuming that there has been no double crossover, what is the expected number of progeny with the genotypes:

(a) $\frac{a^+ \quad b \quad c^+}{a \quad b \quad c}$ (b) $\frac{a^+ \quad b^+ \quad c^+}{a \quad b \quad c}$

(c) $\frac{a^+ \quad b^+ \quad c}{a \quad b \quad c}$

Select the set which shows the correct number of expected progeny. **(DEC 2013)**

Set	(A)	(B)	(C)
(a)	100	50	850
(b)	50	25	425
(c)	100	850	50
(d)	0	425	75

11. When F_1 female *Drosophila* of the genotype $a^+a\ b^+b\ c^+c$ is test crossed, the following progenies were obtained:

Progeny classes*	No. of progenies
$a^+ b^+ c^+$	22
$a^+ b^+ c$	28
$a\ b\ c^+$	26
$a\ b\ c$	24
$a^+ b\ c^+$	230
$a^+ b\ c$	220
$a\ b^+ c^+$	225
$a\ b^+ c$	225
Total	**1000**

*The progeny has been shown as classes derived from the female gamete. Statements A to F as given below are conclusions derived from the above result.

A. Genes *a* and *b* are linked in *cis*.
B. Genes *a* and *b* are linked in *trans*.
C. Genes *a* and *b* are linked in *cis* while *b* and c are linked in *trans*.
D. The genotype of the parents are $a^+a^+b^+b^+$ and *aabb*
E. The genotype of the parents are a^+a^+bb and aab^+b^+
F. Genes *a* and *b* are 10cM apart.

Which of the above statements are correct? **(JUNE 2011)**

(a) C alone. (b) A, E and F
(c) B, E and F (d) A, D and F

12. A virgin *Drosophila* female was crossed with a wild type male. The F_1 progeny obtained had four types of males as shown below.

Phenotype	White eyed	Wild type	Cross veinless	White eyed and crossveinless
Number	50	3	44	3

Assuming that white eye and crossveinless mutations are X-linked and recessive, the following statements were made:

A. F_1 females were also of four types as that of males.
B. The white eyed crossveinless male flies appeared due to independent assortment.

C. The map distance between the genes for white eye and crossveinless is estimated to be 12 cM.

D. The map distance between white eye and crossveinless is estimated to be 6 cM.

E. All F_1 females are expected to be wild type.

F. The F_1 wild type males appeared due to crossing over.

The combination with correct statements is: **(JUNE 2018)**

(a) C, E, F (b) A, B, D
(c) A, D, F (d) B, D, E

13. A Lod score of 3 represents a Recombination Frequency (RF) that is **(JUNE 2019)**

(a) 3 times as likely as the hypothesis of no linkage

(b) 30 times as likely as the hypothesis of no linkage

(c) 100 times as likely as the hypothesis of no linkage

(d) 1000 times as likely as the hypothesis of no linkage

14. A pedigree shown below depicts that the individual I-1 is heterozygous for a dominant disease allele *D* and for molecular markers M1/M2. The paternal molecular markers present in the progeny individuals are indicated in the pedigree.

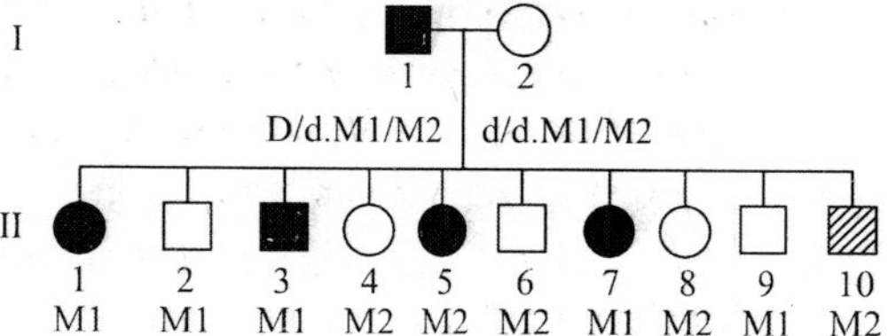

The following statements may be drawn from the above pedigree:

A. The two loci D/d and M1/M2 appears to be linked

B. The recombination frequency between the two loci is 20%

C. If LOD score comes out to be 3, then it ensures that the two loci are independently assorting

D. A LOD score <1 would have ensured that the two Loci are linked

Which combination of the above statements can correctly interpret the depicted pedigree? **(DEC 2019)**

(a) C and D (b) Only C
(c) A and B (d) Only D

Answer Key

1.(b)	2.(a)	3.(a)	4.(d)	5.(b)	6.(a)	7.(a)
8.(b)	9.(a)	10.(b)	11.(c)	12.(a)	13.(d)	14.(c)

7. Tetrad analysis

1. Segregation of alleles can occur at Anaphase I or at Anaphase n of meiosis. With reference to this statement, which one of the following organism is an ideal model system for identifying stage of allelic segregation at meiosis? **(JUNE 2015)**

(a) *Neurospora crassa*

(b) *Saccharomyces cerevisiae*

(c) *Pisum sativum*

(d) *Drosophila melanogaster*

2. In *Neurospora* a cross between the genotypes '*A*' and '*a*' results in an ascus with ascospores of genotypes as shown below:

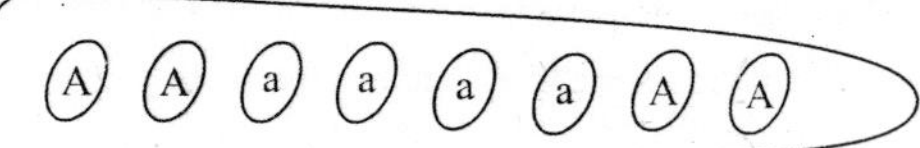

Statements A to D are events that could have occurred during meiosis.

A. Crossing over between the centromere and the gene.

B. Segregation of alleles '*A*' and '*a*' in meiosis I.

C. Segregation of alleles '*A*' and '*a*' in meiosis II.

D. Assortment of alleles '*A*' and '*a*'.

Which of the above events could correctly explain the observation shown in the figure? **(JUNE 2011)**

(a) A followed by C (b) C alone
(b) A followed by B (d) D alone

3. The following table summarizes the result of a cross between two strains of *Neurospora* having the alleles *D* and *d*, respectively. The table shows the different patterns of octad arrangement and the number of ascus observed of each type.

Octads						
	D	d	D	d	D	d
	D	d	D	d	D	d
	D	d	d	D	d	D
	D	d	d	D	d	D
	d	D	D	d	d	D
	d	D	D	d	d	D
	d	D	d	D	D	d
	d	D	d	D	D	d
Number of ascus observed	115	125	14	16	17	13
					Total = 300	

Based on the above, fill in the blanks from the options given below.

"The first two columns are from meiosis with no crossover between locus D and ___[A]___. The pattern for these two columns represent ___[B]___ segregation pattern. The distance between the locus D and the centromere is ___[C]___ map units". **(JUNE 2014)**

	A	B	C
(a)	d allele	first division	10
(b)	centromere	first division	10
(c)	d allele	second division	20
(d)	centromere	second division	10

4. Centromere positions can be mapped in linear tetrads in some fungi. A cross was made between two strains *a b* and a^+ b^+ and 100 linear tetrads were analysed. The genes a and b are located on two arms of the chromosome. The tetrad were divided into 5 classes as shown below.

Class	1	2	3	4	5
	$a\ b$	$a\ b$	$a\ b$	$a\ b$	$a\ b$
	$a^+\ b$	$a\ b^+$	$a\ b$	$a^+\ b^+$	$a^+\ b$
	$a\ b^+$	$a^+\ b$	$a^+\ b^+$	$a\ b$	$a^+\ b$
	$a^+\ b^+$	$a^+\ b^+$	$a^+\ b^+$	$a^+\ b^+$	$a\ b^+$
Linear tetrads	15	29	52	2	2

Based on the above observation the following conclusion were drawn:

(A) Class 1 is a result of a cross over between '*a*' and the centromere.

(B) Class 2 is result of double crossover involving 3 strands between 'a' and the centromere.

(C) Class 5 is a result of a double crossover between '*a*' centromere and '*b*'-centromere.

(D) Class 4 is a result of a double cross over involving all the 4 strands.

Which one of the following options represents all correct statements? **(JUNE 2019)**

(a) A and B (b) A and C

(c) B and D (d) C and D

Answer Key

1.(a)	2.(a)	3.(b)	4.(b)

8. Molecular markers in genetics

1. Which one of the following DNA markers can be used to distinguish between a homozygote and heterozygote? **(JUNE 2019)**

(a) RAPD (b) AFLP

(c) RFLP (d) ISSR

2. Marker-assisted selection (MAS) defined as selection based on molecular markers should have some important criterion for plant breeding activities. Some statements about these criteria are mentioned below:

A. Marker should co-segregate with the desired trait of interest.

B. Marker should not co-segregate with the desired trait of interest.

C. Marker should be un-linked with the desired trait of interest.

D. Marker is used for indirect selection of a genetic determinant or determinants of a trait of interest.

Which onc of abovc combinations is correct?

(JUNE 2014)

(a) A and B (b) B and C

(c) C and D (d) A and D

3. The following are statements above molecular markers in the context of plant breeding

A. Molecular markers can be used for elimination of undesirable traits

B. Molecular markers cannot be used for elimination of genetic contribution of each individual parent in a segregating population

C. Molecular markers are used for mapping of QTLs, which is also possible by conventional techniques.

D. Molecular markers can be used for selection of individuals from a population that are homozygous for the recurrent parent genotype at loci flanking the target locus.

Which of the above statements are TRUE?

(DEC 2011)

(a) A and B (b) A and C

(c) A and D (d) B and C

4. Two homozygous individuals (P1 and P2), were genotyped using dominant DNA markers A and B, as shown below. The F1 progeny obtained was test crossed and frequency of progeny with which different genotypes appear, is given below:

Marker	**P1**	**P2**	**Progeny of test cross and their frequencies**			
			45	45	5	5
A	—		—		—	
B		—		—	—	

Profile of DNA markers following gel electrophoresis

The following conclusions were made:

A. In the F1 markers A and B are linked and in coupling phase (cis)

B. In the F1 markers A and B are linked and in repulsion phase (trans)

C. The distance between A and B is 10 cM

D. The distance between A and a is 5 cM

Which of the above conclusions are correct?

(JUNE 2015)

(a) A and C (b) A and D

(c) B and C (d) B and D

5. A new mutant called early ripening (EL) is identified in a plant. The wild type phenotype is late ripening (LR). Further DNA marker(s) is/are observed to be polymorphic between EL and LR plants. A cross was made between pure lines of LR and ER. The F_1 progeny was test crossed and its progeny was analyzed. The parental, F_1 and progeny of test cross were also analyzed for the DNA marker. The table below summarizes the phenotype of the progeny and the pattern of DNA marker observed in each case:

Generation	Parent		F1	Progeny of test cross			
Phenotype	ER	LR	LR only	LR		ER	
No. of progency				52		48	
Pattern of DNA market on gel	—		—	—		—	
electrophoresis		—	—	—	—	—	—
Number of progeny from each parent				26	27	23	25

Based on the above information, the following statements were made:

A. LR is dominant to ER

B. The DNA marker used is a dominant marker

C. The DNA marker is linked to the phenotype

Which of the above statements are correct?

(DEC 2017)

(a) A only (b) A and B only

(c) A and C only (d) A, B and C only

6. A pair of alleles govern seed size in a crop plant. 'B' allele responsible for bold seed is dominant over 'b' allele controlling small seed. An experiment was carried out to test if an identified dominant DNA marker (5kb band) is linked to alleles controlling seed size. A plant heterozygous for the marker and the alleles was crossed to a small seeded plant lacking the 5kb band. 100 progeny obtained from the cross were analysed for the presence and absence of the DNA marker. The result are tabulated below:

Phenotype	**Plant with bold seed**		**Plant with small seed**	
No. of progeny showing presence or absence of DNA marker	Present	Absent	Present	Absent
	22	23	27	28

Based on the above observations which one of the following conclusions is correct?

(DEC 2016)

(a) The DNA marker assorts independently of the phenotype

(b) The 5kb band is linked to the allele 'B'

(c) The 5kb band is linked to the allele 'b'

(d) The DNA marker assorts independently with bold seed but is linked to the small seed trait.

7. The RFLP pattern observed for two pure parental lines (P1 and P2) and their F1 progeny is represented below. Further, the P1 plant had red flowers while the P2 had white flowers. The F1 progeny was backcrossed to P(b) The result obtained, showing the number of progeny with red and white flowers and their RFLP patterns is also represented below.

	P1	P2	F_1	Progeny of the cross between F_1 and P2			
Phenotype	Red	White	Red	Red		White	
	—		—	—			—
RFLP pattern		— —	— —	— —	— —	— —	— —
		No. of progeny		45	5	45	5

Which one of the following conclusions made is correct? **(DEC 2012)**

(a) The DNA marker and the gene for the flower colour are 10cM apart.

(b) The marker and the gene for the flower colour are 5cM apart.

(c) The marker and the phenotype are independently assorting.

(d) The marker and the gene for the colour segregate from one another.

8. In normal individuals, there are three *MstII* restriction sites, two flanking the β-globin gene and one within the gene. In individuals affected by a disease, a single nucleotide polymorphism in the β-globin gene abolishes the internal *MstII* recognition site. The RFLP pattern for this locus, obtained by hybridization using a probe internal to the flanking *MstII* sites, from three siblings of a family is shown below:

bp (kb)	**Normal son**	**Normal daughter**	**Affected son**
1.35	—		—
1.15	—	—	
0.2	—	—	

Based on the above profile, what is the nature of the genetic disorder? **(JUNE 2017)**

(a) X-linked Recessive

(b) Autosomal Dominant

(c) Autosomal Recessive

(d) X-linked Dominant

9. Sickle cell anemia is a recessive genetic disease caused due to a point mutation in the 6th codon abolishing one of the *MspII* endonuclease digestion site present in the β-globin gene. *MspII* digested DNA from a normal person gives two bands, 1150 bp and 200 bp, in β-globin gene. A family with a proband (based on the disease phenotype) gave the following *MspII* digestion pattern:

bp	Mother	Father	Son (I)	Daughter	Son (II)
1350		—	—	—	
1150	—	—		—	—
200	—	—		—	—

The following conclusions were drawn:

A. Son (I) is the proband and the given mutation is not present in Son (II).

B. The daughter is a carrier for the given mutation.

C. The gene is X-linked and thus Son (I), becomes the proband.

D. The father and daughter are affected

E. A de novo mutation in same site on normal allele has allowed appearance of diseased phenotype in the proband

Which of the following combination of conclusions wilt be the most appropriate for the figure given above? **(JUNE 2015)**

(a) A, B and E (b) A, B and C

(c) B, C and E (d) C and D

10. An analysis of four microsatellite markers was carried out in a family showing a genetic disorder. The results are summarized below

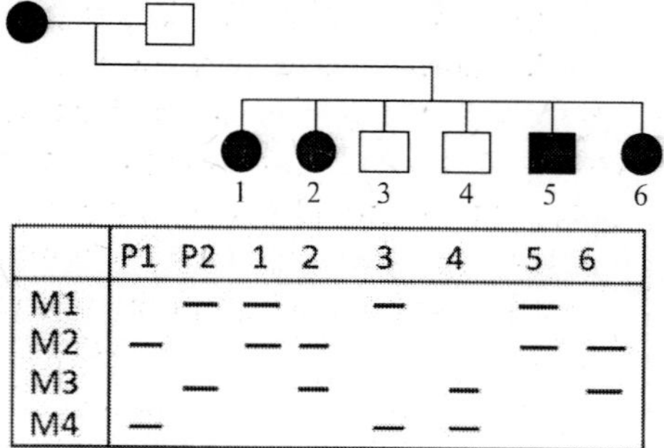

	P1	P2	1	2	3	4	5	6
M1		—	—		—		—	
M2	—		—	—			—	—
M3		—		—		—		—
M4	—				—	—		

Based on the above, which of the markers shows linkage to the disorder? **(DEC 2014)**

(a) M1 (b) M2

(c) M3 (d) M4

11. The pedigree given below follows the inheritance pattern of a late-onset (after age of 30 years) genetic disease that is 100% penetrant. Affected individuals are indicated by a solid circle (woman) or solid ·square (males). RFLP analysis of DNA from each individual is shown below in the pedigree.

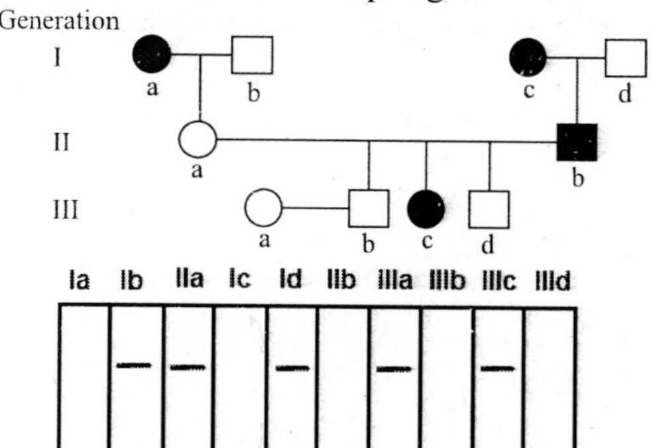

Which grandchildren (IIIb to IIId) will be affected by the disease after attaining the age of 30 years? **(JUNE 2018)**

(a) Only IIIb

(b) Both IIIb and IIIe

(c) Both IIIc and IIId

(d) Both IIIb and IIId

12. Microsatellite are used as marker for identifying individuals via DNA fingerprinting as the alleles may differ in the number of repeals. From the Southern blot shown below identify the progeny (A, B, C and D) for the given parents (M= mother, F= father). **(JUNE 2016)**

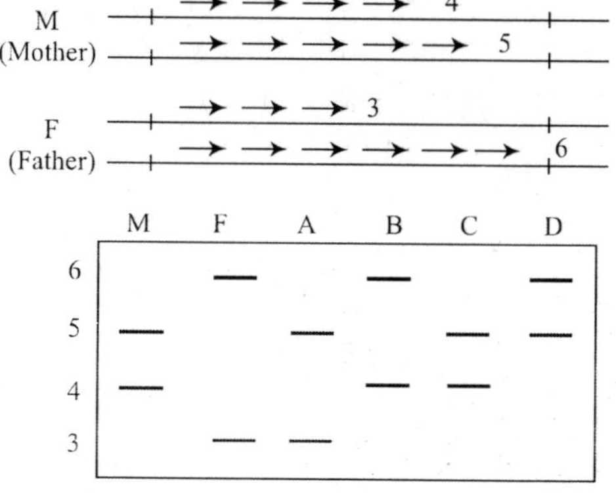

(a) A, B, C and D (b) A, B and D

(c) A and D only (d) B, C and D

13. Polymorphic DNA sequences are used for molecular identification. Short tandem repeats (STRs) and Single Nucleotide Polymorphism (SNPs) are used· as polymorphic markers. The table below summarizes the status of autosomal. SNP, autosomal STR, mitochondrial SNP, Y-linked STR for four individuals related to each other, representing parents and their two children.

	Autosomal		Mt DNA		Y-linked
Individual	SNP	STR	SNP1	SNP2	STR
A	C/C	13/12	C	G	13
B	C/G	13/13	C	A	13
C	C/G	14/13	C	A	-
D	C/C	13/14	C	A	-

Based on the above data, identify the individuals representing the two parents. **(DEC 2018)**

(a) Individuals A and D

(b) Individuals A and C

(c) Individuals B and C

(d) Individuals C and D

14. In the following pedigree three STR loci A, B and C are linked on the long arm of the X-chromosome in the order centromere-A-B-C-telomere. Further in the table, the STR alleles present in each individual is indicated.

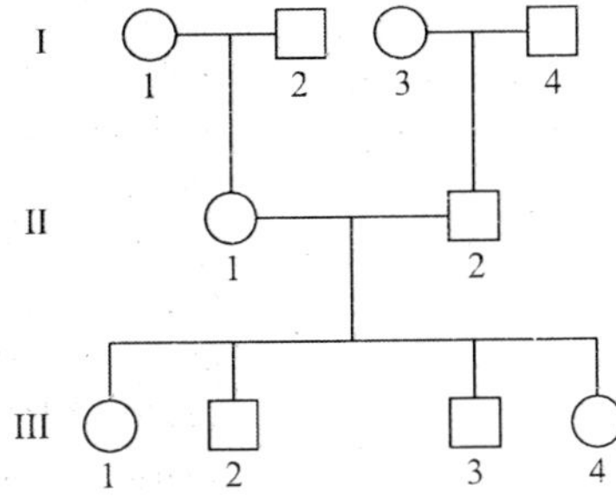

Loci	Gemeratopm									
	I				II		III			
	1	2	3	4	1	2	1	2	3	4
A	5,6	4	5,6	4	4,6	6	6	4	6	6
B	7,8	9	8,9	7	7,9	8	8,9	9	7	7,8
C	1,2	2	1,2	3	1,2	1	1,2	2	2	1

Based on the above, X-chromosome(s) in which of the following individuals are recombinant? **(DEC 2019)**

[Hint: X-chromosome in males will help identify the phase of the alleles]

(a) II - 1, III - 1 and III - 2

(b) II - 2, III - 1 and III - 2

(c) III - 1 and III - 3

(d) III - 2 and III - 4

15. In 'TaqMan' assay for detection of base substitutions (DNA variants), probes (oligonucleotides) with fluorescent dyes at the 5'-end and quencher at 3'-end are used. While the probe is intact, the proximity of the quencher reduces the fluorescence emitted by reporter dye. If the target sequences (wild type or variant) are present, the probe anneals to the target sequence, down stream to one of the primers used for amplifying DNA sequence flanking the position of variants. For an assay two flanking PCR primers, two probes corresponding to the wild type and variant allele and labeled with two different reporter dyes and quencher were used. During extension the probe may be cleaved by the Taq-polymerase separating the reporter dye and the quencher. Three individuals were genotyped using the assay. Sample for individual I shows maximum fluorescence for the dye attached to the wild type probe, sample for individual II shows maximum fluorescence for the dye attached to variant probe and sample for individual III exhibits equal fluorescence for both the dyes. Which of the following is correct? **(DEC 2011)**

(a) Individual I is homozygous for the variant allele.

(b) Individual II is homozygous for the variant allele.

(c) Individual II is homozygous for the wild type allele.

(d) Individual III is homozygous for the wild type allele.

16. To detect mutation (GAG→GTG) allele specific hybridization method is used. Four members of an affected family are investigated. DNA isolated from blood samples of parents and two offsprings are spotted on a membrane after appropriate processing and probed with either TGACTCCTGAGGAGAAGTC (first probe) or TGACTCCTGTGGAGAAGTC (second probe) after labelling. While probed with first oligonucleotide, signals are obtained for the positions where DNA are spotted from parents and offspring II. When probed with second oligonucleotide, signals are obtained at position where DNA from the parents and offspring I are spotted. Results are shown below:

	Father	Mother	Offspring I	Offspring II
First probe	+	+	–	+
Second probe	+	+	+	–

On the basis of the result, which of the following statements is correct? **(JUNE 2011)**

(a) Parents are affected

(b) Offspring I is affected

(c) Offspring II is carrier

(d) Offspring II is affected

Answer Key

1.(c)	2.(d)	3.(c)	4.(c)	5.(a)	6.(a)	7.(a)
8.(c)	9.(a)	10.(b)	11.(d)	12.(b)	13.(b)	14.(c)
15.(b)	16.(b)					

9. Mapping by deletion and somatic cell hybrids

1. A novel enzyme was identified in humans. The following approaches are available to identify the chromosome on which the gene encoding the enzyme is present:

(A) Suppress the activity of enzyme by RNAi.

(B) Identify polymorphism in the population and carry out pedigree analysis to study its inheritance.

(C) Purify the enzyme, decipher its amino acid sequence, predict its DNA sequence and search for its presence in the available human genome sequence.

(D) Create chromosome addition lines by making somatic hybrids between human and mouse cells, identify lines showing the enzyme activity and the human chromosome present in it.

Which of the above approaches can be used? **(JUNE 2013)**

(a) (A) or (B) (b) (B) or (C)

(c) (C) or (D) (d) (A) or (C)

2. The following is a schematic representation of region (showing six bands) of the polytene chromosome of Drosophila, along with the extent of five deletions (Del1 to Del5):

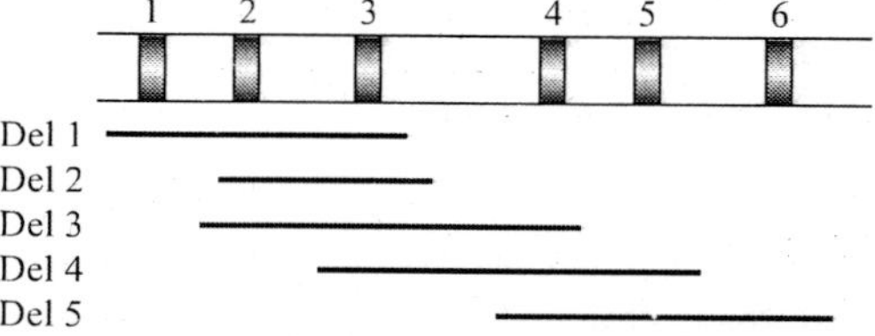

Recessive alleles *a, b, c, d, e* and *f* are known to correspond to each of the bands (1 to 6), but their order is not known. When the recessive alleles are placed against each of these deletions, the following results are obtained. The plus (+) in the table indicates wild type phenotype of the corresponding allele, while a minus (-) indicates the phenotype governed by the corresponding mutant allele.

	a	*b*	*c*	*d*	*e*	*f*
Del 1	+	–	–	–	+	–
Del 2	+	+	–	–	+	+
Del 3	–	+	–	–	+	+
Del 4	–	+	+	–	–	+
Del 5	–	+	+	+	–	–

Which one of the following indicates the correct location of the recessive alleles on the bands of the polytene chromosomes? **(DEC 2011)**

(a) a-3; b-1; c-2; d-4; e-5; f-6.
(b) a-2; b-1; c-3; d-4; e-5; f-6
(c) a-4; b-1; c-2; d-3; e-5; f-6.
(d) a-6; b-2; c-3; d-4; e-1; f-5.

3. The locations of five overlapping deletions have been mapped to a *Drosophila* chromosome as shown below

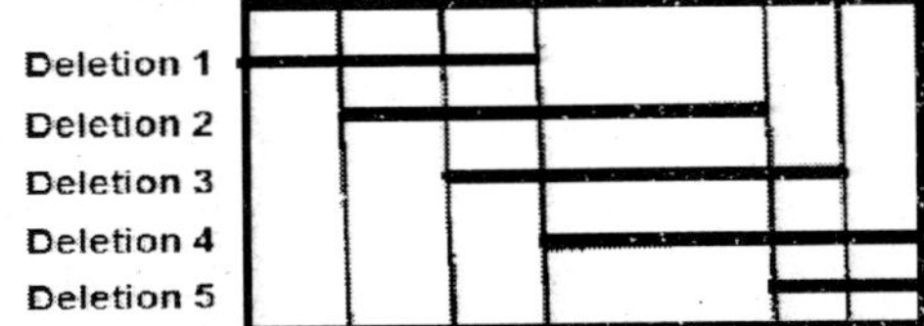

(Horizontal lines in the above figure indicate the deleted regions) Recessive mutations a, b, c, d and e are known to be located within this region, but the order of mutations on the chromosome is not known. When the flies homozygous for the recessive mutations are crossed with flies homozygous for the deletions, the following results are obtained (letter "m" represents mutant phenotype and "+" represents the wild type).

Deletion	Mutation				
	a	b	c	d	e
1	+	m	m	m	+
2	+	+	m	m	+
3	+	+	+	m	m
4	m	+	+	+	m
5	m	+	+	+	m

On the basis of the above data, the relative order of the five mutant genes on the chromosome is **(JUNE 2018)**

(a) b c d e a (b) a b c d e
(c) b c e a d (d) c d b e a

4. The location of six deletion (Shows as solid line underneath the chromosomes) has been mapped to the *Drosophila* chromosome as shown in the diagram given below:

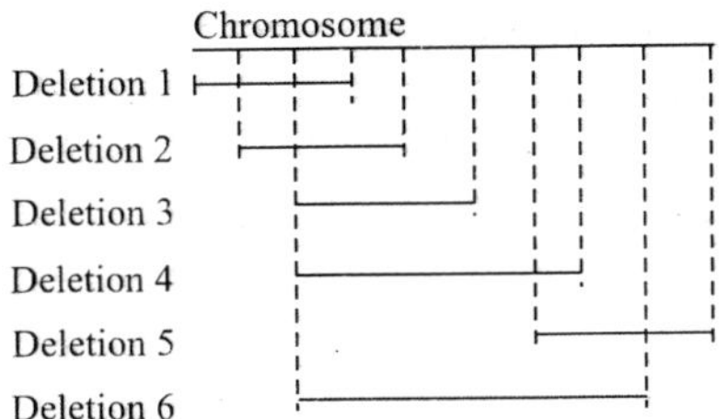

The recessive mutations a, b, c, d, e f and g are known to be located in the region of deletions, but order of mutations on the chromosome is not known. When flies homozygous for the recessive mutations are crossed with flies homozygous for the deletion, the following result were obtained where the letter 'm' represents a mutant phenotype and '+' represents the wild type.

	Mutations						
Deletion	**a**	**b**	**c**	**d**	**e**	**f**	**g**
1	+	M	M	M	+	+	+
2	+	+	M	M	+	+	+
3	+	+	+	M	M	+	+
4	M	+	+	M	M	+	+
5	M	+	+	+	+	M	M
6	M	+	+	M	M	M	+

The relative order of the seven mutant genes on chromosome is : **(JUNE 2019)**

(a) b c e a f g d (b) b c d f g e a
(c) b c d e a f g (d) c d e a g f b

5. The following scheme represents deletions (1-4) in the rII locus of phage T4 from a common reference point:

Deletion 1 ____________
Deletion 2 ____________________
Deletion 3 ____________________________
Deletion 4 ____________________________________

(The bars represent the extent of deletion in each case)

Four point mutations (a to d) are tested against four deletions for their ability (+) or inability (-) to give wild type (rII+) recombinants.

The results are summarized below:

	a	b	c	d
1	+	+	+	+
2	+	+	+	-
3	+	-	+	-
4	-	-	+	-

Based on the above the predicted order of the point mutations is: **(DEC 2016)**

(a) b-d-a-c (b) d-b-a-c

(c) d-b-c-a (d) c-d-a-b

6. Somatic cell hybridization is used to assign a gene to a particular chromosome. When two cell lines from two different species are fused, they form a heterokaryon which tends to Jose chromosomes as they divide, preferentially from one species. A panel of cell lines was created from mouse-monkey somatic cell fusions. Each line was examined for the presence of monkey chromosomes and for the production of a given enzyme.

The following results were obtained:

Cell line	Presence of Enzyme	Presence of Monkey chromosomes									
		1	2	3	4	5	6	7	8	9	10
A	+	+	+	+	+	+	–	+	–	+	+
B	+	–	+	+	–	+	+	+	–	+	–
C	+–	–	+	–	–	–	–	–	+	+	+
D	+	+	+	+	–	–	+	+	+	+	–
E	–	–	–	–	+	+	–	–	–	+	+
F	+	+	+	–	+	+	+	+	–	+	–

On the basis of these results, which chromosome has the gene that codes for the given enzyme? **(JUNE 2017)**

(a) Chromosome 10 (b) Chromosome 7

(c) Chromosome 1 (d) Chromosome 5

7. Three somatic hybrid cell lines, designated as X, Y and Z, have been scored for the presence or absence of chromosomes 1 through 8, as well as for their ability to produce the hypothetical gene product A, B, C and D as shown in the following table:

Hybrid cell lines	Human chromosome present								Gene product expressed			
	1	2	3	4	5	6	7	8	A	B	C	D
X	+	+	+	+					-	+	-	+
Y	+	+			+	+			+	-	-	+
Z	+		+		+		+		+	+	-	+

Which of the following option has most appropriately assigned chromosomes for each of the given genes? **(DEC 2015)**

(a) Gene A on chromosome 5, Gene B on chromosome 3, Gene C on chromosome 8 and Gene D on chromosome 1

(b) Gene A on chromosome 5 and Gene B on chromosome 3 only

(c) Gene D on chromosome 8, Gene C on chromosome 1, Gene B on chromosome 5 and Gene A on chromosome 4

(d) Gene A on chromosome 5, Gene B on chromosome 3 and Gene D on chromosome 1

8. A series of cell lines was created by fusing mouse and human somatic cells. In mouse-human somatic cell hybrids, human chromosomes tend to get lost before becoming a stable cell line. Some hybrid cell lines may carry human chromosome deletions. Each cell line was examined for the presence of chromosomes and for the production of an enzyme.

The following results were obtained:

Cell line	Gene-product	Chromosomal segments									
		1p	1q	2p	2q	3p	3q	4p	4q	5p	5q
A	+	+	+	–	–	+	+	–	–	+	+
B	+	–	–	+	+	–	–	+	+	+	+
C	–	+	+	–	–	+	+	–	+	–	+
D	+	–	–	+	+	+	+	–	–	+	–
E	–	+	+	+	+	–	+	–	–	–	–

Which segment of the chromosome has the gene encoding for the enzyme **(JUNE 2014)**

(a) 1p (b) 5p

(c) 5q (d) 4p

9. The following figure depicts the relationship between a genetic map for four genes (A, B, C and D) and their corresponding physical map:

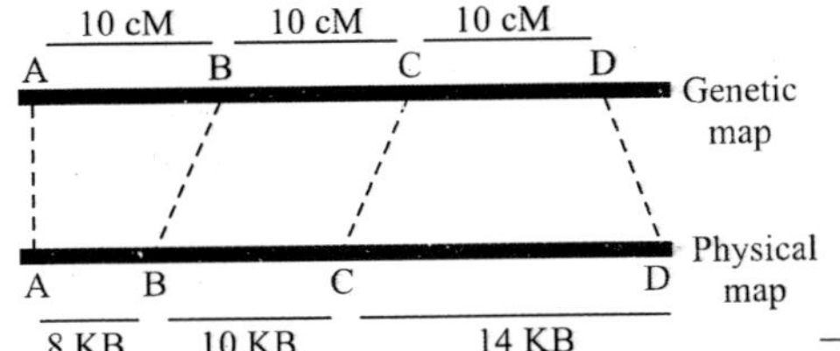

The following statements are made to explain this relationship.

A. More number of recombination events occur between *A* and *B* as compared to *B* and *C*.

B. Lesser number of recombination events occur between *C* and *D* as compared to *B* and *C*.

C. Although the physical distance between *A* and *B* is less than that between *C* and *D*, the region between *A* and *B* is more recombinogenic.

D. The physical distance between *A* and *B* is less than that between *C* and *D*, and thus the region between *A* and *B* is less recombinogenic.

E. Although the physical distance between *C* and *D* is more than that between *B* and *C*, the region between *C* and *D* is less recombinogenic.

F. Although the physical distance between *C* and *D* is more than *B* and *C*, the region between *C* and *D* is more recombinogenic.

Which statements are correct? **(DEC 2011)**

(a) A and B (b) C and E
(c) D and F (d) A, C and E

Answer Key

1.(c)	2.(c)	3.(a)	4.(c)	5.(b)	6.(b)	7.(d)
8.(b)	9.(b)					

10. Extra chromosomal inheritance and Maternal effect

1. In *Neurospora*, the mutant stp exhibits erratic stop-and-start growth. When a female of sip strain is crossed with a normal strain acting as a male, all progeny individuals' showed stp mutant phenotype. However, the reciprocal cross resulted in all normal progeny, individuals. These results can be explained on the basis of

A. maternal inheritance

B. sex limited inheritance

C. sex influenced inheritance

D. stp mutation may be located In mitochondrial DNA

The most appropriate statement or combination of the above statements for explaining the experimental results is: **(DEC 2014)**

* question

(a) A and C (b) C only
(c) A and D (d) B and D

2. Maternal inheritance of coiling of shell in snail (*limnaea peregra*) is well established. The dextral coiling depends on dominant allele *D* and sinistral coiling depends upon recessive allele *d*. A female F1 progeny of dextral(*Dd*) type is crossed with a male sinistral snail. What will be the ratio of heterozygous : homozygous individuals in its F2 progeny? (DEC 2016) * Question

(a) 3:1 (b) 1:1
(c) 1:3 (d) 1:2:1

3. A male snail homozygous for dextral alleles is crossed with a female homozygous for sinistral alleles. All the F_1 individuals showed sinistral phenotype. When F_1 progeny snails were self-fertilized all individuals of F_2 progeny had dextral coiling. This experiment demonstrated **(JUNE 2018)**

(a) dominant epistasis as dextral allele is dominant over sinistral allele

(b) recessive epistasis as in F_2 dextral allele appeared in homozygous condition

(c) maternal effect as the nuclear genotype of the F_1 mother has governed the phenotype of the F_2 individuals.

(d) maternal inheritance as the mitochondrial genes of the F_1 mother has governed the phenotype of the F_2 individuals.

4. A mutant mating type *mt* strain of *Chlamydomonas* that was resistant to the antibiotic kanamycin (kan^r) and herbicide PPT (ppt^r) was crossed to a wild type mating mt^+ kan^s ppt^s strain that was sensitive to kanamycin and PPT. Twenty tetrads of the progeny were analyzed for mating type and resistance/sensitivity to kanamycin and PPT. The following observations were made:

	Type I	Type II	Type III
	$mt\ kan^r\ ppt^r$	$mt\ kan^r\ ppt^s$	$mt\ kan^r\ ppt^r$
	$mt\ kan^r\ ppt^r$	$mt\ kan^r\ ppt^s$	$mt\ kan^r\ ppt^s$
	$mt^+\ kan^r\ ppt^s$	$mt^+\ kan^r\ ppt^r$	$mt^+\ kan^r\ ppt^r$
	$mt^+\ kan^r\ ppt^s$	$mt^+\ kan^r\ ppt^r$	$mt^+\ kan^r\ ppt^s$
Number of each type observed	8	9	3

The following statements were made to explain the observations:

A. *mt* and *ppt* loci are on two different chromosomes

B. Inheritance of mating type and *ppt* resistance/ sensitivity are demonstrating cytoplasmic inheritance

C. Inheritance of kanamycin-resistance/ sensitivity is demonstrating nuclear inheritance

D. Nuclear inheritance is being demonstrated by mating type and *ppt*-resistance/ sensitivity analysis

Which one of the combinations of above statements is correct? **(DEC 2019)**

(a) A and B (b) A and D

(c) B and C (d) C and D

5. Two petite yeasts, *Mat-a* and *Mat-α* are crossed. The diploid is grande. After a few mitotic divisions, the grande diploid is sporulated. The analysis of a large number of tetrads yielded a 2:2 ratio of petite: grande. A few potential scenarios describing the reason for this segregation pattern are stated below:

A. The parental strains had two different mitochondrial rho- mutations

B. One of the parents had a recessive nuclear petite mutation

C. Only one of the mitochondrial petite mutation is inherited in the tetrads

D. The mitochondria inherited are wild type.

Which one of the following options represents a combination of all correct statements?

(DEC 2019 ASSAM)

(a) A, B and C (b) B and D only

(c) A and D only (d) B and C only

Answer Key

1.(*)	2.(*)	3.(c)	4.(b)	5.(b)

11. Microbial Transformation and Transduction

1. A lac^- culture of E. coli was mutagenised. On what media would one spread the mutagenised cells to **select** for lac^+ cells? **(JUNE 2017)**

(a) Minimal media+ lactose.

(b) Rich media + lactose.

(c) Minimal media + glycerol+ IPTG + X-Gal

(d) Rich media+ IPTG + X-Gal

2. DNA from a strain of bacteria with genotype a^+ b^+ c^+ d^+ e^+ was isolated and used to transform a strain of bacteria that was a^- b^- c^- d^- e^-. The transformed cells were tested for the presence of donated genes. The following genes are found be co-transformed

i. a^+ and d^+ ii. b^+ and e^+

iii. c^+ and d^+ iv. c^+ and e^+

The order of genes on the bacterial chromosome is **(DEC 2013)**

(a) a-b-c-d-e (b) a-d-c-e-b

(c) a-c-d-e-b (d) a-d-b-e-c

3. DNA from a bacterial strain with genotype at $b^+c^+d^+e^+f^+$ was isolated and used to transform a strain of bacteria with genotype $a^-b^-c^-d^-e^-f^-$. The transformed cells were tested for the presence of co-transformed genes and the following types of co-transformants were observed:

a^+ and c^+

e^+ and d^+

d^+ and b^+

b^+ and f^+

c^+ and e^+

The correct order of genes will be:

(DEC 2019 ASSAM)

(a) a b c d e f (b) a c e b f d

(c) a c e d b f (d) a b f e d c

4. In a transformation experiment, donor DNA from an *E. coli* strain with the genotypes Z^+Y^+ was used to transform a strain of genotype $Z^- Y^-$. The frequencies of transformed classes were:

Z^+Y^+	200
Z^+Y^-	400
Z^-Y^+	400
Total	1000

What is the frequency (%) with which *Y* locus is co-transformed with the *Z* locus?

(DEC 2012)

(a) 1 (b) 20

(c) 33.3 (d) 40

5. An experiment was designed to find out the relation between DNA uptake and transformation efficiency. ^{32}P-labelled genomic DNA from *Bacillus subtilis* (A) and cold genomic DNA from *Clostridium jejeuni* (B) mixed in various proportions was used to transform *Bacillus subtilis*. The results obtained are tabulated below

Set No	Concentration of DNA used (µg/ml)		Uptake of ^{32}P labeled DNA (%)	Transforming efficiency (CFU)
	A	B		
1	1	1	90	30
2	1	10	82	10
3	1	100	15	5
4	1	1000	5	1

The following interpretations (A to D) could be made

A. The transformation is dependent on recombination between homologous sequences.

B. Cells did not distinguish between homologous or heterologous sequences for uptake of DNA.

C. DNA uptake is based on specific receptors

D. DNA degradation dictates transformation efficiency

Which of the following interpretations are correct? **(JUNE 2017)**

(a) A and B (b) A and C

(c) C and D (d) A and D

6. In a transduction experiment using $a^+b^+c^+$ genotype as a donor and $a^-b^-c^-$ as the recipient, a^+ transductants were selected and screened for *b* and c. The data obtained are shown below.

Genotype	**No. of Recombinant**
a^+ b- c-	573
a^+ b^+ c-	98
a+ b^- c+	11
a^+ b^+ c^+	68

The co-transduction frequencies for a^+b^+ and b^+c^+, respectively, are: **(JUNE 2017)**

(a) 17% and 12% (b) 22% and 9%

(c) 22% and 17% (d) 17% and 9%

7. A co-transduction experiment was performed to decipher the linear order of 4 genes: a, b, c and d. Three sets of experiments were done where transductants were selected for a (Set-1) or b (Set-2) or c (Set-3) and screened for co-transduction of the other markers.

Set-1		
Selected for	Co-transduction	Frequency
a	*b*	31
a	*c*	3
a	*d*	89
Set-2		
Selected for	Co-transduction	Frequency
b	*a*	22
b	*c*	78
b	*d*	68
Set-3		
Selected for	Co-transduction	Frequency
c	*a*	0
c	*b*	69
c	*d*	43

Based on the frequencies shown above, identify the most likely order in the genome.

(DEC 2014)

(a) *a b c d* (b) *b c d a*

(c) *c d a b* (d) *a d b c*

8. Phages are collected from an infected E. coli donor strain of genotype cys$^+$ Leu$^+$ thr$^+$ and used to transduce a recipient of genotype cys$^-$ Leu$^-$ thr$^-$, The treated recipient population is plated on a minimal medium supplemented with leucine and threonine.

   Many colonies grew. Which one of the following combination of genotypes are appropriate for the colonies that grew?

   (DEC 2017)

   (a) cys$^+$ leu$^+$ thr$^+$, cys$^+$ leu$^-$ thr$^+$, cys$^-$ leu$^+$ thr$^-$

   (b) cys$^-$ leu$^+$ thr$^+$, cys$^+$ leu$^-$thr$^-$, cys$^+$ leu$^+$ thr$^-$

   (c) cys$^-$ leu$^-$ thr$^-$, cys$^-$ leu$^-$ thr$^+$, cys$^-$ leu$^+$ thr$^-$

   (d) cys$^+$ leu$^-$ thr$^-$, cys$^+$ leu$^-$ thr$^+$, cys$^+$ leu$^+$ thr$^-$

9. Five bacterial markers were followed for a co-transduction experiment. The following table documents the observations of this experiment. '+' denotes co-transduction and '–'denotes lack thereof; 'ND' stands for not determined.

	arg	leu	str	met
gal	+	–	+	–
leu	ND		+	+
arg		ND	–	ND
str	–	+		ND

Pick the correct order in which the genes are arranged on the bacterial chromosome

(JUNE 2016)

(a) *str-gal-leu-arg-met*

(b) *leu-met-arg-str-gal*

(c) *leu-str-met-gal-arg*

(d) *arg-gal-str-leu-met*

Answer Key

1.(a)	2.(b)	3.(c)	4.(b)	5.(a)	6.(b)	7.(d)
8.(d)	9.(d)					

12. Mapping genes by interrupted mating

1. Conjugation involves transfer of DNA from one cell to another. In *E. coli*, the cells that cannot transfer their DNA into the recipient cell are **(DEC 2013 GU)**

   (a) Hfr (b) F'

   (c) F+ (d) F–

2. Which one of the following is true for cells harbouring F' plasmid? **(DEC 2016)**

   (a) Their F plasmid is non-functional.

   (b) They exhibit increased rates of transfer of all chromosomal genes.

   (c) They are merodiploids.

   (d) They fail to survive as the chromosomal origin of replication is inactivated.

3. The distance between bacterial genes as determined from interrupted mating experiments are measured in units of

   (JUNE 2014)

   (a) cM (b) minutes

   (c) bp (d) micrometers

4. An interrupted mating experiment was performed between *Hfr Str^s a$^+$b$^+$c$^+$* and *F$^-$ Str^r a$^-$ b$^-$c$^-$* strains. The genotype of majority of

streptomycin resistant (*Str* ') exconjugant after 10, 20 and 30 minutes of interrupted mating is given below:

10 min	$a^+ b^- c^-$
20 min	$a^+ b^- c^+$
30 min	$a^+ b^+ c^+$

The most probable gene order would be

(DEC 2014)

(a) *a b c* (b) *c a b*

(c) *b a c* (d) *a c b*

5. The following table shows mapping data from three interrupted mating experiments using three different Hfr strains and an F^- strain:

Appearance of genes in F^- cells

Hfr 1	Genes	e+	f+	c+	d+	b+
	Time	6	24	34	49	54
Hfr 2	Genes	b+	d+	c+	f+	g+
	Time	1	6	21	31	63
Hfr 3	Genes	d+	c+	f+	e+	g+
	Time	4	19	29	47	61

(Time is represented in minutes)

The following answers are derived:

The order of genes is

A. e g b d c f B. f g b d c e

The distance between

C. f and g is 32 min and between f and b is 30 min

D. c and e is 28 min and between band c is 20 min

The correct combination of answers is

(DEC 2017)

(a) A, C and D (b) Band C

(c) A and B (d) B and D

6. Interrupted mating experiments were performed using three different *Hfr* strains (1-3). The three strains have different combinations of selectable markers. The time of entry for markers for each strain is shown is shown in the table given.

Strain	**Time of Entry**				
Hfr #1	*met* (5')	*thr* (17')	*str^r* (25')	*phe* (30')	*pro* (45')
Hfr #2	*str^r* (15')	*pur^r* (28')	*pro* (35')	*his* (45')	*met* (55')
Hfr #3	*pro* (2')	*his* (12')	*met* (22')	*str^r* (42')	*phe* (47')

Using the above data, predict the correct sequence of markers on the *E. coli* chromosome.

(DEC 2016)

(a) *met-thr- str^r -phe-pro-pur^r-his*

(b) *pur^r-pro-his-met-thr- str^r-phe*

(c) *str^r- pur^r-his-met-phe-pro- str^r*

(d) *his-met-phe-thr-pro- str^r- pur^r*

7. Two Hfr strains, Hfr-1 (arg^+leu^+ gal^+str^s) and Hfr-2 (arg^+ his^+ $gal^+pur^+str^s$) were mated with a F^- strain (arg^-leu^- gal^- $his^-pur^-str^s$). The results of the interrupted mating experiment are shown as plots 'a' and 'b', respectively.

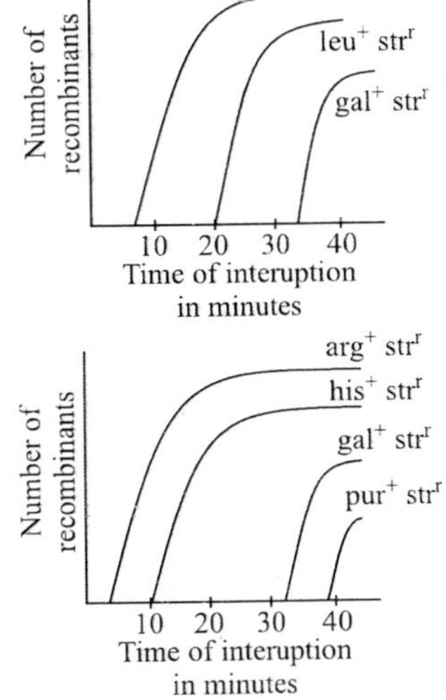

Based on these results, identify which of the options accurately reflects the order of loci?

(JUNE 2018)

(a) arg — leu — his — gal — pur^r

(b) arg — his — leu — gal — pur^r

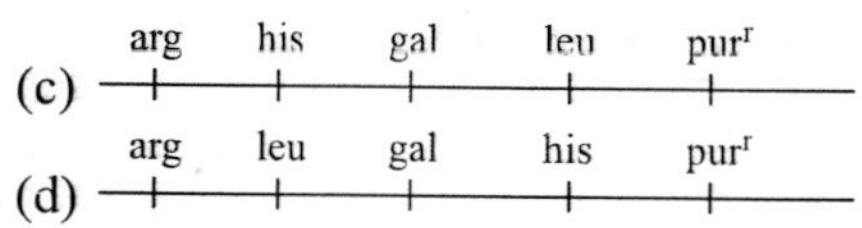

8. A cross was made between *Hfr met$^+$ arg$^+$ leu$^+$ str^S* X F- *met$^-$ arg$^-$ leu$^-$ str^r* in which *leu$^+$* exconjugants are selected. If the linear organization of the genes are *leu$^+$ arg$^+$ met$^+$*, which one of the following genotypes is expected to occur in the lowest frequency?

(JUNE 2015)

(a) *leu$^+$ arg$^-$ met$^-$* (b) *leu$^+$ arg$^+$ met$^-$*

(c) *leu$^+$ arg$^+$ met$^+$* (d) *leu$^+$ arg$^-$ met$^+$*

9. A cross is mode between an Hfr strain that is met$^+$ thi$^+$ pur$^+$ and an F$^-$ that is met- thi$^-$ pur-. The recombinants are selected on medium that contains supplements that satisfy only the requirements for pur and thi and not for met. These recombinants are tested for the presence of the thi$^+$ and pur$^+$ alleles. The following number of colonies are found for each genotype:

met$^+$ thi$^+$ pur$^+$ 200

met$^+$ thi$^+$ pur$^-$ 0

met$^+$ thi$^-$ pur$^+$ 7

met$^+$ thi- pur - 60

What is the order of genes in Hfr strain?

(DEC 2013 GU)

(a) met$^+$ pur$^+$ thi$^+$ (b) met$^+$ thi$^+$ pur$^+$

(c) thi$^+$ met$^+$ pur$^+$ (d) pur$^+$ met$^+$ thi$^+$

10. In *E. coli*, four Hfr strains donate the genetic markers shown in the order shown:

Strain 1: F L K O Z

Strain 2: C Y A Z O

Strain 3: P D W C Y

Strain 4: P F L K O

All the Hfr strains are derived from the same F$^+$ strain. What is the order of these markers on the circular chromosome of the original *E. coli* F$^+$ strain? **(MODELPAPER)**

(a) K O Z P D F L W C Y A K

(b) F L K O Z P D W C Y A F

(c) F L K O Z A Y C W D P F

(d) F L K C Y A Z O W D P F

11. In *E. coli* four Hfr strains donate the following genetic markers in the order shown below:

Strain 1: L Q W X Y

Strain 2: M T A D Y

Strain 3: E C M T A

Strain 4: W Q L E C

Which of the following depicts the correct order of the markers and the site of integration (→) of the F - factor in the four Hfr strains?

(DEC 2011)

(a) Strain 1 Strain 4 Strain 2 Strain 3

L Q W X Y D A T M C E

(b)

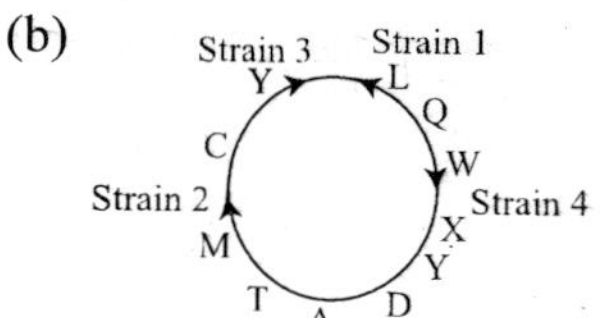

(c)

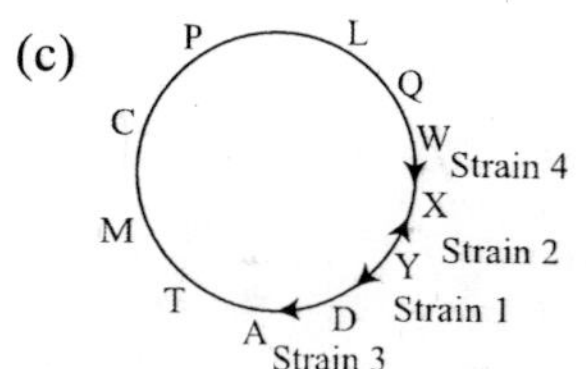

(d) Strain 4 Strain 1 Strain 2 Strain 3

12. Using interrupted mating, four Hfr strains were analysed for the sequence in which they transmitted a number of different genes to a F$^-$ strain. Each Hfr strain was found to transmit its genes in a unique order as summarized in the table [Only the first five genes were scored].

Order of transmission	*Hfr* strain			
	1	2	3	4
First	A	B	M	F
	B	A	K	H
	D	L	J	G
	F	M	G	J
Last	H	K	H	K

Which one of the following correctly represents the gene sequence in the original strain from which the Hfr strains were derived as well as the place of integration and polarity of the F plasmid? **(DEC 2018)**

(a) 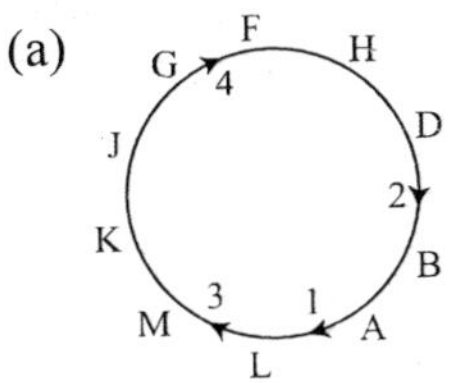

(b) 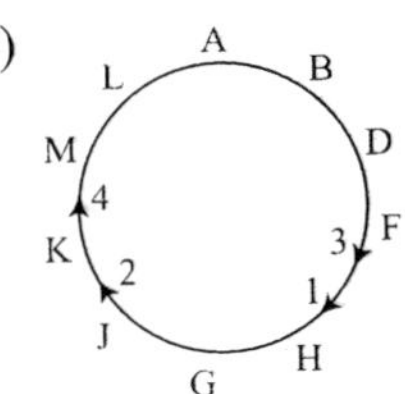

(c) 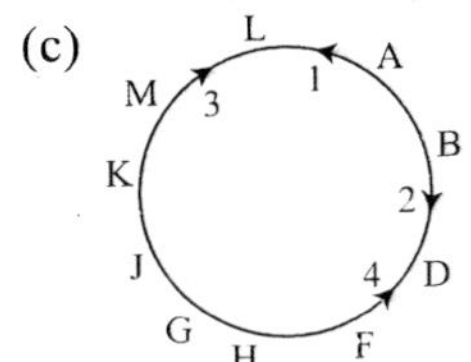

(d) 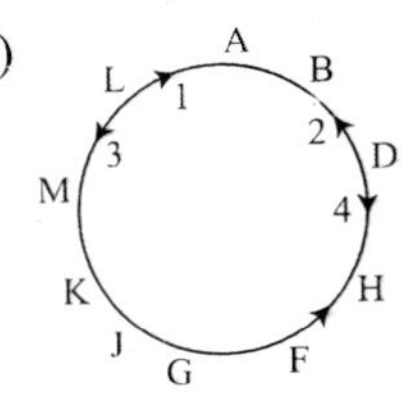

13. A conjugation experiment is carried out between F^+ *his⁺ leu⁺ thr⁺ pro⁺* bacteria and F^- *his⁻ leu⁻ thr⁻ pro⁻* bacteria for a period of 25 minutes. At this time the mating is stopped, and the genotypes of the recipient F^- bacteria are determined. The results are shown below:

Genotype	Number of colonies
his^+	0
leu^+	13
thr^+	26
pro^+	6

What is the probable order of these genes on the bacterial chromosome? **(MODEL PAPER)**

(a) *thr, leu, pro, his*

(b) *pro, leu, thr* and the position of *his* cannot be determined.

(c) *thr, leu, pro*, and the position of *his* cannot be determined

(d) *his, pro, leu, thr*

14. An *E. coli* strain has *metB1* (90 min) and *leuA5* (2 min) mutations. It also has *strA7* (73 min) mutation and *Tn5* transposon which confers streptomycin and kanamycin resistance, respectively, inserted in its chromosome. The mutant strain was crossed with an Hfr strain that is streptomycin sensitive and has a *hisG2* mutation (44 min) that makes it require histidine. After incubation for 100 min, the cells were plated on minimal plate supplemented with leucine, histidine and streptomycin to select the *metB* marker. After purifying 100 of the Met^+ transconjugants, one finds that 15 are His^+, 2 are Leu^+ and 12 are kanamycin sensitive.

The unselected markers are

A. *metB1* and *leuA5* mutation.

B. *leuA5* and *Tn5* insertion mutation.

Which of the above statement is correct and what is the position of transposon insertion? **(DEC 2012)**

(a) A and before 73 min

(b) B and before 44 min

(c) B and before 73 min

(d) A and before 44 min

Answer Key

1.(d)	2.(c)	3.(b)	4.(d)	5.(a)	6.(b)	7.(b)
8.(d)	9.(a)	10.(c)	11.(b)	12.(c)	13.(a)	14.(b)

13. Pedigree analysis

1. Of the following, which one of the individuals will NOT necessarily carry the allele responsible for the mentioned trait? **(DEC 2015)**

(a) A woman in a family where an autosomal dominant trait is segregating and her mother and son are affected.

(b) A daughter of a man who is affected by an X linked dominant trait

(c) A father of a child who is affected with an autosomal recessive trait

(d) A father of a boy affected with X- linked recessive trait

2. The following is the inheritance pattern of a trait under observation:

(i) The trait often skips a generation

(ii) The number of affected males and females is almost equal

(iii) The trait is often found in pedigrees with consanguineous marriages.

The trait is likely to be **(DEC 2013)**

(a) autosomal recessive.

(b) autosomal dominant.

(c) sex-linked recessive.

(d) sex-linked dominant.

3. The inheritance of a given disorder is recorded in three small families shown below:

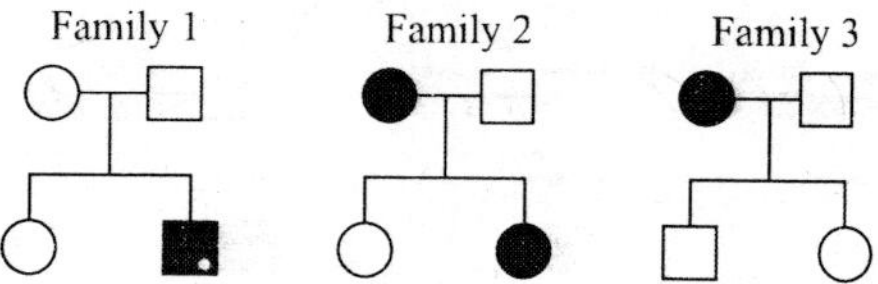

Based on the above limited information, which one of the following inheritance pattern best explains the observations? **(DEC 2017)**

(a) X-linked recessive

(b) X-linked dominant

(c) Autosomal recessive

(d) Autosomal dominant

4. The following pedigree chart shows inheritance of a given trait.

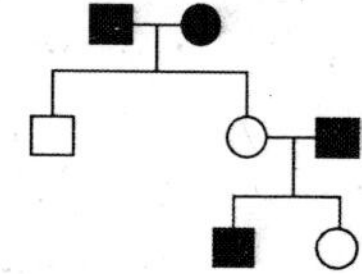

The trait can be called **(DEC 2015)**

(a) autosomal dominant

(b) autosomal recessive

(c) X-linked dominant

(d) sex limited

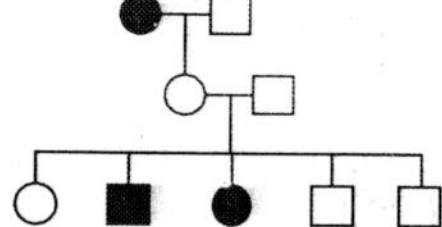

5. The above pedigree shows the inheritance of a rare allele. The allele is: **(JUNE 2015)**

(a) X-linked recessive

(b) autosomal recessive

(c) dominant with incomplete penetrance

(d) autosomal recessive with incomplete penetrance

6. The following pedigree chart shows inheritance of a given trait.

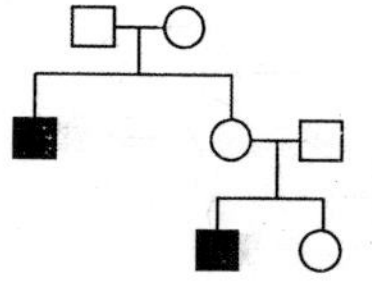

This can be taken as an example of **(DEC 2013 GU)**

(a) Y -linked inheritance.

(b) X-linked recessive as well as autosomal recessive inheritance

(c) Autosomal dominant inheritance.

(d) X-linked dominant inheritance.

7.

Find the pattern of inheritance of the trait showing incomplete penetrance from the figure shown above. **(MODELPAPER)**

(a) Autosomal dominant.

(b) Autosomal recessive.

(c) Mitochondrial inheritance.

(d) X-linked recessive.

8. Following are four modes of inheritance:

(A X-linked recessive

(B) X-linked dominant

(C) Autosomal recessive

(D) Autosomal dominant

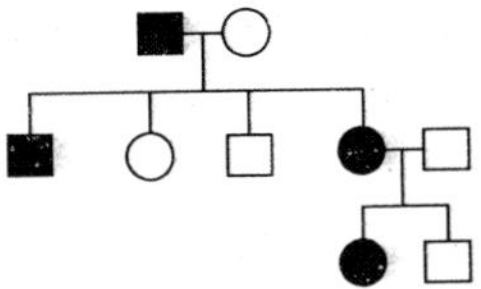

Which of the above modes can explain the inheritance of a common trait shown in the pedigree below? **(DEC 2013 GU)**

(a) B and D only (b) A, C and D only

(c) C only (d) D only

9. Following are four modes of inheritance

A. X-linked recessive

B. X-linked dominant

C. Autosomal recessive

D. Autosomal dominant

Which of the above modes of inheritance can explain the pedigree shown below?

(JUNE 2012)

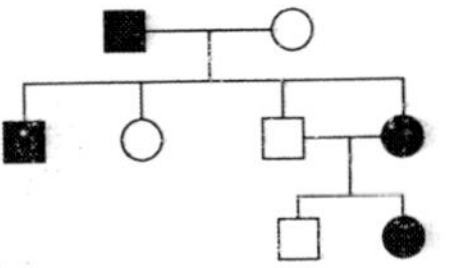

(a) A and C (b) B and C

(c) C and D (d) D only

10. A Family was examined for a given trait which is represented in the pedigree shown below. Further, the degree of expression of the trait is highly variable among members of the family; some are only slightly affected while others developed severe symptoms at an early stage

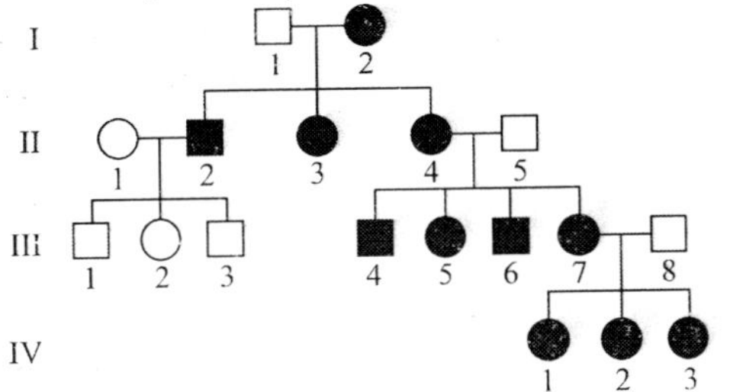

The following statements are made to explain the pattern of inheritance shown in the pedigree.

(A) X-linked dominant mutation

(B) X-linked recessive mutation

(C) Mitochondrial inheritance

(D) Variable expression can be due to heteroplasty

The best possible explanation for this inheritance is **(JUNE 2019)**

(a) A and D (b) C and D

(c) B only (d) A only

11. The following pedigree represents inheritance of a trait in an extended family:

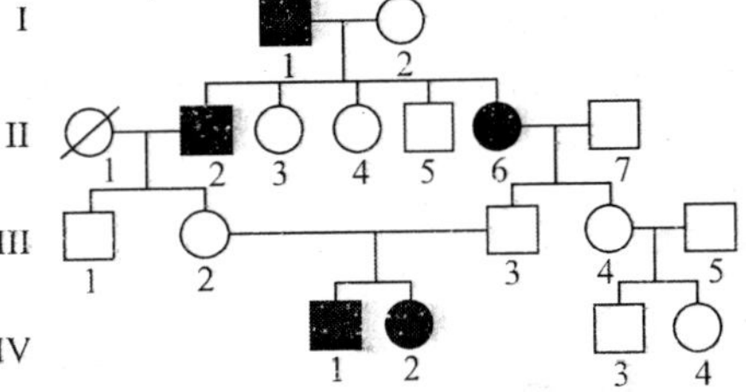

What is the probable mode of inheritance and which individuals conclusively demonstrate this mode of inheritance? **(DEC 2012)**

(a) Autosomal recessive, III-2, 3 and IV- 1, 2 conclusively demonstrate the mode of inheritance.

(b) Autosomal recessive, I-1,2 and II-2 conclusively demonstrate the mode of inheritance.

(c) Autosomal dominant, III- 2, 3 and IV- 1, 2 conclusively demonstrate the mode of inheritance.

(d) X-linked recessive, II- 3, 4 and 5 conclusively demonstrate the mode of inheritance.

12. The given pedigree shows the inheritance of a trait

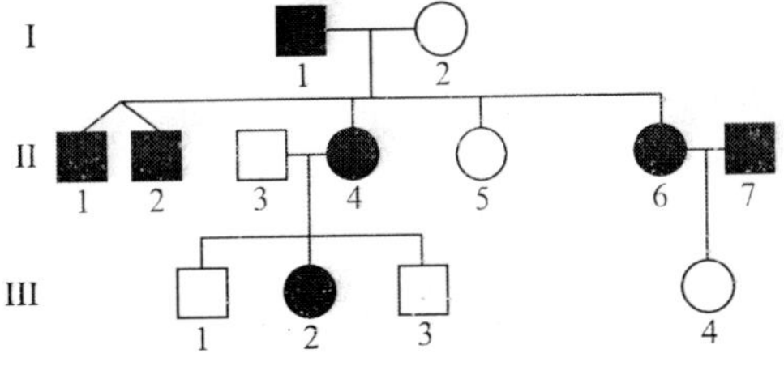

The following derivations are made from the pedigree chart:

The trait can be Y-linked because I-parent produced II-4 child

The trait cannot be X-linked recessive because II-4 parent produced III-1 and III -3 children

The trait can be X-linked dominant because I-1 parent produced II-5 child

The trait is unlikely to be autosomal recessive because II-6 and II-7 parents producedIII-4 child.

The trait may be autosomal dominant

Which one of the following options represents a combination of all correct statements?

(DEC 2019 ASSAM)

(a) B, D, E (b) A, B, D

(c) B, C, D (d) A, C.

13. Affected individuals from the pedigree given below are suffering from albinism, an autosomal recessive disease. Identify the confirmed carrier individuals in this pedigree assuming that the members coming from outside the family are homozygous for the dominant allele. **(JUNE 2011)**

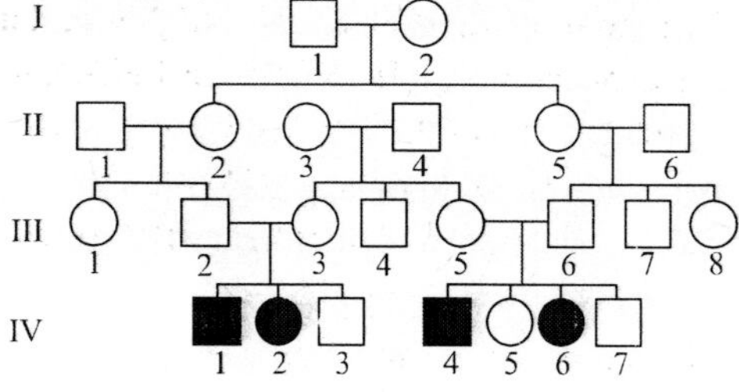

(a) III-2, III-3, III-5, III-6, II-I, II-3 and II-6

(b) III-2, III-3, III-5, III-6, II-2, II-4, II-5 and I-2

(c) III-2, III-3, III-5, III-6, II-2, II-4 and II-5

(d) III-I, III-4, III-7, II-2, II-4 and II-5

14. The following pedigree shows the inheritance pattern of a trait.

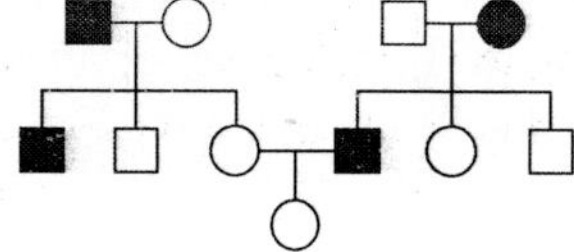

From the following select the possible mode of inheritance and the probability that the daughter in generation III will show the trait.

(DEC 2016)

(a) X-linked recessive, probability is 1/2

(b) X-linked recessive, probability is 1/4

(c) Autosomal recessive, probability is 1/2

(d) Autosomal recessive, probability is 1/3

15. The following pedigree shows the inheritance of a common phenotype controlled by an autosomal recessive allele. The probability of carriers in the population is 1/3

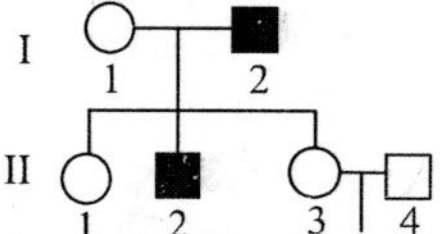

What is the probability that a child from parents II-3 and II-4 will show the phenotype?

(DEC 2014)

(a) 1/12 (b) 1/18

(c) 1/36 (d) 3/16

16. The inheritance pattern of a common trait which shows complete penetrance is shown below:

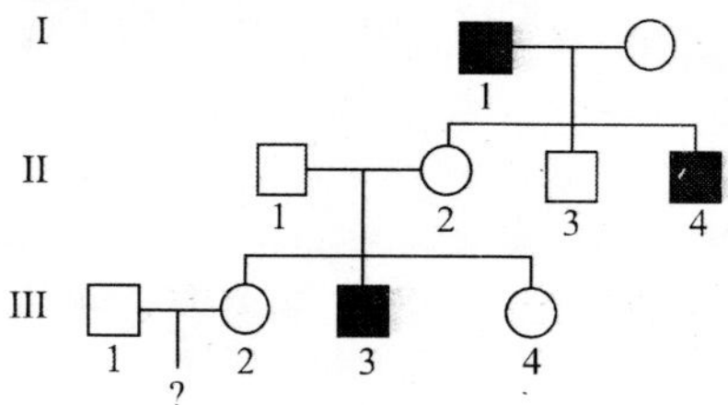

Based on the above pedigree, fill in the blanks from the options given below:

"The trait is __[A]__. The probability that a child from the marriage of individual III-I and III-2 will show the trait is __[B]__ considering that the individual III-I is a carrier of the trait."

(JUNE 2014)

	A	B
(a)	Y-linked	0
(b)	Y-linked	1/2
(c)	Autosomal	1/8
(d)	Autosomal	1/6

17. The following pedigree shows the inheritance pattern of a rare recessive disorder with complete penetrance.

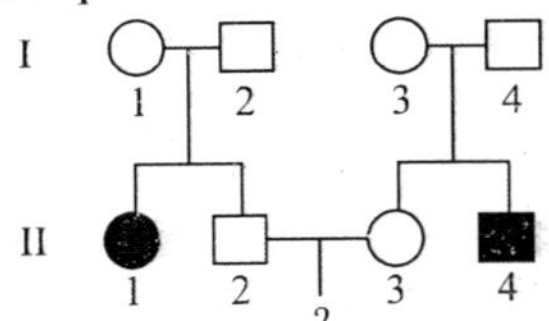

A child from marriage between individuals II-2 & II-3 will show the disorder only if the parents carry the recessive allele. What is the probability that the child will show the disorder? **(DEC 2013)**

(a) 1/9, and the probability of the parents carry the recessive allele is 2/3.

(b) 1/4, and the probability of the parents carry the recessive allele is 3/4.

(c) 1/16, and the probability of the parents carry the recessive allele is 2/3

(d) 1/64, and the probability of the parents carry the recessive allele is 3/4

18. The pedigree below represents the inheritance of an autosomal recessive trait. **(DEC 2018)**

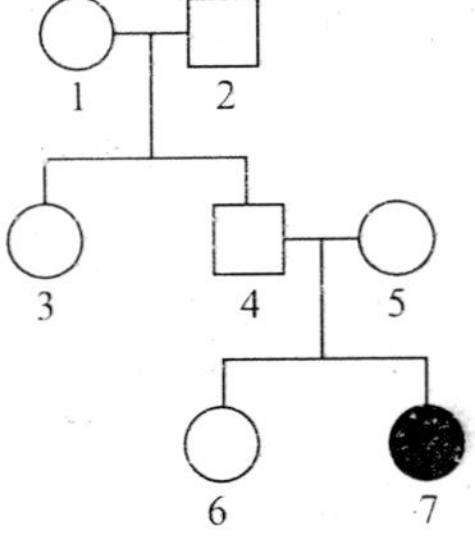

What is the probability that individual '6' is a heterozygote?

(a) 1/4 (b) 1/2

(c) 2/3 (d) 1/3

19. In the following pedigree , individuals with shaded circle or shaded square show presence of a recessive autosomal trait.

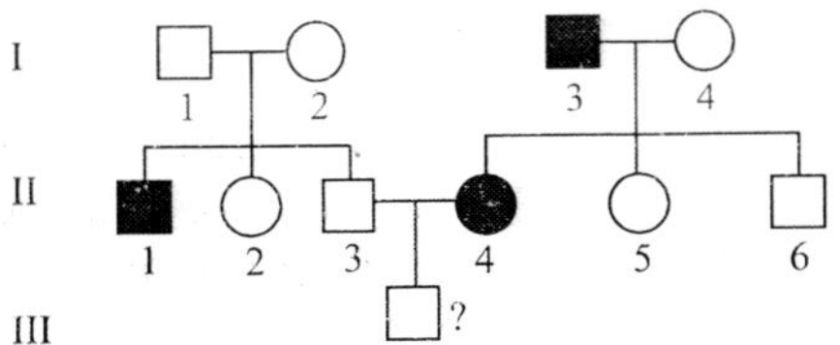

The calculated risk of occurrence of this trait for III-I is **(JUNE 2019)**

(a) 1/2 (b) 1/4

(c) 1/8 (d) 1/3

Answer Key

1.(d)	2.(a)	3.(c)	4.(a)	5.(c)	6.(b)	7.(b)
8.(d)	9.(c)	10.(b)	11.(a)	12.(a)	13.(c)	14.(c)
15.(a)	16.(d)	17.(a)	18.(c)	19.(d)		

14. Quantitative traits and heritability

1. Which one of the following statements is INCORRECT? **(JUNE 2016)**

(a) Quantitative inheritance results in a range of measurable phenotypes for a polygenic trait.

(b) Polygenic traits often demonstrate continuous variation.

(c) Certain alleles of quantitative trait loci (QTL) have an additive effect on the character/trait.

(d) Alleles governing quantitative traits do not segregate and assort independently.

2. A new trait was found to be highly variable in a population. It showed a bell-shaped distribution and is shown to be influenced by environmental factors. The trait can be **(DEC 2019 ASSAM)**

(a) a monogenic trait

(b) a polygenic trait

(c) a non-quantitative trait

(d) due to point mutation in the regulatory region of a gene

3. In an experiment, clones of a plant is grown in a field. The plants were observed to be of different heights. When a graph was plotted for frequency of plants (Y-axis) against different heights (X-axis), a bell-shaped curve was obtained. From the above, it can be concluded that the observed variation in height is due to **(JUNE 2015)**

(a) it being a polygenic trait
(b) environmental effect
(c) variation in genotype.
(d) influence of environment on different genotypes

4. Interacting genes which are involved in producing continuous variation in phenotypes in a population are known as/constitute **(DEC 2019)**

(a) codominant genes
(b) pseudogenes
(c) alleles
(d) QTLs

5. The total variance in a phenotypic character can be split into two components - genetic (V_G) and environmental (V_E). The heritability of a phenotypic trait can be expressed quantitatively as heritability coefficient (h^2) which is calculated as h^2= **(JUNE 2011)**

(a) V_G-V_E (b) V_E/V_G
(c) V_G/V_G+V_E (d) V_G/V_G-V_E

6. Two varieties of maize averaging 48 and 72 inches in height, respectively, are crossed. The F1 progeny is quite uniform averaging 60 inches in height. Of the 500 F_2 plants, the shortest 2 are 48 inches and the tallest 2 are 72 inches. What is the probable number of polygenes involved in this trait? (MODEL PAPER)

(a) Four. (b) Eight.
(c) Sixteen. (d) Thirty two

7. Two pure line of corn have mean cob length of 9 cm and 3 cm, respectively. The polygenes involved in this trait exhibit additive gene action. Crossing these two lines is expected to produce a progeny population with mean cob length of **(JUNE 2012)**

(a) 12.0 cm (b) 7.5 cm
(c) 6.0 cm (d) 2.75 cm

8. Two inbred lines of beans were intercrossed. In F_1 the variance in bean length was measured as 2.0. The F_1 was selfed to obtain F_2 and the variance in bean length in F_2 was 7.0. The broad heritability of bean length in the F_2 population will be:

(a) 0.75 (b) 9.0
(c) 5.0 (d) 0.71

9. A disease-resistant plant was crossed with a susceptible plant and the resultant F_1 plants were disease resistant. The F_1 plant was selfed and the F_2 individuals were analysed for qualitative and quantitative disease resistance. The following statements were hypothesized

A. Qualitative resistance follows Mendelian ratio.
B. In the F_2 individuals demonstrating qualitative resistance, "resistance" is dominant
C. Quantitative resistance is always monogenic
D. Qualitative resistance can be polygenic

Which one of the following combination of statements is correct? **(DEC 2019)**

(a) A, C and D (b) A, B, and C
(c) A, B and D (d) B, C and D

Answer Key

1.(d)	2.(b)	3.(b)	4.(d)	5.(c)	6.(a)	7.(c)
8.(d)	9.(c)					

15. Mutation : Types, causes and detection

1. Mutation in gene 'X' leads to lethality in a haploid organism. Which one of the following is best suited to analyse the function of gene 'X'? **(DEC 2015)**

(a) Pleiotropic mutants
(b) Temperature-sensitive mutants
(c) Recessive mutants
(d) Mutants with low penetrance

2. Aspartic acid (Asp) is specified by the codon GAU and GAC. After mutation, Asp is changed to Alanine represented by GCX, where X may be A, U, C and G. The reversion of the mutation could only be done with reactive oxygen species. The nature of mutation is considered to be (JUNE 2013)

(a) transition
(b) transversion
(c) depurination
(d) either transition or transversion

3. A gene encoding t-RNA undergoes a mutational event in its anticodon region that enables it to recognize a mutant nonsense codon and permit completion of translation. Such a mutation is known as (JUNE 2012)

(a) silent mutation
(b) neutral mutation
(c) reversion
(d) Non sense suppressor

4. A hypothetical gene encodes a protein with the following amino acid sequence:

Phe- Pro- Thr- Ala- Val-Arg- Ser

A mutation of single nucleotide alters the amino acid sequence to

Phe- Leu- Leu- Leu- Leu-

A second single nucleotide mutation occurs in same gene restoring back the amino acid sequence to the original. The following statements were made regarding the nature and location of the first mutation and that of the intragenic suppressor mutation:

A. The first mutation is a deletion in the second codon.
B. The first mutation is an insertion in the second codon.
C. The intragenic suppressor mutation is an insertion in the second codon
D. The intragenic suppressor mutation is a deletion in the third codon

Which combination of the above statements is correct? (JUNE 2014)

(a) A and C (b) A and D
(c) B and C (d) B and D

5. The effect of nonsense mutation could be nullified by reversion as well as suppression. Which of the following processes will help to distinguish between the two kinds of revertants? (DEC 2012)

(a) Complementation
(b) Transgenesis
(c) Test for allelism
(d) Recombination

6. To replace animal use in testing hepatic toxicity of a drug on trial, which one of the following would be used *in vitro* to be closest to the *in vivo* scenario? (JUNE 2012)

(a) Liver cells
(b) Hepatic cell lines
(c) Liver slices
(d) Co-culture of liver parenchymal cells and kupfer cells

7. Ames test is used to evaluate mutagens in the environment. Which of the following statements, about Ames test are true? (DEC 2013)

A. The mutagenic effect of a compound tested using an auxotrophic strain of *Salmonella typhimurium*
B. The mutagenic effect of a compound is tested using His strain of *Escherichia coli.*
C. Using appropriate strains, compounds causing base substitutions and frame shift
D. Liver enzymes are important as they are activated by test compound to evaluate its mutagenicity potential.
E. Many compounds may have to be converted to bioactive metabolites, which is carried out by the enzymes from the liver

(a) A, C and D (b) A, B and D
(c) A, C and E (d) A and E only

8. Which of the following mutagens is most likely to results in a single amino acid change in a gene product? (DEC 2016)

(a) Acridine orange (b) X-rays
(c) EMS (d) Ethidium bromide

9. 5-Bromouracil is a base analog that can cause mutation when incorporated into DNA. Which of the following is the most likely change that 5-Bromouracil induces: **(DEC 2014)**

(a) T:A to C:G (b) T:A to A:T .

(c) G:C to T:A (d) C:G to A:T

10. Which one of the following chemicals is a DNA intercalator? **(DEC 2015)**

(a) 5-Bromouracil

(b) Ethyl methane sulfonate

(c) Acridine orange

(d) UV

11. 2- Aminopurine induces mutation by **(DEC 2012)**

(a) Base pair change.

(b) Frameshift

(c) Duplication

(d) Deletion

12. Deamination of bases is a common chemical event that produces spontaneous mutation. Which one of the following bases will be formed by deamination of 5-methylcytosine? **(DEC 2018)**

(a) Uracil (b) Thymine

(c) Cytosine (d) Guanine

13. The base analog 2-aminopurine pairs with thymine, and can occasionally pair with cytosine. The type of mutation induced by 2-aminopurine is **(DEC 2011)**

(a) transversion (b) Transition.

(c) deletion (d) nonsense.

14. A protein has Gly at position 28 (wild type). On mutagenesis, a point mutation leads to conversion of Gly to Arg at position 28 (mutant 1). When mutant 1 is further mutagenized, four different point mutants (mutants 2 to 5) were isolated where the Arg at position 28 was mutated to Ile, Thr, Ser or Gly as represented below:

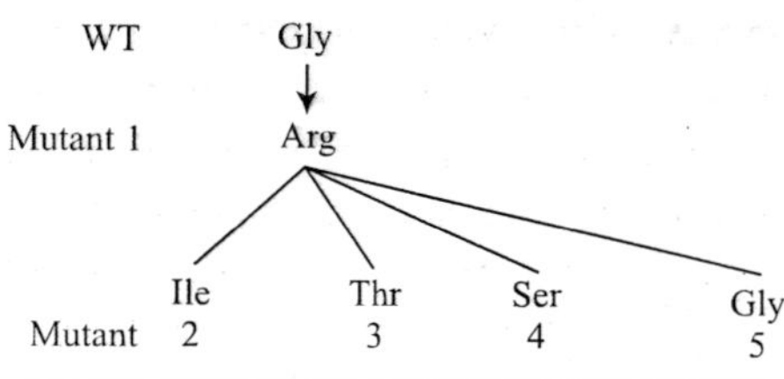

Gly	Thr	Ile	Ser	Arg
GGU	ACU	AUU	AGU	AGA
GGC	ACC	AUC	AGC	AGG
GGA	ACA	AUA		
GGG	ACG			

Based on the codons given in the table above, which one of the following codons codes for Gly at position 28 of the wild type protein? **(DEC 2019 ASSAM)**

(a) GGU (b) GGC

(c) GGA (d) GGG

15. The following is the amino acid sequence of a part of a protein encoded by gene 'X'.

.....Phe Leu Val Pro Ser Tyr Cys.....

A mutant for gene 'X' is isolated following treatment with a mutagen. The amino acid sequence of the same region encoded by the mutant gene is as follows:

.....Phe Leu Phe Arg Arg Ile.....

Which of the following mutagens is most likely to have been used? **(DEC 2014)**

(a) 5-bromouracil

(b) 2-amino-purine

(c) Ethyl methanesulfonate

(d) Acridine orange

16. Three *met E. coli* mutant strains were isolated. To study the nature of mutation these mutant strains were treated with mutagens EMS or proflavins and scored for revertants.

The results obtained are summarized below:

Mutant Strain	Mutagen treatment	
	EMS	Proflavin
A	–	+
B	+	–
C	–	–

(+ stands for revertants of the original mutants and – stands for no revertants obtained).

Based on the above and the typical mutagenic effects of EMS and proflavin, what was the nature of the original mutation in each strain? **(DEC 2016)**

(a) A-Transversion; B- Insertion or deletion of a single base; C- Deletion of multiple bases

(b) A-Transition; B- Transversion; C- Insertion or deletion of a single base

(c) A- Insertion or deletion of a single base; B- Transition; C- Deletion of multiple bases

(d) A-Transition; B- Insertion or deletion of a multiple bases; C- Transversion

17. A chemist synthesized three new chemical compounds, M1, M2 and M3. The compounds were tested for their mutagenic potential and were found to be highly mutagenic. Tests were made to characterize the nature of mutations by allowing the reversion with other mutagens. The following results were obtained:

Mutations Produced by	Reversed by			
	2- Amino Purine	Nitrous acid	Hydroxyl amine	Acridine orange
M1	NO	NO	NO	NO
M2	YES	YES	NO	NO
M3	NO	NO	NO	YES

Which one of the following conclusions drawn regarding the nature of mutations by the compounds is correct? **(DEC 2012)**

(a) M1- transversion, M2 - insertion, M3 - deletion

(b) M1- transition, M2- transversion, M3 - insertion

(c) M1 - insertion, M2 - transition, M3 - transverslion

(d) M1 - transversion, M2 - transition, M3 – insertion

18. A chemist synthesizes three new chemical compounds in the laboratory and names them as X, Y and Z. After analysing mutagenic potential of all these compounds, the geneticist observed that all are highly mutagenic. The geneticist also tested the potential of mutations induced by these compounds to be reversed by other known mutagens and obtained the following results **(JUNE 2018)**

Mutation produced by	**Reversed by**		
	Nitrous acid	**Hydroxyl amine**	**Acridine orange**
X	Yes	some	No
Y	No	No	No
Z	No	No	yes

Assuming that X, Y and Z caused any of the three types of mutations, transition, transversion or single base deletion, what conclusions can you make about the nature of mutations produced by these compounds?

(a) X causes transversion; Y causes transition; Z causes single base deletion

(b) X causes transition; Y causes transversion; Z causes single base deletion

(c) X causes transition; Y causes single base deletion; Z causes transversion

(d) X causes transversion; Y causes single base deletion; Z causes transition

19. Three *E. coli* mutants were isolated which require compound 'A' for their growth. The compounds B, C and D are known to be involved in biosynthetic pathway to A. In order to determine pathway, the mutants were grown in a minimal medium supplemented with ONE OF THE COMPOUNDS, A TO D. The results obtained are summarized below:

Mutant	**Medium supplemented with compound**			
	A	B	C	D
1	+	0	0	0
2	+	0	0	+
3	+	0	+	+

'+' = Growth on medium '0' = No growth

Which of the following equation represents the biosynthetic pathway of A? **(JUNE 2012)**

(a) B→C→D→A (b) C→D→B→A

(c) B→D→C→A (d) A→C→D→B

20. Several mutants (1-4) are isolated, all of which require compound E for growth. The compounds A to D in the biosynthetic pathway to E are known; but their order in the pathway is not known. Each compound is tested for its ability.to support .the growth of each mutant (l-4). In the following table, a plus sign indicates growth and a minus sign indicates no growth.

Mutants	Medium supplemented with compound				
	A	B	C	D	E
1	-	-	-	+	+
2	-	+	+	+	+
3	-	-	+	+	+
4	-	-	-	-	+

What is the order of the compounds (A to E) in the pathway? **(DEC 2018)**

(a) E → D → C → B → A

(b) A → C→ D→ B → E

(c) E→ B→ D→ C → A

(d) A→ B → C →D → E

21. Two auxotrophic strains of *E. coli:* A (met^- bio^- thr^+ leu^+ thi^+) and B (met^+ bio^+ thr^- leu^- thi^+) were incubated together for 18 hours in liquid medium and then ~10^8 cells were plated on minimal medium. Prototrophs were observed at frequency of 1 X 10^{-7} cells. This may have happened by process of genetic recombination between two strains or by mutation of strains. Which of the following control experiment would help rule out the possibility of mutation? **(JUNE 2012)**

(a) Plating strains A and B directly on minimal medium

(b) Growing the mixture of strain A and B for 18 hours and then plating on complete medium

(c) Growing strains A and B individually in a liquid complete medium for 18 hours and then plating on a minimal medium

(d) Growing the obtained prototroph in a liquid complete medium for 18 hours and then plating them on minimal medium

22. Somatic recombination was caused by mild exposure to radiation on flies heterozygous for a given allele during specific stages of development and the individuals were allowed to develop. Such individuals are likely to have

A. clones of homozygous cells in heterozygous body.

B. site specific mutagenesis.

C. twin spots, i.e., patches of mutants cells and homozygous wild type cells in heterozygous body.

D. tissue specific expression of the given allele.

Which of the following combination of answers will be most appropriate? **(JUNE 2015)**

(a) A and B (b) B and C

(c) A and C (d) C and D

Answer Key

1.(b)	**2.**(b)	**3.**(d)	**4.**(a)	**5.**(d)	**6.**(c)	**7.**(c)
8.(c)	**9.**(a)	**10.**(c)	**11.**(a)	**12.**(b)	**13.**(b)	**14.**(c)
15.(d)	**16.**(c)	**17.**(d)	**18.**(b)	**19.**(a)	**20.**(d)	**21.**(c)
22.(c)						

16. Structural alterations of chromosomes

1. Angelman syndrome (AS) and Prader-Willi Syndrome (PWS) have very distinct symptoms. Factors responsible for the occurrence of these syndromes are given below:

A. Microdeletion of 15q 11-13 in paternal chromosome.

B. Uniparental disomy of maternal chromosome 15.

C. Lack of functional maternal copy of ubiquitin ligase E3A.

D. Lack of SNURF-SNRPN transcript, which is produced only from paternal chromosome.

E. Deficiencies of small nucleolar RNAs, which are encoded from the introns of SNURF-SNRPN transcript from paternal chromosome.

Which of the following combination of answers is correct for Angelman and Prader- Willi Syndromes **(DEC 2018)**

(a) PWS - A, c, D; AS - B, E

(b) PWS -B only; AS - A, C, D E

(c) PWS - A, B, D, E; AS – C only

(d) PWS-A, B; AS-C,D,E

2. Inversions are considered as cross-over suppressors because: **(DEC 2016)**

(a) Homozygous inversions are lethal and thus they do not appear in next generation.

(b) Inversion heterozygotes, i.e., one copy having normal chromosome and its homologue having inversion, does not allow crossing over to occur as they cannot pair at all.

(c) Due to inversion present, four chromosomes take part in the pairing and crossing over events and make the structure difficult for separation and gamete formation.

(d) The pairing and crossing overs do occur in inversion heterozygotes but the gametes having cross over products are lethal.

3. The following diagram shows meiotic pairing in an inversion heterozygote and a point where single crossing over has occurred

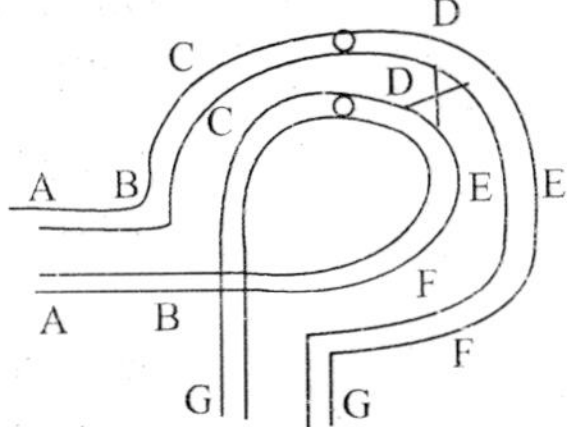

The resulting gametes produced may have

A. the chromosome having normal gene sequence

B. the chromosome having inverted gene sequence

C. a dicentric chromosome with duplication and deletion

D. an acentric chromosome having duplication and deletion

E. the chromosome having duplication and deletion

Which of the following combination will be most appropriate for the diagram shown: **(DEC 2015)**

(a) A, B, C and D (b) A, B and E

(c) B, C, D and E (d) A, C, D and E

4. Following is the picture of an inversion heterozygote undergoing a single crossing-over event

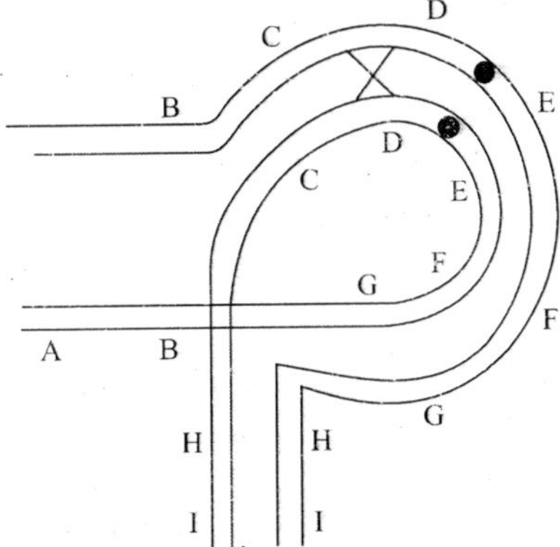

The following statements are given towards explaining the consequences at the end of meiosis.

A. The resultant two chromosomes will have deletions and duplications.

B. A dicentric and an acentric chromosome will be formed.

C. The inversion does not allow crossing over to occur, so even if a crossing over is initiated, it will fail to occur.

D. The crossing over is considered suppressed by inversion as the acentric chromosome will not segregate normally.

E. All the gametes formed with cross-over chromatids at the end of meiosis will be non-viable as they carry large deletion or duplication.

F. The gametes having non-crossover (parental) chromatid will survive.

Which combination of statements is correct?

(DEC 2018)

(a) B and E (b) A and C

(c) B, D and F (d) A, E and F

5. Following is the diagram of a paracentric inversion heterozygote ABCDEFG/ABFEDCG involved in recombination during meiosis I:

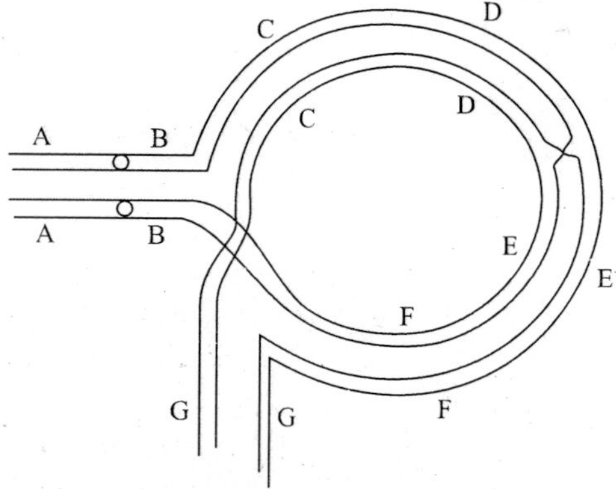

The consequence of this recombination will be the formation of

A. A dicentric and an acentric chromosome in meiosis I as the chiasmata gets terminated.

B. No dicentric or acentric chromosome but appearance of deletion and duplication in both the chromosomes.

C. All non-viable gametes.

D. Non-viable gametes from crossover products.

Which of the above statements are correct?

(DEC 2012)

(a) A and B (b) A and C

(c) A and D (d) B and C

6. An individual is having an inversion in heterozygous condition. The regions on normal chromosome are marked as A, B, C, D, E, F, G while the chromosome having inversion has the regions as a, b, e, d, c, f, g. The diagram given below shows pairing of these two homologous chromosomes during meiosis and the site of a crossing over is indicated:

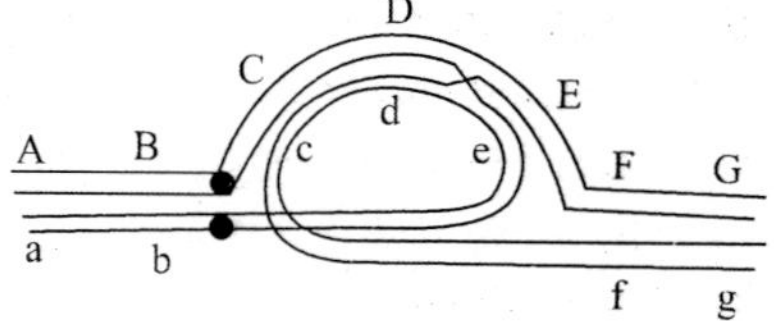

The following statements are given to describe the inversion and the consequence of crossing over shown in the above diagram:

A. This is a pericentric inversion

B. This will generate a dicentric and an acentric chromosome following separation of chromosomes after crossing over

C. This will generate two monocentric recombinant chromosomes following separation of chromosomes after crossing over

D. All the gametes thus formed will have deletion and/or duplication and will be non-viable

E. 50% of the gametes having recombinant chromatid will be non-viable, while 50% gametes having non-recombinant chromatid will survive

F. This is a paracentric inversion

Which combination of the above statements describe the inversion and meiotic consequences correctly? (JUNE 2018)

(a) A, B and C (b) A, C and E

(c) B, E and F (d) C, D and F

7. When gamete formation occurs in individuals heterozygous for an inversion, some of the following possibilities may occur:

A. A peri centric inversion heterozygote results in an acentric and a dicentric chromosome at the end of meiosis I

B. A paracentric inversion heterozygote also results in an acentric and a dicentric chromosome at the end of meiosis I

C. 25% of gametes formed in inversion heterozygotes are non-viable

D. Inversions are considered as cross-over suppressors because the gametes having a cross over product do not survive

Which combination of the above statements is correct to explain the meiotic consequence in inversion heterozygote? **(DEC 2013 GU)**

(a) A and D (b) B and C

(c) A and B (d) B and D

8. An individual is having a paracentric inversion (denoted by the region f-e-d, marked by arrows) in homozygous condition.

↓ ↓

a b c f e d g h i

The meiotic consequences of inversion can be:

A. generation of an acentric and a dicentric chromosome

B. the recombinants will have long deletion or duplication and may be lethal

C. the inversion will suppress crossing over

D. all gametes will have complete genome and will survive normally

Which of the above statement or their combinations will explain the meiotic consequence of the given inversion logically? **(DEC 2019)**

(a) A, B and C (b) A and B

(c) B and C (d) Only D

9. Which of the following illustrations explain the correct pairing preceding recombination between a chromosome (ABC•DEFG/ ABC•DEFG) and its inverted homologue (ABC•DGFE/ABC•DGFE). The dot in genotype represent the centromere. **(MODEL PAPER)**

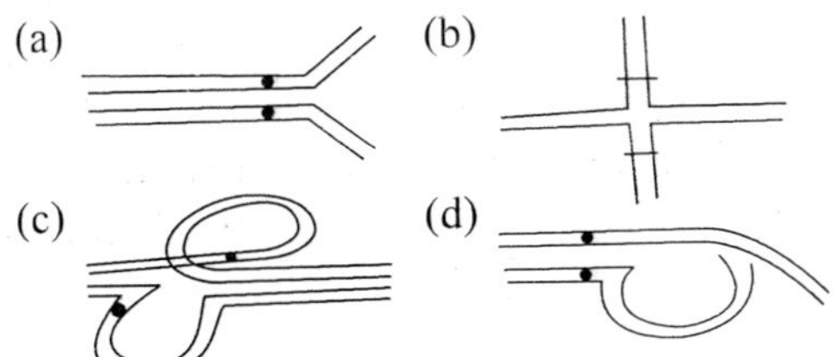

10. A mechanism that can cause a gene to move from one linkage group to another is **(JUNE 2011)**

(a) crossing over. (b) inversion.

(c) translocation. (d) duplication.

11. In Burkitt's Lymphoma a reciprocal translocation between chromosome 8 and 14 is observed. If an individual is heterozygous for this translocation, the consequence in meiosis will be as follows:

A. Four chromosomes, i.e., normal chromosome 8 and 14, and translocated chromosome 8 and 14, pair together

B. The two normal chromosomes 8 and 14, and-two translocated chromosomes pair separately

C. All gametes produced from this meiosis are non-viable as they have deletions and duplications

D. In one of the cross configurations called "alternate segregation" all gametes having normal or translocated chromosomes, survive

E. The gametes having normal chromosomes only survive while all gametes having translocated chromosomes are non-viable, hence the translocations are used as crossover suppressors

Which of the following combinations best describes the meiotic consequences for the translocation described above? **(DEC 2017)**

(a) B and E (b) B, C and E

(c) A, C and E (d) A and D

12. A chromosome aberration leads to change in the order of genes in a genetic map but does not alter its linkage group. This is due to **(JUNE 2013)**

(a) translocation. (b) recombination.

(c) transposition. (d) inversion.

13. A reciprocal translocation heterozygote at the end of meiosis I generate **(DEC 2019 ASSAM)**

(a) An acentric and a dicentric chromosome

(b) Viable gametes with deletions and duplications

(c) Viable gametes with only parental type chromosomes

(d) All non-viable gametes

14. The following statements were made regarding chromosome pairing (shown in the figure below) and subsequent segregation during meiosis-I in the reciprocal translocation heterozygote:

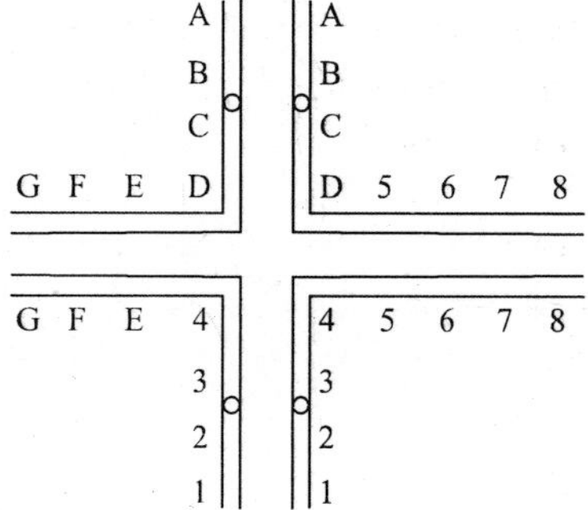

A. Three ways of segregation in Anaphase I would be: adjacent 1 (vertically in the above figure) adjacent 2 (horizontally) and alternate.

B. Gametes resulting from adjacent 1 and adjacent 2 segregation will be non-viable because of deletions and duplications of several genes.

C. All gametes resulting from alternate segregation will be viable as they will carry both normal chromosomes or both chromosomes having trans locations in the two poles, respectively.

D. A dicentric and an acentric chromosome will be generated following alternate segregation.

Which of the following combination of statements will most appropriately explain the consequence? (JUNE 2014)

(a) A, B and D (b) A, B and C

(c) Only A and B (d) Only A and C

15. Which of the following statements are true Robertsonian translocations?

A. The size oft h e non-homologous chromosome involved in translocation will differ.

B. Genes on the chromosome involved in translocation will show linkage with genes with which it normal independently assorts.

C. There will be change in the physical map, but no change in the genetic map.

D. It can be identified by G-banding chromosomes.

E. It can be identified by C-banding chromosomes.

F. It can lead to Down syndrome

Which one of the following combination is correct? **(DEC 2013)**

(a) A, C and D (b) A, D and F

(c) A, B, D and F (d) A, C, E and F

Answer Key

1.(c)	**2.**(d)	**3.**(b)	**4.**(d)	**5.**(c)	**6.**(c)	**7.**(d)
8.(d)	**9.**(a)	**10.**(c)	**11.**(d)	**12.**(d)	**13.**(c)	**14.**(b)
15.(c)						

17. Numerical changes and Polyploidy

1. If non-disjunction occurs in meiosis I, which of the following scenario is most likely to occur? **(DEC 2016)**

(a) Two gametes will be n+1 and two will be n-1

(b) One gametes will be n+1, two will be 'n' and one will be n-1

(c) Two gametes will be normal and two will be n-1

(d) Two gametes will be normal and two will be n+1

2. What kind of aneuploid gametes will be generated if meiotic non-disjunction occurs at first division? ('n' represents the haploid number of chromosomes) **(DEC 2011)**

(a) only n + 1 and n
(b) only n - 1 and n
(c) both n + 1 and n – 1
(d) either n + 1 or n – 1

3. A cell undergoing meiosis produces four daughter cells, two of which are aneuploids, while two are haploid. This can occur due to: **(JUNE 2012)**
(a) Non-dysjunction during first meiotic division only
(b) Non-dysjunctin during second meiotic division only
(c) Non-dysjunction during either first or second meiotic divisions
(d) Non-dysjunction during both first and second meiotic divisions

4. Turner syndrome results from **(DEC 2013 GU)**
(a) Monosomy of sex-chromosome.
(b) monosomy of autosome.
(c) nullisomy of sex-chromosome.
(d) trisomy of autosome.

5. Aneuploid females with only one X chromosome is a characteristic of individuals with **(JUNE 2011)**
(a) *Cri du chat* syndrome
(b) Klinefelter syndrome
(c) Down syndrome.
(d) Turner syndrome.

6. In a family, father is homozygous dominant (*AA)* for a gene A and his wife is homozygous for its recessive allele (*aa*) showing albino phenotype. It was surprising that their child showed the albino phenotype. Which of the following phenomenon can explain the phenotype? **(MODELPAPER)**
(a) Nondisjunction
(b) Uniparental Disomy
(c) Gene conversion
(d) All of the above

7. Autotetraploids arise by the doubling of 2n complement to 4n. There are three different pairing possibilities at meiosis in tetraploids as given below:
A. Two bivalents
B. One quadrivalent
C. One univalent + one trivalent
Which of the above pairings can lead to production of diploid? **(DEC 2014)**
(a) Only A (b) B and C
(c) A and C (d) A and B

8. Two near inbred parental lines P1 and P2 of an angiosperm species are crossed to produce F1 seeds in which, the ploidy of the endosperm is 6N. If plants generated from these F1 seeds are backcrossed with P1, what will be the ploidy of the somatic cells in the next generation? **(JUNE 2018)**
(a) 2N (b) 4N
(c) 5N (d) 6N

Answer Key

1.(a)	2.(c)	3.(b)	4.(a)	5.(d)	6.(d)	7.(d)
8.(b)						

18. Homologous and Site Specific Recombination

1. During an experiment, an investigator found that the cell line used are RecA $^{-/-}$. What could be the probable finding which led him to such observation? **(MODELPAPER)**
(a) Loss of recombination.
(b) Showing aberrant cell morphology.
(c) Cells developed phagocytic properly.
(d) Cells were aggregating.

2. Following statements have been made about recombination in a diploid organism:
A. Recombination could be identified by genotyping parents and offsprings for a pair of loci.

B. Recombination frequency does not exceed 0.5, and therefore, 50cM would be the maximum distance between two loci.

C. Recombination is a reciprocal process. However, a non-reciprocal exchange may cause gene conversion.

D. Occasionally non-homologous recombination happens and this functions as a source of chromosomal rearrangement.

Select the combination with all correct statements. **(DEC 2018)**

(a) A,B,C (b) A, B, D

(c) B,C,D (d) A, C, D

3. To generate a knock-in or a knock-out mouse, it is essential to use antibiotic selection to select the stem cells that have putatively integrated the construct accurately. Each construct will have 5 homology arm and 3' homology arm to help in homologus recombination. A positive antibiotic selection cassette and a negative antibiotic selection cassette are also used. Negative selection cassette helps in de-selection of nonhomologous recombinant stem cells.

Choose one statement indicated below that reflects the correct position of the negative selection cassette. **(MODELPAPER)**

(a) Negative selection cassette is supplied in a separate construct.

(b) Negative selection cassette is placed between the two homology arms.

(c) Negative selection cassette is placed 3' to the 3'-homology arm.

(d) Negative selection cassette is inserted between the promoter Exon I of positive selection cassette.

4. In an effort to produce gene knockout mice, a gene targeted homologous recombination was tried, with the exogenous. DNA containing *neo*^{r} gene (confer G-418 resistance) and tk^{Hsv} gene (confers sensitivity to the cytotoxic nucleotide analog ganciclovir). If the *neo*^{r} gene was inserted within the target gene in the exogenous DNA and considering that both homologous and non-homologous recombination (random integration) is taking place, which one of the following statements is NOT correct about the possible outcome of the experiment?

(JUNE 2016)

(a) Cells with non-homologous insertion will be sensitive to ganciclovir.

(b) Non-recombinant cells will be sensitive towards G-418 and resistant to ganciclovir.

(c) Homologous recombination will ensure that cells will be resistant to both ganciclovir and G-418.

(d) Homologous recombinants will grow in G-418 containing media but will be sensitive towards ganciclovir.

5. A student while constructing knock-out mice isolated mouse embryonic, stem cells and introduced an engineered DNA into the cells. However, none of the mice were transgenic. On checking the cells containing DNA construct, he found that he had made a mistake in constructing the DNA since the cells were resistant to gancyclovir but sensitive to G418. Which one of the following constructs had he

(JUNE 2015)

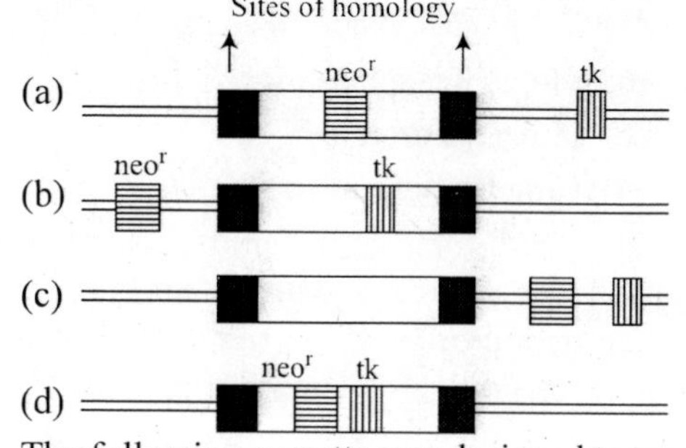

6. The following cassette was designed to create estrogen receptor knock-out mice:

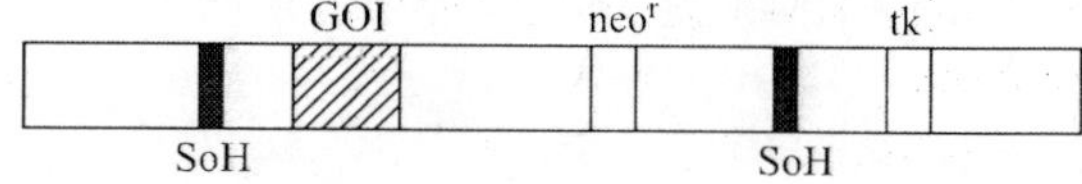

SoH: site of homology; GoI: gene of interest

What would ensure that proper recombination has taken place? **(DEC 2018)**

(a) Cells survive when cultured in presence of only G418

(b) Cells survive when cultured in presence of G418 followed by ganciclovir

(c) Cells die when cultured in presence of G418

(d) Cells survive when cultured with G418 and die when cultured with ganciclovir

7. In the following diagram, segments A and C are copies of 10 base pair repeat DNA sequences, flanking a unique stretch shown as B. A and C are in inverted orientation as indicated by arrows. Intramolecular recombination between A and C leads to which event: **(JUNE 2017)**

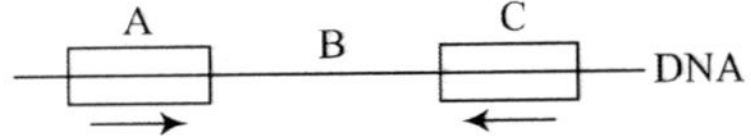

(a) The complete region encompassing A to C will be inverted

(b) Only A and B will be inverted

(c) Only B will be inverted

(d) Only regions A and C will be inverted

8. In transgenic mice, the orientation an location of the loxP sites determine whether *Cre* recombinase induces a deletion, an inversion or a chromosomal translocation. If a researcher wants to put loxP sites in such a manner that only inversion will take place, which one of the following construct best justifies their intension. Gene X is the target gene.

(JUNE 2015)

(a) Gene X — Lox P — Lox P

(b) lox P — Gene X — lox P

(c) lox P — Gene X — lox P

(d) Gene X — lox P — lox P

9. A student was asked to design a knockout cassette for specifically deleting the p53 gene from the prostate gland of mice. Which one of the following pairs of cassettes will ensure deletion of the gene? **(DEC 2014)**

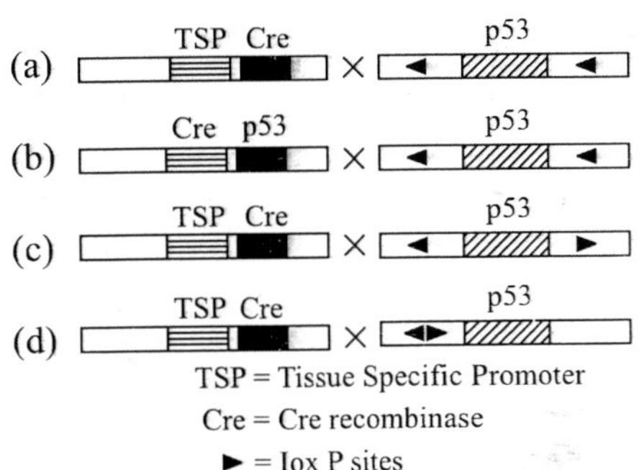

10. In a transgenic mice line, lox P sites are introduced in the target gene A in the following manner

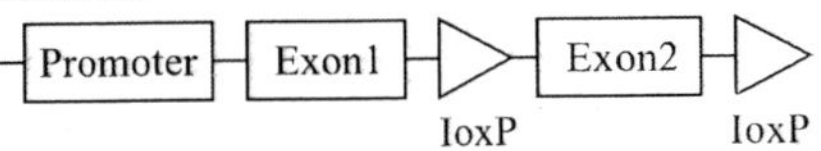

This transgenic mice line was mated with another transgenic mice line where Cre recombinase is expressed only in B cells.

What will be the expression profile of gene A in Cre/lox recombinant mice? **(JUNE 2014)**

(a) Gene will not be expressed in B cells, as orientation of exon 1 will be inverted by Cre.

(b) Gene will not be expressed in B cells, as exon 2 will be deleted by Cre.

(c) Gene will only be expressed in B cells of the recombinant mice where Cre removes the two lox P sites.

(d) Gene will not be expressed in B cells as orientation of exon 2 will be inverted.

11. In order to create an animal whose prostate gland is devoid of the p53 gene, engineered animals were created by cloning specially designed knock-out cassettes. Identify the most appropriate cassette combination for creating the knock out animal by breeding.

(DEC 2013 GU)

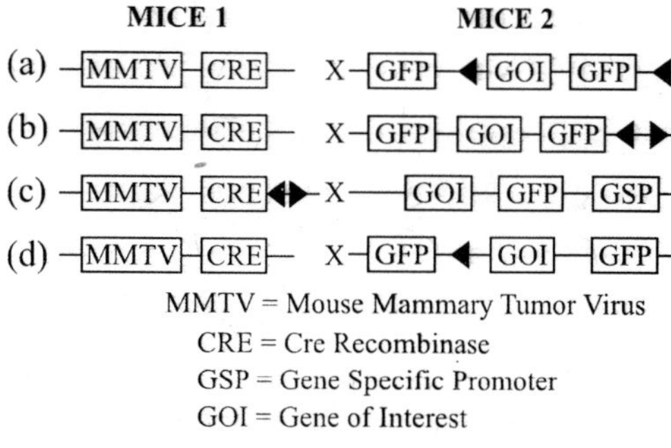

12. Site specific recombination results in precise DNA rearrangements, which is limited to specific sequences. The enzymes that are important to carry out the process are

(JUNE 2012)

(a) Restriction endonuclease and ligase

(b) nuclease and ligase

(c) DNA polymerase and ligase

(d) DNA polymerase and DNA gyrase

13. Cre/loxP system is used by phage P1 to remove terminally redundant sequences that arise during packaging of the phage DNA. Cre-lox system can be used to create targeted deletions, insertion and inversion in genomes of transgenic animals and plants. Consider a series of genetic markers A to K. How should the Lox P sites be positioned in order that Cre recombinase can create an inversion in the EFG segment relative to ABCD and HIJK?

(DEC 2011)

(a) A B C D ➔ E F G ➔ H I J K (Lox P, Lox P)

(b) A B C ➔ D E F G ➔ H I J K (Lox P, Lox P)

(c) A B C D ➔ E F G ← H I J K (Lox P, Lox P)

(d) A B C D ← E F G ← H I J K (Lox P, Lox P)

14. Construction of knockout mice may be performed using the Cre-LoxP system. Eventually, the Cre recombinase of the bacteriophage P1 mediates site-specific recombination at a 34 bp sequence, lox P. From the following statements, choose the INCORRECT event: **(DEC 2018)**

(a) The alteration of the chromosomal copy of the target gene requires a guide RNA.

(b) The *lox*P containing mice should not express *Cre* recombinase prior to mating.

(c) The *Cre* recombinase can be expressed by an inducible promoter.

(d) Induction of the promoter results in the expression of *Cre*, recombination at lox P sites and excision of the sequence in between.

Answer Key

1.(a)	2.(d)	3.(c)	4.(d)	5.(c)	6.(b)	7.(c)
8.(c)	9.(a)	10.(b)	11.(a)	12.(b)	13.(c)	14.(a)

SECTION-9

DIVERSITY IN LIFE FORMS

1. Principles of taxonomy and nomenclature

1. Which of the following is a correct hierarchical sequence for classification of a living organism? **(JUNE 2014)**
   (a) Domain-Kingdom-Phylum-Class-Order-Family-Genus-Species.
   (b) Kingdom-Domain-Phylum-Class-Order-Family-Genus-Species.
   (c) Domain-Kingdom-Phylum-Order-Class-Family-Genus-Species.
   (d) Kingdom-Domain-Phylum-Order-Class-Family-Genus-Species.
2. Which of following shows the correct systematic hierarchy? **(DEC 2013)**
   (a) Kingdom → Phylum → Subphylum → Superclass → Class → Subclass → Cohort → Order → Suborder → Super family→ Family → Subfamily → Genus → Subgenus → Species → Subspecies
   (b) Kingdom → Phylum → Subphylum → Cohort → Superclass → Class → Subclass → Superfamily → Family → Subfamily → Order → Suborder → Genus → Subgenus →Species → Subspecies
   (c) Kingdom → Phylum → Class → Order →Cohort → Family → Genus → Species
   (d) Kingdom → Phylum → Class → Cohort → Family → Order → Genus → Species
3. Which one of the following plants group combinations reflects the correct increasing order of the number of species it has? **(DEC 2013)**
   (a) Gymnosperms, Bryophytes, Algae, Angiosperms
   (b) Angiosperms, Algae, Fungi
   (c) Algae, Bryophytes, Gymnosperms, Angiosperms
   (d) Angiosperms, Gymnosperms, Bryophytes, Algae
4. Species that cannot be diagnosed as separate species based on their morphological characters alone, but can be distinguished by other means are known as **(DEC 2013 GU)**
   (a) typological species.
   (b) Linnaean species.
   (c) phylogenetic species.
   (d) sibling species.
5. Tautonym is an informal taxonomic designation used for animals referring to **(JUNE 2013)**
   (a) same name of genus and species
   (b) same name for species and subspecies
   (c) trinomial nomenclature
   (d) the name of the author for the species
6. As per the of International Code of Botanical Nomenclature, 2006 (Vienna Code), which of the following is a Nothospecies? (JUNE 2013)
   (a) *Polypodium vulgare* subsp. *prionodes* (Asch.) Rothm.
   (b) *Polypogon monspeliensis* (L.) Desf.
   (c) *Agrostis stolonifera* L.
   (d) Agrostis stolonifera L. x Polypogon monspeliensis (L.) Desf.
7. The 'Tribe" refers to a taxonomic group recognized between the **(DEC 2012)**
   (a) Genus and species
   (b) Family and genus
   (c) Order and family
   (d) Class and order
8. A plant species has been descrobe for the fisrt time by a author "x". Later, the species has been transferrred to some other genys by author "y". Then the author citation for the new combination will be **(DEC 2012)**
   (a) x *et* y
   (b) x ex y
   (c) (x) y
   (d) (y) x
9. Two new plant species, A and B, were described in 1872. Subsequently it was found that the type for species A was never designated and for species B there was one specimen designated as type but missing. As per International

Code of Botanical Nomenclature (ICBN), typification should be **(JUNE 2011)**

(a) neotype for A only.

(b) neotype for A and lectotype for B.

(c) neotypes for both A and B.

(d) lectotype for both A and B.

10. Assume a new subspecies *Ficus callosa* subsp. *microcarpa* has been published by Jacobs. The nomenclature of the resulting entities would be **(JUNE 2011)**

(a) *F callosa* and *F. callosa* subsp. *microcarpa* Jacobs

(b) *F. callosa* subsp. *microcarpa* Jacobs and other yet to be named subspecies of *F. callosa*

(c) *F. callosa* subsp. *callosa* Jacobs and *F.callosa* subsp. *microcarpa* Jacobs.

(d) F. callosa subsp. callosa and F.callosa subsp. microcarpa Jacobs.

11. Among the following statements, which is the correct one that refers to a Holotype? **(DEC 2017)**

(a) Specimens collected from different locations and designated as type specimens by the author

(b) A single specimen or illustration upon which name is based and designated as nomenclature type by the author

(c) Specimens collected from different places and desigrated as type specimen

(d) A single specimen designated to serve as nomenclatural type when all of the materials on which the name of the taxon was based is missing

12. Given below are some properties related to botanical and zoological nomenclature.

A. Absence of tautonyms

B. Presence of genus and species ranks only

C. Absence of principle of coordination

D. Presence of only holotype and neotype

Select the correct combination that distinguishes botanical nomenclature from zoological nomenclature system. **(JUNE 2018)**

(a) A, B and D (b) A, B and C

(c) A and C only (d) A, C and D

Answer Key

1.(a)	2.(a)	3.(a)	4.(d)	5.(a)	6.(d)	7.(b)
8.(c)	9.(b)	10.(d)	11.(b)	12.(c)		

2. Classical & quantititative methods of taxonomy

1. As a biologist you want to classify three taxa, A, B and C. You have the information of the three traits, p, q and r. The trait that is ancestral is counted "0" and the trait that is derived is counted as "1". The distribution of the trait found in the three taxa given below:

	A	B	C
p	1	1	0
q	1	1	0
r	0	1	1

Based on the above table the following cladograms were drawn:

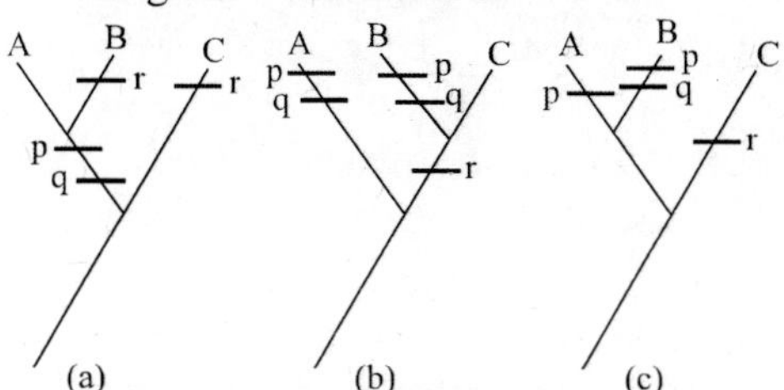

Based on the trait distribution and the principle of parsimony select the correct option: **(DEC 2016)**

(a) Both 'a' and 'b' cladograms are possible

(b) Only 'b' cladogram is possible

(c) Only 'c' cladogram is possible

(d) Only 'a' cladogram is possible

2. Cladistic classification is based on **(JUNE 2013)**
   (a) sequential order in which branches arise from a phylogenetic tree
   (b) the order of sequence divergence
   (c) morphological features and skeleton of individuals
   (d) cellular organization and cytoskeleton
3. Summary of characters between two taxa based on presence (l) and absence (0) data is given below:

		Taxon A	
		1	**0**
Taxon B	**1**	**45**	**10**
	0	**20**	**25**

Based on above, the Jaccard's coefficient will be **(DEC 2013 GU)**
   (a) 0.35. (b) 0.25.
   (c) 0.55. (d) 0.60
4. The following table shows the summary of characters between two taxa based on presence (a) and absence (0) data

		Taxon A	
		1	**0**
Taxon B	**1**	40	18
	0	22	20

Which of the following represents Jaccard's coefficient and Simple matching coefficient respectively? **(DEC 2012)**
   (a) 0.8, 0.5 (b) 0.6, 0.5
   (c) 0.8, 0.6 (d) 0.5, 0.6
5. Character similarity that can be misinterpreted as common descent is called **(DEC 2012)**
   (a) symplesiomorphy
   (b) synapomorphy
   (c) homology
   (d) homoplasy
6. Depicted below is a phylogenetic tree of selected taxa: **(JUNE 2019)**

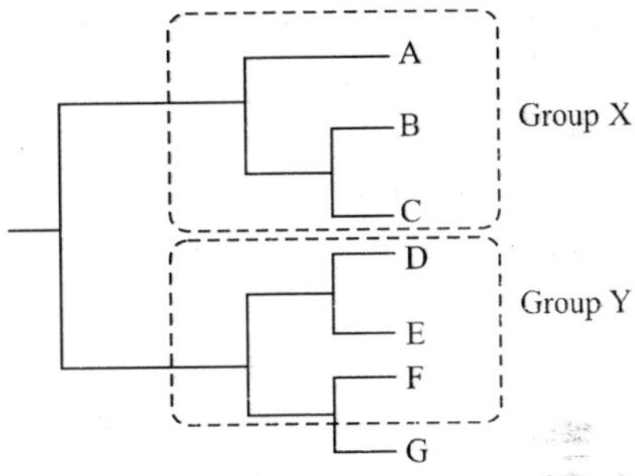

   (a) Group X is monophyletic and group Y is polyphyletic
   (b) Group X is paraphyletic and group Y is monophyletic
   (c) Both group X and Y are monophyletic
   (d) Group X is monophyletic and Group Y is paraphyletic
7. A paraphyletic group **(JUNE 2012)**
   (a) contains unrelated organisms
   (b) includes the most recent common ancestors but not all of its decedents
   (c) includes all the representatives of a clade but not the most recent common ancestor
   (d) contains all the representative of a clade and most recent common ancestor
8. Identify the most appropriate cladogram that can be constructed using the data matrix given below, assuming '0's are pleisomorphic and '1's are apomorphic characters. **(JUNE 2011)**

	1	2	3	4	5
A	0	0	0	0	0
B	0	1	1	0	0
C	0	1	0	0	0
D	0	1	1	0	1

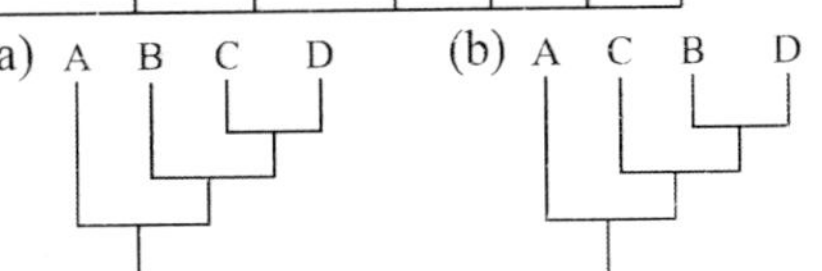

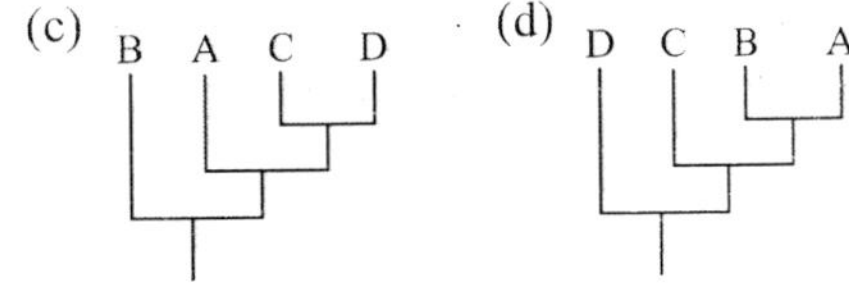

9. Based on Coefficient of Association table given below, which of the following taxonomic phenogram of relationship is correct?

(MODEL PAPER)

	S	T	W	X	Y	Z
S	1.0					
T	0.8	1.0				
W	0.5	0.7	1.0			
X	0.6	0.6	0.9	1.0		
Y	0.3	0.1	0.2	0.2	1.0	
Z	0.2	0.0	0.3	0.4	0.9	1.0

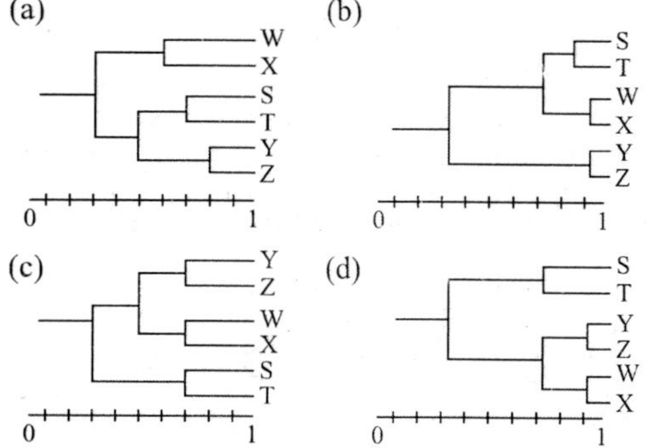

10. Which of the following phylogenetic trees appropriately uses principle of parsimony?

(DEC 2012)

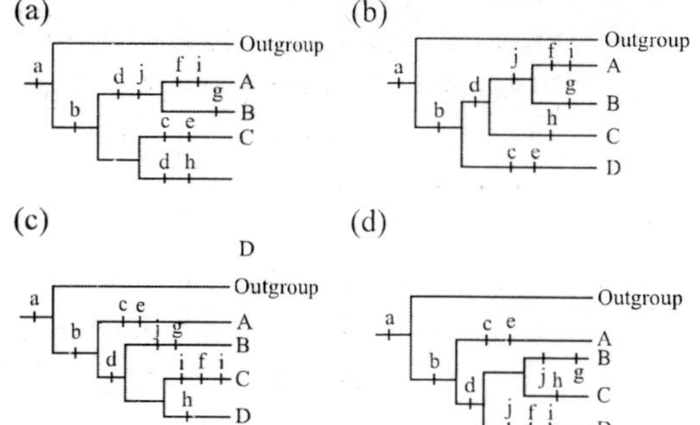

11. The phylogeny below shows evolutionary relationships between 9 extant bird species and whether they display red or blue plumage.

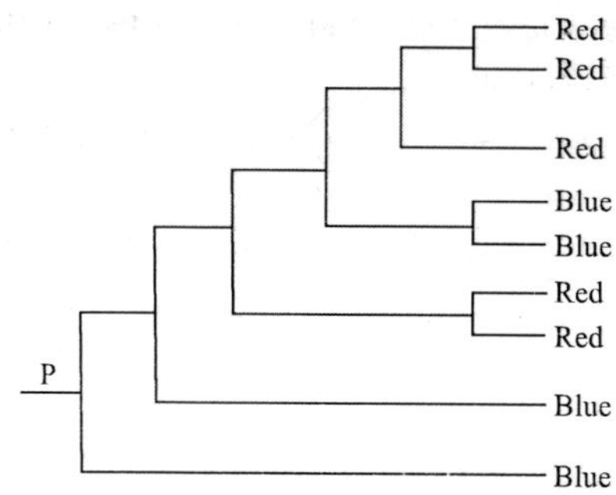

Based on the above phylogeny and the distribution of red and blue character states among the extant species, and using the principle of parsimony, which of the following is the correct inference about plumage colour of the ancestor at the root P? **(DEC 2019)**

(a) Ancestral state at P is blue.

(b) Ancestral state at P is red.

(c) Ancestral state at P is more likely to be red than blue

(d) Ancestral state at P is equally likely to be red or blue.

12. Which one of the following trees represents a fully resolved phylogram? (a) **(DEC 2017)**

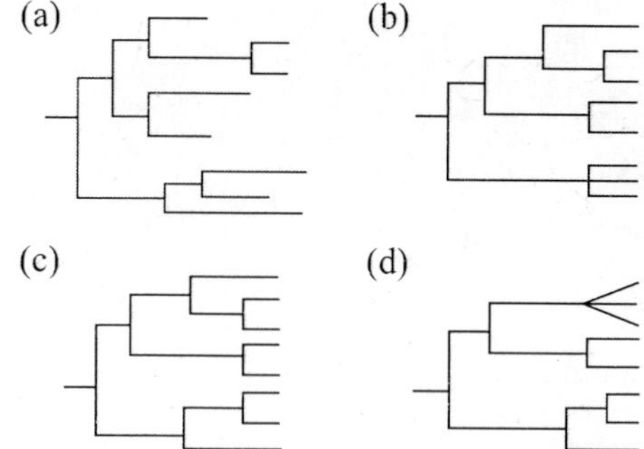

13. Identify 'a', 'b' and 'c' in the figure

(JUNE 2012)

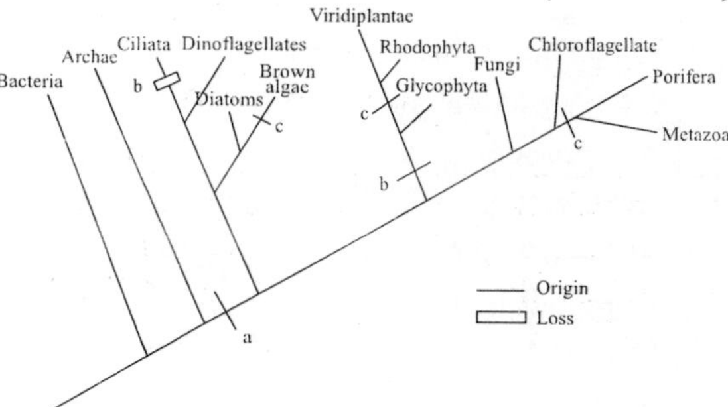

(a) a=mitochondria, b= multicellularity, c=chloroplast
(b) a= mitochondria, b= chloroplast, c=multicellularity
(c) a=chloroplast, b= multicellularity, c= mitochondria
(d) a= chloroplast, b= nucleus, c= multicellularity

Answer Key

1.(a)	2.(a)	3.(d)	4.(d)	5.(e)	6.(d)	7.(b)
8.(b)	9.(b)	10.(b)	11.(a)	12.(a)	13.(b)	

3. Criteria used for classification in animals

1. If you want to divide a human body into dorsal and ventral sections, what plane will you use? **(JUNE 2017)**
(a) Coronal (b) Abdomino-pelvic
(c) Transverse (d) Sagittal

2. In the following columns, certain terms and their descriptions are given in random order. **(JUNE 2017)**

	Column I		Column-II
A	Protostome	i	A fluid filled cavity lying inside the external body wall bathing the internal organs
B	Deuterostome	ii	Mouth forming from the blastopore
C	Pseudocoely	iii	Coelom formed by splitting the mesodermal tissue
D	Schizocoely	iv	Mouth forming from a second opening other than blastopore
E	Enterocoely	v	Coelom formed from pouches pinched off from the digestive tract

Which of the following combination gives correct match for the terms in column I from column II? **(JUNE 2017)**
(a) A-i; B-ii; C-iii; D-iv ; E-v
(b) A-ii; B-iv; C-i; D-v ; E-iii
(c) A-iv; B-ii; C-iv; D-I ; E-iii
(d) A-ii; B-iv; C-i; D- iii ; E-v

3. The following are some important features which are commonly associated with animal development:
A. Position of anus development with respect to blastopore
B. Method of cell division
C. Mechanism of coelom formation
D. Cleavage pattern during egg development
Based on the above, which one of the following combinations differentiate the development of deuterostomes from that of protostomes? **(DEC 2015)**
(a) A, B and C (b) B, C and D
(c) A, C and D (d) A and B

4. Which one of the following advancements in on the animal classification is correct? **(DEC 2013)**
(a) Protostomes → Pseudocoelomates → Deuterostomes → Eucoelomates
(b) Acoelomates → Protostomes →Eucoelomates → Deuterostomes
(c) Pseudocoelomates → Eucoelomates → Protostomes → Deuterostomes
(d) Protostomes → Deuterostomes → Acoelomates → Eucoelomates

5. Coelomates have
A. fluid filled body cavity.
B. a complete lining called peritoneum, derived from mesoderm covering the body cavity.
C. a complete lining called peritoneum, derived from ectoderm covering the body cavity.
D. Round worm as representative of this group.
E. Flat worm as representative of this group.

Select the correct combination **(DEC 2013)**

(a) A, C and D (b) A, C and E

(c) A, B and D (d) A and B

6. Schizocoelous coelom formation, mouth formation from embryonic blastopore, spiral and determinate cleavage are characteristic of **(DEC 2014)**

(a) deuterostomes (b) pseudocoelomates

(c) protists. (d) protostomes

7. Branchiostoma is a **(JUNE 2014)**

(a) deuterostome and schizocoelomate.

(b) protostome and schizocoelomate.

(c) deuterostome and enterocoelomate.

(d) protostome and enterocoelomate.

8. Cnidarians are **(DEC 2019)**

(a) triploblastic animals with bilateral symmetry.

(b) diploblastic animals with medusa as one of the basic body forms.

(c) monoblastic organisms with tube feet.

(d) asymmetric organisms with tentacles containing poison glands.

9. The schematic section given below of an animal indicates that the animal is **(JUNE 2012)**

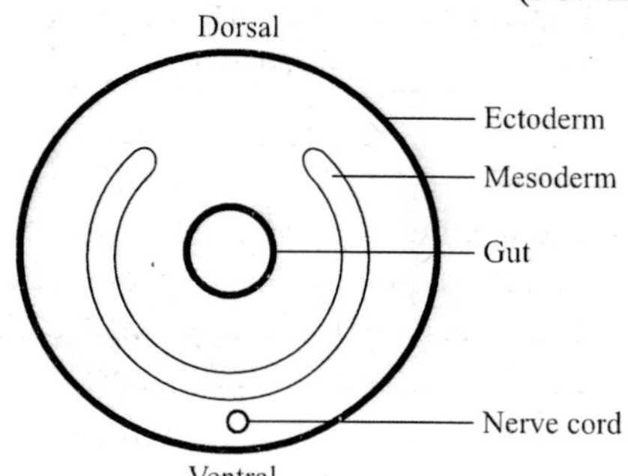

Based on above diagram the organism is

(a) triploblastic, coelomate and invertebrate

(b) triplobastic, acolemate and invertebrate

(c) diploblastic, coelomate and invertebrate

(d) triplobalstic, coelmate and vertebrate

10. Identify the correct match between the animal (flatworm, earthworm, roundworm) and its body cavity type (acoelomate, coelomate, pseudocoelomate): **(DEC 2015)**

(a) Roundworm - pseudocoelomate; Earthworm - acoelomate; Flatworm-coelomate

(b) Roundworm - acoelomate; Earthworm - coelomate; Flatworm-acoelomate

(c) Roundworm - pseudocoelomate; Earthworm - coelom ate; Flatworm-acoelomate

(d) Roundworm -coelomate; Earthworm-pseudocoelomate; Flatworm-acoelomate

11. Given below are some statements related to lower eumetazoans. Select the INCORRECT statement. **(DEC 2018)**

(a) Ctenophores are diploblastic with radial symmetry.

(b) Placozoans, with weakly differentiated tissue layers, are not diploblasts.

(c) Cnidarians are diploblastic with typically two stages in their life cycle

(d) Hydrozoans, a Cnidarian class, often have colonial polyps in their life cycle.

12. A marine biologist dug up a small animal from the ocean floor. The animal was uniformly segmented with short, stiff appendages and soft, flexible skin. It had a complete digestive system and an open circulatory system but no exoskeleton. Based on this description, the animal appears to be a **(JUNE 2013)**

(a) lancelet (b) roundworm

(c) mollusc. (d) crustacean

13. Match the five (A -E) group of organisms with their correct' taxonomic rank (i - v) given below: **(JUNE 2015)**

Group	Taxonomic rank
A. Crustacea	(i) Order
B. Hominidae	(ii) Domain
C. Dermaptera	(iii) Class
D. Ctenophora	(iv) Phylum
E. Archae	(v) Family

(a) A- (iii), B-(i), C- (v), D- (iv), E- (ii)
(b) A- (i), B-(ii), C- (iii), D- (iv), E- (v)
(c) A- (iv), B-(iii), C- (ii), D- (i), E- (v)
(d) A- (iii), B-(v), C- (i), D- (iv), E- (ii)

14. Which of the following is a correct match of the animal with its taxonomic group? **(DEC 2014)**

(a) Hirudinea-Leech; Chelicerata-Horse shoe crab; Cestoda-Tapeworm; Echinoidea-Sea Urchin; Cephalopoda-Octopus; Oligochaeta- Earthworm
(b) Hirudinea-Earthworm; Chelicerata-Horse shoe crab; Cestoda-Octopus; Echinoidea-Tapeworm; Cephalopoda-Earthworm; Oligochaeta- Leech
(c) Hirudinea-Tapeworm; Chelicerata- Leech; Cestoda-Tapeworm. Echinoidea-Horse shoe crab; Cephalopoda-Earthworm; Oligochaeta-Octopus
(d) Hirudinea-:Leech; Chelicerata- Tapeworm; Cestoda-Earthworm Echinoidea-Sea urchins; Cephalopoda-Octopus, Oligochaeta- Horse shoe crab

15. The organs 'radula' and 'clitellum' are found in **(JUNE 14)**

(a) Coelenterata and Echinodennata, respectively.
(b) Echinodennata and Coelenterata, respectively.
(c) Annelida and Mollusca, respectively.
(d) Mollusca and Annelida, respectively.

16. The animal belonging to phylum Onchophora **(JUNE 18)**

(a) havearthropodan characteristics and thus also considered as a class of Arthropoda
(b) have both annelidian and arthropodian characteristics
(c) have both arthroidian and molluscan characteristics
(d) serve as a connecting link between Annelida and mollusca

17. The list below includes names of animal phyla and classes.

A. Echinodermata B. Cephalopoda
C. Annelida D. Mollusca
E. Hirudinea F. Asteroidea
G. Arthropoda H. Crustacea.

For a leech and lobster, the correct classification of the of phylum and class, respectively, is **(JUNE 2014)**

(a) Leech: Phylum - D, Class - B; Lobster: Phylum- A, Class -H
(b) Leech: Phylum - C, Class - B; Lobster: Phylum- D, Class - C
(c) Leech: Phylum– C, Class – E; Lobster: Phylum- G, Class- H
(d) Leech: Phylum- A, Class - G; Lobster: Phylum - C, Class - F

18. In Group I are given 4 orders of class Insecta. Match each one with a common name (Group II) and its diagnostic characters (Group III) **(DEC 2014)**

Group I	Group II
Dermaptera (A)	Ant (E)
Ephemeroptera (B)	Mayfly (F)
Odonata (C)	Grasshopper (G)
Plecoptera (D)	Damselfly
	Stonefly (I)
	Earwig (J)

Group III

(i) Elongate, membranous wings with netlike venation, abdomen long and slender, compound eyes occupy most of head, hemimetabolous metamorphosis
(ii) Elongate chewing mouthparts, threadlike antennae, abdomen with unsegmented, forceps-like cerci, hemimetabolous metamorphosis
(iii) Forewing long, narrow and leathery, hind wing broad and membranous, chewing mouthparts, hemimetabolous metamorphosis

(iv) Elongate abdomen with two or three tail filaments, two pairs of membranous wings with many veins, forewings traiangular, short, bristle-like antennae, hemimetabolous metamorphosis

(v) Adults with reduced mouthparts, elongate antennae, long cerci, nymphs aquatic with gills, hemimetabolous metamorphosis

(vi) Wings membranous with few veins, well developed ovipositiors, sometimes modified into a sting, mouth parts modified for biting and lapping, holometabolous metamorphosis.

(a) A-J-(ii) (b) B-I-(vi)
(c) C-II-(v) (d) D-F-(i)

19. During a field study, three insects with the following characteristics were observed:

A: elongate, membranous wings with netlike venation, long and slender abdomen, large compound eyes

B: small bodied, sucking mouth parts, narrow wings fringed with setae

C: sclerotized forewings, membranous hindwings, chewing mouth parts

They can be identified to their respective orders as **(DEC 2011)**

(a) A - Orthoptera; B-Hemiptera; C-Coleoptera
(b) A - Odonata; B-Coleoptera; C-Hemiptera
(c) A - Orthoptera; B - Odonata; C-Coleoptera
(d) A - Odonata; B-Thysanoptera; C-Coleoptera

20. The table given below provides a list of groups of Arthropods (A-D) and some features (I -V):

A.	Onychophorans	(i)	Includes insects
B.	Trilobites	(ii)	Have cephalothorax and often pincer like appendage
C.	Hexapods	(iii)	Marine arthropods that disappeared in the Permian extinction
D.	Chelicerates	(iv)	Are considered related to arthropods but have unjoined appendages
		(v)	Only arthropod group without antennae

Which one of the following options represents the correct match between the arthropod groups with these features? **(DEC 2018)**

(a) A - (iv) ; B - (iii) ; C - (i) ; D - (v)
(b) A - (i) ; B - (iii) ; C - (iv) ; D - (ii)
(c) A- (iii) ; B - (iv) ; C -(ii) ; D - (i)
(d) A - (iv) ; B - (iii) ; C - (v) ; D - (ii)

21. Which of the following groups have only two wings? **(JUNE 2013)**

(a) Honey bee, beetle, ant
(b) Butterfly, housefly, fruitfly
(c) Dragonfly, butterfly, fruitfly
(d) Housefly, fruitfly, mosquito

22. Which one of the following show complete metamorphosis in all three orders? **(DEC 2019)**

(a) Coleopterans, Dipterans and Hymenopterans
(b) Coleopterans, Hymenopterans and Orthopterans
(c) Dipterans, Lepidopterans and Hemipterans
(d) Hymenopterans, Lepidopterans and Orthopterans

23. Horse shoe crabs belong to the group **(DEC 2011)**

(a) Onychophora (b) Chelicerata
(c) Uniramia (d) Crustacea

24. Some key characteristics of the four classes of phylum Mollusca are listed below:

A. They have two lateral (left and right) shells (valves) hinged together dorsally; they do not have distinct head or radula; they disperse from place to place largely as larvae.

B. They generally creep on their foot; the heads of most of this group have a pair of tentacles with eyes at the end; during embryological development, they undergo torsion

C. They have oval bodies with overlapping calcareous plates; underneath the plates, the body is not segmented; they creep along using a broad, flat foot surrounded by a groove or mantle cavity in which the gills are arranged.

D. They have highly developed nervous system; most members of this class have closed circulatory systems.

The correct match of the above characteristics with the classes of Mollusca is **(JUNE 2014)**

(a) A - Polyplacophora B – Bivalvia C - Gastropoda D - Cephalopoda

(b) A - Cephalopoda B – Polyplacophora C - Bivalvia D - Gastropoda

(c) A- Bivalvia B- Gastropoda C - Polyplacophora D- Cephalopoda

(d) A-Gastropoda B–Bivalvia C - Cephalopoda D – Polyplacophora

25. The terms expressing some of the developmental events or specific body structures are given in column X and the names of animals that are associated with them in column Y:

Column X		Column Y	
A	Torsion	i	Star Fish
B	Metagenesis	ii	Obelia
C	Apolysis	iii	Taenia
D	Pedicellaria	iv	Apple snail

The correct match of the terms in column X with the name of animals in column Y is: **(JUNE 2019)**

(a) A-i ,B-ii,C-iii,D-iv

(b) A-iv ,B-ii,C-iii,D-i

(c) A-iii ,B-i,C-ii,D-iv

(d) A-ii ,B-iv,C-i,D-iii

26. An organism has the following architectural pattern:

(i) multicellular with germ layers

(ii) a coelom derived from the mesoderm

(iii) primary bilateral symmetry with secondary radial symmetry

(iv) presence of endoskeletal plates

Such an organism is most likely to

(A) have mesohyl as its connective tissue.

(B) undergo torsion, whereby the mouth and anus are properly oriented.

(C) be devoid of a brain but have calcareous spicules.

(D) have comb plates to help in locomotion.

Which of the following is true? **(JUNE 2011)**

(a) A and C (b) C only

(c) D only (d) B and C

27. The character that is shared by Arthropoda and Annelida is **(DEC 2013 GU)**

(a) joint appendages.

(b) segmental arrangement of muscles.

(c) exoskeleton made up of chitin.

(d) open circulatory system.

28. Paripatus is an interesting living animal having unjointed legs, nephridia, haemocoel, trachea, dorsal tubular heart, claws, jaws, continuous muscle layers in body wall. This is considered as a connecting link between **(DEC 2016)**

(a) Nematoda and Annelida: continuous muscle layers in body wall, unjointed legs and nephridia being nematode character while haemocoel, trachea and dorsal tubular heart being annelid character.

(b) Annelida and Arthropoda: unjointed legs and nephridia being annelid characters while claws, jaws, haemocoel, trachea and dorsal tubular heart being arthropod characters.

(c) Arthropod and Mollusca: unjointed legs and nephridia being mollusca characters while claws, jaws, trachea, dorsal tubular heart being arthropod characters

(d) Nematoda and Arthropoda: contentious muscles layers, unjointed legs and nephridia being nematode characters while claws, jaws, haemocoel, trachea and dorsal tubular heart being arthropod characters

29. The names of some specialized organs/ structures are given in Column A and their location in organisms are given in Column B.

	Column A		Column B
A.	Crystalline style	i.	Stomach of bivalves
B.	Hastate plate	ii.	Stomach of crustaceans
C.	Organs of Bojanus	iii.	Kidney of gastropods
D.	Mehlis's gland	iv.	Female reproductive system of *Fasciola*

Select the correct matches:

(DEC 2019 ASSAM)

(a) A-ii, B-i, C-iv, D-iii
(b) A-iii, B-iv, C-ii, D-i
(c) A-i, B-ii, C-iii, D-iv
(d) A-iv, B-iii, C-i, D-ii

30. Given below is an evolutionary tree

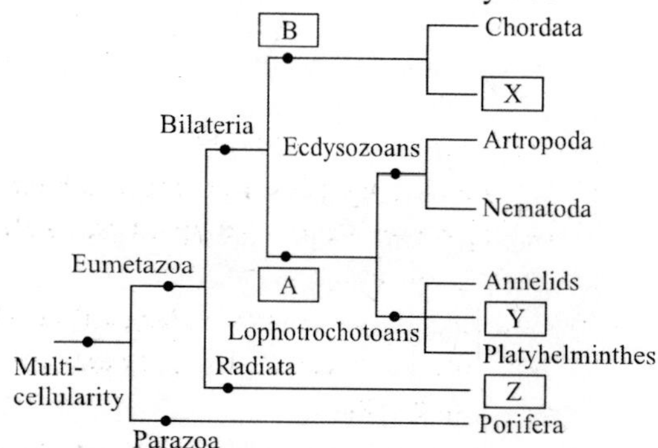

Based on the above, which one of the following combinations is correct? **(DEC 2013)**

(a) A - Protostome; B - Deuterostome; X - Mollusca; Y - Cnidaria; Z- Protozoa
(b) A- Protostome; B - Deuterostome; X - Echinodermata; Y - Mollusca; Z - Cnidaria
(c) A - Deuterostome; B - Protostome; X - Crustacea; Y - Mollusca; Z-Cnidaria
(d) A - Deuterostome; B - Protostome; X - Echinodermata Y- Roundworm; Z - Ctenophora

31. Three anatomical characteristics (A, B and C) of invertebrate nervous system are used to build a generalized cladogram given below. Presence of the anatomical character is indicated by '+' **(JUNE 2018)**

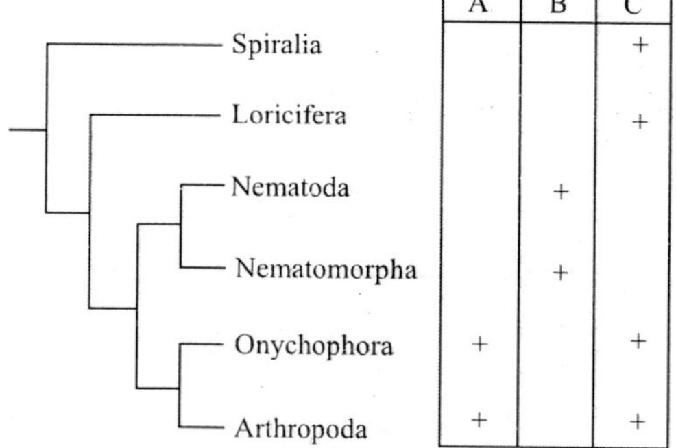

Based on the pattern of character distribution, pick the correct combination that are represented by A, B and C

(a) A-unpaired nerve cord, B-paired nerve cord, C-cephalic ganglia
(b) A-cephalic ganglia, B-unpaired nerve cord, C-paired nerve cord
(c) A-cephalic ganglia, B-paired nerve cord, C-unpaired nerve cord
(d) A-unpaired nerve cord, B-cephalic ganglia, C-paired nerve cord

32. The figure below shows the nervous system of Mollusca with ganglia and the connecting nerves. The connecting nerves are labelled as A, B, C and D.

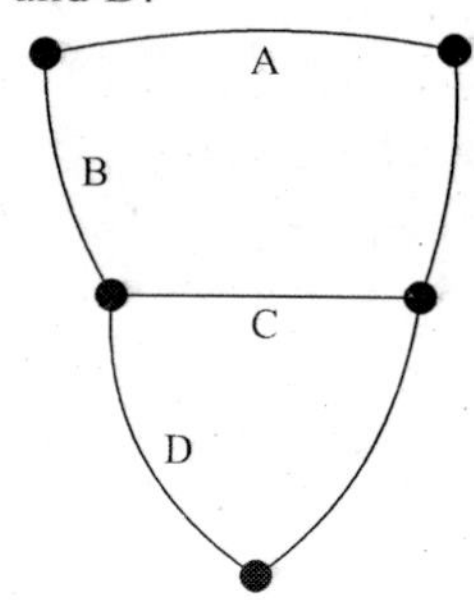

Which one of the following options has correct labelling of A, B, C and D? **(JUNE 2018)**

(a) A-Cerebral commissure; B-Left Cerebro-pedal connective; C-Pedal commissure; D-Left Pedal-visceral connective

(b) A-Cerebral connective; B-Left Cerebropedal commissure; C-Pedal connective; D-Left Pedal-visceral commissure

(c) A-Occipital commissure; B-Occipitopedal connective; C-Pedal commissure; D-Left Pedocaudal connective

(d) A-Cerebral connective; B-Left Cerebropedal commissure; C-Pedal commissure; D-Pedal-caudal connective

33. From among the five animals listed below, match the two attributes - amniotic egg and endothenny, with the correct animal(s): **(DEC 2011)**

(a) fish (b) frog
(c) crocodile (d) pigeon
(e) zebra

(a) Amniotic egg: b, c, d; Endothermy: d, e
(b) Amniotic egg: c, d, e; Endothermy: d, e
(c) Amniotic egg: a, b, c, d; Endothermy: c, d, e
(d) Amniotic egg: b, c, d; Endothermy: c, d, e

34. Which of the following is not a characteristic of phylum Chordata? **(JUNE 2011)**

(a) Pharyngeal slits (b) Amniotic egg
(c) Post anal tail (d) Notochord

35. Given below are some statements on vertebrates. Which one of the following statements is **INCORRECT**? **(JUNE 2017)**

(a) Muscular post-anal tail and pharyngeal slits are derived characters in vertebrates like notochord and dorsal hollow nerve cord

(b) Like echinoderms, vertebrates are deuterosomes

(c) Presence of two or more sets of HOX genes in living vertebrates distinguish them from cephalochordates and urochordates which have only one set

(d) Since adult hagfishes and lampreys lack Vertebral column, they are categorized outside class Vertebrata, but are retained under "chordata" along with Cephalochordates and urochordates.

36. As per the cladistic taxonomy, Archosaurs are a group of diapsid amniotes which include extinct dinosaurs. The living representatives of the group consist of **(JUNE 2017)**

(a) Anurans and Aves
(b) Aves and Crocodilia
(c) Aves and Agnatha
(d) Osteichthyes and Squamata

37. Identify the apomorphic characters marked in the cladogram: **(JUNE 2011)**

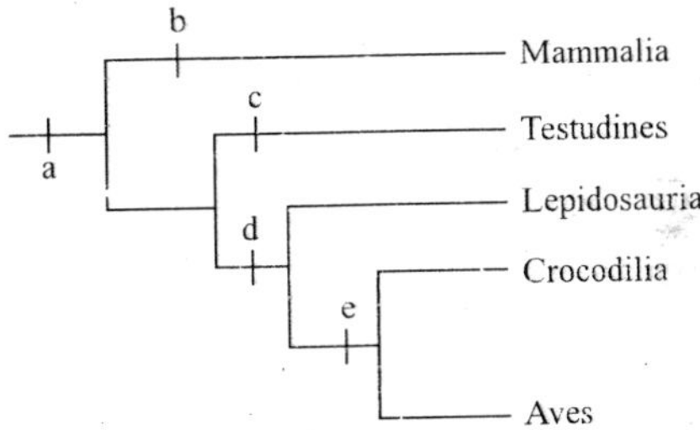

(a) a - amniotic egg; b - 4-chambered heart; c-anapsidian skull; d-diapsidian skull; e-synapsid skull.

(b) a - amniotic egg; b-synapsidan skull; c--4 -chambered heart; d. anapsidan skull; e-diapsidan skull.

(c) a - 4-chambered heart; b-synapsid skull; c-amniotic egg; d- diapsidan skull; e-anapsidan skull.

(d) a - amniotic egg; b - synapsidan skull; c-anapsidan skull, d. diapsidan skull; e-4-chambered heart.

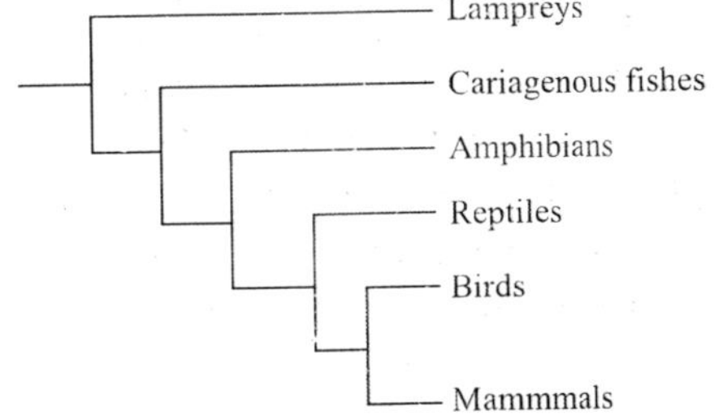

38. With reference to the phylogenetic tree resented above, which of the following statements is true? **(JUNE 2016)**

(a) Amphibians, reptiles, birds and mammals share a common ancestor.

(b) Birds are more closely related to reptiles than to mammals.

(c) Cartilagenous fishes are the ancestors of amphibians.

(d) Lampreys and mammals are not related.

39. The Phylogenetic tree of amniote vertebrates is given in diagram **(DEC 2015)**

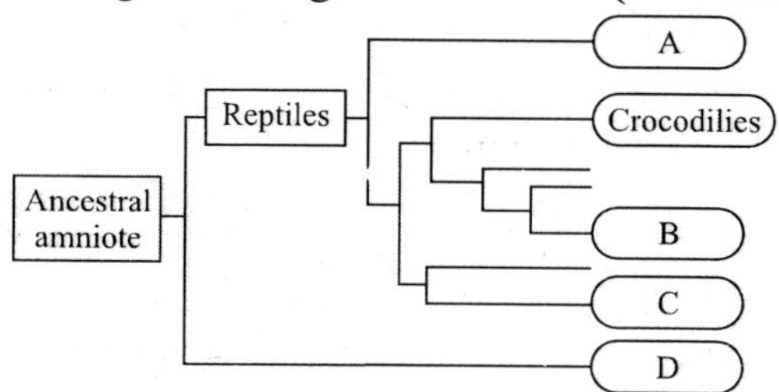

The groups labeled A, B, C, D are

(a) A- Snakes, B- Turtles, C-Birds, D-Mammals

(b) A-Snakes, B- Turtles, C-Mammals, D-Birds

(c) A-Turtles, B-Birds, C-Snakes, D-Mammals

(d) A-Birds, B-Turtles, C-Snakes, D-Mammals

40. Which one of the following phylogenies best represents the evolutionary relationship among whales, dolphins, seals, deer and dogs? **(DEC 2019)**

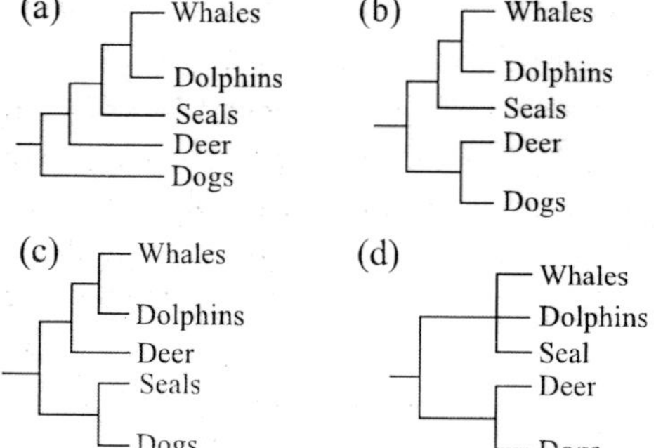

41. In which of the following classes of vertebrates there are group of animals without limbs? **(JUNE 2012)**

(a) Fish, reptiles and mammals

(b) Reptiles only

(c) Amphibians and Reptiles

(d) Amphibian only

42. Given below are names of the animals in column X and the accessory respiratory organs teleost fishes (Column Y).

Column X		Column Y	
A	Anabas	i	Labyrinithine organ
B	Clarias	ii	Air sacs
C	Amphipnous	iii	Supra branchial cavity
D	Channa	iv	Arborescent organ

The correct match of the animals with the accessory respiratory organs they have are: **(JUNE 2019)**

(a) A-iv ,B-iii,C-ii,D-i

(b) A-iii ,B-iv,C-i,D- i

(c) A-i ,B-iv,C-ii,D-iii

(d) A-ii ,B-i,C-iii,D-iv

43. The reptilian order Squamata includes **(MODEL PAPER)**

(a) crocodiles and alligators.

(b) the living fossil 'tuatara'.

(c) turtles and tortoises.

(d) snakes and lizards.

44. Which of the following describes the identification features of non-poisonous snakes? **(JUNE 2019)**

(a) Cylindrical tail and small belly scales

(b) Cylindrical tail, broad transverse belly scales and 4th infralabial scale is the largest

(c) Flat tail, broad transverse scales and 3rd supralabial scale touches eye and nose

(d) Cylindrical tail, broad transverse belly scales and a loreal pit between eye and nostril

45. Given below are certain adaptations which are seen in various groups of animals

A. Ovipary

B. Streamlined body

C. Pouch for carrying eggs
D. Porous egg shell
E. Breast bone as large keel
F. Webbed feet
G. Laterally compressed coccygeal bone
H. Unidirectional pulmonary system to provide large quantities of oxygen
I. Barbules or hooklets on the vanes of each feather

Which combination of the above adaptations facilitate bird flight? **(DEC 2019)**

(a) B, E, H, I (b) A, B, C, G
(c) D, F, G, I (d) B, D, E, F

46. Given below are the characteristics of a few mammalian order. Match the names of the animals with the characteristics of their orders:

A. Hooves with even number of toes on each foot, omnivorous
B. Teeth consisting of many thin tubes cemented together, eats ants and termites
C. Opposable thumbs, forward facing eyes, well developed cerebral cortex, omnivorous.
D. Hooves with an add number of toes on each foot; herbivorous

(i) Tapir (ii) Lemur
(iii) Aardvark (iv) Pig

Choose the correct combination **(JUNE 2015)**

(a) A- (iii) B- (iv) C- (i) D- (ii)
(b) A- (i) B- (iv) C- (ii) D- (iii)
(c) A- (iv) B- (iii) C- (ii) D- (i)
(d) A- (iv) B- (iii) C- (i) D- (ii)

47. During the course of evolution, the jawbones got modified into three ear ossicles in mammals. Which one of the following is a correct match of all three ear ossicles and their jawbones? **(DEC 2017)**

(a) stapes - articular; incus - quadrate; malleus - hyomandibular.
(b) stapes - quadrate; incus - articular; malleus - hyomandibular
(c) stapes- articular; incus - hyomandibular; malleus - quardate
(d) stapes - hyomandibular; incus - quadrate; malleus - articular

48. During the course of vertebrate evolution , the jaw bones got modified into three ear ossicles in mammals .Which one of the following is a correct match of ear ossicle and its ancestral jaw bone ? **(JUNE 2019)**

(a) Stapes–Articular; Incus-Hymoandibular; Malleus-Quadrate
(b) Stapes-Quadrate; Incus-Articular; Malleus-Hymoandibular
(c) Stapes–Quadrate; Incus-Hymoandibular; Malleus- Articular
(d) Stapes-Hymoandibular; Incus- Quadrate; Malleus- Articular

49. The following table shows names of bones (Column A) and specific features (Column B).

	COLUMN A		COLUMN B
A	Axis vertrebra	i	Deltoid Ridge
B	Humerus	ii	Acromion Process
C	Ulana	iii	Odontoid process
D	Pictorial girdle	iv	Sigmoid Notch

Which one of the following options gives the correct match of the bones with their specific features? **(JUNE 2018)**

(a) A - iii; B - i; C - iv; D – ii
(b) A - ii; B - i; C - iii; D - iv
(c) A - iv; B - ii; C - i; D - iii
(d) A - i; B - iii; C - iv; D – ii

50. Based on the type of excretion of nitrogenous waste, animals can be categorized as ammonotelic, ureotelic and uricotelic. Given below are combinations of groups of organisms and type of excretion. Select the correct combination. **(DEC 2018)**

(a) Poriferans, adult amphibians, cartilaginous fishes are ammonotelic.
(b) Ascaris, cockroaches, prawn are uricotelic.

(c) Paramecium, amphibian tadpoles, crocodiles are mainly ammonotelic.

(d) Humans, sharks and aquatic anurans are ureotelic

Answer Key

1.(a)	**2.**(d)	**3.**(c)	**4.**(c)	**5.**(d)	**6.**(d)	**7.**(c)
8.(b)	**9.**(a)	**10.**(c)	**11.**(1/2)	**12.**(b)	**13.**(d)	**14.**(a)
15.(d)	**16.**(b)	**17.**(c)	**18.**(a)	**19.**(d)	**20.**(a)	**21.**(d)
22.(a)	**23.**(b)	**24.**(c)	**25.**(b)	**26.**(b)	**27.**(b)	**28.**(b)
29.(c)	**30.**(b)	**31.**(b)	**32.**(a)	**33.**(b)	**34.**(b)	**35.**(d)
36.(b)	**37.**(d)	**38.**(a)	**39.**(c)	**40.**(b)	**41.**(c)	**42.**(c)
43.(d)	**44.**(a)	**45.**(a)	**46.**(c)	**47.**(d)	**48.**(d)	**49.**(a)
50.(*)						

4. Animal Larvae

1. Given below are larval stages and phylum to which they belong. Select the **INCORRECT** combination. **(DEC 2017)**

(a) Parenchyma and Amphiblastula - Phylum Porifera

(b) Bipinnaria and Auricularia- Phylum Echinodermata

(c) Tornasia and Axolotl- Phylum Hemichordata

(d) Planula and Ephyra- Phylum Cnidaria

2. Match the following larval forms with phyla that they occur in **(JUNE 2016)**

Larva		**Phylum**
(a)	Amphiblastula	(i) Mollusca
(b)	Nauplius	(ii) Echinodermata
(c)	Glochidium	(iii) Porifera
(d)	Bipinnaria	(iv) Arthropoda
		(v) Annelida

(a) a - iii, b - iv, c - i, d - ii

(b) a - iv, b - iii, c - i, d - v

(c) a-ii, b-v, c-iv, d-i

(d) a-ii, b-iv, c-v, d-i

3. For the following invertebrate structures/ organs, identify their major function and the animal group in which they are found:

Nematocyst (A), Protonephridia (B)

Malpighian Tubules (C) Radula (D)

(JUNE 2016)

(a) A - Porifera, Skeletal Support; B - Mollusca, excretion C --Insecta, respiration; D - Anthozoa, prey capture

(b) A - Anthozoa, prey capture; B -Planaria, excretion C -- Mollusca, excretion; D - Insecta, food processing

(c) A - Planaria, excretion; B- Mollusca, respiration; C - Insecta, respiration; D - Porifera, prey capture

(d) A - Anthozoa, prey capture; B-Planaria, excretion; C - Insecta, excretion; D - Mollusca, food processing

4. The following are matches made between adult animals and their larval forms: **(DEC 2015)**

A. Copepods- Nauplius

B. Sea cucumber- Zoea

C. Sea urchin- Echinopleuteus

D. crabs- Auricularia

E. Star fish - Bipinnaria

F. Brittle star- Ophiopleuteus

Which one of the combinations below reflects INCORRECT matches?

(a) A, C, E (b) B and D

(c) B only (d) F only

5. The phylum in which the animals are bilaterally symmetrical in the larval stage and radially symmetrical in the adult stage is (JUNE 2015)

(a) Coelenterata. (b) Nematoda.

(c) Mollusca. (d) Echinodermata.

6. Which of the following is a correct match of the animal with its attribute? **(DEC 2014)**

Animal	**Attributes**
A-Rotifer	(i) Naupliuas larva stage
B-Sea anemone	(ii) Radial symetry
C-Bamacle	(iii) Pseudocoelomate body cavity
D-Seaurchin	(iv) Water vascular system

(a) A- (iii), B-(ii), C- (i), D- (iv)
(b) A-(ii), B- (iii), C- (i),D- (iv)
(c) A- (iii), B- (iv), C -(i), D- (ii)
(d) A- (iv), B- (Hi), C- (ii), D- (i)

7. The special features of 4 different aquatic invertebrates (A, B, C, D) are given below. **(DEC 2013 GU)**

A. It is sessile as adult but free-swimming in larval stage.
B. It reproduces both sexually and asexually.
C. It undergoes a torsion of internal organs during development.
D. Its respiratory organs are called book lungs.

Which of the following names the 4 animals correctly?

(a) A-Barnacle, B-Daphnia, C-Snail, D-Scorpion
(b) A- Snail, B-Daphnia, C-Barnacle, D-Scorpion
(c) A-Daphnia, B-Scorpion, C-Snail, D-Barnacle
(d) A-Snail, B-Barnacle, C-Scorpion, D-Daphnia

8. Select the correct match of larval forms to the given sequence of animals- **(DEC 2013 GU)**
Molluscs, Cnidarians, Crustaceans, Starfishes and Sea urchin.

A. Glochidium, Planula, Zoea, Bipinnaria, Echinopluteus
B. Glochidium, Actinula, Zoea, Brachiolaria, Echinopluteus
C. Ammocoete, Megalopa, Maggot, Bipinnaria, Ophiopleuteus
D. Bipinnaria, Planula, Glochidium, Brachiolaria, Zoea

Which of the above order or combination orders of larval forms match with the given sequence of animals:

(a) B and C (b) A and B
(c) C and D (d) D and A

9. Four Cnidarians with the following characteristics were observed:

A. Asexual polyps and sexual medusae; solitary or colonial; both freshwater and marine.
B. Polyp stage reduced or absent, medusae with velum; solitary; all marine.
C. Polyp stage reduced, bell shaped medusae; solitary; all marine.
D. All polyps, no medusae; solitary or colonial; all marine.

They can be identified to their respective classes: **(DEC 2012)**

(a) A - Scyphozoa B - Anthozoa C - Cubozoa D - Hydrozoa
(b) A - Hydrozoa B - Scyphozoa C - Cubozoa D- Anthozoa
(c) A- Anthozoa B - Cubozoa C- Hydrozoa D - Scyphozoa
(d) A. Cubozoa B - Scyphozoa C - Anthozoa D - Hydrozoa

10. Bipinnaria and brachiolaria are the larval forms of **(JUNE 2018)**

(a) Crustacea
(b) Arthropoda and Mollusca, respectively
(c) Ophiuroidea and Holothuroidea, respectively
(d) Asteroidea

Answer Key

1.(c)	2.(a)	3.(d)	4.(b)	5.(1/4)	6.(a)	7.(a)
8.(b)	9.(b)	10.(d)				

5. Classification of plants

1. The following is a list of reproductive structures found in vascular and non-vascular plants.

A. Archegonia B. Megaspore
C. Capsule D. Fern frond
E. Pollen F. Corolla

Which of the following combinations represents structures primarily associated with the gametophytic life cycle of these plants?

(JUNE 2017)

(a) A, C, F (b) A, B, E

(c) B, D, E (d) C, D, F

2. Following diagram shows phylogeny of plants, fungi, and animals:

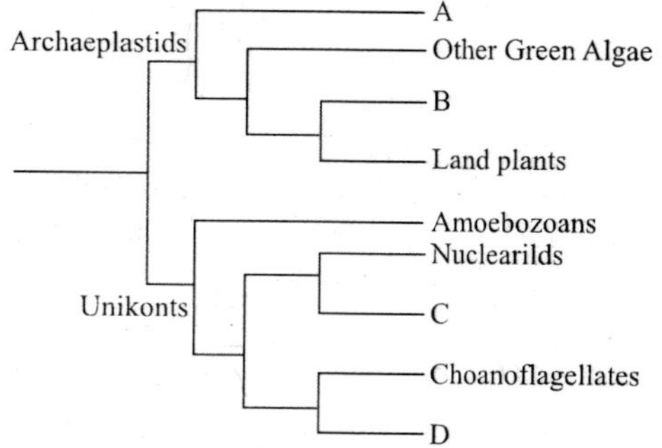

In the above diagram A, B, C and D represent; respectively; **(DEC 2019 ASSAM)**

(a) Fungi, Red algae, Charophytes, Animals

(b) Red algae, Charophytes, Animals, Fungi

(c) Green algae, Red algae, Animals, Fungi

(d) Red algae, Charophytes, Fungi, Animals

3. Which one of the following bryophyte has multicellular rhizoids and its cells mostly contain numerous chloroplasts?

(JUNE 2017)

(a) Anthoceros (b) Sphagnum

(c) Riccia (d) Marchantia

4. Most members of bryophyte phylum Anthocerophyta are characterized by

(DEC 2015)

(a) gametophyte with single chloroplast per cell and multicellular rhizoids; sporophyte without stomata.

(b) gametophyte with single chloroplast per cell and unicellular rhizoids; sporophyte with stomata

(c) gametophyte with multiple chloroplasts per cell and unicellular rhizoids; sporophyte without stomata.

(d) gametophyte with single chloroplast per cell and multicellular rhizoids; sporophyte with stomata.

5. Following table presents bryophyte phyla with their selected characteristics: **(JUNE 2018)**

Characters	Phylum		
	A	B	C
Gametophyte cells with numerous chloroplast	+	+	–
Gametophyte with multicellular rhizoid	+	–	–
Sporophyte body with stomata	+	–	+

\+ **Present; - Absent**

In the above table, phyla A, B and C represent

(a) A - Marchantiophyta, B - Bryophyta, C - Anthocerotophyta

(b) A - Bryophyta, B - Marchantiophyta, C – Anthocerotophyta

(c) A - Anthocerotophyta, B - Marchantiophyta, C - Bryophyta

(d) A - Bryophyta, B - Anthocerotophyta, C – Marchantiophyta

6. In transverse sections of a young stem, if vallecular canals and carinal canals are present, then the plant belongs to **(DEC 2011)**

(a) Lycopodiales (b) Isoetales

(c) Selaginellales (d) Equisetales

7. The following table gives vascular tissue characteristics of four divisions of Tracheophyta.

(JUNE 2013)

	Divisions	Vascular tissue characteristics
(A)	Psilophyta	i. Well-developed tracheid and pits in lateral wall
(B)	Lycopodiophyta	ii. Tracheids
(C)	Sphenophyta	iii. Tracheids, vessels and well-developed phloem
(D)	Pteridophyta	iv. Primitive tracheids and pits in lateral wall

Identify the correct combinations:

(a) (A) - i, (B) - ii, (C) - iii, (D) - iv

(b) (A) - ii, (B) - i, (C) - iv, (D) - iii

(c) (A) - iv, (B) - iii, (C) - ii, (D) - i
(d) (A) - iii, (B) - iv, (C) - i, (D) - ii

8. The table lists characteristic anatomical features and names of plants.

Anatomical features		Plants	
i	Prostele, xylem core surrounded by phloem.	A	Lycopodium
ii	Siphonostele, center pith present or modulated protostele	B	Marsilea rhizome
iii	Eustele, conjoint vasculature on edges of the path.	C	Selaginella species
		D	Equisetum

Choose the option that correctly matches plant with their characteristic features. **(JUNE 2019)**

(a) i-C,ii-B,iii-D. (b) i-A,ii-C,iii-D.
(c) i-C,ii-A,iii-B. (d) i-A,ii-B,iii-D.

9. Which one of the following gymnosperm phyla produces motile sperms, bears ovulate and microsporangiate cones on separate plants and has fleshy, coated seeds? **(DEC 2015)**

(a) Coniferophyta (b) Cycadophyta
(c) Ginkgophyta (d) Gnetophyta

10. Identify the characters shown in the diagram depicting phylogenetic relationships among major groups of ferns and fern allies. **(DEC 2012)**

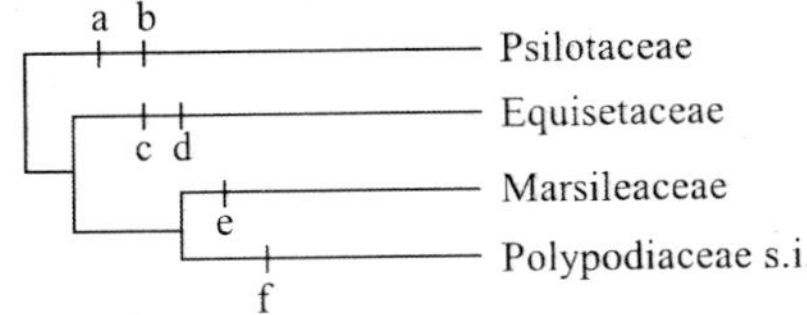

(a) a) Roots absent, b) Sporangiophores, c) Vertical, interrupted annulus, d) Heterospory, e) Leaves scale like, f) Elaters.

(b) a) Roots absent, b) Leaves scale like, c) Sporangiophores, d) Elaters, e) Heteroapory, f) Vertical, interrupted annulus

(c) a) Leaves scale like, b) Sporangiophores, c) Elaters, d) Heterospory, e) Roots absent, f) Vertical, interrupted annulus.

(d) a) Heterospory, b) Roots absent, c) Elaters, d) Sporangiophores, e) Leaves scale like, t) Vertical, interrupted annulus.

11. Which of the following is NOT an advantage to seed-based reproduction? **(JUNE 2013)**

(a) Reserve food material is provided for the developing embryo.

(b) Seed coat protects the embryo and allows it to remain dormant until favourable environmental conditions are available.

(c) The amount of energy spent per female gametophyte is less than that spent on making a spore.

(d) The female gametophyte remains on the sporophyte which provides protection and nourishment

12. Identify the synapomorphies in the following cladogram: **(DEC 2011)**

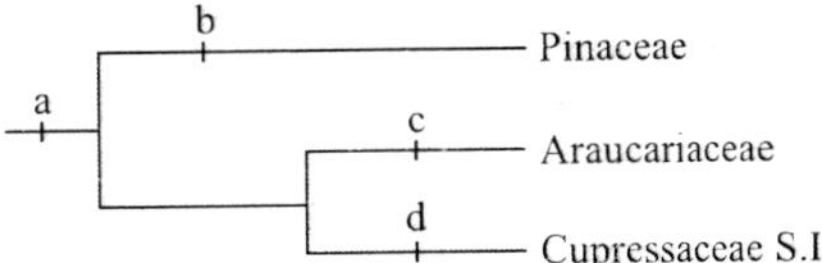

(a) (a) seeds with long terminal wing; (b) ovules 1-20 per scale; (c) resin canals; (d) 1 ovule per scale

(b) (a) resin canals; (b) seeds with long terminal wing; (c) 1 ovule per scale; (d) ovules 120 per scale

(c) (a) resin canals; (b) ovules 1-20 per scale; (c) seeds with long terminal wing; (d) 1 ovule per scale .

(d) (a) seeds with long terminal wing; (b) ovules 1-20 per scale; (c) 1 ovule per scale; (d) resin canals

13. Following is a diagram showing phylogenetic relationship among plant taxa. **(DEC 2013 GU)**

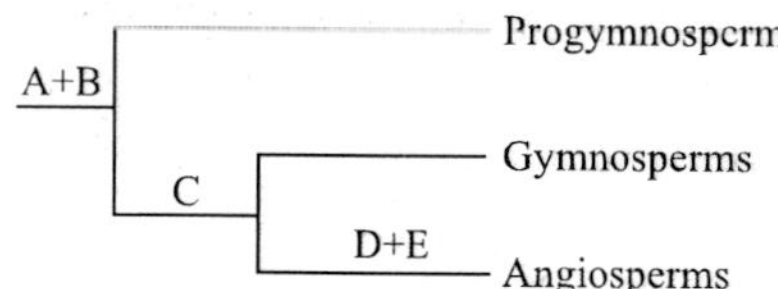

In the above diagram, symbols A, B, C, D and E represent:

(a) A=Cambium, B= Carpel, C = Euphyll D= Seed (ovule), E= Double Fertilization

(b) A=Euphyll, B= Cambium, C = Seed (ovule) D= Carpel, E= Double Fertilization

(c) A= Carpel, B= Euphyll, C = Cambium D= Double Fertilization E= Seed (ovule)

(d) A= Carpel, B= Double Fertilization, C = Euphyll D=Seed (ovule), E= Cambium

14. Following is a cladogram showing phylogenetic relationships among a group of plants;

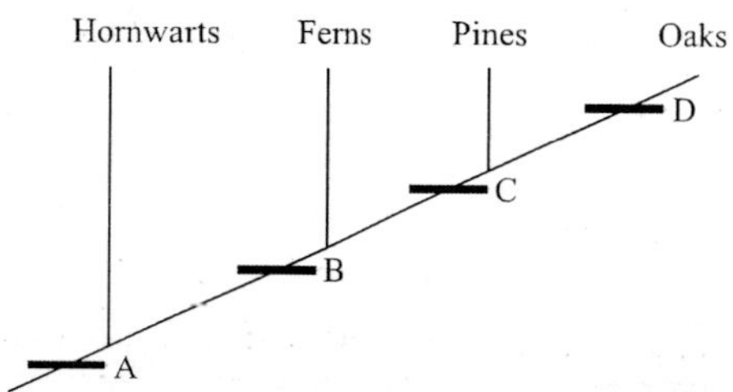

In the above representation. A. B, C and D respectively represent **(JUNE 2016)**

(a) xylem and phloem, embryo. Flower, seed.

(b) embryo. xylem and phloem, seed, flowers.

(c) embryo. xylem and phloem. Flower, seed.

(d) xylem and phloem, flower, embryo, seed.

15. The following table shows selected characters used in analyzing the phylogenetic relationships of four plant taxa:

Taxon	**Characters**			
	Xylem or Phloem	Wood	Seed	Flowers
T1	+	–	–	–
T2	+	+	+	+
T3	+	+	+	-
T4	-	-	-	-

Taxa T1, T2, T3 and T4 are respectively: **(JUNE 2015)**

(a) Ferns, Oaks, Pines, Hornworts

(b) Oaks, Pines, Hornworts, Ferns

(c) Hornworts, Pines, Oaks, Ferns

(d) Ferns, Pines, Oaks, Hornworts

16. Following table shows presence (+) and absence (-) of selected distinguishing characters of different plant taxa:

	CHARACTERS			
TAXON	Xylem and Phloem	Woods	Flowers	Seeds
A	+	–	–	–
B	+	+	+	+
C	–	–	–	–
D	+	+	–	+

Based on the above, which of the following shows correct identity of taxa A, B, C and D? **(DEC 2018)**

(a) A - Homworts; B - Oaks; C - Ferns; D-Pines

(b) A - Fems; B - Oaks; C - Hornworts; D-Pines

(c) A - Homworts; B - Pines; C - Ferns; D-0aks

(d) A - Fems; B - Pines; C - Hornworts; D-0aks

17. Which one of the following plants has this combination of key plant traits: sporophyte dominant in the lifecycle, vascular tissue, lack of seeds? **(DEC 2019)**

(a) Mosses (b) Ferns

(c) Cycads (d) Monocots

18. Which of the following options match the plant tissue type with its correct function in vascular plants? **(JUNE 2015)**

Tissue	**Function**
A. Tracheids	(i) Chief water conducting elements in gymnosperms
B. Vessel elements	(ii) Chief water conducting elements in angiosperms
C. Sieve tube elements	(iii) Food conducting elements in gymnosperms
D. Sieve cell	(iv) Food conducting elements in angiosperms

(a) A-(i) B-(ii) C-(iv) D-(iii)
(b) A-(ii) B-(i) C-(iii) D-(iv)
(c) A-(i) B-(ii) C-(iii) D-(iv)
(d) A-(i) B-(iii)C-(iv) D-(ii)

19. In order to survive in a non-aquatic environment, plants acquired several adaptations with specialized functions. Given below is a list of features/characteristics (Column A) and their potential role (Column B).

Column A		Column A	
A	Waxy cutical	(i)	Mechanical support
B	Thickened or lignified cell walls	(ii)	Protection against excess light
C	Hopmoiohydry	(iii)	Restrict water loss
D	Pigmentation	(iv)	Vascular system

Which one of the following options represents a correct match between the adaptations and their functions? **(JUNE 2018)**
(a) A - (iv), B - (ii), C - (i), D - (iii)
(b) A - (iii), B - (i), C - (iv), D - (ii)
(c) A - (ii), B - (iii), C - (ii), D - (i)
(d) A - (i), B - (iv), C - (iii), D - (iv)

20. Which of the following characteristics make *Amborella* the most basal living angiosperm? **(DEC 2014)**
(a) Carpels fused by tissue connection and absence of vessel elements
(b) Absence of carpels and presence of vessel elements
(c) Carpels free and presence of vessel elements
(d) Presence of carpels and absence of vessel elements

21. The species which is considered sister to all extant angiosperms is **(DEC 2013 GU)**
(a) Amborellaceae. (b) Ceratophyllaceae.
(c) Magnoliaceae. (d) Hydatellaceae.

22. Following is a cladogram of the major taxonomic groups of the angiosperms: **(DEC 2015)**

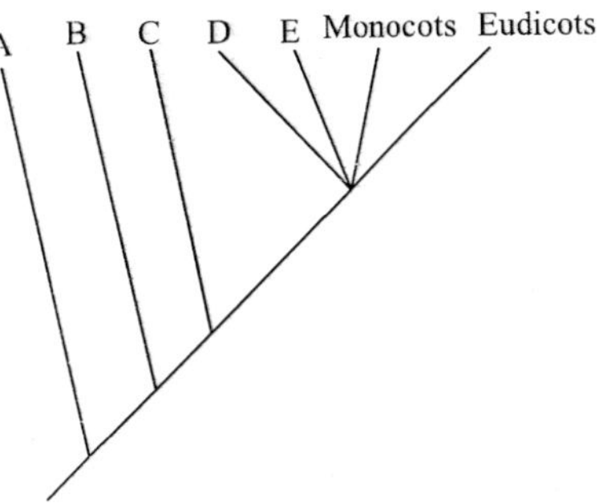

Groups A-E represent respectively:
(a) *Astrobaileyales, Nymphaedales, Amborellales, Chloranthaceae, Magnoliids*
(b) *Amborellales. Astrobaileyales, Nymphaedales, Magnoliids, Chloranthaceae*
(c) *Amborellales, Nymphaedales, Astrobaileyales, Chloranthaceae, Magnoliids*
(d) *Amborellales, Nymphaedales, Chloranthaceae, Magnoliids, Astrobaileyales*

23. Basal angiosperms are NOT represented by the members of: **(DEC 2018)**
(a) Chloranthales (b) Nymphaeales
(c) Austrobaileyales (d) Amborellales

24. Following table shows a list of clades and plants:

Clades	Plants
A. Basal angiosperms	(i) Black pepper
B. Magnolides	(ii) Orchid
C. Monocots	(iii) Star anise
D. Eudicots	(iv) Strawberry

Which one of the following is a correct match for the above? **(JUNE 2018)**
(a) A - (iii), B - (iv), C - (ii), D-(i)
(b) A - (ii), B - (i), C-(iii), D-(iv)
(c) A - (ii), B - (iv), C - (iii), D-(i)
(d) A - (iii), B - (i), C-(ii), D-(iv)

25. The table given below provides a list of female gametophyte features and plant genera

Female Gametophyte		Plant genera	
(i)	Monosporic, 8 nucleate	A	*Allium*
(ii)	Monosporic, 4 nucleate	B	*Oenothera*
(iii)	Bisporic, 8 nucleate	C	*Peperomia*
(iv)	Tetrasporic, 16 nucleate	D	*Polygonum*

Which one of the following options correctly matches the plant genera to female gametophyte features? **(JUNE 2019)**

(a) i-D; ii-C; iii-A; iv-B

(b) i-D; ii-B; iii-A; iv-C

(c) i-A; ii-B; iii-D; iv-C

(d) i-D; ii-B; iii-C; iv-A

26. Which one of the following plants has a bisporic, 8-nucleate, bipolar embryo sac development? **(DEC 2019)**

(a) *Oenothera* (b) *Penaea*

(c) *Plumbago* (d) *Allium*

27. Match the two columns that represent plant organs (I) and parts within these organs (II). **(DEC 2017)**

	Column I		Column II
A	Carpel	(i)	Petals, corolla
B	Perianth	(ii)	Vegetative cell, generative cell
C	Microsporocyte	(iii)	Stigma, style, ovary
D	Megasporocyte	(iv)	Antipodal cells, polar nuclei. synergid cells

(a) A-(i); B-(iii); C-(ii); D-(iv)

(b) A-(iii); B-(iv); C-(i); D-(ii)

(c) A-(i); B-(iii); C-(iv); D-(ii)

(d) A-(iii); B-(i); C-(ii); D-(iv)

28. Which of the following is NOT true for monocots? **(DEC 2016)**

(a) Sieve tube members with companion cells

(b) Vasculature atactostelic

(c) Tricolpate pollen

(d) Vascular cambium absent.

29. Which of the following statements is NOT correct? **(DEC 2014)**

(a) Stomata are present in mosses and hornworts but absent in liverworts.

(b) Only the lycophytes have microphylls and almost all other vascular plants have megaphylls.

(c) Monocot pollen grains have three openings whereas eudicot pollen grains have one opening

(d) Monocots have fibrous root system whereas eudicots have taproot.

30. Following are some generalizations related to wood anatomy of higher plants:

A. The axial system of conifer woods consist mainly or entirely of tracheids.

B. The rays of conifers typically contain only parenchyma cells.

C. The rays of angiosperms typically contain both sclerenchyma cells and tracheids.

D. Angiosperm wood may be either diffuse porous or ring-porous.

Which one of the following options represents all correct statements? **(JUNE 2019)**

(a) A and B only (b) A and D only

(c) B and C only (d) C and D only

31. The following schematic diagram represents secondary growth in angiosperms

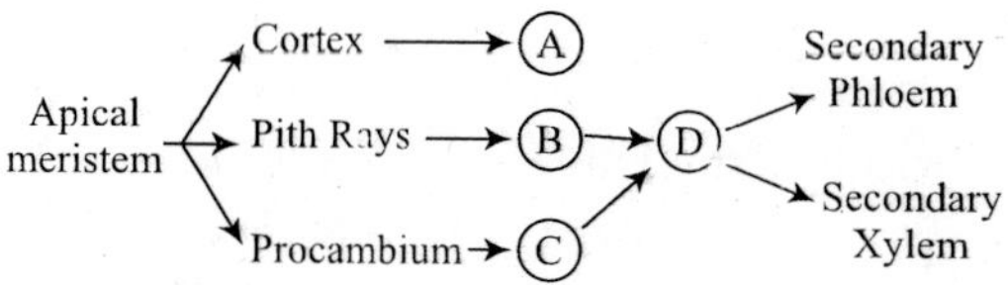

Based on above scheme which of the following option represents correct identity of the cambia labelled as A, B, C and D **(DEC 2016)**

(a) A- Inter-fascicular, B- Fascicular, C- Vascular, D- Cork

(b) A- Fascicular, B- Inter-fascicular, C- - Vascular, D- Cork

(c) A- Cork, B- Inter-fascicular, C- Fascicular, D- Vascular

(d) A- Cork, B- Fascicular, C- Inter-fascicular, D- Vascular

32. Euphorbiaceae generally represents milky sap bearing plants but there are also some non-milky sap bearing plants that belong to this family. Identify the correct combination of the following given plants which belong to family Euphorbiaceae. **(DEC 2017)**

(a) *Terminalia bellirica, Euphorbia hirta, Nerium indicum*

(b) *Mallotus philippense, Ficus religiosa, Ricinus communis*

(c) *Mallotus philippense, Acalypha indica, Emblica officinalis*

(d) *Acalypha indica; Ricinus communis, Mangifera indica*

33. Given below are some of the methods used to assess evolutionary phylogenetic relationships among plant taxa. **(JUNE 2014)**

A. 16S rRNA sequence

B. Mitochondrial microsatellite

C. Biochemical characterization

D. Morphology

Which two of the above methods can best reveal the evolutionary phylogenetic relationships?

(a) A and B (b) B and C

(c) C and D (d) A and D

34. Analyses of nucleotide sequences of ribosomal RNA (rRNA) are particularly useful for evolutionary studies of living organisms because of the following reasons EXCEPT. **(JUNE 2019)**

(a) rRNA is evolutionary ancient

(b) no free – living organisms lacks rRNA

(c) rRNA, since critical for translation, can undergo lateral transfer amongst distant species.

(d) rRNA has evolved slowly over geological time.

35. Which of the following plastid coding region(s) have been recommended as a core barcode by Plant Working Group of the consortium for the Barcode of Life? **(DEC 2018)**

(a) CO1 and rbcL (b) *rbc*L and *mat*K

(c) CO1 and *mat*K (d) *rbc*L only

36. A list of floral formulae and plant families are given in the following table:

Floral Formula	Family
i. ⊕ ⚥ K(5) C(5) A(5) G(2)	A. Brassicaseae
ii. ⊕ ⚥ P(3 + 3) A(3 + 3) G(3)	B. Fabaceae
iii. ⊕ ⚥ K2 + 2 C4 A2 + 4G(2)	C. Liliaceae
iv. † K(5) C(2),3 A10 G1	D. Solanaceae

Which of the following options most appropriately matches given plant families with their representative floral formulae? **(DEC 2018)**

(a) i-D; ii-B; iii-A; iv-C

(b) i - D; ii - C; iii - A; iv - B

(c) i - D; ii - C; iii - B; iv - A

(d) i-A; ii-C; iii-B; iv-D

37. Which of the following is NOT an adaptive modification in a xerophytic plant? **(JUNE 2012)**

(a) Strongly developed Sclerenchyma

(b) Sunken stomata

(c) Sparse stomata

(d) Presence of Lacunar tissues

Answer Key

1.(b)	**2.**(d)	**3.**(b)	**4.**(b)	**5.**(b)	**6.**(d)	**7.**(b)
8.(a)	**9.**(2/3)	**10.**(b)	**11.**(c)	**12.**(b)	**13.**(b)	**14.**(b)
15.(a)	**16.**(b)	**17.**(b)	**18.**(a)	**19.**(b)	**20.**(d)	**21.**(a)
22.(c)	**23.**(*)	**24.**(d)	**25.**(b)	**26.**(d)	**27.**(d)	**28.**(c)
29.(c)	**30.**(b)	**31.**(c)	**32.**(c)	**33.**(*)	**34.**(c)	**35.**(b)
36.(b)	**37.**(d)					

6. Classification of Bacteria

1. Which of the following is **NOT** true for the Anammox bacteria? (JUNE 2017)

(a) They convert nitrate and ammonium into dinitrogen

(b) They are responsible for 30-50% of the dinitrogen gas produced in the ocean

(c) They belong to the bacterial phylum Planctomycetes

(d) Membranes of these bacteria contain ladderane lipids

2. Given below are statements pertaining to organisms belonging to three domains of life. Identify INCORRECT statement.**(DEC 2016)**

(a) Unlike Bacteria and Eukarya, some Archeal membrane lipids contain long chain hydrocarbons connected to Glycerol molecule by ether linkage

(b) Peptidoglycans are absent in cell wall of Archea

(c) Proteobacteria include many species bacteriochlorophyll containing suphur using Photoautotrophs

(d) Mycoplasma, a group of low GC content gram positive bacteria lack the cell wall, belonging to same family of gram positive of mycobacteriaceae

3. Which groups of prokaryotic organisms are mainly classified based on morphology? **(DEC 2013 GU)**

(a) Cyanobacteria and Actinomycetes.

(b) Archaea and Enteric bacteria.

(c) Nitrogen fixing bacteria.

(d) Yeast and Molds.

4. Which one of the following parameters is **NOT** used in phenetic classification of bacteria? **(JUNE 2018)**

(a) trophism

(b) susceptibility of a bacteria to a particular bacteriophage

(c) reaction to a particular antibody

(d) 16S rRNA sequence

5. Which of the following statements is NOT true regarding the closer affinity of Archaea to Eukarya than to Bacteria? **(JUNE 2016)**

(a) Both Archaea and Eukarya lack peptidoglycan in their cell walls.

(b) The initiator amino acid for protein synthesis is methionine in both Archaea and Eukarya.

(c) Histones associated with DNA are absent in both Archaea and Eukarya.

(d) In both Archaea and Eukarya the RNA polymerase is of several kinds.

6. A comparsion of Bacteria, Archaea and Eukarya with respect to a few characteristics is given below.

	Characteristics	**Bacteria**	**Archaea**	**Eukarya**
A	Initiator amino acid for protein synthesis	Formyl Met	Met	Met
B	Histones associated with DNA	Absent	Present in some species	present
C	Response to streptomycin and chloramphenicol.	Growth not Inhibited	Growth not Inhibited	Growth usually Inhibited
D	RNA Polymerase	Three	Three	Three or more
E	Introns in genes	Very rare	Present in some genes	Present in many genes.

Which of the following combinations present a correct comparison of characteristics in the table above? **(JUNE 2017)**

(a) A, B, C and E (b) A, C and D

(c) B, D and E (d) A, B and E

7. A researcher conducts a standard test to identify enteric bacteria (A, B, C) on the basis of their biochemical properties. The result is given in the following table

	Bacteria		
Test	**A**	**B**	**C**
Indole	+	–	–
Methyl Red	+	+	+/–
Voges-Proskauer	–	–	+

Based on the above, the identified bacteria A, B and C are most probably **(DEC 2015)**

(a) *Enterobacter, Salmonella, Escherichia*

(b) *Escherichia, Salmonella, Enterobacter*

(c) *Salmonella, Enterobacter, Escherichia*

(d) *Escherichia, Enterobacter, Salmonella*

8. Which of the following characteristic differentiate Eubacteria from Archaebacteria? **(MODEL PAPER)**
   (a) Circular nature of chromosome.
   (b) Absence of nuclear membrane.
   (c) Presence of 70S ribosomes.
   (d) Presence of murein (peptidoglycan) in cell wall.

9. Which of the following statements best describe archaebacteria? **(DEC 2014)**
   (a) Mostly autotrophic, cell wall contains peptidoglycan, 60S ribosomes, live in extreme environment.
   (b) Divide by fission, not susceptible to lysozyme, live in extreme envioronments, mostly autotropic.
   (c) Not susceptible to lysozyme, contain golgi and linear chromosomes.
   (d) Chitinous cell wall, obligate aerobic, circular chromosomes.

10. Identify the proteobacteria based on the key given below: **(DEC 2012)**

i.	Cause disease in humans	(ii)
i.	Do not cause disease in humans	(iii)
ii.	An obligate intracellular parasite	(A)
ii.	Not all obligate intracellular parasite	(B)
iii.	Live in insects	(C)
iii.	Do not live in insects	(iv)
iv.	Chemoautotrophic	(D)
iv.	Not Chemoautotrophic	(v)
v.	Plant pathogen	(E)
v.	Not a plant pathogen	(vi)
vi.	Fix nitrogen	(vii)
vi.	Do not fix nitrogen	(F)
vii.	Associated with legumes	(G)
vii.	Not associated with legumes	(H)

   (a) A - Rickettsia; B - Brucella; C - Wolbachia; D Nitrobacter; E - Agrobacterium; F - Acetobacter; G - Rhizobium; H - Azospirillum
   (b) A - Rickettsia; B - Wolbachia; C - Brucella; D - Nttrobacter; E - Acetobacter; F – Agrabacterium; G -Rhizobium; H- Azospirillum
   (c) A - Rickettsia; B - Brucella; C- Wolbachia; D-Nitrobacter; E - Agrobacterium; F -Acetobacter; G- Azospirillum; H - Rhizobium
   (d) A- Rickettsia; B - Brucella; C - Wolbachia; D - Nilrobacter; E - Acetobacter;F - Agrobacterium; G- Azospirillum; H-Rhizobium.

11. The diagram below depicts a simplified tree of life with three domains and one of the domains including Whittaker's three major kingdoms.

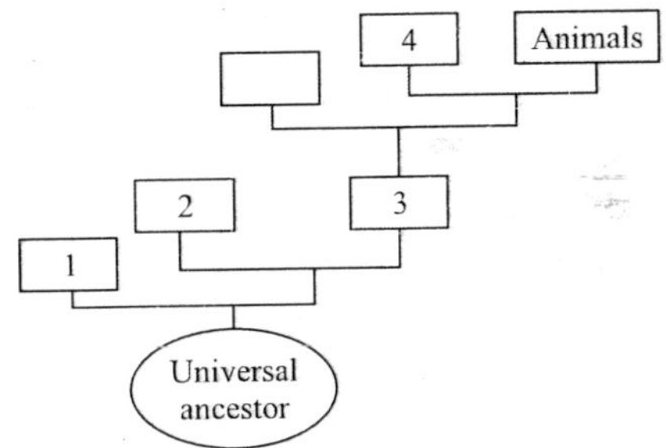

Which of the following is the correct naming of the numbered boxes? **(JUNE 2014)**
   (a) 1- Bacteria 2 - Archaea 3 – Eukarya 4 - Fungi
   (b) 1 - Archaea 2 - Bacteria 3 - Eukarya 4 - Plants
   (c) 1- Eukarya 2 - Bacteria 3 - Archaea 4 - Plants
   (d) 1- Archaea 2 - Bacteria 3 - Eukarya 4 – Fungi

12. Given below is a list of bacteria either functioning as methanogens or methanotrophs:
   A. Methanobacterium sp
   B. Methanococcus sp
   C. Methylomonas sp
   D. Methylosinus sp

   Which of the following options classifies the above list correctly? **(DEC 2018)**

(a) Methanogen - A; Methanotrophs - B, C, D
(b) Methonogens - A, B, C; Methanotroph -D
(c) Methanogens -A, B; Methanotrophs-C, D
(d) Methonogens - A, D; Methanotrophs -B, C

Answer Key

1.(a)	**2.**(d)	**3.**(a)	**4.**(d)	**5.**(c)	**6.**(d)	**7.**(b)
8.(d)	**9.**(b)	**10.**(a)	**11.**(a)	**12.**(c)		

7. Fungus

1. The fungus associated with human oral or vaginal infection is **(DEC 2013)**
(a) *Fusarium.* (b) *Aspergillus*
(c) *Candida.* (d) *Pneumocytis*

2. The fungal group presently classified under protists is **(DEC 2013)**
(a) Zygomycetes. (b) Oomycetes.
(c) Deuteromycetes. (d) Discomycetes

3. The group which is no longer considered under the fungi is **(DEC 2012)**
(a) Ascomycetes (b) Basidiomycetes
(c) Chytridiomycetes (d) oomycetes

4. *Batrachochytrium dendrobatidis*, a fungus, has been implicated in the decline of population of **(DEC 2011)**
(a) fish (b) frogs
(c) pelicans (d) bats

5. 'Imperfect fungi' is a group represented by fungal species which have **(JUNE 2011)**
(a) simple mycelia.
(b) no known mechanisms of sexual reproduction.
(c) unkown phylogenetic relationship.
(d) lost its survival mechanism against harsh environments.

6. Which of the following fungal groups has septate hyphae and reproduces asexually by budding, conidia and fragmentation? **(JUNE 2015)**
(a) Basidiomycota (b) Zygomycetes
(c) Chytrids (d) Glomeromycota

7. The table below list the major fungal groups and their characteristics:

Fungal Groups		**Characteristics**	
A	Asco-mycota	i	Hyphae aseptate, coenocytic, asexual reproduction by sporangiophores
B	Chytrids	ii	Hyphae aseptate, coenocytic, asexual reproduction by zoospores
C	Glomero-mycetes	iii	Hyphae aseptate, coenocytic, no sexual spores
D	Zygo-mycetes	iv	Hyphae septate or unicellular, asexual reproduction by conidia

Which of the following represents the appropriate match between the fungal group and their characteristics? **(DEC 2016)**
(a) A-ii, B-iii, C-i, D-iv
(b) A-iv, B-ii, C-iii, D-i
(c) A- i, B-iv, C-iii, D-ii
(d) A-ii, B-iv, C-iii, D-i

8. Following is a table showing selected characteristics of important fungal groups.

Fungal group	**Characteristic**
A	No regularly occurring septa in thallus
B	Perforated septa
C	Forms arbuscular mycorrhizae on plant roots
D	Have zoospores with flagella

In the above table, the fungal groups A, B, C and D, are, respectively, **(DEC 2014)**
(a) Chytridiomycetes, Ascomycetes, Glomeromycetes, zygomycetes
(b) Zygomycetes, Ascomycetes, Glomeromycetes, Chytridiomycetes
(c) Ascomycetes, Zygomycetes, Glomeromycetes, Chytridiomycetes
(d) Chytridiomycetes, Zygomycetes, Ascomycetes, Glomeromycetes

9. Given below are statements on key characteristics of two fungal lineages.
   (A) Microsporidia do not have true mitochondria
   (B) Most chytrids produce flagellated gametes
   (C) Microsporidia are usually free-living
   (D) Chytrids reproduce sexually without a dikaryon stage
   Choose the combination with statements **(DEC 2017)**
   (a) A, Band C only (b) A, B and D only
   (c) B, C and D only (d) A and B only
10. Match the following taxa with the genus of the microorganism **(DEC 2019)**

	Taxa		Genus
A	Ascomycota	i	*Rhizopus*
B	Basidiomycota	ii	*Erysiphe*
C	Zygomycota	iii	*Pythium*
D	Oomycota	iv	*Ustilago*

   (a) A-ii; B-iv; C-i D-iii
   (b) A-ii; B-ii; C-ii; D-iv
   (c) A-ii; B-iv, C-iii; D-I
   (d) A-i; B-ii; C-iv; D-ii

Answer Key

1.(c)	2.(b)	3.(d)	4.(b)	5.(b)	6.(a)	7.(b)
8.(b)	9.(b)	10.(a)				

8. Algae

1. Which of the following groups represents SAR clade of protists? **(JUNE 2018)**
   (a) Euglenozoans, Red algae, Parabasilids
   (b) Brown algae, Forams, Radiolarians
   (c) Slime moulds, Entamoebas, Diplomonads
   (d) Charophyes, Choanoflagellates, Tubulinids
2. An alga having chlorophyll a, floridean starch as storage product and lacking flagellate cells belongs to the class **(DEC 2016)**
   (a) Phaeophyceae (b) Chlorophyceae
   (c) Rhodophyceae (d) Xanthophyceae
3. Which of the following is the correct match of the algal group with its food reserve? **(DEC 2015)**

	Algal Group		Carbohydrate Reserve
A.	Bacillariophyceae	(i)	Oil
B.	Xanthophyceae	(ii)	Floridean starch
C.	Phaeophyceae	(iii)	Laminarin
D.	Rhodophyceae	(iv)	Chrysolaminarin
		(v)	Starch

   (a) A-(iv), B-(i), C-(iii), D-(ii)
   (b) A-(iii), B-(i), C-(iii), D-(iv)
   (c) A-(iv), B-(i), C-(ii), D-(v)
   (d) A-(i), B-(v), C-(iii), D-(ii)
4. Which of the following is a correct statement? **(DEC 2014)**
   (a) Euglenids have spiral or crystalline rod inside flagella
   (b) Pheophytes have a spiral or crystalline rod inside flagella.
   (c) Euglenids have a hairy and smooth flagella.
   (d) Euglenids and pheophytes both have a spiral or crystalline rod inside flagella.
5. Most biologists agree that seaweeds are protists. Some biologists think that at least some seaweeds should be considered plants, not protists. Which of the following would support the latter one? **(DEC 2013)**
   (a) Certain seaweeds contain several kinds of specialised cells.
   (b) Certain seaweeds have multicellular organization.
   (c) Certain seaweeds are found to be prokaryotic.
   (d) Certain seaweeds undergo sexual and asexual reproduction
6. Chlorophyll pigment composition and carbohydrate food reserves of some algal groups are given below:

Pigments:

(i) Chlorophyll *a* and *b*;

(ii) Chlorophyll *a* and *c*.

Carbohydrate food reserve:

(a) Paramylon; (b) Starch;

(c) Laminarin; (d) Leucosin.

Identify the correct combination of the characters for the given groups. **(DEC 2011)**

(a) Euglenophyta - (i and a); Bacillariophyta - (ii and d); Phaeophyta - (ii and c); Chlorophyta - (i and b).

(b) Euglenophyta - (ii and a); Bacillariophyta - (ii and d); Phaeophyta - (i and c); Chlorophyta - (i and b)

(c) Euglenophyta - (i and a); Bacillariophyta - (ii and b); Phaeophyta - (i and c); Chlorophyta - (ii and d)

(d) Euglenophyta - (i and d); Bacillariophyta - (ii and a); Phaeophyta - (ii and c); Chlorophyta - (i and b)

Answer Key

1.(b)	2.(c)	3.(a)	4.(a)	5.(b)	6.(a)

9. Natural history of Indian subcontinent

1. Tropical semi-evergreen forests are found in the region having rainfall **(MODEL PAPER)**

   (a) > 3000 mm. (b) 2000 – 2500 mm.

   (c) 1200–2500 mm. (d) 800–1200 mm.

2. Which of the following hypothesis best explains the occurrence of Himalayan floral element in Western Ghats of India? **(JUNE 2012)**

   (a) Continental drift theory

   (b) Deccan trap hypothesis

   (c) Himalayan glaciations theory

   (d) Coromondal coast hypothesis

3. Presence of which of the following flora and fauna in Peninsular India is explained by the Satpura hypothesis? **(MODELPAPER)**

   (a) Chinese. (b) Malayan.

   (c) African. (d) Mediterranean.

4. Which of the following groups of species are typical of grassland habitats in India **(DEC 2011)**

   (a) Black buck, wolf, great Indian bustard, lesser florican

   (b) Spotted deer, dhole, peacock, finch- lark

   (c) Sambar, tiger, paradise fly catcher

   (d) Otter, Cormorant, darter, pelican

5. In India, brown antlered deer (sangai) is found only in the floating landmasses of **(JUNE 2011)**

   (a) Wular lake. (b) Sasthamkotta lake.

   (c) Dal lake. (d) Lok Tak lake

6. Three Indian animals -cormorant, lion-tailed macaque and gerbil are to be matched with the ecosystem they inhabit -Wetland (A), Desert (B), Deciduous forest (C), or Rain' forest (D). Which of the following is the correct match of each animal with its habitat? **(JUNE 2014)**

   (a) Cormorant - D; Lion-tailed macaque - C; Gerbil - B

   (b) Cormorant - A; Lion-tailed macaque - C; Gerbil - D

   (c) Cormorant - A; Lion-tailed macaque - D; Gerbil - B

   (d) Cormorant - B; Lion-tailed macaque - C; Gerbil- D

7. Which one of the following sets contains at least one primate species NOT found in India? **(DEC 2019 ASSAM)**

   (a) Lion Tailed Macaque, Slender Loris, Common Langur

   (b) Golden Langur, Hoolock Gibbon, Howler Monkey

   (c) Bonnet Macaque, Stump Tailed Macaque, Slow Loris

   (d) Capped Langur, Rhesus Macaque, Phayre's Leaf Monkey

8. Name the common Indian bird that is generally seen in groups (aggregation) **(DEC 2013)**
   (a) Bulbul (b) Warbler
   (c) Babbler (d) Sun bird

9. The most common vegetation in the Western Ghats of India is tropical moist deciduous forest but that in Deccan plateu is depleted thorn forest. The possible reason is **(DEC 2013)**
   (a) richer soil of Western Ghats compare to Deccan plateau.
   (b) extensive deforestation in Deccan plateau compared to Western Ghats.
   (c) higher rainfall in Western Ghat compared to Deccan plateau.
   (d) higher temperature in Deccan plate compared to Western Ghats.

10. Which of the following biotic provinces are part of Deccan Peninsula biogeographic zone of India? **(DEC 2013)**
   (a) Malabar Coast, Western Plateau, Eastern Plateau
   (b) Western Ghats, Central Plateau, Eastern Plateau
   (c) Central Plateau, Eastern Plateau, chhota Nagapur
   (d) Central Plateau, Malabar Coast, Western Ghats

11. Following table shows plants and animals commonly associated with a biogeographic zone of India:

Biogeographic region	Plant	Animal
A	*Salvadora oleoides*	*Eqqus hemionus*
B	*Tectona grandis*	*Elephas maximus*
C	*Quercus leucotrichophora*	*Ailurus fulgens*

Which of the following represents three biogeographic zones correctly?
(DEC 2013 GU)
   (a) A. Deccan Peninsula, B. Gangetic Plains, C. Himalaya.
   (b) A. Western Ghats, B. Gangetic Plains, C. Trans-Himalaya
   (c) A. Indian Desert, B. Deccan Peninsula, C. Himalaya
   (d) A. Western Ghats, B. Gangetic Plains, C. Northeast India

12. Based on the information given in the table below, which combination is correct? **(DEC 2012)**

Biographic zone	Plant	Animal
A_1 Mediterranean	B_1 *Rhododendron*	C_1 Gibbon
A_2 Indo-Chinese	B_2 *Dipterocarpus*	C_2 Jungle Fowl
A_3 Indo-Malayan	B_3 *Euphorbia*	C_3 Takin
A_4 Peninsular India	B_4 *Deodar*	C_4 Ibex

   (a) $A_1 - B_3 - C_3$ (b) $A_2 - B_1 - C_4$
   (c) $A_3 - B_2 - C_1$ (d) $A_4 - B_4 - C_2$

13. Associate the forest/vegetation type with the plants: **(DEC 2012)**
   a. Grass land
   b. Subalpine forest
   c. Shola forest
   d. Subtropical pine forest
   e. Tropical thorn forest
   f. Tropical dry deciduous forest
   g. Tropical semievergreen forest
   h. Tropical wet evergreen forest
   (a) a - Ilex, b - Dichanthium, c - Abies, d - Pinus, e - Acacia. f - Anogeissus g - Cinnamomum. h - Dipterocarpus
   (b) a . Dichanthium, b . Abies, c . Ilex, d - Pinus, e - Acacia. f. Anogeissus g - Cinnamomum, h - Dipterocarpus
   (c) a - Dichanthium, b - Abies, c - Ilex, d - Pinus, e - Dipterocarpus. f - Cinnamomum, g - Acacia, h - Anogeissus
   (d) a - Anogeissus, b. Dichanthium, c - Ilex, d - Pinus, c - Acacia, f - Abies, g - Cinnamomum, h – Dipterocarpus

14. Following table shows an alphabetical list of certain domesticated crops and places of origin:

	Crop		Place of Origin
i	Barley	A	China
ii	Maize	B	Fertile crescent
iii	Mung Bean	C	India
iv	Rice	D	Southern maxico
v	Wheat		

Based Number of grass species on the above, which one of the following options represent the correct match between crops and their place of origin? **(DEC 2018)**

(a) i-C; ii-D; iii-A; iv-B; v-B

(b) i-B;D; ii-B; iii-A,C; iv-A; v-B

(c) i-C; ii-D; iii-C; iv-B; v-C

(d) i - B; ii- D; iii - C; iv -A, C; v –B

15. The table given below lists animals and primary regions of domestication

Animal		Region of domestication	
A.	Alpaca	(i)	Central Andes
B.	Goat	(ii)	South Asia
C.	Llama	(iii)	Southwest Asia
D.	Zebu cattle	(iv)	China

Which one of the following options represents the best match for animals and the primary region of their domestication?

(DEC 2019 ASSAM)

(a) A-(ii); B-(iv); C-(iii); D-(i)

(b) A-(i); B - (iii); C-(i); D-(iv)

(c) A-(iii); B -(i); C-(iv); D-(iii)

(d) A-(i); B - (iii); C-(i); D-(ii)

Answer Key

1.(b)	**2.**(c)	**3.**(b)	**4.**(a)	**5.**(d)	**6.**(c)	**7.**(b)
8.(c)	**9.**(c)	**10.**(c)	**11.**(c)	**12.**(c)	**13.**(b)	**14.**(d)
15.(d)						

10. Common parasites and pathogens

1. Which one of the following is a fungal disease of plants? **(JUNE 2019)**

(a) Cucumber mosaic

(b) Fire Blight of pear

(c) Crown gall

(d) Apple scab

2. The table given below provides a list of diseases and causal organisms. **(JUNE 2017)**

	Disease		Causal Organism
A	Sleeping sickness in humans	i	*Trypanosoma cruzi*
B	Chagas disease in humans	ii	*Trypanosoma brucei*
C	Blast disease of rice	iii	*Magnaporthe graminis*
D	Powdery mildew of grasses	iv	*Magnaporthe oryzae*
		v	*Blumeria oryzae*
		vi	Blumeria graminis

Which of the following options represent the correct match between disease and the causal organism?

(a) A-i; B- ii; C-v; D-vi

(b) A-ii; B-I; C-iii; D-v

(c) A-i; B-ii; C-vi; D-iv

(d) A-ii; B-I; C-iv; D-vi

3. Which of the following diseases does not leave any paleontological evidence? **(DEC 2013)**

(a) Tuberculosis (b) Arthritis.

(c) Rickets (d) Cholera

4. Which of the following organism is widely used as a biocontrol in organic farming?

(JUNE 2012)

(a) *Rhyzobium tropicii*

(b) *Trichorderma viridis*

(c) *Fusarium oxysporum*

(d) *Nostoc muscorum*

5. Given below are some pathogens and diseases of humans, animals and plants.

	Pathogen		Disease
A	*Bordetella pertusis*	i	Lyme disease of humans
B	*Tilletia indica*	ii	Grain rot in rice
C	*Borrelia burgdorferi*	iii	Karnal blunt of wheat
D	*Anaplasma marginale*	iv	Whooping cough in Humans
E	*Burkholderia glumae*	v	Hemolytic anemia in Cattle

Which of the following is the match between the pathogen and the disease caused?

(DEC 2016)

(a) A- iv, B-iii, C-i, D-v, E-ii
(b) A-iv, B- v, C-i , D-ii, E-iii
(c) A- iii, B-iv, C-v, D-i, E- ii
(d) A-ii, B-v, C-i, D-iii, E-iv

6. Which one of the following viruses cause acute gastrointestinal illness due to contamination of drinking water? **(JUNE 2015)**

(a) Norovirus (b) Poliovirus
(c) Rotavirus (d) Filoviruses

7. The most common cause of blindness induced by microbial infection in humans is:

(DEC 2019 ASSAM)

(a) *Chlamydia psittaci*
(b) *Chlamydia pneumonia*
(c) *Chlamydia muridarum*
(d) *Chlamydia trachomatis*

8. Identify the plant species from which artemisinin, an anti-malarial drug, is extracted.

(DEC 2017)

(a) *Artemisia maritima*
(b) *Artemisia scoparia*
(c) *Artemisia annua*
(d) *Cinchona officinalis*

9. Match the following human diseases with their causal organisms **(JUNE 2016)**

A.	Sleeping Sickness	(i)	*Trypanosoma cruzi*
B.	Chagas disease	(ii)	*Trypanosoma brucei*
C.	Elephantiasis	(iii)	*Borrelia burgdorfei*
D.	Lyme disease	(iv)	*Wuchereria bancrofti*

(a) A - (ii); B - (iv); C - (iii); D- (i)
(b) A - (i); B - (ii); C - (iv); D - (iii)
(c) A- (ii); B - (i); C -(iv); D - (iii)
(d) A - (ii); B - (iv); C - (i); D -- (iii)

10. Which of the following gives the correct human disease- causal microbe match for each?

(JUNE 2015)

Human Disease	Causal Microbe
A. Chronic gastritis	(a) *Borrelia burdorferi*
B. Lyme disease	(b) *Heliobacter pylori*
C. Scarlet fever	(c) *Rickettsia prowazekii*
D. Typhus	(d) *Streptococcus pyrogenes*

(a) A - (ii) B – (i) C - (iv) D - (iii)
(b) A - (ii) B - (iii) C - (i) D - (iv)
(c) A - (iv) B - (i) C - (ii) D - (iii)
(d) A - (iv) B - (iii)C - (i) D- (ii)

11. Sleeping sickness is a disease caused by a protozoan parasite and the following statements pertain to that disease:

A. The vector for this disease is tsetse fly.
B. The vector for this disease is *Trypanosoma brucei*.
C. The parasite's body is covered by a dense coat of variable surface glycoprotein (VSG).
D. There are several thousands of VSG genes, only one of them expressing at a time, which helps the parasite in evading host's immune response.
E. Several thousand copies of VSG genes express concurrently, paralyzing the host immune system.

Which of the following is the correct combination of the statements given above?

(JUNE 2014)

(a) B, C and D (b) A, C and D
(c) A, C and E (d) B, C and E

12. Match the following: **(DEC 2013)**

Disease		Pathogen/ Causative factor	
A	Creutzfeldt- Jakob	(i)	Fungi
B	Pneumocystis	(ii)	Virus
C	Legionnaires disease	(iii)	Prion
D	Rabies	(iv)	Bacteria

(a) A-iv, B - iii, C - ii, D – I
(b) A-iii, B-i, C-,iv, D-ii
(c) A-i, B - ii, C - iii, D – iv
(d) A-ii, B - iv, C - i, D - iii

13. Which one of the following influenza A virus subtypes caused severe avian flu and was responsible for disease outbreak in the year 1997 in Hong Kong ? **(JUNE 2019)**

(a) H1N1 (b) H7N7
(c) H3N2 (d) H5N1

14. Match the following plant diseases with the name of pathogen associated with the disease **(DEC 2019)**

Disease	Pathogen
A. Powdery mildew	i. *Erwinia amylovora*
B Rice blast	ii *Pseudomonas syringae*
C. Bacterial canker	iii. *Magnaporthe oryzae v. Syringae*
D. Fire blight	iv *Erysiphe cichoracearum*

(a) A-ii, B-iii; C-i; D-iv
(b) A-i; B-iv; C-ii; D-iii
(c) A-iv, B-iii; C-ii; D-i
(d) A-iii; B-ii; C-iv; D-i

Answer Key

1.(d)	**2.**(d)	**3.**(d)	**4.**(b)	**5.**(a)	**6.**(1/3)	**7.**(d)
8.(c)	**9.**(c)	**10.**(a)	**11.**(b)	**12.**(b)	**13.**(d)	**14.**(c)

SECTION-10
ECOLOGY

1. Ecology, Habitat, Niche

1. A general increase in the average body mass of animal population within a species with latitude is known as **(JUNE 2018)**
   (a) Allen's rule (b) Bergmann's rule
   (c) Allee effect (d) Hamilton's rule
2. Individuals with greater mass have a smaller surface area to volume ratio, which helps to conserve heat. This is known as **(DEC 2014)**
   (a) Leibig's rule. (b) Cope's rule.
   (c) Gloger's rule (d) Bergmann's rule
3. Individuals occupying a particular habitat and adapted to it phenotypically but not genotypically are known as **(DEC 2016)**
   (a) Ecophenes (b) Ecotypes
   (c) Ecospecies (d) Coenospecies
4. Ecological compression differs from character displacement in that it operates on a **(DEC 2011)**
   (a) shorter timescale and does not involve heritable change.
   (b) longer timescale and does not involve heritable change.
   (c) shorter timescale and involves heritable change.
   (d) longer timescale and involves heritable change.
5. Which of the following statements is the most appropriate example of character displacement? **(MODEL PAPER)**
   (a) Two related species depending on the same prey species avoid competition by feeding at different times of the day.
   (b) The body sizes of two related species are very similar when they are allopatric, but in geographical areas of sympatry, one species is significantly smaller than the other.
   (c) The food niche of a species is generally wider in the absence of competing species than in their presence.
   (d) Closely related species can coexist if their densities are regulated by a predator.
6. The diagram below shows the frequency distribution of three closely related bird species based on their beak sizes, across three different habitat patches A, B and C. All the three species co-exist in habitat A, whereas only species 1 occurs in habitat B and species 2occurs in habitat C.

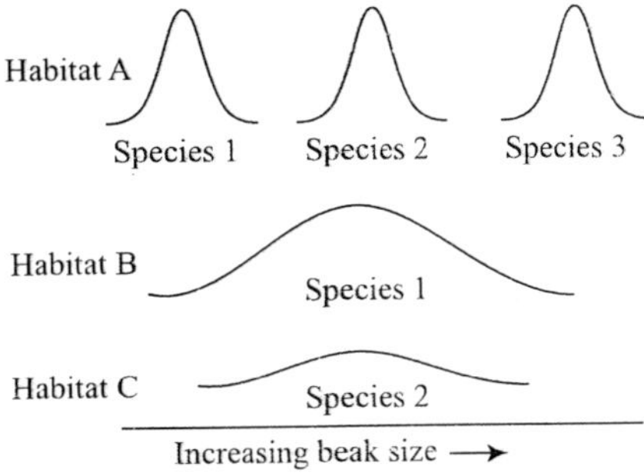

   Which one of the following phenomena most appropriately explains the change in frequency distributions of species 1 and 2 in the figure given above? **(DEC 2019 ASSAM)**
   (a) Migration
   (b) Character displacement
   (c) Colonization
   (d) Succession
7. The Hutchinsonian concept of ecological niche is based on **(DEC 2011)**
   (a) microhabitat occupied
   (b) multidimensional hypervolume
   (c) role played in the ecosystem
   (d) a combination of role played and microhabitat occupied
8. Gause's 'Competitive exclusion' principle states that two species with identical niches cannot coexist indefinitely. Which of the following statements is the most appropriate regarding the validity of the principle? **(JUNE 2011)**
   (a) It depends on how one defines niche.
   (b) There are in nature many instances of continued coexistence of closely related species.

(c) The principle is universally true.

(d) It does not predict the outcome where, both the species are equally strong competitors

9. You observed the two species of barnacles, species 1 and species 2, occupy upper and lower strata of intertidal rocks, respectively. Only when species 2 was removed by you from the lowers strata, species 1 could occupy both the upper and lower strata. From the choice given below what would be your inference from these observations. **(DEC 2016)**

(a) Upper strata of intertidal rocks is the realised niche of the species1

(b) Upper strata of intertidal rocks is the fundamental niche of the species1

(c) Species 1 and species 2 exhibit mutualism

(d) Species 1 can compete out species 2

10. In a 2-species interaction, assume that it leads to competition in habitat A, but to facilitation in habitat B. Which of the following is true about the niches of these species? **(DEC 2013 GU)**

(a) Fundamental niches are larger than realized niches in habitat A but the opposite in habitat B.

(b) Fundamental niches are smaller than realized niches in habitat A, but opposite in habitat B.

(c) Fundamental niche and realized niches are equal in habitat B, but fundamental niches are larger than realized niches in habitat A.

(d) Fundamental niches and realized niches are equal in habitat A, but realized niches are larger than fundamental niches in habitat B.

11. Three bird species with similar habitat and diet preferences co-exist in a habitat. For these species, which one of the following statements is most likely to be correct?

(DEC 2019 ASSAM)

(a) The fundamental and realised niches of these species are same.

(b) Their fundamental niches are greater than their realised niches.

(c) Their realised niches are greater than their fundamental niches.

(d) The fundamental and realised niches both will expand.

12. Followings are the niche characteristics of the constituent species and resource partitioning pattern in different ecosystems. Which of these would lead to competitive exclusion of species? **(JUNE 2013)**

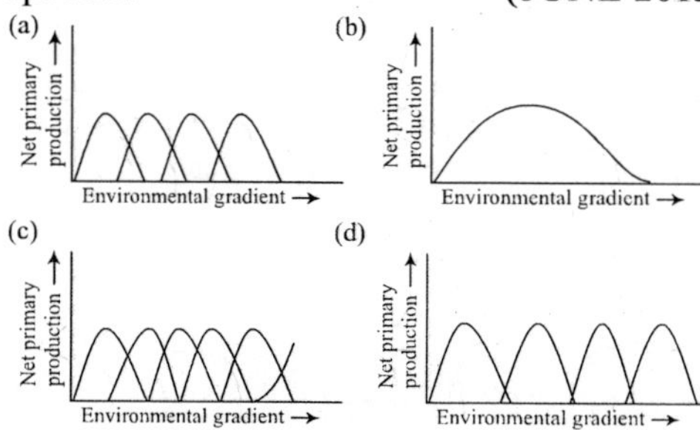

13. Which of the following is NOT a mechanism for species coexistence? **(DEC 2019)**

(a) Niche differentiation

(b) Niche complementarity

(c) Niche overlap

(d) Amount of limiting resources is greater than the number of species

Answer Key

1.(b)	2.(d)	3.(a)	4.(a)	5.(b)	6.(b)	7.(b)
8.(a)	9.(a)	10.(a)	11.(?)	12.(b)	13.(c)	

2. Characteristics of a population

1. Given below are the population pyramids of three different populations A, B and C depicting the relationship between birth and death rates in each.

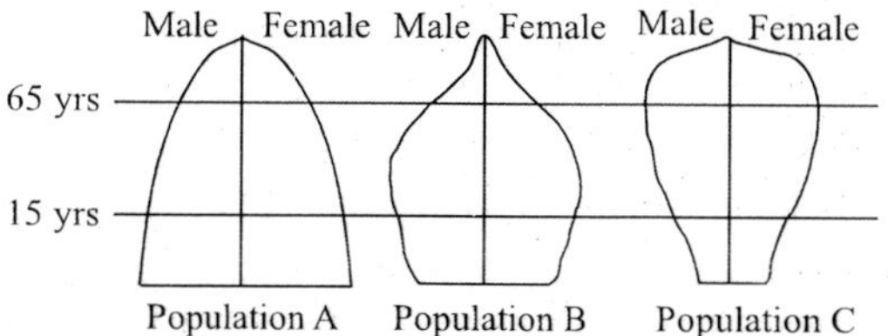

Based on the population pyramids given above, which one of the following is **INCORRECT**? **(DEC 2017)**

(a) Population B has slower growth rate than population A.

(b) Population C has birth rate higher than its death rate.

(c) Population A represent a rapidly growing population.

(d) Population B has the highest death rate among the three populations.

2. In natural system, a species producing large number of offsprings with little or no parental care generally exhibits which one of the following kind of survivorship curve **(DEC 2016)**

(a) % Survivors → / Time →

(b) % Survivors → / Time →

(c) % Survivors → / Time →

(d) % Survivors → / Time →

3. Following is the diagram of three idealized survivorship curves of animals.

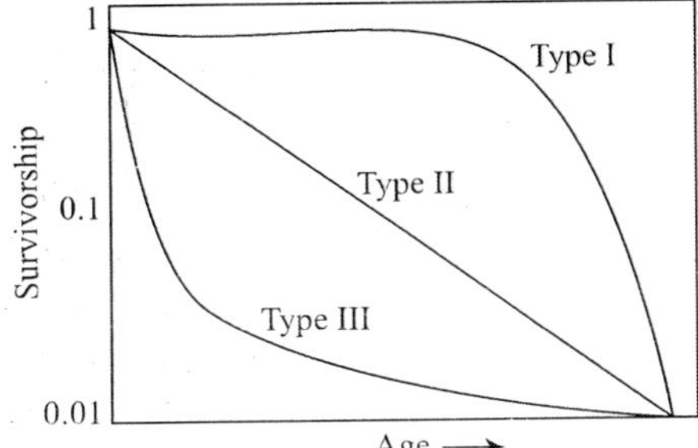

Find the correct match between the group of animals and the respective survivorship curves. **(JUNE 2014)**

(a) Marine pelagic fish and large mammals -III and I, respectively.

(b) Marine pelagic fish and large mammals- I and II, respectively.

(c) Some birds and large mammals - I and III, respectively.

(d) Marine pelagic fish and some birds –I and III, respectively.

4. Given below are the steps to assess the population size of grasshoppers in a given area:

A. 'n' individuals are collected randomly from the study area in a defined period of time.

B. The captured individuals are counted, marked and released at the site of collection. Next day, individuals are captured from the same site for same length of time. Number of marked (n_M) and unmarked (n_U)

C. This capture-release and recapture is continued till one gets 100% marked individuals

D. The size of the population (N) is estimated as follows:

$$\frac{N}{n} = \frac{n_m + n_u}{n_m}$$

E. The size of population (N) is estimated as follows:

$$\frac{N}{n} = \frac{n_u}{n_m}$$

The most appropriate combination of steps for estimating population size using mark-recapture method is: **(JUNE 2019)**

(a) A, B and D (b) A, B and E

(c) A, B. C and D (d) A, B, C and E

5. In a census for a lake fish, 10 individuals were marked and released. In second sampling after a few days 15 individuals were caught, of which 5 individuals were found marked. The estimated population of the fish in the lake will be **(DEC 2012)**

(a) 20 (b) 30

(c) 25 (d) 35

6. In an experiment to determine the number of rats in a field, 80 rats were initially captured, marked and released. After one month, 100 rats were captured in the same field, of which 20 were previously marked ones. Based on the above observation, estimated population size of the rats in the field will be: **(JUNE 2017)**

(a) 160 (b) 200

(c) 400 (d) 1600

7. A grasshopper population is being assessed by capture-mark-release-recapture method. On the first day, 100 grasshoppers were captured from a given area in 1 hour time, marked and released. On the next day during recapture, 10 marked and 90 unmarked grasshoppers could be found in the same time period from same area. What will be the estimated population size in the given area? **(JUNE 2013)**

(a) 80. (b) 100.

(c) 1,000. (d) 10,000.

8. In order to estimate population size of a fish species in a lake, a researcher captures 100 fish from the lake and marks them with coloured tags a week later, the researcher return to the lake and catches 150 fish of the and finds that 25 of them are tagged ones. Assuming no immigration or emigration occurred, thetotal population size of the fish species in the lake will be: **(DEC 2018)**

(a) 17 (b) 38

(c) 600 (c) 86

9. During line-transect sampling of two solitary species of ground mammals the following observations were made:

		Smaller species	Larger Species
A	Transect length (Km)	100	100
B	Number of animals recorded	30	36
C	Mean perpendicular distance from transect line (m)	10	40

Which of the following can be inferred from the data? **(JUNE 2012)**

(a) Smaller species is more abundant but seen less frequently

(b) Smaller species is less abundant but seen less frequently

(c) Larger species is more abundant but seen less frequently

(d) The large species is seen more frequently but its abundance cannot be compared with smaller species

10. The following table shows the mean and variance of population densities of species A, B and C. **(JUNE 2016)**

Statistic	Species A	Species B	Species C
Mean $\overline{x}$	5.3	7.05	5.30
Variance s^2	5.05	0.35	50.5

Based on the above, which of the following statements is correct?

(a) Species A and B show uniform distribution, whereas species C shows clumped distribution.

(b) Species A shows random distribution, species B shows uniform distribution, and species C shows clumped distribution.

(c) Species A and B show clumped distribution, whereas species C shows uniform distribution.

(d) Species A shows clumped distribution, species B shows random distribution, and species C shows uniform distribution.

11. Two species of plants were sampled in 32 quadrats in a forest. The mean and variance for the occurrence of species 1 were 16.2 and 48 and species 2 were 3.6 and 3.2 respectively. Which of the following statements about the distribution of the two species in these quadrats is supported by these findings? **(JUNE 2013)**

(a) Both species are distributed randomly.

(b) Species 1 is distributed randomly and species 2 is clustered.

(c) Species 1 is clustered and species 2 is distributed randomly.

(d) Both species are clustered.

Answer Key

1.(b)	2.(b)	3.(a)	4.(a)	5.(b)	6.(c)	7.(c)
8.(c)	9.(a)	10.(b)	11.(c)			

3. Population growth

1. The net reproductive rate (R_o) is 1.5 for a given population. If N_t, the population of females at generation t, is 500, then what will be the population of females after four generations (N_{t+4})? **(JUNE 2017)**

(a) 1125.000 (b) 2531.250

(c) 1265.625 (d) 3796.875

2. In the following equations

(a) dN /dt = rN

(b) $N_t = N_o e^{rt}$

(c) dN/dt = rN (K-N/K)

(d) dN/dt=rN x N/K

Exponential population growth is described by **(DEC 2015)**

(a) a and b. (b) a only.

(c) e only. (d) b and d.

3. Given below are growth equations where dN/dt is defined as

A. rN/K B. rN

C. rN[(K-N)/N] D. rN[(K-N)/K]

With reference to the above equations, which one of the following statements is correct? **(DEC 2017)**

(a) B represents exponential growth and A represents logistic growth.

(b) B represents exponential growth and D represents logistic growth

(c) B represents zero growth and C represents logistic growth

(d) A represents exponential growth and D represents logistic growth

4. For a population growing exponentially with a growth rate *r*, its population doubling time is **(JUNE 2016)**

(a) ($N_0 \times 2$) (b) ln 2/r

(c) λ ln 2 (d) ln *r* X 2

5. The general relation between generation time (T) and population growth rate (r) is described by the equation **(JUNE 2015)**

(a) *ln*r = *ln*a – b *In*T (b) r = a – bT

(c) *ln*r= lna + b*ln*T (d) r = a + bT

6. If gypsy moth egg density is 160 at time t and 200 at t + 1, what will be its value at time t + 3, assuming that egg density continues to increase at constant rate? **(JUNE 2016)**

(a) 250 (b) 280

(c) 312 (d) 390

7. The population density of an insect species increases from 40 to 46 in one month. If the birth rate during that period is 0.4 What is the death rate? **(DEC 2014)**

(a) 0.25 (b) 0.15

(c) 0.87 (d) 0.40

8. The population size of a bird increased from 600 to 645 in one year. If the per capita birth rate of this population is 0.125, what is its per capita death rate? **(JUNE 2014)**

(a) 0.25 (b) 0.15

(c) 0.05 (d) 0.02

9. A Population is growing logistically with a growth rate (r) of 0.15/week, in an environment with a carrying capacity of 400. What is the maximum growth rate (N_o) of individuals/week) that this population can achieve? **(DEC 2015)**

(a) 15 (b) 30

(c) 22.5 (d) 60

10. The birth rates and death rates of an insect are 0.25 and 0.05 respectively. In a habitat with a carrying capacity of 500, when the population density is 100, by how much does the insect population size increase per unit time if (a)

the growth is exponential and (b) growth is logistic? **(DEC 2013 GU)**

(a) a: 16, b:20 (b) a: 20, b: 20

(c) a: 20, b: 16 (d) a: 25, b: 20

11. For a species having logistic growth, if K = 20,000 and r = 0.15, the maximum sustainable yield will be **(DEC 2016)**

(a) 450 (b) 1500

(c) 3000 (d) 6000

12. A population grows according to the logistic growth equation,

$$\frac{dN}{dt} = \gamma N\left(1 - \frac{N}{K}\right)$$

Where, $\frac{dN}{dt}$ is the rate of population growth, r is the intrinsic rate of increase, N is population size and K is the carrying 'capacity of the environment. According to this equation, population growth rate is maximum at **(DEC 2018)**

(a) K/4 (b) K/2

(c) K (d) 2K

13. In a population growing logistically and approaching Carrying Capacity (K), the change in density (N) per unit time (dN/dt) is maximum when N equals to **(MODEL PAPER)**

(a) K^2 (b) K/2.

(c) K. (d) K/4

14. A small number (approximately 10) of mice are introduced into an uninhabited island. Their population grows exponentially initially and after 3 years, reaches a population size of 520 afterwhich the population becomes stable. At what point would you expect their population to attain their highest growth rate? **(DEC 2019, ASSAM)**

(a) When the mice population was first introduced.

(b) When the population size is 260.

(c) Their population growth rate remains constant throughout.

(d) When the population size reaches 520.

15. A plot of dN/dt as a function of population density yields a **(DEC 2016)**

(a) rectangular hyperbola

(b) negative exponential curve

(c) positive rectilinear curve

(d) bell- shaped curved

16. The following graph is for a logistically growing population, with N_t plotted on the X-axis. What is the parameter plotted on the Y-axis? **(DEC 2013)**

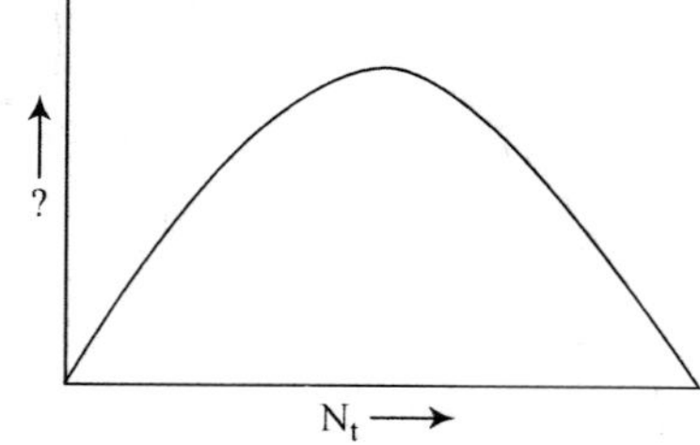

(a) dN/dt (b) N_{t+1}

(c) dN/dt. 1/N (d) K

17. Which of the following represents exponential growth in populations? **(DEC 2019)**

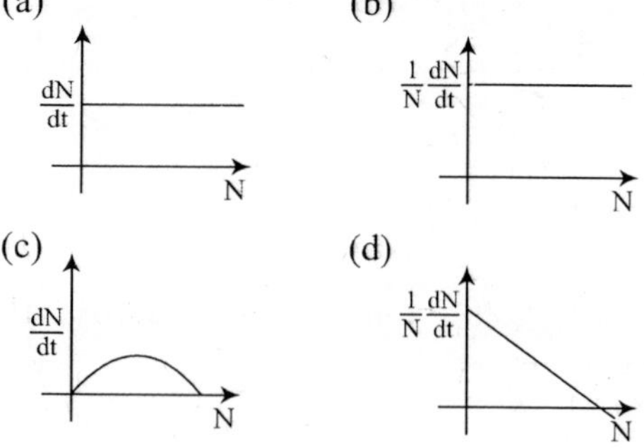

18. The three graphs (A, B, C) show population growth (N) patterns in relation to N or time (t)

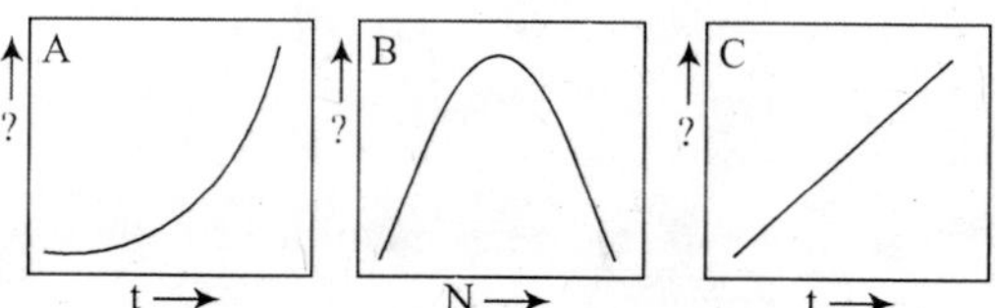

Which of the following is correct with reference to the Y-axis label and the type of population growth? **(JUNE 2014)**

(a) A: Y-axis: N_t, exponential growth
B: Y-axis: dN/dt, logistic growth
C: Y-axis: ln(N), exponential growth

(b) A: Y-axis: dN/dt, exponential growth
B: Y-axis: ln(N), logistic growth
C: Y-axis: N_t, exponential growth

(c) A: Y-axis: ln(N), exponential growth
B: Y-axis: dN/dt, logistic growth
C: Y-axis: N_t, exponential growth

(d) A: Y-axis: dN/dt, exponential growth
B: Y-axis: ln(N), logistic growth
C: Y-axis: N_t, exponential growth

19. What parameter, plotted on Y-axis against generation time, would yield the curve shown in the figure? **(DEC 2014)**

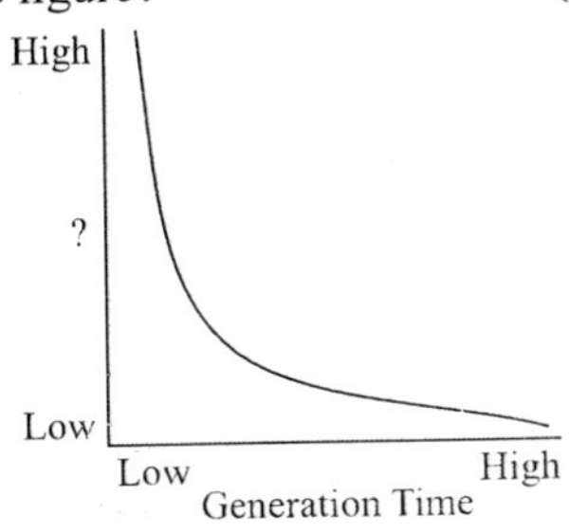

(a) Survivorship
(b) Body size
(c) Lifespan
(d) Intrinsic rate of growth

20. Three important biological parameters – generation time, population growth rate (r) and metabolic rate per gram body weight are a function of the organism's body size.

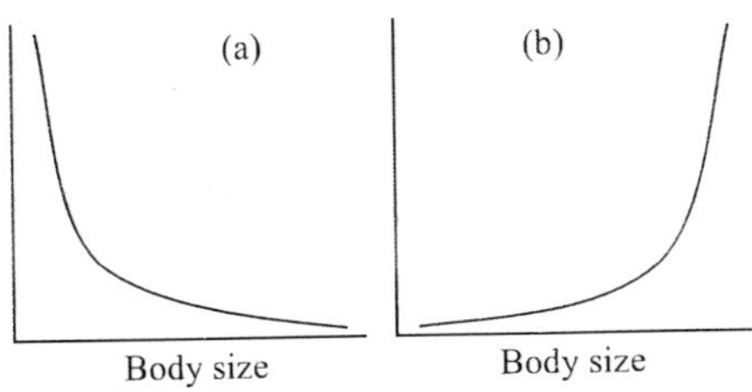

Which of the curves (a) or (b) represents the correct relation of each of the parameters to body size? **(MODELPAPER)**

(a) Generation time – (a); population growth rate – (b); metabolic rate/g bw – (a)
(b) Generation time – (a); population growth rate – (b); metabolic rate/g bw – (b)
(c) Generation time – (b); population growth rate – (a); metabolic rate/g bw – (a)
(d) Generation time – (b); population growth rate – (b), metabolic rate rate/g bw – (a)

21. The graph below shows the relationships of per capita population growth rate *(r)*, fecundity *(b)* and age at first reproduction *(α)* in an animal species.

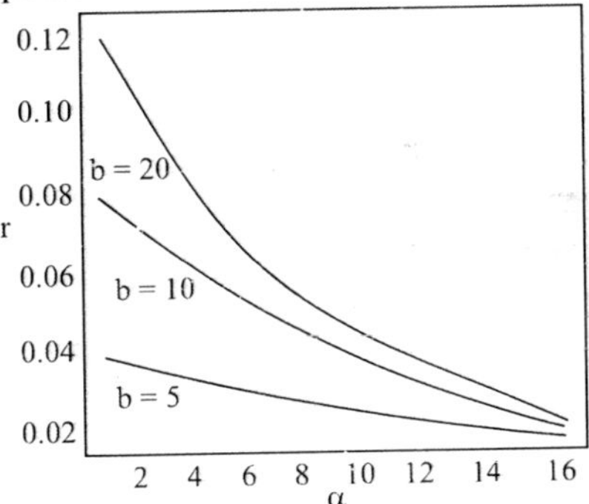

What is the *most important* conclusion to be drawn from the graph? **(JUNE 2011)**

(a) The later the age of first reproduction, the lower is the population growth rate achieved.
(b) The population growth rate decreases as first reproduction is postponed to a later stage, regardless of the fecundity.
(c) At any α, the higher the fecundity, the higher is the population growth rate achieved.
(d) As the age at first reproduction is postponed further, the benefits of increasing fecundity on the population growth rate become progressively negligible.

22. The following table shows survival and fertility data for a seasonally breeding species. Based on above data net reproductive rate (Ro) of the species will be **(DEC 2012)**

Season	Proportion Surviving	Fertility
0	1.0	0
1	0.5	20
2	0.0	–

(a) 1 (b) 5
(c) 10 (d) 20

23. Following is the hypothetical life table for species

Age Class (x)	Number alive (n_x)	Number dying (d_x)	Age specific survivorship (l_x)	Age specific fertility (m_x)
0-5	100	50	--	0.0
5-10	50	--	0.5	0.5
10-15	--	10	--	0.5
15-20	20	10	0.2	1.0
20-25	--	10	--	1.0

Which one of the following is the correct net reproductive rate (R_0) **(DEC 2016)**

(a) 0.0 (b) 0.3
(c) 0.7 (d) 1.5

24. Complete the following hypothetical life table of a species to calculate the net reproductive rate Ro: **(DEC 2014)**

Age class (x)	Number alive (n_x)	Number of dying (d_x)	Age specific survivorship	Age specific Fertility	$l_x m_x$
0–1	1000			0	
1–2	800			0	
2–3		200		0.5	
3–4	300	100		1.0	
4–5		200		1.0	

The calculated Ro will be

(a) 0.75 (b) 1.00.
(c) 0.65. (d) 1.15.

25. Suppose a population has three age classes. Females in the second and third age classes produce four and three offsprings, respectively. While 50% female in the first age class survive into the second age class, only 30% females survive into the third age class. The

Leslie matrix for this population is given by

$$L = \begin{pmatrix} 0 & 4 & 3 \\ 0.5 & 0 & 0 \\ 0 & 0.3 & 0 \end{pmatrix}$$

If there are 10 individuals in each of the three age classes, the number of individuals in the next iteration would be: **(DEC 2019 ASSAM)**

(a) 50 (b) 78
(c) 100 (d) 65

26. At a given time, the age class distribution of a population was as shown in figure:

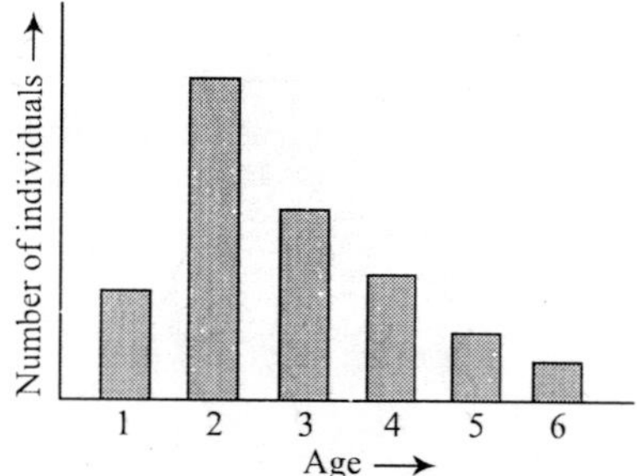

Which of the following can be inferred from the figure? **(JUNE 2012)**

(a) Age class 2 has maximum fecundity
(b) Age class 2 has maximum survival
(c) Age class distribution is at equilibrium
(d) Age class distribution is not at equilibrium

27. Biologist randomly sampled about 300 insects from a newly found island. The distribution of their abundance in the sample was shown in the figure given below.

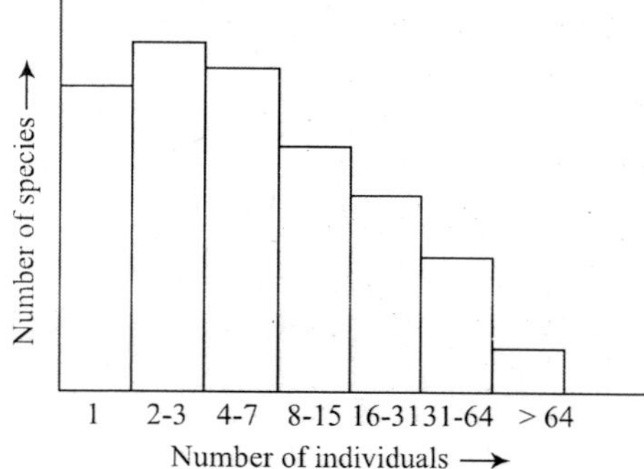

Which of the following can be correctly inferred from the graph? **(JUNE 2012)**

(a) Many species have only one individual each on the island

(b) The bar on the extreme right represents a large number of species with very few individuals

(c) Summation of the height of all columns will be exactly equal to the total number of species present on the island

(d) All species from the island may not be represented in the sample

28. Area of patch 1 is 2000 m^2 with a resource density of 5 units/m^2 Area of patch 2 is 3000 m^2 with a resource density of 10 units/m^2. As per the theory of ideal-free distribution, organisms distribute themselves such that the expected ratio of abundance of organisms in the two patches (patch 1: patch 2) is **(DEC 2019)**

(a) 1:2 (b) 2:3

(c) 1:3 (d) 3:2

29. What will be the approximate effective population size in a panmictic population of 240 with 200 females and 40 polygamous males? **(JUNE 2011)**

(a) 160 (b) 133

(c) 63 (d) 67

30. From population dynamics point of view, what would be the effective population size of a population of 150 breeding females and 50 breeding males? **(DEC 2013 GU)**

(a) 200 (b) 100

(c) 150 (d) 50

Answer Key

1.(b)	**2.**(a)	**3.**(b)	**4.**(b)	**5.**(a)	**6.**(c)	**7.**(a)
8.(c)	**9.**(a)	**10.**(c)	**11.**(b)	**12.**(b)	**13.**(b)	**14.**(b)
15.(d)	**16.**(a)	**17.**(b)	**18.**(a)	**19.**(d)	**20.**(c)	**21.**(d)
22.(c)	**23.**(c)	**24.**(a)	**25.**(b)	**26.**(d)	**27.**(d)	**28.**(c)
29.(b)	**30.**(c)					

4. Population regulation

1. An observation was made on a species experiencing three factors A, B and C in order to infer a density dependent population regulation by a factor. The following graph shows the relationship between the adverse effect of the factors in terms of number and population density.

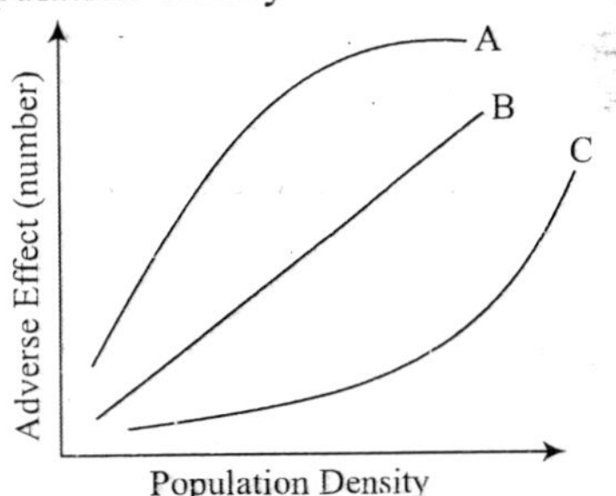

Based on the above observation, which of the following is correct? **(DEC 2013)**

(a) A - Density independent; B = Density dependent; C - Inversely density dependent

(b) A - Inversely density dependent; B- Density independent; C - Density dependent

(c) A - Density dependent; B = Inversely density independent; C - Density independent

(d) A - Density dependent; B = Density independent; C - Inversely density dependent

2. The birth rates (b) and death rates (d) of two - species 1 and 2 in relation to population density (N) are shown in the graph. Which of the following is NOT true about the density dependent effects on birth rates and death rates? **(JUNE 2016)**

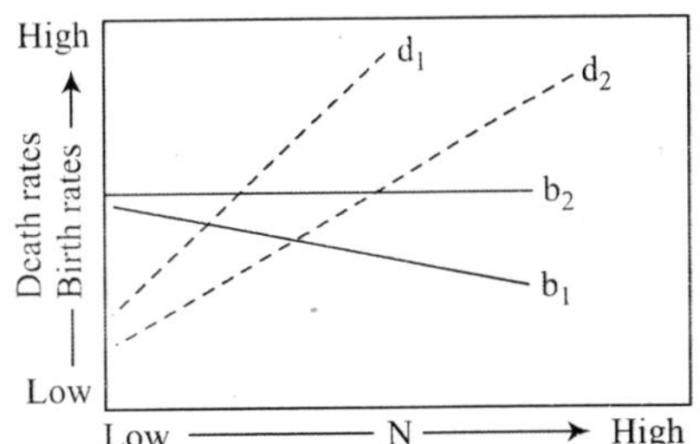

(a) Birth rates are density-dependent in species 1 and density-independent in species 2
(b) Death rates are density-dependent in both the species.
(c) Density-dependent effect on birth rate is stronger in species 1 than in species 2 .
(d) The density-dependent effects on death rates are similar in both the species.

3. Which of the following statements about the birth rates (b_1, b_2) and death rates (d_1, d_2) of species 1 and 2 indicated in the figure is NOT true? **(DEC 2015)**

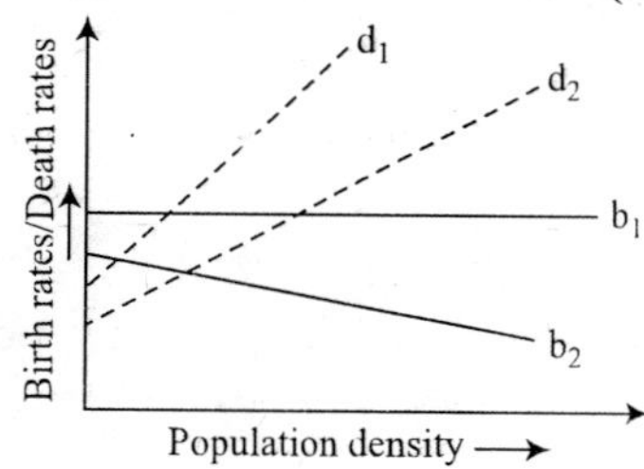

(a) Birth rates of species 1 are density independent.
(b) Death rates of both species are density dependent.
(c) Birth rates of species 2 are density dependent.
(d) Density dependent effects on death rates are similar for both the species.

4. Two lakes (I and II) with a similar trophic structure of phytoplankton-zooplankton, planktivorous fish food chain were chosen.

To understand the 'top-down' effects, some piscivorous fish (those that feed on planktivorous fish) were introduced into Lake I, making it a system with four trophic levels. Lake II was enriched by adding large quantities of nitrates and phosphates to study the 'bottom-up' effects over a period of time. Changes in the biomasses of each trophic level were measured.

The expected major changes in the two lakes are **(JUNE 2015)**

(a) In Lake I zooplankton biomass increases, phytoplankton biomass decreases.
In Lake II both phytoplankton and planktivorous fish biomasses increase.
(b) In Lake I zooplankton biomass decreases, phytoplankton biomass increases.
In Lake II both phytoplankton and planktivorousfish biomasses increase.
(c) In Lake I planktivorous fish biomass and phytoplankton biomass decrease.
In Lake II phytoplankton biomass increases, planktivorous fish biomass decreases.
(d) In Lake I planktivorous fish and zooplankton biomasses increase.
In Lake II both phytoplankton and planktivorus fish biomasses increase.

5. In a lake ecosystem, bottom-up effects (B) refers to control of a lower trophic level by the higher trophic levels and top down effects (T) refer to the opposite. In a lake with three trophic levels - Phytoplankton (P), Zooplankton (Z) and Carnivore (C), **(JUNE 2011)**

(a) P and C are controlled by B, and Z is controlled by T
(b) P, Z and C are all controlled by T
(c) P is controlled by B, Z is controlled by T and C is controlled by B
(d) P is controlled by T, Z is controlled by B and C is controlled by T

In a lake, reducing the population of a fish which feeds on plankton was followed by a decline in the rate of primary productivity. This is consistent with which one of the following hypotheses regarding the regulation of primary productivity? **(DEC 2019)**

(a) Bottom-up control
(b) Eutrophication
(3) Top-down control
(d) Trophic pyramid

6. A small lake has three trophic levels-phytoplankton (autotrophs), Zooplankton (herbivore) and planktivorous fish (primary carnivore). Into this lake, a population of

piscivorous fish (secondary carnivore) was introduced to study the 'top-down' effects". What is the expected long-term consequence of such an introduction to phytoplankton and zooplankton trophic levels? **(JUNE 2014)**

(a) Zooplankton biomass will increase and phytoplankton biomass will decrease.

(b) Zooplankton biomass will decrease and phytoplankton biomass will increase.

(c) The biomasses of both zooplankton and phytoplankton will increase.

(d) The biomasses of both zooplankton and phytoplankton will decrease.

7. The dynamics of any subpopulation within a metapopulation differs from-that of a normal population in that the **(JUNE 2015)**

(a) birth rates are lower than the death rates.

(b) death rates are lower than the birth rates.

(c) immigration and emigration rates are significantly higher.

(d) immigration and emigration rates are negligible.

Answer Key

1.(b)	2.(d)	3.(d)	4.(a)	5.(c)	6.(a)	7.(c)

5. Life history strategies (r and K selection)

1. What do mayflies, Pacific salmon (*Oncorhynchus* spp.) and annual grain crops have in common? They all are **(JUNE 2017)**

(a) semelparous (b) iteroparous

(c) oviparous (d) Viviparous

2. Which of the following is NOT semelparous? **(JUNE 2015)**

(a) Dracena (b) Bamboo

(c) Cicada (d) Mayfly

3. A specialist species has a **(DEC 2011)**

(a) wider niche and high efficiency of niche utilization

(b) narrower niche and high efficiency of niche utilization

(c) wider niche and low efficiency of niche utilization

(d) narrower niche and low efficiency of niche utilization

4. One life history trait that is not characteristic of very small sized organisms is **(MODEL PAPER)**

(a) delayed age at first reproduction.

(b) earlier age at first reproduction.

(c) high population growth rate.

(d) short lifespan.

5. A species whose life history strategies allow for high intrinsic rates of increase (r- strategist) will also exhibit the following **EXCEPT** **(DEC 2017)**

(a) high tolerance for both environmental instability and low-quality resources.

(b) short period of exponential population growth (r)

(c) reproductive strategy that involves random mating, semelparity and little or no parental investment

(d) survivorship that show density-dependent mortality, typically exhibiting Type 1 or 2 survivorship curves.

6. Compared to K-selection, r-selection favours **(JUNE 2015)**

(a) rapid development, smaller body size and early, semelparous reproduction.

(b) rapid development, smaller body size and early, iteroparous reproduction.

(c) slow development, larger body size and late, iteroparous reproduction.

(d) slow development, smaller body size and late, iteroparous reproduction

7. Which one of the following trait set characterizes best a *r* selected species? **(MODELPAPER)**

(a) Usually a type III survivorship curve, short life span and density dependent mortality

(b) Usually a type I survivorship curve, short life span and density dependent mortality

(c) Usually a type I survivorship curve, long life span and density independent mortality
(d) Usually a type III survivorship curve, short life span and density independent mortality

8. A hormone facilitates oogenesis in mammals. The same hormone down regulates the transfer of nutrients from the placenta. The likely effect of the hormone on life history strategies is that it facilitates **(DEC 2013 GU)**
(a) 'r' type of reproduction.
(b) 'K' type of reproduction.
(c) both 'r' and 'K' reproduction.
(d) neither 'r' nor 'K' reproduction.

9. The most important reproductive strategies of big trees in a forest are **(JUNE 2014)**
(a) earlier age at first reproduction and production of a large number of small seeds.
(b) earlier age at first reproduction and production of a small number of large seeds.
(c) later age at first reproduction and production of a large number of small seeds.
(d) later age at first reproduction and production of a small number of large seeds.

10. Which one of the following statements supports the concepts of trade-off in the evolution of life history trades? **(DEC 2016)**
(a) Level of parental care and clutch size are positively correlated.
(b) Animals mature in early tend to live longer
(c) An increase in the seed size is usually associated with the decrease in the seed number.
(d) Allocation of higher energy for reproduction leads to higher population growth

11. In life history evolution there is generally a trade off between the size and number of offsprings produced. Some conditions are listed below:
A. Scarcity of food during the early stages of life
B. Provision of parental care
C. High mortality during early stages of life
D. Predator's preference for large sized prey

What are the above two conditions that would favour the production of a small number of large-sized offspring? **(DEC 2014)**
(a) B and C (b) B and D
(c) A and B (d) A and C

12. Suppose you discovered a new species about which you know only two facts: it is small sized (<10 cm) and short lived (<20 days). Which of the following strategies is most likely to be true for this species? **(DEC 2011)**
(a) Breed early and more than once in life and produces large number of small sized offspring
(b) Breed late and only once in life and produces large number of small sized offspring
(c) Breed early and only once in life and produces large number of small sized offspring
(d) Breed early and only once in life and produces small number of large sized offspring

13. There are three species of frogs - A, B and C. Species A does not provide parental care for its eggs and larvae. Species B is subjected to predation by a predator that selectively feeds only on small-sized larvae. Species C faces progressively decreasing opportunities for breeding with increasing age. Assuming that resources available for reproduction are similar for A, B and C. Which of the following strategies would have been favored? **(JUNE 2014)**
(a) A should produce large number of small-sized offspring; B should produce a small number of large-sized offsprings; C should breed earlier in life.
(b) Species A and B should produce a small number, of large-sized offspring and C should breed earlier in life.

(c) Both species A and B should produce a large number of small-sized offspring and C should breed later in life but increase its clutch size.

(d) Species A should produce a small number of large-sized offspring; B should 'produce a large number of small sized offspring and C should breed earlier in life with a small clutch size.

Answer Key

1.(a)	2.(a)	3.(b)	4.(a)	5.(d)	6.(a)	7.(d)
8.(a)	9.(d)	10.(c)	11.(c)	12.(c)	13.(a)	

6. Species Interactions and competition

1. Given below is a matrix of possible interactions beneficial. (+), harmful (-), Neutral (0) between species 1 and (b) The names of interactions, A, B, C and D, respectively; are **(JUNE 2015)**

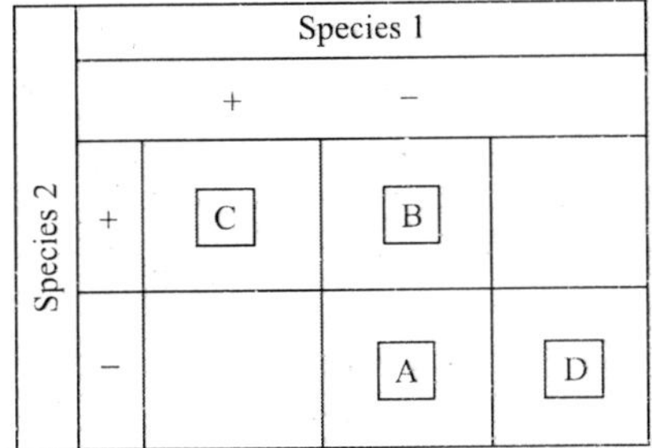

(a) Predation, competition, mutualism, commensalism

(b) Mutualism, competition, amensalism, commensalism

(c) Competition, predation, mutualism, amensalism

(d) Competition, mutualism, commensalism, predation

2. An example of the species interaction called commensalism is **(DEC 2014)**

(a) nitrogen-fixing bacteria in association with legume plant roots.

(b) A microbes in living human gut.

(c) female mosquito deriving nourishment from human blood

(d) orchid plant growing on the trunk of mango tree

3. The following graphs show the population growth of two species P and Q, each growing either alone (a) or in the presence of other species (b). **(MODELPAPER)**

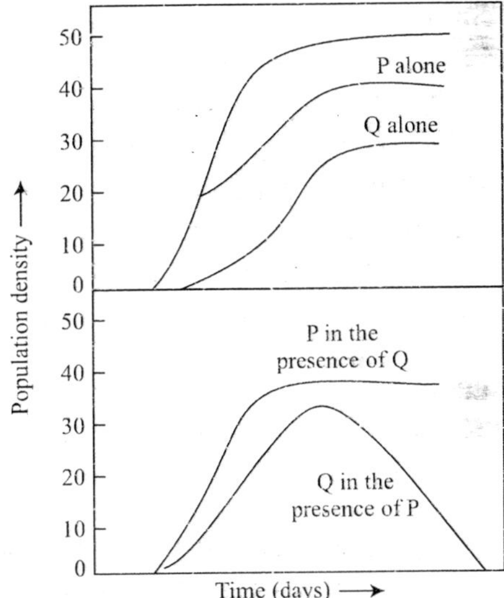

The most important conclusion to be drawn from the graph is

(a) P and Q are equally competitive.

(b) In competition, the growth of both species is adversely affected.

(c) In competition, species P remains unaffected while Q suffers.

(d) There is no evidence of competitive exclusion.

4. Given below in Column A are schematic representations of three types of pairwise species interactions and the name of some interactions are in Column B. **(DEC 2018)**

	Column A		Column B
A	A ⇄ B (−, −)	(i)	Apparent competition
B	Shared Resource; +/− A, +/− B; A ⇄ B (−, −)	(ii)	Interference competition
C	Herbivore; −/+ A, −/+ B; A ⇄ B (−, −)	(iii)	Direct amensalism
		(iv)	Exploitation competition

Select the best match for interaction between column A and B in each schematic

(a) A - (iii); B - (ii); C - (iv)

(b) A -(iv); B -(ii); C - (iii)

(c) A - (ii); B -(iv); C -(i)

(d) A - (iii); B - (i); C - (ii)

5. In column I are given equations and in column II what the equations represent. Column I and II are incorrectly matched.

I.	II.
1. $\frac{dN_1}{dt} = r_1N_1\left(\frac{K_1 - N_1}{K_1}\right)$	(a) Population growth of species 1 in the presence of competing species 2
2. $\frac{dN_1}{dt} = r_1N_1$	(b) Logistic population growth of species 1
3. $\frac{dN_1}{dt} = r_1N_1\left(\frac{K_1 - N_1 - \alpha N_2}{K_1}\right)$	(c) Population grwoth of prey species 1 in the presence of predator species 2.
4. $\frac{dN_1}{dt} = N_1b_1 - d_1N_1N_2$	(d) Exponential population grwoth of species 1.

The correct match in sequence of the four equations from column II is

(MODEL PAPER)

(a) b, d, a and c (b) d, b, c and d

(c) b, d, c and a (d) d, b, a and c

6. Three islands have identical habitat characteristics. On first island rodent species A is present at a density 325/km^2 Second island has only species B at a density of 179/km^2 On the third island, both A and B co-exist with densities 297/km^2 and 150/km^2, respectively. Which of the following can be inferred from this? **(JUNE 2013)**

(a) The two species do not compete with each other.

(b) The intra-species competition is more intense than inter-species competition.

(c) The inter-species competition is more intense than intra-species competition.

(d) The inter and intra species competition are of the same intensity.

7. In Lotka and Voltera's two species competition model:

$$\frac{dN_1}{dt} = r_1N_1\left[\frac{K_1 - N_1 - N_2\alpha_{12}}{K_2}\right] \text{ and}$$

$$\frac{dN_2}{dt} = r_2N_2\left[\frac{K_2 - N_2 - \alpha_{21}N_1}{K_2}\right]$$

Where N represents population size, r growth rate and K maximum carrying capacity for species 1 and (b) The inter-specific competition coefficient α_{12} <1 will mean: **(JUNE 2012)**

(a) Individuals of species 2 have less inhibiting effect on individuals of species 1 than individuals of species 1 on other of their own species

(b) Individuals of species 2 have greater inhibiting effect on individuals of species 1 than individuals of species 1 on other of their own species

(c) Individuals of species 1 have less inhibiting effect on individuals of species 2 than individuals of species 2 on other of their own species

(d) Individuals of species 1 have greater inhibiting effect on individuals of species 2 than individuals of species 2 on other of their own species

8. The Lotka-Volterra model of competition between species A and B is given by the equations

Species A: $dN_A/dt = r_A N_A (K_A - N_A - \alpha N_B / K_A)$

Species B: $dN_B/dt = r_B N_B (K_B - N_B - \beta N_A / K_B)$

Given that species A always wins, which of the following is true according to the model **(DEC 2013 GU)**

(a) $K_2 > K1/\beta$ and $K_1 < K_2/\alpha$

(b) $K_2 > K1/\alpha$ and $K_1 < K_2/\beta$

(c) $K_2 < K1/\beta$ and $K_1 > K_2/\alpha$

(d) $K_2 < K1/\alpha$ and $K_1 > K_2/\beta$

9. The diagram represents competition between species 1 and species 2 according to Lotka-Volterra model of competition.

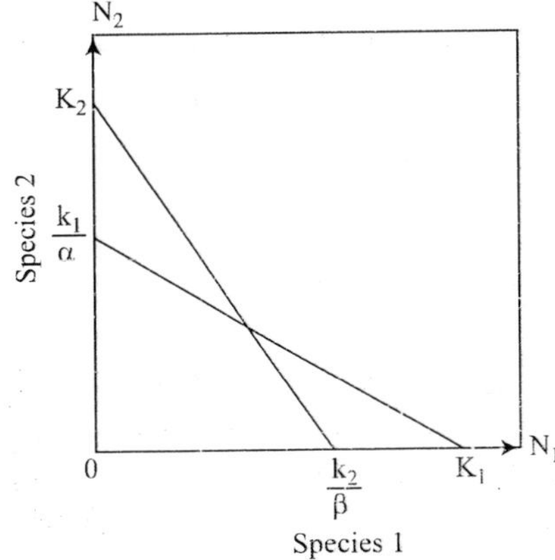

Given the conditions in the diagram, the predicted outcome of competition is **(DEC 2014)**

(a) Unstable coexistence between species 1 and 2 because $K1 > K2/\beta$ and $K2 > K1/\alpha$

(b) Unstable coexistence between species 1 and 2 because $K1 < K2/\beta$ and $K2 < K1/\alpha$

(c) Stable coexistence between species 1 and 2 because $K1 > K2/\beta$ and $K2 > K1/\alpha$

(d) Stable coexistence between species 1 and 2 because $K1 < K2/\beta$ and $K2 < K1/\alpha$

10. For two species A and B in competition, the carrying capacities and competition coefficients are

$K_A = 150$ $K_B = 200$

$\alpha = 1$ $\beta = 1.3$

According to the Lotka-Volterra model of interspecific competition, the outcome of competition will be **(JUNE 2016)**

(a) Species A wins.

(b) Species B wins.

(c) Both species reach a stable equilibrium.

(d) Both species reach an unstable equilibrium.

11. Two species, M and N, occupy the same habitat. Given below is a 'state-space' graph in which the abundance of species M is plotted on the X-axis and abundance of species N is plotted on the Y-axis. For each species, the zero-growth isocline is plotted.

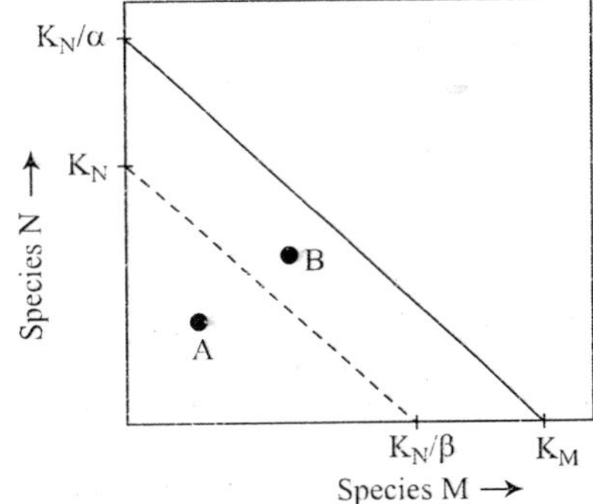

—— zero-growth isocline for species M

_ _ _ zero-growth isocline for species N

K_M = carrying capacity of the habitat for species M -in absence of species N

K_N = carrying capacity of the habitat for species N in absence of species M

α = per capita effect of species N on M

β = per capita effect of species M on N

Based on the above plot some deductions are made. Which one of the following statements is **INCORRECT**? **(JUNE 2017)**

(a) At point A, populations of both the species M and N increase

(b) At point B, population of species M increase while that of species N decreases

(c) At point B, population of species N increase while that of species M decreases

(d) Ultimately species N will be eliminated

Answer Key

1.(c)	2.(d)	3.(c)	4.(c)	5.(a)	6.(b)	7.(a)
8.(d)	9.(a)	10.(b)	11.(c)			

7. Herbivory, Carnivory, Pollination, Symbiosis.

1. Which of the following statements is correct according to the plant vigor hypothesis dealing with herbivore-plant interaction? **(DEC 2013 GU)**

(a) Herbivores prefer to attack slow growing stressed plants which produce leaves that are higher in nitrogen.

(b) Herbivores prefer to attack fast growing plants rather than slow growing, stressed plants.

(c) Small amount of grazing will increase plant growth and fitness rather than cause harm to the plants.

(d) Higher plant growth rates will result in less investment in defensive chemicals.

2. Nearly 25% of all insect species are known to be herbivores. Yet, in spite of such heavy herbivore pressure, globally green plants tend to persist, contributing to a green earth". Which of the following account for the relative success of green plants? **(MODEL PAPER)**

(A) Herbivore insects are inefficient feeders

(B) Herbivore insect densities are kept low by predators

(C) Plants secrete herbivore-deterrent chemicals

(a) (A) and (C) (b) (A) only

(c) (B) and (C) (d) (A), (B) and (C)

3. Three species M, N and O when grown independently in a laboratory, showed typical logistic growth curves. However, when grown in pairs, the following growth curves were observed.

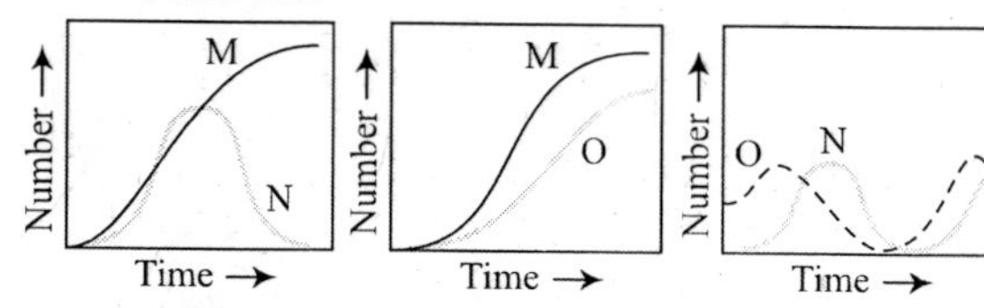

What interpretation regarding the interspecific relationship between M, N and O can be deduced from the above observations? **(DEC 2017)**

(a) N predates over O and therefore can also predate on M.

(b) N is competed out by M and O

(c) N and O possibly have a prey-predator relationship.

(d) M and O exhibit prey-predator relationship.

4. To understand prey-predator relationship, *Didinium* (predator) and Paramecium (prey) were used. *Paramecium* population was grown with sand sediment as hiding place or refuge. To this population, *Didinium* was introduced only once. What would happen to the prey population in the course of time? **(JUNE 2017)**

(a) The population will steadily decrease and vanish

(b) The population will initially increase and then stabilize

(c) The population will initially decrease, then increase and stabilize

(d) The population will steadily increase

5. Incorporating additional ecological factors into the Lotka-Voltera predator-prey model can change the predator isocline. Given below are three state-space graphs (A-C) representing modification of predator isocline due to the ecological factors listed below (i-iii).

(i) Victim abundance acting as predator carrying capacity

(ii) Availability of alternate prey (victim) population

(iii) Predator carrying capacity determined by factors other than victim abundance.

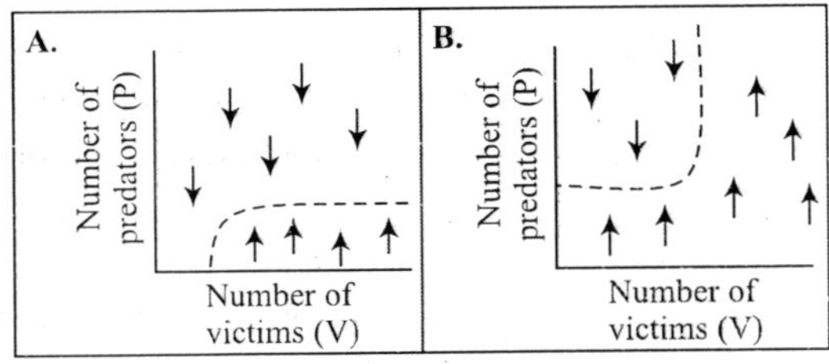

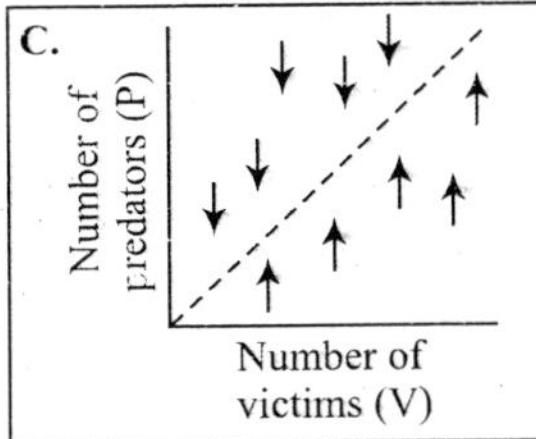

Which one of the following options represents all correct matches of the state space graphs with their ecological factor?

(JUNE 2019)

(a) A — (ii), B — (iii), C —(i)

(b) A — (ii). B — (i), C - (iii)

(c) A — (iii), B — (ii), C — (i)

(d) A — (i). B — (ii), C — (iii)

6. Predator-prey interactions in a community are represented in the figure below.

(DEC 2019 ASSAM)

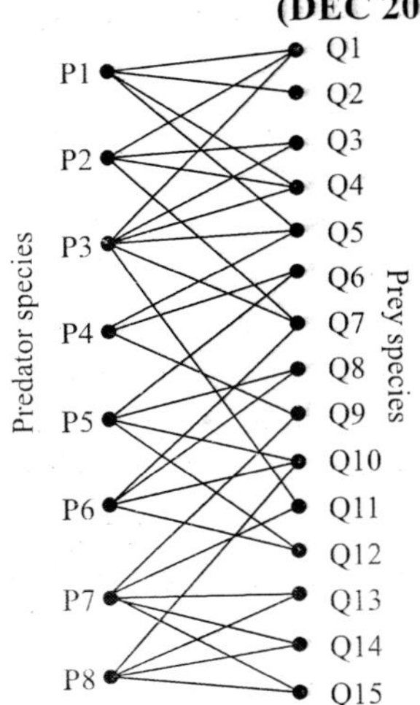

Given below are some statements related to the above figure:

(i) Predators and prey are generalists

(ii) Predators have expanded their prey base over time

(iii) Predators and prey have co-evolved

(iv) There is speciation in prey while predators are prone to extinction

Which one of the following options represents statement/s that can be correctly inferred from the figure?

(a) (i) and (iii) (b) (iii) only

(c) (i) only (d) (ii) and (iv)

7. Following are the main types of defense employed by prey species against predators

Types of defense:

Chemical with aposematic coloration (A);

Cryptic coloration (B);

Batesian mimicry (C);

Intimidation display (D)

Prey Species:

Grasshoppers and seahorses (i);

Hoverflies and wasps (ii);

Bombardier beetles, ladybird beetles, many butterflies (iii);

Frilled lizard, Porcupine fish (iv)

Which one of the following combinations is correct? **(JUNE 2015)**

(a) A-(i) B-(iii) C- (ii) D (iv)

(b) A-(iv) B-(ii) C-(i) D (iii)

(c) A-(iii) B-(i) C-(ii) D (iv)

(d) A-(ii) B-(iii) C- (i) D- (iv)

8. Given below are few traits and related functions:

TRAIT		FUNCTION	
(i)	Aposematism	A	Acquiring food
(ii)	Basking	B	Avoiding predation
(iii)	Cooperative hunting	C	Territory defence
(iv)	Song	D	Thermoregulation

Match the above given traits to their most likely functions. **(JUNE 2019)**

(a) i-C; ii-D; iii-B; iv-A
(b) i-D; ii-B; iii-A; iv-C
(c) i-B; ii-D; iii-A; iv-C
(d) i-C; ii-A; iii-D; iv-B

9. Fish species X and Y feed on mayfly nymphs in their stream habitat. In a laboratory experiment, the predation intensity of X and Y on their prey was tested under dark (D) and light (L) conditions. Thus, the experimental protocol included four aquaria - LX, LY, DX and DY. In each aquarium containing 100 mayfly nymphs, one fish was introduced and allowed to feed for 30 minutes. Then the fish was removed and the number of mayfly nymphs left uneaten in each aquarium was counted. The results are shown graphically below

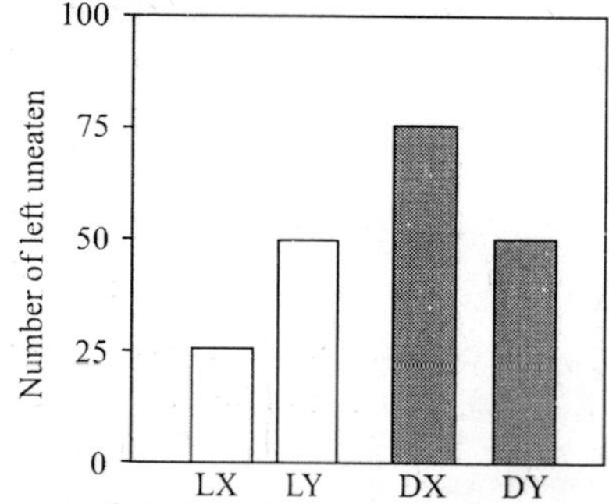

The most significant conclusion from the results is: **(JUNE 2014)**

(a) X is a visual predator, but has less predation impact on the prey than Y.
(b) X is a visual predator and has greater predation impact on the prey than Y.
(c) Y is a visual predator and has greater predation impact on, the prey than X.
(d) Y is a visual predator but has less predation impact on the prey than X.

10. A plant is visited by bats during the night and sunbirds during the day. Given this information, which of the following characters best match this plant? **(JUNE 2017)**

(a) The plant is a herb with saucer shaped white flowers
(b) The plant is a shrub with tubular, red, diurnal flowers
(c) The plant is a liana with tubular cream coloured flowers
(d) The plant is a grass with white coloured fragrant, spikelets

11. Identify the pollinators for the flowers with following pollination syndromes

A. Flowers dull colored, located away from foliage, floral parts turgid,
B. Flowers bright red, crowded, turgid, neclar watery and sucrose rich.
C. Flowers white with pleasant odor, corolla tube long, night blooming. **(DEC 2012)**

(a) (A) Bird: (B) Bat; (C) Butterfly
(b) (A) Bat: (B) Bird: (C) Moth
(c) (A) Bat; (B) Bird; (C) Bee
(d) (A) Bird; (B) Bat; (C) Carrion fly

12. A red coloured tubular flower without any odour is most likely to be pollinated by **(DEC 2015)**

(a) beetles. (b) bees.
(c) butterflies. (d) birds.

13. A plant with blue-coloured flowers was observed to attract a large number of pollinators. However, these flowers were not producing any nectar. Which of the following can be a logical explanation to the observation? **(JUNE 2013)**

(a) There could be another species in the vicinity that has blue flowers and is rich in nectar.
(b) There is no other species with blue flowers in the vicinity so pollinators are compelled to visit this species.
(c) Pollinators may not have blue-colour vision.
(d) Pollinators may be able to see only blue colour.

14. The following matrix shows the relationship between probability of death and duration of species association.

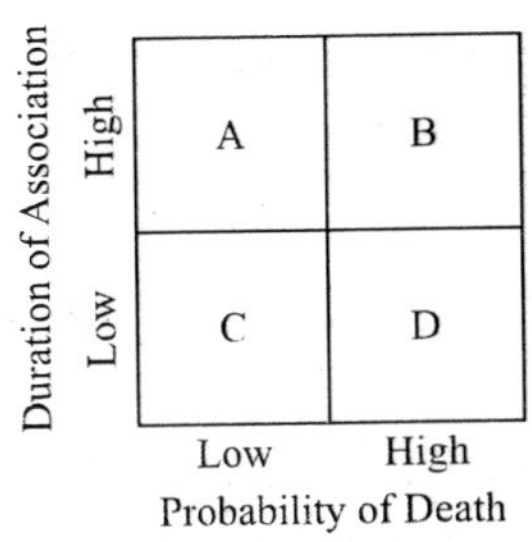

In the above, A, B, C and D are: **(DEC 2013)**

(a) A-Parasites, B - Parasitoids, C - Grazers, D - Predators

(b) A-Carnivores, B - Herbivores, C- Parasites, D -Parasitoids

(c) A-Grazers, B - Parasites, C - Herbivores, D - Parasitoids

(d) A-Predators, B - Parasitoids, C - Parasites, D - Carnivores

Answer Key

1.(b)	2.(c)	3.(c)	4.(c)	5.(c)	6.(c)	7.(c)
8.(c)	9.(b)	10.(c)	11.(b)	12.(d)	13.(a)	14.(a)

8. Community structure and attributes

1. Following are the graphical representation of various hypothesis proposed for explaining the possible relationship between the species richness on (X) axis and community services on (Y) axis. **(DEC 2016)**

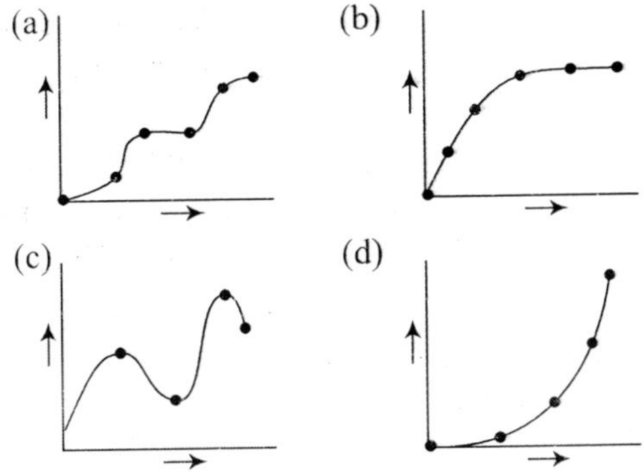

Which of the following is the correct match between the graphical representation and the hypothesis **(DEC 2016)**

(a) a-redundancy, b- keystone, c-rivat, d-idiosyncratic

(b) a- idiosyncratic, b- rivat, c- keystone, d-redundancy

(c) a- rivat, b- redundancy, c- idiosyncratic, d- keystone

(d) a-rivat, b- keystone, c- redundancy, d-idiosyncratic

2. Given below are the species accumulation curves and rarefaction curves measured in an ecological community.

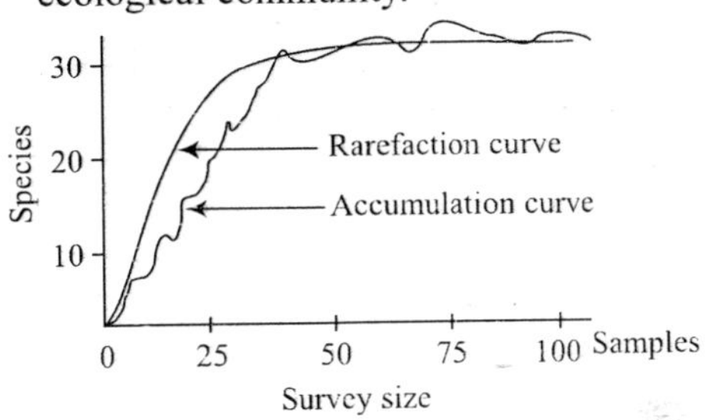

Which one of the following statements is **INCORRECT** about the two curves? **(DEC 2017)**

(a) Species accumulation curve moves from left to right and rarefaction curve moves from right to left.

(b) Species accumulation curve represents the total species richness of the assemblage.

(c) Rarefaction curve represents the mean of repeated resampling of all pooled samples.

(d) Rarefaction curve is the realized accumulation value of the total species in a community.

3. The rank/abundance plot given illustrates three well known abundance curves (a, b, c). **(DEC 2017)**

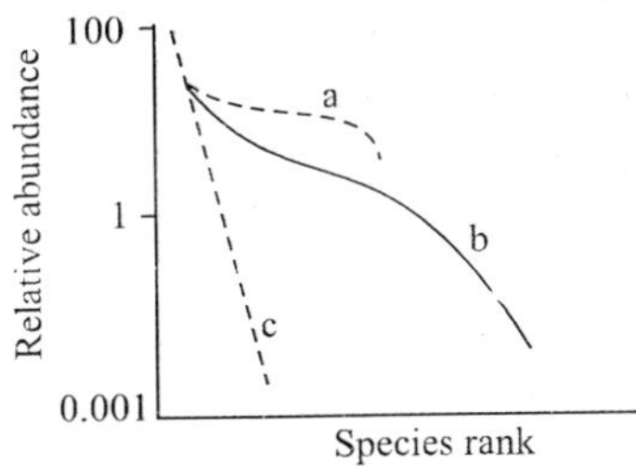

Based on the shapes of the curves select the correct option.

(a) a-exponential, b-log normal, c-geometric series
(b) a-broken stick, b-geometric series, c-log normal
(c) a-broken stick, b-log normal, c-geometric series
(d) a-exponential, b-geometric series, c-log normal

4. For a dominance-diversity relationship curve of a community, species sequence was plotted on the X axis and relative importance on the log Y axis. The relationship resulted in a straight line, which represents the **(DEC 2013 GU)**
(a) geometric model and niche pre-emption hypothesis.
(b) log normal model and random-niche boundary hypothesis.
(c) broken stick model and random-niche boundary hypothesis.
(d) lognormal model and niche pre-emption hypothesis.

5. Which of the following characterizes natural communities under equilibrium? **(DEC 2013 GU)**
(a) Biotic decoupling.
(b) Species independence.
(c) Large stochastic effects.
(d) Density dependence.

6. Annual weeds of arable lands are classified as **(DEC 2013)**
(a) phonerophytes. (b) therophytes.
(c) chamaephytes. (d) geophytes.

7. The table given below lists types of plant communities and types of growth forms.

Plant communities		Growth forms	
(i)	Dry grassland	A	Chamaephytes
(ii)	Semi desert	B	Cryptophytes
(iii)	Tropical forests	C	Hemicryptophytes
(iv)	Tundra	D	Phanerophytes

Which of the following is the best match for the plant communities with most dominant growth form generally present in that community? **(JUNE 2019)**
(a) i - D; ii- A; iii - B; iv -C
(b) i -C; ii- A; iii - D; iv -D
(c) i - B; ii-C; iii -D; iv -C
(d) i - C; ii- A; iii -D; iv – C

8. A matrix of plant strategies based on four possible combinations of the intensities of stress and disturbance is given below:

Disturbance	Stress	
	Low	High
Low	A	B
High	C	D

In this matrix, ruderal (r) plants are represented by **(DEC 2013 GU)**
(a) A (b) B
(c) C (d) D

9. Given below is a graphical representation of plant life histories based on Grime's model in which disturbance and competition are the important factors.

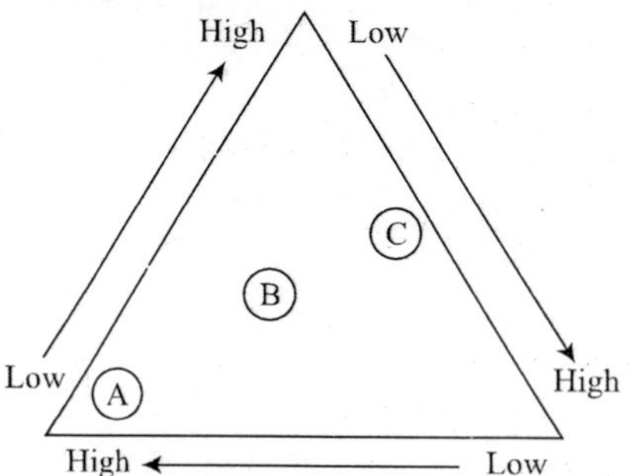

Which of the following options correctly represents A, B and C, respectively in the figure above? **(DEC 2018)**
(a) perennial herbs, trees and shrubs, annual plants.
(b) annual plants, perennial herbs, trees and shrubs.
(c) annual plants, trees and shrubs, perennial herbs.

(d) trees and shrubs, perennial herbs, annual plants.

10. Ruderal species are those which are found in the environments with **(JUNE 2018)**

(a) low disturbance, high competition

(b) high disturbance, low competition

(c) low disturbance, low competition

(d) high disturbance, high competition

11. Following four types of species were observed in a community:

A. Species A has a large effect on community because of its abundance.

B. Species B has a large role in community out of proportion to its abundance.

C. Status of species C provides information on the overall health of an ecosystem

D. Significant conservation resources are allocated to species D which is single, large and instantly recognizable.

According to above description, species A, B, C and D are called respectively **(DEC 2014)**

(a) Dominant, Keystone, Indicator and Flagship

(b) Keystone, Flagship, Dominant and Indicator.

(c) Keystone, Dominant, Indicator and Flagship.

(d) Flagship, Dominant, Keystone and Indicator.

12. Species serving as a proxy for entire communities and ecosystems, so that the entire system is conserved if they are conserved, are known as: **(DEC 2013 GU)**

(a) indicator species. (b) flagship species.

(c) keystone species. (d) umbrella species.

13. Following are the descriptions used by conservation biologists for characterizing species / groups in a community:

A. Species with a disproportionally large effect on its environment relative to its abundance

B. Species defining a trait or characteristics of the environment

C. Species whose conservation leads to direct protection of other species

D. Species which is instantly recognizable and used as the focus of a broader conservation effort

Which of the following combination correctly identifies these species / groups? **(JUNE 2017)**

(a) A - Keystone species, B- Indicator species C - Flagship species, D - Umbrella species

(b) A - Keystone species, B - Indicator species C - Umbrella species, D -Flagship species

(c) A- Indicator species, B - Flagship species C - Umbrella species. D - Keystone species

(d) A - Umbrella species, B - Indicator species C -Keystone species, D - Flagship species

14. Which of the following statements is INCORRECT about a keystone species? **(DEC 2017)**

(a) Species other than consumers can be a keystone species

(b) Keystone species has influence on a community proportionate to its abundance

(c) Removing a keystone species can reduce species richness of a community

(d) Removing a keystone species can effect successive trophic levels causing a trophic cascade

15. Which Of the following X-Y relationships does NOT follow the pattern shown in the graph? **(DEC 2015)**

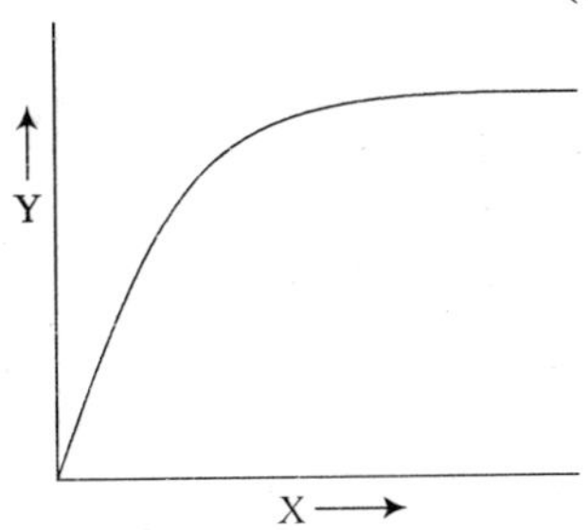

(a) Number Of prey killed (Y) in relation to prey density (X)

(b) Photosynthetic rate (Y) in relation to light intensity (X)

(c) Species richness (Y) in relation to area (X)

(d) Tree species richness (Y) in relation to actual evapotranspiration

Answer Key

1.(c)	2.(d)	3.(c)	4.(a)	5.(d)	6.(b)	7.(d)
8.(c)	9.(b)	10.(b)	11.(a)	12.(d)	13.(b)	14.(b)
15.(d)						

9. Levels of species diversity and its measurement

1. In the context of diversity patterns of species, which one of the following statements is **INCORRECT**? **(DEC 2017)**
   (a) Alpha diversity is diversity within a single community
   (b) Beta diversity is a measure of the change in species composition from one community or habitat to another
   (c) Alpha diversity is the regional diversity found among range of communities in a geographical region
   (d) Gamma diversity is the regional diversity found among range of communities/habitats in a geographical region

2. The possible relationships between levels of disturbance and species diversity in a biological community are that species diversity
   A. is unaffected by disturbance.
   B. is highest at intermediate levels of disturbance.
   C. decreases exponentially with increasing levels of disturbance.
   D. starts decreasing only at higher levels of disturbance.

(a)

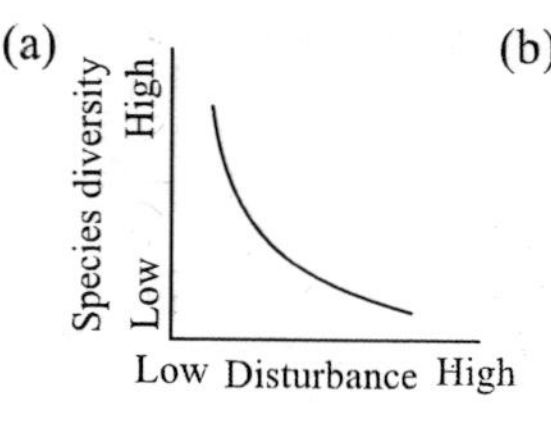

(b)

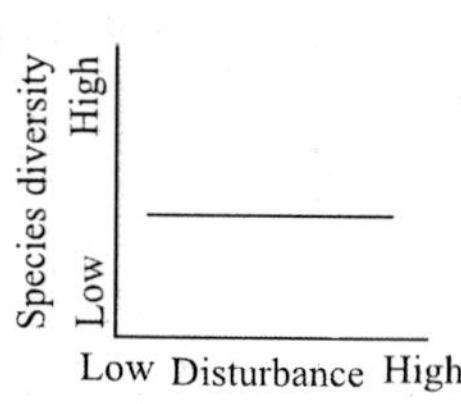

(c)

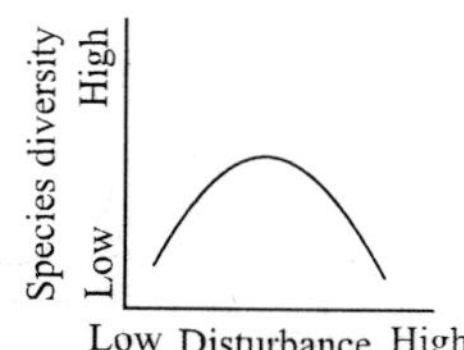

(d)

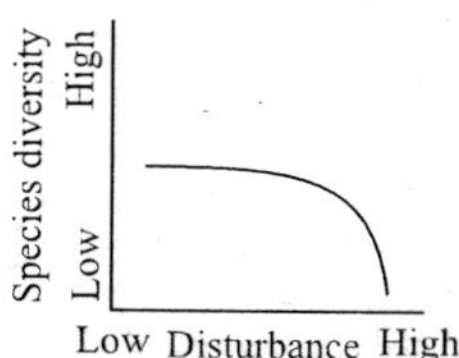

Match each graph with its corresponding statements above: **(DEC 2014)**
   (a) 1-D, 2-C, 3-B, 4-D
   (b) 1-C, 2-D, 3-B, 4-A
   (c) 1-A, 2-B, 3-C, 4-D
   (d) 1-C, 2-A, 3-B, 4-D

3. Which of the following graphs illustrates the current consensus on the role of *disturbance* on the species richness of a community? **(DEC 2011)**

(a)

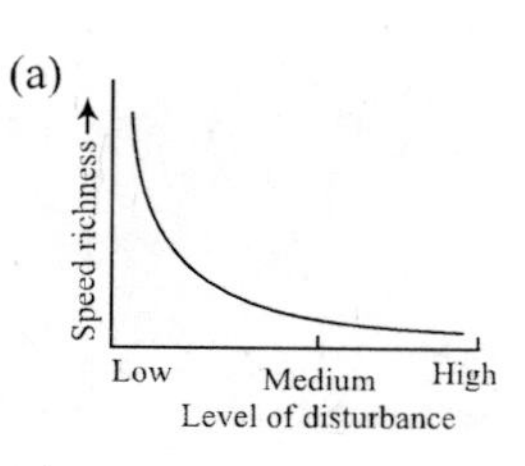

(b)

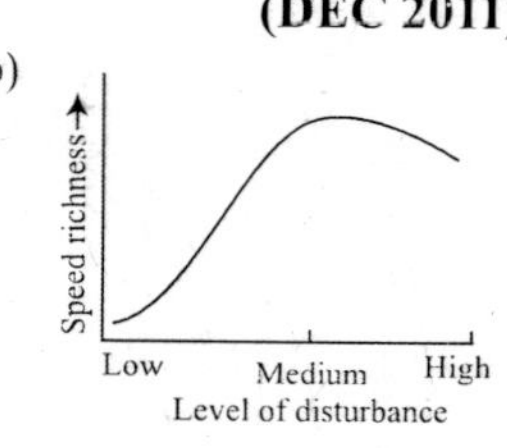

(c)

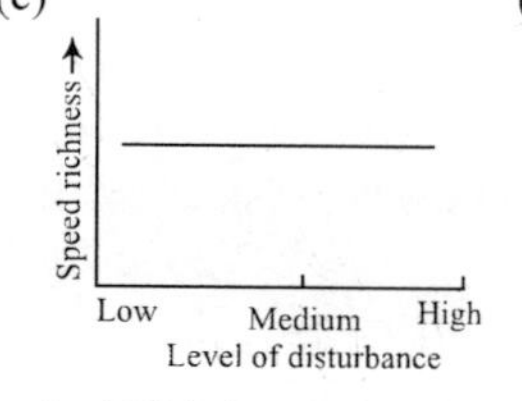

(d)

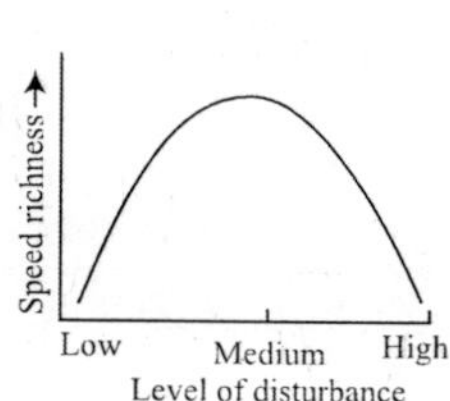

4. Which of the following is/are NOT valid explanation(s) for the observed pattern of species richness? **(JUNE 2015)**
   A. Older communities are more species rich.
   B. Large areas support more species.
   C. Natural enemies promote reduced species richness at local level.

D. Communities in climatically similar habitats may themselves be similar in species richness.

E. Greater productivity permits existence of more species.

(a) B, C and D (b) Only C

(c) Only D (d) A, B and E

5. Following table shows the presence (+) or absence (-) of five species in three communities (A, B, C):

Species	Community		
	A	**B**	**C**
1	+	+	+
2	+	+	–
3	–	–	+
4	+	–	–
5	–	–	+

Based on the above, which of the following is the correct order of similarity between two pairs of communities? **(DEC 2015)**

(a) A and B > B and C > A and C

(b) A and B > A and C > B and C

(c) B and C > A and B > A and C

(d) A and C > A and B > B and C

6. Following table shows the number of individuals of five tree species in a community; **(DEC 2015)**

Tree Species	No. of Individuals
A	50
B	20
C	20
D	05
E	05

Based on the above, the Simpsons diversity (DS) index of the community will be

(a) 0.552. (b) 0.335.

(c) 0.435. (d) 0.345.

7. The following table shows the number of individuals of each species found in two communities: **(JUNE 2015)**

Community	Species			
	A	**B**	**C**	**D**
C1	25	25	25	25
C2	80	05	05	10

(Hint: In values for 0.05, 0.10, 0.25 and 0.80 are -3.0, -2.3, -1.4 and -0.2, respectively)

The calculated Shannon diversity index (H) values for communities C1 and C2, respectively are

(a) 1.4 and 0.69 (b) 1.2 and 0.34

(c) 2.1 and 0.43 (d) 1.8 and 0.37

8. Following table shows the number of individuals of different species in two communities, X and Y.

Species	Community	
	X	Y
1	250	490
2	200	125
3	180	80
4	100	65
5	65	50
6	30	10
7	5	10
Total	830	830

Based on the above data and the Simpson's Diversity Index which one of the following statements is true? **(DEC 2019 ASSAM)**

(a) X has greater alpha diversity than Y

(b) Y has greater alpha diversity than X

(c) X and Y have equal alpha diversities

(d) X has greater species richness than Y

9. While studying the diversity of 3 communities, 5 species and 50 individuals were recorded from each community. The number of individuals under each species was listed as mentioned in the following Table. In which of the following community Pielou's Evenness index (*e*) will be 1? **(JUNE 2012)**

	Community	Species				
		1	2	3	4	5
(a)	A	20	8	7	5	10
(b)	B	10	10	10	10	10
(c)	C	10	12	10	8	10
(d)	D	1	1	1	1	46

10. A field ecologist gathers following data (abundance values) in order to study diversity of species in four plant communities.

Species	Community			
	C1	C2	C3	C4
SP1	90	0	0	40
SP2	60	0	0	65
SP3	15	25	25	40
SP4	0	180	0	0
SP5	25	0	215	45
SP6	65	0	20	55

Based on the above observations, the ecologist draws following conclusions:

A. Plant communities C1 and C4 show strong similarity with each other

B. Plant communities C1 and C4 as well as communities C2 and C3 show strong similarity with each other

C. Plant community C1 is most diverse

D. Plant community C4 is most diverse

Which of the following statements is correct regarding above conclusions? **(JUNE 2019)**

(a) All the conclusions are correct

(b) Only conclusions A and D are correct

(c) Only conclusions A and C are correct

(d) Only conclusions B and D are correct

11. In the graph below, large boxes (denoted by P, Q, R, S) represent a region, whereas smaller boxes represent habitats. Labels S1, S2 above each small box represent species present in the given habitat denoted by that box.

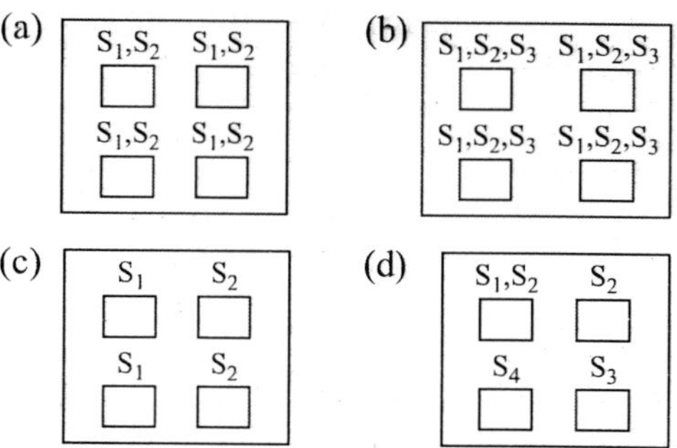

Given the above graphs, choose the option which correctly depicts the regions which show maximum α and maximum β diversity, respectively **(DEC 2019)**

(a) Q and S (b) P and R

(c) S and P (d) Q and R

12. Grassland plots with varying number of grass species were cultivated for 10 years. At the end of the experiment, total plant cover was measured. Soil nitrogen was also measured to assess its utilization by plants. The relationships are shown in the following plots.

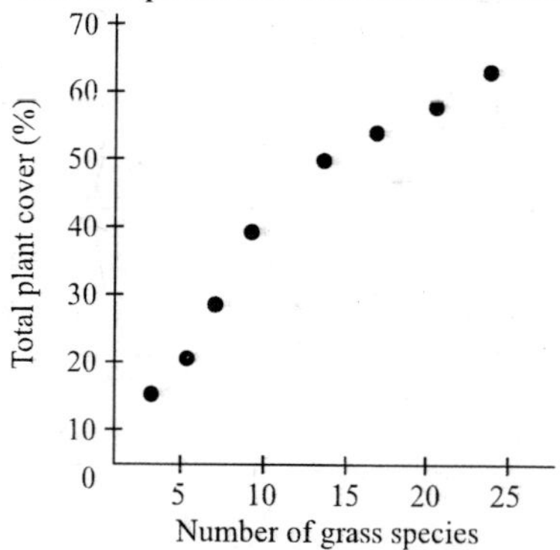

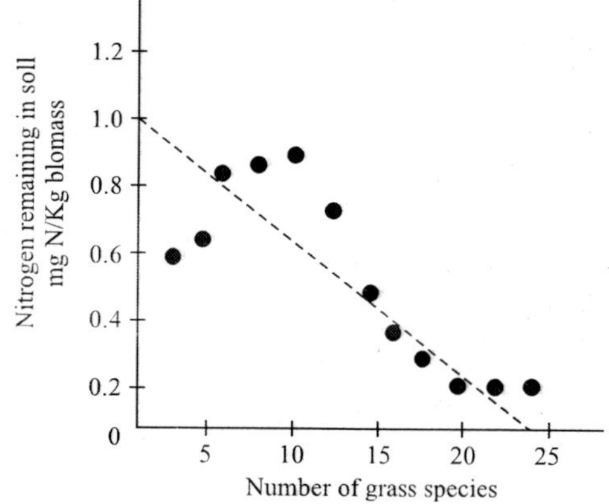

Which one of the following inferences can be drawn from the above experiment? **(DEC 2018)**

(a) Grasses in plots with lower species richness enriched soil nitrogen, thereby increasing the plant cover.

(b) Plots with greater species richness showed greater stability and more efficient soil nitrogen utilization.

(c) Plots with greater species richness utilized nitrogen more efficiently, but would not show increased net primary production.

(d) No correlation can be drawn between species richness, community productivity and nitrogen utilization.

Answer Key

1.(c)	2.(d)	3.(d)	4.(b)	5.(a)	6.(b)	7.(a)
8.(a)	9.(b)	10.(b)	11.(a)	12.(b)		

10. Ecological Succession

1. If the milk is left open, lactose is fermented first to produce acid. This is followed by protelytic bacteria which increases the pH. Ultimately milk fats are degraded to produce rancidity. This is an example of **(JUNE 2012)**

   (a) ecological succession

   (b) Antagonism

   (c) interference competition

   (d) Microevolution

2. Which of the following characteristic of an early community? **(JUNE 2011)**

   (a) Narrow niche specialization.

   (b) High species diversity.

   (c) community production.

   (d) open mineral cycling.

3. Which of the following is NOT a physiological characteristic of early successional plants? **(DEC 2011)**

   (a) High respiration rate

   (b) Inhibition by far-red light

   (c) High transpiration rate

   (d) Low photosynthetic rate

4. Which of the following is NOT a characteristic of late successional forest plant species? **(DEC 2012)**

   (a) Large seed size, high root to shoot ratio.

   (b) Long seed dispersal distance, long seed viability

   (c) Slow growth rate, long maximum life span.

   (d) Low light saturation intensity, high efficiency at low light.

5. In tolerance model of succession, environment modified by early occupants is **(DEC 2013 GU)**

   (a) less suitable for subsequent recruitment of Early Successional Species (ESS), but has little or no effect on the subsequent recruitment of Late Successional Species (LSS).

   (b) less suitable for subsequent recruitment of ESS and more suitable for the recruitment of LSS.

   (c) more suitable for subsequent recruitment of ESS and less suitable for the recruitment of LSS.

   (d) less suitable for the subsequent recruitment of both ESS and LSS.

6. In the following diagram, two models succession are represented. In this diagram A, C and D are species and arrows indicate replaced by **(DEC 2013)**

   (i) 

   (ii) 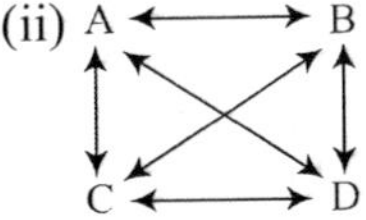

   Based on the above, which statement is correct

   (a) Fig.(i) represents facilitation and Fig.(ii) represents tolerance model

   (b) Fig.(i) represents tolerance model Fig. (ii) represents facilitation model

(c) Fig.(i) represents facilitation m and Fig.(ii) represents inhibition model

(d) Fig.(i) represents tolerance model Fig.(ii) represents inhibition model

7. Given below are two patterns of ecological succession. Four species are represented by A, B, C and D. An arrow indicates "is replaced by".

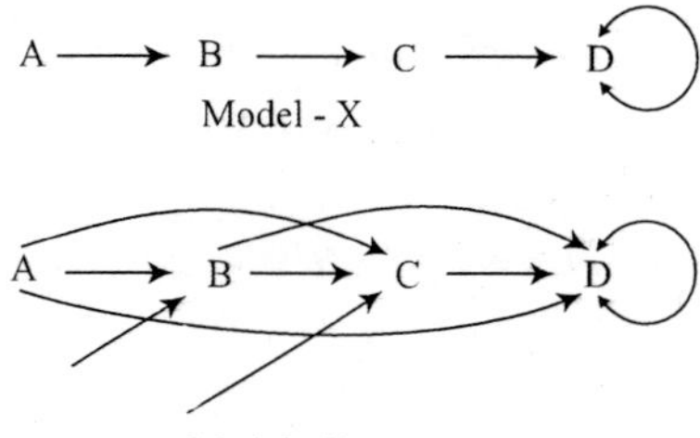

In the context of ecological succession, which of the following statements is **INCORRECT** with respect to the figures given above. **(DEC 2017)**

(a) Model - X represents facilitation model and Model - Y represents tolerance model.

(b) Model - X represents tolerance model and Model - Y represents inhibition model.

(c) As per Model - Y, C can out-compete B but can also invade a habitat in their absence.

(d) As per the Model - X, A makes the environments more suitable for B to invade.

8. Which of the following species replacement sequence depicts tolerance model of community succession? **(MODELPAPER)**

(a) A → B, B ↓ C, D — C

(b) A → B (BCD), B ↓ CD, D — CD

(c) A → B, B ↓ C, D — C

(d) A —//→ B

9. Environmental conditions can influence accumulation of species in successional communities. Curves representing changes in forest species over time are given in the figure below. Which of the following keys is correct for the curves? **(JUNE 2013)**

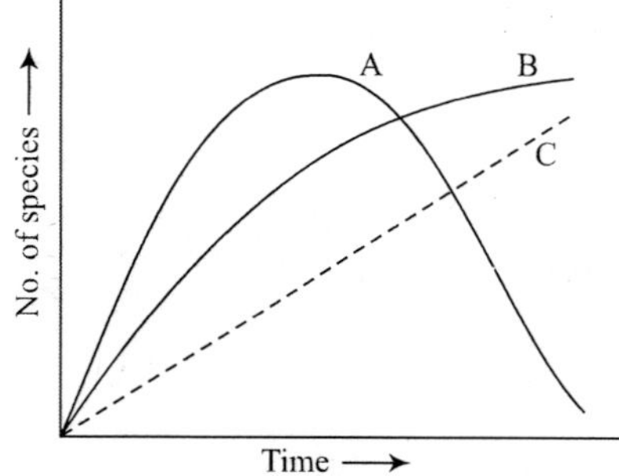

(a) A = xeric, B = mesic, C = intermediate

(b) A = intermediate, B = xeric, C = mesic

(c) A = intermediate, B = mesic, C = xeric

(d) A = mesic, B = intermediate, C = xeric

10. Complete the following sentence with the most appropriate option.

Global analysis of a large number of plant species traits showed that with increase in leaf lifespan **(DEC 2018)**

(a) specific leaf area increases whereas leaf nitrogen and net photosynthesis rate decrease.

(b) specific leaf area, leaf nitrogen and net photosynthesis rate increase.

(c) specific leaf area, leaf nitrogen and net photosynthesis rate decrease.

(d) specific leaf area decreases whereas leaf nitrogen and net photosynthesis rate increase.

Answer Key

1.(a)	2.(d)	3.(d)	4.(b)	5.(a)	6.(c)	7.(b)
8.(b)	9.(d)	10.(c)				

11. Ecosystem structure and ecosystem function

1. Possible explanations for the age related decline in primary productivity of trees are:
   A. As trees grow larger with age, they have more tissues that respire and loose energy and proportionately less leaf area to photosynthesis.
   B. Nutrient limitation by nitrogen due to reduced rate of woody litter decomposition as forest ages.
   C. As trees become larger, water transport to the top canopy leaves becomes limited because cf increased hydraulic resistance. This results in reduced stomatal conductance and reduction in photosynthetic rate.

   Which of the above is/are correct? **(DEC 2012)**
   (a) A, B and C (b) only A
   (c) only A and C (d) only B and C
2. Which of the following is the most appropriate spectral bands for vegetation analysis using remote sensing platforms? **(DEC 2013)**
   (a) Red, Near Infrared
   (b) Infrared, Visible
   (c) Red, Microwave
   (d) Visible, Microwave
3. A researcher collected information from four forest using sensors to assess their green cover. Observed average spectral values from each of the forest are given in the table below:

Forest	spectral value	
	NIR	VIS
A	0.50	0.08
B	0.04	0.30
C	0.50	0.20
D	0.60	0.20

   The forest cover in the order of highest to lowest is **(JUNE 2012)**
   (a) A>C>B>D (b) A>D>C>B
   (c) B>C>D>A (d) D>A>B>C
4. Primary production in aquatic ecosystem is measured using Light-and-Dark-Bottle technique. In this method, as an indirect measure of photosynthetic production, dissolved oxygen concentration of the pond water enclosed in a BOD bottle is measured initially (I) and after a fixed duration of incubation in a light bottle (L) and a dark bottle (D). Then, the gross and net primary productions are estimated is **(JUNE 2011)**
   (a) (L-D) and (L-I), respectively.
   (b) (L-I) and (L--D), respectively.
   (c) (L-I) and (I-D), respectively.
   (d) (L-D) and (I-D), respectively.
5. Aquatic primary production was measured using Light-and-Dark Bottle technique. If the initial oxygen concentration was *I* and the final oxygen concentration in the light bottle is *L* and that in the dark bottle *D*, the gross productivity (in terms of oxygen released) is given as **(DEC 2011)**
   (a) *D-L* (b) *I-D*
   (c) *I-L* (d) *L-D*
6. The most commonly used method of estimating primary productivity of a pond involves measurement of the amount **(JUNE 2011)**
   (a) CO_2 utilized (b) autotroph biomass.
   (c) oxygen released. (d) organic carbon
7. In a field experiment, autotrophs are provided a ^{14}C-labelled carbon compound for photosynthesis. Radioactivity (^{14}C) levels were then monitored at regular intervals in all the trophic levels. In which ecosystem is the radioactivity likely to be detected fastest at the primary carnivore level? **(DEC 2015)**
   (a) open ocean (b) Desert
   (c) Deciduous forest (d) Grassland
8. The complexity of a food web in a community is quantified using certain parameters which are defined below. Which of the following is an **INCORRECT** representation? **(JUNE 2018)**

(a) Chain length

$$= \frac{\text{Avergae number of links between trophic levels}}{\text{Total number of trophic levels}}$$

(b) Connectance

$$= \frac{\text{Actual number of links in food web}}{\text{Potential number of links in a food web}}$$

(c) Potential links in a food web where 'n' species are present

$$= \frac{n(n-1)}{\text{Actual number of links}}$$

(d) Chain length

$$= \frac{\text{Actual number of links in food web}}{\text{Number of species in a food web}}$$

9. Below is a table showing the number of species and actual food web links observed in three different ecosystems.

Ecosystem	No. of species	Observed food web links
A	11	25
B	21	70
C	51	200

By calculating connectance as a measure of relative complexity of food webs, infer which of the following statements is correct? **(DEC 2019 ASSAM)**

(a) Connectance increases as species richness increases.

(b) Connectance decreases as species richness increases.

(c) Connectance is constant regardless of species richness.

(d) Relationship between species richness and connectance is stochastic.

10. In which ecosystem is the detrital pathway of energy flow most important? **(DEC 2014)**

(a) Lakes

(b) Grasslands

(c) Tropical rain forests

(d) Oceans

11. In which ecosystem is the autotroph-fixed energy likely to reach the primary carnivore level in the shortest time? **(JUNE 2016)**

(a) Grassland

(b) Temperate deciduous forest

(c) Ocean

(d) Tropical rain forest

12. Which of the following is likely to contribute to the stability of an ecosystem? **(JUNE 2015)**

(a) High number of specialists

(b) Fewer number of functional links

(c) More omnivores

(d) Linear rather than reticulate food webs

13. Which of the following is the correct decreasing order for the rate of decomposition of litter constituents? **(DEC 2014)**

(a) Hemicellulose, cellulose, lignin, phenol

(b) Cellulose, hemicellulose, phenol, lignin

(c) Hemicellulose, cellulose, phenol, lignin

(d) Lignin, phenol, hemicellulose, cellulose

14. A much greater proportion of energy fixed by autotrophs is transferred to the herbivore level in the open ocean ecosystem than in a forest ecosystem because **(JUNE 2011)**

(a) aquatic autotrophs are small.

(b) aquatic herbivores are more efficient feeders.

(c) terrestrial autotrophs are less efficient feeders.

(d) terrestrial autotrophs have more indigestible tissues.

15. The utilization or consumption efficiency of herbivores is highest in **(JUNE 2016)**

(a) plankton communities of ocean waters.

(b) mature temperate forests.

(c) managed grasslands.

(d) managed rangelands.

16. The approximate P:B (Net Primary Production: Biomass) ratios in four different ecosystems (A, B, C, D) are A - 0.29; B - 0.042, C - 16.48; D - 8.2. The four ecosystems are **(JUNE 2016)**

(a) A- Ocean; B - Lake; C - Grassland; D - Tropical forest
(b) A - Grassland; B - Tropical forest; C - Ocean; D - Lake
(c) A - Tropical forest; B - Ocean; C – Grassland; D- Lake
(d) A -Grassland; B - Ocean; C - Lake; D - Tropical forest

17. Autotrophs in the aquatic ecosystem, unlike their counterparts in the terrestrial ecosystem are mostly microscopic and very low in indigestible (to the herbivores) matter. This explains the fact that compared to the terrestrial ecosystem, in the aquatic ecosystem **(DEC 2011)**

(a) Productivity/Biomass ratios are higher and energy transfer rates to higher trophic levels are faster.
(b) Productivity/Biomass ratios are lower and the energy transfer rates to higher trophic levels are slower.
(c) Productivity Biomass ratios are lower and the energy transfer rate to higher trophic levels are faster.
(d) Productivity/Biomass ratios are higher and the energy transfer rate to higher trophic levels are slower

18. Which of the following set of observations is true with reference to a comparison of aquatic (A) and terrestrial (T) ecosystems? **(JUNE 2014)**

(a) Number of trophic levels is more in A than T.
Productivity/Biomass ratio is higher in T than in A.
Herbivore assimilation efficiency is higher in A than in T.
(b) Number of trophic levels is more in T than in A.
Productivity/Biomass ratio is greater in A than in T.
Herbivore assimilation efficiency is higher in T than in A.
(c) Number of trophic levels is more in T than in A.
Productivity/Biomass ratio is higher in T than in A.
Herbivore assimilation efficiency is higher in T than in A.
(d) Number of trophic levels is more in A than in T.
Productivity/Biomass ratio is greater in A than in T.
Herbivore assimilation efficiency is higher in A than in T.

19. Lindeman's efficiency between trophic levels is depicted by the formula: Efficiency = A/B Where, A and B, respectively are: **(DEC 2013)**

(a) 'assimilation at trophic level n' and 'assimilation at trophic level n - 1'
(b) 'intake at trophic level n' and 'assimilation at trophic level n-1'
(c) 'assimilation at trophic level n' and 'net productivity at trophic level n - 1'
(d) 'intake at trophic level n' and productivity at trophic level n - 1'

20. Which of the following is true about the Digestion Efficiency (DE)-(assimilation/ consumption) and Ecological Efficiency (EE) (production/consumption) of ectotherms and endotherms? **(JUNE 2015)**

(a) Endotherms have a high DE and ectotherms have a high EE. .
(b) Endotherms have a low DE and ectotherms have a high EE.
(c) Endotherms have a high DE and ectothenns have a low EE.
(d) Endotherms have a low DE and ectotherms have a low EE.

21. Following are certain statements regarding energy efficiencies of ectotherms and endotherms:

A. Ectotherms have high assimilation efficiency but low production efficiency.

B. Ectotherms have low assimilation efficiency but high production efficiency.

C. Endotherms have high assimilation efficiency but low production efficiency.

D. Endotherms have low assimilation efficiency but high production efficiency.

Which one of the following represents the combination of correct statements?

(JUNE 2018)

(a) A and B (b) B and C

(c) C and D (d) A and C

22. Following are some of the generalizations regarding energy flow in an ecosystem:

A. Assimilation efficiency of carnivores is higher than herbivores.

B. Consumption efficiency of aquatic herbivores is higher than terrestrial herbivores.

C. Vertebrates have higher production efficiencies than invertebrates.

D. Trophic level transfer efficiency is higher in terrestrial food chains than in marine.

Based on the above, select the correct option.

(JUNE 2019)

(a) Only A and C (b) Only A and B

(c) A, B and C (d) A, C and D

23. Which one of the following statement is **NOT** correct?

(a) Herbivores enhance the productivity of a productive ecosystem and reduce the productivity of an unproductive ecosystem.

(b) Detritus based food chains are longer in more productive ecosystem

(c) Consumption efficiency of herbivores is higher in grasslands than ocean

(d) Production efficiency of carnivores is higher than herbivores. **(JUNE 2017)**

24. Given below is an ecological pyramid.

(DEC 2017)

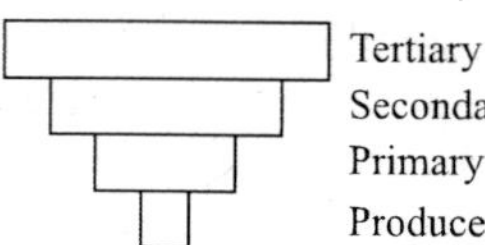

The above pyramid represents:

(a) Pyramid of number of a parasitic food chain and pyramid of biomass of a pond ecosystem

(b) Pyramid of number of a pond ecosystem and pyramid of biomass of a forest ecosystem

(c) Pyramid of energy of a grassland and pyramid of biomass of an open ocean ecosystem

(d) Pyramid of biomass of a grassland and pyramid of number of a tropical forest ecosystem

25. which of the following is a correct ranking of ecosystems based on the roots: shoot ratio of plants? **(DEC 2016)**

(a) Tropical wet forest □ Tropical dry forest □ Temperate grassland □ Tropical grassland

(b) Temperate grass land □ Tropical grassland □ Tropical wet forest□ Tropical dry forest

(c) Tropical dry forest □ Tropical wet forest □ Tropical grassland □ Temperate grassland

(d) Temperate grass land □ Tropical grassland Tropical dry forest □ Tropical wet forest

26. Which of the following is the correct increasing order for the daily net primary productivity (NPP) per unit leaf area in different ecosystems?

(DEC 2018)

(a) Deserts < Temperate forests < Tropical forests

(b) Deserts < Tropical forests < Temperate forests

(c) Temperate forests< Tropical forests< Deserts

(d) Tropical forests < Temperate forests < Deserts

27. Which of the following options lists ecosystems in increasing order of plant productivity per day per unit leaf area? **(JUNE 2018)**

(a) Tropical forests, hot deserts, temperate forests

(b) Hot deserts, temperate forests, tropical forests

(c) Hot deserts, temperate grasslands, tropical forests

(d) Tropical forests, temperate grasslands, hot deserts

28. What is the significance of upwelling zone for marine ecosystems? **(DEC 2018)**

(a) It is responsible for uniformity of temperature in ocean to support the marine life.

(b) It brings nutrients from deeper zones to relatively nutrient poor ocean surface thus increasing marine productivity.

(c) It is responsible for uniform oxygenation of marine waters thus increasing marine productivity.

(d) It helps in circulating decomposers from the bottom of ocean to surface for proper decomposition of dead material on the surface.

Answer Key

1.(d)	2.(a)	3.(b)	4.(a)	5.(d)	6.(c)	7.(a)
8.(c)	9.(b)	10.(c)	11.(c)	12.(c)	13.(a)	14.(d)
15.(a)	16.(b)	17.(a)	18.(d)	19.(d)	20.(a)	21.(b)
22.(b)	23.(c)	24.(a)	25.(d)	26.(c)	27.(d)	28.(b)

12. Major terrestrial biome and Island biogeography

1. Average annual precipitation and temperature are two important determinants of world's major biomes. Which of the following combinations are correct? **(JUNE 2012)**

	Temperature and Precipitation		Biome
A	25°C and 255 cm	i.	Temperate Forest
B	15°C and 300 cm	ii.	Savannah
C	15°C and 100 cm	iii.	Temperate rain forest
D	25°C and 255 cm	iv.	Tropical rain forest

(a) A-iv; B-iii; C-i; D-ii

(b) A-iii; B-ii; C-iv; D-i

(c) A-ii; B-i; C-iii; D-iv

(d) A-i; B-iv; C-ii; D-iii

2. Match the correct local names of temperate grasslands with their geographical range **(DEC 2016)**

Geographical range		Local name of grassland	
i	Asia	A	Pampas
ii	North America	B	Prairies
iii	South America	C	Steppes
iv	South Africa	D	Veldt

(a) i-C, ii-B, iii- D, iv-A

(b) i- C, ii-B, iii-A, iv-D

(c) i- D, ii-B, iii-A, iv-C

(d) i-B, ii-C, iii-A, iv-D

3. Which one of the following is **NOT** correct? **(JUNE 2017)**

(a) Island ecosystems are less prone to biological invasion because of their distance from mainland

(b) Invasive species have greater phenotypic plasticity compared to native species

(c) Invasive species have high dispersal ability

(d) At a large scale, diversity rich ecosystems are generally more prone to invasion

4. Following figure shows McArthur and Wilson's equilibrium model of biota on a single island. **(JUNE 2011)**

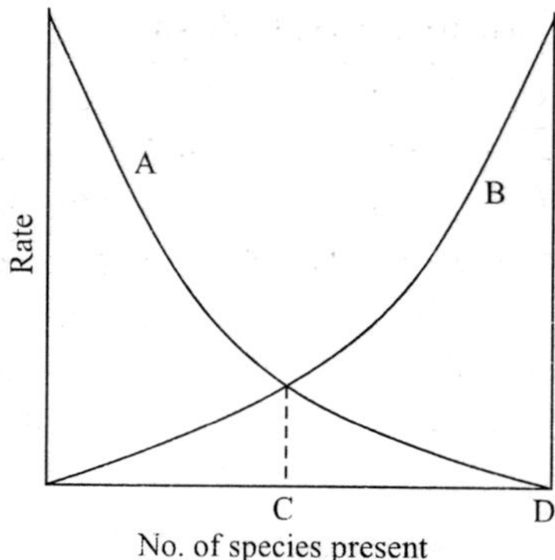

In this figure, terms A, B, C and D in order are

(a) extinction. immigration, equilibrium number of species, size of species pool.
(b) immigration. extinction, equilibrium number of species, size of species pool.
(c) extinction. immigration, size of species pool, equilibrium number of species.
(d) immigration, extinction, size of species pool, equilibrium number of species.

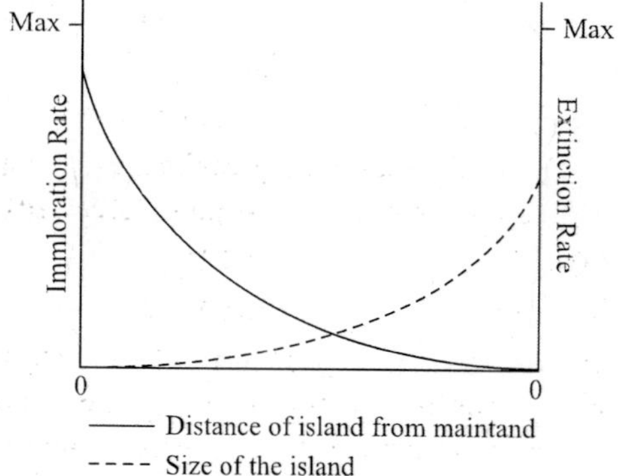

5. The above graph illustrates two lines that represent the immigration and extinction rates for an island based on its distance from mainland (solid line) and its size (dotted line). Which of the following is true for this island? **(JUNE 2018)**

(a) It is close to the mainland and is very small.
(b) It is far from mainland and is very large.
(c) It is close to the mainland and is very large.
(d) It is far from the mainland and is very small.

6. According to MacArthur and Wilson's equilibrium theory, which of the following is true? **(JUNE 2011)**

(a) Larger islands and islands closer to continent are expected to have more species than smaller and isolated islands.
(b) Smaller islands and islands far from the continent are expected to have more species than larger and isolated islands.
(c) Smaller islands and islands closer to the continent are expected to have more species than far away smaller and isolated islands.
(d) More species are expected on all islands irrespective of their size and distance from the continent.

7. Which of the following is NOT a prediction arising out of Wilson-MacArthur's -Theory of Island Biogeography? **(JUNE 2016)**

(a) The number of species on an island should increase with its size/area.
(b) The number of species should decrease with increasing distance of the island from the source pool.
(c) The turnover of species should be common and frequent.
(d) Species richness on an island should be related to its average distance to the neighbouring islands.

8. The equilibrium model of island biogeography proposed by MacArthur and Wilson assumes that the number of species on an island represents a balance between **(DEC 2018)**

(a) resource consumption rate and predation rate.
(b) birth rate and death rate.
(c) colonization rate and extinction rate.
(d) speciation rate and hybridization rate

9. Two islands (A and B) located approximately 180 km and 100 km respectively from the mainland. Island A is 160 sq. km, while Island

B is 220 sq. km in area. Which one of the following statements is correct? **(DEC 2019 ASSAM)**

(a) Island A is likely to have higher extinction and lower colonization rates than Island B.

(b) Island B is likely to have higher extinction and lower colonization rates than Island A.

(c) Island A is likely to have lower extinction rates and higher colonization rates than Island B.

(d) Islands A and B are likely to have equal rates of extinction and colonization

10. Forest fragments in an agricultural landscape can be viewed as islands of habitat in an ocean of non-habitat. MacArthur and Wilson's island biogeography model can be used to predict patterns of species richness in these forest fragments which are represented ill the graphs below.

(a)

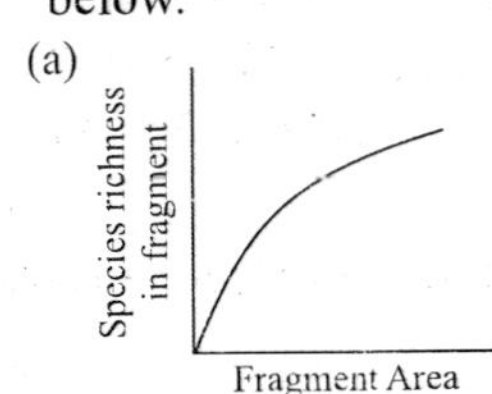

(b)

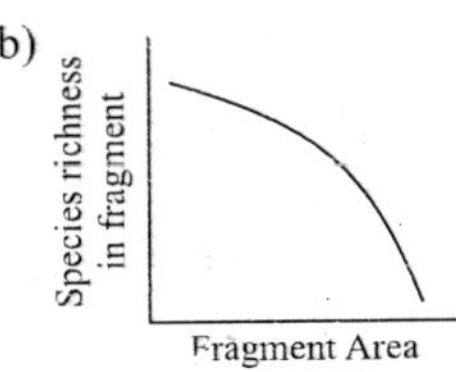

(c)

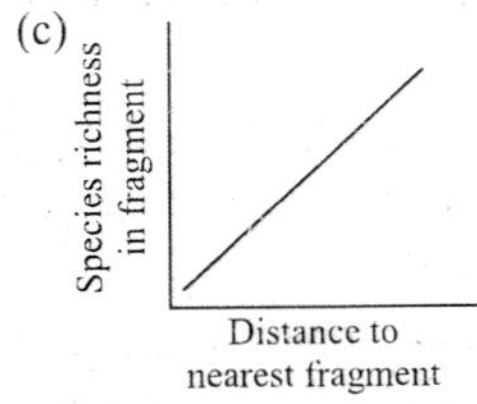

(d)

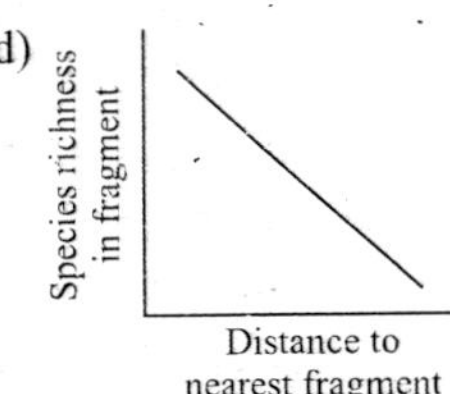

Which one of the following combinations of the graphs correctly represents predictions from the model? **(DEC 2018)**

(a) A and C (b) A and D

(c) B and C (d) B and D

Answer Key

1.(a)	2.(b)	3.(d)	4.(b)	5.(c)	6.(a)	7.(c)
8.(c)	9.(a)	10.(b)				

13. Environmental pollution

1. Which one of the following is in the correct decreasing order for the major reservoirs of carbon on Earth? **(JUNE 2017)**

(a) Terrestrial soils > Terrestrial vegetation > Atmospheric CO_2 > Large lake sediments

(b) Terrestrial soil > Large lake sediments > Terrestrial vegetation > Atmospheric CO_2

(c) Atmospheric CO_2 > Large lake sediments >Terrestrial soils > Terrestrial vegetation

(d) Large lake sediments > Terrestrial soils > Atmospheric CO_2 >Terrestrial vegetation

2. Which gas does NOT contribute to global warming through its greenhouse effect? **(DEC 2015)**

(a) Nitrous oxide (b) Methane

(c) Carbon dioxide (d) Nitric oxide

3. Based on per molecule, which of the following gas has the most powerful greenhouse effect? **(JUNE 2012)**

(a) CO_2 (b) CH_4

(c) N_2O (d) CFCs

4. Which of the following greenhouse gases has got highest atmospheric life time? **(DEC 2012)**

(a) CO_2 (b) CH_4

(c) N_2O (d) CFCS

5. As per national air quality standard for India, which one of the following options gives correct concentration limits (μg m^{-3}, annual) of various gaseous air pollutants for a residential area? **(JUNE 2017)**

(a) SO_2-100, NO_2-40, O_3-40, CO-50

(b) SO_2 - 50, NO_2 - 40; O_3 - 100, CO - 02

(c) SO_2 -40, NO_2 - 50, O_3 - 50, CO - 10

(d) SO_2 - 50, NO_2 - 100, O_3 -40, CO- 02

6. The losses of ozone over Arctic are significantly lower than that over Antarctica because **(MODEL PAPER)**

(a) polar vortex over Arctic is not as tight as over Antarctic.

(b) Arctic stratosphere warms slower in the spring.

(c) concentration of chlorine in the atmosphere over Arctic is less than over Antarctic.

(d) freezing of NO_2 and CH_4 are slower over Arctic than over Antarctic.

7. The present global warming trend is expected to result in an increased incidence of malaria in temperate countries. The supposed underlying mechanism is that **(JUNE 2014)**

(a) higher temperatures make temperate country people more vulnerable to diseases

(b) malarial parasite grows better at higher temperatures.

(c) the vector mosquito species requires warmer temperatures for reproduction.

(d) anti-malaria drugs are less effective in temperate countries.

8. Based on the selected physicochemical features of four water bodies given in the following table, which of the following statements is correct? **(DEC 2013 GU)**

Parameter	Water Bodies			
	A	B	C	D
Total dissolved solids (ppm)	25	75	300	400
NO_3–N(ug/L)	200	300	400	500
PO_4-P(ug/L)	1	15	20	50

(a) Water bodies A and B are oligotrophic whereas water bodies C and D are eutrophic.

(b) Only water body A is oligotrophic and all other water bodies are eutrophic.

(c) Only water body D is eutrophic and all other water bodies are oligotrophic.

(d) All the four water bodies are eutrophic.

9. Which one of the following statements is true for trends of the dissolved oxygen (DO) and biological oxygen demand in water stream receiving pollutants from point source.

(DEC 2016)

(a) In septic zone, both DO and BOD levels remains stationary

(b) In recovery zone, both DO and BOD levels increase rapidly

(c) In decomposition zone, DO levels drop rapidly where as BOD level remains more or less stable

(d) In septic zone, DO levels decrease and BOD level increase whereas in recovery zone DO increase and BOD decrease

10. In a lake subjected to progressive eutrophication, temporal changes in the magnitude of selected parameters (A, B, C, D) are shown in the graph **(JUNE 2016)**

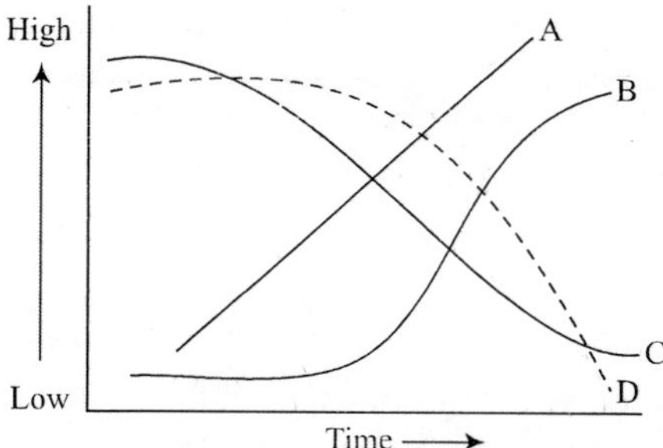

The parameters A, B, C, D are

(a) A-Green algal biomass, B – Cyano-bacterial biomass, C - Dissolved Oxygen concentration, D - Biological Oxygen Demand

(b) A- Biological Oxygen Demand, B -Cyanobacterial biomass, C- Dissolved Oxygen concentration, D - Green algal biomass

(c) A- Biological Oxygen demand, B -Green algal biomass, C - Cyanobacterial biomass, D - Dissolved Oxygen concentration

(d) A- Cyanobacterial biomass, B - Biological Oxygen Demand, C -Green algal biomass, D - Dissolved Oxygen concentration

11. Following table gives a list of international environmental agreements and areas covered.

	Agreement		Area covered
A	Basal convention	(i)	Biosafety
B	Cartagena protocol	(ii)	Control of trans-boundary movements of hazardous wastes and their disposal
C	Kyoto protocol	(iii)	Greenhous gas emission reductions
D	Stockholm convention	(iv)	Persistent organic pollutants

Which of the following is correct combination?

(JUNE 2018)

(a) A - (i), B - (ii), C - (iv), D - (iii)
(b) A - (ii), B - (i), C - (iii), D - (iv)
(c) A - (iv), B - (i), C - (iii), D - (ii)
(d) A - (ii), B - (iv), C - (iii), D - (i)

Answer Key

1.(d)	2.(d)	3.(d)	4.(d)	5.(b)	6.(a)	7.(c)
8.(b)	9.(c)	10.(b)	11.(b)			

14. Biodiversity: Status

1. Following is a coded matrix representing threatened plants and animals of India.

	Endangered	Vulnerable
Plants	A	B
Animals	C	D

In the above matrix, letters A, B, C and D represent **(DEC 2013 GU)**

(a) A = *Capparis pachyphylla*,
B = *Myristica malabarica*
C = *Macaca silenus*,
D = *Equus hemionus*

(b) A = *Saraca asoca*,
B = *Madhuca diplostemon*
C = *Ursus thibetanus*,
D = *Ailursus fulgens*

(c) A = *Plerocarpus santalinus*,
B = *Berberis nilghiriensis*
C = *Elephas maximus*,
D = *Viverra civettina*

(d) A = *Cupressus cashmeriana*,
B = *Dipterocarpus grandiflorus*
C = *Ursus thibetanus*,
D = *Sus salvanius*

2. The table given below lists species and conservation status

	Species		Conservation status
A	White bellied Heron	i	Critically endangered
B	Ganges river dolphin	ii	Endangered
C	Gaur	iii	Vulnerable
D	Clouded leopard		

Which one of the following is the correct pairing between Indian animal species and their conservation status? **(JUNE 2017)**

(a) A-i; B-i; C-ii; D-iii
(b) A-ii; B-ii; C-iii; D-ii
(c) A-i; B-ii; C-iii; D-iii
(d) A-iii; B-iii; C-ii, D-ii

3. In the following table, a list of threat categories and animals of India is given in an alphabetical order:

Animals		Threat category	
i.	Bengal florican	A.	Critically endangered
ii.	Ganga river Dolphin	B.	Endangered
iii.	Indian Rhinoceros	C.	vulnerable
iv.	Indian Vulture		

Which of the following options show correct combination of animals and their threat category as per Red Data list of IUCN?

(DEC 2018)

(a) i - A ; ii - A ; iii - B; iv- C
(b) i-A; ii-B; iii-C; iv-A
(c) i-B; ii-C; iii-B; iv-A
(d) i-C; ii-A; iii-a; iv-B

4. Which one of the following was recently reported to be the first mammal to have become extinct as a result of climate change? **(DEC 2019)**

(a) Bramble Cay melomys – *Melomys rubicola*

(b) Gangetic river dolphin – *Platanista gangetica*

(c) Malaga giant rat – *Hypogeomys antimena*

(d) Tapanuli orangutan – *Pongo tapanuliensis*

5. Which of the following global hotspots of biodiversity has the highest number of endemic plants and vertebrates? **(JUNE 2016)**

(a) Sundaland

(b) Tropical Andes

(c) Brazil's Atlantic Forest

(d) Mesoamerican forests

6. Given below are biodiversity hotspots in decreasing order of endemic plant species recorded in them. Select the correct order. **(JUNE 2018)**

(a) Western Ghats and Sri Lanka> Indo Burma>Sundaland> Philippines

(b) Philippines>Sundaland> Indo-Burma > Western Ghats and Sri Lanka

(c) Sundaland> Indo-Burma > Philippines > Western Ghats and Sri Lanka

(d) Western Ghats and Sri Lanka>Sundaland> Philippines> Indo-Burma

7. According to 2014 IUCN Red List, which of the following vertebrate classes has the largest percentage of threatened species? **(DEC2015)**

(a) Mammals (b) Birds

(c) Reptiles (d) Amphibians

8. Which of the following is NOT true for a critically endangered species? **(DEC 2011)**

(a) Reduction of population breeding ability due to increased relatedness through the action of incompatibility mechanisms in plants or behavioural difficulties in animals.

(b) The individuals of the species which have declined to low numbers are still a genetically open system.

(c) Loss of some alleles from the species causing loss of genetic diversity with consequent inability to respond rapidly to selection

(d) Expression of deleterious alleles and increased homozygosity increases mortality of young, and inbreeding depression leads to reduced offspring fitness.

9. A species has the following population characteristics:

A. Reduction in population size ≥ 90 % over last 10 years or 3 generations

B. Geographical range: Extent of occurrence <100 Km^2 and Area of occupancy: <10km^2

C. Population size less than 50 mature individuals

D. Probability of extinction in wild at least 50 % within the next 10 years or 3 generations

To which of the following categories the species will be assigned according to IUCN categorization of threatened species (version 3.1)? **(JUNE 2012)**

(a) Endangered

(b) Vulnerable

(c) Critically endangered

(d) Extinct in wild

10. Which of the following set of conditions will qualify a species to be considered as endangered (EN) as per IUCN criteria? **(DEC 2019)**

(a) 80% reduction in population size, <100 km2 area of occupancy, at least 50% estimated extinction risk in five generations.

(b) <2,500 individuals with declining population, <250 mature individuals, at least 20% estimated extinction risk in 10 years.

(c) <10,000 individuals with declining population, <1000 mature individuals, at least 10% estimated extinction risk in 10 years.

(d) 75% reduction in population size, <500 km^2 area of occupancy, at least 20% estimated extinction risk in five generations

11. Based on the table given below, which of the following option represents the correct match? **(JUNE 2016)**

	Category	Plant species
A.	Critically endangered	*(i) Chromolaena odorata*
B.	Vulnerable	*(ii) Dipterocarpus grandiflorus*
C.	Extinct	*(iii) Euphorbia mayuranathanii*
D.	Invasive	*(iv) Saraca ashoka*

(a) A - (i); B - (iv); C - (iii); D - (ii)

(b) A - (ii); B - (iii); C - (iv); D - (i)

(c) A - (i); B - (iv); C - (ii); D - (iii)

(d) A - (ii); B - (iv); C - (iii); D - (i)

12. The following figure is a "risk-graph" that illustrates the percent risk a species faces towards extinction.

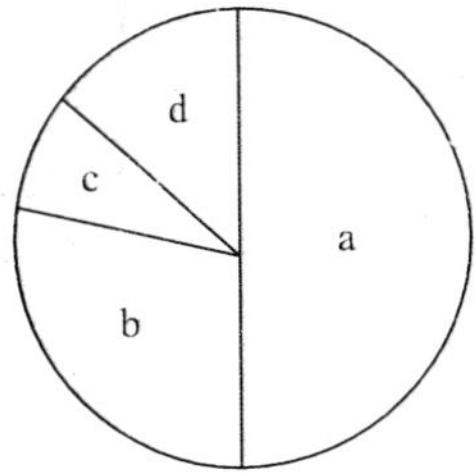

The following are ranks assigned according to IUCN's red-list category:

(i) Critically endangered

(ii) Near threatened

(iii) Vulnerable

(iv) Least concern

Which one of the following is the most appropriate match between the percent-risk and their assigned rank? **(JUNE 2018)**

(a) a- (i), b - (iii), c - (iv), d - (ii)

(b) a -(i), b- (iv), c- (iii), d - (ii)

(c) a- (iii), b- (ii), c - (i), d - (iv)

(d) a- (iv), b - (iii), c - (ii), d - (i)

13. Complete the following sentence using options given below the sentence as a, b, c, d and e. "Species are critically endangered when it is not endangered but is facing ________ risk of extinction in the wild in the ________ future". **(MODELPAPER)**

(a) high (b) very high

(c) extremely high (d) near

(e) immediate

(a) a, a (b) b, c

(c) c, e (d) c, d

14. Tropical regions may have more species diversity because of the following possible reasons, EXCEPT **(JUNE 2019)**

(a) tropical regions have had more limit to diversify under relatively stable climatic conditions than temperate regions.

(b) tropical regions have high spatial heterogeneity

(c) greater biological competition in the tropics leads to narrower niches

(d) lower predation intensity in the tropics allows survival of more prey species

15. A road is constructed, through a wet tropical forest, following which the population of a species of forest butterfly declines. Which of the following is NOT a possible explanation for the road causing a decline in the forest butterfly population? **(JUNE 2019)**

(a) Road facilitates immigration of gap-loving species which compete with the forest species.

(b) Road facilitates increased movement of the forest butterfly within the forest, which reduces genetic diversity.

(c) Road internally fragments the habitat and negatively affects important micro-habitat conditions for the forest butterfly.

(d) Road facilitates invasion by non-native plants that displace native host and nectar plants of the forest butterfly.

16. Which of the following is NOT an attribute of a species that makes it vulnerable to extinction? **(JUNE 2016)**

(a) Specialized diet
(b) Low dispersal ability
(c) Low trophic status
(d) Variable population density

17. Which of the following attributes increase the extinction risk for a species? **(DEC 2013 GU)**

(a) High trophic status, limited geographical distribution, specialized niche
(b) Low trophic status, limited geographical distribution, specialized niche
(c) High trophic status, wide geographical distribution, specialized niche
(d) Low trophic status, wide geographical distribution, specialized niche

18. Following are the characteristics of species that make them more or less prone to extinction:

Rare – a
Common – b
Good dispersal rate – c
Poor dispersal rate – d
Low specialization – e
High specialization – f
High variability – g
Low variability – h
Low trophic status – i
High trophic status – j
Long life span – k
Short life span – l
High reproductive output – m
Low reproductive output – n

Which of the following is the correct combination of characteristics that makes the species more prone to extinction? **(JUNE 2014)**

(a) a d f g j l n (b) a e f h i k m
(c) b d e g i l n (d) b c f h j k m

19. Species characteristics that make them more prone to extinction are listed below:

A. High degree of specialization
B. High sexual dimorphism
C. High trophic status
D. Short life span

Which of the following is the correct combination? **(DEC 2012)**

(a) A, B and C (b) A, C and D
(c) A, B and D (d) B, C and D

20. Following table shows attributes of selected species A, B, C and D:

	Species			
Attributes	**A**	**B**	**C**	**D**
Dispersal ability	High	Low	High	Low
r Selected	Yes	Yes	No	No
Predominant reproduction mode	Asexual	Sexual	Asexual	Sexual
Competitive ability	High	Low	High	Low

Based on the above information, which of the following is most likely to become invasive if climate matches between its site of origin and site of colonization? **(DEC 2018)**

(a) Species A and D
(b) Species A only
(c) Species D only
(d) Species B and C

21. Match the following invasive plants to the likely habitats in which they are expected to occur: **(JUNE 2019)**

Invasive plant		**Habitat (s) that they invade**	
A	*Eichornia crassipes*	(i)	Arid and semi-arid habitats
B	*Lantana camara*	(ii)	Dry and moist tropical forests
C	*Prosopis juliflora*	(iii)	Wetlands

(a) A – ii, B – i, C – iii
(b) A – i, B – iii, C - ii
(c) A – iii, B – ii, C – i
(d) A – iii, B – i, C – ii

22. In a meta study, percentage of species affected by different ecological threats in different habitats was found to be as follows: **(DEC 2019 ASSAM)**

Habitat	Over-exploitation	Invasive species	Pollution
A	20	22	7
B	19	20	44
C	86	5	38

Based on the above information, habitats A, B and C will be respectively:

(a) Terrestrial, Fresh water, Marine
(b) Marine, Fresh water, Terrestrial
(c) Fresh water, Terrestrial, Marine
(d) Marine, Terrestrial, Fresh water

Answer Key

1.(a)	2.(c)	3.(b)	4.(a)	5.(b)	6.(c)	7.(d)
8.(b)	9.(c)	10.(d)	11.(d)	12.(a)	13.(d)	14.(d)
15.(b)	16.(c)	17.(a)	18.(a)	19.(b)	20.(b)	21.(c)
22.(a)						

15. Conservation Biology

1. Given below are names of some of the National Parks of India and their key protected animals.

	Name of the National Park		Key protected animal
A	Dibru-Saikhowa National Park, Assam	(i)	Indian, Rhinoceros
B	Jaldapara National Park, West Bengal	(ii)	Hangul
C	Mukurthi National Park, Tamil Nadu	(iii)	Feral Horse
D	Dachigam National Park, Jammu and Kashmir	(iv)	Asiatic Lion
E	Gir Forest National Park, Gujarat	(v)	Nilgiri Tahr

Based on the table given above, which of the following options represents the correct match? **(DEC 2017)**

(a) A-(iii); B-(i); C-(iv); D-(ii); E-(v)
(b) A-(i); B-(ii); C-(v); D-(iii); E-(iv)
(c) A-(i); B-(ii); C-(iii); D-(iv); E-(v)
(d) A-(iii); B-(i); C-(v); D-(ii); E-(iv)

2. Which of the following National parks has the highest density of tigers among protected area in the world? **(JUNE 2016)**

(a) Jim Corbett (b) Kaziranga
(c) Keoladeo Ghana (d) Manas

3. Which one of the following is the most appropriate match for the protected areas of India? **(JUNE 2015)**

Category	Protected Area
A. Biosphere Reserve	(i) Chambal
B. National Park	(ii) Loktak
C. Ramsar Site	(iii) Nanda Devi
D. Wildlife Sanctuary	(iv) Rajaji
	(v) Sunderbans

(a) A: (iii) B- (iv) C-(ii) D-(i)
(b) A: (ii) B- (iv) C-(iii) D-(i)
(c) A: (i) B- (v) C-(iii) D-(i)
(d) A: (iii) B- (ii) C-(v) D-(iv)

4. Wetlands are conserved internationally through an effort called as **(DEC 2011)**

(a) Basel Convention
(b) Rio Convention
(c) Montreal Convention
(d) Ramsar Convention

5. Which of these programs is used to conserve a species facing extinction? **(JUNE 2013)**

(a) Captive breeding.
(b) Natural resources.
(c) Sustainable use.
(d) Edge effects.

6. India has currently 17 biosphere -reserves representing different ecosystems. These conservation areas significantly differ from the conventional protected areas of the country. Identify the correct combination of attributes (A to G) that best .explains the concept of biosphere reserve. **(JUNE 2013)**

(A) Conservation,

(B) Education,

(C) Human habitation allowed,

(D) Human habitation not allowed,

(E) Strong legal back-up,

(F) No supporting act,

(G) Research.

(a) (A), (B), (C), (F), (G)

(b) (A), (B), (D), (F), (G)

(c) (A), (B), (C), (E), (G)

(d) (A), (B), (E), (G)

7. Which of the following combinations is good for setting up a nature reserves: **(DEC 2012)**

(a)

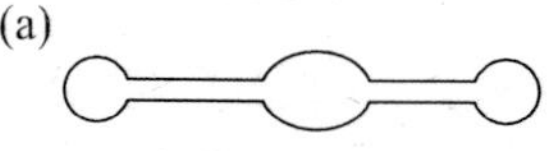

Linked reserves

(b)

Large, compact shape

(c)

High edge-area ratio

(d)

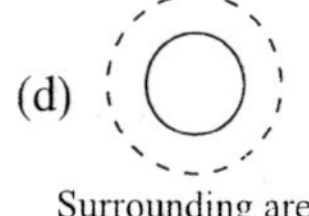

Surrounding area of same ecosytems

(a) i, ii, iii

(b) ii, iii, iv

(c) i, iii, iv

(d) I, ii, iv

8. The prominent mammal species found in four different protected areas are listed below:

Area A: Tiger, Wild dog, Leopard, Elephant

Area B: Common langur, Barking deer, Wild dog, Elephant

Area C: Tiger, Indian rhinoceros, Pygmy hog, Wild pig

Area D: Blackbuck, Indian gazelle, Hyena, Indian wolf

The area with the greatest phylogenetic diversity is **(JUNE 2019)**

(a) A

(b) B

(c) C

(d) D

Answer Key

1.(d)	2.(b)	3.(a)	4.(d)	5.(a)	6.(a)	7.(d)
8.(b)						

SECTION-11

EVOLUTION

1. Lamarck; Darwin–concepts

1. Column A lists names of evolutionary biologists and column B lists descriptions of evolutionary mechanisms proposed by them in random order. **(JUNE 2017)**

	Column-A		Column-B
A.	Jean Beptiste Lamarck	(i)	Variation at the molecular level is selectively neutral
B.	Charles Darwin	(ii)	Inheritance of acquired characters
C.	Motoo kimura	(iii)	Differential reproduction of genotypes
D.	Sewall Wright	(iv)	Changes in allele frequency due to random genetic drift

(a) A -(i), B -(ii), C -(iv), D –(iii)
(b) A-(ii), B-(iii), C-(i), D-(iv)
(c) A-(iii), B- (i), C-(ii), D-(iv)
(d) A - (ii), B - (iii), C -(iv), D - (i)

2. Fossils of the same species of fresh water reptiles have been found in South America and Africa. Based on the current understanding. Which of the following is the best possible explanation for this pattern? **(JUNE 2016)**
(a) The same species originated and evolved independently in these two places.
(b) Species migrated from Africa to establish new populations in South America. ,
(c) Species migrated from South America to establish new populations in Africa.
(d) South America and Africa were joined at some point in Earth's history.

3. The mean (μ) and standard deviation (σ) of body size in a Drosophila population are 8.5 and 2.2 mm, respectively. Under natural selection over many generations the μ and σ of body size change to 8.5 and 0.8 mm, respectively. The type of natural selection responsible for the change is called **(JUNE 2015)**
(a) directional. (b) neutral.
(c) disruptive. (d) stabilizing.

4. Two kinds of natural selection (A and B) acting on a trait are shown in the figure below. In each, the top graph shows the trait frequency before and the bottom graph frequency after the action of natural selection.

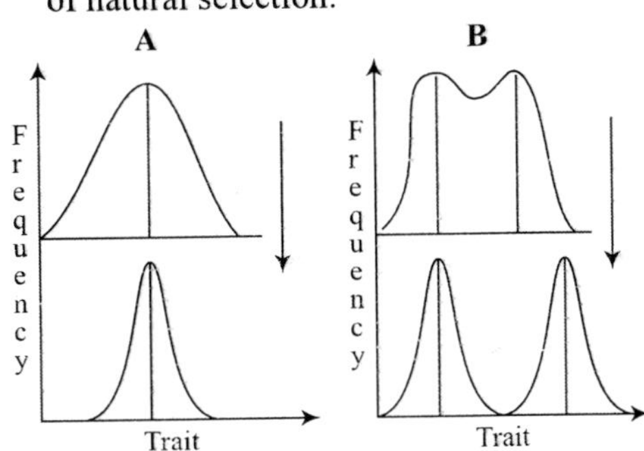

The kind of natural selection in A and B are **(DEC 2014)**
(a) A- Directional, B-Disruptive
(b) A- Neutral, B- Disruptive
(c) A- Stabilizing, B- Disruptive
(d) A-Disruptive, B- Stabilizing

5. The body weight of adult female of a strain of Drosophila is 1.8±0.45 mg (mean ± standard deviation). In a laboratory experiment, each of the 3 groups (A, B, C) of this strain was subjected to a different type of selection pressure having influence on the female body weight. After many generations of experimental selection pressure, the body weight changed as follows:

Group A: Body weight distribution - Normal, 1.8 ± 0.08 mg

Group B: Body weight distribution - Bimodal at 1.4 and 2.2 mg

Group C: Body weight distribution - Normal, 2.2 ± 0.08 mg

Which of the following correctly gives the types of selection that have occurred in the three groups? **(JUNE 2014)**

(a) Group A: Directional; Group B: Stabilizing; Group C: Disruptive
(b) Group A: Disruptive; Group B: Directional; Group C: Stabilizing
(c) Group A: Stabilizing; Group B: Disruptive; Group C: Directional
(d) Group A: Directional; Group B: Disruptive; Group C: Stabilizing

6. Selection in which two extreme phenotypes leave more offspring than the intermediate phenotype is called **(DEC 2013 GU)**
(a) directional selection.
(b) disruptive selection.
(c) stabilizing selection.
(d) neutral selection.

7.

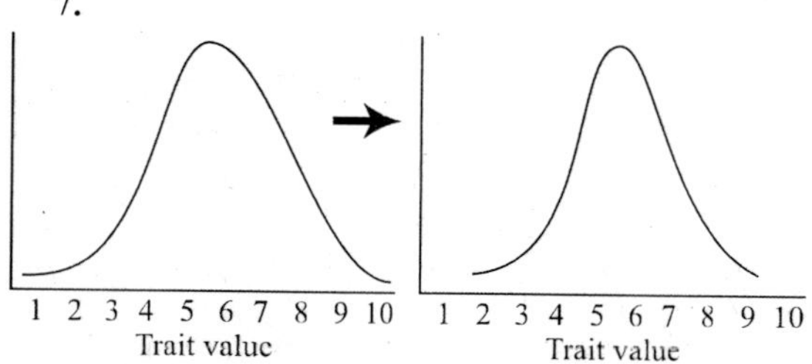

The change in a trait with time as a result of natural selection is shown above. This type of natural selection is **(MODELPAPER)**
(a) directional. (b) disruptive.
(c) stabilizing. (d) random.

8. It is found that people with the genetic disease called sickle cell anaemia are resistant to malaria. Which of the following best describes the underlying mechanism? **(MODEL PAPER)**
(a) Frequency-dependent selection
(b) Superiority of heterozygotes/Heterozygote Advantage
(c) Transient polymorphism
(d) Balanced polymorphism

9. In pre-industrial period in England, peppered moths had light coloration which effectively camouflaged' them against light coloured trees and lichens. During industrial revolution, many lichens died out and trees became blackened by soot from factories and interestingly, dark coloured moths were predominantly seen. This happened due to **(JUNE 2013)**
(a) natural selection of dark coloured moths which were initially present in fewer numbers.
(b) new mutation which arose due to environmental pollution.
(c) macroevolution occurring due to environmental change.
(d) natural selection of the camouflaging mechanism or the moths.

10. A moth species occurs as two distinct morphs based on wing colour – pale and dark. In the forest there are trees with dark coloured trunks as well as those with light coloured trunks and the moths can rest on either tree. Birds capture the resting moths and eat. In a field experiment, the proportion (%) of dark and pale morphs captured from dark and light trunk trees was recorded.

	Month	
Tree Trunk Colour	**Dark**	**Pale**
Dark	45	55
Light	48	52

The most plausible conclusion to be drawn from the results is **(MODELPAPER)**
(a) Natural selection favours dark morphs in forests where trees with dark trunks are dominant.
(b) Birds can detect dark morphs better than light morphs.
(c) Pale morphs prefer to rest on light coloured trunks
(d) Birds detect the moths by cues other than their wing colour.

11. The Galapagos finches were an important clue to Darwin's thinking about the origin of species. These finches are believed to have descended from a single ancestral species that colonized the Galapagos archipelago, America, over a short period of time. The Galapagos finches differ in their beak shape and size. Different species feed on seeds that vary in size and hardness.

Which of the following is the most likely explanation for these patterns? **(JUNE 2012)**

(a) The finches represent an example of directional trend in beak size from small to big.

(b) Beak shapes changed in response to different seed types and these changes were inherited by subsequent generations.

(c) The ancestral finch already had all the beak variations and different lineages formed that were specialized to eat different seed

(d) The finches represent an example of adaptive radiation in which beak variation was generated by mutation followed by selection by different seed types.

12. Inspite of its two-fold cost, sexual reproduction is the most dominant mode of reproduction among the living organisms. Which of the following reasons might account for this? **(MODEL PAPER)**

(A) Sexual reproduction generates genetic heterogeneity through recombination

(B) Sexual reproduction helps in purging deleterious mutations

(C) Sexual reproduction evolved to stay evolutionarily ahead of fast evolving internal parasites.

(a) (A) only (b) (A) and (B)

(c) (C) only (d) (A), (B) and (C)

13. The flightless birds ostrich, rhea and emu are distributed on different continents. What is the most plausible explanation that is given by an evolutionary biologist for this discontinuous observation? **(MODELPAPER)**

(a) The birds were able to fly earlier, but lost their flight ability later

(b) Prehistoric humans transported these birds to different continents

(c) The birds, although flightless, may have used drifting logs to cross the ocean and reach other continents

(d) All the continents used to be one single supercontinent earlier and the flightless birds were isolated after the break up of the landmass.

14. How is genetic variation in a population related to inbreeding and finite population size? **(DEC 2013 GU)**

(a) Both finite population size and inbreeding lead to decrease in genetic variation.

(b) Only inbreeding leads to a decrease in genetic variation, population size has no effect.

(c) Both inbreeding and finite population size lead to an increase in genetic variation.

(d) Neither of the two factors has any effect on genetic variation in the population.

15. Which one of the following statements is TRUE for positive-frequency dependent selection? **(JUNE 2018)**

(a) Fitness of a genotype increases as it becomes less common.

(b) Fitness of a genotype increases as it becomes more common.

(c) Fitness of a genotype decreases as it becomes less common.

(d) Fitness of a genotype decreases as it becomes common and gets fixed.

Answer Key

1.(b)	**2.**(d)	**3.**(d)	**4.**(c)	**5.**(c)	**6.**(b)	**7.**(c)
8.(b)	**9.**(a)	**10.**(d)	**11.**(d)	**12.**(d)	**13.**(d)	**14.**(a)
15.(b)						

2. The evolutionary synthesis

1. Which of the following statements about evolution is NOT true? **(DEC 2015)**

(a) Evolution is the product of natural selection.

(b) Evolution is goal-oriented.

(c) Prokaryotes evolve faster than eukaryotes.

(d) Evolution need not always lead to a better phenotype.

2. The rate of mutation in *E. coli* from *lac*$^-$ to *lac*$^+$ are determined using medium containing lactose, as the only sole source of energy. The principle of spontaneity can be said to be violated if: **(JUNE 2013)**

(a) the rate of mutation increase during starvation

(b) in the presence of lactose the rate of mutation from *lac*$^-$ to *lac*$^+$ increases but overall rate of mutation is not

(c) The rate of mutation in lac gene is always greater than in other genes

(d) The rate of mutation in lac gene is always less than in other genes

3. In an experiment that has continued for more than 50 years, corn has been propagated by breeding only from plants with the height amount of oil in the seeds. The average oil content is now much greater than any of the plants in the original population. The following hypotheses were proposed as explanations for this observation.

(A) Mutations occurred that increased the oil content in seeds.

(B) Plants with high oil content were stimulated to produce offspring with more oil in their seeds.

(C) The breeding led to increased frequency of alleles at multiple loci, so that new combinations of genes for even higher oil content were formed.

Which of the following represents a combination of correct statements? **(JUNE 2012)**

(a) (A) and (B) only (b) (A) and (C) only

(c) (B) and (C) only (d) (A), (B) and (C)

4. According to which evolutionary theory, there are long periods without significant evolutionary changes interrupted by short episodes of rapid evolution? **(JUNE 2015)**

(a) Punctuated equilibrium

(b) Saltation

(c) Mutation

(d) Neutrality

5. Plasmids are self-replicating small circular DNA elements in bacterial cells that can be said to have a stable symbiotic existence with the host cell. They often carry genes useful to the host. Which of the following is a potential threat to the evolution and stability of the symbiotic coexistence? **(JUNE 2012)**

(a) 'Copy-up' mutations that increase the rate of plasmid replication per host cell cycle.

(b) Reversible integration of plasmid DNA into the host DNA

(c) Transfer of plasmids to new cells by conjugation

(d) Spontaneous curing of plasmids in a small proportion of host cells

6. Which of the following events/examples represent observation of evolutionary changes on a human time scale? Here human time scale is defined by an average human age of approximately 70-80 years.

(DEC 2019 ASSAM)

(a) Complete eye degeneration in cave fishes

(b) Human bipedalism

(c) Evolution of flowering plants

(d) Drug resistance in HIV

7. Following table contains some of the generalizations of evolutionary biology:

Column I		Column-II	
A	Copes Rule	i	Population lineages tend to increase in body size over evolutionary time
B	Dollo's Rule	ii	There is constant probability of extinction in family of related organism.
C	Ockham's Principle	iii	Complex charcters once lost are not regained.
D	Van Valen's Law	iv	Accept the simplest theory that works

Which of the following is correct match between Column I and II? **(JUNE 2018)**

(a) A - (i), B - (ii), C - (iv), D - (iii)

(b) A - (i), B - (iii), C - (iv), D - (ii)

(c) A - (ii), B - (iii), C - (i), D - (iv)

(d) A - (iv), B - (iii), C - (i), D - (ii)

Answer Key

1.(b)	2.(b)	3.(b)	4.(a)	5.(a)	6.(d)	7.(b)

3. Origin of Life

1. Complex eukaryotic cells may have evolved from simpler prokaryotic cells because complexity of organization increases the **(JUNE 2012)**
   (a) growth rate
   (b) efficiency of energy utilization
   (c) tolerance to starvation
   (d) Ability to attain large size
2. The first living beings on earth were anaerobic because **(DEC 2011)**
   (a) There was no oxygen in air
   (b) oxygen damages proteins
   (c) oxygen interferes with the action of ribozymes
   (d) they evolved in deep sea
3. Recent studies on Archaea suggest that life could have originated **(JUNE 2011)**
   (a) extra terrestrially and seeded through meteorite impacts.
   (b) in shallow coastal areas.
   (c) in deep hydrothermal vents..
   (d) in hot, terrestrial habitats.
4. Among the following events in history of life
   a. prokaryotic cell
   b. eukaryotic cell
   c. natural selection
   d. organic molecules
   e. self replicating molecule

   Which is the correct chronological order? **(JUNE 2012)**
   (a) d$\rightarrow$e$\rightarrow$c$\rightarrow$a$\rightarrow$b
   (b) d$\rightarrow$e$\rightarrow$a$\rightarrow$b$\rightarrow$c
   (c) e$\rightarrow$d$\rightarrow$a$\rightarrow$c$\rightarrow$b
   (d) d$\rightarrow$e$\rightarrow$a$\rightarrow$c$\rightarrow$b
5. Which one of the following statements regarding 'Endosymbiotic hypothesis of origin of eukaryotes' is **INCORRECT**? **(JUNE 2017)**
   (a) Mitochondria arose from an α-proteobacterium and plastids arose from cyanobacteria.
   (b) The event of engulfinent of a photosynthetic cyariobacterium by a host cell was primitive to engulfinent of an α- proteobacterium during the eukaryotic origin.
   (c) Protists chlorarachniophytes, likely evolved when a heterotrophic eukaryote engulfed green alga, exemplifying secondary cndmymbiosis.
   (d) One of the membranes of the engulfed double-membraned cyanobacteria was lost in some of the hosts that eventually led to red and green algae descendants.

Answer Key

1.(d)	2.(a)	3.(c)	4.(d)	5.(b)

4. Paleontology and Evolutionary History

1. The correct order of periods from Palaeozoic to Mesozoic era is **(JUNE 2017)**
   (a) Triassic $\rightarrow$ Jurassic $\rightarrow$ Cretaceous $\rightarrow$ Cambrian $\rightarrow$ Ordovician $\rightarrow$ Silurian $\rightarrow$ Devonian $\rightarrow$ Carboniferous $\rightarrow$ Permian
   (b) Palaeocene $\rightarrow$ Eocene $\rightarrow$ Oligocene $\rightarrow$ Miocene $\rightarrow$ Pliocene $\rightarrow$ Pleistocene $\rightarrow$ Holocene
   (c) Cambrian $\rightarrow$ Ordovician $\rightarrow$ Silurian $\rightarrow$ Devonian $\rightarrow$ Carboniferous $\rightarrow$ Permian $\rightarrow$ Triassic $\rightarrow$ Jurassic $\rightarrow$ Cretaceous

(d) Pliocene → Eocene → Oligocene → Silurian → Devonian → Carboniferous → Triassic → Jurassic → Cretaceous

2. During which geological period did the greatest diversification of life on earth occur? **(MODEL PAPER)**

(a) Permian (b) Jurassic
(c) Cambrian (d) Ordovician

3. During which geological period was there an explosive increase in the number of many marine invertebrate phyla? **(DEC 2014)**

(a) Ordovician (b) Devonian
(c) Permian (d) Cambrian

4. Which of the following plant groups evolved during the Silurian period? **(MODEL PAPER)**

(a) Bryophyta (b) Psilophyta
(c) Lycophyta (d) Spherrophyta

5. Which of the following periods is known as "Age of fishes"? **(DEC 2016)**

(a) Devonian (b) Jurassic
(c) Cambrian (d) Carboniferous

6. During which of the following major mass extinction events, over 95% of the marine species disappeared from the planet Earth? **(JUNE 2016)**

(a) Ordovician (b) Devonian
(c) Permian (d) Triassic

7. Which of the following geographical periods is characterized by the first appearance of mammals? **(DEC 2016)**

(a) Tertiary (b) Cretaceous
(c) Permian (d) Triassic

8. The origin and diversification of Angiosperms was during which geological period? **(DEC 2015)**

(a) Permian (b) Triassic
(c) Jurassic (d) Cretaceous

9. The first vertebrate animal appeared in which of the following geological ages?

A) Paleozoic era
B) Mesozoic era
C) Ordovician period
D) Cretaceous period
E) Mississippian epoch
F) Paleocene epoch

Which of the following combinations give the best answer? **(DEC 2013)**

(a) A, C and F (b) A and F
(c) B, D and E (d) A and C

10. Match major events in the history of life with Earth's geological period. **(JUNE 2016)**

	Event		Geological period
A.	First reptiles	(i)	Quaternary
B.	First mammals	(ii)	Tertiary
C.	First humans	(iii)	Cretaceous
D.	First amphibians	(iv)	Triassic
		(v)	Carboniferous
		(vi)	Devonian

(a) A- (v); B-(i); C-(ii); D-(v)
(b) A - (v); B - (iv); C -(i); D - (vi)
(c) A - (vi); B - (iv); C -(ii); D -(vi)
(d) A - (iii); B - (i); C -(vi); D - (v)

11. The following geological eras mark the advent of important events in the history of earth - origin of terrestrial plants, origin of mammals and break up of the land mass Pangaea:

(a) Early Cambrian (b) Late Devonian
(c) Late Cretaceous (d) Late Jurassic

Identify the correct match of the events with the geological era **(JUNE 2011)**

(a) Origin of terrestrial plants - (c); Origin of mammals (d); Break-up of Pangaea - (a)
(b) Origin of terrestrial plants - (b); Origin of mammals (d); Break up of Pangaea –(c)
(c) Origin of terrestrial plants - (d); Origin of mammals (c); Break up of Pangaea –(b)
(d) Origin of terrestrial plants - (b); Origin of mammals (c); Break up of Pangaea –(d)

12. A few events in the history of life on earth are given below

A. Radiation of mammals and bird; Flourishing of insects and angiosperms

B. Primitive plants and fungi colonize land; Diversification of echinoderms.

C. Seed plants appear; Fishes and Trilobites abundant; earliest amphibians and insects.

D. Earliest birds and Angiosperms appear; Gymnosperms dominant

E. Invasion of land by primitive land plants and Arthropods

F. Mass marine extinctions; Reptiles radiate; Amphibians decline.

Which of the following is a correct match of the above events with the geological period during which they had occurred?

(JUNE 2014)

(a) A: Ordovician; B: Tertiary; C: Permian; D: Silurian E: Devonian; F: Jurassic

(b) A: Permian; B: Devonian; C: Silurian; D: Ordovician; E: Tertiary; F: Jurassic

(c) A: Tertiary; B: Ordovician; C: Devonian; D: Jurassic; E: Silurian; F: Permian

(d) A: Permian; B: Devonian; C: Jurassic; D: Tertiary; E: Silurian; F: Ordovician

13. Some important events in the history of life on Earth are given below.

A. First vertebrates (jawless fishes); first plants.

B. Forest of ferns and conifers; amphibians arise; insects radiate.

C. Conifers dominant; dinosaurs arise; insects radiate.

D. Flowering plants appear; climax of dinosaurs followed by extinction.

E. Radiation of flowering plants, most modern mammalian orders represented.

F. Ice Ages, Modern humans appear

Match the above with the geological time periods and choose the correct combination,

(DEC 2012)

(a) A- Silurian; B- Permian; C- Triassic; D- Jurassic; E- Cretaceous; F- Tertiary

(b) A- Ordovician; B- Carboniferous; C- Triassic; D- Cretaceous; E- Tertiary; F- Quaternary

(c) A- Cambrian; B- Ordovician; C- Silurian; D- Devonian; E- Permian; F- Tertiary

(d) A- Devonian; B- Permian; C- Triassic; D- Cretaceous; E- Tertiary; F- Quaternary

14. Following is the list of some important events in the history of life and the names of the epochs of Cenozoic era.

Events

A. Angiosperm dominance increases; continue radiation of most present day mammalian orders

B. Major radiation of mammals, birds and pollinating insects

C. Origins of many primate groups

D. Origin of genus *Homo*

E. Appearance of bipedal human ancestors

F. Continued radiation of mammals and angiosperms, earliest direct human ancestors

Epochs I

(i) Paleocene (ii) Pleistocene

(iii) Oligocene (iv) Pliocene

(v) Eocene (vi) Miocene

Which one of the following is the correct match of events with the epochs? **(JUNE 2015)**

(a) A-(v) B-(ii) C-(i) D-(iii) E-(iv) F-(vi)

(b) A-(vi) B-(i) C-(ii) D-(iv) E-(iii) F-(v)

(c) A-(v) B-(i) C-(iii) D-(ii) E-(iv) F-(vi)

(d) A-(iv) B-(i) C-(ii) D-(iii) E-(v) F-(vi)

15. Match the geological time period with the extinction or diversification events associated with them: **(DEC 2019, ASSAM)**

Geological time	Event
A. Cenozoic	i. Angiosperm diversification
B. Cretaeous	ii. Modem fauna diversification (bivalves, gastropods, bryozoans, malacostracan crustaceans)
C. Paleozoic	iii. Mega fauna extinction
D. Quaternary	iv. Mammal diversification

(a) A-ii, B-i, C-iii, D-iv
(b) A-iv, B-i, C-ii, D-iii
(c) A-ii, B-ii, C-i, D-iv
(d) A-iv, B- iii, C-i, D-iv

16. Several distinct time periods and different routes might explain the entrance of marsupials into Australia.

(i) Late Jurassic - early therians arrived in Antarctica - Australia where the marsupials subsequently evolved.

(ii) Early to middle Cretaceous - early marsupials arrived in Australia from northern regions and then radiated in isolation.

(iii) Paleocene - marsupials entered Australia from South-East Asia.

(iv) Eocene - chance dispersal of marsupials into Australia. **(DEC 2011)**

Which of the following is the correct combination?

(a) (i) (ii) (iii) (b) (i) (iii) (iv)
(c) (ii) (iii) (iv) (d) (i) (ii) (iv)

17. According to fossil record, the earliest fossils of liverworts are found in late Devonian, of mosses in early cretaceous, and vascular plants in the later Silurian/early Devonian. *Anthoceros* (hornworts) fossils have not been recovered. Reading fossil records we would say that vascular plants appeared first and then liverworts.

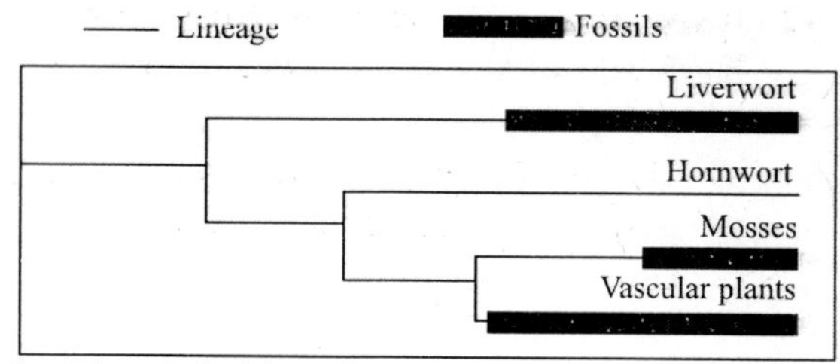

However phylogenetic relationship (shown in figure) suggest otherwise. It may be that

(A) evolutionary history can be read directly from fossil record

(B) The moss lineage goes back to at least early Silurian/early Devonian

(C) Fossil can only set a maximum age for a lineage

(D) Fossil can only sets a minimum age for a lineage

(E) the divergence between liverworts and rest of land plants goes back to at least early Ordovician

Which of the following statements is correct? **(JUNE 2013)**

(a) A, B, C and E (b) B, D and E
(c) A, B, D and E (d) B, C and E

18. In the evolutionary tree given below terms A, B, C, D and E represent respectively **(JUNE 2015)**

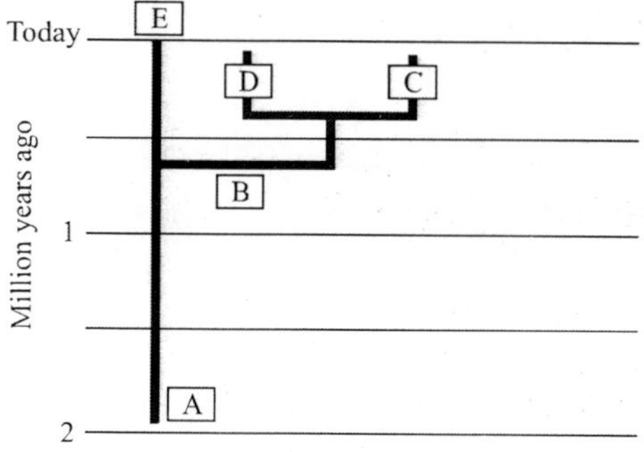

(a) *Homo erectus*, *Homo heidlebergensis*, Neanderthal, Denisovan and *Homo sapiens*.

(b) *Homo heidelbergerisis, Homo erectus,* Denisovan, Neanderthal and *Homo sapiens.*

(c) *Homo heidelbergensis,* Denisovan, Neanderthal, *Homo erectus* and *Homo sapiens.*

(d) *Homo heidelbergensis, Homo sapiens,* Derrisovan, Neanderthal, and *Homo erectus.*

19. A small-bodied hominid whose fossil was discovered very recently, ventured into the islands of South-East Asia from mainland Asia about 67,000 years ago. This species is believed to be closely related to another Homo species discovered previously from an island in Indonesia. The given scientific name of the newly discovered species is: **(DEC 2019 ASSAM)**

(a) *Homo floresiensis*

(b) *Homo luzonensis*

(c) *Homo naledi*

(d) *Homo habilis*

20. Following is a diagrammatic representation of human evolutionary tree.

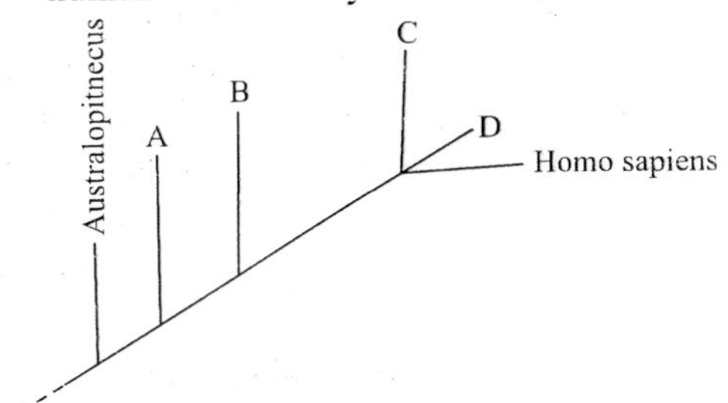

In the above diagram A, B, C and D respectively represent: **(JUNE 2019)**

(a) Denisovan, *Homo habilis, Homo erectus, Homo neanderthalensis*

(b) *Homo habilis, Homo erectus, Homo neanderthalensis,* Denisovan

(c) *Homo erectus, Homohabilis, Homo neanderthalensis,* Denisovan

(d) *Homo erectus,* Denisovan, *Homo neanderthalensis, Homo habilis*

21. Given below are statements related to the two competing hypotheses on the origin of modern humans: the Out-of-Africa hypothesis and the multi-regional hypothesis. Which of the following statements is **INCORRECT?** **(JUNE 2018)**

(a) Both the hypotheses support that *Homoerectus* originated in Africa and expanded to Eurasia.

(b) Mitochondrial DNA (mtDNA) and Y chromosome DNA evidence support the 'Out-of-Africa' hypothesis.

(c) The principal conflict between the two hypotheses is that multi-regional hypothesis does not support African origin of *Homo erectus.*

(d) The multi-regional hypothesis states that independent multiple origins occurred in the million years since *Homo erectus* came out of Africa.

22. A group of palaeontologists digging in an area discovers a pre-historic human burialsite. The same group, while exploring a nearby area, discovered fossil remains of what appeared to be more than 100 million year old dinosaur bones. Which of the following combinations of modern radiometric dating techniques should they use to calculate the age of these fossils most accurately? **(JUNE 2018)**

(a) ^{14}C dating for human remains and ^{235}U dating for dinosaur remains

(b) ^{87}Rb dating for both human and dinosaur remains

(c) ^{14}C dating for both human and dinosaur remains

(d) ^{129}I dating for human remains and ^{129}Xe for dinosaur remains

Answer Key

1.(c)	2.(c)	3.(d)	4.(b)	5.(a)	6.(c)	7.(d)
8.(d)	9.(d)	10.(b)	11.(b)	12.(c)	13.(b)	14.(c)
15.(b)	16.(d)	17.(b)	18.(a)	19.(b)	20.(b)	21.(c)
22.(a)						

5. Molecular Evolution

1. Which one of the following statements is **NOT TRUE** about the Neutral Theory as proposed by Motoo Kimura? **(DEC 2017)**
   (a) Except for advantageous mutations, most alleles are under neutral selection
   (b) The rate of evolution for most genes will be equal to the neutral mutation rate
   (c) Advantageous mutations are exceedingly rare
   (d) At the level of DNA sequences, genetic drift dominates evolution

2. Which of the following is a prediction of neutral theory of molecular evolution that is supported by data? **(JUNE 2013)**
   (a) Human and chimps differ more in DNA sequences of psuedogenes than in coding regions of functional genes
   (b) Human and chimps differ more in DNA sequences of coding regions of functional genes than of pseduogenes
   (c) Human have a faster evolution of most of their DNA sequences than chimps
   (d) The more advanced species have more number of functional genes

3. If the ratio of the number of non-synonymous to synonymous substitutions per site in protein coding gene is greater than one, it is an evidence of selection that is **(JUNE 2011)**
   (a) positive. (b) negative.
   (c) neutral. (d) random.

4. A gene 'X' in *Drosophila* contributes to inviability of hybrids. The phylogeny below shows the evolutionary history of gene 'X' and on each branch the numbers indicate the non-synonymous/synonymous substitutions.

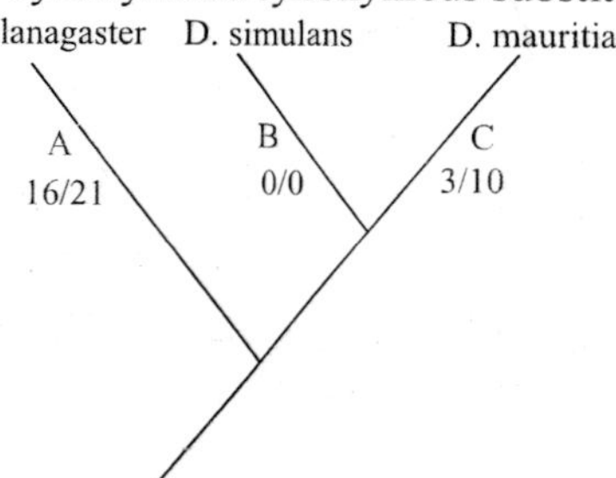

Based on the above information, select the correct statement from the choices below: **(DEC 2017)**
   (a) High proportion of non-synonymous changes at A indicates evolution by natural selection.
   (b) High proportion of synonymous changes at A indicates absence of natural selection.
   (c) Equal proportion of non-synonymous and synonymous changes at B indicates deleterious selection.
   (d) Low proportion of non-synonymous changes at C indicates positive selection.

5. Using molecular clock, it was estimated that two species A and B must have diverged from their common ancestor about 9 x 10^6 years ago. If the rate of divergence per base pair is estimated to be 0.0015 per million years, what is the proportion of base pairs that differ between the two species now? **(JUNE 2011)**
   (a) 0.0270 (b) 0.0135
   (c) 0.00017 (d) 0.0035

6. Scientists discovered two new plant species. "A" and "B" that look similar except that, species "A" bears flowers and leaves that are twice the size of those in species "B". Which method should the scientists use to appropriately investigate if species "A" is a result of gene duplication in species "B"? **(JUNE 2018)**
   (a) Sequence similarity, gene structure and gene size.
   (b) Plant size, physical proximity of gene and genome size.
   (c) Sequence similarity, physical proximity of gene, genome size.
   (d) Sequence length, gene structure and chromosome count.

Answer Key

1.(a)	**2.**(a)	**3.**(a)	**4.**(a)	**5.**(a)	**6.**(c)

6. Molecular tools in phylogeny

1. Similarities in sequence and function of two proteins indicate that they are members of a family that share a common ancestor. If they are from different species, they are called **(MODEL PAPER)**

(a) homologs. (b) orthologs.
(c) paralogs. (d) proteologs.

2. Mayfair genes (HYPOTHETICAL) consist of a superfamily of transcription factors. They are found in 4 clusters in mammals; in 2 clusters in insects; and in a single cluster in an ancestor to insects. These data consistent with all of the following explanations EXCEPT: **(DEC 2016)**

(a) Two successive genome duplication event occurred between ancestral organisms and vertebrates
(b) the first duplication may have taken place before divergence of vertebrates
(c) exon shuffling exclusively produced such cluster
(d) whole genome duplication could lead to such observation.

3. The most commonly used molecular tool for phylogentic analysis involves sequencing of **(JUNE 2015)**

(a) mitochondrial DNA.
(b) mitochondrial RNA.
(c) ribosomal RNA.
(d) nuclear DNA.

4. Which species concept utilizes morphological and molecular characters to distinguish between species? **(DEC 2014)**

(a) Evolutionary
(b) Ecological
(c) Biological
(d) Phylogenetic

5. For which, of the following characters, convergent evolution is most unlikely? **(DEC 2013 GU)**

(a) Locomotor organs.
(b) Secondary metabolites.
(c) Conditionally expressed genes.
(d) base sequence of a gene.

6. Evolution of multi-gene family occurs by **(JUNE 2013)**

(a) only gene duplication.
(b) only unequal crossing-over.
(c) random mutations.
(d) both duplication and unequal crossing over.

The table below shows the number of unrooted and rooted trees generated when the number of taxa are three and four, respectively. Number of taxa Number of unrooted trees Number of rooted trees?

Number of taxa	Number of unrooted trees	Number of rooted trees
3	1	3
4	3	15
5	15	:?

The number of rooted trees generated when there are five taxa is: **(DEC 2019 ASSAM)**

(a) 75 (b) 78
(c) 105 (d) 126

7. Following diagrams represent various ways in which a character may evolve.

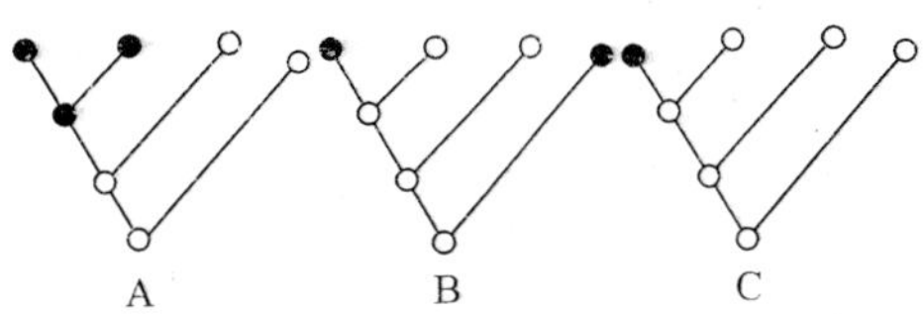

Which of the following is the correct definition for the character evolution patterns shown above? **(JUNE 2017)**

(a) A- Autapomorphy, B - Synapomorphy, C -Homoplasy
(b) A-Autapomorphy, B - Homoplasy, C - Synapomorphy
(c) A- Synapomorphy, B-Autapomorphy, C - Homoplasy
(d) A- Synapomorphy, B - Homoplasy, C – Autapomorphy

8. Which of the following processes interferes in sequence-based phylogeny? **(DEC 2011)**
   (a) Horizontal gene transfer
   (b) Adaptive mutations
   (c) DNA repair
   (d) Reverse transcription

9. Following tree represents phylogenetic relationships among species of a moth family. Circles represent species having eye spots on the wings. Other species do not have eye spots.

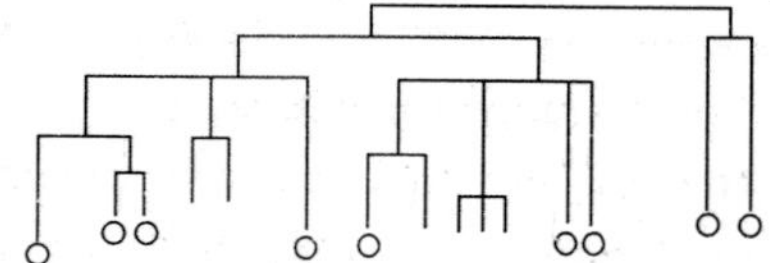

The following interferences were made by different researchers:

Eye spots were present in the ancestors and some species lost them.

Eye spots were not present in the ancestors.

Eye spots were lost more than once in evolution of the family.

Eye spots were gained only once while evolving from ancestors without them.

Which of the interferences are correct? **(DEC 2012)**

(a) A and B (b) C and D
(c) A and C (d) B and D

10. The phylogeny below shows evolutionary relationships between 9 extant bird species and whether they display red or blue plumage.

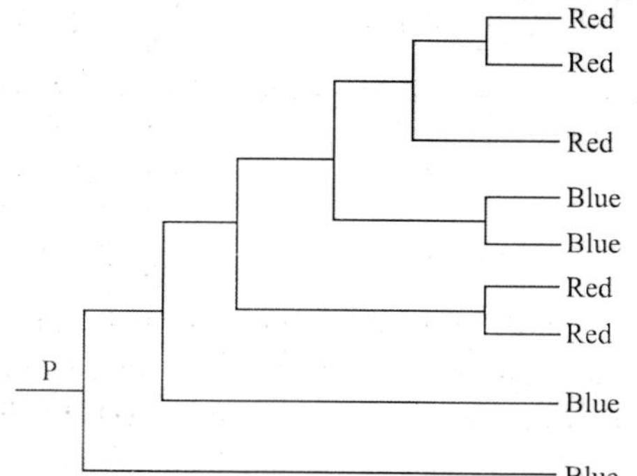

Based on the above phylogeny and the distribution of red and blue character states among the extant species, and using the principle of parsimony, which of the following is the correct inference about plumage colour of the ancestor at the root P? **(DEC 2019)**
   (a) Ancestral state at P is blue.
   (b) Ancestral state at P is red.
   (c) Ancestral state at P is more likely to be red than blue
   (d) Ancestral state at P is equally likely to be red or blue.

11. The phylogenetic tree below shows evolutionary relationships among 8 species. Males of these species are either blue (b) or red (r) in colour, the colour being indicated next to each species name. **(DEC 2018)**

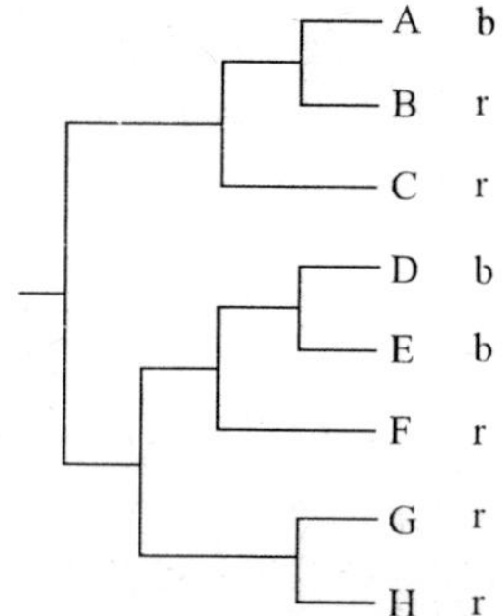

Based on the principle of parsimony, which of the following statements best represents, the evolution of male body colour in this set of species?
   (a) The most recent common ancestor of all 8 species was blue; red evolved independently 5 times.
   (b) The most recent common ancestor of all 8 species was blue; red evolved independently 4 times.
   (c) The most recent common ancestor of all 8 species was red; blue evolved independently 3 times.
   (d) The most recent common ancestor of all 8 species was red; blue evolved independently 2 times.

12. *Knox* gene code transcriptionl factor important for regulation of indeterminate growth in plant

shoots. These genes also regulate patterns of development of plant organs such as leaves and flowers. The figure represents a phylogenetic tree of the multigene family in some land plants. The circles represents the genes that act to maintain shoot apical meristem (equivalent to stem cells). Orthologues are genes that duplicate due to speciation and paralogues are genes that duplicate within a species.

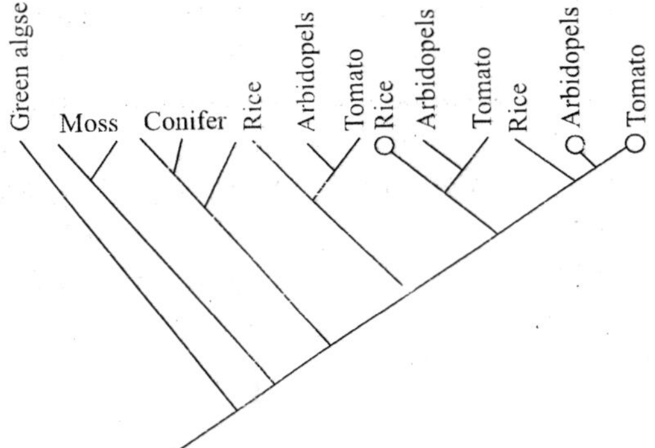

From the figure; the following inferences were made.

(A) Multiple gene duplication occurred in vascular plants.

(B) Gene duplications may have enabled shoot diversification in vascular plants.

(C) Shoot apical meristems are regulated by orthologous genes in vascular plants.

(D) Shoot apical meristems are regulated by paralogous genes in vascular plants.

Which of the following represents a combination of correct inferences? **(JUNE 2012)**

(a) (A), (B) and (D)

(b) (A), (B) and (C)

(c) (B) and (C) only

(d) (B) and (D) only

13. In the phylogenetic tree above, branch-lengths are drawn proportional to the number of changes along a lineage.

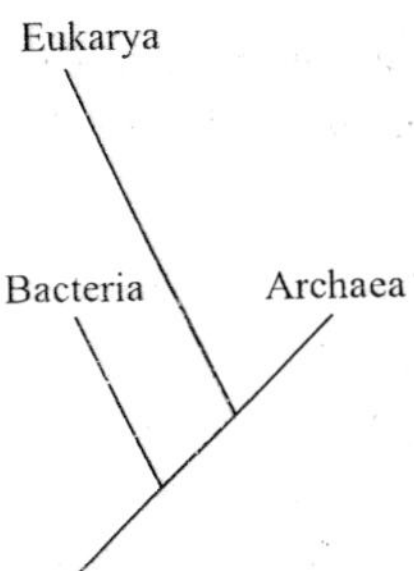

The following inferences were made from this tree. **(JUNE 2013)**

(A) Bacteria are more closely related to Eukarya than to Archaea.

(B) Bacteria and Archaea are more similar to each other than either is to Eukarya.

(C) Archaea and Eukarya diverged from each other after their common ancestor diverged from bacteria.

Which of the following represents a combination of correct inferences?

(a) (A), (B) and (C) (b) (A) and (B) only

(c) (B) and (C) only (d) (A) and (C) only

14. When James R Brown and W Ford Doolittle (1997) reconstructed the tree of life using a variety of different genes, they found that different genes gave fundamentally different phylogenies as shown below. Note that the genes are unique to the specific trees.

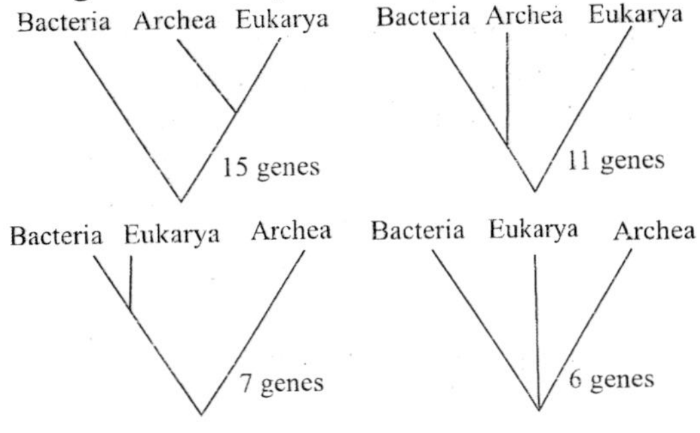

From the given options select the process that best explains the observed discrepancies between the trees. **(DEC 2017)**

(a) polyploidization
(b) horizontal gene transfer
(c) allopatric speciation
(d) localized extinctions

15. Given are some statements with reference to the use of genes in plant molecular systematics.
A. mtDNA are not preferred over cpDNA or rDNA because they generally show slow rate of sequence evolution and fast rate of structural evolution.
B. cpDNA are not preferred because of their haploidy, uniparental inheritance, and absence of recombination among cpDNA molecules.
C. rDNA such as ITS are preferred for their higher evolutionary rates as well as shorter sequence length.
D. rDNA and cpDNA cannot be used simultaneously in molecular systematics since they represent conflicting patterns of inheritance.
Which of the above statements are **INCORRECT?** **(JUNE 2018)**
(a) A, C and D (b) A, B and C
(c) A and C only (d) B and D only

Answer Key

1.(b)	2.(c)	3.(1/4)	4.(d)	5.(d)	6.(d)	7.(d)
8.(a)	9.(c)	10.(a)	11.(d)	12.(a)	13.(c)	14.(b)
15.(d)						

7. Population genetics

1. Which of the following is NOT an assumption of the Hardy-Weinberg model? **(DEC 2016)**
(a) Population mates at random with respect to the locus in question
(b) Selection is not acting on the locus in question.
(c) One allele is dominant and other is recessive at this locus
(d) The population is effectively infinite in size

2. Which of the following assumption support the Hardy-Weinberg Equilibrium?
(MODEL PAPER)
(a) Presence of Natural Selection
(b) Random mating.
(c) Genetic Drift.
(d) Assortative mating

3. The Hardy-Wienberg principle comes from considering what happens when Medelian genes act on population. The model predicts that there will be no change in allele frequencies when **(JUNE 2012)**
(a) Migration into the population occurs at a steady rate
(b) The population suffers a bottle neck
(c) a rare new mutation is associated with a sharp increase in fitness
(d) no evolutionary process is at work

4. The Hardy-Weinberg principle states that allele frequencies in a population will remain constant over generations if certain assumptions are met.
A. Random mating
B. Mate choice
C. Small population size
D. Large population size
E. Lack of mutations
F. Directional selection
Which of the above factors will cause changes in allele frequencies overgenerations?
(JUNE 2019)
(a) A, D and F (b) B, D and F
(c) A, C and E (d) B, C and F

5. The frequency of M-N blood types in a population of 6129 individuals is as follows:

Blood type	Genotype	Number of individuals
M	L^ML^M	1787
MN	L^ML^N	3039
M	L^NL^N	1303

The frequency of L^N allele in this population is **(DEC 2016)**

(a) 0.4605 (b) 0.2121
(c) 0.5395 (d) 0.2911

6. The following genotypes were observed in a population

Genotype	Number
HH	90
Hh	60
hh	50

Which of the following is the correct frequency of H allele and what will be the expected number of HH in the given population? **(DEC 2013)**

(a) 0.60 and 72 (b) 0.80 and 96
(c) 0.50 and 32 (d) 0.30 and 90

7. In a population, the genotype frequencies are: $f(A_1A_1) = 0.59$; $f(A_1A_2) = 0.16$; $f(A_2A_2) = 0.25$. What are the frequencies of the two alleles at this locus? **(JUNE 2015)**

(a) $A_1 = 0.59$ $A_2=41$ (b) $A_1=0.75$ $A_2=25$
(c) $A_1 = 0.67$ $A_2=33$ (d) $A_1=0.55$ $A_2=44$

8. There are 'n' numbers of alleles at a given locus in a diploid population. The proportion of all homozygotes in the population **(JUNE 2013)**

(A) All alleles are equal abundant
(B) All alleles are not in equal abundant

(a) $1/n$ and $<1/n$
(b) $1/n$ and $>1/n$
(c) $1/n^2$ and $<1/n^2$
(d) $1/n^2$ and $>1/n^2$

9. Red hair is a recessive trait in human. In a randomly mating population in Hardy-Weinberg equilibrium approximately 9% of individuals are red- haired. What is the frequency of the heterozygotes? **(DEC 2016)**

(a) 81% (b) 49%
(c) 42% (d) 18%

10. in a random sample of 400 individuals from a population with allele of trait in Hardy-Weinberg equilibrium, 36 individuals are homozygous for allele *a*. How many individuals in the sample are expected to carry atleast one allele A? **(DEC 2014)**

(a) 36 (b) 168
(c) 364 (d) 196

11. In a population, alleles p and q are known to be in a ratio of 0.7p: 0.3q. At Hardy- Weinberg equilibrium how many heterozygotes (pq) can be expected in a sample of 60? **(DEC 2013 GU)**

(a) 25 (b) 42
(c) 49 (d) 9

12. The frequencies of alleles 'A' and 'a' in a population at Hardy- Weinburg equilibrium are 0.7 and 0.3, respectively. In a random sample of 250 individuals taken from the population, how many are expected to be heterozygous? **(JUNE 2011)**

(a) 112 (b) 81
(c) 105 (d) 145

13. Consider a single locus with 2 alleles which are at Hardy-Weinberg equilibrium. If the frequency of one of the homozygous genotypes is 0.64, what is the frequency of heterozygotes in the population? **(DEC 2018)**

(a) 0.16 (b) 0.20
(c) 0.32 (d) 0.36

14. The frequency of homozygotes in a diploid population is 0.68. Assuming that the population is in Hardy-Weinberg equilibrium, the frequencies of the two alleles are **(DEC 19)**

(a) 0.1 and 0.9 (b) 0.2 and 0.8
(c) 0.4 and 0.6 (d) 0.5 and 0.5

15. In a sample from a population there were 65 individuals with BB genotype, 30 individuals with Bb genotype and 15 individuals with bb genotype. The frequency of the '*b*' allele is **(JUNE 2018)**

(a) 0.59 (b) 0.27
(c) 0.41 (d) 0.73

16. In a· population of 2000 individuals of a plant species, genetic difference at a single locus leads to different flower colours. The alleles are incompletely dominant. The population has 100 individuals with the genotype *rr* (white flowers), 800 individuals with the genotype *Rr* (pink flower) and the remaining have. genotype*RR* (red flowers). What is the frequency of the r allele in the population?

(DEC 2018)

(a) 0.25 (b) 0.50
(c) 0.75 (d) 1.00

17. An autosomal recessive condition affects 1 newborn in 10,000 in a random mating population without any disruptive acting force. What is the approximate expected frequency of carriers in this population?

(DEC 2018)

(a) 1 in 1000 newborns
(b) 1 in 500 newborns
(c) 1 in 100 newborns
(d) 1 in 50 newborns

18. If a given gene in a randomly mating population has three alleles a, b and c in the ratio of 0.5, 0.2 and 0.3 respectively, what is the expected frequency of genotypes ab and bc in the population at equilibrium?

(MODEL PAPER)

(a) 0.1 and 0.06 (b) 0.2 and 0.15
(c) 0.2 and 0.12 (d) 0.04 and 0.09

19. Fruit colour of wild *Solanum nigrum* is controlled by two alleles of a gene (A and a). The frequency of A, p=0.8 and a, q=0.(b) In a neighbouring field a tetraploid genotype of *S. nigrum* was found.

After critical examination five distinct genotypes found; which are AAAA, AAAa, AAaa, Aaaa and aaaa. Following Hardy Weinberg principle and assuming the same allelic frequency as that of diploid population the numbers of phenotypes calculated within a population of 1000 plants are close to one of the following: **(JUNE 2016)**

AAAA : AAAa : AAaa : Aaaa ; aaaa

(a) 409 :409:154:26:2
(b) 420: 420: 140: 18: 2
(c) 409:409:144:36:2
(d) 409: 420 144: 25: 2

Answer Key

1.(c)	2.(b)	3.(d)	4.(d)	5.(a)	6.(a)	7.(c)
8.(a)	9.(c)	10.(c)	11.(a)	12.(c)	13.(c)	14.(b)
15.(b)	16.(a)	17.(d)	18.(c)	19.(a)		

8. Microevulution

1. Micro-evolution is the term used for changes in allele frequencies that occur over time.
   A) Within a population at species level
   B) within a community at genus level
   C) due to appearance of new genes infections
   D) due to mutation, natural selection, flow and genetic drift

   Which of the following combinations is NOT appropriate? **(DEC 2013)** *

   (a) A and C (b) A and D
   (c) B and C (d) B and D
2. Which one of the following does NOT contribute to micro-evolutionary change?

   (JUNE 2019)

   (a) Mutation (b) Random mating
   (c) Genetic drift (d) Natural selection
3. Which one of the following will have the least impact on allele frequencies in small populations? **(DEC 2018)**

   (a) Inbreeding (b) Random mating
   (c) Genetic drift (d) Outbreeding
4. Which one of the following statements is INCORRECT? **(JUNE 2016)**

   (a) Loss of genetic variation occurs within a small population due to genetic drift. .

(b) The number of deleterious alleles present in the gene pool of a population is called the genetic load.

(c) Genetic erosion is a reduction in levels of homozygosity.

(d) Inbreeding depression results from increased homozygosity for deleterious alleles.

5. The frequencies of two alleles p and q for a gene locus in a population at Hardy-Weinberg equilibrium are 0.3 and 0.7, respectively. After a few generations of inbreeding, the heterozygote frequency was found to be 0.28, The inbreeding coefficient will be **(DEC 2011)**

(a) 0.42 (b) 0.28
(c) 0.33 (d) 0.67

6. In a plant species, a segregating line (one that contains both homozygotes and heterozygotes at a locus) can be made homozygous by repeated selfing for several generations. What is the level of remaining heterozygosity after three generations of selting, if the level of heterozygosity in generation '0' is denoted as 1? **(JUNE 2013)**

(a) 0.5 (b) 0.25
(c) 0.125 (d) 0.0625

7. Consider an autosomal locus with two alleles A_1 and A_2 at frequencies of 0.6 and 0.4 respectively. Each generation, A_1 mutates to A_2 at a rate of $\mu = 1 \times 10^{-5}$ while A_2 mutates to A_1 at a rate of $\upsilon = 2 \times 10^{-5}$. Assume that the population is infinitely large and no other evolutionary force is acting. . The equilibrium frequency of allele A_1 is **(JUNE 2016)**

(a) 1.0. (b) 0.5.
(c) 0.67 (d) 0.33

Answer Key

1.(*)	2.(b)	3.(b)	4.(c)	5.(c)	6.(c)	7.(c)

9. Random genetic drift

1. Northern elephant seal had been reduced to about 20 in 1800s. Biologist studied variation in protein in the species. They found no genetic differences in the protein among individuals. This lack of variation is due to **(JUNE 2013)**

(a) the fact that elephant seal lives in constant environment where there was no need for genetic variation

(b) population bottle neck and genetic drift

(c) natural selection resulting in a single best genotype

(d) a very low rate of mutation

2. Twenty small populations of a species, each polymorphic for a given locus (T, t) were bred in captivity. In 10 of them the population size was kept constant by random removal of individuals, while other 10 were allowed to increase their population size.

After several generations it was observed that in 7 of the size restricted populations only T was present, in the remaining 3 only t was observed. The experiment illustrates **(DEC 2012)**

(a) Genetic drift which is more likely in large populations.

(b) Genetic drift which is more likely in small populations

(c) Density dependant selection against T.

(d) Density dependant selection against t.

3. Which of the following statements is NOT correct regarding effect of genetic drift? **(DEC 2014)**

(a) It alters allele frequency substantially only in small population.

(b) It can cause allele frequencies to change at random.

(c) It can lead to a loss of genetic variation within populations.

(d) It can cause harmful alleles to become eliminated.

4. Following are key points about the effect of genetic drift:

A. Genetic drift is significant in small populations.

B. Genetic drift can cause allele frequencies to change in a pre-directed way.

C. Genetic drift can lead to a loss of genetic variation within populations.

D. Genetic drift can cause harmful alleles to become fixed.

Which one of the following combination of the above statements are true? **(JUNE 2018)**

(a) A and B only (b) A and C only

(c) A, B and C (d) A, C and D

5. In. very small populations, genetic variation is often lost through genetic drift. If the population size of a mammal on an isolated island is 50, what percentage of its genetic variation is lost every generation

(JUNE 2014)

(a) 0.01 (b) 0.5

(c) 0.1 (d) 0.05

6. One hundred independent populations of Drosophila are established with 10 individuals in each population, of which, one individual is of Aa genotype and the other nine are of AA genotype. If random genetic drift is the only mechanism acting on these populations, then, after a large number of generations, the expected number of populations fixed for the "a" allele is **(JUNE 2016)**

(a) 75 (b) 50

(c) 25 (d) 5

7. In several populations, each of size N =20, if genetic drift results in a change in the relative frequencies of alleles,

A. What is the rate of increase per generation in the proportion of populations in which the allele is lost or fixed?

B. What is the rate of decrease per generation in each allele frequency class between 0 and 1?

The correct answer for A and B is:

(DEC 2015)

(a) A-0.25, B-0.125

(b) A-0.025, B-0.0125

(c) A-0.0125, B-0.025

(d) A-0.125, B-0.25

8. In a population of effective Population size N_e with rate of neutral mutation μ_o, the frequency of heterozygotes per nucleotide per site at equilibrium between mutation and genetic drift is calculated as **(DEC 2015)**

(a) $\frac{2N_e\mu_0}{4N_e\mu_0+1}$ (b) $\frac{4N_e\mu_0}{4N_e\mu_0+1}$

(c) $\frac{N_e\mu_0}{4N_e\mu_0+1}$ (d) $\frac{4N_e\propto_0}{(4N_e\propto_0 \quad 1)}$

Answer Key

1.(b)	2.(b)	3.(d)	4.(d)	5.(a)	6.(d)	7.(c)	8.(b)

10. Convergent evolution/ Divergent evolution

1. Flufftails in mainland Asia show high variation in tail colour. However, in the far out Pacific island, the flufftails show very little variation in tail colour. This variation in tail colour can be explained by all of the following EXCEPT

(JUNE 2017)

(a) founder effect

(b) homologous evolution

(c) genetic drift

(d) frequency dependent selection

2. In order to study evolutionary relatedness between two given individuals, homology and analogy are considered. By this study we may observe the following possible information:

A. Presence of homologous structures indicate adaptive radiation.

B. Presence of analogous structure indicate convergent evolution.

C. Homology also refers to divergent evolution.

D. Analogous structures develop due to the fact that different unrelated groups of organism evolved under similar environmental conditions.

Which of the above possibilities correctly explain the evolutionary process? **(DEC 2013 GU)**

(a) A, C and D (b) A and B
(c) B, C and D (d) A, B, C and D

3. Which one of the following statements is **NOT TRUE** about homologous characters? **(DEC 2017)**

(a) Similar traits may not be homologous
(b) Similar traits may be a result of convergent evolution
(c) A homologous trait that is derived and shared from a common ancestor is called an autapomorphy
(d) Homologous characters may show structural similarity but functional diversity

4. Fore limb of human and flippers of whale are embryologically homologous structures. What does the study of homologous structures tell us about evolution?

A. This is the example of adaptive radiation, occurred due to similar group of organisms inhabiting different environments
B. This is the example of divergent evolution, occurred due to similar group of organisms inhabiting different environments
C. Similar group of organisms with mutations and variations getting naturally selected in different environments
D. This is the example of convergent evolution, occurred due to similar group of organisms inhabiting different environments

Which of the following is the correct combination? **(DEC 2013)**

(a) A, B and C (b) A and D
(c) B and D (d) Only D

5. The wings of insects and the wings of bats represent a case of **(DEC 2014)**

(a) divergent evolution
(b) convergent evolution
(c) parallel evolution.
(d) neutral evolution.

6. An alphabetical list of tropical rainforest mammals from South America and Africa is given below:

	South America		Africa
i	Agouti	A	Royal Antelope
ii	Giant Armadilo	B	Pangolin
iii	Paca	C	Chevrotain
iv	Three toed sloth	D	Bosman's Potto

Pair the species in the list to demonstrate the concept of convergent evolution between the two continents. **(DEC 2018)**

(a) i- D; ii- C; iii - A; iv-B
(b) i - A; ii - C; iii - D; iv - B
(c) i - A; ii - B; iii- D; iv: C
(d) i - D; ii - C; iii - B; iv – A

Answer Key

1.(b)	2.(d)	3.(c)	4.(a)	5.(b)	6.(d)
2	4	3	1	2	4

11. Speciation: Allopatricity and Sympatricity

1. The speciation in which a population splits into two geographically isolated populations experience dissimilar selective pressure and genetic drift is known as **(JUNE 2013)**

(a) sympatric speciation.
(b) parapatric speciation.
(c) peripatric speciation.
(d) allopatric speciation.

2. A species of grass grows around a mine area having patches of heavy metal contaminated soil. Some of the populations of the species grew selectively on the soil that was contaminated with heavy metals. Over a period of time, though the tolerant and non- tolerant grass populations were continuously distributed and not separated by geographical barriers, they eventually evolved different flowering times and became different species. What kind of speciation would you call this? **(JUNE 2017)**

(a) Allopatric speciation

(b) Sympatric speciation

(c) Parapatric speciation

(d) Bottle-neck effect

3. Which one of the following statement is correct for the process of speciation? **(JUNE 2019)**

(a) Allopatric speciation occurs between adjacent populations.

(b) Parapatric speciation may occur between geographically separated populations.

(c) Sympatric speciation occurs within one continuously distributed population.

(d) Sympatric speciation occurs when continuously distributed populations are fragmented.

4. Wolbachia are obligate intracellular bacteria, many different strains of which are abundantly present in insects. They induce mating incompatibility in host, i.e. males infected with one strain can only fertilize females infected with the same strain. No other pathological effects are observed in host. A possible evolutionary consequence of this phenomenon would be: **(DEC 2012)**

(a) Extinction of many insect species.

(b) Termination of sexual reproduction in many insect species.

(c) Co-extinction of host and parasite.

(d) Reproductive isolation leading to rapid speciation in insects.

5. Orchids of the genus *Cryptostylis* are known to maintain reproductive isolation because their flowers look and smell like females of the wasps of genus *Lissopimpla*. When the male wasp visits and attempts to mate with the flower, the shape of anther and stigma allows correct placement and transfer of pollen to the wasp, which then transfers the pollen to species specific flower that it next attempts to mate with. This prezygotic barrier that prevents inter-species cross-pollination in *Cryptostylis* is best explained by: **(DEC 2017)**

(a) behavioural isolation through mimicry

(b) mechanical isolation through mimicry

(c) temporal isolation

(d) habitat isolation

6. Temporal isolation in breeding seasons between closely related species leads to reproductive isolation. Given below are breeding seasons of different species of frogs.

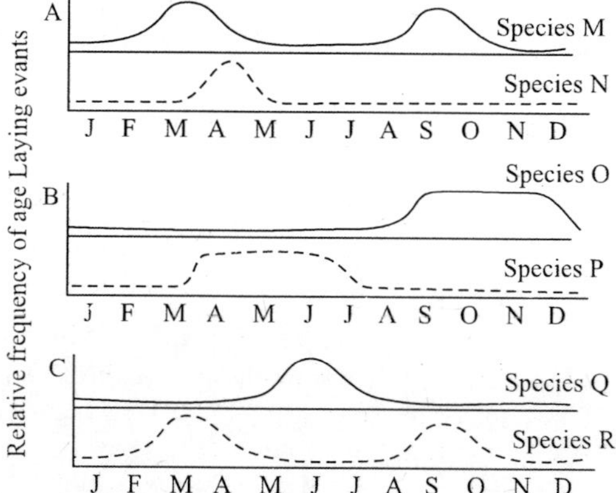

Which of the above plots represents temporal isolation in breeding seasons among closely related sympatric species? **(DEC 2017)**

(a) Plot A (b) Plots A and B

(c) Plots B and C (d) Plots A and C

7. In a study of sexual isolation in a species of salamander, scientists brought together males and females from different populations and from the same population. They observed the frequency of mating and calculated a sexual isolation index. One graph shows the relationship between mating frequency and genetic distance, and the other shows the relationship between sexual isolation index and geographic isolation.

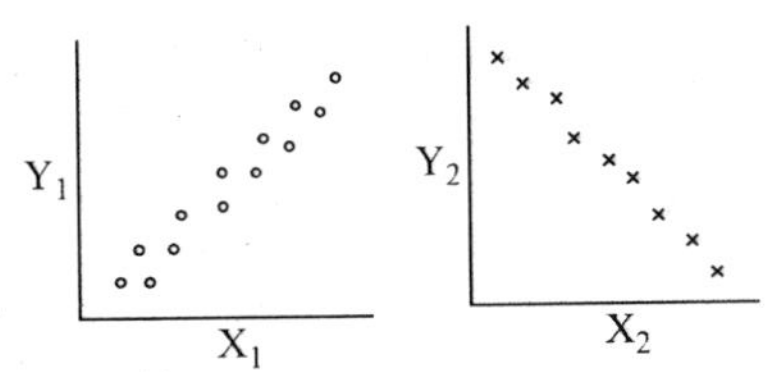

Choose the appropriate terms for of X_1, Y_1, X_2 and Y_2 in the figures, above. **(JUNE 2013)**

(a) X_1 = Geographic distance, Y_1 = Sexual isolation index; X_2 = Genetic distance, Y_2 = mating frequency

(b) X_1 = Geographic distance; Y_1 = mating frequency; X_2 = Genetic distance, Y_2 = Sexual isolation index

(c) X_1 = Genetic distance; Y_1 = mating frequency; X_2 = Sexual isolation index; Y_2 = Geographic distance

(d) X_1= Genetic distance; Y_1 = Geographic distance; X_2 = Sexual isolation index; Y_2 = mating frequency

8. Four different species concepts are given below:

A. Species separate based on their use of different ecological niches and their presence in different habitats and environments.

B. Differences in physical characteristics or molecular characteristics are used to distinguish species.

C. Species are distinct if they are reproductively isolated.

D. Phylogenetic trees and analyses of ancestry serve to differentiate species.

Which of the following gives the correct names of the above concepts? **(JUNE 2014)**

(a) A: Biological; B: Phylogenetic; C: Evolutionary; D: Ecological

(b) A: Ecological; B: Phylogenetic; C: Biological; D: Evolutionary

(c) A: Evolutionary; B: Ecological; C: Biological; D: Phylogenetic

(d) A: Phylogenetic; B: Evolutionary; C: Ecological; D: Biological

9. The given below is the graphical representation of the changes in morphological features over a period of the geological time scale, where population A accumulates heritable morphological features and give rise to distinct species B. Population B splits in to a distinct species B2

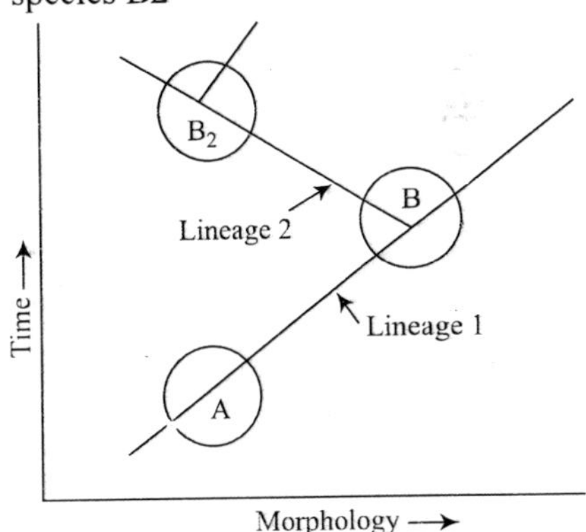

Which of the following lineage represent the pattern of speciation by cladogenesis? **(DEC 2016)**

(a) Lineage 1

(b) Both lineage 1 and 2

(c) Lineage 2

(d) Neither of the lineage 1 and 2

10. If the number of new species evolving is directly proportional to the number of existing species and the probability of extinction of any species is inversely proportional to the number of existing species, the number of species present at a time during evolution will follow a curve given by: **(DEC 2012)**

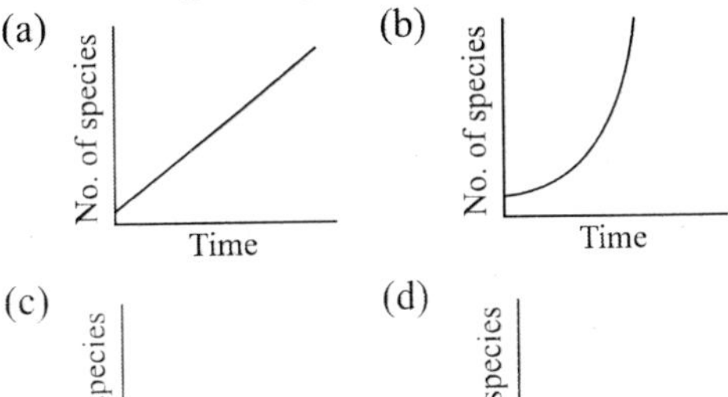

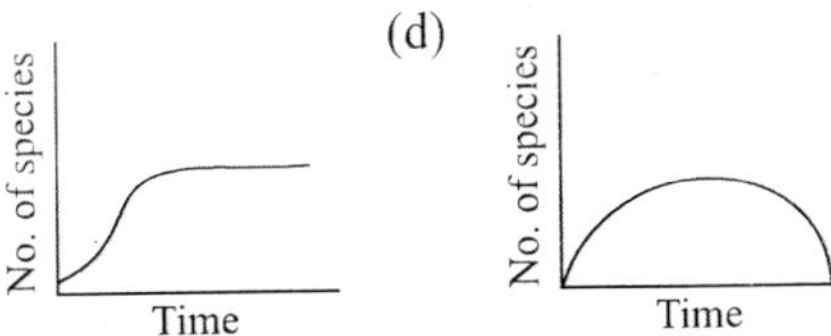

Answer Key

1.(d)	2.(c)	3.(c)	4.(d)	5.(b)	6.(c)	7.(a)
8.(b)	9.(c)	10.(b)				

12. Sexual selection and Co-evolution

1. The peacock's tail is an example of **(DEC 2011)**
   (a) natural selection
   (b) diversifying selection
   (c) sexual selection
   (d) group selection
2. In order to demonstrate that the long tails of males attracted females in a bird species, experimenters captured and cut the tails of 'n' number of males and monitored the number of females mated by each male.
   They had two types of controls in the experiment.
   (i) 'n' males that were not captured
   (ii) 'n' males that were captured, had their tails cut and then stitched back to attain the original size.
   The males with cut tails mated with a significantly smaller number of females than both the controls. Which of the following alternative explanations is NOT ruled out by the, experiment? **(JUNE 2012)**
   (a) The stress of cutting tails affected the performance of males.
   (b) The time wasted in the capture reduced mating opportunities of males.
   (c) Females avoided any deviation from normal.
   (d) Females chose males randomly.
3. Sexual selection results in variation in the reproductive success of males, often due to female choice with particular phenotypes. This type of sexual selection is because **(JUNE 2012)**
   (a) Males cannot compete with other males
   (b) cost of breeding is higher for females as compared to males
   (c) inappropriate mating results in a similar reduction in fitness of females and males
   (d) males are a limiting resource for females
4. The long feather train of a peacock is quoted as an example supporting **(JUNE 2014)**
   (a) Hamilton's rule.
   (b) Zahavi's handicap principle
   (c) The Red Queen hypothesis.
   (d) Haldane's rule.
5. The "Red Queen Hypothesis" is related to **(DEC 2015)**
   (a) the mating order in the harem of a Polygamous male.
   (b) the elimination by deleterious mutations by sexual reproduction.
   (c) mate selection process by a female in a lek.
   (d) the evolutionary arms race between the host and the parasite
6. Which of the following evolutionary processes played an important role in the evolution of immune system? **(DEC 2012)**
   (a) Reproductive isolation
   (b) Adaptive radiation
   (c) Neutral evolution
   (d) Co-evolution
7. Beak shape in birds has evolved in response to their diet.
   The table listing bird species and food type is given below:

	BIRD SPECIES		FOOD TYPE
i	Barn swallow	A	Fruits
ii	Great hornbill	B	Insects
iii	House sparrow	C	Nectar
iv	Purple sunbird	D	Seeds

   Match the bird species shown above to their main food resource. **(DEC 2018)**
   (a) i-D; ii-A; iii-C; iv-B
   (b) i-B; ii-D; iii-A; iv-C
   (c) i-C; ii-B; iii-D; iv-A
   (d) i-B; -ii-A; iii-D; iv-C

Answer Key

1.(c)	2.(b)	3.(b)	4.(b)	5.(d)	6.(d)	7.(d)

13. Brain, Behavior and Evolution

1. The Western honey bee (*Apis mellifera*) collects nectar and pollen from flowers. The following are few hypotheses proposed to explain this behaviour in *A. mellifera*:
   A. In the past, those individuals that fed on nectar and pollen left more descendants than those who preferred only nectar or only pollen
   B. The sensory stimulus from taste receptors in the honey bees lead to a positive reinforcement to look for more of the same food
   C. The honey bee's nervous system is predisposed to like the sweet taste
   D. The ancestor of honey bee was dependant on some sugar and protein rich diet and the honey bees have inherited the same taste perception

   Which of the following combination of ultimate hypotheses best explains the bee's feeding behaviour? **(JUNE 2017)**
   (a) A and B (b) B and C
   (c) A and D (d) B and D

2. After gull nestlings hatch, the parents remove the egg-shells from the nest. This behaviour is to **(DEC 2013)**
   (a) clean the area.
   (b) reduce infection
   (c) make more space in the nest.
   (d) minimise nest detection by predators.

3. Number of trials required for rates to learn a task when they were exposed to various conditions were as follows: **(DEC 2011)**

Experimental conditions	Observation
A. Light : Dark cycle- 12 h: 12h	N- trials
B. Bright light- 24 h	Significantly more trails than 'N'
C. Bright light-24 h + continuous physical disturbances	Significantly more trails than 'N'
D. Dark light-24 h + continuous physical disturbances	Significantly more trails than 'N'

   Which of the following inferences is most appropriate?
   (a) Continuous light enhanced learning
   (b) Continuous darkness inhibited learning
   (c) Physical activity inhibited learning
   (d) Learning was reduced by sleep loss

4. Assume a male sparrow (species X) is hatched and reared in isolation and allowed a critical imprinting period to hear the song of male of another sparrow (species Y). Now after the isolation, what kind of behaviour wills species X show? **(DEC 2011)**
   (a) It will sing the song of species Y that it had heard in the critical period
   (b) It will sing the song of its own species X
   (c) It will not sing at all
   (d) It will sing a song not sung by either X or Y

5. A 20-week old infant was exposed to the following stimuli and the responses were measured. **(DEC 2017)**

Stimuli	Mean No. of responses elicited
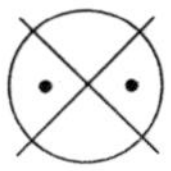	100
	15
	25
	10

Based on the response patterns shown above to the given stimuli, select the correct theory that best describes the observed responses.

(a) Heterogenous summation

(b) Gestalt principle

(c) Supernormal stimuli

(d) Sign stimulus

6. A particular behavioural variant affects fitness of an organism. The relationship between the frequency of the variant in the population and fitness are plotted below. In which of these cases is the behavioural variant most likely to reach a frequency of 1? **(JUNE 2016)**

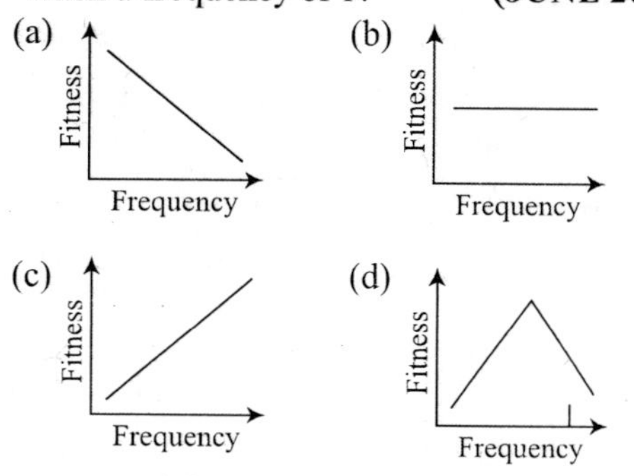

(a) only b (b) only c

(c) b and d (d) a and d

7. Shown in the graph below are the fitness costs and benefits of four alternatives behavioural phenotypes (P,Q,R,S). Given sufficient evolutionary time, which phenotype(s) is likely to evolve as an adaptation? **(JUNE 2011)**

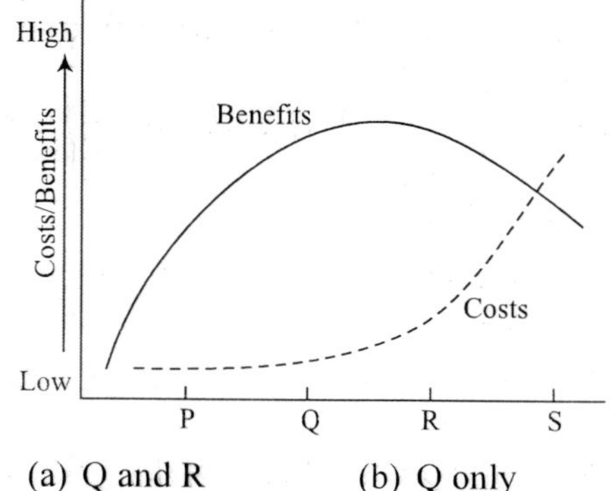

(a) Q and R (b) Q only

(c) P, Q and R (d) S only

8. When the electrical response of the two receptors A1 and A2 in a noctuid moth that was exposed to a variety of sounds was measured, it produced the following patterns: **(DEC 2017)**

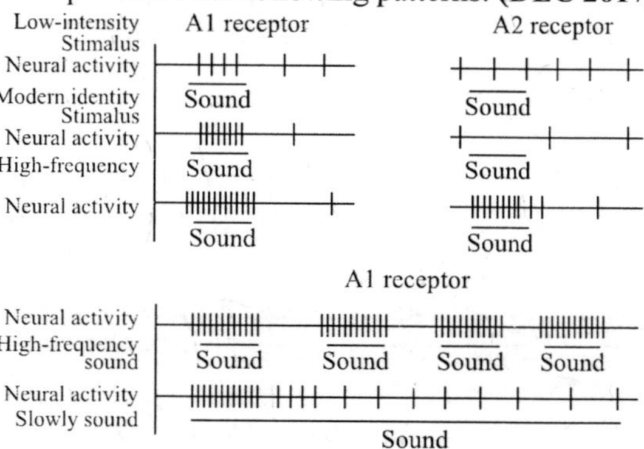

Given this, which one of the following statements is **INCORRECT**: **(DEC 2017)**

(a) The A1 receptor is sensitive to sound of low to high intensity.

(b) A2 receptor begins to produce action potentials only when a sound is loud.

(c) Both the receptors have similar response to high intensity sound.

(d) The Al receptor fires much more frequently to steady, uninterrupted sounds than to high frequency pulses of sound.

9. A study tested the importance of learning mechanisms in the development of antipredator escape responses in tadpoles of a frog species. Tadpoles hatched from eggs in the lab were kept individually either with predator chemical cues (PRIOR EXPOSURE) or without predator chemical cues (NAIVE) for 1 week. These individuals were tested for their escape response when exposed to a live predator. They were tested either alone or together with 3 older experienced tadpoles. The graph below shows the escape response of the test individuals in the four different treatments.

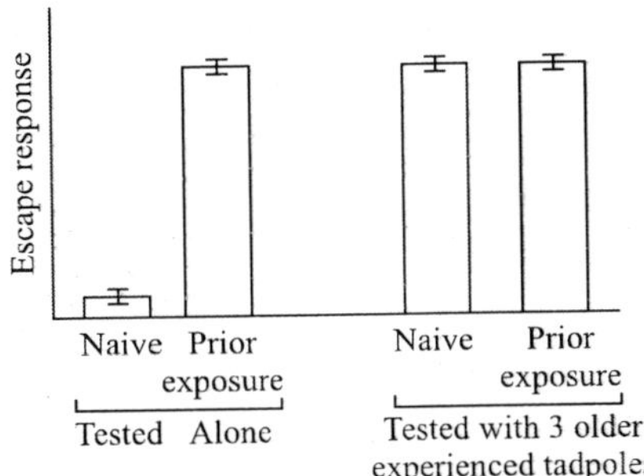

Some of the inferences drawn are given below:

A. Prior exposure to predator cues is necessary for the development of escape response.

B. Prior exposure to predator cues positively influences the development of escape response.

C. The presence of older experienced individuals is necessary for the development of escape response.

D. The presence of older experienced individuals positively influences the development of escape response.

E. An individual with prior exposure and with older and experienced individuals showed the strongest escape response.

Which one of the following combination of statements represents the correct inference from the experiment? **(DEC 2018)**

(a) A and C (b) B and D

(b) A and D (d) A, C and E

10. Antelopes are proposed to form groups to reduce the risk of predation. A researcher measured the predation of individuals in groups of different sizes. She found that per capita mortality risk decreased with increasing group size for males (solid line) but remained unchanged for female (dashed line). Furthermore, males in all groups experienced greater per capita mortality risk than females. Identify the graph below that best depicts the above findings: **(JUNE 2019)**

(a)

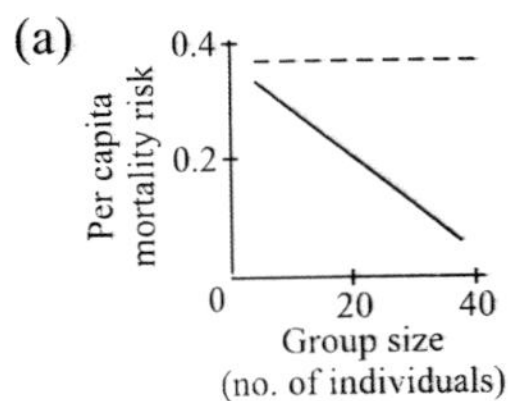

(b)

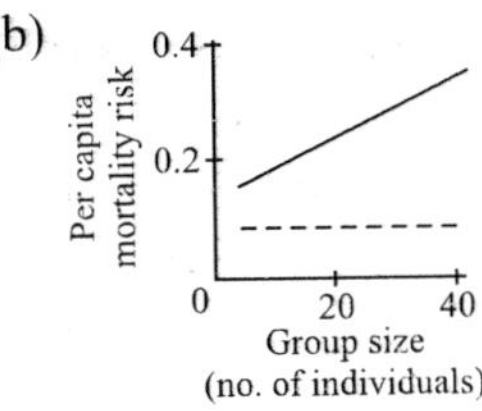

(c)

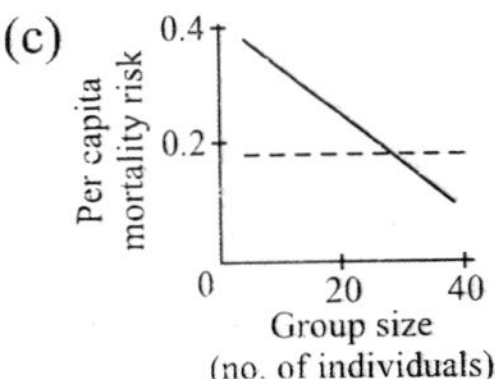

(d)

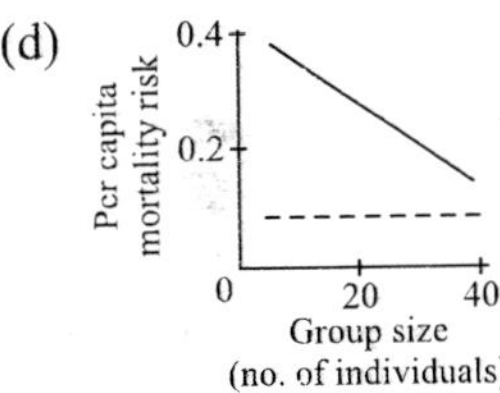

11. Birds in a population show two foraging phenotypes: A and B. Birds of phenotype A search, attack and capture prey while birds of phenotype B steal prey from birds of phenotype A. A and B are maintained in the population through negative frequency-dependent selection. The graph below shows the fitness of A (broken line) and B (solid line) at different relative frequencies of A (frequency of B = 1- frequency of A). **(JUNE 2019)**

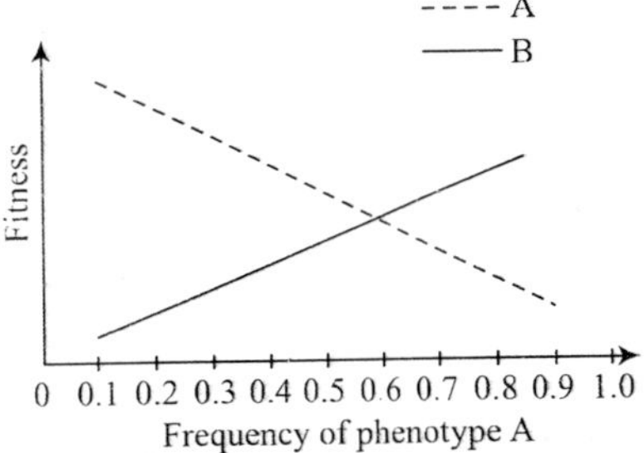

Which of the following statements does the graph support?

(a) A outcompetes B; at equilibrium. A goes to fixation.

(b) B outcompetes A: at equilibrium. B goes to fixation.

(c) A and B are both maintained in the population; the equilibrium frequencies are A= 0.6, B= 0.4.

(d) A and B are both maintained in the population; the equilibrium frequencies are A= 0.9, B= 0.1.

12. Given below are examples of animals and their behaviours.
    A. Elephant: Innate and learnt behaviour
    B. Ant: Innate behaviour only
    C. Whale: Learnt behaviours only
    D. Bee: Innate and learnt behaviours

    Which one of the following options represents a correct combination of true and false?
    (DEC 2019 ASSAM)
    (a) A: True B: False C: False D : True
    (b) A: True B: True C: True D: False
    (c) A: False B: True C: True D : False
    (d) A: True B: True C: True D : True

Answer Key

1.(c)	2.(d)	3.(d)	4.(d)	5.(a)	6.(b)	7.(b)
8.(d)	9.(b)	10.(d)	11.(c)	12.(a)		

14. Group selection, Kin selection, Altruism

1. In an altruistic act, if a donor sacrifices 'C' offspring which helps the recipient to gain 'B' offspring and the donor is related to the recipient by a coefficient γ, under which condition would kin selection favour this altruistic trait? **(JUNE 2011)**
   (a) B>C. (b) $B > \gamma C$
   (c) $\gamma B - C = O$ (d) $\gamma B - C > O$
2. According to Hamilton's rule, altruistic behaviour can evolve when rb>c, where b is the extra benefit gained by the recipient as a result of the altruistic act, c is the cost to the actor arising from performing the altruistic act and r is the relatedness between the:
   (JUNE 2019)
   (a) individual performing the altruistic act and the offspring of the recipient
   (b) individual performing the altruistic act and the recipient
   (c) recipient and the offspring of the individual performing the altruistic act
   (d) individual performing the altruistic act and the member of its population.
3. The idea that an altruistic gene will be favored if r>C/B, where r is the coefficient of relatedness, B is the benefit to the recipient of the altriusm, and c is the cost incurred to the donor, is known as **(DEC 2012)**
   (a) red queen hypothesis
   (b) handicap principle
   (c) Hamilton rule
   (d) Competitive exclusion principle
4. Individual A can derive fitness benefit of 160 units by helping Individual B, but incurs a fitness cost of 50 units in doing so following Hamilton's rule, A should help B ONLY if B is his **(DEC 2015)**
   (a) brother or Sister. (b) first Cousin only.
   (c) cousin or uncle. (d) nephew or niece.
5. Brothers A and B have the same father but different mothers. B wants A to help him, which involves both benefits (b) and costs (c) for A. If A incurs a cost of 30 'Darwinian fitness units' in that act, under what condition, should he help B, following Hamilton's rule?
   (JUNE 2015)
   (a) only if b>30 (b) only if b>60
   (c) only if b>120 (d) only if b>240
6. The degree of genetic relatedness between the offspring and their parents is **(DEC 2014)**
   (a) higher than that between sister and brother.
   (b) lower than that between sister and brother.
   (c) the same as that between sister and brother.
   (d) dependent on the number of siblings.
7. Assume that in terms of 'genetic fitness' the 'benefit' of performing an altruistic act to a relative is 500 units and the 'cost' involved is 150 units. Following Hamilton's Rule the act should be performed if the relative is a
   (JUNE 2014)

(a) only brother.
(b) nephew or niece.
(c) brother or step-sister.
(d) only step-sister.

8. For an individual A to help his relative B or C, the fitness benefits and costs are 50 and 20 units, respectively. However, following Hamilton's Rule, A should help B only but not C. How are B and C genetically related to A? **(DEC 2013 GU)**

(a) B-Son, C-Nephew
(b) B- Son, C-Brother
(c) B- Sister, C- Nephew
(d) B- Father, C-Mother

9. Assume that individual A wants to do an altruistic act to individual B and that benefit and cost of doing this act are, in 'fitness' units, 40 and 12, respectively. According Hamilton's Rule, A should perform the altruistic act only if B is his **(DEC 2014)**

(a) nephew.
(b) grandson or granddaughter.
(c) niece.
(d) daughter or son

10. Inclusive fitness of an animal can be measured as a sum of direct fitness and indirect fitness. Imagine you have 10 off springs. Through diligent parental care, 5 survive to reproduce. You give your life in a heroic deed to save a total of 5 of your nieces and nephews. What is your inclusive fitness? **(JUNE 2018)**

(a) 15 (b) 12.5
(c) 7.5 (d) 3.75

11. The genetic relatedness (r) of an individual to his nephew is 0.25. The alleles that cause uncles to care for nephews will spread, according to Hamilton's Rule, only if the fitness benefit is **(DEC 2011)**

(a) equal to cost of care
(b) more than the cost of care by 25%
(c) double the cost of care
(d) four times the cost of care

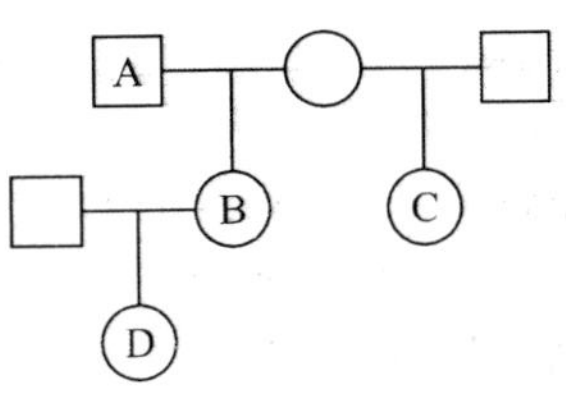

12. The coefficient of relatedness between individuals A and B, A and D, and between D and C is **(JUNE 2016)**

(a) 0.5, 0.25, 0.125 respectively.
(b) 0.5, 0.5, 0.25 respectively.
(c) 0.5, 0.25, 0.75 respectively.
(d) 0.125, 0.5, 0.5 respectively.

13. Worker bees, instead of themselves reproducing, help the queen reproduce. This behaviour is explained as an example of **(DEC 2014)**

(a) kin selection
(b) group selection
(c) sexual selection
(d) natural selection

14. In some species of new world monkeys, only one female reproduces in a group. One or more younger females have suppressed reproduction and asssit the reproductive female. This is an example of **(DEC 2012)**

(a) Sexual selection
(b) Group selection
(c) Kin selection
(d) Reciprocal altruism

15. In eusocial insects, males develop from unfertilized eggs while females develop from fertilized eggs. The ultimate consequence of this difference is that **(JUNE 2015)**

(a) in any colony there are always more males than females.
(b) a female is genetically more closely related to her sister than to her own offspring.
(c) females are behaviorally more dominant than the males.
(d) in any colony there are always more females than males.

16. In hymenoptera insects, males are haploid and females are diploid. All fertilized eggs give rise to female and unfertilized eggs give rise to males. As a result, if a female mates with a single male, the females in progeny are related to 75 %. But if the mother had mated with males, the mean genetic relatedness of female progeny is correctly represented by **(JUNE 2013)**

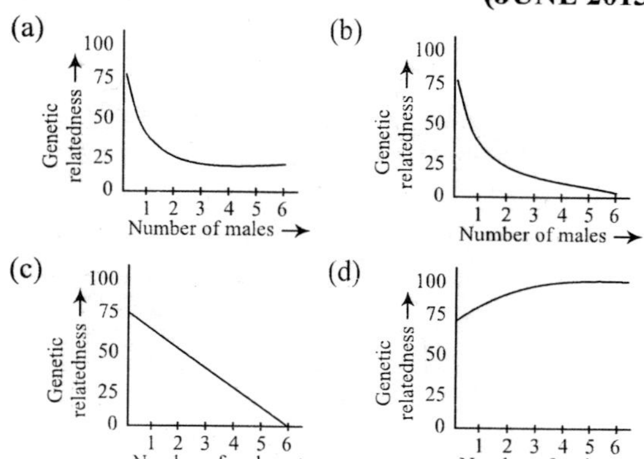

17. Altruism describes a behaviour performed by animals that may be disadvantageous to self while benefitting others. Which one of the following statements is **INCORRECT** about altruism? **(JUNE 2018)**

(a) It is the net gain of direct fitness when sociality is facultative.

(b) It is under positive selection via indirect fitness benefits that exceed direct fitness costs.

(c) It generates indirect benefit by enhancing survivorship of kin.

(d) It is favoured when rb - c > 0 where c is fitness cost to altruist, b is fitness benefit to recipient; and r is genetic relatedness.

18. Individual A performs to another individual a behavioral act which has a fitness consequence. Match the behavioral acts (a to e) with the correct fitness consequence (i) to (iv) **(DEC 2014)**

Behavioral act	**Fitness consequence to A**
Cooperation (a)	Gains direct fitness but after delay (i)
Adapative altruism (b)	Looses inclusive fitness (ii)
Spite (c)	Gain indirect fitness (iii)
Deciet and manipulation (d)	Gain direct fitness but immediately (iv)
Reciprocity (e)	

(a) a-(iv),; b-(iii); c(ii); d-(ii); e-(i)

(b) a-(i); b(ii); c(ii); d(iii); e(iv)

(c) a-(i); b(iii); c(ii); d(ii); e(iv)

(d) a-(i); b(ii); c(iii); d(i); e(iv)

19. An interaction where the actor and the recipient both suffer a cost is referred to as **(DEC 2019)**

(a) Altruism (b) Cooperation

(c) Mutualism (d) Spite

Answer Key

1.(d)	**2.**(b)	**3.**(c)	**4.**(a)	**5.**(c)	**6.**(a)	**7.**(c)
8.(a)	**9.**(d)	**10.**(d)	**11.**(d)	**12.**(a)	**13.**(a)	**14.**(c)
15.(b)	**16.**(a)	**17.**(a)	**18.**(a)	**19.**(d)		

15. Biological rhythm

1. The photoreceptor commonly involved In light entrainment of the biological clock in flies, moulds and plants is **(JUNE 2011)**

(a) phytochrome. (b) rhodopsin

(c) carotenoid. (d) Cryptochrome.

2. An animal was first maintained in a constant environmental condition for several days until a consistent biological rhythm (B) was established. The animal was then exposed to an experimental physical rhythm (E).which modulates the phase and period of B. However, upon withdrawal of E, the B gradually regained its pattern of pre-exposure condition.

From these observations which one or more of the following should be the most logical inference?

A) E is a Zeitgeber.

B) E is a masking agent.

C) E causes entrainment of B.

D) B is a conditioned to E.

The correct answer is **(JUNE 2011)**

(a) A and C (b) B and D

(c) B only (d) D only

3. In circadian rhythm studies, following may be possible generalizations for the effectiveness of light entrainment to the day/ night cycle:

A. Shorter exposures have a greater effect than longer exposures

B. Bright light exposures have a greater effect than dim light

C. Intermittent light exposures have a greater effect than consistent exposures

D. Dim light can affect entrainment relative to darkness

Which combination of the above statements is correct? **(JUNE 2017)**

(a) B and C only (b) B and D only

(c) A, C and D (d) A, B and D

4. Following are the plots representing biological rhythms at different time points depicted as: SR = Sunrise ; N= Noon ; SS = Sunset ; MN = Midnight

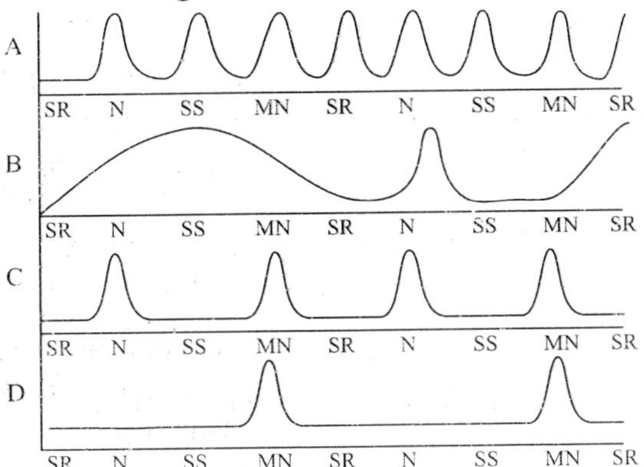

Which of the plot(s) represents the ultradian biological rhythm(s)? **(DEC 2016)**

(a) Plot B (b) Plots A and C

(c) Plots C and D (d) Plot D

5. Given below are some examples of biological rhythms in humans:

(a) REM and non-REM sleep cycle

(b) Menstrual cycle

(c) Sleep-wake cycle

Which one of the following is a correct representation of the type of biological rhythms given above?

(DEC 2019 ASSAM)

(a) (a) Ultradian cycle; (b) Infradian cycle; (c) Circadian cycle

(b) (a) Circadian cycle; (b) Ultradian cycle; (c) Infradian cycle

(c) (a) Infradian cycle; (b) Ultradian cycle; (c) Circadian cycle

(d) (a) Infradian cycle; (b) Circadian cycle; (c) Ultradian cycle

6. Four drugs (A,B,C,D) were used to disrupt a biological rhythm in experimental animals. The changes in the pattern of the biological rhythm as compared to untreated are shown below. The solid line represents the biological rhythm of the untreated and broken line represents that of the treated animals.

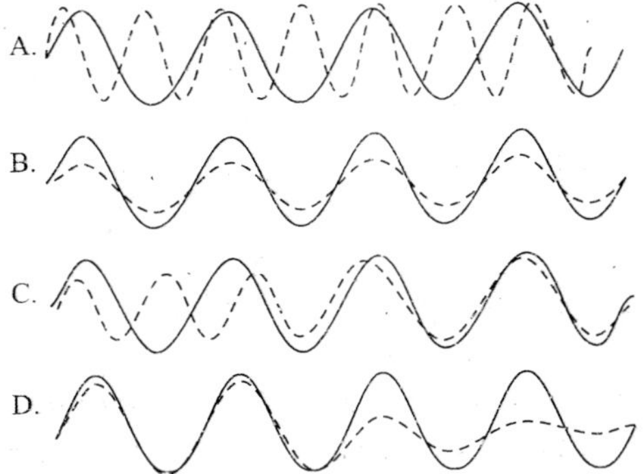

Which of the following interpretations from the above experiment is INCORRECT?

(DEC 2018)

(a) Drug A can be used to reduce the period length of the rhythm.

(b) Drug B can be used for sustained lowering of amplitude of the rhythm without changing its period.

(c) Drug C can be used for sustained lowering of amplitude and period of the rhythm.

(d) Drug D can be used to reduce the robustness and dampen out the rhythm.

Answer Key

1.(d)	2.(a)	3.(b)	4.(b)	5.(a)	6.(c)

16. Social dominance, Use of space and territoriality

1. A barn owl sits on its perch 10m above ground. It hears a mouse underneath on the ground at an angle α as shown in the figure below.

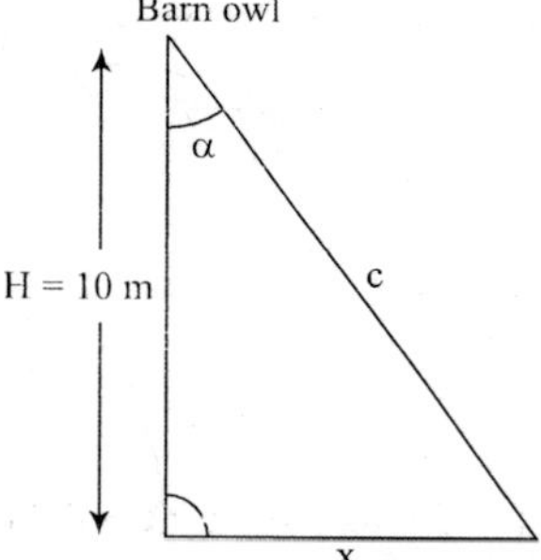

The error range with which it can locate the mouse on the ground is given by **(DEC 2017)**

(a) tan α = X/H (b) Cos α = X/H

(c) sin α = X/H (d) Cos α = X/C

2. An extraordinary sensory ability that elephants possess is. **(JUNE 2015)**

(a) emission and detection of ultra high frequency sounds.

(b) emission and detection of ultra low frequency sounds.

(c) detection of changes in earth's magnetic field.

(d) possession of ultraviolet vision.

3. With reference to the graph given below, identity the optimal territory size. **(JUNE 2016)**

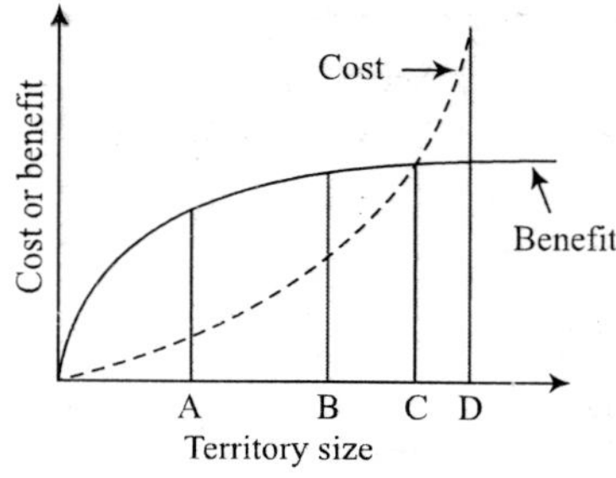

(a) A (b) B

(c) C (d) D

4. To understand the singing behaviour in songbirds, the following three characters were measured as shown in the graph:

A. Territoriality rate

B. Female fertility rate

C. Song rate

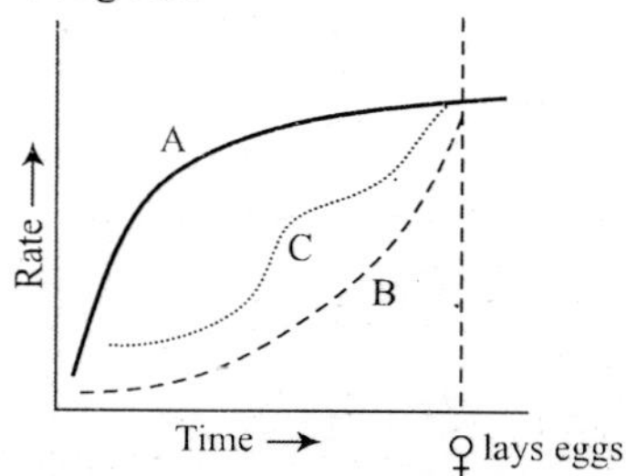

Which one of the following conclusions is most appropriate? **(JUNE 2017)**

(a) Male birds sing as a display of strength to rivals and to attract females

(b) Male birds sing to display parental care behaviour

(c) Male birds sing only to display that females are sexually receptive

(d) Male birds sing only to deter other male rivals from competing for territories

5. A few males and females of a species were introduced to a new island. Their population was monitored over several .generations and followed a pattern shown in the figure.

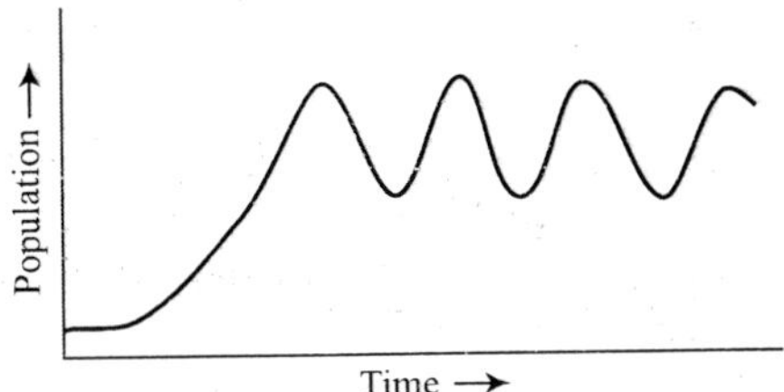

Which of the characteristics of the species does NOT explain the pattern? **(JUNE 2012)**

(a) Skewed sex ratio (more females than males)

(b) Large litter size

(c) Delayed sexual maturity

(d) Effects of intra-uterine development on fecundity

Answer Key

1.(a)	2.(b)	3.(a)	4.(a)	5.(a)

17. Parental investment and Reproductive success

1. In which of the following mating systems there is likely to be NO conflict of interest over reproductive success between the sexes? **(DEC 2016)**

(a) polyandry (b) monogamy

(c) promiscuity (d) polygamy

2. Which one of the following conditions is NOT likely to favour male monogamy? **(DEC 2015)**

(a) When the male has to guard his mate against mating by another male.

(b) When the male wants to spend more time for foraging.

(c) When the male has to assist the mate in brood and nestling care.

(d) When the female guards her mate against seeking other females to mate.

3. Which of the following is NOT a benefit for the female adopting polyandry? **(DEC 2014)**

(a) Greater probability of getting all her eggs fertilized.

(b) Ability to receive more resources from the males.

(c) Ability to produce more offspring than normal.

(d) Improved chances of genetic compatibility with her own DNA.

4. In bird species where both parents contribute equally to parental care, generally **(DEC 2012)**

(a) males are larger than females

(b) females are more colorful than males

(c) females are larger than males

(d) both sexes are morphologically similar

5. Evolutionarily, with which of the following could parental care in animals be associated? **(MODEL PAPER)**

(a) Smaller clutch size.

(b) Polygamy.

(c) Greater longevity.

(d) Semelparity.

6. Reproductive success of individuals in a population is likely to be skewed under all the following conditions, EXCEPT when **(DEC 2019 ASSAM)**

(a) Females are choosy.

(b) male-male competition is intense.

(c) pair-bonding occurs in the species.

(d) Females prefer males with larger territories

7. In a bird species A, the male alone builds the nest, incubates the eggs and feeds the nestling. In bird species B, it is the female that does all that. In bird species C, both sexes contribute equally to the above activities. In species A and B, the uninvolved partner may fly away and mate again. Which sex among A, B and C is most likely to develop colourful plumage during breeding season? **(MODELPAPER)**

(a) Male in species A and B, both sexes in species C.

(b) Female in species A, male in species B and C.

(c) Female in species A and B, neither in species C.

(d) Female in species A, male in species B, neither in species C.

8. The following situations might lead to the evolution of monogyny in birds:

A. Male has to assist the female in rearing the offspring.

B. Male guards the female against other males trying to mate with her.

C. One male may not produce enough sperm required to fertilize all the eggs produced by the female.

Which of the above is/are correct?

(JUNE 2014)

(a) Only A (b) Only B

(c) A and B (d) A and C

9. Competition for mates and variance in fitness is higher among females than among males in which of the following animal mating systems?

(DEC 2018)

(a) Monogamy

(b) Polygyny

(c) Polyandry

(d) Sequential monogamy

10. The ratio of variance in male mating success (V_m) to variance in female mating success (V_f) is strongly male biased ($V_m>V_f$) in species P, strongly female biased in species Q ($V_f>V_m$) and similar in species R ($V_m=V_f$). All else being equal, which one of the following matches between species and mating systems is most likely? **(DEC 2019)**

(a) P-monogamy; Q-polyandry, R-polygyny

(b) P- polyandry; Q-polygyny, R-monogamy

(c) P-polygyny; Q-polyandry, R- monogamy

(d) P-monogamy; Q-polygyny, R-polyandry

11. In a large, healthy, polygynous population of an ungulate species with distinct, short seasonal mating, the operational sex ratio (Male: Female) is likely to resemble which one of the following conditions:

(DEC 2019 ASSAM)

(a) Male > Female (b) Female ≥ Male

(c) Male = Female (d) Female > Male

12. If the relationship between life time reproductive success and body size for males and females of a species as shown in figure below:

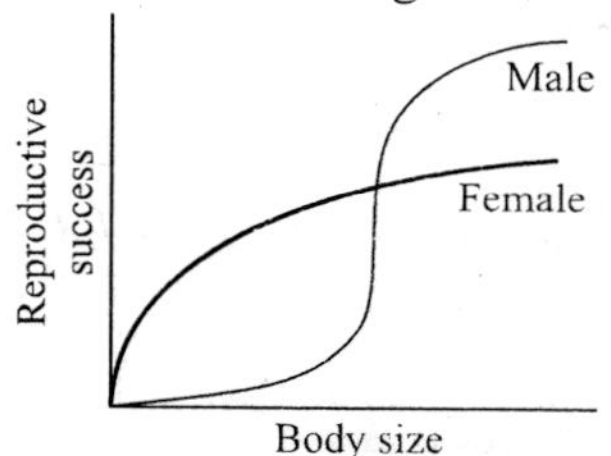

The species is most likely to evolve

(DEC 2012)

(a) Sexual dimorphism

(b) Asexual reproduction

(c) Polyandry

(d) Obligate monogamy

13. Greater male investment in the care of offspring is most likely to lead to **(DEC 2013)**

(a) a lek system.

(b) stronger female choice.

(c) Reverse sexual dimorphism

(d) run-away selection.

14. The top panel (graphs a-c) represents trends of number of sperms produced per mating season with respect to number of mates, while the bottom panel (graphs i-iv) represents trends of time invested in paternal care with respect to number of mates in birds.

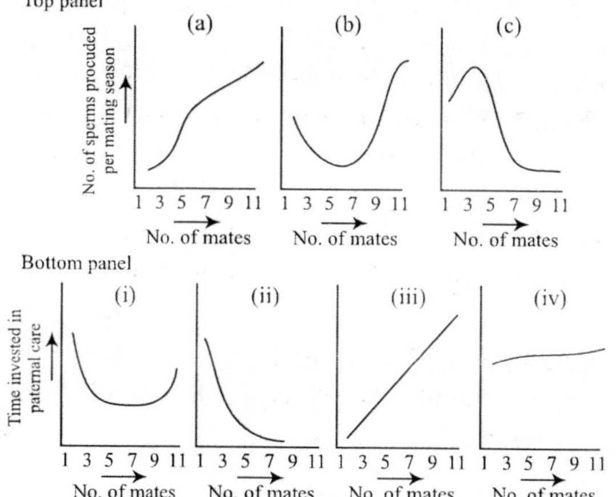

Select the correct trend from each panel. (DEC 2019)

(a) c, iv (b) a, ii
(c) b, iii (d) a, i

15. In rats, after the delivery of the offspring, mother shows the following behaviors. Which one of the following behaviors is NOT maternal? **(JUNE 2014)**
(a) Licking the pups
(b) Huddling above the pups to access the ventrum
(c) Lordosis of mother rat
(d) Bringing back to the nest pups that wander away from it

16. Which of the following behavioural changes are expected in a rat when its nucleus accumbens is experimentally ablated? **(DEC 2011)**
(a) Aggressive behaviour increases
(b) Exploratory behaviour decreases
(c) Nest-building activity increases
(d) Level of parental care drops

Answer Key

1.(b)	2.(b)	3.(c)	4.(d)	5.(a)	6.(c)	7.(d)
8.(c)	9.(c)	10.(c)	11.(a)	12.(a)	13.(c)	14.(b)
15.(c)	16.(a)					

18. Mimicry, Migration

1. The Population of the Non-poisonous butterflies have the same the color pattern as some highly poisonous butterflies. Assume that the population of non-poisonous butterflies is higher than the poisonous butterflies. Given this, what will be the impact of this mimicry on the fitness of the population of the poisonous butterflies in the presence of the predator? **(DEC 2016)**
(a) It will lower the fitness, that is, fitness of the mimic is negatively frequency – dependent
(b) It will increase the fitness, that is, fitness of the mimic is positively frequency dependent
(c) It will not affect the fitness, that is, fitness of the mimic is frequency independent
(d) It will increase fitness, that is, fitness of the mimic is negatively frequency dependent

Condition	Abundance	Predators learn	Resource overlap
A	Mimic>Model	Yes	Yes
B	Model>Mimic	Yes	No
C	Mimic=Model	No	Yes
D	Model > Mimic	No	No

2. Which among the above sets of conditions are best suited for mimicry to be successful? **(DEC 2019, ASSAM)**
(a) Condition A (b) Condition B
(c) Condition C (d) Condition D

3. Homing pigeons, when released at a place far away from their home, use earth's magnetic field or the sun as navigational cues and choose the right direction to fly. To test the hypotheses, two experiments were conducted. In Experiment I, one group of pigeons (Test) were equipped with Helmholtz coils (which disrupt magnetic field detection), while the second group (Control) were not. Both groups were released on a sunny day. Experiment II used the same Control and Test groups of pigeons, but they were released on a cloudy, overcast day. The expected results, if the hypotheses is true, would be **(JUNE 2015)**
(a) In Experiment I, both Control and Test groups fly in the right direction, but in Expt. II, only Control group does.
(b) In both experiments, Test groups fail to choose the right direction.
(c) In Experiments I and II; Test groups fly in the right direction.
(d) In Experiment I both groups fly in the right direction but in Expt. II both groups fail to choose the right direction

4. The mechanism by which ants find their way back to the nests after searching and finding food is: **(DEC 2019 ASSAM)**

(a) Migration

(b) Topographical mapping

(c) Piloting

(d) Path integration

Answer Key

1.(a)	**2.**(b)	**3.**(a)	**4.**(d)

SECTION-12

APPLIED BIOLOGY

1. Microbial fermentation

1. The word "fermentation" is used in biochemistry and Microbial technology to denote different phenomenon. If the former is called C and latter is called T. Which of the following statement is true? **(JUNE 2012)**
   (a) All C is T but all T is not C
   (b) All T is C but C is not T
   (c) T is always a product of genetic engineering while C is not
   (d) C is always an aerobic process, while T can be aerobic or anaerobic

2. Which one of the following statements CANNOT be included while defining the fermentation process? **(DEC 2019)**
   (a) Alcohol is formed from sugar residues
   (b) Requires an electron transport system.
   (c) Utilizes an organic compound as the final electron acceptor
   (d) Produces lactic acid in oxygen deprived muscle

3. Which of the following microbial fermentations are anaerobic? **(DEC 2013)**
   (a) Ethanol and acetone-butanol.
   (b) Citric acid and propionic acid.
   (c) Penicillin and vitamin B_{12}
   (d) Streptomycin and rifampicin.

4. During the production of alcohol by fermentation using budding yeast, oxygen supply is kept 'limited. Why? **(DEC 2014)**
   (a) Budding yeasts are obligate anaerobes and cannot tolerate oxygen.
   (b) Budding yeasts lose mitochondria in the absence of oxygen.
   (c) Budding yeasts are facultative anaerobes
   (d) Alcohol is oxidized further in the presence of oxygen.

5. Novel groups of polyketides can be synthesized in *Actinomycetes* by **(DEC 2013 GU)**
   (a) gene rearrangement.
   (b) gene amplification.
   (c) combinatorial biosynthesis.
   (d) metabolic engineering.

6. *Coliforms* are used as indicator organisms of sewage pollution because they **(DEC 2013 GU)**
   (a) are pathogens.
   (b) ferment lactose.
   (c) are abundant in human intestines.
   (d) grow within 24 hours.

7. Enzymes are nowadays used extensively in bio-processing industries.
   Enzymes 1 is used for treatment of hides to provide a finer texture, in leather processing and manufacture of glue.
   Enzyme 2 is used for clarification of fruit juices.
   Identify Enzymes 1 and 2 **(DEC 2011)**

	Enzyme 1	Enzyme 2
(a)	Amylase	Pectinase
(b)	Protease	Amylase
(c)	Protease	Pectinase
(d)	Pectinase	Amylase

8. Electrons are transferred from reduced c enzymes like NADH, NADPH to pyruvic acid its derivatives during fermentation. Those final electron acceptors are reduced to the end-product for example latic acid, propionic acid, etc. end products depend on the particular microorganism and the substrate

Organism		End product (s)	
A	*Streptococcus*	a.	Butanediol, formic acid
B	*Clostridium*	b.	Lactic acid
C	*Salmonella*	c.	Butyric acid and iso proanol
D	*Enterobacter*	d.	Succinic acid, acetic acid

The correct match is **(DEC 2013)**
   (a) A - a, B - b, C - c, D – d
   (b) A - d, B - c, C - a, D -c
   (c) A - b, B - c, C - d, D - a
   (d) A - c, B - d, C - a, D – b

9. Industrial products in which bacteria are employed for production are shown in the following table: **(JUNE 2011)**

I. List of products	II. Microorganims
A. 2,3-Butane diol	i) *Leuconostoc*
B. Dextran	ii) *Brevibacterium*
C. Glutamic acid	iii) *Bacillus polymyxa*
D. Cobalamine	iv) *Propionibacterium*

The correct combinations are

(a) A - iii; B – i; C – ii; D - iv

(b) A - i; B - ii; C - iii; D - iv

(c) A - iii; B - ii; C - iv; D – I

(d) A - ii ; B - iii; C - iv; D - i

10. The following table lists some of the enzymes of fungi and bacteria having wide variety of industrial applications, including alcoholic beverages, food, detergents and pharmaceuticals, along with their microbial original **(MODELPAPER)**

Enzyme	Microorganism
A. Amylase	E. *Azotobacter vinelandii*
B. Asparginase	F. *Serratia marcescens*
C. Lipase	G. *Aspergillus aureus*
D. Pectinase	H. *Aspergillus oryzae*

The correct combinations are

(a) A and H (b) B and G

(c) C and E (d) D and F

11. Which one of the following microorganisms is used for the fermentation of coffee beans? **(DEC 2019 ASSAM)**

(a) *Candida krusei*

(b) *Erwinia dissolvens*

(c) *Lactobacillus delbrueckii*

(d) *Lactobacillus plantarum*

12. Column A lists a set of nutrients and Column B lists some sources for these nutrients that are used in industrial fermentation.

	Column A		Column B
A	Nitrogen	(i)	Soya meal
B	Carbon	(ii)	Yeast extract
C	Micronutrient	(iii)	Corn liquor
		(iv)	Molasses
		(V)	Tryptone

Choose the option that matches the nutrient with its most common source.

(DEC 2019 ASSAM)

(a) A: (i), (ii); B :(iii), (iv); C: (v)

(b) A: (i), (v); B: (iii), (iv); C: (ii)

(c) A: (ii), (v); B: (iii), (iv); C: (i)

(d) A: (v); B: (i), (iii); C: (ii), (iv)

13. Which of the following curves correctly represents the process of ethanol production by yeast? **(JUNE 2013)**

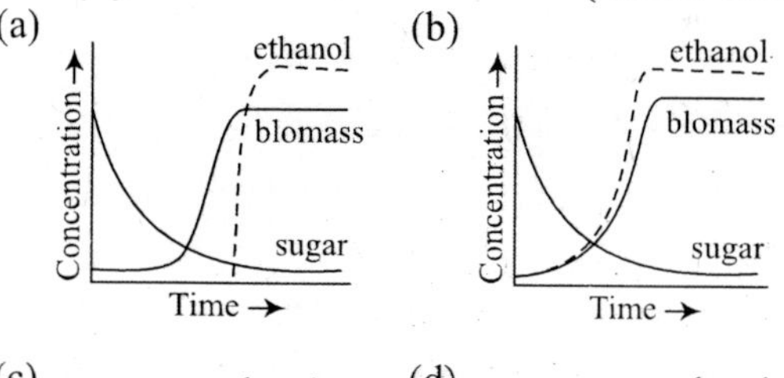

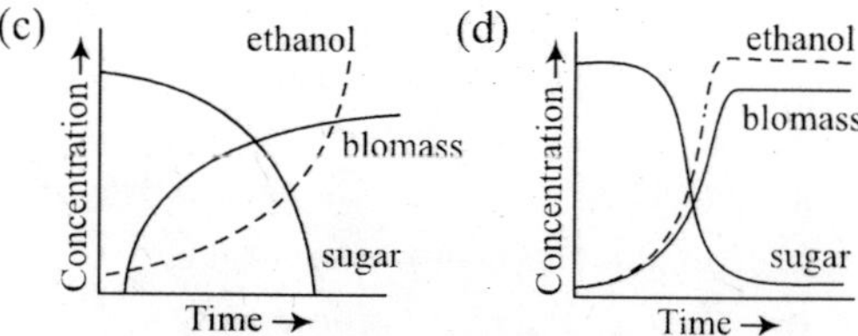

14. Which of the following curve correctly represents the antibiotic production by *Streptomyces* species? **(JUNE 2012)**

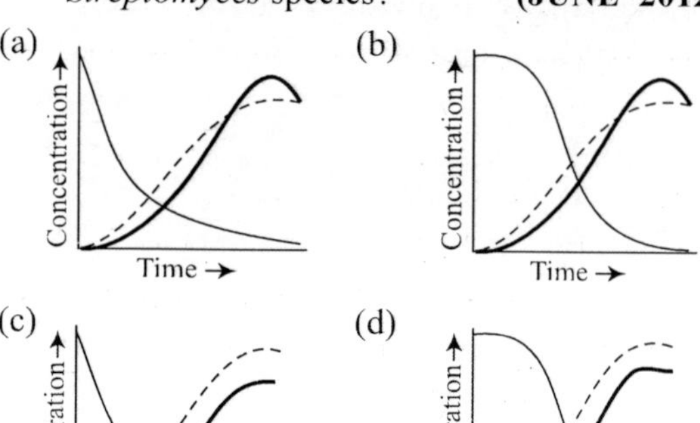

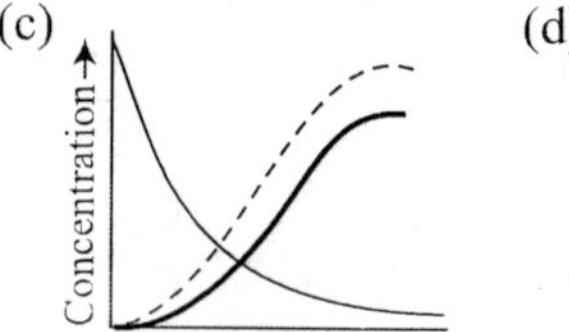

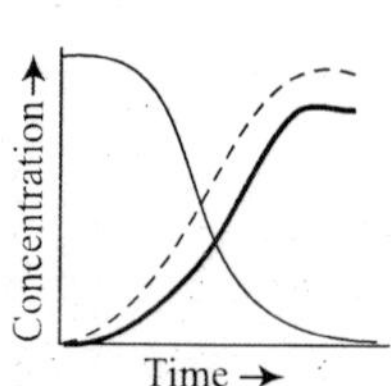

15. Microbes produce either primary or secondary metabolites during fermentation. A metabolite production curve is shown below:

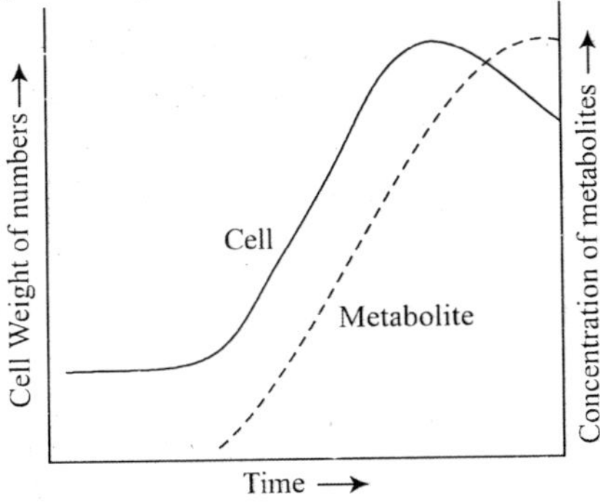

The following statements refer to the above figure:

A primary metabolite has a production curve that lags behind the line showing cell growth.

A primary metabolite is produced after the Trophophase is completed.

A secondary metabolite is produced mainly during Idiophase.

The curve shows the production of Penicillin from mold.

Which of the above statements are correct? **(DEC 2012)**

(a) A and B (b) C and D

(c) A and C (d) B and D

Answer Key

1.(a)	2.(b)	3.(a)	4.(c)	5.(c)	6.(c)	7.(c)
8.(c)	9.(a)	10.(a)	11.(b)	12.(b)	13.(d)	14.(b)
15.(b)						

2. Tissue and cell culture methods for plants

1. Which one of the following will be observed when auxin to cytokinin ratio is increased in the culture medium during organogenesis from tobacco pith callus? **(JUNE 2016)**

(a) Adventitious roots will form.

(b) Adventitious shoot will form.

(c) There will be no root formation.

(d) There will be no shoot formation.

2. Following are few statements for regeneration of plants from explants/tissues.

(A) Cytokinin is required for shoot development.

(B) Auxin is required for shoot development.

(C) Auxin to cytokinin ratio is very important.

(D) Jasmonic acid is required for both root and shoot development.

Which of the following combinations of above statements is true? **(JUNE 2013)**

(a) (A) and (C) (b) (B) and (D)

(c) (A) and (D) (d) (B) and (C)

3. Given below is a table comprising various terms associated with plant tissue culture in Column A and Column B.

	Column A		Column B
A	Auxin	i	Embryogenesis
B	Protoplast culture	ii	6-Furfuryl amino purine
C	Cytokinin	iii	Pectinase and Cellulase
D	Microspore culture	iv	Indole-3-acetic acid

Which one of the following options represents the most appropriate match between all the terms of Column A and Column B? **(DEC 2018)**

(a) A-ii; B-i; C-iv; D-iii

(b) A-iv; B-iii; C-ii; D-i

(c) A-ii; B-iv; C-i; D-iii

(d) A-iv; B-i; C-ii; D-iii

4. Somatic embryogenesis is an important exercise in micropropagation and genetic engineering of plants. The following steps are considered as critical for achieving somatic embryogenesis:

A. Reducing the concentration of sucrose in the medium by half.

B. Addition of the hormone, 2,4-dichlorophenoxyacetic acid to induce somatic embryos.

C. Reduce agar concentration to 0.6% (w/v)

D. Use maltose in place of sucrose as a carbon source

Which one of the following combinations is correct? **(JUNE 2015)**

(a) A and C (b) B and D

(c) A and B (d) C and D

5. Encasing of which of the following plant eel in a gelatinous matrix is referred as artificial seed? **(DEC 2013)**

(a) Microcalli (b) Somatic embryos

(c) Root tips (d) Shoot tips

6. The following statements are related to plant tissue culture

A. Friable callus provides the inoculum to form cell-suspension cultures.

B. The process known as 'habituation' refers to the property of callus loosing the requirement of auxin and/or cytokinin during long term culture.

C. Cellulase and pectinase enzymes are usually used for generating protoplast cultures.

D. During somatic embryo development, torpedo stage embryo is formed before heart stage embryo.

Which one of the following combinations of above statements is correct? **(DEC 2015)**

(a) A, B and C (b) A, B and E

(c) A, C and D (d) B, C and D

7. Following are certain statements regarding somatic hybridization, a technique used for plant improvement.

A. Protoplasts of only sexually compatible plant species can be fused.

B. Hybrids are produced with variable and asymmetric amounts of genetic material of parental species

C. Protoplast fusion permits transfer of gene block or chromosomes.

D. Genes to be transferred need to be identified and isolated.

Which one of the following combinations of the above statements is correct? **(DEC 2014)**

(a) A and C (b) B and C

(c) A and D (d) B and D

8. Protoplast fusion is used in plant tissue culture for various applications. In protoplast fusion:

(A) naked plant cells are used.

(B) transfer of organelles is not possible.

(C) partial genome transfer is involved.

(D) cells from two different plants can be mixed together and forced to fuse.

Which one of the following combinations of the above statements is correct? **(JUNE 2013)**

(a) (A), (B) and (C) (b) (A), (C) and (D)

(c) (A), (B) and (D) (d) (B), (C) and (D)

9. *Arabidopsis thaliana* seeds were planted on Murashige Skoog (MS) plates with or without a hormone added to the medium. Seeds were found to germinate late in the hormone containing MS plates as compared to MS plates without hormone. Identity the hormone.

(JUNE 2014)

(a) Jasmonic acid (b) Cytokinin

(c) Auxin (d) Abscisic acid

Answer Key

1.(a)	**2.**(a)	**3.**(b)	**4.**(b)	**5.**(2/4)	**6.**(a)	**7.**(b)
8.(b)	**9.**(d)					

3. Transgenic plants

1. Which of the following GM crops is the most widely cultivated globally?

(MODEL PAPER)

(a) Herbicide resistant soybean.

(b) Insect resistant cotton.

(c) Insect resistant brinjal.

(d) delayed ripening tomato

2. Which of the following features is NOT shown by glyphosate, a broad spectrum herbicide?

(DEC 2011)

(a) Little residual soil activity

(b) Ready translocation in phloem

(c) Inhibition of a chloroplast enzyme catalyzing the synthesis of aromatic amino acids

(d) Inhibition of early steps in the biosynthesis of branched chain amino acids

3. The following genes have been genetically engineered to develop herbicide resistance in plants:

A. Resistance to glyphosate using the 5-enolpyruvyl shikimate-3-phosphate synthase gene

B. Bialaphos resistance using the bar gene

C. Sulfonyl urea resistance using the acetolactate synthase gene

D. Atrazine resistance using the glutathione S-transferase gene

In which of the above two cases the mechanism is based on abolition of herbicide binding to the enzyme? **(JUNE 2015)**

(a) A and D (b) B and C

(c) C and D (d) A and C

4. Which of the following transgenic crops have been approved for commercial cultivation in India? **(DEC 2013)**

(a) Cotton

(b) Brinjal

(c) Cotton and Brinjal

(d) Cotton, *Brassica*, Brinjal

5. Which genes have been introduced in Bollgard II cotton to get resistance against cotton bollworm, tobacco bollworm and pink bollworm? **(DEC 2016)**

(a) cry 1Ab + cry 1Ac

(b) cry 1Ac + cry 2 Ab

(c) cry 1Ab + cry 2 Ab

(d) cry9 C cry 1Ab

6. Neomycin phospho-transferase gene, frequently used as a selection marker during plant trans formation, inactivates which one of the following antibiotics? **(DEC 2015)**

(a) Hygromycin (b) Ampicillin

(c) Streptomycin (d) Kanamycin

7. Which of the following genes was engineered in the "Flavr Savr" transgenic tomato variety? **(JUNE 2013)**

(a) 1-Amino cyclopropane-1-carboxylic acid synthase.

(b) 1-Amino cyclop'ropane-1-carboxylic acid oxidase.

(c) Expansin.

(d) Polygalacturonase.

8. Which of the following methods of plant transformation can be used to introduce a gene into chloroplast genome? **(JUNE 2011)**

(a) Agrobacterium-mediated transformation

(b) Particle delivery system

(c) Permeabilization

(d) Electroporation

9. Which one of the following is the most appropriate definition of 'Gene Pyramiding' in plants? **(JUNE 2019)**

(a) Introducing different genes for resistance to a specific pest in different genotypes.

(b) Introducing a single gene for resistance to a particular pest in different genotypes.

(c) Introducing different genes for resistance to a single pest in a single genotype.

(d) Introducing a single gene for resistance to multiple pests in different genotypes.

10. Given below are four statements regarding genetic transformation of plants in the laboratory:

A. Plants incapable of sexual reproduction cannot be transformed by *Agrobacterium tumifaciens*

B. Integration of transgene in organellar (chloroplast) genome occurs primarily by homologous recombination

C. An enhancer trap construct used in Agrobacterium-mediated transformation would contain a functional coding sequence of a reporter gene and a minimal promoter

D. A T_0 transgenic plant containing two unlinked copies of a selection marker gene (*hpt*) and one copy of the passenger gene (*gfp*) would segregate in a 1 : 1 ratio for hygromycin resistance: sensitivity in the backcrossed progeny grown on selection media

Which one of the combinations of above statements are correct? **(JUNE 2017)**

(a) A and D (b) B and C

(c) A and C (d) B and D

11. A T_0 transgenic plant containing a transgene for herbicide resistance shows two bands on Southern blot analysis using a probe that is internal to the restriction sites used for genomic DNA digestion. However, it segregates in a 3: 1 ratio for herbicide resistance: sensitivity in the T_1 progeny obtained by self-pollination. Which one of the following statements is correct? **(JUNE 2017)**

(a) The T_0 plant is a single-copy event

(b) The T_0 plant is a double-copy event and the two transgene copies are tightly linked

(c) The T_0 plant is a double-copy event and the two transgene copies are integrated in two different chromosomes

(d) The T_0 plant contains two unlinked copies of the transgene, both of which are truncated versions of the herbicide resistance gene.

12. For the generation of transgenic plant in crop improvement, one important regulatory gene 'X' was overexpressed in a crop plant. Out of 30 transgenic rice plants generated, 22 showed high levels of gene 'X' expression. However, rest 8 lines displayed low level of expressionff. One explanation of such observation may be: **(JUNE 2012)**

(a) Suppression effect of the transgene

(b) Knock down effect of gene X

(c) Gene silencing effect

(d) Co-suppression effect of transgene

13. *Agrobacterium* mediated transformation was used to generate transgenic plants using a construct with a selection marker gene "X" and a passenger gene "Y". Expression levels of "Y" protein in eight independent transgenic plants are given below:

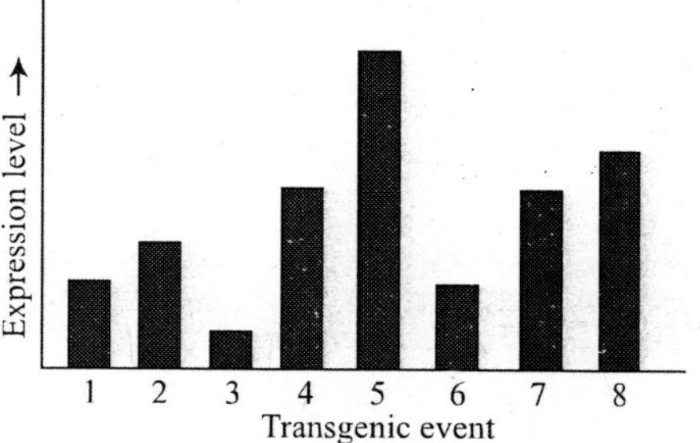

The following could represent probable reasons for the observed variability in transgene expression levels

A. Position effects on passenger gene

B. Transgene silencing of the marker gene

C. Variation in copy number of passenger gene

D. mRNA instability of marker gene

Which one of the following combinations of above statements is correct? **(JUNE 2017)**

(a) A and C (b) C and D

(c) B and D (d) A and B

14. In a gene stacking experiment, a transgenic plant generated using bar as a selection marker was crossed with another transgenic line containing hpt as the selection marker. Analysis of the F_1 progeny of this cross showed that 50% of the progeny was resistant to both selection agents (Basta and hygromycin) while the remaining 50% of the progeny was resistant to only hygromycin. Which one of the following statements is a possible explanation of the above results? **(DEC 2019 ASSAM)**

(a) Both the transgenic lines used as parents for the cross are single-copy events that are homozygous for the transgenes.

(b) The Basta-resistant transgenic plant is a single-copy, homozygous event and the hygromycin-resistant plant is a double-copy event that is homozygous for two unlinked copies of the transgene.

(c) The bar containing transgenic plant is a hemizygous single-copy event and the hpt containing transgenic plant is a homozygous, single-copy event

(d) Both the transgenic plants used as parents for the cross are hemizygous, single-copy

15. Engineering of metabolic pathways in plants can be achieved by introduction and overexpression of appropriate candidate gene(s) using transgenic technology. The figure given below represents a biochemical pathway in plants where a precursor molecule 'A' is converted in to products 'T' and 'X' through a series of enzymatic reactions. Enzymes 1-5 are involved in this pathway. Scientists attempted to increase the levels of 'X' by introducing additional copy of the gene for enzyme '5' under transcriptional control of a strong constitutive promoter. However, the developed transgenic plants did not display a proportionate increase in the level of 'X'.

$$A \xrightarrow{'1'} C \xrightarrow{'2'} D \xrightarrow{'3'} E \quad D \xrightarrow{'4'} T \quad E \xrightarrow{'5'} X$$

The following statements were proposed for explaining the above results:

A- Enzyme '4' has greater affinity for D than enzyme '3'

B- Feedback inhibition of enzyme '5' by compound X

C- Substrate limitation for enzyme '5'

Which of the above statement could represent probable reasons for NOT obtaining proportionate increase in the amount of 'X' in the transgenic plants? **(DEC 2016)**

(a) Only C (b) Only A and B

(c) Only A (d) A, B and C

16. In an attempt to increase the yield of a commercially important enzyme from natural isolate several strategies were adopted as follows:

A. Genome was selectively modified to increase yield.

B. Reappraisal of culture requirements of the modified organism to increase yield.

C. Induced mutants were screened and selected for organism synthesizing improved levels of the enzyme.

D. Organism was genetically modified, so that it produces a factor that enhances stability of the enzyme.

Which one of the following options represents strategies that are appropriate for the purpose? **(DEC 2018)**

(a) A, B, C and D (b) B and C only

(c) A, C and D only (d) A and B only

17. Which one of the following statements regarding use of hybrid nucleases for plant genome engineering is **INCORRECT**? **(DEC 2017)**

(a) For gene knock out experiments, the nuclease is expressed without a donor DNA template.

(b) Hybrid nucleases typically comprise two protein subunits that dimerize at their nuclease domain.

(c) Zinc finger nucleases (ZFNs) can efficiently target all nuclear genes with equal efficiency.

(d) Both ZFNs and TALENs are fused with cleavage domains of FokI endonuclease.

Answer Key

1.(a)	**2.**(d)	**3.**(d)	**4.**(a)	**5.**(b)	**6.**(d)	**7.**(d)
8.(b)	**9.**(c)	**10.**(b)	**11.**(a)	**12.**(c)	**13.**(a)	**14.**(c)
15.(d)	**16.**(a)	**17.**(c)				

4. Agrobacterium Tumefaciens

1. The T-DNA region of the Ti plasmid of *Agrobacterium tumefaciens* harbours two genes: X and Y. Mutation of gene 'X' produces a rooty tumour while mutation of gene 'Y' produces shoots in the tumor. Based on the above information, which one of the following statements is correct? **(DEC 2019)**

(a) Gene 'X' encodes auxins and gene 'Y' encodes cytokinins

(b) Gene 'X' encodes cytokinins and gene 'Y' encodes auxins

(c) Gene 'X' and gene 'Y' both encode auxins

(d) Gene 'X' encodes opines while gene 'Y' encodes cytokinins

2. *Agrobacterium tumefaciens* is frequently used as a vector to create transgenic plants. Under laboratory conditions *Agrobacterium* - mediated plant transformation does not require **(DEC 2019)**

(a) host plant genes

(b) bacterial type IV secretion system

(c) vir genes

(d) opine catabolism genes

3. Some of the following transgenic approaches could be used for functional characterization of endogenous genes in plants:

A. Transformation using a binary vector containing a strong enhancer element and lacking the right border of T-DNA

B. Transformation using a binary vector containing a promoter-less reporter gene sequence and a selection maker gene cassette within the T-DNA.

C. Transformation using a binary vector containing only a strong enhancer element and a selection marker gene cassette within the T-DNA.

D. Transformation using a binary vector lacking a reporter gene as well as both the left and right borders of T-DNA.

Which one of the following combinations can be used? **(DEC 2017)**

(a) A and B only (b) B and C only

(c) C and D only (d) A and D only

4. Given below are four statements related to Agrobacterium-mediated transfer of T-DNA into plant cells:

A. Production of single-stranded T-DNA by VirD1 and VirD2 proteins.

B. Interaction of VirE2 with VIP1 and VirE3.

C. Use of VirB/VirD4 type IV secretion system.

D. Activation of VirA-VirG complex.

The correct sequence of events (from earliest to latest) is: **(DEC 2018)**

(a) A-B-D-C (b) B-C-A-D

(c) C-A-B-D (d) D-A-C-B

5. Match the following Vir proteins with their correct function during *Agrobacterium* mediated transfer of T-DNA to plant cells. **(DEC 2017)**

	Vir protein		**Function**
A	Vir G	(i)	Nucleus targeting of T-DNA
B	Vir D2	(ii)	Component of membrane structure (transfer apratus) for T-DNA transfer
C	Vir B1	(iii)	Proteosome mediated destruction of proteins coating T-DNA complex
D	Vir F	(iv)	Induction of Vir genes

(a) A-(i), B-(ii), C-(iii), D-(iv)

(b) A-(iv). B-(i). C-(ii), D-(iii)

(c) A-(i), B-(iii), C-(iv), D-(ii)

(d) A-(iii); B-(ii), C-(i), D-(iv)

6. *Agrobacterium* Ti plasmid vectors are used to generate transgenic plants. The following are examples of vir gene encoded proteins that are important for transfer of T-DNA in to plants.

A. vir E, single stranded DNA binding protein

B. virD_2 that generates T-strands

C. vir A that sense plant phenolic compounds

D. vir F which directs T complex proteins for destruction in proteasome

Which one of the following combination of the proteins functions inside the plant cell?

(DEC 2016)

(a) Only A and C (b) A, B and C

(c) Only B and C (d) A, B and D

7. A student noted the following points regarding *Agrobacterium tumefaciens*:

A. *A. tumefaciens* is a gram-negative soil bacterium.

B. Opine catabolism genes are present in T-DNA region of Ti-plasmid.

C. Opines are synthesized by condensation of amino acids and α-ketoacids or amino acids and sugars.

D. A callus culture of crown gall tissue caused by *A. tumefaciens* in plants can be multiplied without adding phytohormones.

Which one of the combinations of above statements is correct? **(DEC 2015)**

(a) A, B and C (b) A, B and D

(c) B, C and D (d) A, C and D

8. Following are certain statements regarding the use of *Agrobacterium* in plant transformation.

A. *A. tumefaciens* causes crown gall disease and *A. rhizogenes* causes hairy root disease.

B. Region A in Ti plasmid is responsible for replication.

C. Region D in Ti plasmid is responsible for virulence.

D. Oncogenic (one) region in T-DNA is responsible for unusual amino acid synthesis.

Which one of the following combinations of above statements is correct? **(JUNE 2014)**

(a) A and B (b) C and D

(c) A and C (d) B and D

9. While designing an experiment for *Agrobacterium* mediated plant transformation, a student noted down the following points:

A. Ti and Ri plasmids induce crown gall and hairy root disease, respectively

B. Enzymes octopine synthase and nopaline synthase involved in the synthesis of octopine and nopaline, respectively are encoded by T-DNA.

C. All the six vir genes, vir A, vir B, vir C, vir D, vir E and vir G are absolutely required for virulence.

D. Almost perfect 25 bp direct repeat sequences flanking all Ti and Ri plasmids in the T-DNA region is essential for T-DNA transfer.

Which one of the following combinations of the above statement is correct? **(DEC 2013)**

(a) A, B and C (b) B, C and D

(c) A, C and D (d) A, B and D

10. Following are some statements related to raising transgenic plants

A. T-DNA of Agrobacterium contains oncogenes and opine synthesis genes

B. All 8 operons present in vir region of Agrobacterium is essential for virulence

C. The CaMV 35S RNA promoter is a very strong constitutive promoter used in plant transformation

D. Conjugative transfer (oriT or tra) region functions in conjugative transfer of the Ti-plasmids

Which one of the following combinations of above statements is correct? **(DEC 2013 GU)**

(a) B, C and D (b) A, C and D

(c) A, B and D (d) A, B and C

11. *Agrobacterium tumefaciens*, also known as natural genetic engineers, causes crown-gall disease in plants. However, when the same bacteria are used to raise transgenic plants with improved agronomic traits, no such tumor (disease) is observed. This is due to:

A. Vir D_2 gene is mutated in Ti plasmid.

B. Diseased Ti plasmid is generally used.

C. Heat-shock during transformation destroys virulence.

D. Oncogenes have been removed.

Which one of the following combination of above statements is correct? **(DEC 2012)**

(a) A and C (b) A and D

(c) B and C (d) B and D

12. A set of virulence genes (*vir* genes), located in the *Agrobacterium* Ti-plasmid, is activated by **(MODEL PAPER)**

(a) octopine (b) nopaline

(c) acetosyringone (d) auxin

13. Which of the following statement is NOT true during infection of plant cell by *Agrobacterium*? **(JUNE 2012)**

(a) The protein products of virulence genes *Vir A* and *VirG* perceives acetosyringone

(b) The VirB protein forms a connection between *Agrobacterium* and the plant cell and facilitates T-DNA transfer into the plant

(c) The T-DNA is excised and bound to VirD2 protein

(d) The T-DNA, after becoming coated with VirF binds to phosphorylated VIP1, which allows the complex to enter the plant's nuclues

14. A transposon carrying a promoterless β-galactosidase *(lac* Z) was used to create insertional mutation in the *vir* region of Ti-plasmid of *Agrobacterium tumefaciens.* All the mutants in which lac Z fusion was in frame were divided into the following three groups:

A. The virulence of the bacteria was completely lost and the *lac* Z was induced by acetosyringone.

B. The virulence of the bacteria was reduced and the *lac* Z was induced by acetosyringone.

C. The virulence ofthe bactèria was completely lost and *lac* Z was not induced by acetosyringone.

Which of the following assumptions are valid about these mutants? **(JUNE 2011)**

(a) In group A, the insertion could be in *vir B,* C, or D; in group B the insertion could be in *virC* or *E;* and in group C the insertion could be in *virA* or G.

(b) In group A, the insertion could be in *virA, B,* C or *D;* in group B, the insertion could be in either *virC* or *D;* and in group C the insertion could be in *virG.*

(c) In group A, the insertion could be in *virA;* in group B, the insertion could be in *vir B;* and in group C, the insertion could be in *vir* C.

(d) In group A, the insertion could be in *virG;* in group B, the insertion could be in *virB, D* and *E;* in group C, the insertion could be in *virA.*

15. Following are some statements about *Agrobacterium* – plant interactions

(A) *Agrobacterium* transfers a part of its chromosome into plant cell.

(B) Agrobacterium transfers a part of one of its plasmid DNA into plant cell.

(C) All the virulence genes of *Agrobacterium* are inducible.

(D) All the virulence genes product of *Agrobacterium* are functional only inside the bacterial cells.

(E) Some of the virulence genes of *Agrobacterium* are inducible.

(F) Some of the virulence genes products of *Agrobacterium* are functional both in bacterial and plant cells.

Which of the following combination of statements is true? **(MODELPAPER)**

(a) (A), (C) and (D) (b) (B), (D) and (E)

(c) (C), (D) and (E) (d) (B), (E) and (F)

16. Given below are a few statements on *Agrobacterium* mediated transformation of plants. Which one of the following statements is CORRECT? **(JUNE 2018)**

(a) T-DNA transfer occurs from left border to right border.

(b) The *gfp* reporter gene can never be used for selection of transgenic plants.

(c) Transformation frequencies will decrease on over-expression of *virulence* genes.

(d) Host plant genes play an important role in influencing transformation frequencies.

17. Following statements were given regarding factors influencing variation in expression levels of transgene in transgenic plants:

A. Difference in restriction enzyme sites within the T-DNA.

B. Difference in copy number of the transgene.

C. Variations in site of integration of the T-DNA within the plant genome.

D. Presence of multiple promoters within the T-DNA region.

Which one of the following options represents a combinations of statements that would NOT lead to variations in transgene expression levels in transgenic plants generated using the same T-DNA/ binary vector? **(JUNE 2019)**

(a) A and C only (b) B only

(c) C and D only (d) A and D only

Answer Key

1.(b)	2.(d)	3.(b)	4.(d)	5.(b)	6.(d)	7.(*)
8.(c)	9.(d)	10.(b)	11.(d)	12.(c)	13.(d)	14.(a)
15.(d)	16.(d)	17.(d)				

5. Molecular approaches to transgene diagnosis

1. During transgenesis, the location of the genes and their number integrated into the genome of the transgenic animal are random. It is often necessary to determine the copy number of genes and their tissue-specific transcription. The following are the possible methods used for the determination.

A. Polymerase Chain Reaction(PCR)

B. Southern blot hybridization

C. Reverse Transcriptase PCR

D. Western blot

Choose the correct set of combinations. **(DEC 2012)**

(a) A and B (b) B and C

(c) B and D (d) A and D

2. A single copy homozygous transgenic plant containing the transgene 'A ' for fungal resistance was subsequently re-transformed with another gene ' B' for conforming resistance to salt stress. The selection marker genes used for both the transformation experiment were different.

Transgenic plant obtained following the retransformation experiment were screened for salt stress resistance and single copy event were identified by southern hybridization. These single copy event were self - pollinated. In the event of the two T-DNA (containing the A and B transgenes) getting integrated in unlinked location in all transgenic plants, the phenotypic ratios among the T_1 progeny would be: **(DEC 2016)**

(a) 3 (fungal resistant + salt- stress resistant): 1(fungal resistant)

(b) 1 (fungal resistant): 2 (fungal resistant + salt resistant): 1(salt- stress resistant)

(c) 3 (salt -stress resistant): 1 (fungal resistant)

(d) 1 (fungal resistant) :1(salt- stress resistant): 1(fungal resistant+ salt stress resistant)

3. A T_0 transgenic plant contains two unlinked copies of the T- DNA of which one is functional and the other is silenced. Segregation of the transgenic to non transgenic phenotype would occur in a

(i) Ratio in progeny obtained by backcrossing and in a (ii) ration in F_1 progeny obtained by self pollination.

Fill in the blanks with the correct combination of (i) and (ii) from the options given below: **(JUNE 2019)**

(a) (i)— 3:1 and (ii)— 15 : 1

(b) (i)— 1:1 and (ii) -- 3 : I

(c) (i)—3:1 and (ii) — 3 : 1

(d) (i) —1:1 and (ii) -15 : 1

4. A transgenic plant is developed with the following T-DNA construct

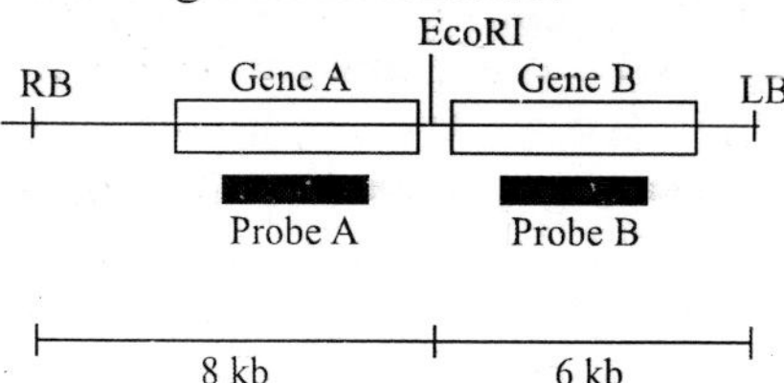

In order to analyze the nature of integration, genomic DNA digested with EcoRI was used for Southern hybridization using either probe A or B. The result obtained is as

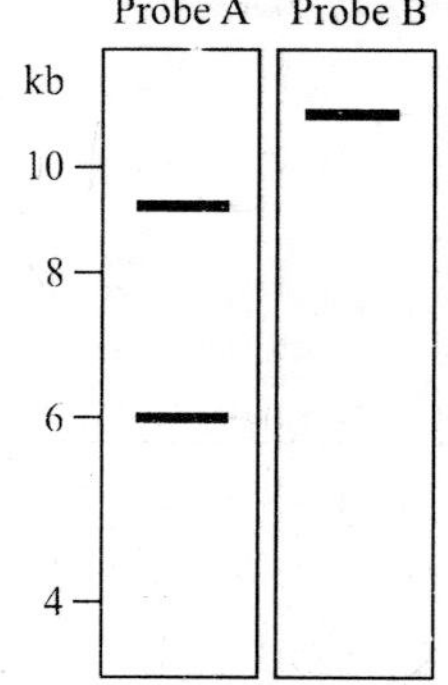

The following conclusions were made:

A. There are two copies of the T-DNA cassette integrated at one loci and a third copy at another loci.

B. There are two copies of the T-DNA cassette integrated at one loci.

C. Complete T-DNA cassette has been integrated in all cases.

D. In one T-DNA cassette there is a truncation towards the RB

E. In one T-DNA cassette there is a towards the LB

F. The arrangement of the T-DNA cassettes integrated at the same loci is

RB LB LB RB

G. The arrangement of the T-DNA cassette integrated at the same loci is

RB LB LB RB

Which of the above are correct? **(DEC 2013)**

(a) B, E and G (b) B, D and F

(c) A, D and F (d) B, C and F

5. "In *Agrobacterium*-mediated transformation of a plant using a binary vector construct,"

(DEC 2019 ASSAM)

Complete the above statement with the correct option.

(a) All transgenic plants generated would show similar levels of transgenic expression".

(b) The Agrobacterium cells lose the binary vector after the transfer of T-DNA".

(c) Some transgenic plants generated may contain partial or truncated versions of the T-DNA".

(d) Proteins of the host plant do not play any role".

6. Genomic DNA of transgenic plants (P1, P2 and P3) obtained by transforming with binary vector A whose map is depicted below, was digested with BamH I and Sal I and hybridized with genomic fragment X

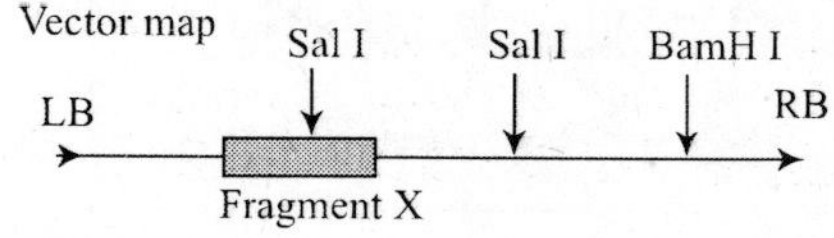

P1		P2		P3	
BamH I	Sal I	BamH I	Sal I	BamH I	Sal I
	—	—		—	—
—	—		—		
—	—		—		—

Pattern obtained in Southern hybridization is shown below

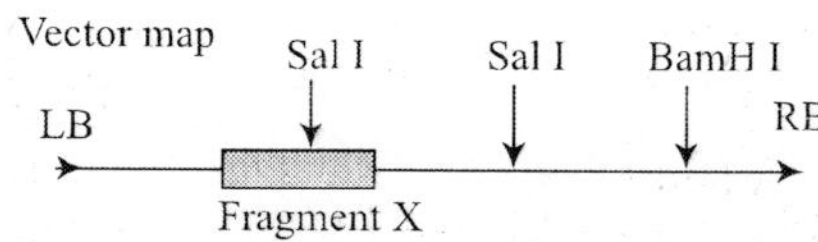

P1		P2		P3	
BamH I	Sal I	BamH I	Sal I	BamH I	Sal I
	—	—		—	—
—	—		—		
—	—		—		—

Based on the above, which of the following interpretation is correct: **(DEC 2011)**

(a) All the plants (P_1, P_2 and P_3) contains two copies of the transgene

(b) P_1, P_3 contains one and P_2 contains two copies of the transgene

(c) P_1 contains two, whereas P_2 and P_3 contains one copy of transgene each

(d) P_1 and P_2 contains two and P_3 contains one copy of the trangene

7. Given below are schematic representations of the T-DNA regions of four constructs that are to be used for *Agrobacterium* – mediated transformation to silence an endogenous plant gene represented as XYFP that is expressed constitutively in the plant.

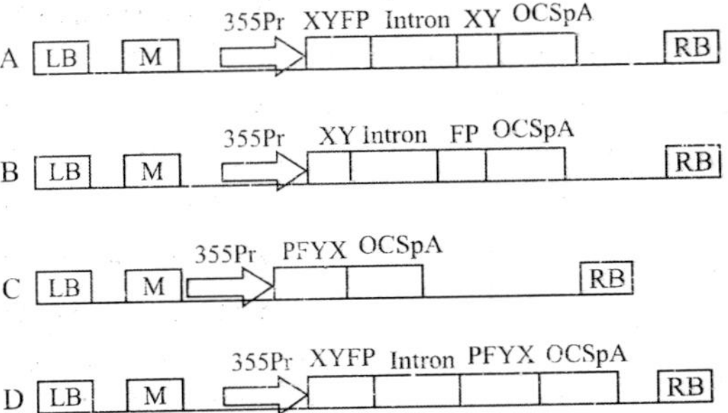

M: Selectable marker gene expression cassette

LB Left Border

RB: Right Border

Which of the four constructs depicted above could be used to silence the target gene XYFP? **(DEC 2019)**

(a) A and B only (b) B and D only

(c) A and C only (d) C and D only

8. In a breeding experiment; two homozygous parental lines (P1 and P2) were crossed to produce F_1 hybrids. Due to an experimental error, seeds of these hybrids got mixed up with the seeds of two other germplasm lines (P3 and P4) and hybrid seeds derived from them.

A marker-based fingerprinting exercise was performed using six randomly selected seeds (F1-F6) from the mixed material and the four parental lines. Results of this analysis are shown below:

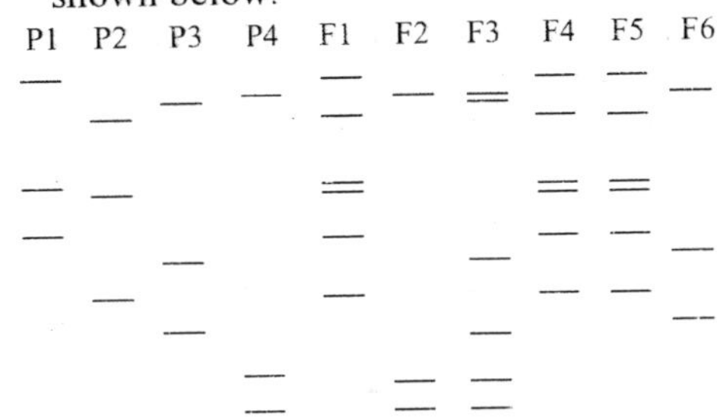

Based on the above data, which one of the following options represents the correct set of parents and their F_1 progeny? **(JUNE 2017)**

(a) P1 X P2 = F3 (b) P3 X P4 = F2

(c) P1 X P2 = F1 (d) P3 X P4 = F6

9. TILLING is a reverse genetics approach used in functional genomics. Which one of the following is used for TILLING? (JUNE 2016)

(a) T-DNA tagging by Agrobacterium-mediated transformation.

(b) Transposon tagging using Ac/Ds elements.

(c) Mutagenesis with ethylmethane sulphonate.

(d) Protoplast transformation by electroporation.

Answer Key

1.(b)	2.(a)	3.(b)	4.(b)	5.(c)	6.(c)	7.(d)
8.(c)	9.(c)					

6. Animal Cell Culture and Transgenic Animals

1. Application of gene therapy in clinical trials did NOT succeed due to **(DEC 2015)**
   (a) poor integration of a gene in the host genome
   (b) lack of expression of integrated gene in cells
   (c) degradation of gene inside the cell
   (d) activation of oncogenes consequent to integration of the gene
2. For sustainable expression of a trangene in the successive generation of a cell line in culture, the ideal gene transfer can be obtained using **(DEC 2012)**
   (a) lentiviral vector
   (b) adenoviral vector
   (c) plasmid DNA containing the transgene
   (d) only transgenic DNA
3. Which of the following host systems is best suited to express large amounts of glycosylated protein for structural studies? **(DEC 2017)**
   (a) GST fusion protein in *E. coli*
   (b) His-tagged protein in *E. coli*
   (c) Native protein in baculovirus
   (d) Native protein in *Pseudomonas fluorescens*
4. The genes whose promoters are extensively used for production of pharmaceutical proteins in transgenic dairy cattles are **(MODEL PAPER)**
   (a) lactalbumin and ovalbumin.
   (b) lactoglobulin and casein.
   (c) lactoferrin and transferrin.
   (d) casein and ovalbumin.
5. For developing transgenic mice, embryonic stem cells are engineered to express the transgene. These cells are selected by **(JUNE 2013)**
   (a) novobiocin. (b) neomycin.
   (c) tetracycline. (d) penicillin.
6. To keep them in a totipotent state, embryonic stem cells need to be maintained in a medium supplemented with **(DEC 2011)**
   (a) growth hormone
   (b) leukemia inhibiting factor
   (c) nestin
   (d) insulin
7. Which one of the following techniques is generally used to produce transgenic animals? **(JUNE 2014)**
   (a) Processed mRNA containing only exons are introduced into blastocyst stage embryo.
   (b) Entire foreign nucleus is introduced in enucleated unfertilized egg.
   (c) Desired DNA is micro injected into fertilized eggs followed by implantation of embryo in a foster mother.
   (d) c-DNA of desired gene is introduced into animal embryos and implanted in a foster mother
8. The following statements are made regarding developing a transgenic mouse:
   A. The transgenic mouse thus born will be a homozygous transgenic animal and can be maintained by crossing with another transgenic animal.
   B. The fertilized transgenic eggs are allowed to develop in vitro.
   C. The desired gene is preferably microinjected in male pronucleus after sperm entry in oocyte.
   D. For best efficiency, the desired gene is always microinjected in the male gametes and then they are allowed to fertilize the female gametes.
   E. Blastocyst stage embryos are transferred to the uterus of hormonally prepared mother.
   F. The fertilized eggs are collected from specific strain of mouse.
   G. The female mouse of specific strain is super ovulated, oocytes are collected and allowed to fertilize in vitro.

Choose the combination of statements arranged in the correct sequence for developing transgenic mouse. **(DEC 2019)**

(a) G→C→E (b) F→ B→A

(c) G→ D→ A (d) D→F→B→A

9. Which one of the following statements is NOT correct for propagation and maintenance of mammalian cells in vitro? **(JUNE 2014)**

(a) Transformed cell lines do need external supply of serum to grow.

(b) The cells that are obtained directly from the organism is a primary culture.

(c) Trypsin is added to cell culture media to maintain cell's health.

(d) HEPES buffer is generally used to maintain pH of the culture media.

10. To prepare individual tissue cells from a primary culture, the cell-cell and cell-matrix interaction must be broken. To achieve this would NOT use: **(JUNE 2019)**

(a) EDTA (b) Trypsin

(c) Collagenase (d) Separase

11. Which is best method for checking mycoplasma contamination in a mammalian cell line? **(JUNE 2012)**

(a) Southern Hybridization

(b) ELISA

(c) PCR

(d) Western Hybridization

12. Animal cell cultures are frequently used for production of therapeutic proteins. NIH3T3 (a fibroblast cell line) and CHO (Chinese hamster ovarian cell line) are some of the popular cell lines used. Choose the best combination of cell line (for transfection) and starting material for purification of human growth hormone, a secretary protein **(MODELPAPER)**

(a) NIH3T3 cell pellet

(b) CHO and cell pellet

(c) NIH3T3 and culture medium

(d) CHO and culture medium

13. Inbreeding for 5 generations led to production of homozygous transgenic mice. However, these homozygous males or females were infertile. Which of the following approach is most preferable and economical to obtain heterozygous transgenic animals continuously? **(JUNE 2013)**

(a) More transgenic founder (1st animal) should be generated.

(b) Crossing (breeding) of transgenic mice with wild type mice in earlier generations should be done for continued production of transgenic heterozygous offspring.

(c) Inbreeding should be avoided after 5th generation.

(d) Homozygous transgenic mice should be mated with wild type mice for continued production of transgenic heterozygous offspring.

14. The muscle tone was increased after electrolytic lesion of the caudate nucleus in a cat. The muscle tone decreased within seven days. The following explanations were given by the researcher.

A. The functional recovery was due to plastic changes of nervous system.

B. The brain tissue surrounding the lesioned area was non-functional due to circulatory insufficiency immediately after surgery which led to the greater functional loss.

C. The circulatory status in surrounding tissue recovered with time resulting in partial functional recovery.

D. The degenerating nerve fibres were regenerated which underlie functional recovery.

Which one of the following is correct? **(DEC 2012)**

(a) A and B (b) B and C

(c) C and D (d) A and D

15. Stem cells are widely used for their regenerative property and capacity to differentiate into different lineages. A person with a damaged liver approaches a stem cell therapist. Which of the following therapeutic strategy would be safest? **(JUNE 2012)**

(a) Transplanting adult liver cells from healthy donor and grafting them into patient

(b) Transforming skin cells from patient into iPS cells and using them for further differentiation and grafting in liver

(c) Injecting embryonic stem cells into the damaged liver

(d) Injecting cord blood cells into the liver directly

16. Following are the different ways of obtaining human stem cells-

(A) Cells from morula are dispersed and cultured to give rise pluripotent stem cells.

(B) Isolated inner cell mass of a blastocyst are cultured and they become pluripotent stem cells.

(C) The primordial germ cells from a fetus are collected and cultured, which grow into pluripotent stem cells.

(D) Stem cells are collected from adult tissues and grown in specific manner to develop into pluripotent stem cells.

Which of the above cell type(s) is extensively used for somatic cell gene therapy in human? **(MODEL PAPER)**

(a) (A) and (B) (b) (B) only
(c) (B) and (D) (d) (D) only

17. Stem cell therapies are being used in regenerative medicine like forming new adult bone, which usually does not regrow to bridge wide gaps. Successful attempts have now been made in this area because the same paracrine and endocrine factors were found to be involved in both endochondral ossification and fracture repair. Few methods to achieve the above are given below.

A. Develop a collagen gel containing plasmids carrying the human parathyroid hormone gene and place in the gap between the ends of broken leg.

B. Develop a gel matrix disc containing genetically modified stem cells stem to secrete BMP4 and VEGF-A and implant it at the site of wound.

C. Make scaffold of material that resembles normal extracellular matrix that could be molded to form the shape of a bone needed and seed them with bone marrow stems cells

D. Develop a collagen gel containing plasmids carrying the human bone marrow cells and place them between ends of bones.

Which of the above methods would you employ to develop a new functional bone in patients with severely fractured bones? **(DEC 2011)**

(a) A and B (b) A, B and C
(c) A and C (d) C and D

18. Factor IX is essential for blood clotting. Deficiency of Factor IX could be corrected by delivering factor IX gene using viral vectors.

In an experiment the gene for Factor IX was cloned appropriately into Adenovirus (AV), Adeno-associated virus (AAV) and Retrovirus (RV). Retrovirus integrate the gene into the dividing cells. AAV integrates partially into non-dividing cells. AV does not integrate the gene but transfects both dividing and non-dividing cells.

Following expression profile for Factor IX was observed when the three vectors were injected' intra-muscularly into three groups of experimental mice. Which one of the following outcomes is correct? **(JUNE 2015)**

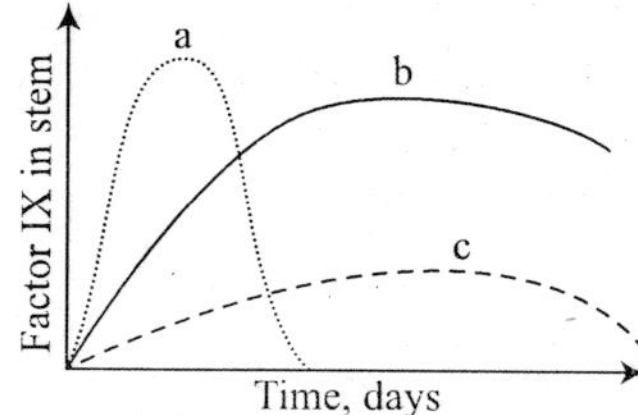

(a) a by RV; b by AV; c by AAV
(b) a by AV; b by AAV; c by RV
(c) a by AA V; b by RV; c by AV
(d) a by AV; b by RV; c by AAV

Answer Key

1.(d)	2.(a)	3.(c)	4.(b)	5.(b)	6.(b)	7.(c)
8.(a)	9.(c)	10.(d)	11.(c)	12.(d)	13.(b)	14.(b)
15.(b)	16.(d)	17.(b)	18.(d)			

7. Breeding in plants and animals

1. Which one of the following options is an INCORRECT match? **(DEC 2019 ASSAM)**

1	F_2 population	Selfing or sib-mating F_1 hybrids
2	$F_{2:3}$ population	immortal population
3	Backcross population	Difference in segregation patterns of dominant and recessive markers
4	F_1 doubled haploids	immortal population

2. Given below are a few statements on use of plant breeding to develop improved varieties of a crop plant

   A. Genotypic/phenotypic variation in the desired trait should be available in the germplasm resources of the crop plant.

   B. Availability of molecular markers linked to the trait of interest would decelerate the process of trait introgression into desired varieties.

   C. Breeding procedures to improve plant varieties are generally more successful among sexually compatible species as compared to sexually incompatible species.

   D. Co-dominant molecular markers cannot be used for selection of plants with the desired trait

   Which of the above statement(s) is/are INCORRECT? **(DEC 2019)**

   (a) A and C (b) B only
   (c) C and D (d) B and D

3. Given below are four statements regarding the use of marker assisted selection (MAS) in plant breeding.

   A. Markers that are not linked to target traits cannot be used to test accumulation of the desired recurrent parent's genetic background in a backcross breeding program.

   B. Co-dominant markers are more effective for marker assisted backcrossing since they facilitate selection of heterozygous progeny.

   C. MAS can be used to facilitate backcrossing of recessive genes that influence traits of interest.

   D. Use of tightly linked flanking makers for a trait of interest increases linkage drag in a breeding program.

   Which one of the following options represents a combination of all INCORRECT statements? **(DEC 2019 ASSAM)**

   (a) A and B (b) B and C
   (c) A and D (d) C and D

4. Which one of the following statements regarding crop improvement programs using molecular breeding approaches is **INCORRECT**? **(JUNE 2017)**

   (a) Allelic diversity for traits of interest should be available in the naturally occurring crossable germplasm
   (b) The genes of interest cannot be derived from a sexually incompatible organism
   (c) Availability of markers and linkage maps would facilitate the breeding program
   (d) The crop plant should necessarily have an optimized robust system for production of doubled haploids

5. Which of the following is wild relative of wheat? **(JUNE 2016)**

   (a) *Triticum monococcum*
   (b) *Triticum compactum*
   (c) *Triticum vulgare*
   (d) *Triticum boeoticum*

6. For the aquaculture farming of Indian major carps several techniques are used. Which one of the following is NOT used for this purpose: **(JUNE 2015)**
   (a) Induced breeding
   (b) Selective breeding
   (c) Inbreeding
   (d) Composite fish farming
7. Use of double haploids in plant breeding helps to **(DEC 2012)**
   (a) reduce generation time while interogressing recessive traits
   (b) reduce generation time while interogressing dominant traits
   (c) develop somatic hybrids
   (d) interogress transgenic traits
8. Cultivated bananas are sterile because **(MODEL PAPER)**
   (a) male flower-bearing plants are very rare.
   (b) they lack natural pollinators in the crop plants.
   (c) they are triploid and therefore seeds are not set.
   (d) they are a cross of two unrelated species.
9. The rattans and canes that we use in furniture belong to **(DEC 2011)**
   (a) bamboos (b) palms
   (c) arborescent lilies (d) legumes

Answer Key

1.(b)	2.(d)	3.(c)	4.(d)	5.(a)	6.(c)	7.(a)
8.(c)	9.(b)					

8. Bioremediation and phytoremediation

1. In bioremediation by microorganisms detailed below, choose INCORRECT option? **(JUNE 2019)**
   (a) The organic contaminants provide a source of carbon
   (b) The bacteria do not get energy by degrading contaminants.
   (c) Bacteria can produce oxydized and reduced species that can cause metals to precipitate.
   (d) Bacteria act on contaminants by aerobic and anaerobic respiration.
2. Microbial leaching involves the process of dissolution of metals from ore breaking rocks using microorganisms. Which one of the following bacteria helps in leaching copper from its ore? **(JUNE 2013)**
   (a) *Acidithiobacillus ferroxidans.*
   (b) *Pseudomonas putida.*
   (c) *Deinococcus radiodurans.*
   (d) *Rhodopseudomonas capsulate.*
3. *Desulphovibrio desulfuricans* (A) and *Pseudomonas species* (B) are involved in mercury bioremediation. Which of the statement below is correct? **(DEC 2012)**
   (a) A converts methyl mercury to mercuric ion, B converts mercury to methyl mercury
   (b) A converts mercury to methyl mercury, B converts mercury to mercuric ions
   (c) A converts mercury to methyl mercury, B converts methyl mercury to mercuric ions
   (d) A converts methyl mercury to mercuric ions, B converts mercury to mercury ions
4. The following are certain facts regarding bioremediation:
   A. Biodegradable plastics are made using polyhydroxy alkanoates (PHAs) such as polyhydroxybutyrate (PHB).
   B. Pseudomonas putida F1 bacterial strain is involved in degradation of aromatic hydrocarbon.
   C. The bacterium *Deinococcus radiodurans* consume and digest toluene and ionic mercury from highly radioactive nuclear waste.
   D. Bioaugmentation is a process of improving the microorganisms already existing in the system for degradation of xenobiotic compound.

Which one of the following combination of above statements is correct? **(DEC 2013)**

(a) A, B and C (b) A, B and D

(c) A, C and D (d) B, C and D

5. Chinese Brake fern (*Pteris vittata*) is hyperaccumulator of: **(DEC 2013)**

(a) Cadmium. (b) Arsenic.

(c) Lead. (d) Chromium.

6. Column A lists names of microorganisms and column B lists the pollutants that are degraded by the microbes.

	Column A		Column B
I	*Pseudomonas putida*	i	Methyl tert-butyl ether
II	*Methylicibium petroleiphilum*	ii	Heavy metal detoxification
III	*Alconivorax borkumensis*	iii	Toluene, naphthalene
IV	*Deinococcus radiodurans*	iv	Fuel hydrocarbon

Which one of the following options correctly matches the organism to the pollutant it most commonly degrades? **(DEC 2019 ASSAM)**

(a) 1 - (ii); II - (iv); III - (i); IV - (iii)

(b) 1 - (iii); II - (iv); III - (ii); IV - (i)

(c) 1 - (iii); II - (i); III - (iv); IV - (ii)

(d) 1 - (iv); II - (iii); III - (i); IV -(ii)

7. Microorganisms having nitrilase activity can be enriched from soil samples by supplying nitriles as the sole nitrogen or carbon source in the growth medium. A problem with this approach is that false-positive strains may be isolated, which use nitrites to produce amide with the help of a different enzyme activity such as **(DEC 2013 GU)**

(a) phosphorylases (b) phosphatases

(c) hydratases (d) phosphoestarases

8. The process of phytoremediation where complexation and immobilization of toxin takes place within the soil is called **(DEC 2013 GU)**

(a) phytoextraction.

(b) phytodegradation

(c) phytovolatilization

(d) phytostabilization.

9. Release of nutrients, oxidants or electron donors into the environment to stimulate naturally occurring microorganisms to degrade a contaminant, is referred to as **(JUNE 2011)**

(a) biostimulation. (b) phytoremediation.

(c) bioaugmentation. (d) bioremediation.

10. A plot of soil contaminated with diesel oil was inoculated with oyster mushrooms. After 4 weeks, more than 95% of the polycyclic aromatic hydrocarbons had been reduced to non-toxic compounds. This process is called **(JUNE 2013)**

(a) phytoremediation.

(b) chemoremediation.

(c) mycoremediation.

(d) zooremediation.

11. Which of the following does not represents strategy for phytoremediation? **(JUNE 2012)**

(a) Phytodegradation

(b) Phytomining

(c) Continous removal through hyper accumulators

(d) Chelate mediated extraction of pollutants

12. Secondary sewage treatment involves **(JUNE 2012)**

(a) Physical removal of solids from polluted water by filtration and sedimentation

(b) removal of chemical remains by precipitation

(c) removal of dissolved organic compounds by activated sludge or trickling filter

(d) Removal of microbial pathogens by chlorination or ozonization

Answer Key

1.(b)	2.(a)	3.(c)	4.(b)	5.(b)	6.(c)	7.(b)
8.(d)	9.(a)	10.(c)	11.(d)	12.(c)		

9. Biosensors

1. Lower limits of detection by sensors is important. Which method of detection is more sensitive than. Glass electrode used for pH measurement? **(DEC 2014)**
   (a) Absorption spectroscopy
   (b) Refractive index
   (c) Circular dichroism
   (d) fluorescence spectroscopy

2. Routinely used glucose biosensor estimates blood glucose level by sensing the concentration of **(JUNE 2011)**
   (a) glucose. (b) oxygen.
   (c) δ-giuconolactone. (d) H_2O_2.

3. Performance of biosensor is evaluated by their response to the presence of an analyte. The physiological relevant concentration of analyte is between 10μM and 50μM. Which among the following biosensor responses is best? **(JUNE 2016)**

(a)

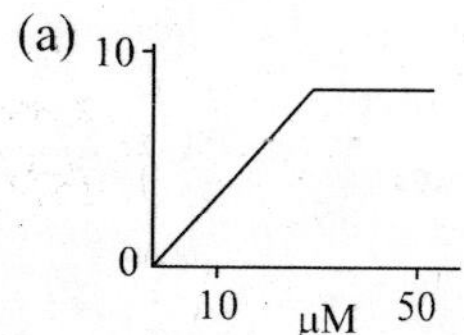

(b)

10
0
10 μM 50

(c)

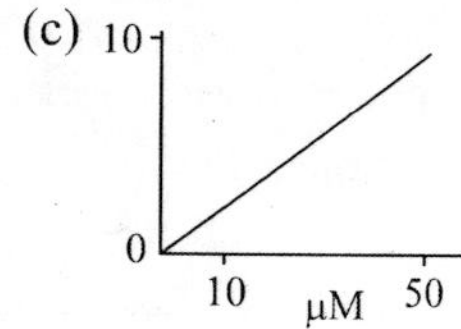

(d)

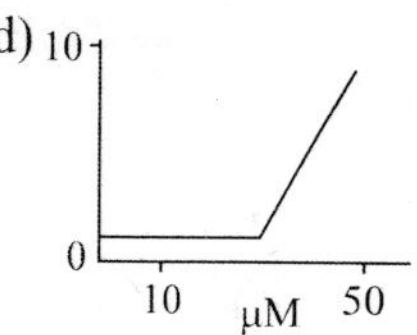

4. Glucose in the blood is detected by four different methods (a, b, c and d). The sensitivity and range of detection of glucose by these four methods is shown below. Clinically relevant concentration of glucose in blood is between 80 – 250 mg/dL

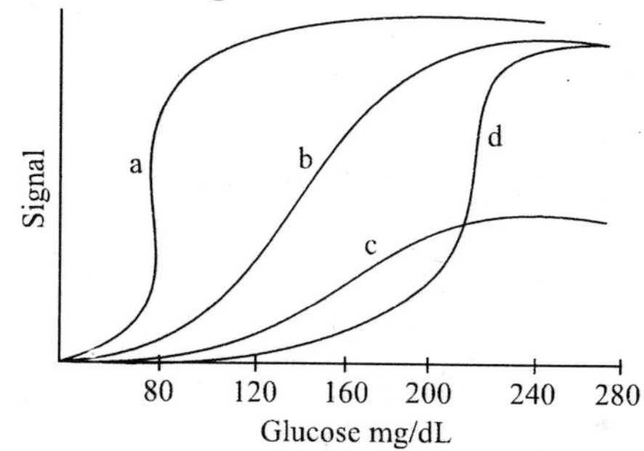

Which of the following method is most appropriate? **(DEC 2015)**
   (a) a (b) b
   (c) c (d) d

Answer Key

1.(d)	2.(b)	3.(c)	4.(b)

SECTION-13

METHODS IN BIOLOGY

1. Molecular cloning and restriction digestion

1. Among existing technologies, which of the following vector system would you prefer to use for generating a library for 140 kb eukaryotic genomic DNA fragments, while giving due consideration to size as well as stability of insert? **(JUNE 2012)**
   (a) Phage (b) Cosmid
   (c) BAC (d) YAC
2. Yeast artificial chromosome (YAC) vectors contain selectable markers. Loss of which marker at the cloning site distinguishes the religated YACs from the original vector marker? **(JUNE 2011)**
   (a) TRP 1 (b) SUP4
   (c) URA3 (d) CEN
3. Which of the following statements about YACs is **FALSE**? **(DEC 2017)**
   (a) YACs can carry DNA fragments of 1 Mb
   (b) A single yeast may contain more than one YAC
   (c) Rearrangements in YACs are rare because recombination is poor in yeast
   (d) YACs have to be transferred to bacteria for subsequent DNA manipulations
4. The average size of human genome is about 3.2×10^9 base pairs. Considering 60% of it is made up of A- T base pairs, what will be the frequency of the following restriction sites to be present in human genome in high to low order? **(DEC 2013 GU)**
   X= 5'-GAATTC-3'
   Y= 5'-ACCGGT-3'
   Z= 5'-CGCG-3'
   (a) X > Y > Z (b) Z > X > Y
   (c) Y > Z > X (d) Z > Y > X
5. A gene encoding for protein X was cloned in an expression vector under the T7 RNA polymerase promoter and lac operator. Cells were incubated by the addition of 1 mM IPTG at 37^0 c for 6 h. Cells were lysed and fractionised into insoluble bodies. Which one of the following strategies would you use to express protein X in the soluble fraction (cell free supernatant)? **(DEC 2016)**
   (a) Increase the duration of induction with 1mM IPTG
   (b) Grow cells at lower temperature after induction with 1mM IPTG
   (c) Increase the concentration of IPTG
   (d) Grow cells at higher temperature after induction with 1 mM IPTG
6. In a typical gene cloning experiment, by mistake a researcher introduced the DNA of interest within ampicillin resistant gene instead of *lac Z* gene. The competent cells were allowed to take up the plasmid and then plated in the media containing ampicillin, X-gal and IPTG and subjected to blue-white screening. Considering all plasmids were recombinant which one of the following statements correctly describes the outcome of the experiment? **(JUNE 2016)**
   (a) The bacteria which took up the plasmids would grow and give blue colonies. .
   (b) The bacteria which took up the plasmids would not grow.
   (c) The bacteria which took up the plasmids would form white colonies.
   (d) All of the bacteria would grow and give white colonies.
7. In order to clone an eukaryotic gene in pBR322 plasmid vector, the desired DNA fragment was produced by *PstI* cleavage and incubated with *PstI* digested pRR322 (*PstI* cleavage site lies within the ampicillin resistant gene) and ligated. Mixtures of ligated cells were used to transform *E. coli* and plasmid containing bacteria were selected by their growth in tetracycline containing medium. Which type of plasmid/s will be found? **(DEC 2012)**

(a) Circular pBR322 plasmid containing the target gene and resistant to only tetracycline.

(b) Circular pBR322 plasmid containing the target gene and resistant to tetracycline only and re-circularized pBR322 plasmid resistant to both ampicillin and tetracycline.

(c) Circular pBR322 plasmid containing the target gene and resistant to only tetracycline, re-circularized pBR322 resistant to both ampicillin and tetracycline and concatemerised pBR322 resistant to both ampicillin and tetracycline.

(d) Circular pBR322 plasmid containing the target gene and resistant to both ampicillin and tetracycline.

8. Polynucleotide kinase (PNK) is frequently used for radio labeling DNA or RNA by phosphorylating 5' end of non-phosphorylated polynucleotide gene. Which of the following statements about PNK is NOT true?

(DEC 2016)

(a) PNK catalyse the transfer of alpha phosphate from ATP to 5' end of polypeptide chains of (DNA or RNA)

(b) PNK has three phosphatase activity

(c) PNK is inhibited by small amount of ammonium ions

(d) PNK is T4 bacteriophage encoded enzyme

9. For 5' end labelling of DNA, the following reaction are carried out sequentially as indicated

5' dephosphorylated DNA $\rightarrow$[γ^{32}] dATP + T_4 polynucleotide kinase (T4PNK) and incubated for 2 hours$\rightarrow$Ammonium acetate$\rightarrow$Tris EDTA$\rightarrow$Ethanol

If the trace amount of ammonium ion (NH_4^+) is present in initial DNA mix, which of the following statement would most likely be true?

(JUNE 2012)

(a) NH_4^+ ion activates T4PNK, thereby increasing the labeling efficiency

(b) NH_4^+ ion inhibits T4PNK, therefore should not be present in DNA mix

(c) NH_4^+ ion does not have any effect on T4PNK

(d) NH_4^+ ion dephosphosphorylates DNA, thereby increasing labeling efficiency

10. A gene from genomic library is screened by using hybridization technique. After hybridizing the probe, usually a stringent washing step is given. The following statements are given to explain the stringent washing step:

A. Stringent washing takes care of removing unincorporated and non-specifically hybridized probe molecules

B. Stringent washing is done in solution having high salt concentration and lower temperature

C. Stringent washing is done in solution having low salt concentration keeping higher temperature

D. Salt present in washing solution supports hybrids to stay intact by shielding the interference of water molecules

E. Salt reacts with DNA molecules and allows easy dissociation of hybrids

F. Stability of hybrid is directly proportional to the temperature

Which combination of the above statements is most appropriate for stringent washing, step?

(JUNE 2015)

(a) A, B and D (b) A, C and D

(c) A, B and F (d) C, E and F

11. You are inserting a gene of 2 kb length in to a vector of 3kb to make a GST fusion protein the gene is being inserted at the *E. coli* site and the insert has a Hind III site 500bp downstream of the first codon. You are screening for the clone with the correct orientation by a restriction digestion of the plasmid using Hind III plus Bam HI (H+P) and Hind iii plus PSTI (H+P) the map of the relevant region of the vector is shown below

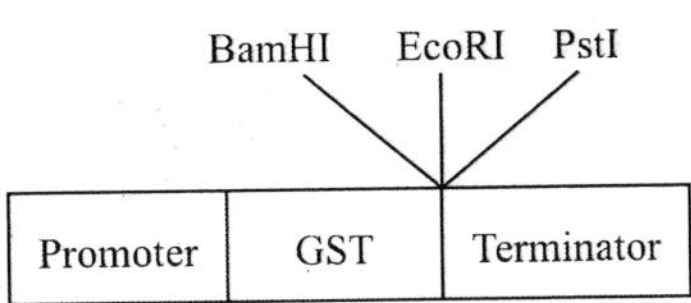

Which of the following given below is the pattern following the restriction digestion of a plasmid isolated from four independent clones. (A, B, C or D)

	A		B		C		D	
	H+B	H+P	H+B	H+P	H+B	H+P	H+B	H+P
5 —								
4 —								
3 —								
2 —								
1.5 —								
1 —								
500 —								

Which of the plasmid shown above represents the clone in the correct orientation? **(DEC 2016)**

(a) A (b) B
(c) C (d) D

12. The emission maximum of tryptophan fluorescence in a protein is ~ 335 nm. This suggests that tryptophan **(JUNE 2019)**

(a) is in a hydrophobic environment.
(b) occurs in a helical segment
(c) has proximal cysteine residues
(d) is oxidized.

13. A protein D is encoded by a gene, which is 5 Kb long and has three Hind III restriction enzyme sites. The first one is 0.5 Kb from the transcription start site, the second one is 2.5 Kb from the first site and the third one is 0.5 Kb internal to the stop codon. The second site is polymorphic. In order to find out whether fetal cells contain the normal or the mutated gene, total genomic DNA from fetal cells was isolated, completely digested with Hind III, separated in an agarose gel, transferred to membrane and detected by a probe against the region between the second and third restriction site. Which one of the following band patterns will be obtained if the fetal cell is heterozygous? **(DEC 2013)**

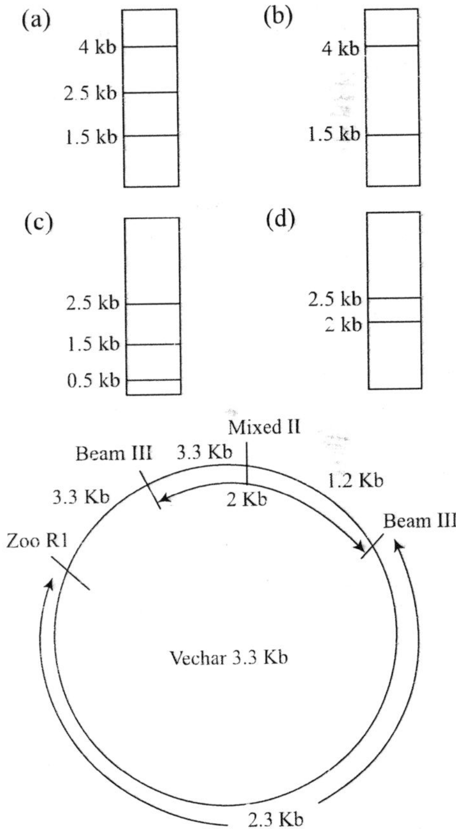

14. The above diagram represents a 2 Kb insert successfully introduced between two BamH1 sites of a 3.8 Kb vector in desired orientation. The HindIII site on the insert and EcoRl site on the vector is also indicated. If the insert was introduced in the opposite orientation, which one of the following statements is INCORRECT? **(JUNE 2014)**

(a) Digestion with EcoRl will linearize the 5.8 Kb plasmid.
(b) Digestion with BamH1 will yield 3.8 and one 2.0 Kb fragment.

(c) Double digestion with EcoRl and HindIII will produce 1.7 Kb and 4.1 Kb segments.

(d) Double digestion with EcoRl and HindIII will produce 2.1 Kb and 3.7 kb fragments.

15. A researcher attempted to clone a 630bp coding sequence of a gene downstream to a bacterial promoter (500bp in length) for over expression and purification of the encoded protein. The gene sequence was isolated as a Smal fragment (Recognition sequence: CCC ↓↓GGG; arrow indicates the site of restriction) and cloned at a Smal site located downstream to the promoter. The gene sequence contained a single site for EcoRI located 30bp downstream to the start codon. A schematic representation of the plasmid along with locations(s) of some restriction enzyme sites of the vector is given below.

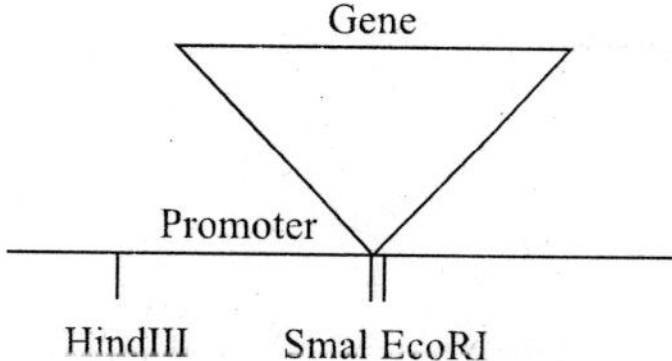

The researcher screened the obtained colonies by a double digestion using *Hind III* and *Eco RI.* Which one of the following restriction digestion patterns would represent the restriction profile of the desired clone that could be used for overexpression? **(DEC 2017)**

(a) ~530bp + ~600bp + vector backbone

(b) ~1100bp + ~30bp + vector backbone

(c) ~500bp + vector backbone

(d) ~1130bp + vector backbone

16. In an experiment, a 1 kb fragment with a single BamHI site (as shown below in figure "A") is to be cloned in the Smal (CCC ↓ GGG) site of a cloning vector of 3kb length (figure 'B'). None of the other enzymes of the multiple cloning sites are present in the fragment to be cloned.

(a) 5' BamHI 3' 0.2 0.8 (b) P Xhol HindII Smal BamHI EcoRI EcoRV

Based on the information given above, a series of digestions were set up for the potential clones and their fragment profiles are given below:

A. BamHI : 200bp + 3.8 kb

B. BamHI : 800bp + 3.2 kb

C. HindIII + EcoRI : ~1 kb +3kb

D. XhoI + BamHI : ~200bp + ~800bp + ≈3kb

Which one of the above digestion profiles confirms successful cloning of the fragment in the vector in an orientation wherein the 5' end of the cloned fragment is towards P?

(DEC 2019)

(a) A only (b) B only

(c) A and C (d) C and D

17. A circular plasmid of 4 kb produces fragments on digestion with different enzymes as given below:

BamHI	- 4 kb
BamHI + HindiIII	- 1 kb + 3 kb
BamHI + EcoRI	- 2.5 kb + 1.5 kb

Based on the above information, the enzymes that are present as unique cutters in the plasmid are: **(DEC 2019 ASSAM)**

(a) BamHI and HindIII only

(b) BamHI only

(c) HindIII and EcoRI only

(d) BamHI, HindIII and EcoRI

18. Which one of the following statements related to molecular cloning procedures is INCORRECT? **(DEC 2019)**

(a) 5' overhangs of restricted DNA fragments can be blunt-ended by Klenow polymerase but not by DNaseI.

(b) A DNA fragment obtained as an XhoI fragment (C↓TCGAG) may be ligated at the Sallsite (G↓TCGAC) in a vector.

(c) To prevent self-ligation of a vector digested with KpnI (GGTAC↓C), alkaline phosphatase enzyme is used to remove 3'-PO_4 groups from the ends of fragments.

(d) α-complementation/blue-white screening may produce blue coloured recombinant colonies (containing cloned fragments) in case of translational fusion with the β-galactosidase gene.

19. Which one of the following statements regarding restriction/modifying enzymes used in recombinant. DNA technology is correct? **(DEC 2018)**

(a) Endonucleases remove nucleotides, one at a time, from the ends of a sequence.

(b) Type II class of restriction enzymes do not recognise palindromic sequences.

(c) Mung bean nuclease acts on double stranded DNA or RNA termini.

(d) Type II class of restriction enzymes can generate either "sticky" (staggered) or "blunt" ends

Answer Key

1.(c)	2.(b)	3.(c)	4.(b)	5.(b)	6.(b)	7.(c)
8.(a)	9.(b)	10.(b)	11.(c)	12.(a)	13.(b)	14.(c)
15.(a)	16.(b)	17.(d)	18.(c)	19.(d)		

2. Methods for analysis of gene expression

1. A gene expressing a 50 kD protein from an eukaryote was cloned in an E. coli plasmid under the lac promoter and operator. Upon addition of IPTG, the 50 kDa protein was not detected. Which one of the following explains the above observation? **(DEC 2015)**

(a) The cloned sequence lacked the Kozak sequence

(b) E. coli does not make proteins larger than 40 kDa

(c) Differences in codon preference

(d) 50 kDa protein contains a nuclear localization signal

2. You have transiently expressed a new protein (for which no antibody is available) in a cell line to establish structure function relationship. Which one of the following strategies is the most straight forward way to examine the expression profile of this new protein? **(DEC 2015)**

(a) By metabolic labelling using ^{35}S labelled amino acids

(b) Making a GFP fusion protein with this new protein

(c) Immunoprecipitating this protein with the help of another protein for which antibody is available

(d) Running SDS-PAGE and identify the protein

3. In an attempt to detect protein expression profile in a cell, Western blot technique is employed. Expression of two new proteins is to be followed by probing with respective high affinity antibodies (raised in rabbit). Unfortunately, the two proteins were found to co-migrate in SDS-PAGE profile. Under this situation, using one dimensional SDSPAGE and by Western blot, which one of the following is the best way to demonstrate the presence of both the proteins? **(DEC 2014)**

(a) Develop Western blots with their antibodies in the same gel.

(b) Prior to doing SDS-PAGE/Western blot, one protein could be removed by immuno precipitating in the cell extracts.

(c) Silencing the expression of one protein at a time by siRNA and performing Western blotting.

(d) Subjecting the technique of stripping/ re-probbing of the gel after transferring to nitrocellulose membrane while doing Western blotting.

4. In an attempt to follow up the time dependent expression of a new protein "X" following transfection of its gene into an animal cell in culture, total protein was extracted at various time points. Next, it was decided to run the protein mixture in a SDS - PAGE followed

by western blotting to assess its expression (time dependent). A trend of increase in band intensity was observed in the blot. To confirm this increase, one important aspect was missing.

Identify this aspect from the list below.

(DEC 2013 GU)

(a) Equal loading controls

(b) Use of appropriate primary antibody

(c) Use of appropriate secondary antibody conjugated to enzyme

(d) Use low affinity secondary antibody conjugated to enzyme

5. In order to check whether a protein has been phosphorylated during treatment with a drug, you would perform **(JUNE 2017)**

(a) Southern hybridization

(b) Western blot analysis

(c) ChIP assay

(d) RFLP

6. For identification of three proteins moving together (as a single band) upon loading in a single lane of a SDS-PAGE gel the best method is: **(DEC 2015)**

(a) one step Western blot

(b) NMR spectroscopy

(c) Western blot followed by stripping and reprobing

(d) ESR spectroscopy

7. You want to purify a recombinant protein of your interest. You can use affinity chromatography to purify as you have nickel columns available in the laboratory. With what molecule will you tag the protein to purify using those columns?

(DEC 2014)

(a) GST (b) Histidine

(c) Histamine (d) Proline

8. Three students (P, Q, R) in a research lab were trying to identify proteins that interact with a transcription factor X. P performed gel filtration experiments and identified that X was found along with proteins A, B, C and D. Q performed coimmunoprecipitation experiments using antibodies to X and identified A, B and C. R did a yeast-2-hybrid screen and identified only B.

The following are likely conclusions that may explain all the results:

(i) A, B, C and D are in a complex with X.

(ii) X directly interacts with B.

(iii) Only A, Band C are in complex with X.

(iv) D is probably weakly associated with X.

Which of the above conclusions best explains all the results? **(JUNE 2018)**

(a) (i), (ii) and (iii) (b) (i), (ii) and (iv)

(c) (i), (iii) and (iv) (d) (ii), (iii) and (iv)

9. Three proteins *Blm 1, Blm 2, Blm 3* were shown to be involved in repair of DNA double strand breaks. A chromatin immunoprecipitation experiment was performed for the three proteins. The pattern of results obtained is shown below:

	0 Min (before break)			30 Min (after break)		
	Input	0 Kb from break	1 Kb from break	Input	0 Kb from break	1 Kb from break
Blm 1	—	—	—	—	▬	—
Blm 2	—	▬	▬	—	—	—
Blm 3	—	—	—	—	▬	▬

Based on the above figure, choose the option that correctly interprets the data. **(DEC 2018)**

(a) Blm 1, Blm 2, Blm 3 bind to DNA break sites

(b) Blm 1 binds to the break site; Blm 3 binds to the break site and beyond

(c) Blm 2 remains bound to DNA after the break is induced

(d) Blm 3 binds to DNA irrespective of the break

10. A chromatin iminuno precipitation (ChIP) assay was performed to determine specific transcription factor binding sites on the promoter of a gene. Pull down was done using either IgG or anti-bodies against *c-myc*. A DNA containing c-myc binding regions was used as a control for PCR amplification (input). Which one of the following PCR representations of DNA is correct? **(DEC 2014)**

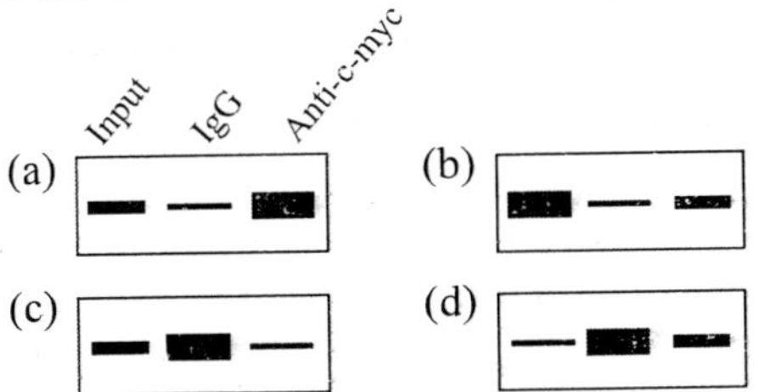

11. A researcher attempted to clone two genes (*X* and *Y*) independently in a plasmid vector for over expression and purification in *E. coli*. All attempts to clone gene *X* were unsuccessful whereas gene '*Y*' could be cloned easily. When the, researcher attempted to clone gene '*X*' in the plasmid clone containing gene '*Y*', gene '*X*' could be cloned. The following statements were proposed to explain the above results.
    A. Protein encoded by gene '*Y*' is not lethal to the cell.
    B. Gene '*X*' has introns, which prevents its expression in *E. coli*
    C. Expression of '*X*' protein is lethal to the cell.
    D. The '*Y*' gene product inhibits the activity of '*X*' protein

    Which one of the following options represents a combination of correct statements to explain the observations? **(JUNE 2018)**
    (a) only A and B (b) B, C and D
    (c) only A and D (d) A, C and D

12. In an experiment to detect a new protein in fixed cells, no secondary antibody tagged with fluorescence dye is available. What should be the best choice out of the following to detect the protein? **(DEC 2014)**
    (a) Protein A-FITC
    (b) Protein A-Sepharose
    (c) Biotin-FITC
    (d) Avidin-FITC

13. A gene is regulated by a novel transcription factor. The following techniques may be used to identify the cis- regulatory element in the 1 kb promoter sequence of the gene where the novel transcription factor binds:
    A. Bioinformatics analysis.
    B. Cell based reporter assay.
    C. S 1 nuclease assay
    D. Electrophoretic mobility shift assay.
    E. DNAse-I foot-printing analysis

    Which one of the following can help to identify the cis element? **(DEC 2013)**
    (a) A and B (b) C and E
    (c) D only (d) E only

14. Based upon phenotypic observation, it was concluded that an unknown gene responsible for an agronomically important trait is present in a particular plant. In order to identify the gene, a researcher proposes to use the following strategies:
    A. PCR amplification of the gene.
    B. Map based cloning of the gene.
    C. Subtractive DNA hybridization
    D. Genome sequencing
    E. Develop molecular markers linked to the trait

    Which one of the following options is most suitable for identifying unknown gene? **(JUNE 2019)**
    (a) A and C (b) B and E
    (c) C only (d) A and D

15. During an experiment; a student found increased activity of a protein, for which there were three possible explanations, viz., increased expression of the protein, increased phosphorylation, or increased interaction with other effector proteins. After conducting several experiments, the student concluded that increased activity was due to increased

phosphorylation. Which one of the following experiments will NOT support/provide the correct explanation drawn by the student?

(DEC 2015)

(a) Western blot analysis
(b) Analysis of transcription rate
(c) Mass spectroscopy
(d) Phospho amino acid analysis

16. Which one of the following techniques will you use to identify more than 1000 differentially expressed genes in and tumor tissues in one experiment? **(DEC 2015)**

(a) RAPD
(b) Genome sequencing
(c) ChiP assay
(d) Transcriptome analysis

17. The presence and distribution of specific mRNAs within a cell can be detected by

(DEC 2016)

(a) Northern blot analysis
(b) RNase protection assay
(c) In situ hybridization
(d) Real-time PCR

18. Which one of the following techniques/ approaches CANNOT be used to analyse expression pattern of a gene?

(DEC 2019 ASSAM)

(a) Western blotting
(b) qRT-PCR
(c) Northern blotting
(d) Southern blotting

19. From the various techniques listed below, which one **CANNOT** be used to precisely map the transcription start-site of a gene?

(JUNE 2018)

(a) S1 Mapping
(b) Sequencing the region downstream of promoter
(c) 5' RACE
(d) Primer Extension Method

20. What will happen if DNA is labeled by nick translation while doing DNA foot- printing?

(JUNE 2015)

(a) Nick translation will facilitate better analysis because entire DNA will be labeled and proteins binding at any region of DNA can be demarcated with precision.
(b) This will allow arranging the DNA fragment in the desired order.
(c) Labeling by random priming may be advantageous as it generates smaller fragments which can penetrate tissue easily.
(d) The linear order of fragments from 5' → 3' end of DNA cannot be arranged.

21. Diagnosis of influenza virus infections can be done using some of the following techniques:

A. Western blot and Southern blot
B. Northern blot and western blot
C. ELISA and RT-PCR
D. PCR and electron microscopy

Choose the combination of techniques that correctly lists the detection methods.

(DEC 2017)

(a) A and B only (b) C and D only
(c) B and C only (d) A and D only

22. Match the technique with its appropriate application/use from the list of options given below **(DEC 2017)**

	Technique		Application
A	ChIp	(i)	Analysis of methylation sites
B	Bisulphite sequencing	(ii)	Identification of 5' and/or 3' ends of transcripts
C	ELISA	(iii)	Identification of binding sites of transcription factors
D	RLM-RACE	(iv)	Quantification of trans gene expression

(a) A-(iii), B-(i), C-(iv), D-(ii)
(b) A-(ii), B-(iii), C-(i), D-(iv)
(c) A-(iv), B-(i), C-(iii), D-(ii)
(d) A-(iii), B-(iv), C-(ii), D-(i)

23. Given below are two sets of terms related to various methods used in biological science.

Column A	Column B
A. RACE	(i) DNA-protein interactions
B. South-Western blotting	(ii) FAM
C. Recursive PCR	(iii) Determining the ends of mRNA
D. TaqMan	(iv) Construction of synthetic DNA

Which one of the following options correctly matches terms of Column A and Column B? **(DEC 2019)**

(a) A-(iv); B-(iii); C-(i); D-(ii)
(b) A-(iii); B-(i); C-(iv); D-(ii)
(c) A-(ii); B-(iv); C-(i); D-(iii)
(d) A-(ii); B-(i); C-(iv); D-(iii)

24. Several fusion constructs were developed to purify heterologous protein in *E. coli*. The table below lists fusion partners and ligands.

	Partner		Ligand
i	Maltose binding protein	a	Specific monoclonal antibody
ii	Streptavidin	b	Nickel
iii	Glutathione-S-transferase	c	Glutathione
iv	Flag-tag	d	Amylose
v	6-Histiden tag	e	Biotin

Which one of the following is the correct match off fusion partner with the ligand? **(JUNE 2017)**

(a) i- b, ii-d, iii-c, iv-a, v-e
(b) i-d, ii-b, iii-e, iv-c, v-a
(c) i-d, ii-e, iii-c, iv-a, v-b
(d) i-c, ii-d, iii-a ,iv-b,v-e

25. If one wishes to design a microarray chip for whole genome expression analysis of an eukaryotic system, which region of the gene should be preferred for selection of unique target sequences? **(JUNE 2012)**

(a) Any region of coding DNA sequence
(b) 3' region from coding DNA sequence (CDS) and 3' untranslated region (UTR)
(c) 5' region of CDS and 5 'UTR
(d) 1st intron only

26. Which one of the following indicate steps involved in DNA foot-printing to monitor interaction of DNA with a protein? **(MODEL PAPER)**

(a) DNA + protein→treat with restriction enzymes → run agarose gel → stain with ethidium bromide and view under UV light.
(b) Label one of the DNA strands with a radiolabel→treat one portion of labelled DNA with the protein of interest followed by reaction with DNAse→treat second portion of DNA with only DNAse. Run both the treated DNA samples on a sequencing gel.
(c) Analyze fragments obtained from the experiments described in (b) by MALDI mass spectrometry.
(d) DNA + protein → treat with DNAse followed by restriction enzymes. Run the sample on SDS page and visualize by commassie blue staining.

27. One aims to find out the role of a gene product in macrophages by using a transgenic mouse expressing the genes under a promoter. Which of the following is the most appropriate promoter? **(JUNE 2013)**

(a) Actin promoter.
(b) MHC Class II promoter.
(c) Mac-1/CD llb promoter.
(d) IL-2 promoter.

28. One of the methods for finding common regulatory motifs present in a set of co-regulated genes is **(DEC 11)**

(a) Prosite
(b) MEME
(c) Mat Inspector
(d) PSSM

29. A bioinformatics tool used to find the sequence similarity in the subunits of hemoglobin is **(JUNE 2014)**

(a) FASTA (b) BLAST
(c) HUMMER (d) PSI:PLOT

30. Choose the correct sequence of events in a next generation sequencing technology-based whole genome sequencing project. **(JUNE 2013)**

(a) DNA extraction → shearing → library preparation → sequencing →assembly → finishing → annotation → submission to Genbank.

(b) DNA extraction → library preparation → sequencing → assembly → annotation → finishing → submission to Genbank.

(c) DNA extraction → shearing → adapter ligation → library amplification → sequencing → assembly→ finishing → annotation → submission to Genbank.

(d) DNA extraction → adapter ligation → library amplification → shearing → sequencing → finishing → assembly →annotation → submission to Genbank.

31. Which one of the following set of essential components are required for Sanger method of DNA sequencing in a required buffer containing $MgCl_2$ and Tris-HCl? **(JUNE 2018)**

(a) DNA template, a primer, 4 deoxyribonucleotides, 4 labelled dideoxyribonucleotides, DNA polymerase

(b) DNA template, a primer, 4 labelled dideoxy ribonucleotides, DNA polymerase, DNA ligase

(c) DNA template, 4 deoxyribonucleotides, 4 labelled dideoxy ribonucleotides, DNA polymerase, DNA ligase

(d) DNA template, a primer, 4 labelled dideoxy ribonucleotides, DNA polymerase, telomerase

32. Figures A and B respectively represent the dideoxy sequencing gels obtained for partial sequences from 5' ends of a bacterial gene and its mutant (with a point mutation). **(DEC 2011)**

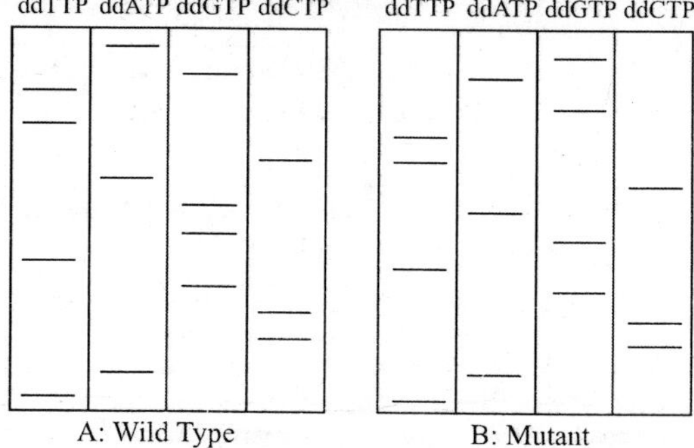

What type of mutation has occurred in the gene?

(a) Nonsense (b) Missense
(c) Frameshift (d) Transversion

33. Given below are four statements on various methods used in molecular biology and recombinant DNA studies:

A. Orientation of a cloned DNA fragment in a plasmid vector can be determined by using primers specific to the cloned fragment.

B. Transcriptome sequencing cannot be used to identify differential expression profiles of mRNA in an organism.

C. A DNA fragment with 5' overhangs can be converted into a blunt-ended fragment by using a nuclease but not by a DNA polymerase

D. Oligonucleotide probes can be labelled at their 5 and 3' ends by using T4 polynucleotide kinase and terminal transferase, respectively

Which one of the following options consists of only correct statement/s? **(DEC 2019 ASSAM)**

(a) A and C only (b) B only
(c) C and D only (d) D only

34. Duschenne muscular dystrophy (DMD) is caused by a defective dystrophin gene and can be cured by removing a specific faulty exon. If a CRISPR/Cas9 based genome editing approach is to be designed to remove the faulty exon, some of the following conditions have to be met.

A. Two guide RNA molecules to match the flanking regions of the faulty exon
B. An efficient non-homologous end joining system in the host
C. Two separate, specific Cas9 endonucleases to target both ends of the exon
D. A single guide RNA that targets the intron preceding the exo

Choose the option that includes all the correct conditions. **(DEC 2019 ASSAM)**

(a) A, B and C only (b) A and B only
(c) C and D only (d) A and C only

Answer Key

1.(c)	2.(b)	3.(d)	4.(a)	5.(b)	6.(c)	7.(b)
8.(b)	9.(b)	10.(b)	11.(d)	12.(a)	13.(d)	14.(b)
15.(b)	16.(d)	17.(c)	18.(d)	19.(b)	20.(d)	21.(c)
22.(a)	23.(b)	24.(c)	25.(b)	26.(b)	27.(c)	28.(b)
29.(b)	30.(c)	31.(a)	32.(c)	33.(d)	34.(b)	

3. Isolation, separation and analysis of carbohydrate and lipid molecules

1. A researcher wants to obtain complete chemical information, i.e., head groups and fatty acids of phospholipids from liver tissues. Phospholipids have fatty acids of different lengths and unsaturation and also the head groups are of different chemistries. Which of the following combination of techniques would provide complete chemical description of phospholipids? **(JUNE 2016)**
   (a) Only thin layer chromatography (TLC)
   (b) TLC and gas chromatography
   (c) Paper and thin layer chromatography
   (d) Only paper chromatography
2. Which among the following is the simplest method to estimate the concentration of glycerol in an aqueous solution of glycerol? **(DEC 2015)**
   (a) UV absorption spectroscopy
   (b) Gas chromatography
   (c) pH measurement
   (d) Viscosity measurement
3. Small molecular weight compounds affect the activity of luciferase differently. The oil/water solubility of various compounds is one property important for its effect on luciferase. The straight line in the graph was obtained by plotting the activity data with 50 different compounds. The luciferase activity of two new derivatives of Benzene (A and B) are shown below:

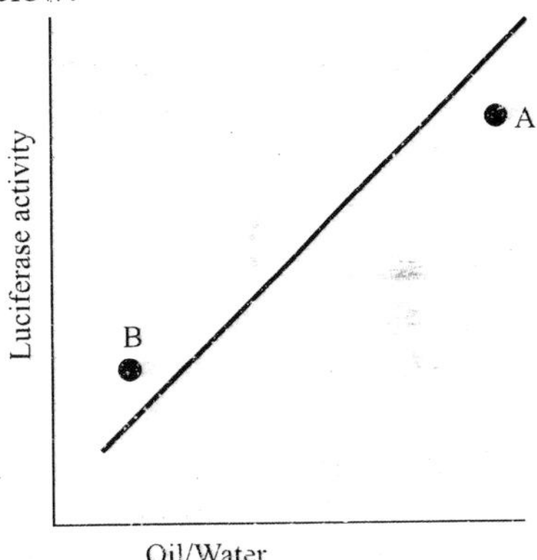

Which of the following statements is correct? **(DEC 2014)**

(a) A is phosphate and B is amine
(b) A is methyl and B is amine
(c) A is methyl and B is propyl
(d) A is amine and B is phosphate

Answer Key

1.(b)	2.(d)	3.(b)

4. RFLP, RAPD and AFLP techniques/ PCR

1. In an experiment designed to clone a PCR-amplified fragment in a cloning vector digested with XhoI (C/TCGAG) and SmaI (CCC/GGG), which one of the following combinations

of restriction enzymes can be used in the PCR primer to generate compatible ends for cloning?

(' /' indicates the site of cleavage within the recognition sequence) **(JUNE 2017)**

(a) XbaI (T/CT AGA) and SpeI (A/CT AGT)
(b) EcoRI (G/ AA TIC) and SmaI (CCC/GGG)
(c) SalI (G/TCGAC) and EcoRV (GAT/ATC)
(d) HindIII(A/AGCTT) and PvuII (CAG/TAG)

2. Molecular polymorphic markers are already known with respect to tobacco mosaic virus (TMV) resistance in tobacco. Among these, which marker system you will select that will be simple, economic and less time consuming: **(JUNE 2016)**

(a) RAPD (b) RFLP
(c) AFLP (d) EST-SSR

3. Which one of the following statements is correct for amplified-fragment length polymorphism (AFLP)? **(JUNE 2016)**

(a) PCR using a combination of random and gene-specific primers.
(b) PCR amplification followed by digestion with restriction enzymes.
(c) Digestion of DNA with restriction enzymes followed by one PCR step.
(d) Digestion of DNA with restriction enzymes followed by two PCR steps.

4. Which one of the following molecular marker types uses combination of both restriction enzyme and PCR techniques? **(JUNE 2014)**

(a) SSR (b) AFLP
(c) SNP (d) RAPD

5. Combination of molecular markers with their classification based on either dominant or co-dominant types are shown below:

A. SSR and RFLP: co-dominant.
B. SSR and RAPD: co-dominant.
C. RAPD and RFLP: dominant.
D. AFLP and RAPD: dominant.

Which one of the following is the correct combination? **(JUNE 2014)**

(a) A and B (b) B and C
(c) C and D (d) A and D

6. PCR based DNA amplification is an essential feature of which of the following combination of molecular markers? **(MODELPAPER)**

(a) RFLP, AFLP and SSR.
(b) AFLP, SSR and RAPD.
(c) RFLP, RAPD and SSR.
(d) RAPD, RFLP and SSR.

7. Which of the following statement about RFLP (restriction fragment length polymorphism) and AFLP (amplified fragment length polymorphism) is NOT correct? **(DEC 2013 GU)**

(a) Southern hybridization is used only for RFLP.
(b) AFLP is a co-dominant molecular marker.
(c) AFLP can generate DNA profiles regardless of the complexity or origin of the DNA molecule.
(d) only RFLP requires prior information about DNA sequence.

8. A student wrote following statements regarding comparison of Restriction Fragment Length Polymorphism (RFLP), Random Amplified Polymorphic DNA (RAPD), Amplified Fragment Length Polymorphism (AFLP) and Simple Sequence Repeats (SSRs) techniques used for generating molecular markers in plants:

A. All these techniques can be used for fingerprinting.
B. Detection of allelic variation can be achieved only by RFLP and SSRs.
C. Use of radioisotopes is required in RFLP and RAPD only.
D. Polymerase chain reaction is required for all the techniques.

Which one of the following combination of above statements is correct? **(DEC 2012)**

(a) A and B (b) B and C
(c) C and D (d) D and A

9. RFLP and RAPD markers are commonly used in plant breeding and diversity analysis. Which of the following combination of statement about these molecular makers are correct? **(MODEL PAPER)**

A. RFLP is co dominant
B. RAPD is co dominant
C. Both the markers are ubiquitous
D. Only RFLP can detect heterozygote
E. Only RAPD can detect heterozygote
F. RAPD cannot detect allelic variation

(a) A, C, D, E (b) B, C, E, F
(c) A, C, D, F (d) C, D, E, F

10. Given below are various types of molecular markers in Column A and properties of these markers in Column B.

	Column A		Column B
A.	RFLP	(i)	Single locus
B.	SSR	(ii)	Multi-allelic
C.	AFLP	(iii)	Co-dominant
D.	RAPD	(iv)	Single-allelic
		(v)	Multi-locus
		(vi)	Dominant

Which one of the options given below correctly matches the molecular markers with their properties? **(DEC 2019)**

(a) A-(vi), B-(i), C-ii), D-(v)
(b) A-(v), B-(ii), C-(iv), D-(iii)
(c) A-(i), B-(ii), C-(v), D-(vi)
(d) A-(ii), B-(iii), C-(i), D-(ii)

11. Given below is a marker profile for two parental lines (P_1 and P_2) and their derived F_1 progeny: **(DEC 2018)**

P_1 P_2 P_1

The marker that is represented in the above figure is most likely to be

(a) RFLP or SSR (b) SSR only
(c) SSR or RAPD (d) RAPD only

12. Molecular beacons (MB) and Taqman (TQ) are used as probes in Real time PCR experiments. Both these probes are based on the principle of FRET and employ a fluorophor (F) and a quencher (Q). However the mechanisms by which they function are different as illustrated below **(JUNE 2011)**

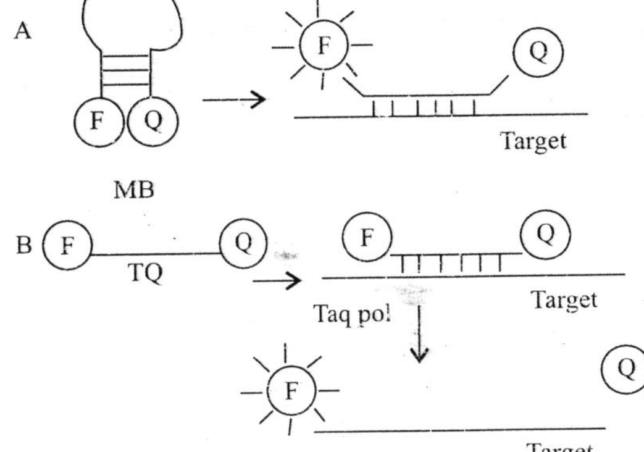

At what stage of the PCR we would be able to detect fluorescence?

(a) Annealing step for both.
(b) Extension step for both.
(c) Annealing for A and Extension for B.
(d) Extension for A and Annealing for B.

Answer Key

1.(c)	2.(d)	3.(d)	4.(b)	5.(d)	6.(b)	7.(b)
8.(a)	9.(c)	10.(c)	11.(d)	12.(c)		

5. ELISA, RIA, western blot, FISH and GISH.

1. Which one of the following analytical techniques does NOT involve an optical measurement? **(DEC 2016)**

(a) ELISA
(b) Microarray

(c) Flow cytometry
(d) Differential Scanning Calorimetry

2. Given below are a few factors that influence the sensitivity of immunoassays:
A. Affinity of the antibody used
B. Epitope density on the antigen
C. Epitope distribution on the antigen
Which one of the following options represents the correct factor(s)?
(DEC 2019 ASSAM)
(a) A only (b) B only
(c) B and C only (d) A, B and C

3. Radioimmuno assay (RIA) can be employed for the detection of insulin in blood plasma. For this, ^{125}I labelled insulin is mixed and allowed to bind with a known concentration of anti-insulin antibody. A known volume of patients' blood plasma is then added to the conjugate and allowed to compete with the antigen binding sites of antibody. The bound antigen is then separated from unbound ones and the radioactivity of free antigen is then measured by gamma counter. Following are some of the statements made about this assay.
(i) The ratio of radioactive count for unbound antigen to the bound one is more at the end of reaction.
(ii) The ratio of radioactive count for unbound antigen to the bound one is less at the end of reaction.
(iii) For a diabetic patient, the radioactive count for free antigen is less than that for a normal individual.
(iv) For a diabetic patient, the radioactive count for free antigen is more than that for a normal individual.
Which of the above statements are true?
(JUNE 2016)
(a) (i) and (iii) (b) (i) and (iv)
(c) (ii) and (iii) (d) (ii) and (iv)

4. In Radio Immuno Assay (RIA) for glucocorticoid hormone, glucocorticoid (tritiated) is added to RIA cocktail. When the amount of bound hormone was measured no radioactive counts were observed. The following possible explanation(s) were proposed:
A. The radioactive hormone was insufficient
B. The radioactive tag to the hormone completely dissociated during storage
C. Antibody was not added to the cocktail
D. The specific activity of tritium was low
Choose the correct options(s): **(JUNE 2012)**
(a) A only (b) C only
(c) B and C (d) A and D

5. ELISA assay uses **(JUNE 2011)**
(a) an enzyme which can react with secondary antibody.
(b) an enzyme which can react with the antigen.
(c) a substrate which gets converted into a coloured product.
(d) a radiolabelled secondary antibody

6. A researcher would like to monitor changes in the level of a serum protein for, which an antibody is available. Which one of the following methods would be best suited for the purpose? **(JUNE 2015)**
(a) Immunofluorescence microscopy
(b) Fluorescence in situ hybridization
(c) Enzyme linked immunosorbent assay.
(d) Fluorescence activated cell sorting

7. In order to most efficiently quantitate an antigen available at low concentration in a biological fluid, which of the following ELISA would be preferred? **(DEC 13 GU)**
(a) Coating the plate with the fluid containing the antigen, followed by antigen-specific antibody-enzyme conjugate.
(b) A sandwich ELISA, capturing the antigen using an antibody specific for the antigen.
(c) A sandwich ELISA, capturing the antigen using an antibody and detecting the antigen with an antibody, which is biotinylated.

(d) Plating the biological fluid on an ELISA plate coated with the antigen, followed by regular steps of ELISA

8. The use of biotinylated secondary antibody in ELISA? **(JUNE 2012)**

(a) increase the sensitivity of assay but compromises the specificity

(b) increases the sensitivity of assay without compromising the specificity

(c) does not alter either sensitivity or specificity

(d) decreases both sensitivity and specificity

9. In order to detect minor variations in antigen concentration, the following procedures were suggested. Which one will likely be the best option? **(JUNE 2018)**

(a) Antigen coated microtitre well → add antibody → add enzyme conjugated secondary antibody → add substrate and measure colour.

(b) Antibody coated microtitre well → add antigen → add enzyme conjugated secondary antibody → add substrate and measure colour.

(c) Preincubate antigen with fixed amount of antibody → add to antigen coated well → add enzyme conjugated secondary antibody → add substrate and measure colour.

(d) Preincubate antigen with fixed amount of antibody → add to antibody coated well → add enzyme conjugated secondary antibody → add substrate and measure colour.

10. For ELISPOT measurements of interferon γ [IFN-γ] secretion by mouse NKT cells, the following steps were performed in an ELISA plate:

A. Addition of purified spleenocytes

B. Stimulation of spleenocytes by peptide pulsed antigen presenting cells (APCs)

C. Addition of biotin-conjugated anti IFN-γ antibody

D. Addition of capture antibody specific to IFN-γ

E. Addition of avidin-conjugated colour changing substrate.

Which one of the following options lists the correct sequence of steps performed?

(DEC 2019 ASSAM)

(a) A→D→B→C→E

(b) D→B→A→C→E

(c) D→ A→ B→C→E

(d) B→A→D→C→E

11. Indirect immunofluorescence involves fluorescently labeled **(DEC 2011)**

(a) Immunoglobulin-specific antibodies

(b) antigen-specific antibodies

(c) hapten-specific antibodies

(d) carrier-specific antibodies

12. A student hypothesized that protein Yfg100 interacted with protein Yfr70 in a tissue-specific manner. To test this, an immunoprecipitation experiment was performed using antibodies to Yfg 100 and the samples were subjected to western blot using antibodies to both Yfg100 and Yfr70. The results obtained are shown below

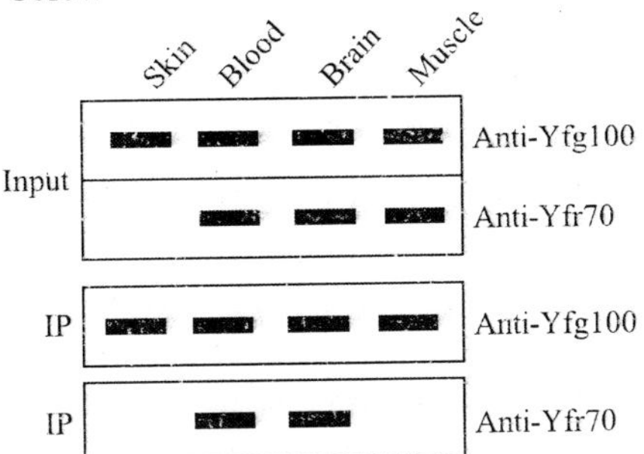

Based on the results shown, identify the INCORRECT statement

(DEC 2019 ASSAM)

(a) Protein Yfg100 is expressed in all tested tissues

(b) Yfg100 does not interact with Yfr70 in muscle

(c) Yfg100 interacts with Yfr70 in muscle but not in other tissues.

(d) Yfr70 is not expressed in skin

13. As cancer progresses, several genome rearrangements including translocation, deletion, duplications etc. Occur. If these rearrangements are to be identified, which of the following techniques would be most suitable? **(DEC 2015)**

(a) RAPD

(b) Microarray

(c) Multi-colour FISH

(d) Flow cytometry

14. Which of the following statements is INCORRECT for fluorescence in situ hybridization (FISH) technique? **(DEC 2014)**

(a) A fluorescence or confocal microscope is used for detection of signal

(b) A labeled sequence of nucleotides is used

(c) Specific fluorescence tagged antibodies are used

(d) A stringent washing step is essential to remove appearance of non-specific signal.

15. Given below are the experimental protocols to find out the exact location of repetitive DNA sequence in mitotic chromosome by FISH (fluorescence in situ hybridization). Which one of the protocols will give the correct result? **(DEC 2013)**

(a) Mitotic chromosomes were fixed on glass slide → incubated with biotinylated telomeric DNA → denatured → incubated with fluorescently labeled avidin → localization observed under fluorescence microscope.

(b) Mitotic chromosomes were fixed on glass slide → denatured → incubated with FITC labeled unrelated non- repetitive DNA sequence → counterstained with propidium iodide -> localization observed under fluorescence microscope.

(c) Mitotic chromosomes were fixed on glass slide → denatured → incubated with biotinylated satellite DNA→ incubated with fluorescently labeled avidin → localization → observed under fluorescence microscope.

(d) mitotic chromosomes were fixed on glass slide → incubated with repetitive DNA sequence binding protein → denatured → FITC labeled antibody against the protein → localization observed under fluorescence microscope.

Answer Key

1.(d)	**2.**(d)	**3.**(a)	**4.**(c)	**5.**(c)	**6.**(c)	**7.**(c)
8.(b)	**9.**(c)	**10.**(c)	**11.**(a)	**12.**(c)	**13.**(c)	**14.**(c)
15.(c)						

6. UV/visible spectroscopy

1. Which one of the following would contribute to intrinsic fluorescence to a protein? **(DEC 2015)**

(a) aromatic amino acids

(b) disulfide bonds

(c) charged amino acids

(d) branched chain amino acids

2. Which one of the methods listed below is the most sensitive label-free quantification method fr proteins **(JUNE 2015)**

(a) UV spectroscopy

(b) Infra-red spectroscopy

(c) Raman spectroscopy

(d) ^{13}C content of protein

3. A protein has one tryptophan and one, tyrosine in its sequence. Assume molar extinction coefficients at 280 nm of tryptophan and tyrosine as 3000 and 1500 $M^{-1}cm^{-1}$, respectively. What would be the molar concentration of that protein if its absorption at 280 nm is 0.90? **(JUNE 2015)**

(a) 2 mM (b) 0.4 mM
(c) 0.2 mM (d) 0.02 mM

4. Molar absorption coefficient of phenylalanine is 200 $M^{-1}cm^{-1}$ at 257 nm. What concentration (g/L) of this amino acid will give an absorption of 1 in a cell of 0.5-cm path length at 257nm? **(JUNE 2013)**

(a) 3.30 (b) 0.33
(c) 1.65 (d) 0.17

5. Optical density of a 400 base pair long 1 ml DNA solution was found to be 0.05(b) How many DNA molecules are present in the solution? [1 base pair=650 dalton, optical density of 1.0 D corresponds to 50 μg DNA/ml] **(DEC 2012)**

(a) 6.023×10^{12} (b) 6.023×10^{13}
(c) 4.633×10^{18} (d) 5.2×10^{13}

6. A fluorophore when transferred from solvent A to solvent B results in an increase in the number of vibrational states in the ground state without any change in the mean energies of either the ground or excited state. What would be the change seen in the fluorophore's emission spectrum? **(DEC 2012)**

(a) An increase in emission intensity.
(b) An increase in emission bandwidth.
(c) An increase in emission wavelength.
(d) A decrease in emission wavelength.

7. A protein contains 2 Trp and 4 Tyr residues. The molecular mass of the protein is 17000 D and that of Trp and Tyr are 204 and 180 D respectively. Values of $E^{1\%}_{1cm}$, the absorption coefficient of 1% (g/v) solutions of Trp and Tyr in 1-cm cell at 280 nm, are 269.60 and 83.33, respectively. The absorption of 1 mg/ml protein solution in 1 cm-cell at 280 nm will be: **(JUNE 2012)**

(a) 0.1 (b) 1.0
(c) 0.7 (d) 1.7

8. The values of molar absorption coefficient (ε) of Trp and Tyr at 240 nm and 280 nm are the following:

Wavelength (nm)	E_{Tyr} (M^{-1} cm^{-1})	ε_{Trp} (M^{-1} cm^{-1})
240	11,300	1,960
280	1,500	5, 380

A 10-mg sample of a protein is hydrolyzed to its constituent amino acids and diluted to 100 ml. The absorption of this solution in a 1-cm path length, is 0.717 at 240 nm and 0.239 at 280nm.

The estimated content of Trp and Tyr in μmol / g protein respectively are **(MODEL PAPER)**

(a) 586 and 28.1. (b) 58.6 and 281.
(c) 586 and 281. (d) 58.6 and 28.1.

9. A solution contains NADH and NAD^+, BOTH AT 0.1 mM concentration. If NADH has a molar extinction co efficient of 6220 and that of NAD^+ is negligible, the optical density measured in a cuvette of 5mm path length will be **(JUNE 2019)**

(a) 0.62 (b) 0.062
(c) 0.31 (d) 0.031

10. The emission maximum of tryptophan fluorescence in a protein is ~ 335 nm. This suggests that tryptophan **(JUNE 2019)**

(a) is in a hydrophobic environment.
(b) occurs in a helical segment
(c) has proximal cysteine residues
(d) is oxydized.

11.

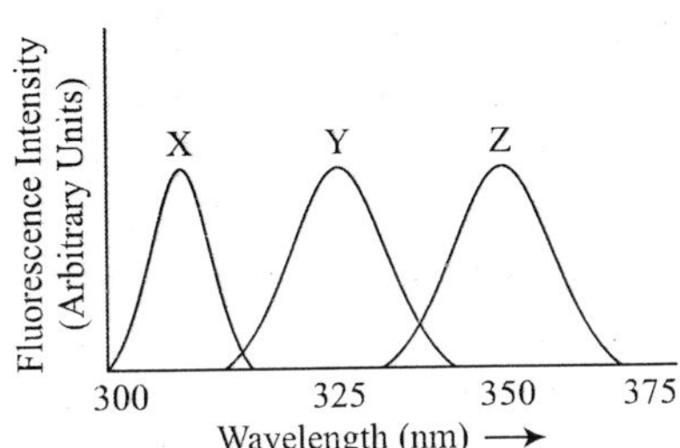

The above figure shows the fluorescence emission spectra of three different proteins; Protein (X), Protein (Y), and Protein (Z) excited at 280 nm. Which one of the following statements gives the correct interpretation?

(a) Proteins (Y) and (Z) have tryptophan while protein (X) has only phenylalanine.

(b) Protein (X) has only tyrosine and protein(Y) has tryptophan on the surface while protein (Z) has tryptophan buried inside.

(c) Protein (X) has tryptophan buried inside while proteins (Y) and (Z) have tryptophan on the surface.

(d) Protein (X) has only tyrosine and protein (Y) has tryptophan buried and protein (Z) has tryptophan on the surface.

12. Absorption spectra of L-tyrosine in acidic (continuous line) and basic (dotted line) medium was estimated and plotted on a graph as depicted below:

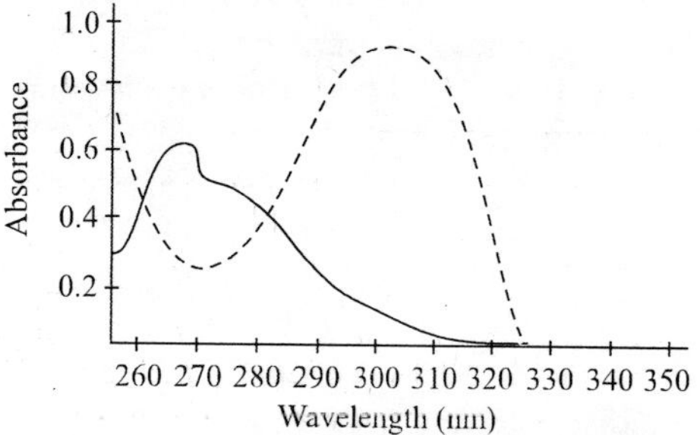

Following interpretations were made:

A. Change in the pH from acidic to basic results in shift in the lowest energy absorption maximum and decrease in the molar absorptivity.

B. Shifting of the absorption band to longer wavelength signifies a shift to lower energy, also known as red shift

C. Shifting of the absorption band to shorter wavelength signifies a shift to higher energy, also known as blue shift.

D. Wavelength shift is always accompanied by change in intensity of the absorption band.

Select the combination with correct interpretations. **(DEC 2019)**

(a) A and B (b) A and C

(c) B and C (d) B and D

Answer Key

1.(a)	2.(a)	3.(c)	4.(c)	5.(a)	6.(2)	7.(a)
8.(c)	9.(c)	10.(a)	11.(d)	12.(c)		

7. NMR, ESR, SPR spectroscopy

1. Point group symmetry operations such as inversion and mirror plane are not applicable to protein crystals. This is because **(JUNE 2017)**

(a) protein molecules assemble in highly ordered fashion

(b) protein molecules have handedness.

(c) protein molecules form a lattice plane that do not diffract X-rays

(d) hydrogen atoms in proteins diffract weakly.

2. A 30-residue peptide containing Phe, Tyr and Trp is dissolved in D_2O and the high field proton NMR is recorded after 24 hours. The resonances that are unlikely to be present are **(DEC 2019 ASSAM)**

(a) aromatic protons (b) C^{α} protons

(c) aliphatic protons (d) amide protons

3. The Nuclear Magnetic Resonance (1D and 2D) spectrum of a 30-residue peptide were recorded at 25°C. The following observations were made.

A. The NH and $C^{\alpha}H$ resonances were well resolved

B. The NOESY spectra showed extensive N_i - N_{i+1} connectivities

C. The NH resonances showed slow exchange with deuterium

The spectra indicates that the peptide adopts **(JUNE 2017)**

(a) Helical conformations

(b) Anti-parallel β-strand conformations

(c) Polyproline conformation

(d) β -turn conformation with four amino acids participating in the tum. Rest of the amino acids are unstructured

4. The hydrogen atoms in the δ (delta) methylene group of lysine will give the following splitting pattern in the ^{1}H-NMR spectra of lysine **(JUNE 2014)**
   (a) Triplet of triplets.
   (b) Quintet.
   (c) Doublet of triplets
   (d) Triplet of a doublet
5. Which of the following atomic nuclei cannot be probed by nuclear magnetic resonance spectroscopy? **(JUNE 2013)**
   (a) ^{1}H. (b) ^{31}P.
   (c) ^{18}O. (d) ^{15}N.
6. The conformation of a 30-residue peptide is studied by NMR spectroscopy. The JNH for most of the amide protons is 4Hz. The 2D NOESY spectrum shows prominent Ni-Ni^{+1} connectivities. The conformation of the peptide is **(MODELPAPER)**
   (a) anti-parallel β sheet.
   (b) parallel β sheet.
   (c) helix-like.
   (d) unordered.
7. Detailed NMR spectra of a 20-residue peptide were recorded using a 600 MHz instrument. If the peptide adopts an αhelical conformation, which one of the following statements is correct? **(DEC 2018)**
   (a) Prominent NH_i - NH_{i+1} NOE peaks would be observed along with $^3J_{NH}$ - HA coupling constants 8.5 Hz
   (b) Prominent $C_\alpha H_i$ - NH_{i+1} NOE peaks would be observed along with $^3J_{NH}$ - HA coupling constants 4.8 Hz
   (c) Prominent $C_\alpha H_i$ - NH_{i+1} NOE peaks with $^3J_{NH-HA}$ coupling constants 8.5 Hz
   (d) Prominent NH_i - NH_{i+1} NOE peaks along with $^3J_{NH-HA}$ coupling constants 4.8Hz
8. Following are statements related to spectroscopic investigation of proteins.
   A. Tryptophan fluorescence in a protein is not sensitive to its environment.
   B. Observation of a large number of N_i - N_{i+1} connectivity's in the NOESY spectrum of a protein suggests the presence of helical conformation.
   C. Only proteins with masses less than 5000 daltons can be identified by MALDI massspectrometry.
   D. Protein conformation can be investigated by ESI mass spectrometry.

   Which one of the following options consists of both correct statements? **(DEC 2019 ASSAM)**
   (a) A and C (b) B and D
   (c) A and B (d) B and C
9. The structure of a protein with 100 residues was determined by X-ray analysis at atomic resolution and NMR spectroscopy. The following observations are possible.
   A. The dihedral angles determined from the X-ray structure and NMR will be identical.
   B. The dihedral angles determined from the X-ray structure will be more accurate.
   C. β-turns can be determined only by NMR.
   D. β-sheets can be more accurately determined from the X-ray structure.

   Indicate the combination with ALL correct answers
   (a) A and C (b) B and D
   (c) B and C (d) A and D
10. Which one of the following can be analysed using Surface Plasmon Resonance method? **(JUNE 2016)**
    (a) Radiolabelled DNA probes.
    (b) Protein structure.
    (c) Optical density of a solution.
    (d) Label-free bimolecular interaction.

Answer Key

1.(b)	2.(d)	3.(a)	4.(a)	5.(c)	6.(c)	7.(d)
8.(b)	9.(b)	10.(d)				

8. Statisitcal Methods

1. Which one of the following statements regarding normal distribution is NOT correct? **(JUNE 2019)**
   (a) It is symmetric around the mean
   (b) It is symmetric around the median
   (c) It is symmetric around the variance.
   (d) It is symmetric around the mode.
2. A weed is assumed to be dispersed randomly in a meadow. What statistical distribution will describe the dispersion correctly? **(DEC 2013)**
   (a) Binomial (b) Negative Binomial
   (c) Poisson (d) Normal
3. A researcher samples n individuals randomly from a population of blackbuck and identifies their sex. The number of females in the sample follows **(DEC 2019)**
   (a) an exponential distribution
   (b) a binomial distribution
   (c) a Poisson distribution
   (d) a normal distribution
4. Given below are names of statistical distribution (Column I) and their characteristic features (Column II)

	Column I		Column II
A	Binomial distribution	i	Each observation represents one of two outcomes (success or failure)
B	Poisson distribution	ii	Probability that is symmetric about the mean
C	Normal distribution	iii	Probability of a given number of events happening in a fixed interval of time

   Which one of the following represents a correct match between columns I and II? **(DEC 2018)**
   (a) A - (ii) ; B - (i) ; C - (iii)
   (b) A - (i) ; B - (ii) ; C - (iii)
   (c) A - (i); B - (iii) ; C - (ii)
   (d) A - (iii) ; B - (ii); C - (i)
5. Choose the correct answer from the statements indicated below: **(DEC 2018)**
   (a) Chi square test is parametric.
   (b) Non-parametric test assumes normal distribution.
   (c) Results can be significantly affected by outliers in a parametric test.
   (d) Non-parametric test is more powerful as compared to parametric test.
6. Following are statements related to statistical methods:
   A. An outlier can be defined as a value in a data set that lies more than three standard deviations from the mean.
   B. Measures of central tendency and dispersion are independent of the presence of outliers in a data set.
   C. Standard deviation is a measure of dispersion.
   D. Mean, median and mode are not measure of central tendency.

   Which one of the following options is a combination with both INCORRECT statements? **(DEC 2019 ASSAM)**
   1. A and B 2. B and C
   3. A and C 4. B and D
7. In the following statement taken from a research paper, what does p in the parenthesis stands for? **(DEC 2011)**
   "The mean temperature of this region now is significantly higher than the one 50 years ago ($p<0.05$, t-test)"
   (a) Ratio of the mean temperature of the two times periods tested
   (b) Probability of the error of rejecting a true null hypothesis
   (c) Probability of the error of accepting a false null hypothesis
   (d) Probability of the *t*-test being effective in detecting significant difference in the mean annual temperatures of the two periods
8. Given below are a set of statistical methods/ parameters (Column A) and their potential applications/utility in biological, research (column B), in a random manner.

Column A		Column B	
A	Variance	(i)	Measure strength of association between two variables
B	Correlation coefficient	(ii)	Prediction of value of a dependent variable based on known value of an associated variable
C	Regression analysis	(iii)	Calculation of deviation between observed and expected values
D	Chi-square analysis	(iv)	Calculate the spread of a distribution

Which of the following options is a correct match of entries in Column A and B? **(JUNE 2017)**

(a) A - (ii), B -(iv), C - (i), D - (iii)
(b) A - (iii), B - (ii), C - (iv), D - (i)
(c) A -(iv), B-(i), C-(ii), D-(iii)
(d) A - (i), B - (iii), C - (ii), D - (iv)

9. The use of Kruskal Wallis test is most appropriate in which of these cases? **(JUNE 2016)**

(a) There are more than two groups and each group is normally distributed.
(b) There are more than two groups and the distribution in each group is not normal.
(c) There are two groups and each group is normally distributed.
(d) There are two groups and the distribution in each group is not normal.

10. Two groups (Control, Treated) are to be compared to test the effect of a treatment. Since individual variability is high in both groups, the appropriate statistical test to use is **(JUNE 2015)**

(a) Analysis of variance.
(b) Kendall's test.
(c) Student's t-test.
(d) Mann-Whitney U-test.

11. If 'r' denotes correlation coefficient and 'm' denotes slope of regression line, interchanging X and Y axes would **(JUNE 2012)**

(a) change 'm' but not 'r'
(b) change 'r' but not 'm'
(c) changes both 'r' and 'm'
(d) nor change 'r' or 'm'

12. Given below are sampling techniques and their features.

Sampling technique		Features	
A.	Straitified sampling	(i)	Population is divided into groups/clusters and a sample of groups/clusters is chosen using a probability method
B.	Systematic sampling	(ii)	Population is divided into groups/clusters and within each group/cluster, a p[robability sample is selected from it.
C.	Opportunity sampling	(iii)	Only participants available and willing to participate are used.
D.	Clustered sampling	(iv)	Samples are selected at regular intervals from the p[opulation

Which one of the following options correctly matches sampling techniques with their features? **(DEC 2019 ASSAM)**

(a) A-(ii); B-(i); C-(iv); D-(iii)
(b) A-(ii); B-(iv); C-(iii); D-(i)
(c) A-(i); B-(iv); C-(iii); D-(ii)
(d) A-(i); B-(iv); C-(ii); D-(iii)

13. The frequency distribution of tree heights in two forest areas with different annual rainfall are given

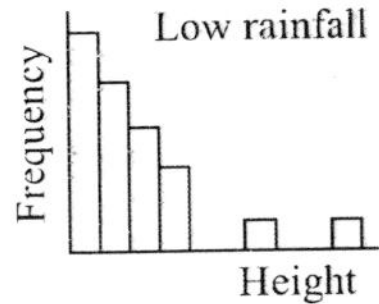

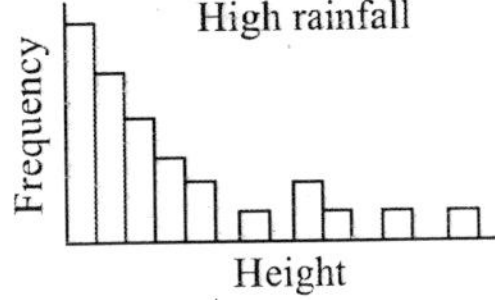

Which of the following statistical analysis will you choose to test whether rainfall has an effect on tree heights? **(JUNE 2013)**

(a) t test for comparison of means.
(b) A non-parametric comparison of the two groups

(c) Correlation analysis of rainfall and mean tree heights.

(d) Regression of tree heights on rainfall.

14. Following are statements to depict relationship among measures of central tendency in a skewed dataset

A. In positively skewed datasets, mean > median > mode

B. In positively skewed datasets, mode >median > mean

C. In negatively skewed datasets, mean> median > mode

D. In negatively skewed datasets, mode> median> mean

Which of the above statements are TRUE?

(JUNE 2018)

(a) A and B (b) A and C

(c) B and D (d) A and D

15. Female fiddler crabs prefer male fiddler crabs with larger claws over males with smaller claws. If the selection pressure exerted is strong resulting in a skewed distribution of claw size, which of the following statements is true about the population's mean, median and mode? **(DEC 2019, ASSAM)**

(a) Mean > Median > Mode

(b) Mean < Median < Mode

(c) Mean = Mode < Median

(d) Mean = Median = Mode

16. The pH of a solution is 7.4 ± 0.02 where 0.02 is standard deviation obtained from eight measurements. If more measurements were carried out, the % of samples whose pH would fall between pH 7.38 and 7.42 is **(JUNE 2017)**

(a) 99.6 (b) 95.4

(c) 68.2 (d) 99.8

17. The μ and σ of wing length (a normally distributed parameter) in a population of fruitflies are 4 and 0.2 mm. respectively. In a random sample of 400 fruitflies, how many individuals are expected to have wing lengths greater than 4.4 mm? **(JUNE 2011)**

(a) 20 (b) 64

(c) 10 (d) 336

18. The mean and standard deviation of serum cholesterol in a population of senior citizens are assumed to be 200 and 24mg/dl, respectively. In a random sample of 36 senior citizens, what values of cholesterol (to the nearest whole number) should lead to rejection of the null hypothesis at 95% confidence level?

(JUNE 2015)

(a) above 224

(b) above 248

(c) below 176 and above 224

(d) below 192 and above 208

19. It is hypothesized that the mean (μ_0) dry weight of a female in a Drosophila population is 4.5 mg. In a sample of 16 female with $\bar{Y} = 4.8$ mg and $s = 0.8$ mg, what dry weight values would lead to rejection of the null hypothesis at $p = 0.05$ level? (take $t_{0.05}$=2.1) **(JUNE 2016)**

(a) Values lower than 4.0 and values higher than 5.6

(b) Values lower than 3.20 and values higher than 6.40

(c) Values lower than 4.38 and values higher than 5.22

(d) Values lower than 3.22 and values higher than 6.48

20. It is claimed that the mean (μ) arsenic concentration in the ground water of village is 20µg/L with $\sigma = 3$. In a random sample of 16 measurements what values of arsenic concentration should lead to rejection of the claim with 95% confidence?

(MODEL PAPER)

(a) values lower then 18.53 and values higher that 21.47

(b) values higher than 21.47

(c) values lower than 18.53

(d) values lower than 18.07 and values higher than 21.93

21. The number of seeds in the fruit of a plant species, Ho : μ=30. A random sample of 9 fruits gives the mean number of seeds as 24 with a standard deviation of 6.12.
    (a) What are the confidence limits for the sample mean?
    (b) Would your reject or accept the null hypothesis at 95% confidence level? **(DEC 2014)**

(a) (a) 18 and 30, (b)reject the hypothesis
(b) (a) 20 and 28, (b) reject the hypothesis
(c) (a) 20 and 28, (b) accept the hypothesis
(d) (a) 18 and 30, (b) accept the hypothesis

22. In the population of an insect species, 50% are known to be female. How many females should turn up in a random sample of 40 insects to reject the null hypothesis (χ^2 value for rejection is $\geq$3 .84)? **(DEC 2013 GU)**

(a) 6 (b) $\leq$26 or $\leq$8
(c) 25 (d) < 7 or > 27

23. The Triver-Willard hypothesis states that the physiological state of a female can bias the sex ratio of offspring. In an experiment in the bird species a group of females were fed a diet 30% lower in calories than the control females. After allowing both the groups to mate and breed freely, the offspring of control 1 group were 22 males and 18 females. The diet restricted females laid a total of 40 eggs.
What should be the minimum deviation from the control to conclude that they have significantly female biased offspring sex ratio. (Chi sq [0.05] df=1 is 3.84) **(DEC 2012)**
(a) 18 male 22 female
(b) 20 male 20 female
(c) 15 male 25 female
(d) 10 male 30 female

24. An ant moving straight, upon encountering an obstacle, may turn either right or left and continue moving. To test the hypothesis that the direction chosen by the ant is random, the most appropriate statistical test is **(JUNE 2014)**
(a) Student's t-test.
(b) X^2-test of independence.
(c) X^2-test of goodness of fit.
(d) correlation test

25. Bacteria often acquire genes by the process of lateral or horizontal transfer. Such 'foreign' genes, if acquired in recent past, may be identified by their atypical GC content, as compared to 'native' genes. Suppose the genomic GC content of a bacterium is 40%. Gene A of this organism contains 1000 bases with 225 G and 215 C. Another gene B of length 800 bases contains 160 G and 140 C. Which one of the following would be the most acceptable hypothesis (given that x^2= 3.841 at 0.05 significance level)? **(JUNE 2011)**
(a) A: native, B: Foreign
(b) A: Foreign, B: Native
(c) A: Foreign, B: Foreign
(d) A: Native, B: Native

Answer Key

1.(c)	2.(c)	3.(b)	4.(c)	5.(c)	6.(d)	7.(b)
8.(c)	9.(b)	10.(d)	11.(a)	12.(b)	13.(b)	14.(d)
15.(b)	16.(c)	17.(c)	18.(d)	19.(c)	20.(d)	21.(b)
22.(d)	23.(c)	24.(c)	25.(b)			

9. Radiolabeling techniques

1. If a metabolically active cell is challenged with ^{55}Fe radioisotope label, which of the following proteins **CANNOT** be detected by autoradiography? **(DEC 2017)**
(a) Aconitase and lipoic acid synthase
(b) Cytochrome C and DNA primase
(c) Calmodulin and Calcineurin
(d) Myoglobin and Homoaconitase

2. Which isotope below is best suited for metabolic labelling of glyceraldehyde-3-phospho-dehydrogenase? **(DEC 2015)**

(a) ^{14}C (b) ^{125}I
(c) ^{32}P (d) ^{131}I

3. The average energy absorbed by 10 gm of tissue from ^{32}P radiation is 14.9 $Jkg^{-(a)}$ The average dose in rads is **(JUNE 2014)**
(a) 1490 (b) 1.49
(c) 14900 (d) 149

4. t½ of an irreversible first order reaction, S→P is 1 hour. The time (in hours) required to reach 75% completion is **(JUNE 2013)**
(a) 1.5. (b) 2.0.
(c) 2.5. (d) 3.0.

5. If you inject a mouse with radioactive material of current activity of 256 Bq, what will be the activity after completion of 6 half-lives? **(DEC 2019)**
(a) 4 Bq (b) 8 Ba
(c) 16 Ba (d) 24 Bq

6. A sample counted for one minute shows a count rate of 752 cpm. For how many minutes should it be counted to have 1 % probable error? **(DEC 2011)**
(a) 13 (b) 5
(c) 2 (d) 75

7. A sample of ^{32}P disintegrates at a rate of 30,120 dpm. The radioactivity in microcuries (1 curie = 3.7×10^{10}dps) is **(MODELPAPER)**
(a) 15,060. (b) 1.36×10^{-2}.
(c) 81.6×10^{-2}. (d) 1.36×10^{-8}.

8. Given below is a table with information on isotopes, their half-life and type of particle(s) they emit.

Symbol and atomic weight		Half life		Type of particle(s) emitted	
(a)	^{11}C	(i)	5700 yrs	(X)	β
(b)	^{14}C	(ii)	15.1 hours	(Y)	□
(c)	^{24}Na	(iii)	20 min		

Choose the correct combination from the options given below. **(JUNE 2018)**
(a) (a) - (iii) - (y); b - (ii) - (x), (y); (c) - (i) - (y)
(b) (a) - (iii) - (x); b - (i) - (x); (c) - (ii) (x), (y)
(c) (a) - (ii) - (x), (y); b - (iii) - (x), (y); (c) - (i) - (x)
(d) (a) - (i) - (x); b - (ii) - (x); (c) - (iii) - (x),(y)

9. Some of the radioisotopes commonly used in biology are listed in column A and their properties are listed in Column B.

	Column A		Column B
I	^{35}S	i	High energy β
II	^{32}P	ii	α particle
III	^{238}U	iii	γ emission
IV	^{125}I	iv	Low energy β

Choose the option that correctly matches the isotope with the property listed. **(DEC 2019 ASSAM)**
(a) I: iv; II: iii; III: ii; IV: i
(b) I: iii; II: ii; III: iv; IV: i
(c) I: ii; II: i; III: iv; IV: iii
(d) I: iv; II: i; III: ii; IV: iii

Answer Key

1.(c)	2.(a)	3.(a)	4.(b)	5.(a)	6.(a)	7.(c)
8.(b)	9.(d)					

10. Microscopy and flow cytometer / FACS/FRET

1. Specimens for light microscopy are commonly fixed with a solution containing chemicals that crosslink/ denature cellular constituents. Commonly used fixatives such as formaldehyde and methanol could act in various ways as described below:
A. Formaldehyde cross-links amino groups on adjacent molecules and stabilizes protein-protein and protein-nucleic acid interactions.
B. Methanol acts as a denaturing fixative and acts by reducing the solubility of protein molecules by disrupting hydrophobic interactions.
C. Formaldehyde cross-links lipid tails in biological membranes.

D. Methanol acts on nucleic acids. Cross links nucleic acids with proteins and thus stabilizes protein-nucleic acid interactions.

Which one of the following combinations represents all correct statements?

(a) A and C (b) B and C

(c) B and D (d) A and B

2. Which one of the following microscopes would you use to visualize a protein fused to an appropriate reporter in a living cell? **(DEC 2017)**

(a) Fluorescence, microscope

(b) Scanning electron microscope

(c) Differential interference contrast microscope

(d) Phase contrast microscope

3. The movement of a single cell was required to be continually monitored during development. This cell was marked with a reporter gene. To visualize this movement one would use **(DEC 2012)**

(a) phase contrast microscopy

(b) bright field microscopy

(c) florescence microscopy

(d) atomic force microscopy

4. Co-localization of two fluorescently labeled proteins in an organelle in cells is usually visualised by **(DEC 2013)**

(a) interference-contrast microscopy.

(b) scanning electron microscopy.

(c) confocal microscopy.

(d) atomic force microscopy.

5. Which one of the following is used as a source of excitation in a confocal microscope? **(DEC 2018)**

(a) Lasers (b) Electron beam

(c) Mercury lamp (d) Masers

6. You wish to localise a given gene product at subcellular levels following immune-fluoroscence staining. Routine microscopy could not resolve whether the gene product is localized inside the nucleus or on the nuclear membrane.

Which of the following will resolve this unambiguously?

A. Sectioning of cell followed by phase contrast microscopy.

B. A simulation of 3D picture following confocal microscopy.

C. Optical sectioning and observing each section.

D. Freeze fracturing followed by Scanning Electron Microscopy. **(DEC 2012)**

(a) A and B (b) B and C

(c) C and D (d) A and C

7. In order to visualize the intracellular organization of a cell, one can utilize various microscopy-based techniques. These include:

A. Differential interference contrast (DIC) microscopy

B. Phase contrast microscopy

C. Dark field microscopy

D. Epifluorescence microscopy

E. Scanning electron microscopy

F. Transmission electron microscopy

G. Confocal microscopy

Which of the above mentioned microscopes can be used to study the intracellular dynamics using live cell imaging? **(JUNE 2018)**

(a) A, B, E, F, G (b) A, B, C, D, G

(c) A, D, E. F, G (d) C, D, E, F, G

8. In electron microscopy, to detect specific macromolecule or structure such as spindle pole body (SPB), the frequently used procedure is to couple secondary antibody with **(DEC 2019)**

(a) Alexa 568 (b) Cy5

(c) Gold particle (d) Osmium tetraoxide

9. Proteins in cells can be visualized by the following methods:

A. Express the gene (coding for the said protein) as a fusion with the green fluorescence protein (GFP) and directly visualize under a fluorescence microscope.

B. Express the gene (coding for the said protein) as a fusion with the β-galactosidase gene (*lac Z*) and directly visualize under a phase contrast bright field microscope.

C. A fluorescence tagged antibody raised against the said protein could be used for visualization in a fluorescence microscope.

D. Over express the protein and directly visualize it under a scanning electron microscope.

Which of the following methods you would choose to visualize a protein in a living cell? **(JUNE 2011)**

(a) A only (b) A and C only
(c) A and B only (d) D only

10. Sub-cellular fractionation-based assays have been used to identify various organelles in the mammalian cells. In order to characterize such organelles in a living mammalian cell, which of the following microscopy-based method would be the most accurate? **(JUNE 2018)**

(a) use of fluorescent probes specific for organelles.
(b) use of organelle specific fluorescent probes followed by microinjection of fluorescent antibodies against organelle specific protein.
(c) use of fluorescent probes in permeabilized cells.
(d) use of organelle specific fluorescent probes followed by cryo-electron microscopy

11. In resting cells, proteins X and Y are localized in the cytosol. Upon stimulation with lipopolysaccharide (LPS), both of them are phosphorylated and translocate to the nucleus. You have used antibodies against phosphorylated forms of proteins X and Y which are conjugated to either red, or green or blue dye. Keeping optical aberration of light in mind, which one of the following will be the best for visualizing X and Y In the nucleus by fluorescence microscopy? **(DEC 2013)**

(a) Anti green X and anti red Y
(b) Anti red X and anti green Y
(c) Anti red X and anti blue Y
(d) Anti blue X and anti green Y

12. To spatiotemporally detect the localization of two proteins A and B using a confocal microscope, an investigator plans to utilize a few fluorescent proteins (X, Y, Z and T). The excitation and emission spectra of these fluorescent proteins are shown below.

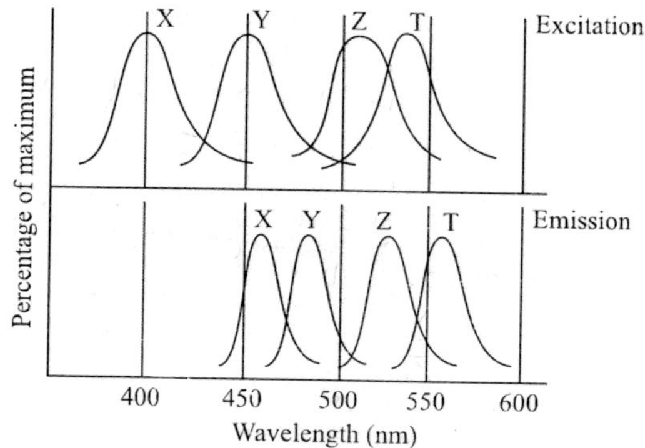

Which one of the following options comprises of a fluorescent pair that is best suited to conduct such an experiment? **(DEC 2019 ASSAM)**

(a) X and Y (b) Y and Z
(c) X and T (d) Z and T

13. Electron microscopes have much higher resolution than any type of light microscope because **(JUNE 2014)**

(a) of their higher magnification.
(b) the lenses used are of much higher quality.
(c) of very short wave length of electrons.
(d) the images are viewed on screen rather than directly using an eye-piece or occular lens.

14. The most important property of any microscope is its power of resolution, which is numerically equivalent to D; the minimum distance between two distinguishable objects. D depends on three parameters namely, the angular aperture, α, the refractive index, N, and wavelength, λ, of the incident light. Below are given few options to increase the resolution of the microscope.

A. Decrease the value of λ or increase either N or α to improve resolution.

B. Moving the objective lens closer to the specimen will decrease sin α and improve the resolution

C. Using a medium with high refraction index between specimen and the objective lens to improve the resolution.

D. Increase the wavelength of the incident light to improve resolution

Which of the following combinations of above statement are correct? **(DEC 2011)**

(a) A and C (b) B and C

(c) A and D (d) C and D

15. The most important property of any microscope is its resolution (D) and can be calculated from the formula

$$D = \frac{0.61\lambda}{N \sin\alpha}$$

where D is, minimum distance between two distinguishable objects, A. is the wave length of incident light, a is, the angular aperture and N is the refractive index of the medium.

Given below are several suggestions to improve the resolution of a microscope:

A. decrease the wave length of incident light

B. increase the wave length of incident light

C. use oil which has a higher refractive index

D. use oil because of its lower refractive index

Which one is the correct suggestion? **(JUNE 2015)**

(a) A and C (b) Only B

(c) Only D (d) B and D

16. In an oil immersion microscope, a drop of oil must be placed between the tip of objective lens and the specimen. The oil

A. reduces the numerical aperture of the objective lens.

B. increases the numerical aperture of the objective lens.

C. prevents light from bending (refraction) as it passes through the specimen.

D. improves the resolving power by a factor of 1/n (n being the refractive index of oil).

Which of the above statement/s is/are true for an oil-immersion microscope? **(DEC 2013 GU)**

(a) B, C and D (b) A and B

(c) Only A (d) Only B

17. An observation was being made under a compound microscope using light of 500 nm wave length. The refractive index of the observation medium was 1.5. The sinα of the lens used was 0.94. The limit of resolution (μm) of the microscope using this lens will be **(DEC 2013 GU)**

(a) 0.216. (b) 216.

(c) 0.94. (d) 1.41.

18. To achieve a best resolution using a fluorescence microscope, what combination of wavelength of emitted light (λ), refractive index and the angle (2^{θ}) by which light enters into the microscope would be the best choice for the user: **(DEC 2017)**

(a) λ = 405; refractive index=1.33; 2^{θ} = 90°

(b) λ = 420; refractive index=1.51; 2^{θ} = 180°

(c) λ = 520; refractive index=1.51; 2^{θ} = 90°

(d) λ = 405; refractive index=1.51; 2^{θ} = 180°

19. The most important property of any microscope is its resolution (D). Which one of the following wavelengths (nm) would be used to achieve the best resolution using a light microscope with lenses having numerical aperture (NA) of 1.4? **(DEC 2018)**

(a) 450 (b) 480

(c) 560 (d) 700

20. Fluorescence recovery after photobleaching in live cells is used to determine **(DEC 2015)**

(a) co-localization of proteins

(b) distance between two organelles

(c) diffusion of proteins

(d) nucleic acid compactness

21. A certain protein has been assumed to play an indispensable role in the survival of an intracellular parasite inside the host cells. Which one of the following techniques will best prove the assumption to be correct?

(JUNE 2018)

(a) Treat the parasite-infected host cells with an inhibitor of the protein and check the number of parasites per host cell under the microscope.

(b) Check the expression of the protein in parasite-infected host cells.

(c) Check the activity of the protein in parasite-infected host cells.

(d) Treat the parasite-infected host cells with an activator of the protein and check the number of parasites per host cell under the microscope.

22. A researcher was repeating a FACS experiment but somehow got confused with the labeling of the tubes. There are four tubes, one control, C (with no fluorescent label), one standard 1, S1 (with FITC label), one standard 2, S2 (with PE label) and the last one test, T (which should be FITC positive). Given below is the result of the FACS experiment.

(a)

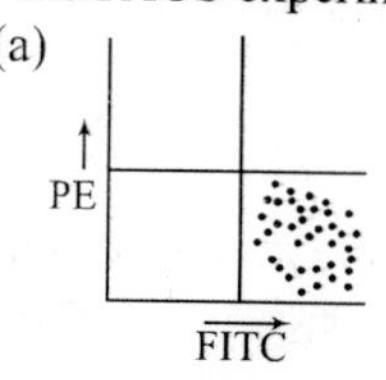

(b)

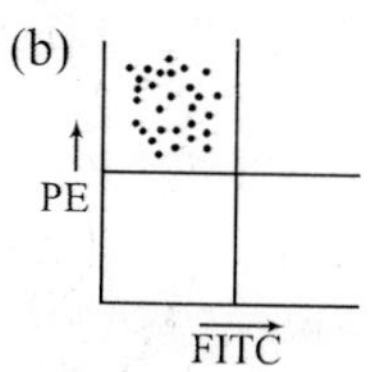

(c)

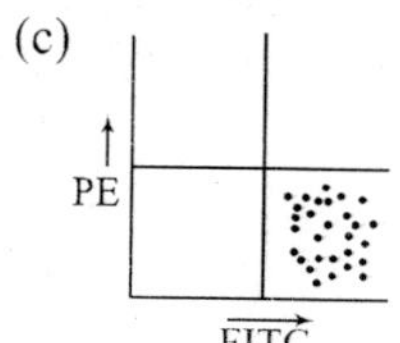

(d)

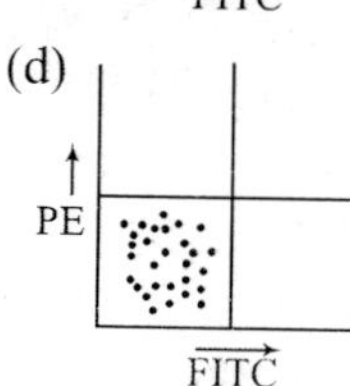

What should be the correct labeling?

(DEC 2014)

(a) a,S2; b,S1; c,T; d,C

(b) a,S1; b,T; c,C; d, S2

(c) a,S2; b,S1/T; c, C; d,S1/T

(d) a,S1/T; b,S2; c, s1/T; d,C

23. A mixed cell population was stained with two antibodies, one specific for cell surface antigen A and the other specific for cell surface antigen B. Anti-A antibody was labelled with fluorescein and anti-B antibody was labelled with Rhodamine. The cell population was then analysed for the presence of antigens by flow cytometry. Which one of the following is the correct outcome for this cell population? **(DEC 2019)**

(a)

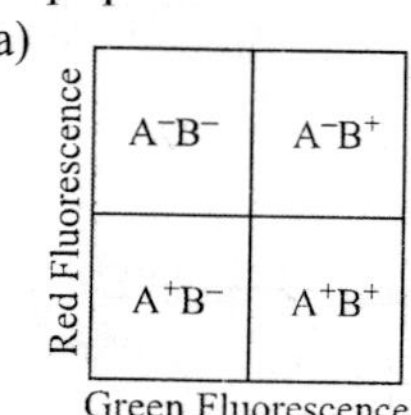

(b)

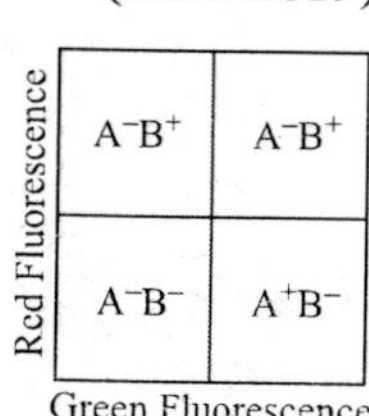

(c)

Red Fluorescence

A^+B^-	A^-B^+
A^-B^-	A^+B^+

Green Fluorescence

(d)

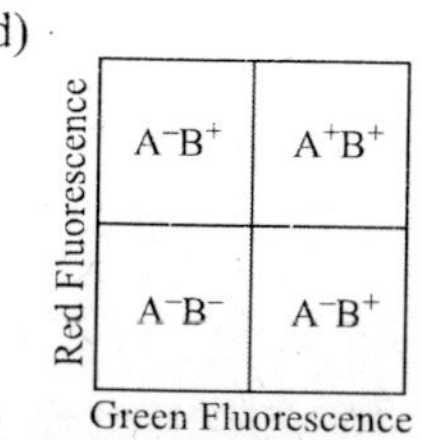

24. T cell proliferation *in vivo* is to be analyzed. The cells are labeled with CFSE (a fluorescent probe) and injected in CD86-deficient mice and BALB/c mice along with the required antigens. Three days later, the cells are recovered and analyzed by flow cytometry. Which one of the following is logically correct? **(DEC 2011)**

— BALB/c

--- CD86-deficient

(a)

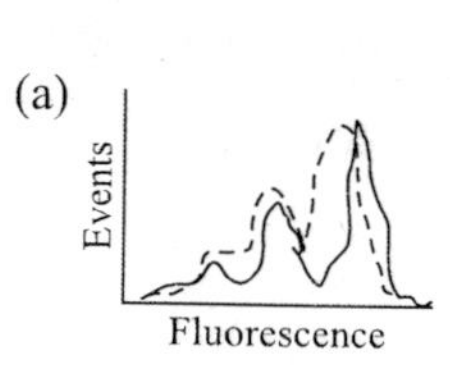

(b)

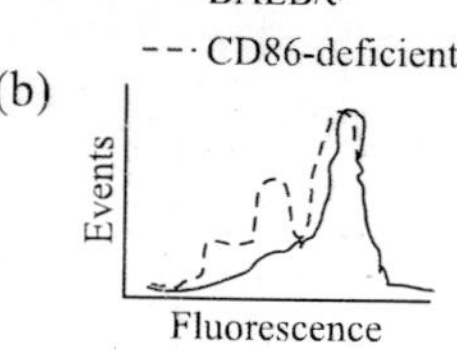

(c)

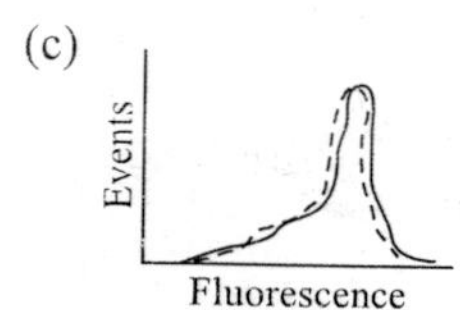

(d)

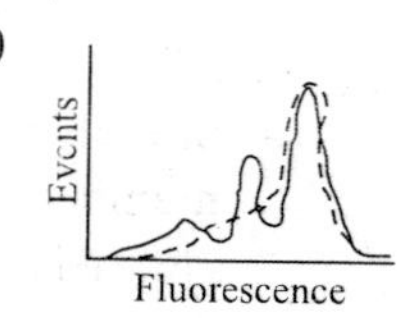

25. To test the impact of cAMP on protein kinase A conformation in cells, an investigator made FRET biosensor by fusing two fluorescent protein at the N and C terminus of protein Kinase A. In the absence of cAMP in the cellular milieu, no FRET signal was detected. However, upon c-AMP addition, a strong emission at 530 nm was observed. What could be the best configuration of fluorophores that were used by the investigator? **(JUNE 2019)**
   (a) Green fluorescent protein (GFP) and Red fluorescent protein (RFP).
   (b) CYAN fluorescent protein (CFP) and Yellow fluorescent protein (YFP)
   (c) Yellow fluorescent protein (YFP) and Red fluorescent protein (RFP)
   (d) Red fluorescent protein (RFP) and CYAN fluorescent protein (CFP)

Answer Key

1.(d)	2.(a)	3.(c)	4.(c)	5.(a)	6.(b)	7.(b)
8.(c)	9.(a)	10.(b)	11.(c)	12.(c)	13.(c)	14.(a)
15.(a)	16.(a)	17.(a)	18.(d)	19.(a)	20.(c)	21.(a)
22.(d)	23.(c)	24.(d)	25.(b)			

11. Patch-clamp, EEG, PET, MRI, FMRI, CAT

1. In an intact cell patch-clamp experiment, **(DEC 2017)**
   (a) two electrodes are inserted into the cytosol but at different depths.
   (b) one electrode is applied to the plasma membrane in a region containing only lipids and one into the cytosol.
   (c) two electrodes are applied to the plasma membrane, one in a region containing only lipids and the other in a region containing one or few ion channels.
   (d) one electrode is applied to the plasma membrane containing one or few ion channels and one electrode inserted into the cytosol.

2. A neurophysiologist was interested in using the patch-clamp technique. Following statements are related to this technique:
   A. Intracellular movement of ion channels.
   B. Post–translational modification of the ion channel protein
   C. Ligand that controls the opening or closing of ion channels.
   D. Change in current flow in a single ion channel.

   Which one of the following combinations will be achievable using the patch-clamp technique? **(JUNE 2019)**
   (a) A and B (b) B and C
   (c) C and D (d) D and A

3. If a researcher intends to identify a specific brain area activity linked to a cognitive function in human subjects, which one of the following techniques should be used? **(DEC 2014)**
   (a) CAT (b) MRI
   (c) fMRI (d) Patch-clamp

4. Measurement and mapping with spatial resolution the membrane potential of a cell, which is too small for microelectrode impalement, is done using **(DEC 2011)**
   (a) radioisotope
   (b) voltage-sensitive dye
   (c) pH sensitive chemical
   (d) vital dyes

5. Nichrome coated stainless steel electrodes were implanted in a rat brain for chronically recording the electrical activity of deep brain structures. During a study of 3 months the intensity of electrical signals gradually decreased.

   The following statements may explain the cause of this observation.
   A. The deposition of metallic iron from the electrode tips caused degeneration of some neurons.

B. The gradual accumulation of microglia at the electrode tips increased the resistance of electrodes.

C. The neurons at the electrode tips were hyperpolarized gradually.

D. The threshold for firing action potential in the neurons at the electrode tips was increased due to prolonged presence of electrodes.

Which one of the following is correct?

(DEC 2014)

(a) A only (b) A and B

(c) C only (d) C and D

6. An EEG was recorded and its power spectrum analyses were done in rats with implant electrode for a long time. The power of the EEG waves decreased two months after electrode implantation. This observation may be due to the following:

A. Glial cells accumulate surrounding exposed tips of electrodes.

B. Degeneration of neurons occur surrounding the electrode tips due metal ion deposition.

C. Coating of electrodes are destroyed with time.

D. The microsocket becomes loose time.

Which one of the following is true?

(DEC 2013)

(a) Only A (b) A and B

(c) Only C (d) C and D

7. Given below are spatial and temporal techniques (column I) used to detect brain activity (column II)

	Column I		**Column II**
A	Functional magnetic resonance imaging	(i)	Uses short-lived radioactive material to map functional processes in brain
B	Positron emission tomography	(ii)	Measure the fluctuation of dipole voltage in neurons of brain
C	Computed tomography	(iii)	Detects changes in blood oxygenation and flow due to neural activity
D	Electroencephalogram	(iv)	Images of brain obtained by differential absorption of X-rays

Select the correct set of combination

(DEC 2017)

(a) A-(i), B-(ii), C-(iii), D-(iv)

(b) A-(iii), B-(i), C-(iv), D-(ii)

(c) A-(iv), B-(iii), C-(ii), D-(i)

(d) A-(ii), B-(iv), C-(i), D-(iii)

8. In an animal experiment;

(i) Electrical stimulation of an area in the brain (A) increase a function (F) which was prevented by systemic injection of adrenergic antagonistic, prazosin.

(ii) Injection of carbachol (cholinergic agonist) into A also increased function F which was, however, not prevented by systemic injection of adrenergic antagonistic, prazosin.

The results are likely to be due to the stimulation of **(DEC 2011)**

(a) Noradrenergic and cholinoceptive neurons

(b) Cholinergic and nor-adrenoceptive neurons

(c) Adrenergic terminals in 'A'

(d) Both neurons and fibres passing through 'A'

9. Membrane potential in mitochondria critical for oxidative phosphorylation and is monitored by **(JUNE 2017)**

(a) Patch clamping

(b) Measuring internal sodium ions after lysing the mitochondria

(c) Measuring distribution of labelled ions across the mitochondrial membrane

(d) Measuring the consumption of ATP

Answer Key

1.(d)	**2.**(c)	**3.**(c)	**4.**(b)	**5.**(b)	**6.**(b)	**7.**(b)
8.(a)	**9.**(c)					